Efficiency in Natural Product Total Synthesis

Efficiency in Natural Product Total Synthesis

Edited by

Pei-Qiang Huang
College of Chemistry and Chemical Engineering, Xiamen University, Xiamen, Fujian, China

Zhu-Jun Yao
School of Chemistry and Chemical Engineering, Nanjing University, Nanjing, China

Richard P. Hsung
Division of Pharmaceutical Sciences, School of Pharmacy, University of Wisconsin-Madison, Madison, WI, USA

WILEY

This edition first published 2018

Registered Office
John Wiley & Sons, Inc., 111 River Street, Hoboken, NJ 07030, USA

Editorial Office
111 River Street, Hoboken, NJ 07030, USA

For details of our global editorial offices, customer services, and more information about Wiley products visit us at www.wiley.com.

Wiley also publishes its books in a variety of electronic formats and by print-on-demand. Some content that appears in standard print versions of this book may not be available in other formats.

Library of Congress Cataloging-in-Publication Data

Names: Huang, Peiqiang, 1962– editor. | Yao, Zhu-Jun, editor. | Hsung, Richard P., editor.
Title: Efficiency in natural product total synthesis / edited by Pei-Qiang Huang, Zhu-Jun Yao, Richard P. Hsung.
Description: First edition. | Hoboken, NJ : John Wiley & Sons, 2018. | Includes bibliographical references and index. |
Identifiers: LCCN 2018008623 (print) | LCCN 2018023632 (ebook) | ISBN 9781118940204 (pdf) | ISBN 9781118940211 (epub) | ISBN 9781118605400 (cloth)
Subjects: LCSH: Natural products–Synthesis. | Organic compounds–Synthesis.
Classification: LCC QD415 (ebook) | LCC QD415 .E38 2018 (print) | DDC 547/.2–dc23
LC record available at https://lccn.loc.gov/2018008623

Cover Design: Wiley
Cover Image: © Monty Rakusen/Getty Images

Set in 10/12pt Warnock by SPi Global, Pondicherry, India

Printed in the United States of America

10 9 8 7 6 5 4 3 2 1

Contents

Contributors

Stellios Arseniyadis
Queen Mary University of London
London, UK

Erica Benedetti
Université Paris Descartes
Paris, France

Cyril Bressy
Aix Marseille Université
Marseille, France

Xin Chen
School of Pharmacy
Presbyterian College
Clinton, SC, USA

Bin Cheng
Department of Chemistry
Lanzhou University
Lanzhou, China

Lichao Fang
China Novartis Institutes for BioMedical Research
Shanghai, China

Fang Fang
Department of Chemistry
Lanzhou University
Lanzhou, China

John B. Feltenberger
Medicinal Chemistry Center
School of Pharmacy
University of Wisconsin-Madison
Madison, WI, USA

Peng Gao
Department of Chemistry
Lanzhou University
Lanzhou, China

Richard P. Hsung
Division of Pharmaceutical Sciences
School of Pharmacy
University of Wisconsin-Madison,
Madison, WI, USA

Hanwei Hu
Department of Chemistry
Lanzhou University
Lanzhou, China

Pei-Qiang Huang
Fujian Provincial Key Laboratory of Chemical Biology
Department of Chemistry and iChEM
Xiamen University
Xiamen, China

Yun Li
Department of Chemistry
Lanzhou University
Lanzhou, China

Zhiqiang Ma
Department of Chemistry
South China University of Technology
Guangzhou, China

Zhi-Xiong Ma
Division of Pharmaceutical Sciences
School of Pharmacy
University of Wisconsin-Madison
Madison, WI, USA

Wai-Lung Ng
Department of Medicine
Harvard Medical School; and
Department of Cancer Biology
Dana-Farber Cancer Institute
Boston, MA, USA; and
Department of Chemistry
The Chinese University of Hong Kong
Shatin, New Territories
Hong Kong SAR, China

Tony K. M. Shing
Department of Chemistry
Keio University, Tokyo, Japan; and
Department of Chemistry
Chinese University of Hong Kong
Hong Kong, China

Michael Smietana
Université de Montpellier
Montpellier, France

Wan-Guo Wei
Shanghai Advanced Research Institute
Chinese Academy of Sciences
Shanghai, China

Anthony W. H. Wong
Department of Chemistry
Chinese University of Hong Kong
Hong Kong, China

Xiao Yu Wu
Department of Chemistry
Shanghai University
Shanghai, China

Zhu-Jun Yao
School of Chemistry and Chemical Engineering
Nanjing University
Nanjing, China

Zheng Qing Ye
Shanghai Institute of Organic Chemistry
Chinese Academy of Sciences
Shanghai, China

Shouyun Yu
School of Chemistry and Chemical Engineering
Nanjing University
Nanjing, China

Hongbin Zhai
School of Chemical Biology and Biotechnology
Shenzhen Graduate School of Peking University
Shenzhen, China

Weihe Zhang
Department of Chemistry
Lanzhou University
Lanzhou, China

Gang Zhao
Shanghai Institute of Organic Chemistry
Chinese Academy of Sciences
Shanghai, China

Foreword

It is my prodigious pleasure to be invited to write this Foreword for the book entitled *Efficiency in Natural Product Total Synthesis*, co-edited by three of my good friends Pei-Qiang Huang, Zhu-Jun Yao, and Richard P. Hsung. Although I do not rather regard myself as a chemist engaging in total synthesis of natural products in a full-time manner, I do from time to time participate in synthetic studies of structurally intriguing naturally occurring molecules. In fact, usually after presenting painstaking steps on my way towards the total synthesis of natural products, I was asked time and again, "What would be the future of natural products synthesis?" This is not a trivial question, especially knowing that even *palytoxin* and *ciguatoxin*, two exceedingly structurally complicated and eye-catching molecules, have both been conquered. Then, what are the synthetic challenges if we are to contemplate the future of natural product total synthesis? Perhaps my quick responses are those modern synthetic procedures involving biomimetic synthesis; procedure-economical synthesis; cascade synthesis; metal-, organometallic-, and organocatalytic synthesis; and multi-component synthesis. These underlying synthetic endeavors would indeed improve the synthetic efficiency, pose new challenges, and even rejuvenate our interests in natural products total synthesis in the years to come.

This book, co-edited by Pei-Qiang, Zhu-Jun and Richard, serves by all means as a timely treatise, summarizing the relevance of those aforementioned reactions in natural product total synthesis. In Chapter 1, entitled *Principles for Synthetic Efficiency and Expansion of the Field*, Pei-Qiang Huang gives an excellent account on concepts of efficiency in organic synthesis, biomimetic synthesis, and expansion of the field toward chemical biology and chemical genetics. At the end of this chapter, Pei-Qiang also addresses the threats we may face in the near future. Pei-Qiang also wrote Chapter 2, which is a comprehensive summary of *Selected Procedure-Economical Enantioselective Total Syntheses of Natural Products*. In this chapter, total syntheses employing 1–15 steps are described. Richard Hsung and his coauthors prepared Chapter 3, which is on *Diels–Alder Cascades in Natural Product Total Synthesis*. Richard, being an expert in this synthetic area, gives a concise account on cascade pathways making use of Diels–Alder cascades with his coauthors. Chapter 4 is concerned with *Organometallics-Based Catalytic (Asymmetric) Synthesis of Natural Products,* and was written by Hong-Bin Zhai and his coauthors, who provide a balanced coverage of Au-, Ag-, Pt-, Co-, Cu-, Cr-, Fe-, Mn-, Ni-, Pd-, Rh-, and Ru-catalyzed reactions, used as pivotal steps in natural product synthesis. Hong-Bin and two coauthors also contributed to Chapter 5, which is on the *C–H Activation-Based Strategy for Natural Product Synthesis.* Stellios Arseniyadis and three coauthors are the contributors of Chapter 6, which reports *Recent Applications of Kagan's Reagent (SmI_2) in Natural Product Synthesis,* exemplifying by several natural products. Organocatalysis is the main theme of Chapter 7, whose title is *Asymmetric Organocatalysis in the Total Synthesis of Complex Natural Products,* and whose authors are Gang Zhao, Zheng-Qing Ye, and Xiao Yu Wu. Gang used naturally occurring molecules as examples to showcase the significance of organocatalysis in total synthesis. Chapter 8 was also written by Stellios Arseniyadis and three coauthors, and deals with *Multicomponent Reactions in Natural Product Synthesis.* Tony Shing, my friend and colleague at the Chinese University of Hong Kong, provided Chapter 9, whose subject matter is on *Renewable Resource-Based Building Blocks/Chirons for the Total Synthesis of Natural Products.* This chapter describes the deployment of renewable chirons such as amino acids, carbohydrates, and terpenes as building blocks in total syntheses of natural products. Chapter 10 is dedicated to applications of natural products, and its title is *Natural Product Synthesis for Drug Discovery and Chemical Biology.* Last but not the least, Chapter 11 delineates some *Modern Technologies in Natural Product Synthesis,* that is, chemical and biological space, natural product-like chemistry, enzymatic synthetic biology, solid phase synthesis, and flow chemistry. Chapter 12 is the *Concluding Remarks and Perspectives.*

All in all, each chapter in this book involves a special synthetic theme, and the common goal is to improve efficiency in natural product total synthesis. Although lamentably it is rather unlikely that we will be able to rival enzymes in achieving 100% chemical and optical yields in the near future, we nonetheless still desire to accomplish at least a certain synthetic efficiency that would help to upgrade chemical yields and stereoselectivities. Therefore, I enthusiastically recommend this book to all practitioners in the area of natural product total synthesis.

December 23, 2017

Henry N. C. Wong
The Chinese University of Hong Kong

Preface

In the past decade, the endeavor of natural product synthesis has lost its pole position as the most dominant research topic in the field of organic chemistry. In fact, the entire discipline of organic chemistry seems to be vanishing into the background after serving as an indispensable engine that drives discoveries in materials and medicines for the most part of twentieth century as well as during the turn of this century. Although no eulogy has been given yet, it has been ordained with the kiss-of-death label of "matured science." While such perception could not be far from the truth, like it or not, perception is often the reality if not perceived as the truth altogether. Scientists in this field are experiencing increasingly more difficult time to obtain financial support from funding agencies. On many continents, funding for organic chemistry in general, let alone natural product synthesis, has been paltry for most and approaching the bottom for some with veer increasing fallout from the grandeur of economic globalization. Consequently, the last 10 years would mark the most frenzied time in the history of organic chemistry to scramble for new identities, new justifications, and new audiences and at times to marginalize one area of research just to uphold another in a bid for limited funds. The past decade also saw an incredible amount of fascination with publications metrics to quantify one's worth and the level of significance therein. These insane measures have been created once again for vying for limited resources. When we know full well that none of these exercises is good for Science spelled with a capital "S," why would any of it be good for the discipline of organic chemistry?

But it gets worse. Major pharmaceutical companies, who had been the main beneficiaries of innovative research in organic synthesis and the resulting well-trained individuals, have completely abandoned natural product chemistry from drug discovery efforts. There are many reasons contributing to this drastic loss of interest - be that an astute recognition that natural product synthesis remains a formidable challenge, or be that greed. Yes, greed began to proliferate among pharmaceutical industries when the drug development process, which is full of unpredictable scientific and human elements, was equated with mathematical models that were at best conducive for Wall Street wizards or Las Vegas odds makers. Nevertheless, despite some of the best-selling drugs are becoming small-molecule and macrobiomolecule based (or biologics), depending upon the source of analysis, somewhere between 60 and 80% of drugs currently on the market remain either a natural product or derived from a natural product.

It is true that as a field advances, it must learn how to adapt and evolve. Thus, many of recent adaptations and interdisciplinary pursuits are not just necessary but also have been simply life-saving to organic chemistry. One could also accept that certain areas can exert a greater influence than others, as long as it is not driven by egos and wills of individuals to survive at all cost in the face of dwindling resources. That said, for any given field of research to truly succeed, diversity is quintessential. Yes, diversification is the common denominator for the survival of any evolutionary process. In this regard, natural product synthesis must be funded and allowed to live on. There is no reason to believe otherwise because its existence has served as an inspiration to nearly all discoveries in organic chemistry and many of those discoveries have played an important role in our everyday life for well over a century, and are continuing to do so.

It is within this context we decided to pursue this major undertaking. We are fully aware that we are not alone in our philosophical thinking; and thus, we would like to join hands with other like-minded folks to pronounce that organic chemistry is far from being mature. One needs to look no further than conceding to the fact that how little we know fundamentally in comparison to the best organic chemist known to date – Mother Nature. More specifically in the context of this book, we would like to proclaim that natural products chemistry and their syntheses should continue to hold a prominent position and serve as a driving force behind endeavors important to humans including drug discovery. It is not too difficult to agree and accept this proclamation. One needs to look no further than recognizing the fact how inefficient, uneconomical, flawed and imperfect, we still are in our pursuit of natural product synthesis. Again compared to Mother Nature, we are still in the Stone Age – and that actually would qualify as a compliment to the current state of the art in organic synthesis. Therefore, this book will strive to reflect the current status in total syntheses of natural products and illustrate principles for achieving synthetic efficiency.

In the Introduction, a brief review of the evolution of the total synthesis of natural products allowed us to identify the problem: the synthetic efficiency. In Chapter 1, the basic concepts and principles for synthetic efficiency are presented. In view of illustrating the concepts and principles of synthetic efficiency presented in Chapter 1, in Chapter 2, we analyzed 39 the shortest enantioselective total syntheses of natural products and compiled key information of 50 ones. From Chapter 3 to Chapter 9, selected efficient synthetic methodology-based total syntheses are presented. Chapters 10 and 11 are devoted to natural product synthesis for drug discovery and chemical biology, and modern technologies in natural product synthesis, respectively. Brief concluding remarks and perspectives are described in the last chapter. We hope that this book will be helpful, not only for researchers in the field of total synthesis, in particular, the graduate students, but also for those in related fields including medicinal chemists, and in pharmaceutical industries. Last but not least, comprehensive literature is given with a focus on review articles, which can be used as further reading materials for a more specific theme.

Introduction

Pei-Qiang Huang

Fujian Provincial Key Laboratory of Chemical Biology, Department of Chemistry and iChEM, Xiamen University, Xiamen, China

Humans rely on the Earth's resources for survival. In modern society, almost no resource except air can be used directly. The mission of chemists is to transfer natural resources from the Earth into usable materials. This task requires the understanding of structure and reactivity, and features the production/creation of natural or unnatural molecules/substances. Plants, animals, microorganisms, and fungal kingdoms both on land and in the sea are rich sources of small organic compounds called secondary metabolites or natural products.[1] Very often, natural products exist in organisms in only minuscule amounts. Moreover, natural products display widespread structural diversity and important bioactivities or are potentially bioactive, which rend them ideal sources for discovering and developing medicines, agrochemicals, and many other useful chemicals.[2] Thus, the chemical synthesis of natural products constitutes an important branch of science. In this context, the synthesis of structurally diverse (including the skeleton, functional group, and stereochemistry) natural products is often a challenging yet a fascinating field that has attracted numerous talented chemists for generations. Among the many books and reviews dedicated to the total synthesis of natural products,[3] *Selected Organic Synthesis* (I. Fleming), *Art in Organic Synthesis* (N. Anand), *The Logic of Chemical Synthesis* (E. J. Corey), and *Classics in Total Synthesis I–III* (K. C. Nicolaou) are representative. Different from all previous literature, in this book we intend to focus on one theme: The efficiency of total synthesis.

Efficiency in Natural Product Total Synthesis, First Edition. Edited by Pei-Qiang Huang, Zhu-Jun Yao, and Richard P. Hsung.

1 The Golden Age of the Total Synthesis of Natural Products: The Era as a Dominant Field

The first chemical synthesis of a natural product can be traced back to 1828 when F. Wöhler synthesized urea.[4] In an attempt to make ammonium cyanate by combining cyanic acid with ammonia, urea was obtained instead (Scheme 1a). This synthesis is the first example of the artificial production of an organic substance from inorganic materials. This serendipitous discovery not only marked the birth of organic chemistry as a branch of science, but is also celebrated as a refutation of vitalism; the hypothesis that living things are alive because of some special "vital force." According this hypothesis, "organic" compounds, so-called animal substances, could be made only by living things.

The first total synthesis of a natural product by a designed synthetic route was H. Kolbe's synthesis of acetic acid reported in 1845[5] (Scheme 1b). This represented a singular example that an organic compound was synthesized from inorganic elements and minerals. It is this synthesis that finally put to rest the *vis vitalis*. Moreover, the term "synthesis" was coined for the first time to describe the process of assembling a chemical compound from other substances.

In 1965, 120 years later, Chinese scientists achieved the chemical synthesis of crystallized bovine insulin (Figure 1),[6] a protein consisted of 51 amino acid residues with a molecular weight of 5733.53 Da. This work marked the first

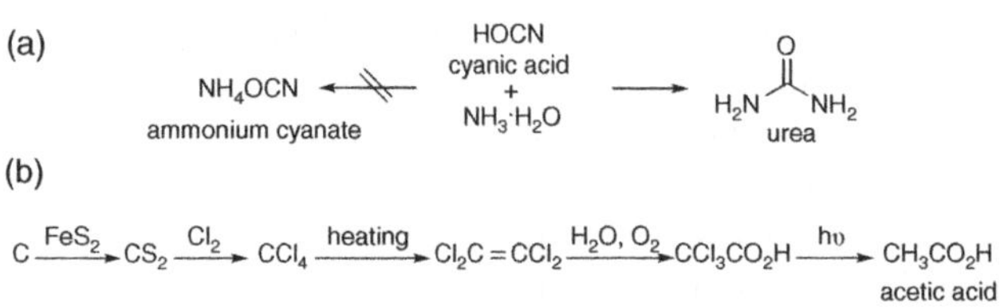

Scheme 1 Wöhler's synthesis of urea and Kolbe's synthesis of acetic acid.

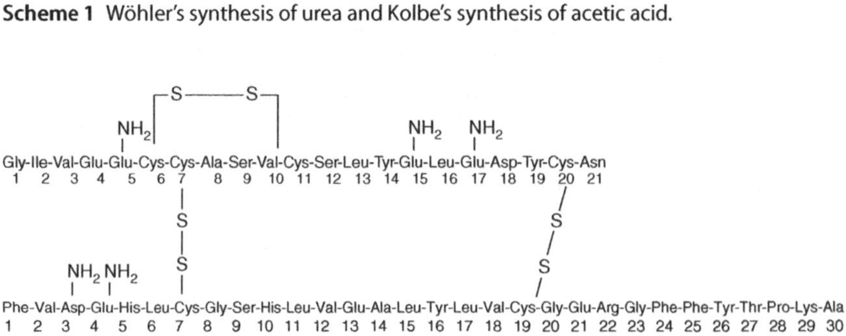

Figure 1 Structure of bovine insulin.

chemical total synthesis of a zoetic protein. In addition, it constituted a significant breakthrough in the field of life science and has had a substantial impact on human endeavors in finding out the secrets of life. This historic accomplishment relied on a state major science and technology program in China that started in 1958. Thanks to the cooperation of Ying-Lai Wang from the Shanghai Institute of Biochemistry (SIBC, CAS) and with the joint forces of research groups directed by Jing-Yi Niu, Cheng-Lu Zou (both from SIBC), You Wang (SIOC, CAS), and Qi-Yi Xing (Pekin University), the ambitious goal was finally achieved after 8 years' work. Note that, before this, V. du Vigneaud was awarded the 1955 Nobel Prize in Chemistry for his work on the first synthesis of polypeptide hormone oxytocin (MW 1007.19 Da.) and, by 1958, the longest peptide synthesized was a 23-residue fragment of adrenocorticotrophic hormone.

Since the structural determination of cholesterol and cholic acid by Windaus and Wieland in 1932, interest in the total synthesis of steroids has been widespread (e.g., **Progesterone** in Figure 2), and augmented as the importance of steroids in medicine and in animal physiology has grown. In particular, with the discovery of the steroids as active hormonal compounds in the 1950s, steroid chemistry blossomed. These longstanding efforts have resulted in numerous partial and total syntheses of steroids.[7] Moreover, the joint efforts of many disciplines together with the pharmaceutical industry on several aspects of steroid sciences have yielded a series of steroid hormone drugs such as cortisone, cholesterol, prednisone, as well as lovastatin for the treatment of hypercholesterolemia in 1987.[8] Thus, in the 1970s, when international society recognized the population explosion as one of the three challenges that

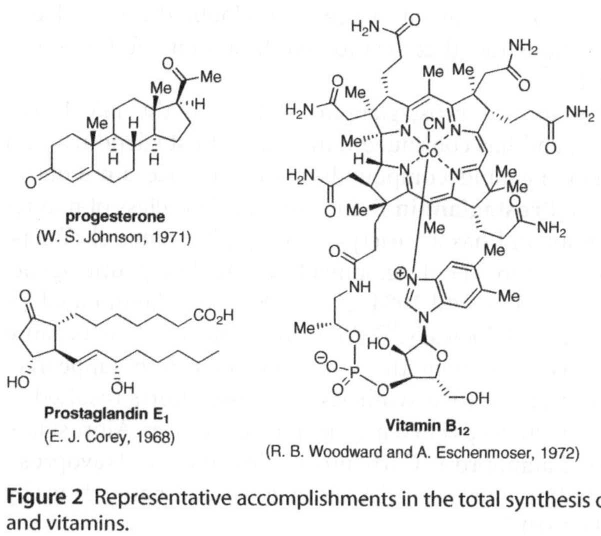

Figure 2 Representative accomplishments in the total synthesis of steroids, prostaglandins, and vitamins.

threatened the survival of the human race, and hormonal contraception was the best method available for birth control,[9] for once synthetic organic chemists were able to say that, with the synthetic methods developed, there was no longer any shortage of contraceptive steroids. Perhaps steroids are the sole class of natural products that have attracted the interest of so many great organic chemists and for so long (several decades). Research in this area has yielded abundant accomplishments (see Table 1 later) including a classical radical C–H functionalization reaction (the Barton reaction,[10] Scheme 2a) and the first enantioselective organocatalytic reaction (Hajos–Parrish–Eder–Sauer–Wiechert reaction,[11] Scheme 2b).

Through the close collaboration between R. B. Woodward's group at Harvard (in Cambridge) and the group at the Eidgenossische Technische Hochschule (ETH, in Zurich), the total synthesis of **vitamin B_{12}** (see Figure 2) was accomplished in 1972.[12] It took a period of 12 years and the efforts of more than 100 experimental chemists to achieve this challenging goal.[13] This is a landmark achievement as commented on by J. W. Cornforth: "The quality of the effort needed for this extension of man's control over matter can quite fittingly be compared to the better known (and much more costly) feat of placing men on the moon…."[14] Moreover, the combined efforts with R. Hoffmann to understand an unexpected stereochemical outcome of a reaction observed during the total synthesis resulted in the well-known Woodward–Hoffmann rules.[15] This brought about a totally unexpected major advance in theoretical organic chemistry, which led Hoffmann and K. Fukui to be honored by the 1981 Nobel Prize in Chemistry. Unfortunately, the premature death of the Woodward prevented the possibility of him receiving a second Nobel Prize together with Hoffmann for the Woodward–Hoffmann rules.[16] The total synthesis of vitamin B_{12} itself and the achievements it triggered undoubtedly provided a marvelous example to demonstrate that the total synthesis of natural products is indeed a dominant field.[17]

A synthetic campaign toward prostaglandins started in the late-1960s, covered the whole 1970s, and has continued until today. Prostaglandins are a group of naturally occurring lipid compounds found in trace amounts in animals and humans (e.g., **Prostaglandin E_1** in Figure 2). This class of natural products acts as mediators and has a variety of strong physiological effects, making them very promising for the development of new therapeutic agents for a number of diseases.[18,19] Up until 1984, over 5000 prostaglandin analogs were synthesized and tested biologically.[19a] From the 1930s until 2010, more than 158,000 publications containing the word "prostaglandin" appeared. Among them, 890 are related to "total synthesis."[19d] Those efforts resulted in the discovery of more than 20 drugs currently in clinical use (e.g., Alprostadil, Misoprostol, Limaprost, Latanoprost, Carboprost, Bimatoprost, Travoprost, and Tafluprost),[19b,c] including four in the list of top 200 brand-name drugs by total US prescriptions in 2010.[20]

Table 1 Selected examples of the impact of total synthesis on science and societies.

Synthetic Target/ Contributor	Achievements
Quinine W. H. Perkin (1856)	Serendipitous discovery and commercialization of first industrial dye, mauveines
Tropanone R. Robinson (1917)	The conception of biogenetic synthesis Biosynthesis of alkaloids
Steroids Many distinguished synthetic organic chemists and their collaborators (biologists)[8]	• Control of population • Steroid industry • Many hormonal drugs, including anti-inflammatories • Conformational analysis[32] • The Stork–Eschenmoser hypothesis for the stereospecific biosynthesis of steroids from squalene[33] • Robinson Annulation reaction[34] • Birch reduction • Barton reaction: the photocleavage of nitrite esters (an early example of C–H functionalization) (Scheme 2a)[10] • Hajos–Parrish–Eder–Sauer–Wiechert reaction (the first enantioselective organocatalysis) (Scheme 2b)[11]
Vitamin B_{12} Woodward Eschenmoser	Woodward–Hoffmann rules[15] and in part the 1981 Nobel Prize in Chemistry to R. Hoffmann Eschenmoser sulfide contraction reaction (Scheme 2c)[35]
Prostaglandins Many scientists	>20 drugs in clinical use
Peptides V. du Vigneaud R. B. Merrifield and many others	The 1955 Nobel Prize in Chemistry to du Vigneaud Solid-phase synthesis concept and technology[36] the 1984 Nobel Prize in Chemistry to Merrifield A significant contribution to modern life sciences including biology, pharmacology, and medicine
A number of natural products E. J. Corey	The 1990 Nobel Prize in Chemistry to Corey Theory (retrosynthetic analysis) of organic synthesis[3c,25,37] Many named reactions, synthetic methods, and reagents

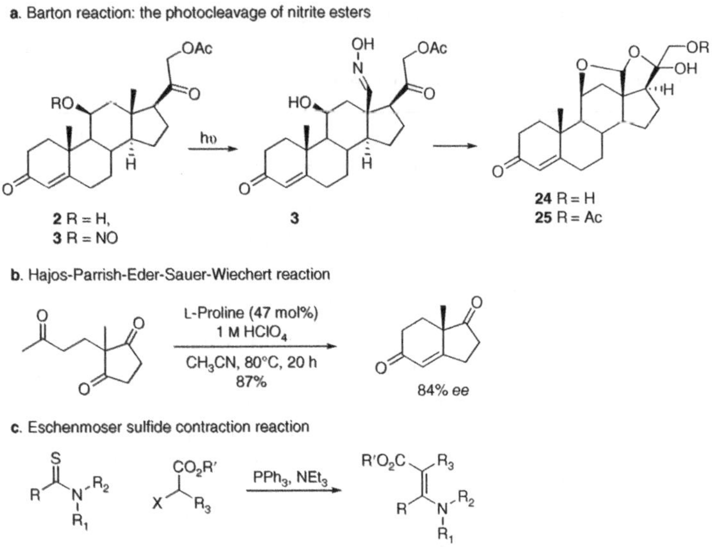

Scheme 2 Selected named reactions invented/developed during total syntheses.

At the end of the 1970s, synthetic organic chemists began to challenge the enantioselective synthesis of alicyclic and macrocyclic molecules possessing many stereogenic centers. The curtain of the campaign was pulled in 1978 by the Corey team's enantioselective total synthesis of erythronolide B (the aglycone of the antibiotic erythromycin B) (Figure 3).[21] Erythronolide B features a 14-membered macrolide containing 10 chiral centers. Subsequently, the elegant enantioselective total syntheses of monensin by Y. Kishi in 1979[22] and by W. C. Still in 1980,[23] and erythromycin by Woodward and coworkers in 1981,[24] have been achieved. It is remarkable that one is able to synthesize the natural stereoisomer out of 2^{22} (131 072 for monensin) and 2^{23} (for erythromycin) possible ones! While most of Woodward's synthetic activities have been conducted in the era of racemic synthesis, his tenacity in embracing the era of enantioselective total synthesis and becoming one of the pioneers was even more respectable. The first enantioselective total synthesis of erythromycin became Woodward's last contribution to the art of organic synthesis and was published posthumously. By the end of this decade and the beginning of 1990s, the enantioselective total syntheses of ginkgolide (Figure 3) by Corey,[25] FK-506 by T. K. Jones (Merck),[26a] S. L. Schreiber (1990),[26b] S. J. Danishefsky (1990),[26c] Sih, C. J. (1990),[26d] A. B. Smith III (1994),[26e] Ireland, R. E. (1996),[26f] and cytovaricin by D. A. Evans[27] (Figure 3) were significant achievements.

The glorious 1980s concluded with the enantioselective total synthesis of palytoxin carboxylic acid (PTC, Figure 4) (1989)[28a,b] by the team directed by Y. Kishi at Harvard. Its conversion to palytoxin (PTX) was accomplished 5 years later.[29]

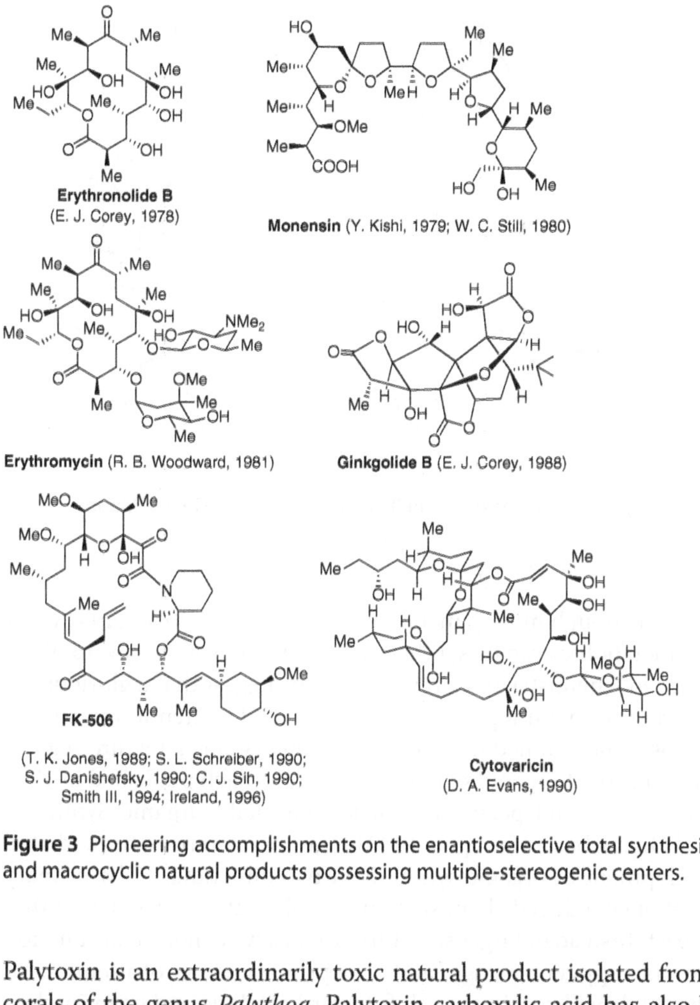

Figure 3 Pioneering accomplishments on the enantioselective total synthesis of alicyclic and macrocyclic natural products possessing multiple-stereogenic centers.

Palytoxin is an extraordinarily toxic natural product isolated from marine soft corals of the genus *Palythoa*. Palytoxin carboxylic acid has also been isolated from the same source in approximately one-third of palytoxin and its physiological activity is similar to palytoxin,[28b,c] Palytoxin, which has a molecular weight of 2680 daltons ($C_{129}H_{223}N_3O_{54}$) and contains 64 chiral centers and seven in-skeleton olefinic bonds, has more than one sextillion ($2^{71} = 1.2 \times 10^{21}$) possible stereoisomers. Moreover, 29 of the chiral centers are in the acyclic portions, and the configurations of 27 of them were unknown. Thus, palytoxin was regarded as a "Mount Everest" of chemistry. Undoubtedly, the enantioselective total synthesis of palytoxin should be a giant project full of challenges if not being an impossible mission.

With the conquest of palytoxin, yet another milestone in the history of the total synthesis, synthetic organic chemistry is conceived as a relatively "mature"

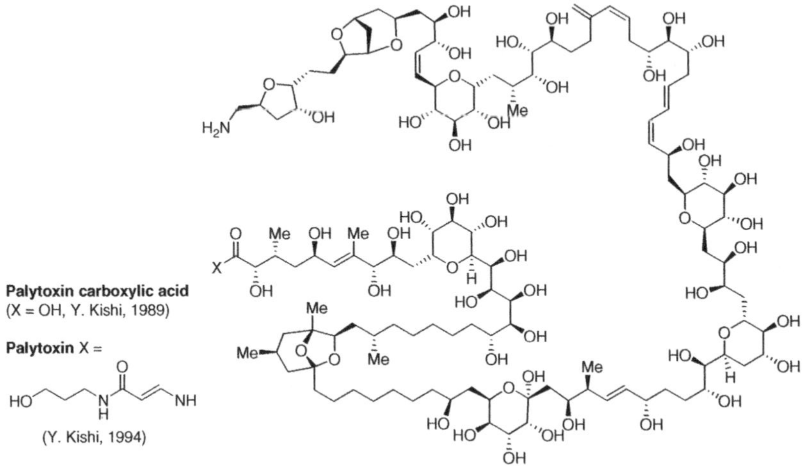

Figure 4 Structures of palytoxin carboxylic acid (PTC) and palytoxin (PTX): A "Mount Everest" of chemistry.

discipline: by given enough money, manpower and time, any structurally complex secondary metabolites can be synthesized.[30,31] Moreover, the award of the 1990 Nobel Prize in Chemistry to Corey seemed to suggest that a glorious era of the total synthesis of natural products ended. With the intention to change the minds of those who claimed that organic chemistry was a mature science and to shape the future of organic synthesis, Dieter Seebach published in 1990 the well-known review and perspective article entitled "Organic Synthesis: Where Now?"[31] On the basis of roughly 1000 literature references, he predicted the future development of organic synthesis in two directions: "The primary motivations that once induced chemists to undertake natural product syntheses no longer exist. Instead of target structures themselves, molecular function and activity now occupy center stage," and "The primary center of attention for all synthetic methods will continue to shift toward catalytic and enantioselective variants." Finally, he concluded that "I consider the most important message: that organic synthesis continues to react forcefully and with vitality to new challenges, still ready to purse old dreams...chemistry including organic synthesis – today more than ever before – is 'the central science.'"[31]

The impact of the achievements of total synthesis is far beyond the synthesis itself. Some discoveries made during total syntheses and their impact on science and society are displayed in Table 1.

As early as 1856, the 18-year-old W. H. Perkin tried to synthesize quinine from coal tar. The "naive," unsuccessful attempt led Perkin to discover serendipitously aniline purple (mauveine). Subsequently, he commercialized this very first synthetic dye, launching thus the synthetic dye industry.

From Wöhler's serendipitous discovery of the synthesis,[4] to Perkin's legendary synthesis of mauveine, to the establishment of the Woodward–Hoffmann rules, and many others from Table 1, these all reveal that the total synthesis of natural products is a fertile field for scientific discovery and invention.

2 1991–2000: A Contrasting Decade

The following decade 1991–2000 is full of contrast for the field of total synthesis led by K. C. Nicolaou.[38a] On one hand, academic research continues to thrive, many extremely challenging, or even impossible target molecules have been conquered (Figure 5).[38b] The prosperity stemmed from including the impressive development of marine natural products, the approval of the natural product Taxol® as an anticancer drug, and the subsequent supply crisis; to the discovery of several natural products as star molecules for their interesting biological profiles and unique structures such as calicheamicins[39], epothilones,[40] and vancomycin[41] (Figure 5); as well as from the advances in spectral analysis and separation techniques.

On the other hand, the traditional natural product-based strategy for drug R&D became less attractive because of the emergence of combinatorial chemistry and high-throughput screening technology.[42] The non-rational design of drug become the dominant strategy for most pharmaceutical companies. However, the perceived supply problem could be the biggest obstacle.[43] Providing materials of drug candidates for preclinical and clinical studies should have provided the opportunity for organic synthesis to demonstrate its power. However, in academic-style research, the synthetic routes for natural products are usually

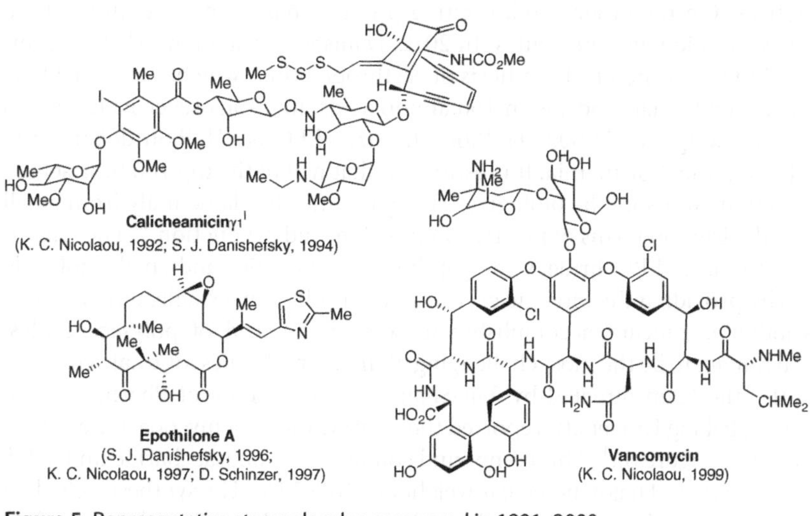

Figure 5 Representative star molecules conquered in 1991–2000.

lengthy and low yielding, making it difficult to scale them up. In addition, these syntheses frequently lack the flexibility to rapidly access analogs. It may be safe to conclude that the efficiency of the total synthesis of natural products cannot meet the needs of pharmaceutical industry.

A consequence is that the society is giving up the total synthesis of natural products as can be seen from the dramatic decline in funding for total synthesis in the USA. In fact, within this branch of science, efficiency has not been taken into consideration as a priority. Only the conquest of structurally complex molecules by the first total synthesis has been taken into consideration. This kind of race for the molecule "summit" is really harmful.[17b] In an era of globalization, increasing the efficiency of total synthesis is key to prevent the pursuit being given up by society.

3 Total Synthesis in the Twenty-First Century

Catalysis has become a dominant field in the twenty-first century, as can be seen by being honored three times by Nobel Prizes for Chemistry[44] (Knowles and Noyori for their metal-catalyzed asymmetric hydrogenation reactions, Sharpless for catalytic asymmetric oxidation reactions in 2001; Chauvin, Grubbs, and Schrock for the metathesis reaction in 2005; Heck, Negishi, and Suzuki for palladium-catalyzed cross-couplings in 2010). By taking advantage of the tremendous progress on synthetic methodologies, the total synthesis of many challenging natural products has been achieved. The enantioselective total syntheses of marine alkaloid sarain A (Figure 6) by L. E. Overman,[45] and of plant natural product azadirachtin by S. V. Ley[46] are two outstanding representatives. On the other hand, exciting progress has been made in the total synthesis of biomacromolecules. In 2012, Danishefsky and coworkers accomplished the total chemical synthesis of erythropoietin,[47] which is a crucial hormone in the human body, with 166 amino acids, four oligosaccharides, and a molecular weight of 34,000 Da. Biosynthesized EPO is a "billion-dollar" drug for the treatment of anemia. It is estimated that half of the top 100 best-selling medications will soon be biologics. Biologics differ fundamentally from small molecule drugs in terms of purity, composition, and production.[48] The chemical synthesis of EPO bridges the gap between biologics and small-molecule drugs by providing the biologic erythropoietin (EPO) in a single, pure form.

Another significant accomplishment was in the field of polysaccharides, which remain to be the most challenging biomacromolecules for total synthesis. In 2017, the team directed by Xin-Shan Ye at the School of Pharmaceutical Sciences, Peking University, reported the total synthesis of mycobacterial arabinogalactan (Figure 6).[49] This compound contains 92 monosaccharide units, 398 chiral centers, and has a molecular weight of 13185.5 Da. The synthesis relied on the preactivation-based one-pot glycosylation protocols developed previously,[50]

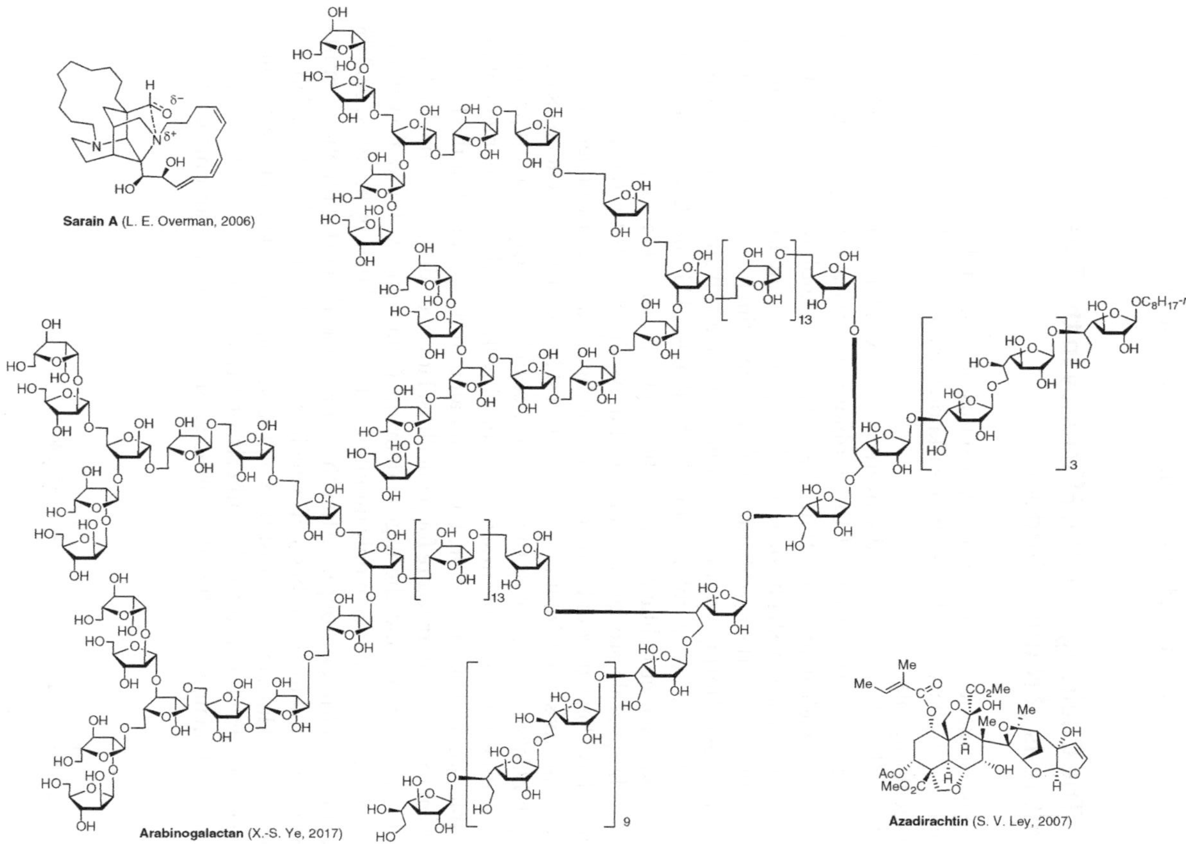

Figure 6 Structures of representative naturally occurring molecules conquered in the twenty-first century.

which allowed a highly convergent [31 + 31 + 30] coupling reaction to establish 92 chiral centers in a highly stereoselective manner. Arabinogalactan is an essential structural constituent of mycobacterial cell wall, which plays critical roles in the infectivity and pathogenicity of *Mycobacterium tuberculosis*.[51]

4 The Challenges of the Efficiency in the Total Synthesis of Natural Products

To have some idea about the current status of total synthesis in terms of efficiency, let us have a look at some well-known natural product-based drugs or medicinal agents.

Penicillin (Figure 7) was discovered by Alexander Fleming in 1928. The remarkable *in vivo* activity of the antibiotic against a variety of pathogenic organisms made it, for a long time, the sole efficient antibiotic against bacterial infections caused by staphylococci and streptococci, and for infections such as pneumonia, gonorrhea, or rheumatic fever. Penicillin heralded the dawn of the antibiotic age. During World War II, due to the need for substantial amounts of the antibiotic to treat soldiers on the battlefield, US and UK governments launched a secret military project (the Anglo–American penicillin project)[52] for the large-scale production of penicillin. As a part of that project, extensive efforts have been devoted to determine the chemical structure of penicillin and to chemically synthesize it in large quantities. More than 1000 chemists working at 39 laboratories across the Atlantic Ocean have engaged in this project.[52] The team was comparable in size to the one involved in the development of the atomic bomb. The joint effort has culminated in the proposal of the fused β-lactam thiazolidine structure for penicillins. However, despite the huge efforts, high expectations, and the urgent demand from society, the total synthesis afforded by penicillin G in only minute yields (0.03–0.08% by bioassay, 0.008% isolated by a Merck group).[53] After the World War II, most synthetic chemists abandoned the total synthesis of penicillin.[52c] In 1948, J. C. Sheehan's laboratory at the Massachusetts Institute of Technology embarked on a substantial program aimed at the total synthesis of the penicillins and their simpler structural analogs.[54a] After the heroic efforts of Sheehan and his coworker

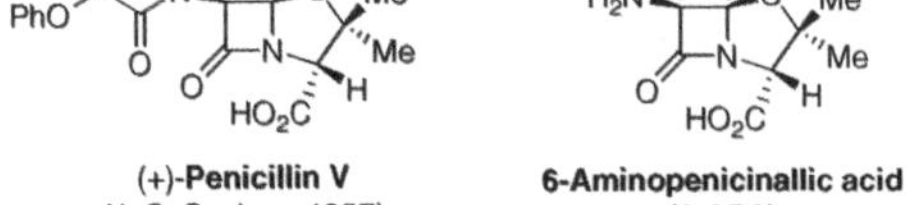

(+)-**Penicillin V**
(J. C. Seehan, 1957)

6-Aminopenicinallic acid
(6-APA)

Figure 7 Penicillin V and its semisynthetic precursor 6-APA.

K. R. Henery-Logan, the total synthesis of penicillin V was finally achieved in 1957,[54] 12 years after the end of the World War II! It is worth mentioning that penicillins are produced commercially by fermentation or by semi-synthesis from 6-aminopenicinallic acid (6-APA), which is in turn obtained from fermentation.

Penicillin is a molecule that has changed the world.[2] The discovery and development of penicillin was designated in November 19, 1999 by American Chemical Society and Royal Society of Chemistry as a historic chemical landmark.[52c]

Taxol® is the second example (Paclitaxel, Figure 8), which was first isolated in 1971 from the Pacific yew and approved by the Food and Drug Administration (FDA) for the treatment of ovarian and breast cancers in 1991 and 1993, respectively.[55] The approval of Taxol as an anticancer drug resulted in the supply crisis.[56] At that time, Taxol was extracted from the bark of the Pacific yew (*Taxus brevifolia*) in a process that inevitably killed the tree. Moreover, yews grow slowly and each plant yields little bark. A 100-year-old tree provides only a gram of the compound; about half the amount needed for a single treatment. The high demand for Taxol from cancer patients' families once caused an eco-environmental incident.[55b] More than 30 research groups have engaged in the total synthesis of Taxol. Although these efforts have resulted in seven total syntheses and three formal syntheses,[57] the synthetic routes were unable to provide sufficient quantities of the natural product for medical testing or therapeutic use.

To address the sourcing problem of Taxol, in the late 1970s, French chemist Pierre Potier at the Institut de Chimie des Substances Naturelles of the Centre National de la Recherche Scientifique (CNRS), in Gif-sur-Yvette, France, investigated an alternative approach. Potier discovered that it was feasible to obtain relatively large quantities of 10-deacetylbaccatin from needles of the European yew, *Taxus baccata*, planted on the ICSN campus at Gif-sur-Yvette. In collaboration with French pharmaceutical company Rhône-Poulenc Rorer (now Sanofi-Aventis), they discovered in 1986 Taxotère (Docetaxel),[58] which

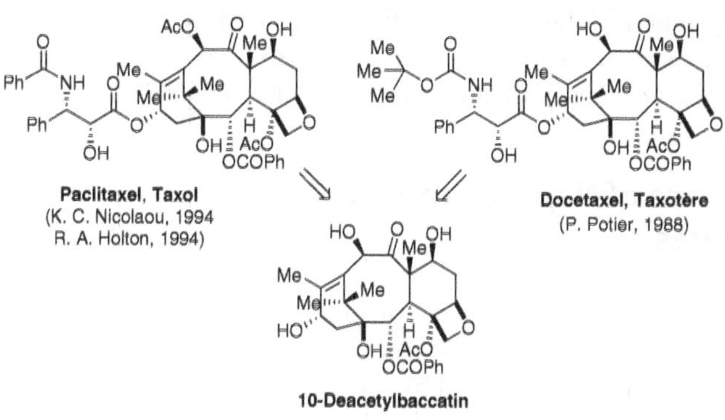

Figure 8 Two anticancer drugs and their common semisynthetic precursors.

was approved in 1995 for the treatment of a number of types of cancer. Another semisynthetic process, commercialized by Bristol Myers Squibb, involves the extraction of paclitaxel (Taxol®) intermediates baccatin III or 10-deacetylbaccatin III from the needles of *Taxus* spp.[59] In addition, plant cell fermentation technology was used by Phyton Biotech, Inc. for the production of paclitaxel.[59] Both Taxol and Taxotère are on the World Health Organization's List of Essential Medicines. Annual sales of Taxotère in 2010 were 2.122 billion euros ($US3.1 billion) and that of paclitaxel US$1.6 billion.

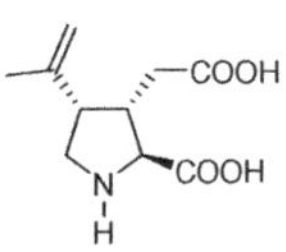

Figure 9 (−)-α-Kainic acid.

(−)-*α*-Kainic acid: Besides the structurally complex natural products, the structurally simpler natural products may also suffer from a supply problem. (−)-*α*-Kainic acid (Figure 9), which is a marine natural product extracted from a red seaweed called digenea, represents a typical example. Called the "drug from the sea," *α*-kainic acid has been used as an anti-worming agent and for fighting parasites in humans and animals for centuries. Moreover, this compound has become indispensable as a useful molecular probe for research in neuroscience.[60] In 2000, it was reported that a shortage of *α*-kainic acid had hampered neuroscience research.[61a,b] Although it was available again a year later for research, the compound was once again obtained from natural sources instead from organic synthesis.[61c] Ironically, by 2001, a total of 53 laboratory syntheses of α-kainic acid have been reported.[62]

Azadirachtin: Agrochemicals form another important field of application for natural products.[63] Agrochemicals refer to a broad range of pesticides, including insecticides, herbicides, fungicides and nematicides, plant hormones, and chemical growth agents, as well as insect pheromones. Many plant species produce molecules to protect themselves against feeding insects. Such compounds are natural insecticides and are named "feeding deterrents." Azadirachtin is found in the neem tree *Azadirachta indica*, and reported to be able to prevent the feeding of more than 200 insect species.[64] The pursuit of the total synthesis of this structurally complex natural product by the team led by Ley turned out to be a 22-year "long-march." The target was finally reached in 2007 with the endeavor of more than 35 coworkers. But the synthetic route contains 71 steps (longest linear sequence 48 steps) with an overall yield of 0.00015%![46c]

5 The Renaissance of Natural Products as Drug Candidates

Since 2005, "The renaissance of natural products as drug candidates" began to appear in the titles and/or abstracts of many perspective and review articles.[65] This change was caused by the failure of high-throughput screening-combinatorial chemistry-based strategies in meeting initial expectations. Indeed, since the shift of drug-discovery strategy from a natural product-based approach to that relied on synthetic compounds libraries (generated by combinatorial chemistry) and HTS, there was a decline in new small-molecule drugs approved by the FDA. According to Newman's recent review,[66] only one new chemical entity from synthetic libraries has been approved by the FDA in the last 30 years, while during the same period several drugs developed from natural products have been approved for clinical uses.[67]

On the other hand, people began to think why a limited number of natural products (those isolated and tested for bioactivities) won the campaign against a huge number of synthetic compounds generated at random from combinatorial chemistry and diversity-oriented synthesis. The answer resides in how they are formed.[68] Natural products that are secondary metabolites, are formed by living organisms via natural evolutionary selection – nature's own high-throughput screening process for the optimization of biologically active compounds.

The third reason for the renaissance in natural product-based drug discovery is due to the longstanding joint effort of outstanding scientists from both academic and pharmaceutical companies. The discovery and successful development of eribulin mesylate (Halaven®, Figure 10) as an anticancer drug for the treatment of patients with late-stage metastatic breast cancer[69] will be helpful for illustrating the main strategies that have been successfully employed for providing sufficient amounts of structurally complex molecules on a commercial scale.[70] Although Halaven® retains only about two-thirds of halichondrin B[71] that contains 32 stereogenic centers and a 54-carbon backbone, this molecule features a complex ring system and 19 stereocenters. The production of such a structurally complex molecule on a commercial scale through total synthesis presented formidable challenges for the pharmaceutical industry. Attracted by Halaven's potency and unique biological profile, Eisai managers and scientists have made a heroic decision to embrace the challenges. Finally, they have succeeded in securing a stable supply of eribulin mesylate at a

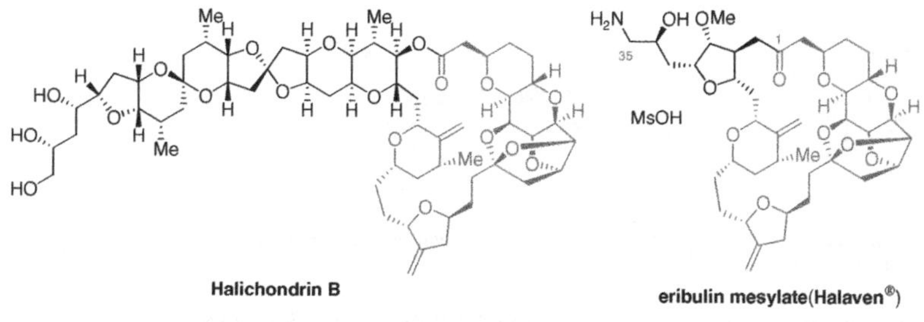

Figure 10 Structures of halichondrin B and Halaven®.

consistent quality on a commercial scale through a 62-step total synthesis (see Chapter 10 for a discussion on the synthesis).[69,70] This represents the most structurally complex drug on the market today that is produced by total synthesis, and reflects the state-of-the-art of the pharmaceutical industry in the implementation of a lengthy total synthesis in commercial manufacturing.[70a] This achievement constitutes a milestone and will encourage the pharmaceutical industry to consider the total synthesis as a reliable avenue for sourcing structurally complex natural products or natural product derivatives.

6 Recent Recognition of the Contribution of Natural Product-Based Drugs to Society

On October 5, 2015, the Nobel Prize in Physiology or Medicine 2015 was awarded with one-half jointly to William C. Campbell and Satoshi Ōmura "for their discoveries concerning a novel therapy against infections caused by roundworm parasites" and the other half to Youyou Tu "for her discoveries concerning a novel therapy against malaria." This award attracted widespread attention from the scientific community and has been universally welcomed by the international health community.[72]

Scientifically, Tu discovered artemisinin (qinghaosu, Figure 11); a drug that has significantly reduced the mortality rates for patients suffering from malaria.[73] Campbell and Ōmura discovered avermectins (Figure 11),[74] the derivatives of which (ivermectin)[75] have radically lowered the incidence of river blindness and lymphatic filariasis, and showed efficacy against an expanding number of other parasitic diseases.

This award can be distinguished from previous ones in three ways. First, the two drugs target two neglected tropical diseases (NTDs: major global health problems affecting hundreds of millions of poor people in poor regions of the world that are not profitable for the pharmaceutical industry), malaria, and river blindness. According to the report published by World Health

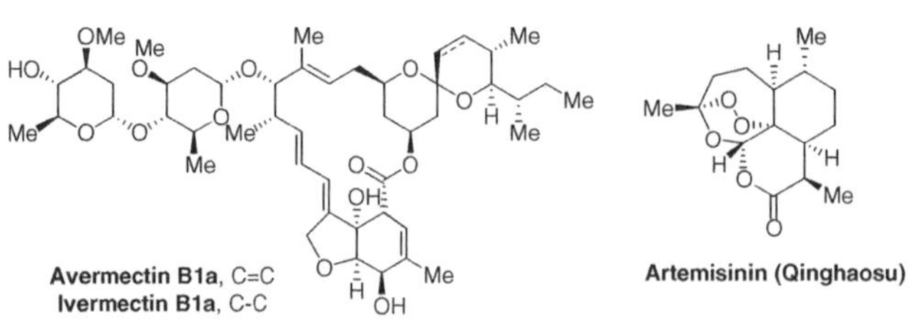

Figure 11 The natural products recognized by the Nobel Prize in Physiology or Medicine 2015.

Organization (WHO), the 128 million cases of malaria caused 584,000 deaths in 2013. The significant of these two drugs in the human history was evidenced by the Nobel committee's conclusion "The discoveries have revolutionized therapy for patients suffering from devastating parasitic diseases. The global impact of their discoveries and the resulting benefit to mankind are immeasurable."[72a–d] According to Bill Gates, "The discovery of artemisinin is one of the most significant breakthroughs in tropical medicine in the 20th century and has saved millions of lives which would otherwise have been lost to malaria."[72e]

Second, the two drugs are natural products, avermectin and artemisinin (**3**) (Chinese: qinghaosu). Antiparasitic avermectin was discovery by Japanese microbiologist Ōmura and colleagues in 1979 from *Streptomyces* strains isolated from soil samples in Japan.[74] From the cultures of Ōmura, Campbell, and an American expert in parasite biology discovered that the modified ivermectin was highly active against roundworms such as those that produce river blindness (onchocerciasis) and lymphatic filariasis (elephantiasis).[75] In addition, avermectins and their derivatives doramectin, selamectin, and abamectin, semisynthetic AVM (C22, C23 dihydroAVM B1) have also become commercial veterinary drugs for treating several types of diseases caused by nematodes and arthropod parasites. These compounds have also found application in the treatment of pathogen-caused diseases on farms.[74,75]

The discovery of artemisinin (qinghaosu) involved a legendary story.[73] During the Vietnam War, tropical epidemics, in particular falciparum malaria, caused heavy casualties for both sides; the Army of the Republic of Vietnam and the US military. During 1967–1970, hundreds of thousands of US soldiers were affected by malaria. To support allied troops in North Vietnam, and at the request of Chi Minh Ho, in May 23, 1967 the Chinese government launched a secret military-oriented state research project known as Project 523 to find an effective anti-malaria drug. The project lasted for 13 years. More than 60 research institutions, more than 500 full-time researchers, and a total of more than 2000 part-time researchers and supporting staff were engaged. Among the researchers, the team from the Academy of Traditional Chinese Medicine headed by Youyou Tu tested more than 640 medicinal plants and remedies in Traditional Chinese Medicine (TCM) against various chills and fevers. Inspired by a note in the *Handbook of Prescriptions for Emergency Treatments* by Ge Hong over 2000 years ago, and after extensive experimentations, Tu discovered that an ethereal extraction of *Artemisia annua L.* (Sweet wormwood) leaves (qinghao) was highly potent with 100% inhibition for *Plasmodium*. The active ingredient isolated as crystals in 1973 was named qinghaosu. The challenging structural determination of qinghaosu was headed by the late synthetic organic chemist Wei-Shan Zhou. Synthetic organic chemist Yu-Lin Wu and medicinal chemist Ying Li contributed to the elucidation of the key endoperoxide bridge substructure.[73d] The revelation of the peroxide bridge and sesquiterpene lactone structure caused a change in the perception that only nitrogen-containing molecules are active against malaria. Although the structure determination of qinghaosu was accomplished in 1976, its structure remained hidden in the literature until 1977.[73b]

Third, originated from TCM, qinghaosu presents the first TCM-based scientific discovery to be awarded the Nobel Prize. This award will open a door for drug discovery from TCM, a major medicinal system that has been used over 4000 years by the largest population in the word.

In short, this Nobel Prize award reveals once again the high potential of natural products in drug discovery. The syntheses of avermectin/ivermectin,[76,77] and artemisinin[78] have attracted considerable attention, which culminated in several total syntheses including gram-scale synthesis of artemisinin.[78a] However, considering that global demand for artemisinin is 180 tons per year, one has to admit that the total synthesis of natural products is just at its infancy or perhaps adolescence! According to Baran's view, "this gap in efficiency is therefore an exciting opportunity for invention and exploration in chemical synthesis."[57] The supply of avermectin/ivermectin is provided by biotechnology-based fermentation,[74] while for plant-originated artemisinin, most of it is still extracted from the sweet wormwood plant, and one-third by semisynthetic production (see Chapter 10). The endeavor for alternative production methods of artemisinin by state-of-the art technologies will be presented in Chapter 12 of this book.

Acknowledgements

Financial support from the National Key R&D Program of China (grant No. 2017YFA0207302), the National Natural Science Foundation of China (21332007, 21472153, and 21672176) and the Program for Changjiang Scholars and Innovative Research Team in University of Ministry of Education, China gratefully acknowledged.

We thank Ms. Yan-Jiao Gao for her assistance in the preparation of this manuscript, and thank graduate students Xiong-Zhi Huang, Hui Geng, and Hang Chen for their help in drawing schemes/figures.

References

1 (a) For Wikipedia's definition of a natural product, see: https://en.wikipedia.org/wiki/Natural_product; (b) for a discussion on the expansion of the concept of "natural," see: Editorial, *Nat. Chem. Biol.* **2007**, *3*, 351. Naturally occurring organic compounds are produced by organisms, which can be divided into two major classes, the primary and secondary metabolites. Primary metabolites

have an intrinsic function that is essential to the survival of the organism that produces them. Using as hormones/pheromones, regulators, chemical messages, or weapons for both attack and defense (repellants, venoms, or toxins), secondary metabolites (natural products by narrow definition) serve to increase the competitiveness of the organism within its environment. Because of their ability to modulate biochemical pathways, many secondary metabolites exhibit useful medicinal profiles.

2 Nicolaou, K. C.; Montagnon, T. *Molecules That Changed the World: A Brief History of the Art and Science of Synthesis and Its Impact on Society*; New York: John Wiley & Sons, Inc., **2008**.

3 (a) Fleming, I. *Selected Organic Syntheses*: New York: John Wiley & Sons, Inc., **1973**; (b) Anand, N.; Bindra, J. S.; Ranganathan, S. *Art in Organic Synthesis*, 2nd Edn; New York: John Wiley & Sons, Inc., **1988**; (c) Corey, E. J.; Cheng, X.-M. *The Logic of Chemical Synthesis*; New York: John Wiley & Sons, Inc., **1989**; (d) Nicolaou, K. C.; Sorensen, E. J. *Classics in Total Synthesis: Targets, Strategies, Methods*; Weinheim: Wiley-VCH, **1996**; (e) Nicolaou, K. C.; Snyder, S. A. *Classics in Total Synthesis II: More Targets, Strategies, Methods*; Weinheim: Wiley-VCH, **2003**; (f) Nicolaou, K. C.; Chen, J. S. *Classics in Total Synthesis III: Further Targets, Strategies, Methods*; Weinheim: Wiley-VCH, **2011**; For an excellent review on the art and science of total synthesis at the dawn of the twenty-first century, see: (g) Nicolaou, K. C.; Vourloumis, D.; Winssinge, N.; Baran, P. S. *Angew. Chem. Int. Ed.* **2000**, *39*, 44.

4 Wöhler, F. *Ann. Phys. Chem.* **1828**, *12*, 253.

5 Kolbe, H. *Annalen* **1845**, *54*, 145.

6 (a) Du, Y. C.; Jiang, R. Q.; Tsou, C.-L. *Sci. Sin. (Engl. Ed.)* **1965**, *14*, 229; (b) Kung, Y. T.; Du, Y. C.; Huang, W. T.; Chen, C. C.; Ke, L. T.; Hu, S. C.; *et al. Sci. Sin. (Engl. Ed.)* **1965**, *14*, 1710; (c) collective work (Institute of Biochemistry, Chinese Academy of Sciences; Department of Chemistry, Beijing University; Institute of Organic Chemistry, Chinese Academy of Sciences), *Kexue Tongbao (Chin. Sci. Bull.)* **1966**, *17*, 241; (d) Kung, Y.-T.; Du, Y.-C.; Huang, W.-T.; *et al. Kexue Tongbao (Sci. Bull.)* **1966**, *15*, 544; for a historic review, see: (e) Tsou, C. L. *Trends Biochem. Sci.* **1995**, *20*, 289; for a recent commemorative article, see: (f) Ye, Y.-H. *Chin. Bull. Life Sci.* **2015**, *27*, 648; for a recent review, see: (g) Moroder, L. Musiol, H.-J. *Angew. Chem. Int. Ed.* **2017**, *56*, 10656.

7 (a) Wiechert, R. *Angew. Chem. Int. Ed. Engl.* **1970**, *9*, 321; (b) Hanson, J. R. *Nat. Prod. Rep.* **2007**, *24*, 1342; (c) Nising, C. F., Bräse S. *Angew. Chem. Int. Ed.* **2008**, *47*, 9389.

8 Hirschemann, R. *Angew. Chem. Int. Ed. Engl.* **1991**, *30*, 1278.

9 Wiechert, R. *Angew. Chem. Int. Ed. Engl.* **1977**, *16*, 506.

10 (a) Barton, D. H. R.; Beaton, J. M.; Geller, L. E.; Pechet, M. M. *J. Am. Chem. Soc.* **1960**, *82*, 2640; (b) Barton, D. H. R.; Beaton, J. M.; Geller, L. E.; Pechet, M. M. *J. Am. Chem. Soc.* **1961**, *83*, 4076; (c) Barton, D. H. R. *Pure Appl. Chem.* **1968**, *16*, 1–15; (d) Barton, D. H. R. *Aldrichim. Acta* **1990**, *23*, 3.

11 (a) Hajos, Z. G.; Parrish, D. R. *Asymmetrische Synthese Polycyclischer Organischer Verbindungen*, **1971**, 2102623; (b) Eder, U.; Wiechert, R.; Sauer, G. *Verfahren Zur Herstellung Optisch Aktiver Bicycloalkan-Derivate*, **1971**, 2014757; (c) Eder, U.; Sauer, G.; Wiechert, R. *Angew. Chem. Int. Ed.* **1971**, *10*, 496; (d) Hajos, Z. G.; Parrish, D. R. *J. Org. Chem.* **1974**, *39*, 1612; (e) Hajos, Z. G.; Parrish, D. R. *J. Org. Chem.* **1974**, *39*, 1615. For a recent review, see: (f) Gallier, F.; Martel, A.; Dujardin, G. *Angew. Chem. Int. Ed.* **2017**, *56*, 12424.

12 (a) Woodward, R. B. *Pure Appl. Chem.* **1973**, *33*, 145; (b) Woodward, R. B. in *Vitamin B_{12}*, eds. Zagalak, B.; W. Friedrich, New York: de Gruyter:, **1979**, p. 37. (c) Eschenmoser, A.; Winter, C. E. *Science* **1977**, *196*, 1410.

13 Battersby, A. R. *Nat. Prod. Rep.* **2000**, *17*, 507.

14 Todd, A. R.; Cornforth, J. W. *Biogr. Mem. Fellows R. Soc.* **1981**, *27*, 629.

15 (a) Hoffmann, R.; Woodward, R. B.; *Acc. Chem. Res.* **1968**, *1*, 17; (b) Woodward, R. B.; Hoffmann, R. *Angew. Chem. Int. Ed. Engl.* **1969**, *8*, 781; (c) Seeman, J. I. *J. Org. Chem.* **2015**, *80*, 11632.

16 Seeman, J. I. *Nat. Chem.* **2017**, *9*, 925.

17 (a) Breslow, R. *Chemistry: Today and Tomorrow – The Central, Useful, and Creative Science*, Washington, D.C.: ACS, **1996**; (b) Service, R. F. *Science* **1999**, *285*, 184.

18 Marks, F.; Furstenberger, G. *Prostaglandins, Leukotrienes and Other Eicosanoids: From Biogenesis to Clinical Application*. Weinheim: Wiley-VCH, **1999**.

19 (a) Noyori, R.; Suzuki, M. *Angew. Chem., Int. Ed. Engl.* **1984**, *23*, 847; (b) Collins, P. W.; Djuric, S. W. *Chem. Rev.* **1993**, *93*, 1533; (c) Das, S.; Chandrasekhar, S.; Yadav, J. S.; Grée, R. *Chem. Rev.* **2007**, *107*, 3286; (d) Peng, H.-H.; Chen, F.-E. *Org. Biomol. Chem.* **2017**, *15*, 6281.

20 McGrath, N. A.; Brichacek, M.; Njardarson, J. T. *J. Chem. Edu.* **2010**, *87*, 1348. (http://njardarson.lab.arizona.edu/content/top-pharmaceuticals-poster)

21 (a) Corey, E. J.; Trybulski, E. J.; Melvin, L. S., Jr.; Nicolaou, K. C.; Secrist, J. A.; Lett, R.; *et al. J. Am. Chem. Soc.* **1978**, *100*, 4618. (b) Corey, E. J.; Kim, S.; Yoo, S.; Nicolaou, K. C.; Melvin Jr., L. S.; Brunelle, D. J.; *et al. J. Am. Chem. Soc.* **1978**, *100*, 4620. For the synthesis of erythronolide A, see: (c) Corey, E. J.; Hopkins, P. B.; Kim, S.; Yoo, S.; Nambiar, K. P.; Falck, J. R. *J. Am. Chem. Soc.* **1979**, *101*, 7131.

22 Fukuyama, T.; Akasaka, K.; Karanewsky, D. S.; Wang, C.-L. J.; Schmid, G.; Kishi, Y. *J. Am. Chem. Soc.* **1979**, *101*, 262.

23 (a) Collum, D. B.; McDonald III, J. H.; Still, W. C. *J. Am. Chem. Soc.* **1980**, *102*, 2117; (b) Collum, D. B.; McDonald III, J. H.; Still, W. C. *J. Am. Chem. Soc.* **1980**, *102*, 2118; (c) Collum, D. B.; McDonald III, J. H.; Still, W. C. *J. Am. Chem. Soc.* **1980**, *102*, 2120.

24 (a) Woodward, R. B.; Logusch, E.; Nambiar, K. P.; Sakan, K.; Ward, D. E.; Au-Yeung, B.-W.; Balaram, P.; Browne, L. J.; Card, P. J.; Chen, C. H.;

Chênevert, R. B.; Fliri, A.; Frobel, K.; Gais, H.-J.; Garratt, D. G.; Hayakawa, K.; Heggie, W.; Hesson, D. P.; Hoppe, D.; Hoppe, I.; Hyatt, J. A.; Ikeda, D.; Jacobi, P. A.; Kim, K. S.; Kobuke, Y.; Kojima, K.; Krowicki, K.; Lee, V. J.; Leutert, T.; Malchenko, S.; Martens, J.; Matthews, R. S.; Ong, B. S.; Press, J. B.; Rajan Babu, T. V.; Rousseau, G.; Sauter, H. M.; Suzuki, M.; Tatsuta, K.; Tolbert, L. M.; Truesdale, E. A.; Uchida, I.; Ueda, Y.; Uyehara, T.; Vasella, A. T.; Vladuchick, W. C.; Wade, P. A.; Williams, R. M.; Wong, H. N.-C. *J. Am. Chem. Soc.* **1981**, *103*, 3210; (b) ibid, *J. Am. Chem. Soc.* **1981**, *103*, 3213; (c) ibid, *J. Am. Chem. Soc.* **1981**, *103*, 3215; See also: (d) Martin, S. F.; Hida, T.; Kym, P. R.; Loft, M.; Hodgson, A. *J. Am. Chem. Soc.* **1997**, *119*, 3193 (asymmetric synthesis of erythromycin B); (e) Kim, H. C.; Kang, S. H. *Angew. Chem. Int. Ed.* **2009**, *48*, 1827 (azithromycin); (f) Andrade, R. B. *Synlett* **2015**, *26*, 2199 (desmethyl macrolide antibiotics).

25 (a) Corey, E. J. *Chem. Soc. Rev.* **1988**, *17*, 111; (b) Corey, E. J.; Gavai, A. V. *Tetrahedron Lett.* **1988**, *29*, 3201.

26 (a) Jones, T. K.; Mills, S. G.; Reamer, R. A.; Askin, D.; Desmond, R.; Volante, R. P.; *et al. J. Am. Chem. Soc.* **1989**, *111*, 1157; For the subsequent total syntheses, see: (b) Nakatsuka, M.; Ragan, J. A.; Sammakia, T.; Smith, D. B.; Uehling, D. E.; Schreiber, S. L. *J. Am. Chem. Soc.* **1990**, *112*, 5583; (c) Jones, A. B.; Villalobos, A.; Linde, R. G.; Danishefsky, S. J. *J. Org. Chem.* **1990**, *55*, 2786; (d) Sih, R. L.; Sih, C. J. *Tetrahedron Lett.* **1990**, *31*, 3287; (e) Smith III, A. B.; Chen, K.; Robinson, D. J.; Laakso, L. M.; Hale, K. J. *Tetrahedron Lett.* **1994**, *35*, 4271; (f) Ireland, R. E.; Gleason, J. L.; Gegnas, L. D.; Highsmith, T. K. *J. Org. Chem.* **1996**, *61*, 6856.

27 Evans, D. A.; Kaldor, S. W.; Jones, T. K.; Clardy, J.; Stout, T. J. *J. Am. Chem. Soc.* **1990**, *112*, 7001.

28 (a) Armstrong, R. W.; Beau, J.-M.; Cheon, S. H.; Christ, W. J.; Fujioka, H.; Ham, W.-H.; *et al. J. Am. Chem. Soc.* **1989**, *111*, 7525; (b) Armstrong, R. W.; Beau, J.-M.; Cheon, S. H.; Christ, W. J.; Fujioka, H.; Ham, W.-H.; *et al. J. Am. Chem. Soc.* **1989**, *111*, 7530; (c) Kishi, Y. *Pure Appl. Chem.* **1989**, *61*, 313.

29 Suh, E. M.; Kishi, Y. *J. Am. Chem. Soc.* **1994**, *116*, 11205.

30 Kishi, Y. *Pure Appl. Chem.* **1998**, *70*, 339.

31 Seebach, D. *Angew. Chem., Int. Ed. Engl.* **1990**, *29*, 1320.

32 (a) Barton, D. H. R. *Experientia* **1950**, *6*, 316; see also: (b) Barton D. H. R. (ed.), *Reason and Imagination*, New Jersey: World Scientific, **1996**.

33 (a) Stork, G.; Burgstahler, A. W. J. *J. Am. Chem. Soc.* **1955**, *77*, 5068; (b) Eschenmoser, A.; Ruzicka, L.; Jeger, O.; Arigoni, D. *Helv. Chim. Acta* **1955**, *38*, 1890; (c) Stadler, P. A.; Eschenmoser, A.; Schinz, H.; Stork, G. *Helv. Chim. Acta* **1957**, *40*, 2191; at the occasion of the Eschenmoser's 1955 *Helvetica Chimica Acta* paper's hemi-centennial, the English version (translated by Lucy Stark) entitled: "Revisited after 50 Years: The 'Stereochemical Interpretation of the Biogenetic Isoprene Rule for the Triterpenes'" appeared: (d) Eschenmoser, A.; Arigoni, D. *Helv. Chim. Acta* **2005**, *88*, 3011.

34 Rapson, W. S.; Robinson, R. *J. Chem. Soc.* **1935**, 1285.

35 (a) Roth, M.; Dubs, P.; Gotschi, E.; Eschenmoser, A. *Helv. Chim. Acta* **1971**, *54*, 710; (b) Gotschi, E.; Hunkeler, W.; Wild, H.-J.; Schneider, P.; Fuhrer, W.; Gleason, B. W.; Eschenmoser, A. *Angew. Chem. Int. Ed. Engl.* **1973**, *12*, 910.

36 Merrifield, R. B. *J. Am. Chem. Soc.* **1963**, *85*, 2149.

37 Corey, E. J. *Pure Appl. Chem.* **1967**, *14*, 19.

38 (a) Hall, N. *Chem. Commun.* **2003**, 661; (b) Comprehensive lists of conquered challenging natural products can be seen from K. C. Nicolaou's kooks and reviews: refs. 3(d–g).

39 (a) Nicolaou, K. C.; Hummel, C. W.; Pitsinos, E. N.; Nakada, M.; Smith, A. L.; *et al. J. Am. Chem. Soc.* **1992**, *114*, 10082; (b) Nicolaou, K. C. *Angew. Chem., Int. Ed. Engl.* **1993**, *32*, 1377; (c) Hitchcock, S. A.; Boyer, S. H.; Chu-Moyer, M. Y.; Olson, S. H.; Danishefsky, S. J. *Angew. Chem., Int. Ed. Engl.* **1994**, *33*, 858.; (d) Danishefsky, S. J.; Shair, M. D. *J. Org. Chem.* **1996**, *61*, 16.

40 (a) Balog, A.; Meng, D.; Kamenecka, T.; Bertinato, P.; Su, D.-S.; Sorensen, E. J.; Danishefsky, S. J. *Angew. Chem., Int. Ed. Engl.* **1996**, *35*, 2801; (b) Yang, Z.; He, Y.; Vourloumis, D.; Vallberg, H.; Nicolaou, K. C. *Angew. Chem., Int. Ed. Engl.* **1997**, *36*, 166; (c) Schinzer, D.; Limberg, A.; Bauer, A.; Böhm, O. M.; Cordes, M. *Angew. Chem., Int. Ed. Engl.* **1997**, *36*, 523.

41 Nicolaou, K. C.; Mitchell, H. J.; Jain, N. F.; Winssinger, N.; Hughes, R.; Bando, T. *Angew. Chem. Int. Ed.* **1999**, *38*, 240.

42 Li, J. W. H.; Vederas, J. C. *Science* **2009**, *325*, 161.

43 Newman, D. J. *Pharmac. Therap.* **2016**, *162*, 1.

44 Nobel Prize Home Page, Available online at: nobelprize.org/nobel_prizes/chemistry/laureates (accessed February **2018**).

45 (a) Garg, N. K.; Hiebert, S.; Overman, L. E. *Angew. Chem. Int. Ed.* **2006**, *45*, 2912; (b) Becker, M. H.; Chua, P.; Downham, R.; Douglas, C. J.; Garg, N. K.; Hiebert, S.; *et al. J. Am. Chem. Soc.* **2007**, *129*, 11987, correction: *J. Am. Chem. Soc.* **2018**, *140*, 5319.

46 (a) Veitch, G. E.; Boyer, A.; Ley, S. V. *Angew. Chem. Int. Ed.* **2008**, *47*, 9402; (b) Veitch, G. E.; Beckmann, E.; Burke, B. J.; Boyer, A.; Maslen. S. L.; Ley, S. V. *Angew. Chem. Int. Ed.* **2007**, *46*, 7629; for a highlight, see: (c) Jauch, J. *Angew. Chem. Int. Ed.* **2008**, *47*, 34.

47 (a) Wang, P.; Dong, S. W.; Brailsford, J. A.; Iyer, K.; Townsend, S. D.; Zhang, Q.; *et al. Angew. Chem. Int. Ed.* **2012**, *51*, 11587; (b) Wang, P.; Dong, S. W.; Shieh, J.-H.; Peguero, E.; Hendrickson, R.; Moore, M. A. S.; Danishefsky, S. J. *Science* **2013**, *342*, 1357.

48 Hsieh-Wilson, L. C.; Griffin, M. E. *Science* **2013**, *342*, 1332.

49 Wu, Y.; Xiong, D. C.; Chen, S. C.; Wang, Y. S.; Ye, X. S. *Nat. Commun.* **2017**, *8*, 14851.

50 (a) Huang, X.; Huang, L.; Wang, H.; Ye, X.-S. *Angew. Chem. Int. Ed.* **2004**, *43*, 5221; (b) Wang, C.; Wang, H.; Huang, X.; Zhang, L.-H.; Ye, X.-S. *Synlett* **2006**, 2846.

51 Brennan, P. J. *Tuberculosis* **2003**, *83*, 91.

52 (a) Clarke, H. T.; Johnson, J. R.; Robinson, R. *The Chemistry of Penicillin*, Princeton, NJ: Princeton University Press, **1949**; (b) Sheehan, J. C. *The Enchanted Ring: The Untold Story of Penicillin*, Cambridge, MA: MIT Press, **1982**; (c) International Historic Chemical Landmark: Discovery and Development of Penicillin, Commemorative Booklet. Available online at: www.acs.org/content/acs/en/education/whatischemistry/landmarks/flemingpenicillin.html (accessed February2018).

53 Folkers, K. In *Perspectives in Organic Chemistry*, Sir A. Todd (ed.), New York: John Wiley & Sons, Inc., **1956**, p. 409.

54 (a) Sheehan, J. C.; Henery-Logan, K. R. *J. Am. Chem. Soc.* **1957**, *79*, 1262; (b) Sheehan, J. C.; Henery-Logan, K. R. *J. Am. Chem. Soc.* **1959**, *81*, 3089.

55 (a) Potier, P. *Chem. Soc. Rev.* **1992**, *21*, 113; (b) Nicolaou, K. C.; Dai, W.-M.; Guy, R. K. *Angew. Chem. Int. Ed. Engl.* **1994**, *33*, 15; (c) Nicolaou, K. C.; Guy, R. K.; Potier, P. *Sci. Am.* **1996**, *272*, 84.

56 Cragg, G. M.; Schepartz, S. A.; Suffness, M.; Grever, M. R. *J. Nat. Prod.* **1993**, *56*, 1657.

57 Yuan, C.-X; Jin, Y.-H.; Wilde, N. C.; Baran, P. S. *Angew. Chem. Int. Ed.* **2016**, *55*, 8280.

58 (a) Chauvière, G.; Guénard, D.; Picot. F.; Sénilh, V.; Potier, P. C. R. *Seances Acad. Sci., Ser. 2*, **1981**, *293*, 501; (b) Colin, M.; Guénard. D.: Guéritte-Voegelein, F.; Potier, P. *French Patent* 2601 676, **1986**; (c) for the first synthesis of Taxol from 10-deaceytlbaccatin III, see: Denis, J.-N.; Greene, A. E.; Guénard, D.; Guéritte-Voegelein, F.; Mangatal, L.; Potier, P. *J. Am. Chem. Soc.* **1988**, *110*, 5917; (d) Guénard, D.; Guéritte-Voegelein, F.; Potier, P. *Acc. Chem. Res.* **1993**, *26*, 160; See also: (e) Borman, S. *Chem. Eng. News Archive* **1991**, *69* (35), 11; (f) Denis, J. N.; Green, A. E.; Guénard, D.; Guéritte-Voegelein, F.; Potier, P. *Chem. Eng. News Archive* **2003**, *81* (49), 4.

59 (a) Liu, W. C.; Gong, T.; Zhu, P. *RSC Adv.* **2016**, *6*, 48800; (b) Howat, S.; Il Suk Oh, B. P.; Jin, Y.-W. Lee, E.-K.; Loake, G. *J. New Biotech.* **2014**, *31*, 242.

60 Hollmann, M.; Heinemann, S. *Annu. Rev. Neurosci.* **1994**, *17*, 31.

61 (a) Tremblay, J.-F. *Chem. Eng. News Archive* **2000**, *78* (1), 14; (b) Tremblay, J.-F. *Chem. Eng. News Archive* **2000**, *78* (10), 31; (c) Tremblay, J.-F. *Chem. Eng. News Archive* **2001**, *79* (5), 19.

62 For a recent review on kainoid syntheses, see: (a) Stathakis, C. I.; Yioti E. G.; Gallos, J. K. *Eur. J. Org. Chem.* **2012**, 4661; For a recent example, see: (b) Suzuki, J.; Miyano, N.; Yashiro, S.; Umezawa, T.; Matsudam, F. *Org. Biomol. Chem.* **2017**, *15*, 6557.

63 (a) Cantrell, C. L.; Dayan, F. E.; Duke, S. O. *J. Nat. Prod.* **2012**, *75*, 1231; (b) Nising, C. F.; Hillebrand, S.; Rodefeld, L. *Chem. Commun.* **2011**, *47*, 4062; (c) Dayan, F. E.; Cantrell, C. L.; Duke, S. O. *Bioorg. Med. Chem.* **2009**, *17*, 4022; (d) Peng, J.-N.; Shen, X.-Y.; El Sayed, K. A.; Dunbar, D. C.; Perry, T. L.; Wilkins, S. P.; *et al. J. Agric. Food Chem.* **2003**, *51*, 2246.

64 Morgan, E. D. *Bioorg. Med. Chem.* **2009**, *17*, 4096.

65 (a) Paterson, I.; Anderson, E. A. *Science* **2005**, *310*, 451; (b) Molinski, T. F. *J. Org. Chem.* **2014**, *79*, 6765; (c) Pawar, S. V.; Ho, J. C. H.; Yadav, G. D.; *et al. Curr. Top. Med. Chem.* **2017**, *17*, 251

66 (a) Newman, D. J.; Cragg, G. M. *J. Nat. Prod.* **2016**, *79*, 629; See also: (b) Newman, D. J.; Cragg, G. M. *J. Nat. Prod.* **2012**, *75*, 311; (c) Newman, D. J.; Cragg, G. M. *J. Nat. Prod.* **2007**, *70*, 461.

67 Ojima, I. *J. Med. Chem.* **2008**, 2587.

68 Wender, P. A.; Miller, B. L. *Nature* **2009**, *460*, 197.

69 (a) Yeung, B. K. S. *Curr. Opin. Chem. Biol.* **2011**, *15*, 523; (b) Yu, M. J.; Kishi, Y.; Littlefield, B. A. Discovery of E7389, A fully synthetic macrocyclic ketone analog of halichondrin B, in *Anticancer Agents from Natural Products*, 2nd Edn, Cragg, G. M.; Kingston, D. G. I.; Newman, D. J. (eds), Boca Raton, FL: CRC Press, **2012**, pp. 317–345; for discovery of eribulin, see (c) Zheng, W.; Seletsky, B. M.; Palme, M. H.; Lydon, P. J.; Singer, L. A.; Chase, C. E.; *et al. Bioorg. Med. Chem. Lett.* **2004**, *14*, 5551.

70 For a review, see: (a) Yu, M. J.; Zheng, W. J.; Seletsky B. M. *Nat. Prod. Rep.* **2013**, *30*, 1158; For process development, see: (b) Chase, C. E.; Fang, F. G.; Lewis, B. M.; Wilkie, G. D.; Schnaderbeck, M. J.; Zhu, X. J. *Synlett* **2013**, *24*, 323; (c) Austad, B. C.; Benayoud, F.; Calkins, T. L.; Campagna, S.; Chase, C. E.; Choi, H.-W.; *et al. Synlett* **2013**, *24*, 327; (d) Austad, B. C.; Calkins, T. L.; Chase, C. E.; Fang, F. G.; Horstmann, T. E.; Hu, Y. B.; *et al. Synlett* **2013**, *24*, 333; see also: (e) Fukuyama, T.; Chiba, H.; Kuroda, H.; Takigawa, T.; Kayano, A.; Tagami, K. *Org. Process Res. Dev.* **2016**, *20*, 503. For a recent convergent synthesis, see: (f) Lee, J. H.; Li, Z. J.; Osawa, A.; Kishi, Y. *J. Am. Chem. Soc.* **2016**, *138*, 16248.

71 For a recent reviews on the total synthase of halichondrins, see: (a) Jackson, K. L.; Henderson, J. A.; Phillips, A. *J. Chem. Rev.* **2009**, *109*, 3044; for a discussion on the synthetic aspects, see Chapter 10. For recent total syntheses of halichondrins, see: (b) Kim, D.-S.; Dong, C.-G.; Kim, J.-T.; Guo, H.; Huang, J.; Tiseni, P. S.; Kishi, Y. *J. Am. Chem. Soc.* **2009**, *131*, 15636; (c) Yamamoto, A.; Ueda, A.; Brémond, P.; Tiseni, P. S.; Kishi, Y. *J. Am. Chem. Soc.* **2012**, *134*, 893 (Halichondrin C); (d) Ueda, A.; Yamamoto, A.; Kato, D.; Kishi, Y. *J. Am. Chem. Soc.* **2014**, *136*, 5171 (Halichondrin A); (e) Yahata, K.; Ye, N.; Ai, Y. R.; Iso, K.; Kishi, Y. *Angew. Chem. Int. Ed.* **2017**, *56*, 10796. See also references 75–78 in Chapter 10.

72 (a) Hertweck, C. *Angew. Chem. Int. Ed.* **2015**, *54*, 14622; (b) McKerrow, J. H. Nat. *Prod. Rep.* **2015**, *32*, 1610; (c) Efferth, T.; Zacchino, S.; Georgiev, M. I.; Liu, L.; Wagner, H.; Panossian, A. *Phytomedicine* **2015**, *22*, A1-A3; (d) Molyneux, D. H.; Ward, S. A. *Trends in Parasitology* **2015**, *31*, 605; (e) Gates, B. *Kexue Tongbao* **2017**, *62*, 1905.

73 (a) Tu, Y.-Y. *Angew. Chem. Int. Ed.* **2016**, *55*, 10210; (b) Qinghaosu Antimalarial Coordinating Research Group. *Kexue Tongbao* **1977**, *22*, 142; (c) Qinghaosu Antimalarial Coordinating Research Group. *Chin. Med. J.* **1979**,

12, 811; (d) Liu, J.-M.; Ni, M.-Y.; Fan, J.-F.; Tu, Y.-Y.; Wu, Z.-H.; Wu, Y.-L.; Chou, W.-S. *Huaxue Xuebao* **1979**, *37*, 129 (structural determination). See also: (e) Kong, L.-Y.; Tan, R.-X. *Nat. Prod. Rep.* **2015**, *32*, 1617; (f) Hsu, E. *Br. J. Clin. Pharmacol.* **2006**, *61*, 666; (g) Li, Y.; Wu, Y. L. *Med. Trop.* **1998**, *58*, 9; (h) Klayman, D. L. *Science* **1985**, *228*, 1049.

74 Ōmura, S. *Angew. Chem. Int. Ed.* **2016**, *55*, 10190.

75 Campbell, W. C. *Angew. Chem. Int. Ed.* **2016**, *55*, 10184.

76 Thuan, N. H.; Pandey, R. P.; Sohng, J. K. *Appl. Microbiol. Biotech.* **2014**, *98*, 7747.

77 For selected total syntheses, see: (a) Yamashita, S.; Hayashi, D.; Nakano, A.; Hayashi, Y.; Hirama, M. *J. Antibiot.* **2016**, *69*, 31; (b) Li, M. S.; O'Doherty, G. A. *Org. Lett.* **2006**, *8*, 3987; (c) Bailey, S.; Helliwell, M.; Teerawutgulrag, A.; Thomas, E. J. *Org. Biomol. Chem.* **2005**, *3*, 3654; (d) Crimmins, M. T.; Alawar, R. S.; Vallin, I. M.; Hollis, G. W.; O'Mahony, R.; Lever, J. G.; Bankaitis-Davis, D. M. *J. Am. Chem. Soc.* **1996**, *118*, 7513; (e) White, J. D.; Bolton, G. L.; Dantanarayana, A. P.; Fox, C. M. J.; Hiner, R. N.; Jackson, R. W.; *et al. J. Am. Chem. Soc.* **1995**, *117*, 1908; (f) Férézou, J.-P.; Julia, M.; Li, Y.; Liu, L. W.; Pancrazi, A. *Bull. Soc. Chim. Fr.* **1995**, *132*, 428; (g) Hirama, M., Noda, T., Yasuda, S.; Itô, S. *J. Am. Chem. Soc.* **1991**, *113*, 1830; (h) Ley, S. V.; Armstrong, A.; Diez-Martin, D.; Ford, M. J.; Grice, P.; Knight, J. G.; *et al. J. Chem. Soc., Perkin Trans. 1* **1991**, 667; (i) White, J. D.; Bolton, G. L. *J. Am. Chem. Soc.* **1990**, *112*, 1626; (j) Danishefsky, S. J.; Armistead, D. M.; Wincott, F. E.; Selnick, H. G.; Hungate, R. *J. Am. Chem. Soc.* **1987**, *109*, 8117; (k) Hanessian, S.; Ugolini, A.; Dube, D.; Hodges, P. J.; Andre, C. *J. Am. Chem. Soc.* **1986**, *108*, 2776; (l) Schow, S. R.; Bloom, J. D.; Thompson, A. S.; Winzenberg, K. N.; Smith, A. B. *J. Am. Chem. Soc.* **1986**, *108*, 2662.

78 For reviews, see: (a) Cook, S. P. *Synlett.* **2014**, *25*, 751 and references cited therein; (b) Zhou, W. S.; Xu, X. X. *Acc. Chem. Res.* **1994**, *27*, 211; (c) Zhou, W.-S. *Pure. Appl. Chem.* **1986**, *58*, 817.

1

Principles for Synthetic Efficiency and Expansion of the Field

Pei-Qiang Huang

Fujian Provincial Key Laboratory of Chemical Biology, Department of Chemistry and iChEM, Xiamen University, Xiamen, China

Today, the field of natural product total synthesis, which was once a dominant sub-discipline of organic chemistry, has lost its prime position. Even with the renaissance of natural products as drug candidates, and the recent recognition of the value and contribution of natural product-based drugs to societies by the Nobel Prize in Physiology or Medicine 2015, the situation has changed. As we have mentioned in the Introduction, "it is estimated that half of the top 100 best-selling medications will soon be biologics." According to Dirk Trauner, "the field is currently under fire, at least in societies that have traditionally supported it."[1a] To address these challenges, we need to significantly improve efficiency of the synthesis,[1b] expand the field of total synthesis, and address the concerns of society. In this chapter, the efforts by chemists from these dimensions will be summarized with a focus on concepts.

1.1 Concepts for Efficiency in the Total Synthesis of Natural Products

As we have pointed out in the previous sections, the total synthesis of natural products is facing with many challenges. The key to tackle these problems is to improve the synthetic efficiency, which involves selectivity, economy of synthesis, and green chemistry. Over the last five decades, many concepts have been advanced by the leading scientists in the field. It is expected that the new concepts will become new criteria for total synthesis.

Efficiency in Natural Product Total Synthesis, First Edition. Edited by Pei-Qiang Huang, Zhu-Jun Yao, and Richard P. Hsung.

1.1.1 Ideal Synthesis

1.1.1.1 Hendrickson's Definition (1975)

The concept of ideal synthesis was first advanced by J. B. Hendrickson in 1975:

> The synthesis would start from available small molecules so functionalized as to allow constructions linking them together directly, in a sequence only of successive construction reactions involving no intermediary refunctionalizations, and leading directly to the structure of the target, not only its skeleton but also its correctly placed functionality. If available, such a synthesis would be the most economical, and it would contain only construction reactions.[2]

1.1.1.2 Wender's Definition (1996)

A more comprehensive definition of ideal synthesis was given by P. A. Wender in 1996.[3a]

> An ideal (the ultimate practical) synthesis is generally regarded as one in which the target molecule (natural or designed) is prepared from readily available, inexpensive starting materials in one simple, safe, environmentally acceptable, and resource-effective operation that proceeds quickly and in quantitative yield.[3]

Wender also indicated two general ways[3a] for approaching the ideal synthesis (i.e., achieving maximum relevant complexity increase while minimizing step count):

1) The use of strategy-level reactions such as the Diels–Alder reaction or multistep processes such as tandem and domino sequences that allow for a great increase in target-relevant complexity in one operation.
2) The design and development of new reactions and reaction sequences that allow for a great increase in target-relevant complexity.

1.1.1.3 Baran's Quantification of the Ideal Synthesis

To furnish a numerical expression for Hendrickson's conception of an ideal synthesis, P. S. Baran gave the following metric definition: "ideality is the percentage of the sum of construction reactions and strategic redox reactions divided by number all synthetic steps."[4]

Construction reactions are skeletal bond (C–C and C-heteroatom) forming reactions; and **strategic redox reactions** are functionality installation reactions (direct introduction of the correct functionality found in the final product) that belong to the construction reaction.

1.1.2 Selectivity

Although the importance of selectivity in organic synthesis is well known, B. M. Trost was the first to give a systematic analysis and take it as a goal for achieving synthetic efficiency.[5] He categorized selectivity according to chemical reactivity (chemoselectivity), orientation (regioselectivity), and spatial arrangement (diastereoselectivity and enantioselectivity), and indicated the increased role of main group and transition metals in enhancing selectivity.[5a] **Chemoselectivity** refers to the preferential reaction of a chemical reagent with one of two or more different functional groups within a molecule. The term is also applied to reacting molecules or intermediates that exhibit selectivity toward chemically different reagents. The controlled reaction of a functional group (e.g., ester, amide) to a product of an intermediate oxidation state (e.g., ketone/aldehyde, ketone/aldehyde/imine) constitutes another type of chemoselective reaction.

While P. S. Baran views chemoselectivity as both the greatest obstacle to complex molecule synthesis and the mother of invention in total synthesis,[6a] A. K. Yudin, states that achieving high levels of chemoselectivity has been the Achilles heel of chemical synthesis.[6b]

Chemoselectivity is highly depend on reagents. Kagan's reagent (samarium(II) diiodide)[7] is a mild single-electron reductant enabling many chemoselective transformations.[7] The Kagan reagent-mediated efficient total syntheses are presented in Chapter 6.

Amides are the least reactive carboxylic acid derivatives. Performing the chemoselective reaction at an amide group in the presence of an ester group is challenging. The pioneering work of B. Ganem showed that Schwartz's reagent (Cp_2ZrHCl)[8] can serve as a chemoselective reducing agent to convert a secondary amide/lactam group in the presence of an ester group to an imine.[9] Using this methodology, they have developed a concise asymmetric total synthesis of (−)-α-kainic acid (**4**, Scheme 1.1a).[9b] The exceptional chemoselectivity of the amide-to-imine reduction was applied to the synthesis of **6** by the selective removal of the amide acyl side chains from the highly functionalized substrates **5a–f**, which are a mixture of primary taxanes extracted from the yew trees growing on Michigan's Upper Peninsula (Scheme 1.1b).[9c] This legendary chemoselective reduction paved a way for efficient commercial semisynthesis production of paclitaxel through semisynthesis by Natural Pharmaceuticals, Inc. (NPI).[9c, 10]

Recently, Chida extended Ganem's chemistry to the reductive functionalization of secondary amides/lactams,[11] leading to (Scheme 1.1c) the most concise and efficient total synthesis of (±)-gephyrotoxin (**9**) to date (14 steps, 9.4 overall yield).[11c]

Chemoselectivity is not only critical for total synthesis but also an important issue in chemical biology,[12a–c] and in protein chemical synthesis.[12d]

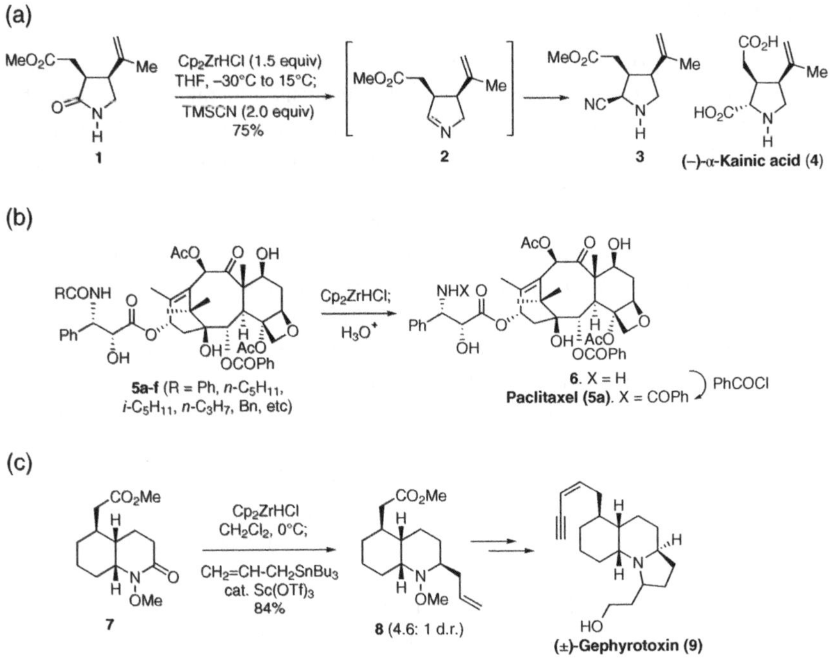

Scheme 1.1 Ganem's and Chida's chemoselective transformations based on Schwartz's reagent.

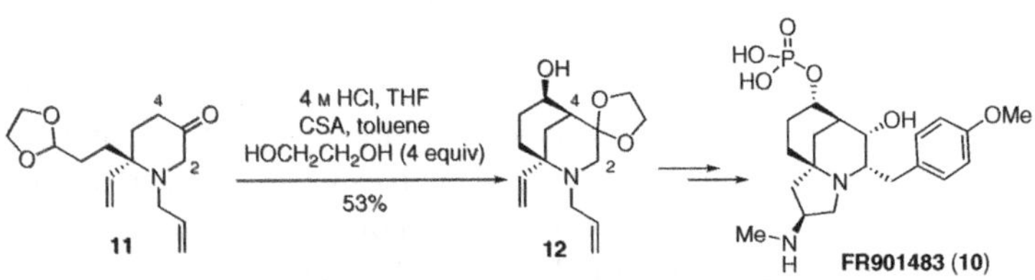

Scheme 1.2 Key C4 regioselective aldol reaction in the total synthesis of FR901483.

Regioselectivity refers to the preferential reaction of a chemical reagent at one direction (position) over all other possible directions (positions) of a molecule. In Huang's enantioselective synthesis of the potent immunosuppressant FR901483 (**10**), a regioselective aldol reaction at C4 of **11** was required.[13a] Under carefully defined conditions, the key aldol reaction proceeded regioselectively at C4 to deliver compound **12** as the sole regio- and diastereoisomer (Scheme 1.2).[13a] The regioselective (and enantioselective) enzymatic $\Delta^{2,3}$ epoxidation of polyene squalene (**46**) to yield (3*S*)-2,3-oxidosqualene (**47**) (see Scheme 1.14 later) is an excellent model of regioselective reaction. Sometimes, reactivity among the same kind of functional groups is referred as site-selectivity, which can be achieved by a variety of methods.[13b–d]

Stereoselectivity includes diastereoselectivity and enantioselectivity. In a diastereoselective reaction one diastereoisomer is formed as preferential to other possible ones. An enantioselective reaction refers to a reaction that leads to preferent formation of one enantioisomer (enantiomers) over another one.

The Sharpless asymmetric epoxidation (of allylic alcohols, AE) and asymmetric dihydroxylation (of alkenes, AD)[14] (Scheme 1.3) are enantioselective reactions. Because of their reliable and predictable enantioselectivity, these two reactions have found widespread applications in the total synthesis of natural products.

R. Noyori and coworkers have developed several BINAP – Ru(II)-based chiral catalysts for enantioselective reduction of alkenes[15a] (Scheme 1.4a) and ketones.[15b] When chiral piperidinyl β-ketone ester **15** was employed as a substrate, the Noyori reaction proceeded diastereoselectively to give the secondary alcohol **16** (Scheme 1.4b).[16]

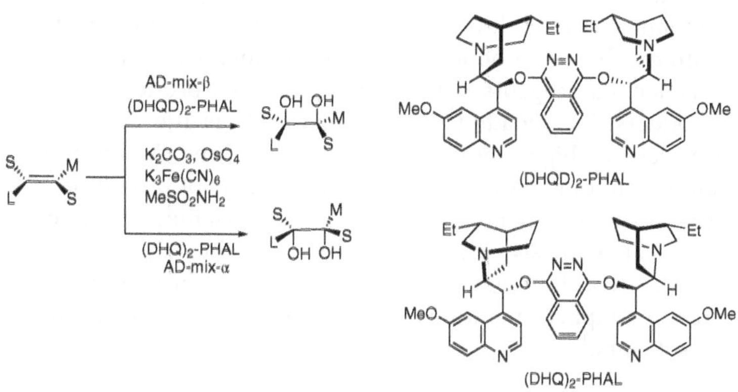

Scheme 1.3 Sharpless enantioselective (asymmetric) dihydroxylation of alkenes.

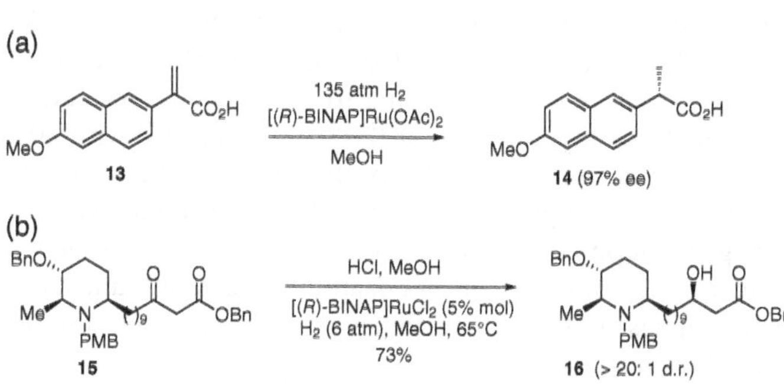

Scheme 1.4 Enantioselective (a) and diastereoselective (b) reductions by Noyori's catalysts.

1.1.3 Green Synthesis

In 1991, the concept of green chemistry was put forward with the "Twelve Principles of Green Chemistry" as a scientific approach for environmental protection.[17a–f] One of the principles states that it is better to prevent waste rather than to treat or clean up waste after it is formed. Anastas and coworkers have also demonstrated that catalysis can be used as a primary tool for achieving the principles of green chemistry. For chemical process in pharmaceutical industry, seven important elements[17g] and eight criteria[17h] that define a good chemical manufacturing process have been suggested.[17i,j] In parallel, sustainable practices in medicinal chemistry have been reviewed and some potential future developments have been recently highlighted.[17k]

1.1.4 Atom Economy

The concept of atom economy[18, 19] was advanced by B. M. Trost in 1991.[18a] He suggested that in addition to selectivity,[5] efficient synthetic methods should be economical at atom level with maximum incorporation of atoms of reactants into the desired products. This concept focuses on minimizing both the use of chemical reagents and additives, and waste production in the form of by-products.[18] Cycloaddition reactions such as the Diels–Alder reaction are a class of highly atom-economical reactions. Methods that involve combining building blocks and catalytic amounts of promotors can reach high atom economy. He also suggested the approaching of the selective and economical synthesis by using transition metal catalysis. In 2008, C.-J. Li and B. M. Trost drew our attention to the design of new synthetic processes that can simplify operations in chemical productions and to the use of greener solvents.[18d]

1.1.5 E Factors

In 1992, Sheldon introduced the concept of the E(nvironmental) Factor[20] to relate the mass of waste to the mass of product formed (E Factor = kg waste/kg product).[19b] In that article, Sheldon also provided a figure for the E factors of the main chemical industrial sectors. According to the list, pharmaceutical industry is the sector with the highest E factor (25 ~ 100 compared with <0.1 for the petroleum refining sector). This reflects that most of the synthetic transformations currently used in pharmaceutical industry are multistep syntheses and employing stoichiometric instead of catalytic amount of reagents. In addition, Sheldon also suggested the concept of atom utilization[20a] and atom efficiency,[20c] which emphasized the importance of catalytic processes in achieving high atom efficiency.

1.1.6 Step Economy

Since the first introduction of the concept of step economy by P. A. Wender in 2006,[21] this brief term has become increasingly popular among the synthetic community. As an approach toward the ideal synthesis, step economy is the drive to increase the efficiency of a synthesis by minimizing synthetic steps, which should cover redox-economy (*vide infra*), quick generation of molecular complexity, and protect group-free synthesis (*vide infra*). However, the intention of Wender was dual.[21c] On one hand, he called for the development of new reactions and new synthetic strategies that allow for shorter routes to a (complex) target, and on the other hand, the design and synthesis of less complex targets while maintaining or enhancing molecular function, namely, function-oriented synthesis (FOS, *vide infra*).[21c–e] The synthesis of the potent vesicant cantharidin (**17**) by Dauben[22] demonstrates the importance of novel synthetic technologies in achieving step economy. Retrosynthetically, the intermediate **18** could be synthesized by a Diels–Alder reaction between furan and dimethyl maleic anhydride (**19**) (Scheme 1.5a). However, this ideal synthesis is difficult to achieve due to the low reactivity of the diene and the dienophile.

Dauben and coworkers overcame these difficulties by employing the high pressure technique and the use of 2,5-dihydrothiophene-3,4-dicarboxylic anhydride (**20**) as a surrogate of the dimethyl maleic anhydride (**19**). Thus, the Diels–Alder reaction of furan with **20** (prepared in three steps from commercially available compound) at 15 kbar pressure yielded quantitatively adducts **21** and **22** in a ratio of 85:15. The former was treated with Raney nickel to

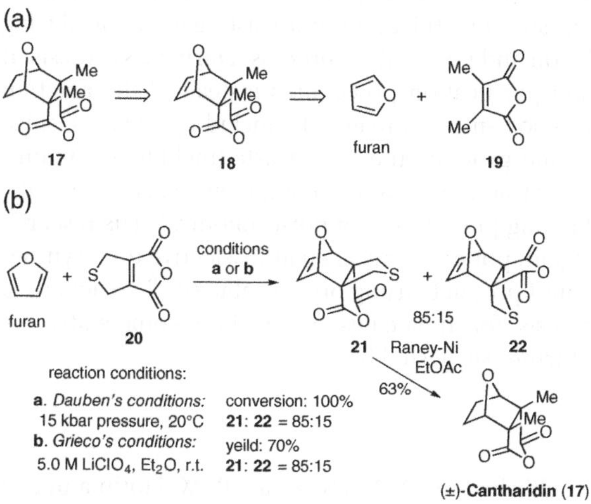

Scheme 1.5 Step-economical synthesis of racemic cantharidin (**17**).

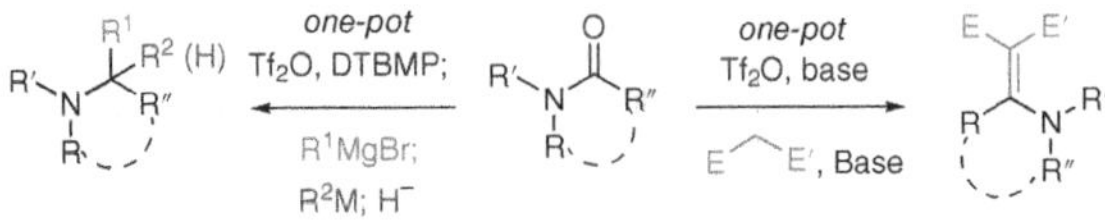

Scheme 1.6 Step-economic transformations of common amides.

produce cantharidin (**17**) in 63% (Scheme 1.5a).[22a] Further improvement by the same group allowed the reaction of >10 g of dihydrothiophene anhydride **20**.[22b] Alternatively, by taking advantage of dramatic rate accelerations of Diels–Alder reactions in 5 M lithium perchlorate-diethyl ether, the Diels–Alder reaction of furan with **20** proceeded smoothly at ambient temperature and pressure to give diastereomeric cycloadducts **21** and **22** in an 85:15 ratio with a 70% combined yield (Scheme 1.5b).[22c]

Amides are a class of versatile compounds found widespread applications in the total synthesis of alkaloids. However, due to their high stability, multistep methods[23] are used for their transformations to compounds at a lower oxidation state with C-C bond formation. By *in situ* amide activation with triflic anhydride (Tf_2O), P.-Q. Huang and coworkers developed a series of step-economic methods for the direct transformations of common amides (Scheme 1.6).[24]

1.1.7 Pot Economy and PASE (Pot, Atom, and Step Economy)

The concepts of both pot economy and its combination with atom and step economy (PASE) were proposed by Clarke in 2007.[25] Although the concept of pot economy overlaps to some extent with step economy, the former addresses the problem of minimizing solvent utilization and waste generation during work-up and product isolation and purification process. During a synthesis, it is the product isolation and purification procedures instead of the reaction itself that consume the most solvents and materials (including solvents, silica gel, or related substances), and generate the most waste (including contaminated aqueous waste from extraction, cleaning equipment, glassware, etc.). Yujiro Hayashi is an outstanding practitioner of pot-economy.[26a] His research group has achieved the elegant, pot economical enantioselective total syntheses of several bioactive natural products using organocatalysis.[26] To achieve a significant "greener" synthesis for molecules of medium complexity, the combination of PASE was further suggested by Clarke.[25]

1.1.8 Redox Economy

In a tutorial review article published in 2009, P. S. Baran, R. W. Hoffmann and N. Z. Burns put forward the concept of redox economy.[27] This concept, "which is an ignored strategy," points out the importance of minimizing the use of redox steps in achieving highly efficient multistep organic synthesis. Redox economy can be achieved by two approaches: through the design of redox economical reactions, and redox economical synthetic strategies. Several total syntheses of high redox economy will be illustrated in Chapter 2.

Alcohols are routinely synthesized by addition of organometallic reagents to aldehydes. The latter in turn are synthesized by oxidation of simpler alcohols. Thus, direct use of alcohols as the surrogates of aldehydes constitute a class of redox-economical reactions. However, such reactions are challenging. In this context, the conceptually novel "C–C bond forming hydrogenations," "C–C bond forming transfer hydrogenations" (Scheme 1.7a), and related reactions (Scheme 1.7b),[28] invented by M. J. Krische and coworkers, represent important breakthroughs in both concept and synthetic methodology. In addition to being redox economical, the reactions also displayed excellent regio-, diastereo-, and enantio-selectivities. Moreover, as can be seen from Scheme 1.7b, the reaction took place chemoselectively at the primary alcohol in the presence of free secondary alcohols allowing a protecting group-free synthesis (*vide infra*). The power of these revolutionary methodologies has been demonstrated by the efficient enantioselective total syntheses of several structurally complex natural products[28a–f] including cyanolide A, which will be discussed in Chapter 2. This chemistry has also been applied by A. Fürstner to the total synthesis of mandelalide A (the proposed structure).[28g]

To overcome the problem of over *N*-alkylation with alkyl halides, the traditional method for the synthesis of tertiary/secondary amines from the corresponding secondary/primary amines consists of reductive *N*-alkylation of secondary/primary amines with aldehydes. Direct use of an alcohol as an alkylating reagent for the *N*-alkylation represents another type of redox-economical method.[27] The Pd/C or $Pd(OH)_2$/C-catalyzed *N*-methylation reactions with

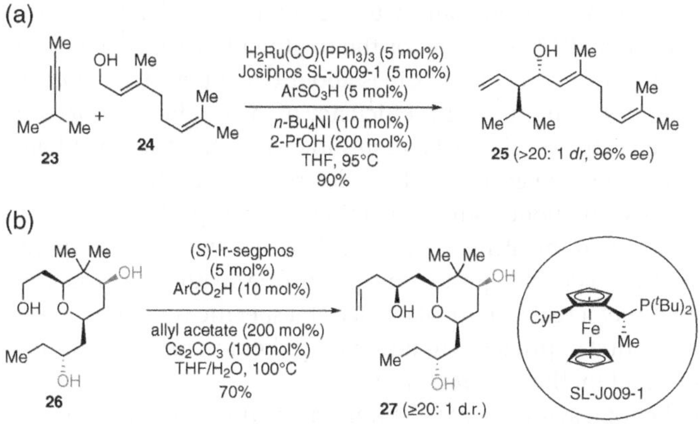

Scheme 1.7 Krische reactions for the redox-economical, diastereo-, and enantioselective synthesis of functionalized alcohols.

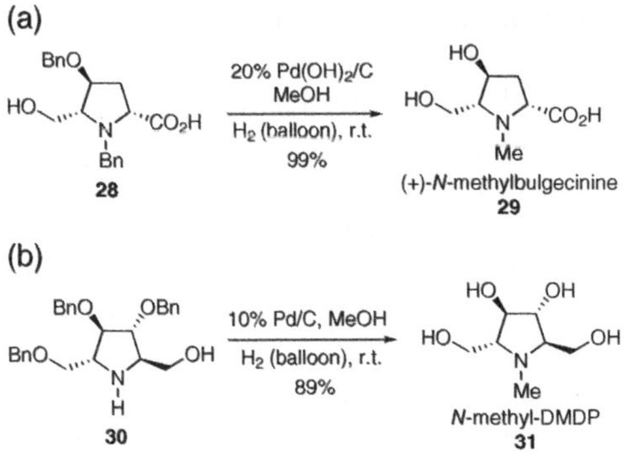

Scheme 1.8 Huang's redox-economical and chemoselective amine *N*-methylation.

methanol outlined in Scheme 1.8 turned out to be a synthetically useful redox-economical reaction.[29] Under mild reaction conditions (at r.t., 1 atm H_2), *N*-,*O*-debenzylations and *N*-methylation took place sequentially and chemoselectively in one pot, and in the presence of unprotected hydroxyl and carboxyl groups.[29a]

1.1.9 Protecting-Group-Free Synthesis

Natural products are generally multifunctionalized compounds. Thus the use of cumbersome protecting groups are indispensable for achieving selective transformations.[30] The semisynthesis of ingenol angelate (**32**) developed by the scientists of Leo Pharma in Denmark is a typical example (Scheme 1.9).[31a] The protection of the two hydroxyl groups in the form of a cyclic acetal ensued the selective esterification of the allylic hydroxyl group at the cyclopentene ring. This semisynthesis provided enough material for clinical trials, and was the basis for the approval of ingenol angelate (**32**) by the US Food and Drug Administration (FDA) in 2012 as a drug for the treatment of AK.[31b] However, the use of a protecting group generally brings about two additional steps to a synthesis, which reduces dramatically the efficiency of the synthesis. Moreover, at the last step of the synthesis, chemists often suffer from frustration at being unable to cleave the protecting groups without destroying the whole molecule. Thus, organic synthesis without using protecting groups (protecting group-free synthesis)[32] has become an attractive approach to achieving step economy.

There are two tactics to develop protecting group-free syntheses. Protecting groups are used to direct a reaction to take place at a specific functional group or position. Thus, the first approach is to developing chemo-, regio-, and stereoselective reactions and methods (*vide supra*).

The last step of the total synthesis of (−)-himalensine A (**37**) (Scheme 1.10a), reported very recently by D. J. Dixon and coworkers,[33a] provides an example of this approach. Amide (lactam) is a highly stable functional group that serves as

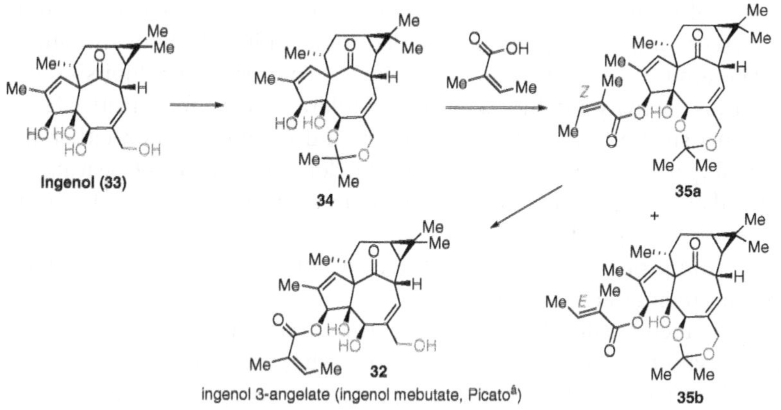

Scheme 1.9 Leo Pharma's semisynthesis of Picato (**32**).

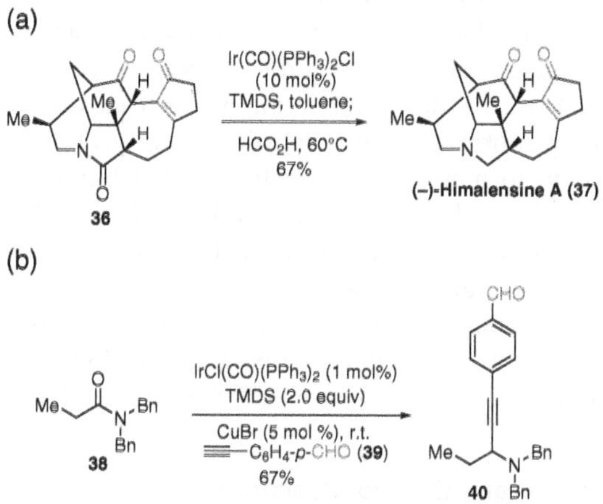

Scheme 1.10 On substrate and on nucleophile protecting-group-free syntheses.

the protected form of the corresponding amine. The last step requires the selective reduction of the lactam in compound **36** in the presence of two more reactive ketone and enone functional groups. The conventional method resides in the protection of the more reactive ketone and enone groups before the reduction of the less reactive lactam. And after lactam reduction, it needs to cleave the two protecting groups. Instead, Dixon employed a new chemoselective method that consisted of partial reduction of the lactam group with Vaska's catalyst/TMDS (1,1,3,3-tetramethyldisiloxane) followed by further reduction in hot formic acid to deliver (−)-himalensine A (**37**). This protecting-group-free reduction not only significantly increased the synthetic efficiency, but also ensued the success of the total synthesis.

By means of Ir and Cu(I) bis-metal sequential catalysis, reductive alkynylation of tertiary amides (e.g., **38**) proceeded under mild conditions to afford propargylic amines (e.g., **40**, Scheme 1.10b).[33b] The reaction exhibits exceptional chemoselectivity and functional group tolerance even for sensitive functional groups such as aldehyde, cyano, ester, and nitro groups on either the amide or alkyne partners. For example, acetylene derivative **39** bearing an unprotected aldehyde group underwent smooth reaction with **38** to afford the desired amino aldehyde **40**.

Now, even unprotected polyhydroxylated carbohydrates can serve as starting materials in organic synthesis.[34]

The second tactic to develop protecting-group-free syntheses is the strategic design of a synthetic route to avoid situations where selectivity is a problem. Many highly efficient total syntheses are protecting group free, and some will be discussed in Chapter 2.

If the use of protecting groups is unavoidable, one should try to economize one step from either protection or deprotection via a one-pot reaction. The reaction showcased in Scheme 1.8 can illustrate this tactic. Because *N,O*-deprotection and *N*-methylation took place in one pot, no additional step is required for the deprotection.[29a] Another tactic for the manipulation of less reactive carbonyl groups in the presence of more reactive aldehyde/ketone groups is through the *in situ* protection of the former. Reagents such as diethylaluminum benzenethiolate (Et_2AlSPh),[35a] the combination HN(OMe)Me·HCl/*n*-BuLi/Me_3Al,[35b] and PPh_3/trifluoromethanesulfonic acid trialkylsilyl esters (R_3SiOTf) have been employed by Markó, Colby, and Fujioka, respectively, for the discrimination of different carbonyl compounds/groups (Scheme 1.11a, b). Notably, H. Fujioka's PPh_3/R_3SiOTf system is able to undergo conjugate addition with α,β-enones, leading to the *in situ* protection of α,β-enones[35d, e] (Scheme 1.11c).

1.1.10 Multicomponent Reactions and One-Pot Reactions

In a multicomponent reaction (MCRs) at least three different starting materials react in a programed sequence to yield, in one pot, the final product, which incorporated most of the atoms of the starting materials.[36] MCRs is a highly efficient methodology because at least two bonds are formed in one pot. Many classical MCRs are named reactions, including the Mannich reaction (M-3CR), Biginelli reaction (B-3CR), and Ugi reaction (U-4CR). Up to eight component reactions have been reported.[37] The high convergence of the MCRs rend them particularly useful for the construction of compound libraries. Merging MCRs with organocatalysis paves an avenue for the asymmetric multicomponent reactions (AMCRs).[38] In this context, Enders and coworkers achieved the control of four stereocenters in a diphenylprolinol silyl ether-mediated asymmetric three-component reaction (Scheme 1.12a).[39a] After this breakthrough, organocatalytic

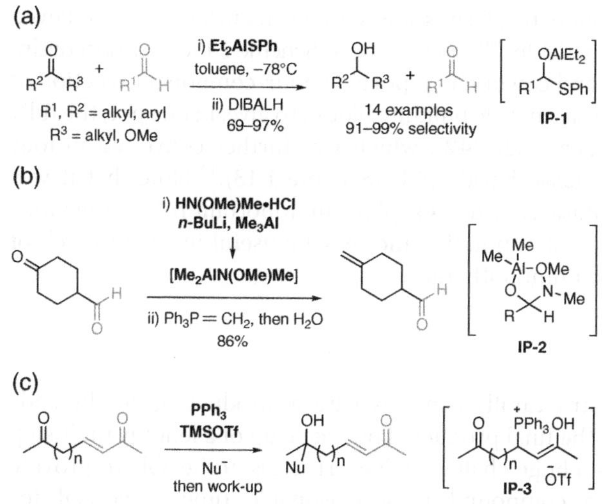

Scheme 1.11 *In situ* protection of aldehydes (a and b) and α,β-enone (c).

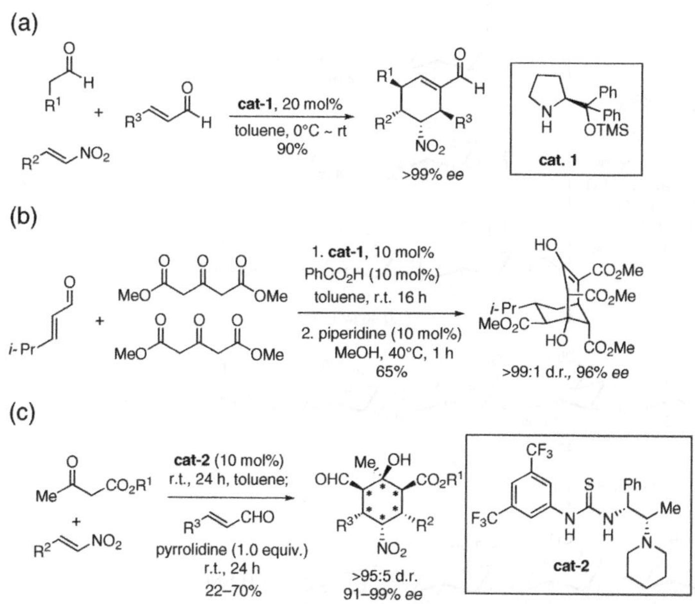

Scheme 1.12 Organocatalytic, asymmetric multicomponent reactions.

one-pot formation of 1 out of 64 (2^6) possible stereoisomers have been achieved by K. A. Jørgensen (Scheme 1.12b)[39b] and D. Enders (Scheme 1.12c),[39c] respectively.

In 2004, Xin-Shan Ye and coworkers reported a four-component reaction of easily available common saccharide building blocks to give, in one pot, the fully protected α-Gal pentasaccharide (**42**), which was further converted in four steps to yield α-Gal pentasaccharide (**43**) (Scheme 1.13).[40] Note that it was reported that α-Gal pentasaccharide (**43**) plays an important role in the interaction with human anti-Gal antibodies and may be useful in the research of xenotransplantation and immunotherapy.

1.1.11 Scalability

The total synthesis of structurally complex natural products generally ends with only milligrams of the final product. Thus scale up of a lengthy multistep total synthesis presents a huge challenge. Nevertheless, to be able to provide sufficient amount of a compound in a reasonable time is crucial for

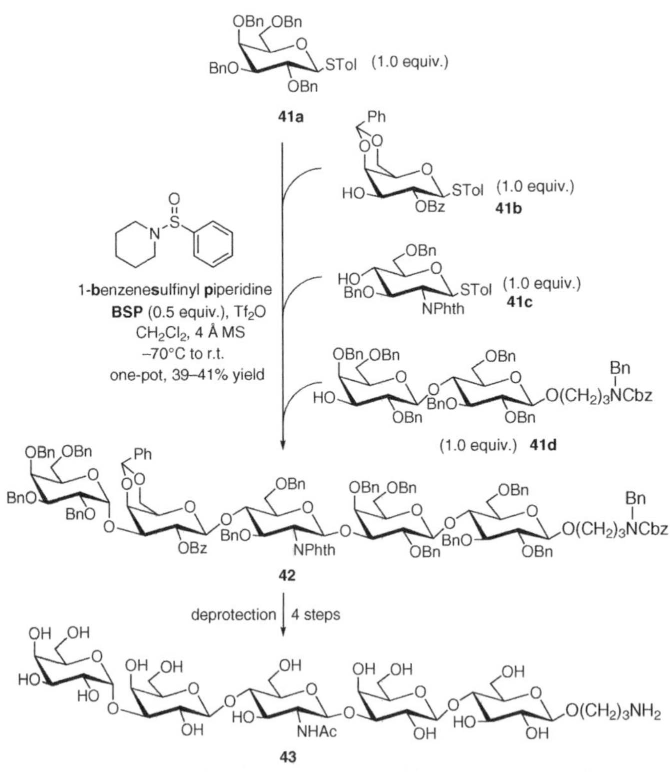

Scheme 1.13 Four-component one-pot synthesis of α-Gal pentasaccharide (**43**).

investigating the function of the molecule. Thus the scalability become an important issue not only for the total synthesis of natural products, but also for organic reactions and synthetic methods. This issue began to attract the attention of synthetic organic chemists. In a recent review article entitled: "Natural Product Synthesis in the Age of Scalability,"[41a] P. S. Baran and coauthors illustrated the gram-scale total syntheses of natural products and kilogram syntheses in pharmaceutical company disclosed in recent years, and emphasized the importance to develop simple, scalable synthetic routes.

Very recently, Allred and coauthors reviewed the state-of-the-art of the production of structurally complex natural product-based drug candidates on scales sufficient to drive human drug trials.[41b]

1.1.12 Convergent Synthesis

Convergent synthesis is a classical yet efficient strategy in total synthesis.[42a,b] Many highly efficient total syntheses employing this strategy will be presented in Chapter 2. The building-block-based convergent strategy can forge practical synthesis for medicinal chemistry. This has been nicely demonstrated by Andrew G. Myers in the context of antibacterial drug discovery (see Section 1.4). In addition, recently, M. Inoue,[42c] J. Šenda,[42d] and coworkers have independently outlined many elegant examples employing convergent strategies for efficient synthesis of structurally complex natural product.[42] Recently, M. Inoue summarized the radical-based convergent strategies for total syntheses of densely oxygenated natural products.[42e]

1.2 Biomimetic Synthesis

In the field of total synthesis of natural products, the structure of the target is usually complex with structure containing multifunctionality and multistereocenters. Hence, multistep synthesis is common, which makes ideal total synthesis and fully atom-economical total synthesis difficult to achieve. A more appropriate approach would be one that imitates nature. Indeed, we admire the biosynthesis of natural products of living organism, which proceeds in an efficient, selective, economical, and environmental friendly manner.

In the literature, when a new natural product is reported, very often the authors suggest a plausible biosynthetic route. Such proposals are generally not confirmed by experimental data and can be incorrect. Even Robinson and Woodward made mistakes on this issue.[43a] In fact, only a small portion of biosynthetic routes have been elucidated. For example, although the biosynthesis of tropane alkaloids has been investigated since the time of Robinson, its exact biogenesis (biosynthetic route) remained unsolved for 100 years.[43b] Biogenetic speculative seems to be a general methodology according to Robert Thomas in a review entitled: "Biogenetic Speculation and Biosynthetic Advances."[43a] He indicated that: "A prerequisite of the experimental investigation of any biosynthetic pathway is the formulation of a hypothetical scheme for the transformation of candidate precursors based on plausible reaction mechanisms" and emphasized the importance of biogenetic speculation.

Moreover, as can be seen from the following examples, in so-called "biomimetic synthesis" only a key strategy, a key intermediate, or a key step but not the whole synthetic sequence is imitated under nonenzymatic conditions.[44] As such, it may be more appropriate to define this type of synthesis as *bio-inspired synthesis*.[45] Thus, it is important to be aware of the basic principles and strategies that nature follows. Although the biosynthesis of many natural products of different origin remains a "black box," general schemes of biosyntheses in plants are known.[46] From these known schemes the following basic principles, key features, and basic strategies of a biosynthesis can be figured out.

1.2.1 Basic Logic of Biosynthesis

1.2.1.1 Basic Principles of Biosynthesis

1) Economy (of materials, energy, time, etc.)
2) Rapid generation of molecular complexity
3) Environmentally benign and sustainable
4) Function-oriented synthesis (of bioactive molecules)

1.2.1.2 Key Features and Basic Strategies of Biosynthesis

1) Aqueous phase synthesis
2) The building blocks strategy
3) The cascade reaction strategy
4) The C–H functionalization strategy
5) The divergent synthesis, collective synthesis
6) The (multi)enzyme catalysis (multidomain protein)
7) The highly selective (chemo-, regio-, and stereoselective) synthesis
8) The protecting-group-free synthesis
9) The small quantity synthesis
10) The umpolung tactic
11) The oligomerization tactic
12) The repetitive tactic

1.2.2 Tandem, Cascade, and Domino Reactions – One-Pot Reactions

Among the biosynthetic strategies that fascinate chemists the most are undoubtedly the one-pot cascade reactions/sequences because they allow rapid generation of molecular complexity, leading to high synthetic efficiency in an elegant manner. Depending on the complexity of the molecule, it may involve a tandem or a cascade/ domino reaction sequence.

The first one-pot reaction can be traced back to the well-known Robinson's tropinone synthesis developed 100 years ago (see Chapter 2, Section 2.1.1). This synthesis is inspirational for synthetic efficiency. The synthesis was achieved through a multicomponent reaction, which involves tandem (sequential) Mannich reactions, and bis-decarboxylation in one pot.

A general view[47] of terpene and sterol biosynthesis is depicted in Scheme 1.14. Central to this scheme is the regio- and enantioselective epoxidation of squalene (**46**) to give (3*S*)-2,3-oxidosqualene (**47**) and the subsequent oxidosqualene cyclase-catalyzed polycyclization to yield a specific product. A plant triterpenoid is built through an all-chair folding, whereas a sterol precursor adopts a chair-boat-chair folding (see **47**) (followed by generation of a chair-boat-chair proto structure, which then undergoes methyl-hydrogen migration and proton loss). The chemical principles of these polyolefin cyclizations are Barton's conformational analysis-based Stork–Eschenmoser hypothesis (rationalization on stereoelectronic grounds). It is fascinating that nature is able to convert the achiral molecule squalene, in just two steps, into a single stereoisomer with 4–5 rings and no less than seven asymmetric centers out of 128 possible stereoisomers!

In light of the Stork–Eschenmoser hypothesis, in 1971 W. S. Johnson and coworkers achieved the biomimetic synthesis of (±)-progesterone (**56**).[48]

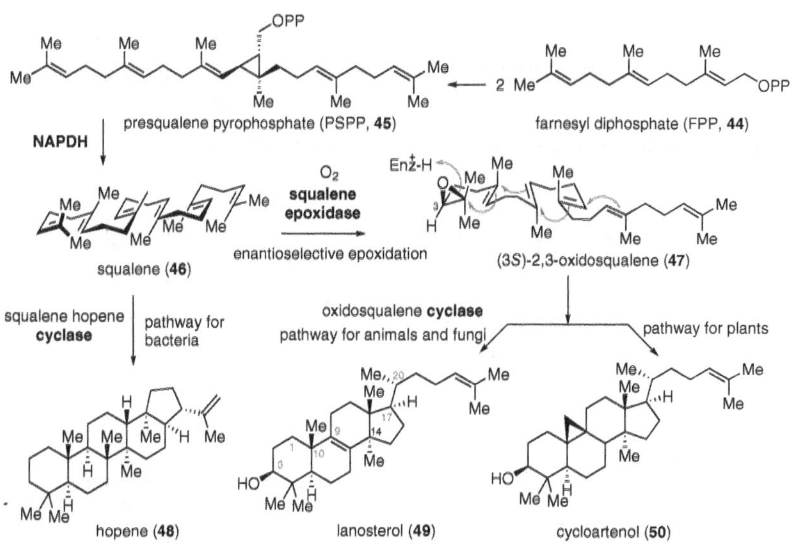

Scheme 1.14 Key steps in terpene and sterol biosynthesis.

The nonenzymatic, stereospecific (except C17), cationic polyolefin cyclization of cyclopentenol derivative **51** afforded, in a one-pot, polycyclic framework **55** with an all-*trans* ("natural") configuration (Scheme 1.15).

Heathcock's biomimetic synthesis of (±)-dihydroprotodaphniphylline (**59b**) (Scheme 1.16) represents another legendary classic in the history of biomimetic synthesis.[49a–c] It was inspired by a possible biosynthesis of one of the alkaloids.[49e] For the key biomimetic step, reaction of *E*-dialdehyde **58** with ammonia/NEt_3-HCl followed by treating with HOAc resulted in the formation of pentacyclization product **59a** at a 13% yield. However, by replacing ammonia with methylamine, (±)-dihydroprotodaphniphylline (**59b**) was obtained at a 65% yield! Interestingly, this spectacular result stemmed from a serendipitous discovery during an early investigation where a bottle of "ammonia" was actually methylamine. In this marvelous reaction,

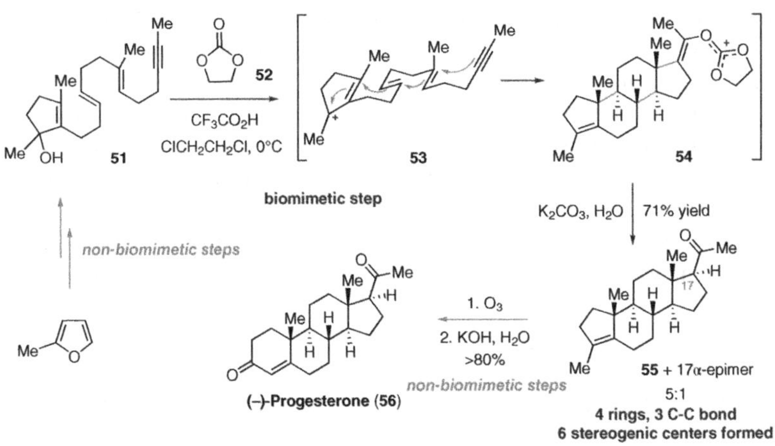

Scheme 1.15 Johnson's biomimetic synthesis of (±)-progesterone (**56**).

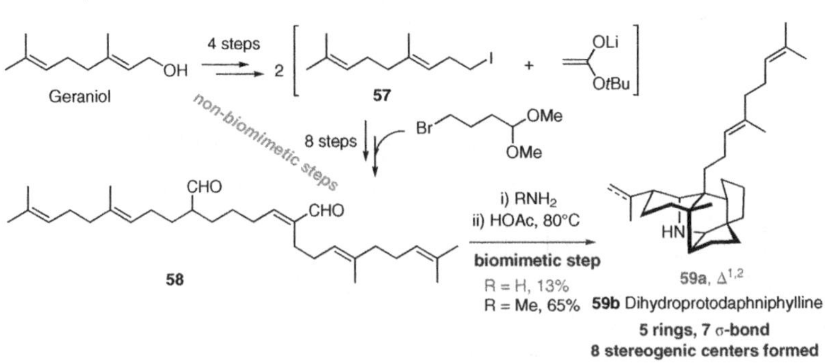

Scheme 1.16 Heathcock's biomimetic synthesis of (±)-dihydroprotodaphniphylline (**59b**).

pentacyclization occurred in a stereospecific manner to form five rings and seven σ-bonds including four carbon-carbon bonds, two carbon-nitrogen bonds, and one carbon-hydrogen bond.[49]

The success of this synthesis led to a concrete proposal about the biosynthesis of the alkaloids.[49b] It is worth mentioning that, in this 13-step synthesis, only the last step is biomimetic.[49b]

After these landmark accomplishments, many cascade polycyclizations have been developed for the total synthesis of natural products,[50] and the one-pot reaction methodology including tandem reactions, cascade reactions, and domino reactions is gaining popularity within the synthetic comminity.[50,51]

In recent years, organocatalytic asymmetric cascade reactions have emerged as a new tool in total synthesis.[52] In this regard, in 2010 D. W. C. MacMillan and S. Rendler developed SOMO catalysis to accomplish polyene tricyclization. Application of this concept to polyenal **60** resulted in the enantioselective formation of a hexacyclization adduct **61** as a single diastereomer at a 62% yield, which translates to an average yield of 92% per bond formed (Scheme 1.17). The level of enantiocontrol was assumed to be similar to those observed for the lower homologs of **61** (92% *ee*). In this remarkable chiral imidazolidinone catalyzed cascade reaction sequence, a total of 11 contiguous stereocenters including five all-carbon quaternary centers, were formed from the acyclic starting material **60**.[52e]

One-pot reaction methodology, in particular cascade polycyclizations, constitutes an important strategy for the rapid generation of molecular complexity.[53] In recent years, diverse strategies providing rapid access to molecular complexity have appeared.[54] In this context, Kagan's reagent (SmI_2) turned out to be a versatile reagent for a range of transformations, which will be illustrated in Chapter 6. Recently, David J. Procter has summarized the development of complexity-generating cascades by Sm(II)-mediated electron transfer to carboxylic acid derivatives.[54d]

A. H. Jackson's approach to the pentacyclic indole derivative **64**, which was related to *Aspidosperma* alkaloids, represents a typical example of rapid generation of molecular complexity from easily available starting materials (**62** and **63**) and under simple reaction conditions (Scheme 1.18a).[55a] Scheme 1.18(b) highlights C. D. Vanderwal's complexity-generating transformation to access

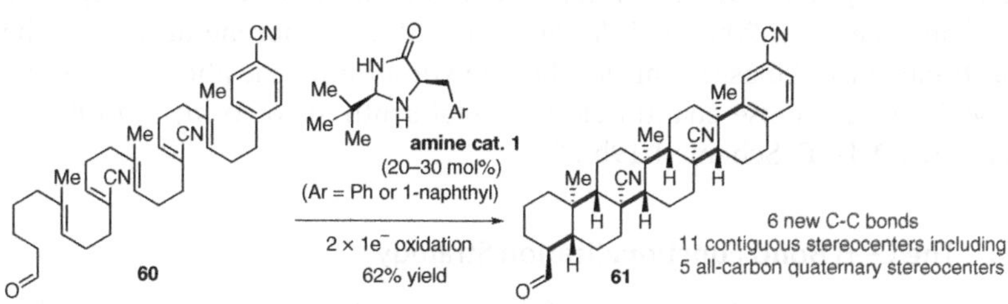

Scheme 1.17 MacMillan's enantioselective polyene cyclization via organocatalysis.

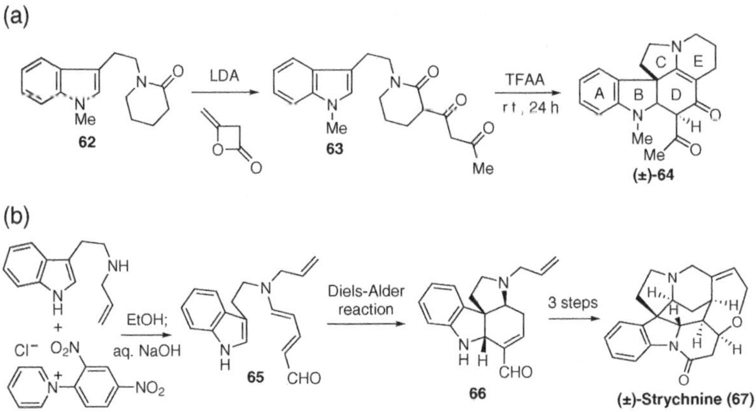

Scheme 1.18 Jackson's and Vanderwal's complexity-generating transformations.

the tetracyclic core (**66**), which is the key intermediate for the short racemic syntheses of norfluorocurarine (five steps), dehydrodesacetylretuline (six steps), valparicine (seven steps), and strychnine (**67**, six steps).[55b]

The assessment of molecule complexity would be helpful for planning efficient and convergent synthetic routes.[56a–e] In 2015, Jun Li and Martin D. Eastgate developed an approach for generating a unique complexity index, which is reflective of both intrinsic molecular complexity and extrinsic synthetic complexity. This approach allows for a direct comparison between molecules, the analysis of trends within research programs, and so on.[56f]

1.2.3 Site and Stereoselective Reactions

The beauty and elegance of biosynthesis resides in the highly regio-, chemo-, and stereoselective reaction at a specific functional group among many other reaction sites. The biosynthesis of D-myoinositol-1-phosphate (**D-I-1P**, Scheme 1.19a) provides an example. In the cell, kinases catalyze the transfer of a phosphoryl group (PO_3^{2-}) from an adenosine triphosphate (ATP) to a specific hydroxyl group among several others in a substrate to yield a specific phosphate ester.[57] In 2001, S. J. Miller and B. R. Sculimbrene developed the **peptide cat. 1** as a kinase mimic for the catalytic asymmetric phosphorylation, and achieved a concise enantioselective total synthesis of D-*myo*-inositol-1-phosphate (**D-I-1P**, Scheme 1.19b).[58]

1.2.4 The C—H Bond Functionalization Strategy

The site selective functionalization of an inert C–H bond among many other reaction sites represents another strategy in biosynthesis. In recent years,

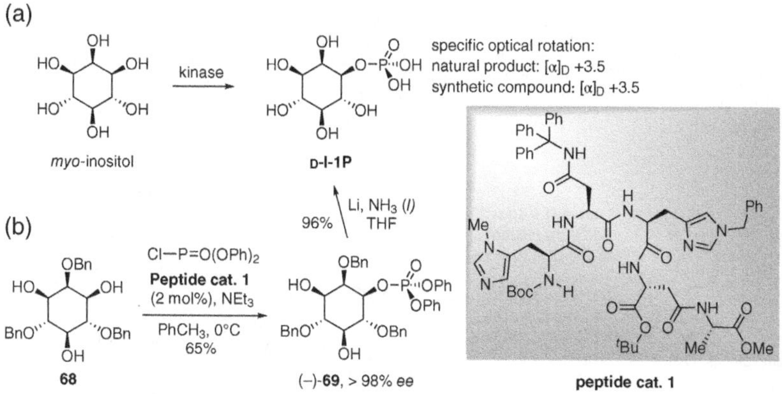

Scheme 1.19 The biosynthetic (a) and S. J. Miller's biomimetic (b) site and enantioselective synthesis of D-*myo*-inositol-1-phosphate (**D-I-1P**).

significant progress has been made on chemical C–H bond functionalization, which is emerging as a powerful tool for organic synthesis.[59] Many applications of C–H bond functionalization to the total synthesis of natural products will be discussed in Chapter 5. Significantly, C–H bond functionalization can also serve as a unique tool for regioselective late-stage modification of agrochemicals such as tetrahydrogibberellic acid analog (+)-**71a** (Scheme 1.20a),[60a] medicinal agents such as the antileukaemic and antitumour alkaloid (+)-camptothecin (**72**) (Scheme 1.20b),[60b] and simaomicin α (**75**), "the most potent natural-occurring anticoccidial agent reported"[60c] (Scheme 1.20c). It is worth mentioning that in 1989, simply by irradiation with medium pressure Hg lamp (quartz), T. Ross Kelly and coworkers achieved the one-pot transformation of **76** to (±)-cervinoymcin A_2 (**77**) at a 36% yield (Scheme 1.20d).[60d] The one-pot transformation involves a cascade of events including regioselective cyclization, cleavage of the MOM ethers, and oxidation.

D. L. Boger's tandem intramolecular Diels–Alder/1,3-dipolar cycloaddition cascade of 1,3,4-oxadiazoles (e.g., **78**) represents an elegant methodology for rapidly accessing molecule complexity (Scheme 1.21).[61a] Combining this methodology with late-stage C–H functionalization forged a short route to anticancer alkaloid vinblastine analog (**81**), which proved to be 10 times more potent than vinblastine, displaying an IC_{50} of 600–700 pM in the cell growth inhibition assays.[61b,c]

1.2.5 The Building-Block Strategy

The building block is an old concept. Even Stephen Hanessian's well-known concept of the "chiron" (chiral building-block, chiral synthon)[62] has seldom appeared in current literature. The building-block-based strategy may seem to be a low

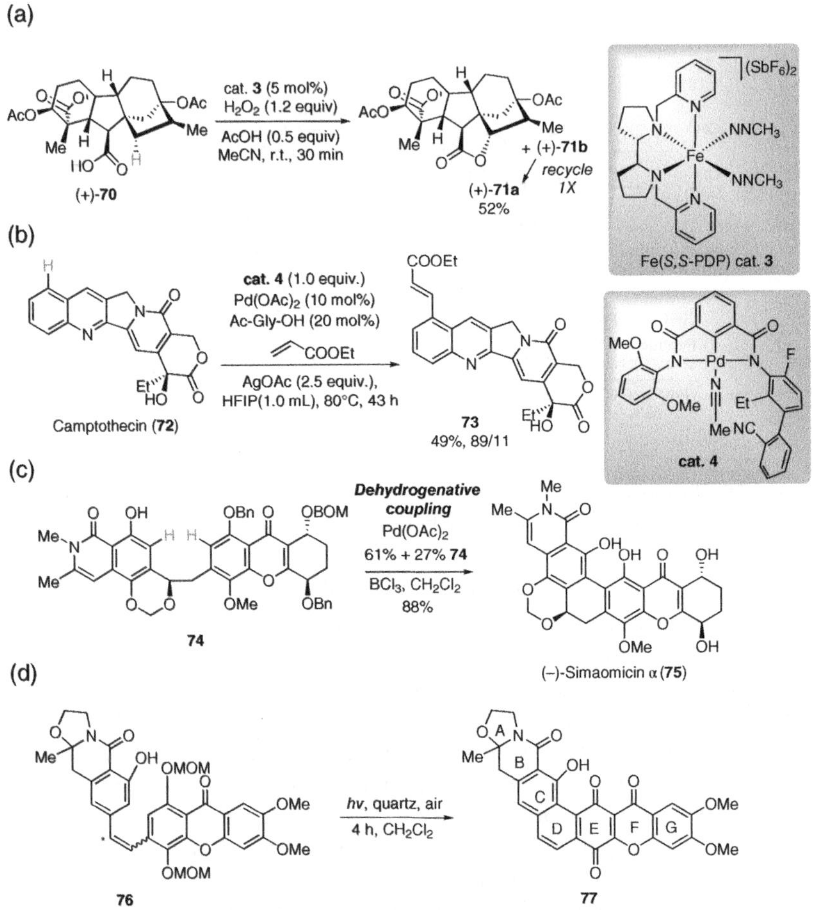

Scheme 1.20 White's, Yu's, Ready's, and Kelly's late-stage selective C–H bond functionalizations.

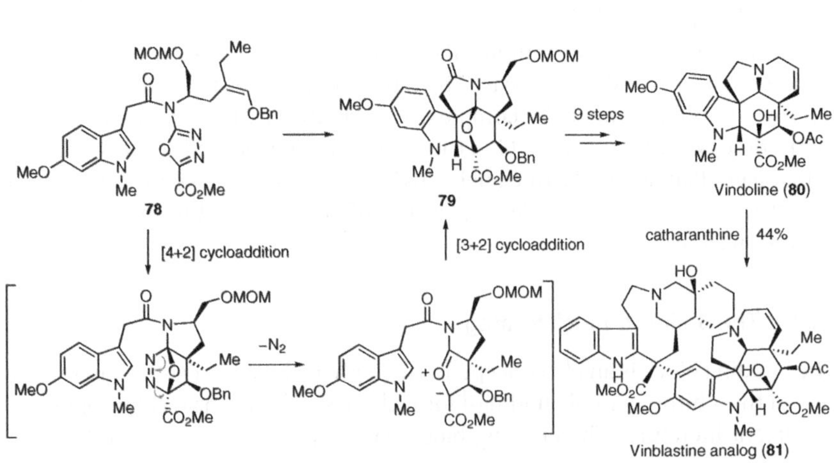

Scheme 1.21 Boger's cascade methodology and its combination with late-stage C–H functionalization.

efficient approach and out of fashion. However, the building block is the essential strategy that nature adopted for almost all biosyntheses! From proteins to nucleic acids, and from polysaccharides to secondary metabolites, all are synthesized starting from a limited number of simple (mono-, bi-, and multifunctional) building blocks such as amino acids for nonribosomal peptides, acyl-CoA thioesters for polyketides, isoprenyl diphosphates for terpenes, monosugars (and four bases) for polysaccharides, and DNA by iterative coupling.[63]

Surprisingly, an analysis of the 39 shortest enantioselective total syntheses, discussed in Chapter 2, revealed that chiral building-block-based approach is still the most commonly adopted (18 out of 39) and reliable strategy. The building-block-based approach is also a biomimetic approach. For these reasons, Chapter 9 is devoted to the total synthesis of natural products based on renewable resource-based building blocks/chirons. In this regard, Thomas J. Maimone and coworkers have very recently illustrated the power of this strategy in modern total synthesis.[64]

In 2014, Martin D. Burke and coworkers reported that it is possible to synthesize most polyene natural product motifs using just 12 building blocks and one coupling reaction.[63a] Their strategy relied on the identification of substructural motifs that are prevalent in natural products and development of suitable bifunctional building blocks and assembly method.[63a] The implementation of this strategy resulted in the first total syntheses of the polyene natural products asnipyrone B, physarigin A, and neurosporaxanthin β-D-glucopyranoside.

1.2.6 The Collective Synthesis Strategy

It is a common phenomenon that secondary metabolites exist collectively as a mixture of many compounds. Thus, the development of synthetic strategies to allow accessing different natural products is another way to increase synthetic efficiency. Many elegant collective syntheses will be discussed in other chapters of this book. Recently, Yong Qin and coworkers disclosed the efficient, enantioselective, collective syntheses of 33 monoterpenoid indole alkaloids belonging to four families[65a] (Scheme 1.22). The method relies on photocatalytic generation of a nitrogencentered radical that leads to umpolung of the reactivity of the nucleophilic amine and triggers radical cascade reactions. The method exhibited excellent chemo-, regio-, and diastereoselectivity.

Recent examples of collective syntheses include bioinspired syntheses of iboga-type indole alkaloids (X. G. She),[65b] metathesis-cascade reactions-based synthesis (C. C. Li),[65c,e] synthesis of (−)-mersicarpine and related alkaloids (J. P. Zhu),[65d] organocatalytic [4+3] cycloaddition reaction-based synthesis of englerin A and B and related natural terpenes (B.-F. Sun, G.-Q. Lin),[65f] and synthesis of lycopodium alkaloids (X. G. Lei; [65g–i] M. D. Shair[65j]).

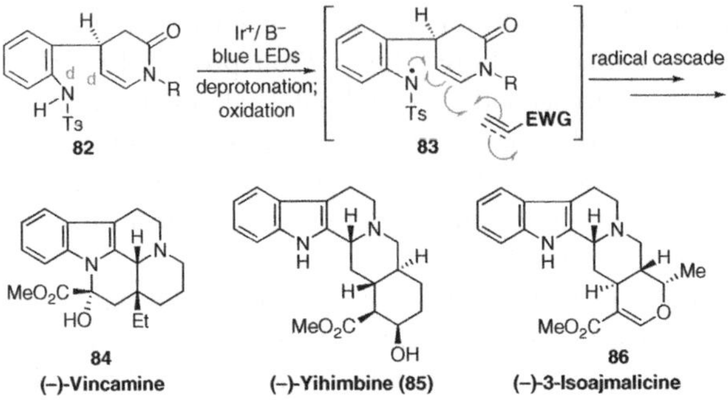

Scheme 1.22 Qin's radical cascade-based collective syntheses of indole alkaloids.

1.2.7 The Oligomerization Tactic

Oligomerization (including dimerization and trimerization) is a tactic used by nature to build complexity in short steps. Thus, identifying the inherent yet hidden symmetry (monomer) and developing methods for assembling monomers are both crucial to simplifying synthetic routes for the synthesis of oligomeric natural products. In 2011, Scott A. Snyder and coworkers developed a programmable resveratrol oligomer synthesis based on regioselective reactions.[66] In the same year, they reviewed synthetic approaches to oligomeric natural products.[67a] In 2014, X. Lei provided an overview for the biomimetic syntheses of oligomeric sesquiterpenoids.[67b] By identifying dehydrozaluzanin C (**87**) as the structure motif, they established a four-step transformation into very complicated trimers ainsliatrimer A (**88**) and ainsliatrimer B (**89**) (Scheme 1.23).[68] Very recently, Wen-Ju Bai and Xiqing Wang published an updated review focusing on symmetry.[67c]

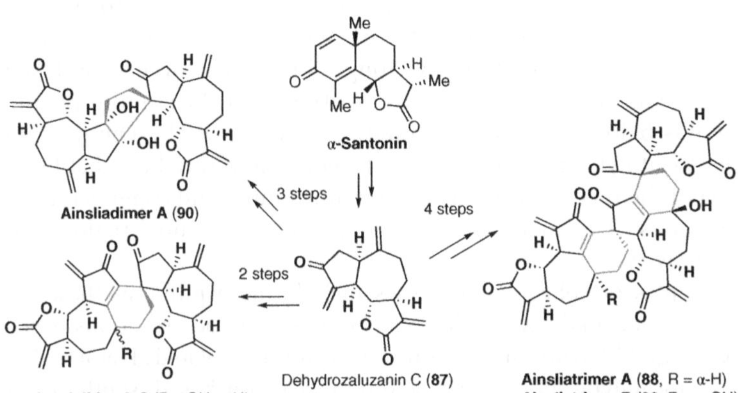

Scheme 1.23 Lei's collective synthesis of oligomeric sesquiterpenoids.

1.3 The Expansion of the Field: Chemical Biology/Chemical Genetics

Chemical biology and chemical genetics are terms used to describe a field that employs the methods of chemistry to study biology. Chemical genetics, which uses small molecules as chemical probes/tools to perturb the function of gene products and allow the systematic dissection of biological processes, and identify small molecules with the ability to induce a biological phenotype or to interact with a particular gene product, is an emerging tool for lead generation in drug discovery.[69]

The studies of FK-506 and rapamycin by S. L. Schreiber and collaborators represent a classical approach of chemical biology in investigating the mechanism of information transfer or signaling through the cytoplasm of the cell, which is one of the great mysteries of cell biology. Cyclosporine A, FK-506, and rapamycin are natural fungal products that possess potent immunosuppressant activity and are capable of specifically inhibit cellular processes. Through chemical biological studies by means of organic synthesis, conformational analysis, and chemical analysis, and in combination with modern techniques in biology such as flow cytometry, they are able to decipher related biological processes at a molecular level.[70] On the other hand, the joint efforts of medicinal chemists, pharmacologists, and chemists from several pharmaceutical companies and universities have also been fruitful, resulting in several approved immunosuppressant and anticancer drugs from rapamycin.[71] Moreover, rapamycin and rapalogs have been shown to increase lifespan and improve other markers of aging in a range of organisms.[71] Table 1.1 shows an overview of these drugs.

1.3.1 Diversity-Oriented Synthesis (DOS)

DOS[72] was conceived as a novel conceptual alternate of combinatorial chemistry for the construction of libraries to study the chemical genetics. The tasks of DOS include the development of efficient pathways to a large amount of skeletal and stereochemical diverse small molecules with defined coordinates in chemical space. In order to achieve the highest levels of structural diversity: The building blocks, stereochemistry, functional groups, and, most importantly, the molecular framework must be varied.[72c]

1.3.2 Function-Oriented Synthesis (FOS)

In view of the difficulty in the efficient synthesis of structurally complex natural products and the uncertainty of the molecule's medicinal profile, the concept of FOS was advanced by Paul A. Wender to achieve function with simple synthetic mimetics.[73] Although such strategy has previously been employed in the

Table 1.1 Drugs developed from immunosuppressant natural products.

Natural Products	Approved Year, Trade Name	Medical Uses
Cyclosporine	1983, Neoral, Sandimmune,	to prevent organ rejection in transplant patients
FK-506	1994, Tacrolimus.	to stave off organ rejection in liver transplants (and other types of organ transplants)
Rapamycin	1999, Rapamune (sirolimus)	to prevent organ transplant rejection;
	2007, Torisel (temsirolimus)	to treat kidney cancer;
	2009, Afinitor (everolimus)	to treat advanced kidney cancer
Approvals for other cancers and for use as an immunosuppressant to prevent rejection of transplanted organs followed.		

pharmaceutical industry, its introduction to academic research is helpful for emphasizing the importance of function,[21c–e] Since simplified targets can be accessed in a step-economic fashion, natural product-based drug discovery can be accelerated. In connection with FOS, very recently, K. Gademann and E. A. Crane reviewed an approach based on natural product derived fragments that can successfully address some of the current challenges in drug discovery.[74] Examples from various stages of the drug development process up to the clinic have been presented.

1.3.3 Biology-Oriented Synthesis (BIOS)

BIOS[75] is a concept advanced by Herbert Waldmann as a structure-based approach to analyze biologically relevant chemical space in view of the use in the development of small molecules for chemical biology and medicinal chemistry research. BIOS is based on structural analysis of the protein and the small-molecule world as well as the combination of structural conservatism and diversity in nature.

1.3.4 Lead-Oriented Synthesis (LOS)

LOS[76] is a concept introduced by Ian Churcher to capture the specific problem of preparing diverse small molecules with lead-like molecular properties[75a] The realization of lead-oriented synthesis requires the development of new synthetic methods and approaches that can deliver large numbers of diverse, lead-like small molecules.

1.4 Addressing the Threats that Humans May Face in the Near Future

A re-recognition by both the scientific community and society of the key role the total synthesis of natural products can play in both science and technology requires tremendous efforts from scientists in the field. Prevising and addressing the major challenges that human may have to face is not only an effective approach but also a duty for scientists.

1.4.1 A. G. Myers' Endeavor

In this context, the longstanding efforts of Andrew G. Myers and his team at Harvard University is very respectable. Eco-environmental problems, resource problems, and resistance of bacteria to various antibiotics may be three major threats that humans will face in the near future. In A. G. Myers and coworkers' recent comprehensive review on the 100-year history of antibiotics discovery and development, they presented their deep concern about the risk of humanity returning to a pre-antibiotic era due to the fact that many major pharmaceutical companies have abandoned antibacterial R&D. Importantly, they revealed the essential and evolving role of chemical synthesis throughout the history of antibiotics, and make the point that this is the clearest path forward to discover future generations of life-saving medicines.[77]

> If the funding system for antibiotics research is not strengthened, if the attitude that academia is not the place for practical innovations persists, and if pharmaceutical companies (and venture capitalists) refuse to prime the pump independently, then the consequences for society could be dire.

This is the last sentence of their review, which presented once again their deep concern on the potential social crisis.

Prior to this review article, in the mid-1990s, A. G. Myers initiated a program aimed at the development of a practical synthetic platform for accessing fully synthetic tetracycline analogs. The motif of this project was to break through the limit of classical approach to antibiotics by semisynthesis or chemical modification of natural antibiotics obtained from fermentation. Via the latter approach, <10 tetracycline antibiotics have been approved over last 60 years, while within the same period of time, for structurally simpler quinolone and β-lactam antibiotics, the approved numbers are >40 and >50, respectively.[78a] Moreover, fully synthetic tetracycline analogs may be helpful in overcome the problems of stability and resistance. After a 12-year of efforts, Myers and coworkers disclosed in 2005 a practical enantioselective synthetic route to a diverse

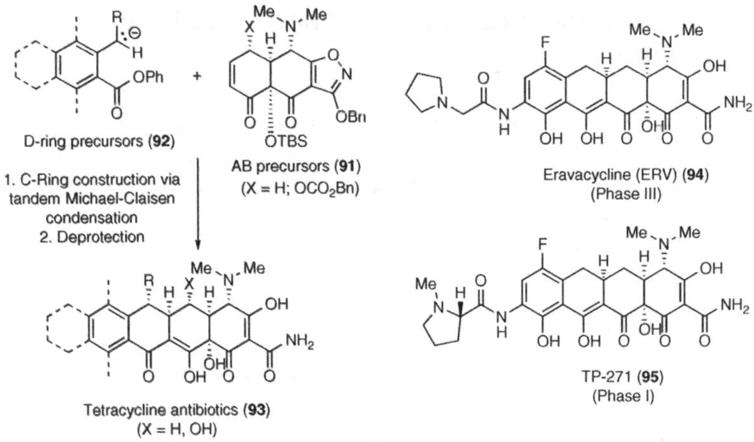

Scheme 1.24 Myers' *de novo* convergent approach to antibiotic tetracycline analogs.

range of fully synthetic 6-deoxytetracycline antibiotics (**93**),[78b,c] which are inaccessible via the conventional semisynthesis approach (Scheme 1.24). Their method consists of separately constructing an AB-ring precursor (**91**) and a D-ring precursor (**92**) containing much of the essential functionality for binding to the bacterial ribosome. The coupling of D-ring precursors (**92**) with an AB precursor (**91**) via tandem Michael–Claisen condensation reactions proceeded diastereoselectively to form a C-ring and thus forged a convergent approach to an unprecedented series of tetracycline analogs (**93**). On the basis of this first generation technology, more than 3000 fully synthetic tetracycline analogs including the clinical candidates eravacycline (**94**, Phase III) and TP-271 (**95**, Phase I), and a preclinical candidate (TP-6076), have been synthesized.[78c] On the other hand, a company, Tetraphase Pharmaceuticals, was founded in 2006 to commercialize the tetracycline synthetic platform. The continuing efforts of Myers' team at Harvard resulted in the development of two entirely different approaches to the key AB enone (**91**).[78a,c] The second-generation synthesis of AB enone was improved and scaling up by the chemists at Tetraphase Pharmaceuticals to allow the synthesis of >100-kg of the AB enone, which enabled clinical development of eravacycline, TP-271, and TP-6076.[78a]

The success inspired and stimulated the development of safer and more effective anti-infective agents of other antibiotic classes. Since the discovery of erythromycin in 1949 by scientists at Eli Lilly, macrolide antibiotics have proven to be safe and effective for use in treating several human infectious diseases. However, due to their structural complexity, in spite of the accomplishments in both the chemical total synthesis and modified biosynthetic routes,[79] all members of this class approved or in clinical development for use in humans have

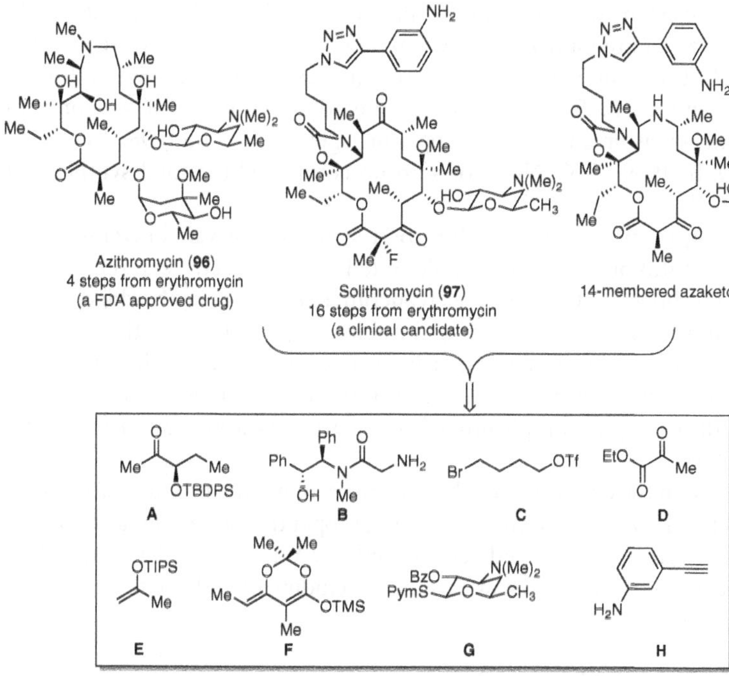

Scheme 1.25 Myers' convergent platform to macrolide antibiotics of the erythromycin class.

been manufactured by semisynthesis from erythromycin and steps for the semisynthesis of new analogs increased over the years.[76a] In 2016, Myers and coworkers developed an ingenious approach to macrolide antibiotics of erythromycin class with rich molecular topological (ring size) and functional group diversity.[80] Their synthetic strategy relies simply on the convergent assembly of simple building blocks **A–H** (Scheme 1.25). Through this convergent platform, more than 300 new macrolide antibiotic candidates, including the clinical candidate solithromycin, have been synthesized. Such fully synthetic and functionality diverse analogs are not accessible by traditional semisynthetic approaches. The majority of those molecules exhibited antibiotic activity, some are efficacious against strains resistant to macrolides in current use.[80a]

1.4.2 D. L. Boger's Endeavor

With the increasingly serious problem of antibiotic resistance, vancomycin (see Figure 5 in the Introduction), aglycopeptide antibiotic, isolated in 1956, and approved as an antibiotic by the FDA in 1958, was considered the last weapon made by human beings to fight against bacteria. However, with the discovery of vancomycin-resistant strains of methicillin-resistant *Staphylococcus aureus* in 1997, there was an urgent need to develop new antibiotics effective against the superbugs. In this connection, the research groups led by D. A. Evans,[81] K. C. Nicolaou,[82] and D. L. Boger[83] achieved the total syntheses of vancomycin aglycon, and its sister antibiotic teicoplanin aglycon,[83a,84] in 1998, and 1999, respectively. In 1999, K. C. Nicolaou's group completed the total synthesis of vancomycin.[85]

Boger's long term endeavor on the total syntheses of vancomycin-related glycopeptide antibiotics and key analogs led to discovery, very recently, of peripheral modified [$\psi[CH_2NH]Tpg^4$]vancomycin analogs with added synergistic mechanisms of action providing durable and potent antibiotics.[86a] Being "over 25,000 times more potent than its' predecessors in activity, the super antibiotic has been hailed as the answer to antibiotic resistance by scientists."[87] This groundbreaking accomplishment is regarded as a victory in the battle against bacteria.[86b]

The progress in the chemistry and biochemistry of natural product-based antibiotics in general,[88] and vancomycin-related glycopeptide antibiotics and analogs in particular,[89] has been reviewed very recently by William M. Wuest and D. L. Boger, respectively, which serves well as the concluding remarks of this chapter.

Acknowledgements

Financial support from the National Key R&D Program of China (grant No. 2017YFA0207302), the National Natural Science Foundation of China (21332007, 21472153, and 21672176) and the Program for Changjiang Scholars and Innovative Research Team in University of Ministry of Education, China gratefully acknowledged.

We thank Ms. Yan-Jiao Gao for her assistance in the preparation of this manuscript, and thank graduate students Xiao-Gao Wang, Xiu-Ning Hu, Yi Lin, Shu-Ren Wang, and Qian He for their help in drawing schemes/figures.

References

1 (a) Trauner, D. *Nat. Prod. Rep.* **2014**, *31*, 411; (b) Mulzer, J. *Nat. Prod. Rep.* **2014**, *31*, 595.
2 Hendrickson, J. B. *J. Am. Chem. Soc.* **1975**, *97*, 5784.
3 (a) Wender, P. A. *Chem. Rev.* **1996**, *96*, 1; (b) Wender, P. A.; Handy, S. T.; Wright, D. L. *Chem. Ind. (London)* **1997**, 765; See also: (c) Wender, P. A.; Bi, F. C.; Gamber, G. G.; Gosselin, F.; Hubbard, R. D.; Scanio, M. J. C.; *et al. Pure Appl. Chem.* **2002**, *74*, 25; (d) Wender, P. A.; Miller, B. L. *Organic Synthesis: Theory and Applications, Vol, 2*; Hudlicky, T. (Ed.) Greenwich: JAI, **1993**, p. 27.

4 Gaich, T.; Baran, P. S. *J. Org. Chem.* **2010**, *75*, 4657.

5 (a) Trost, B. M. *Science* **1983**, *219*, 245; (b) Bartmann, W.; Trost, B. M. (Eds), *Selectivity: A Goal for Synthetic Efficiency*; Weinheim: Verlag Chemie, **1984**.

6 (a) Shenvi, R. A.; O'Malley, D. P.; Baran, P. S. *Acc. Chem. Res.* **2009**, *42*, 530; (b) Afagh, N. A.; Yudin, A. K. *Angew. Chem. Int. Ed.* **2010**, *49*, 262.

7 (a) Szostak, M.; Spain, M.; Procter, D. *J. Chem. Soc. Rev.* **2013**, *42*, 9155; (b) Szostak, M.; Spain, M.; Parmar, D.; Procter, D. J. *Chem. Commun.* **2012**, *48*, 330; (c) Szostak, M.; Procter, D. J. *Angew. Chem. Int. Ed.* **2012**, *51*, 9238.

8 Schwartz, J.; Labinger, J. A. *Angew. Chem., Int. Ed. Engl.* **1976**, *15*, 333.

9 (a) Schedler, D. J. A.; Godfrey, A. G.; Ganem, B. *Tetrahedron Lett.* **1993**, *34*, 5035; (b) Xia, Q.; Ganem, B. *Org. Lett.* **2001**, *3*, 485; (c) Ganem, B.; Franke, R. R. *J. Org. Chem.* **2007**, *72*, 3981.

10 Murray, C. K.; Zheng, Q. Y.; Cheng, X.; Peterson, S. K. U.S. Patent 5,679,807; *Chem. Abstr.* **1996**, *125*, 222220.

11 (a) Oda, Y.; Sato, T.; Chida, N. *Org. Lett.* **2012**, *14*, 950; (b) Nakajima, M.; Wada, T.; Yoritate, M.; Minamikawa, R.; Sato, T.; Chida, N.; *et al. Chem. Eur. J.* **2014**, *20*, 17565; (c) Shirokane, K.; Wada, T.; Yoritate, M.; Minamikawa, R.; Takayama, N.; Sato, T.; *et al. Angew. Chem. Int. Ed.* **2014**, *53*, 512; for a review, see: (d) Sato, T.; Chida, N. *Org. Biomol. Chem.* **2014**, *12*, 3147.

12 (a) Li, X. C.; Lam, H. Y.; Zhang, Y. F.; Chan, C. K. *Org. Lett.* **2010**, *12*, 1724; (b) Tung, C. L.; Wong, C. T. T.; Fung, E. Y. M.; Li, X. C. *Org. Lett.* **2016**, *18*, 2600; (c) Algar, W. R.; Dawson, P.; Medintz, I. L. (eds). *Chemoselective and Bioorthogonal Ligation Reactions. Concepts and Applications: Concepts and Applications. 2* Volumes; Weinheim: Wiley-VCH, **2017**; (d) Zhang, Y. F.; Xu, C.; Lam, H. Y.; Lee, C. L.; Li, X. C. *Proc. Natl. Acad. Sci. U.S.A.* **2013**, *110*, 6657.

13 (a) Huo, H. H.; Xia, X. E.; Zhang, H. K.; Huang, P.-Q. *J. Org. Chem.* **2013**, *78*, 455; (b) Huang, Z. X.; Dong, G. B. *Acc. Chem. Res.* **2017**, *50*, 465; (c) Hartwig, J. F. *Acc. Chem. Res.* **2017**, *50*, 549; (d) Toste, F. D.; Sigman, M. S.; Miller, S. *J. Acc. Chem. Res.* **2017**, *50*, 609.

14 (a) Kolb, H. C; Van Nieuwenhze, M. S.; Sharpless, K. B. *Chem. Rev.* **1994**, *94*, 2483; (b) Hentges, S. G.; Sharpless, K. B. *J. Am. Chem. Soc.* **1980**, *102*, 4263; (c) McKee, B. H.; Gilheany, D. G.; Sharpless, K. B. *Org. Synth. Coll. Vol.* **1998**, *9*, 383; For Noyori's Nobel Prize lecture, see: (d) Sharpless, K. B. *Angew. Chem. Int. Ed.* **2002**, *41*, 2024.

15 (a) Ohta, T.; Takaya, H.; Kitamura, M.; Nagai, K.; Noyori, R. *J. Org. Chem.* **1987**, *52*, 3174; (b) Noyori, R.; Ohkuma, T.; Kitamura, M.; Takaya, H.; Sayo, N.; Kumobayashi, H.; *et al. J. Am. Chem. Soc.* **1987**, *109*, 5856. For Noyori's Nobel Prize lecture, see: (c) Noyori, R. *Angew. Chem. Int. Ed.* **2002**, *41*, 2008; see also: (d) Noyori, R.; Ohkuma, T. *Angew. Chem. Int. Ed.* **2001**, *40*, 40.

16 Xiao, K. J.; Wang, Y.; Huang, Y. H.; Wang, X. G.; Huang, P.-Q. *J. Org. Chem.* **2013**, *78*, 8305.

17 (a) Anastas, P. T. *Aldrichim. Acta* **2015**, *48*, 3; (b) Anastas, P. T.; Kirchhoff, M. M. *Acc. Chem. Res.* **2002**, *35*, 686; (c) Anastas, P. T. *Green Chem.* **2003**, *5*, G29; (d) Anastas, P. T.; Lankey, R. L. *Green Chem.* **2000**, *2*, 289. (e) Anastas, P. T.; Heine, L. G.; Williamson, T. C. Eds. *Green Chemical Syntheses and Processes*; Washington DC: American Chemical Society, **2000**; (f) Anastas, P. T.; Warner, J. C. *Green Chemistry: Theory and Practice*; Oxford University Press: Oxford, **1998**; (g) Leahy, D. K.; Tucker, J. L.; Mergelsberg, I.; Dunn, P. J.; Kopach, M. E.; Purohit, V. C. *Org. Process Res. Dev.* **2013**, *17*, 1099; (h) Dach, R.; Song, J. H. J.; Roschangar, F.; Samstag, W.; C. H. Senanayake *Org. Process Res. Dev.* **2012**, *16*, 1697; (i) Federsel, H.-J. *Acc. Chem. Res.* **2009**, *42*, 671; (j) Andraos, J. *Org. Process Res. Dev.* **2005**, *9*, 149, correction: *Org. Process Res. Dev.* **2005**, *9*, 519; (k) Bryan, M. C.; Dillon, B.; Hamann, L. G.; Hughes, G. J.; Kopach, M. E.; Peterson, E. A.; *et al. J. Med. Chem.* **2013**, *56*, 6007.

18 (a) Trost, B. M. *Science* **1991**, *254*, 1471; (b) Trost, B. M. *Angew. Chem. Int. Ed. Engl.* **1995**, *34*, 259; (c) Trost, B. M. *Acc. Chem. Res.* **2002**, *35*, 695; (d) Trost, B. M.; Frederiksen, M. U.; Rudd, M. T. *Angew. Chem. Int. Ed.* **2005**, *44*, 6630; (e) Li, C. J.; Trost, B. M. *Proc. Natl. Acad. Sci. U.S.A.* **2008**, *105*, 13197.

19 For a review dealing with diverse economies of synthesis, see: (a) Newhouse, T.; Baran, P. S.; Hoffmann, R. W. *Chem. Soc. Rev.* **2009**, *38*, 3010; and (b) Eissen, M.; Mazur, R.; Quebbemann, H.-G.; Pennemann, K.-H. *Helv. Chim. Acta* **2004**, *87*, 524.

20 (a) Sheldon, R. A. *Chem. Ind. (London)* **1992**, 903; (b) Sheldon, R. A. *Chemtech* **1994**, 38; (c) Sheldon, R. A. *Chem. Ind. (London)* **1997**, 12; (d) Sheldon, R. A. *Pure Appl. Chem.* **2000**, *72*, 1233; (e) Sheldon, R. A. *Green Chem.* **2005**, *7*, 267; (f) Sheldon, R. A. *Green Chem.* **2007**, *9*, 1273; (g) Sheldon, R. A. *Chem. Commun.* **2008**, *29*, 3352; (h) Sheldon, R. A. *Green Chem.* **2017**, *19*, 18.

21 (a) Wender, P. A.; Croatt, M. P.; Witulski, B. *Tetrahedron* **2006**, *62*, 7505; (b) Wender, P. A.; Handy, S. T.; Wright, D. L. *Chem. Ind. (London)* **1997**, 765; (c) Wender, P. A.; Verma, V. A.; Paxton, T. J.; Pillow, T. H. *Acc. Chem. Res.* **2008**, *41*, 40; (d) Wender, P. A.; Loy, B. A.; Schrier, A. J. *Israel J. Chem.* **2011**, *51*, 453; (e) Wender, P. A. *Tetrahedron* **2013**, *69*, 7529.

22 (a) Dauben, W. G.; Kessel, C. R.; Takemura, K. H. *J. Am. Chem. Soc.* **1980**, *102*, 6893; (b) Dauben, W. G.; Gerdes, J. M.; Smith, D. B. *J. Org. Chem.* **1985**, *50*, 2576; (c) Grieco, P. A.; Nunes, J. J.; Gaul, M. D. *J. Am. Chem. Soc.* **1990**, *112*, 4595.

23 (a) Mateo, P.; Cinqualbre, J. E.; Mojzes, M. M.; Schenk, K.; Renaud, P. *J. Org. Chem.* **2017**, *82*, 12318; (b) Lee, A. S.; Liau, B. B.; Shair, J. *Am. Chem. Soc.* **2014**, *136*, 13442; (c) Abels, F.; Lindemann, C.; Schneider, C. *Chem.–Eur. J.* **2014**, 1964; (d) Kropf, J. E.; Meigh, I. C.; Bebbington, M. W. P.; Weinreb, S. M. *J. Org. Chem.* **2006**, *71*, 2046.

24 (a) Xiao, K. J.; Luo, J. M.; Ye, K. Y.; Wang, Y.; Huang, P.-Q. *Angew. Chem., Int. Ed.* **2010**, *49*, 3037; (b) Xiao, K. J.; Wang, Y.; Ye, K. Y.; Huang, P.-Q. *Chem. Eur. J.* **2010**, *16*, 12792; (c) Xiao, K. J.; Wang, A. E; Huang, P.-Q. *Angew. Chem., Int. Ed.* **2012**, *51*, 8314; (d) Huo, H. H.; Zhang, H. K.; Xia, X. E.; Huang, P.-Q. *Org.*

Lett. **2012**, *14*, 4834; (e) Huang, P.-Q.; Huang, Y. H.; Xiao, K. J.; Wang, Y.; Xia, X. E. *J. Org. Chem.* **2015**, *80*, 2861; (f) Huang, S. Y.; Chang, Z.; Tuo, S. C.; Gao, L. H.; Wang, A. E; Huang, P.-Q. *Chem. Commun.* **2013**, *49*, 7088; (g) Huang, P.-Q.; Huang, Y. H.; Geng, H.; Ye, J. L. *Sci. Rep.* **2016**, *6*, 28801; (h) Huang, P.-Q.; Ou, W. *Eur. J. Org. Chem.* **2017**, 582; see also: (i) Lindemann, C.; Schneider, C. *Synthesis* **2016**, *48*, 828; (j) Guérot, C; Tchitchanov, B. H.; Knust, H.; Carreira, E. M. *Org. Lett.* **2011**, *13*, 780; for reviews, see: (k) Pace, V.; Holzer, W.; Olofsson, B. *Adv. Synth. Catal.* **2014**, *356*, 3697; (l) Seebach, D. *Angew. Chem., Int. Ed.* **2011**, *50*, 96.

25 Clarke, P. A.; Santos, S.; Martin, W. H. C. *Green Chem.* **2007**, *9*, 438.

26 For reviews, see: (a) Hayashi, Y. *Chem. Sci.* **2016**, *7*, 866; (b) Vaxelaire, C.; Winter, P.; Christmann, M. *Angew. Chem., Int. Ed.* **2011**, *50*, 3605; for recent examples,see: (c) Hayashi, Y.; Koshino, S.; Ojima, K.; Kwon, E. *Angew. Chem. Int. Ed.* **2017**, *56*, 11812; (d) Umemiya, S.; Sakamoto, D.; Kawauchi, G.; Hayashi, Y. *Org. Lett.* **2017**, *19*, 1112. For illustrated examples, see, Chapter 2.

27 Burns, N. Z.; Baran, P. S.; Hoffmann, R. W. *Angew. Chem., Int. Ed.* **2009**, *48*, 2854.

28 For accounts and reviews, see: (a) Kim, S. W.; Zhang, W.; Krische, M. J. *Acc. Chem. Res.* **2017**, *50*, 2371; (b) Feng, J.; Holmes, M.; Krische, M. *J. Chem. Rev.* **2017**, *117*, 12564; (c) Feng, J.; Kasun, Z. A.; Krische, M. J. *J. Am. Chem. Soc.* **2016**, *138*, 5467; (d) Nguyen, K. D.; Park, B. Y.; Luong, T.; Hiroki S.; Garza, V. J.; Krische, M. J. *Science* **2016**, *354*, aah5133; (e) Ketcham, J. M.; Shin, I.; Montgomery, T. P.; Krische, M. J. *Angew. Chem., Int. Ed.* **2014**, *53*, 9142; (f) Dechert-Schmitt, A.-M. R.; Schmitt, D. C.; Gao, X.; Itoh, T.; Krische, M. *J. Nat. Prod. Rep.* **2014**, *31*, 504; (g) Willwacher, J.; Fürstner, A. *Angew. Chem., Int. Ed.* **2014**, *53*, 4217.

29 (a) Xu, C. P.; Xiao, Z. H.; Zhuo, B. Q.; Wang, Y. H.; Huang, P.-Q. *Chem. Commun.* **2010**, *46*, 7834; (b) Wang, R.; Bojase, G.; Willis, A. C.; Paddon-Row, M. N.; Sherburn, M. S. *Org. Lett.* **2012**, *14*, 5652; (c) Chen, J.; Ferreira, A. J.; Beaudry, C. M. *Angew. Chem. Int. Ed.* **2014**, *53*, 11931.

30 (a) Schelhaas, M.; Waldmann, H. *Angew. Chem. Int. Ed. Engl.* **1996**, *35*, 2056; (b) Wang, C. C.; Lee, J. C.; Luo, S. Y.; Kulkarni, S. S.; Huang, Y. W.; Lee, C. C.; Chang, K. L.; Hung, S. C. *Nature* **2007**, *446*, 896; (c) Greene, T. W.; Wuts, P. G. M. *Greene's Protective Groups in Organic Synthesis*, 4th Edn; New York: John Wiley & Sons, Inc., **2006**; (d) Kocienski, P. J. Protecting Groups, 3rd Edn: Stuttgart: Thieme, **2003**.

31 (a) Liang, X., Grue-Sorensen, G., Petersen, A. K., Hogberg, T. *Synlett* **2012**, *23*, 2647; (b) Newman, D. J. *Pharmac. Therap.* **2016**, *162*, 1.

32 (a) Hoffmann, R. W. *Synthesis* **2006**, 3531; (b) Baran, P. S.; Maimone, T. J.; Richter, J. M. *Nature* **2007**, *446*, 404; (c) Hili, R.; Yudin, A. K. *Chem. Eur. J.* **2007**, *13*, 6538; (d) Young, I. S.; Baran, P. S. *Nature Chem.* **2009**, *1*, 193; (e) Li, X. Q.; Liu, A. *Prog. Chem.* **2010**, *22*, 81–90R; (f) Roulland, E. *Angew. Chem., Int. Ed.* **2011**, *50*, 1226; (g) Saicic, R. N. *Tetrahedron* **2014**, *70*, 8183.

33 (a) Shi, H.; Michaelides, I. N.; Darses, B.; Jakubec, P.; Nguyen, Q. N. N.; Paton, R. S.; *et al. J. Am. Chem. Soc.* **2017**, *139*, 17755; (b) Huang, P.-Q.; Ou, W.; Han, F. *Chem. Commun.* **2016**, *52*, 11967.

34 For a review on unprotected carbohydrates as starting material in chemical synthesis, see: Saloranta, T.; Leino, R. *Synlett* **2015**, *26*, 421.

35 (a) Bastug, G.; Dierick, St.; Lebreux, F.; Markó, I. E. *Org. Lett.* **2012**, *14*, 1306; (b) Barrios, F. J.; Springer, B. C.; Colby, D. A. *Org. Lett.* **2013**, *15*, 3082; (c) Yahata, K.; Minami, M.; Watanabe, K.; Fujioka, H. *Org. Lett.* **2014**, *16*, 3680; (d) Morita, K.; Ohta, R.; Aoyama, H.; Yahata, K.; Arisawa, M.; Fujioka, H. *Chem. Commun.* **2017**, *53*, 6605; for a review, see: (e) Ohta, R.; Fujioka, H. *Chem. Pharm. Bull.* **2017**, *65*, 10.

36 (a) Dömling, A.; Wang, W.; Wang, K. *Chem. Rev.* **2012**, *112*, 3083; (b) Cioc, R. C.; Ruijter, E.; Orru, R. V. A. *Green Chem.* **2014**, *16*, 2958; (c) Posner, G. H. *Chem. Rev.* **1986**, *86*, 831.

37 Elders, N.; Born, D. van der; Hendrickx, L. J. D.; Timmer, B. J. J.; Krause, A.; Janssen, E.; de Kanter, F. J. J.; Ruijter, E.; Orru, R. V. A. *Angew. Chem., Int. Ed.* **2009**, *4*, 5856.

38 (a) Guillena, G.; Ramón, D. J.; Yus, M. *Tetrahedron: Asymmetry* **2007**, *18*, 693; (b) Enders, D.; Grondal, C.; Hüttl, M. R. M. *Chem. Int. Ed.* **2007**, *46*, 1570.

39 (a) Enders, D.; Hüttl, M. R. M.; Grondal, C.; Raabe, G. *Nature* **2006**, *441*, 861; (b) Bertelsen, S.; Johansen, R. L.; Jørgensen, K. A. *Chem. Commun.* **2008**, 3016; (c) Enders, D.; Urbanietz, G.; Cassens-Sasse, E.; Keeß, S.; Raabe, G. *Adv. Synth. Catal.* **2012**, *354*, 1481.

40 (a) Wang, Y. H.; Huang, X. F.; Zhang, L. H.; Ye, X. S. *Org. Lett.* **2004**, *6*, 4415; for a perspective, see: (b) Wang, Y. H.; Ye, X. S.; Zhang, L. H. *Org. Biomol. Chem.* **2007**, *5*, 2189.

41 (a) Kuttruff, C. A.; Eastgate, M. D.; Baran, P. S. *Nat. Prod. Rep.* **2014**, *31*, 419; (b) Allred, T. K.; Manoni, F.; Harran, P. G. *Chem. Rev.* **2017**, *117*, 11994.

42 (a) Velluz, L.; Valls, J.; Mathieu, J. *Angew. Chem., Int. Ed. Engl.* **1967**, *6*, 778; (b) Velluz, L.; Valls, J.; Nomine, G. Ibid. **1965**, *4*, 181; (c) Urabe, D.; Asaba, T.; Inoue, M. *Chem. Rev.* **2015**, *115*, 9207; (d) Hill, N.; Paruch, K.; Švenda, J. *Tetrahedron* **2016**, *72*, 3345; (e) Inoue, M. *Acc. Chem. Res.* **2017**, *50*, 460.

43 (a) Thomas, R. *Nat. Prod. Rep.* **2004**, *21*, 224; (b) Humphrey, A. J.; O'Hagan, D. *Nat. Prod. Rep.* **2001**, *18*, 494.

44 (a) Poupon, E.; Nay, B. Eds: *Biomimetic Organic Synthesis*; Weinheim: Wiley-VCH, **2011**, *2*, 956; (b) Poupon, E.; Gravel, E. *Chem. Eur. J.* **2015**, *21*, 10604; (c) Gravel, E.; Poupon, E. *Nat. Prod. Rep.* **2010**, *27*, 32; (d) Bulger, P. G.; Bagal, S. K.; Marquez, R. *Nat. Prod. Rep.* **2008**, *25*, 254; (e) Gravel, E.; Poupon, E. *Eur. J. Org. Chem.* **2008**, 27; (f) Beaudry, C. M.; Malerich, J. P.; Trauner, D. *Chem. Rev.* **2005**, *105*, 4757; (g) De la Torre, M. C.; Sierra, M. A. *Angew. Chem. Int. Ed.* **2004**, *43*, 160; (h) Scholz, U.; Winterfeldt, E. *Nat. Prod. Rep.* **2000**, *17*, 349.

45 (a) Hugelshofer, C. L.; Magauer, T. *Org. Biomol. Chem.* **2017**, *15*, 12; (b) Wang, X.; Ma, Z. Q.; Wang, X. L.; De, S.; Ma, Y. Y.; Chen, C. *Chem. Commun.* **2014**,

50, 8628; (c) Razzak, M. De Brabander, J. K. *Nat. Chem. Biol.* **2011**, *7*, 865; (d) Justicia, J.; De Cienfuegos, L. Á.; Campaña, A. G.; Miguel, D.; Jakoby, V.; Gansäuer, A.; Cuerva, J. M. *Chem. Soc. Rev.* **2011**, *40*, 3525; (e) Ciufolini, M. A.; Canesi, S.; Ousmer, M.; Braun, N. A. *Tetrahedron* **2006**, *62*, 5318; (f) Sorensen, E. J.; Theodorakis, E. A. *Tetrahedron* **2006**, *62*, 5169.

46 (a) Dewick, P. M. *Medicinal Natural Products, A Biosynthetic Approach*, 3rd Edn; Chichester, UK: John Wiley & Sons, Ltd, **2009**; (b) Williams, R. M. *J. Org. Chem.* **2011**, *76*, 4221.

47 Yoder, R. A.; Johnston, J. N. *Chem. Rev.* **2005**, *105*, 4730.

48 For a review, see: (a) Johnson, W. S. *Angew. Chem., Int. Ed. Engl.* **1976**, *15*, 9; (b) Johnson, W. S.; Gravestock, M. B.; McCarry, B. E. *J. Am. Chem. Soc.* **1971**, *93*, 4332; see also: (c) Van Tamelen, E. E. *Acc. Chem. Res.* **1975**, *8*, 152.

49 (a) Piettre, S.; Heathcock, C. H. *Science* **1990**, *248*, 1532; (b) Heathcock, C. H. Angew. *Chem., Int. Ed. Engl.* **1992**, *31*, 665; (c) Heathcock, C. H. *Proc. Natl. Acad. Sci. U. S. A.* **1996**, *93*, 14323; (d) Dean J. Tantillo *Org. Lett.* **2016**, *18*, 4482; for recent reviews on the synthesis of *Daphniphyllum* alkaloids, see: (e) Chattopadhyay, A. K.; Hanessian, S. *Chem. Rev.* **2017**, *117*, 4104; (f) Kang, B.; Jakubec, P.; Dixon, D. J. *Nat. Prod. Rep.* **2014**, *31*, 550.

50 (a) Ardkhean, R.; Caputo, D. F. J.; Morrow, S. M.; Shi, H.; Xiong, Y.; Anderson, E. A. *Chem. Soc. Rev.* **2016**, *45*, 1557; (b) Li, X. W.; Nay, B. *Nat. Prod. Rep.* **2014**, *31*, 533 (transition metal-promoted); (c) Pellissier, H. *Chem. Rev.* **2013**, *113*, 442; (d) Pellissier, H. ed. *Asymmetric Domino Reactions*; London: Royal Society of Chemistry, **2013**. Tietze, L. F.; Düfert, S.-C.; Hierold, J.; *Domino Reactions in the Total Synthesis of Natural Products*, In *Domino Reactions: Concepts for Efficient Organic Synthesis*, **2014**, pp. 523–578; (e) Tietze, L. F.; Stewart, S. G.; Düfert, A. *Domino Reactions in the Enantioselective Synthesis of Bioactive Natural Products, in Modern Tools for the Synthesis of Complex Bioactive Molecules*, **2012**, pp. 271–334. Anderson, E. A. *Org. Biomol. Chem.* **2011**, *9*, 3997; (f) Nicolaou, K. C.; Chen, J. S. *Chem. Soc. Rev.* **2009**, *38*, 2993; (g) Nicolaou, K. C.; Edmonds, D. J.; Bulger, P. G. *Angew. Chem., Int. Ed.* **2006**, *45*, 7134; (h) Kim, J.; Movassaghi, M. *Chem. Soc. Rev.* **2009**, *38*, 3035; (i) Toure, B. B.; Hall, D. G. *Chem. Rev.* **2009**, *109*, 4439; (j) Juhl, M.; Tanner, D. *Chem. Soc. Rev.* **2009**, *38*, 2983; (k) Vilotijevic, I.; Jamison, T. F. *Angew. Chem., Int. Ed.* **2009**, *4*, 5250; (l) Tietze, L. F. *Chem. Rev.* **1996**, *96*, 115; (m) Bunce, R. *Tetrahedron*, **1995**, *51*, 13103.

51 (a) Broadwater, S. J.; Roth, S. L.; Price, K. E.; Kobašlija, M.; McQuade, D. T. *Org. Biomol. Chem.* **2005**, *3*, 2899; for definition of tandem reactions, cascade reactions, and domino reactions etc., see ref. 26a, and references cited therein.

52 (a) Volla, C. M. R.; Atodiresei, I.; Rueping, M. *Chem. Rev.* **2014**, *114*, 2390; (b) Pellissier, H. *Adv. Synth. Catal.* **2012**, *354*, 237; (c) Vaxelaire, C.; Winter, P.; Christmann, M. *Angew. Chem., Int. Ed.* **2011**, *50*, 3605; (d) Albrecht, Ł.; Jiang, H.; Jørgensen, K. A. *Angew. Chem. Int. Ed.* **2011**, *50*, 8492; (e) Grondal, C.;

Jeanty, M.; Enders, D. *Nat. Chem.* **2010**, *2*, 167; (f) Rendler, S.; MacMillan, D. W. C. *J. Am. Chem. Soc.* **2010**, *132*, 5027.

53 For a review on short pathways to complexity generation in biosynthesis, see: (a) Wang, P.; Gao, X.; Tang, Y. *Curr. Opin. Chem. Biol.* **2012**, *16*, 362; for selected examples, see: (b) Walsh, C. T.; Haynes, S. W.; Ames, B. D.; Gao, X.; Tang, Y. *ACS Chem. Biol.* **2013**, *8*, 1366; (c) Haynes, S. W.; Gao, X.; Tang, Y.; Walsh, C. T. *ACS Chem. Biol.* **2013**, *8*, 741.

54 For reviews, see: (a) Chauhan, P.; Mahajan, S.; Enders, D. *Acc. Chem. Res.* **2017**, *50*, 2809; (b) Williamson, A. E.; Ngouansavanh, T.; Pace, R. D. M.; Allen, A. E.; Cuthbertson, J. D.; Gaunt, M. *J. Synlett* **2016**, *27*, 116; (c) Echavarren, R. D. M. *Chem. Rev.* **2015**, *115*, 9028; (d) Just-Baringo, X.; Procter, D. J. *Acc. Chem. Res.* **2015**, *48*, 1263 (SmI_2); (e) Li, J.; Eastgate, M. D. *Org. Biomol. Chem.* **2015**, *13*, 7164; (f) Yoshimitsu, T. *Chem. Rec.* **2014**, *14*, 268; (g) Nicolaou, K. C.; Hale, C. R. H.; Nilewski, C.; Ioannidou, H. A. *Chem. Soc. Rev.* **2012**, *41*, 5185; (h) Ruijter, E.; Scheffelaar, R.; Orru, R. V. A. *Angew. Chem. Int. Ed.* **2011**, *50*, 6234; (i) Davies, H. M. L.; Sorensen, E. *J. Chem. Soc. Rev.* **2009**, *38*, 2981; (j) Walji, A. M.; MacMillan, D. W. C. *Synlett* **2007**, 1477.

55 (a) Jackson, A. H.; Shannon, P. V. R. Wilkins, D. J. *Tetrahedron Lett.* **1987**, *28*, 4901; (b) Martin, D. B. C.; Nguyen, L. Q.; Vanderwal, C. D. *J. Org. Chem.* **2012**, *77*, 17.

56 (a) Bertz, S. H. *J. Am. Chem. Soc.* **1982**, *104*, 5801; (b) Bertz, S. H.; Sommer, T. J. *Chem. Commun.* **1997**, 2409; (c) Chanon, M.; Barone, R.; Baralotto, C.; Julliard, M.; Hendrickson, J. B. *Synthesis* **1998**, 1559; (d) Barone, R.; Chanon, M. *Tetrahedron* **2005**, *61*, 8916; (e) Proudfoot, J. R. *Bioorg. Med. Chem. Lett.* **2017**, *27*, 2014; (f) Li, J.; Eastgate, M. D. *Org. Biomol. Chem.* **2015**, *13*, 716.

57 (a) Berridge, M. J.; Irvine, R. F. *Nature* **1989**, *341*, 197. (b) Potter, B. V. L. *Nat. Prod. Rep.* **1990**, *7*, 1.

58 (a) Sculimbrene, B. R.; Miller, S. J. *J. Am. Chem. Soc.* **2001**, *123*, 10125; for a highlight, see: (b) Gani, D. *Nature* **2001**, *414*, 703.

59 (a) Chen, D. Y.-K.; Youn, S. W. *Chem. Eur. J.* **2012**, *18*, 9452; (b) Yamaguchi, J.; Yamaguchi, A. D.; Itami, K. *Angew. Chem. Int. Ed.* **2012**, *51*, 8960; (c) T. Brückl, R. D. Baxter, Y. Ishihara, P. S. Baran, *Acc. Chem. Res.* **2012**, *45*, 826; (d) Newhouse, T.; Baran, P. S. *Angew. Chem. Int. Ed.* **2011**, *50*, 3362.

60 (a) Chen, M. S.; White, M. C. *Science* **2007**, *318*, 783; (b) Zhang, Z. P.; Tanaka, K.; Yu, J. Q. *Nature* **2017**, *543*, 538; (c) Wang, Y.; Wang, C.; Butler, J. R.; Ready, J. M. *Angew. Chem. Int. Ed.* **2013**, *52*, 10796; (d) Kelly, T. R.; Jagoe, C. T.; Li, Q. *J. Am. Chem. Soc.* **1989**, *111*, 4522.

61 (a) Sears, J. E.; Boger, D. L. *Acc. Chem. Res.* **2016**, *49*, 241; (b) Allemann, O.; Brutsch, M.; Lukesh, J. C., III; Brody, D. M.; Boger, D. L. *J. Am. Chem. Soc.* **2016**, *138*, 8376; (c) Sears, Justin E.; Boger, Dale L. *Acc. Chem. Res.* **2015**, *48*, 653.

62 For selected reviews, see: (a) Hanessian, S. *Total Synthesis of Natural Products: The "Chiron Approach;"* Oxford: Pergamon, **1984**; (b) Hanessian, S. *Aldrichimica Acta* **1989**, *22*, 3; (c) Hanessian, S. *Pure Appl. Chem.* **1993**, *65*,

1189; For recent developments, see: (d) Hanessian, S.; Giroux, S.; Merner, B. L.; *Design and Strategy in Organic Synthesis: From the Chiron Approach to Catalysis*; Wiley-VCH: Weinheim, **2012**. (e) Hanessian, S. *J. Org. Chem.* **2012**, *77*, 6657.

63 (a) Woerly, E. M.; Roy, J.; Burke, M. D. *Nat. Chem.* **2014**, *6*, 484; (b) Wang, A. E; Huang, P.-Q. *Pure Appl. Chem.* **2014**, *86*, 1227.

64 Brill, Z. G.; Condakes, M. L.; Ting, C. P.; Maimone, T. J. *Chem. Rev.* **2017**, *117*, 11753.

65 (a) Wang, X. B.; Xia, D. L.; Qin, W. F.; Zhou, R. J.; Zhou, X. H.; Zhou, Q. L.; *et al. Chem* **2017**, *2*, 803; (b) Zhao, G. Y.; Xie, X. G.; Sun, H. Y.; Yuan, Z. Y.; Zhong, Z. L.; Tang, S. C.; *et al. Org. Lett.* **2016**, *18*, 2447; (c) Han, J. C.; Li, C. C. *Synlett* **2015**, *26*, 1289; (d) Xu, Z. R.; Wang, Q.; Zhu, J. P. *J. Am. Chem. Soc.* **2015**, *137*, 6712; (e) Han, J. C.; Li, F. Z.; Li, C. C. *J. Am. Chem. Soc.* **2014**, *136*, 13610; (f) Wang, J.; Chen, S.-G.; Sun, B.-F.; Lin, G.-Q.; Shang, Y.-J. *Chem. Eur. J.* **2013**, *19*, 2539; (g) Wang, X. M.; Li, H. H.; Lei, X. G. *Synlett* **2013**, *24*, 1032; (h) Li, H. H.; Wang, X. M.; Hong, B. K.; Lei, X. G. *J. Org. Chem.* **2013**, *78*, 800; (i) Zhang, J.; Wu, J.; Hong, B.; Ai, W.; Wang, X.; Li, H.; Lei, X. *Nat. Commun.* **2014**, *5*, 4614; (j) Lee, A. S.; Liau, B. B.; Shair, M. D. *J. Am. Chem. Soc.* **2014**, *136*, 13442.

66 Snyder, S. A.; Gollner, A.; Chiriac, M. I. *Nature* **2011**, *474*, 461.

67 (a) Snyder, S. A.; El Sohly, A. M.; Kontes, F. *Nat. Prod. Rep.* **2011**, *28*, 897; (b) Li, C.; Lei, X. G. *J. Org. Chem.* **2014**, *79*, 3289; (c) Bai, W. J.; Wang X. *Nat. Prod. Rep.* **2017**, *34*, 1345.

68 (a) Li, C.; Yu, X.; Lei, X. *Org. Lett.* **2010**, *12*, 4284; (b) Li, C.; Dian, L.; Zhang, W.; Lei, X. *J. Am. Chem. Soc.* **2012**, *134*, 12414; (c) Li, C.; Dong, T.; Dian, L.; Zhang, W.; Lei, X. *Chem. Sci.* **2013**, *4*, 1163.

69 (a) Schreiber, S. L. *Chem. Eng. News* **2003**, *81*, 51; (b) Strausberg, R. L.; Schreiber, S. L. *Science* **2003**, *300*, 294; (c) Stockwell, B. R. *Nature* **2004**, *432*, 847; (d) Schreiber, S. L. *Nat. Chem. Biol.* **2005**, *1*, 64; (e) Walsh, D. P.; Chang, Y. T. *Chem. Rev.* **2006**, *106*, 2476; (f) O'Connor, C. J.; Laraia, L.; Spring, D. R. *Chem. Soc. Rev.* **2011**, *40*, 4332; (g) Spring, D. R. *Chem. Soc. Rev.* **2005**, *34*, 472; (h) Carlson, E. E. *ACS Chem. Biol.* **2010**, *5*, 639; (i) Thiel, P.; Kaiser, M.; Ottmann, C. *Angew. Chem. Int. Ed.* **2012**, *51*, 2012; (j) Aeluri, M.; Chamakuri, S.; Dasari, B.; Guduru, S. K. R.; Jimmidi, R.; Jogula, S.; Arya, P. *Chem. Rev.* **2014**, *114*, 4640; (k) Szpilman, A. M.; Carreira, E. M. *Angew. Chem. Int. Ed.* **2010**, *49*, 9592.

70 (a) Choi, J. W.; Chen, J.; Schreiber S. L.; Clardy, J. *Science* **1996**, *273*, 239; (b) Schreiber, S. L.; Albers, M. W.; Brown, E. J. *Acc. Chem. Res.* **1993**, *26*, 412; (c) Rasen, M. K.; Schreiber, S. L. *Angew. Chem., Int. Ed. Eng.* **1992**, *31*, 384.

71 Halford, B. *C&EN Global Enterp.* **2016**, *94*, 26.

72 (a) Schreiber, S. L. *Science* **2000**, *287*, 1964; (b) Arya, P.; Chou, D. T. H.; Baek, M.-G. *Angew. Chem. Int. Ed.* **2001**, *40*, 339; (c) Tan, D. S. *Nat. Chem. Biol.* **2005**, *1*, 74; (d) Schreiber, S. L. *Nature* **2009**, *457*, 153; (e) Galloway, W. R. J. D.;

Bender, A.; Welch, M.; Spring, D. R. *Chem. Commun.* **2009**, 2446; (f) Dandapani, S.; Marcaurelle, L. A. *Curr. Opin. Drug Disc. Dev.* **2010**, *14*, 362; (g) Galloway, W. R. J. D.; Isidro-Llobet, A.; Spring, D. R. *Nat. Commun.* **2010**, *1*, 80; (h) O'Connor, C. J.; Beckmann, H. S. G.; Spring, D. R. *Chem. Soc. Rev.* **2012**, *41*, 4444; (i) Collins, I.; Jones, A. M. *Molecules* **2014**, *19*, 17221; (j) Lenci, E.; Guarna, A.; Trabocchi, A. *Molecules* **2014**, *19*, 16506.

73 (a) Wender, P. A. *Nat. Prod. Rep.* **2014**, *31*, 433; (b) Wender, P. A.; Quiroz, R. V.; Stevens, M. C. *Acc. Chem. Res.* **2015**, *48*, 752; see also: (c) Ichikawa, S. *Chem. Rec.* **2016**, 1106.

74 Crane, E. A.; Gademann, K. *Angew. Chem. Int. Ed.* **2016**, *55*, 3882.

75 (a) Wetzel, S.; Bon, R. S.; Kumar, K.; Waldmann, H. *Angew. Chem. Int. Ed.* **2011**, *50*, 10800; (b) Narayan, R.; Potowski, M.; Jia, Z. J.; Antonchick, A. P.; Waldmann, H. *Acc. Chem. Res.* **2014**, *47*, 1296.

76 (a) Keserü, G. M.; Makara, G. M. *Nat. Rev. Drug Discov.* **2009**, *8*, 203; (b) Nadin, A.; Hattotuwagama, C.; Churcher, I. *Angew. Chem. Int. Ed.* **2012**, *51*, 1114; (c) Doveston, R.; Marsden, S.; Nelson, A. *Drug Discov. Today* **2014**, *19*, 813.

77 (a) Wright, P. M.; Seiple, I. B.; Myers, A. G. *Angew. Chem. Int. Ed.* **2014**, *53*, 8840; For related reviews, see: (b) Ng, V.; Chan, W. C. *Chem. Eur. J.* **2016**, *22*, 12606; (c) Blaskovich, M. A. T.; Zuegg, J.; Elliott, A. G.; Cooper, M. A. *ACS Infect. Dis.* **2015**, *1*, 285; (d) O'Connell, K. M. G.; Hodgkinson, J. T.; Sore, H. F.; Welch, M.; Salmond, G. P. C.; Spring, D. R. *Angew. Chem. Int. Ed.* **2013**, *52*, 10706; (e) Stoilova, T.; Colombo, L.; Forloni, G.; Tagliavini, F.; Salmona, M. *J. Med. Chem.* **2013**, *56*, 5987; (f) Nussbaum, F.; Brands, M.; Hinzen, B.; Weigand, S.; Häbich, D. *Angew. Chem. Int. Ed.* **2006**, *45*, 5072; for a review on the chemistry and biology of naturally occurring antibiotics, see: (g) Nicolaou, K. C.; Chen, J. S.; Edmonds, D. J.; Estrada, A. A. *Angew. Chem. Int. Ed.* **2009**, *48*, 660.

78 For a review, see: (a) Liu, F.; Myers. A. G. *Curr. Opin. Chem. Biol.* **2016**, *32*, 48; (b) Charest, M. G.; Lerner, C. D.; Brubaker, J. D.; Siegel, D. R.; Myers, A. G. *Science* **2005**, *308*, 395; (c) Sun, C.; Wang, Q.; Brubaker, J. D.; Wright, P. M.; Lerner, C. D.; Noson, K.; *et al. J. Am. Chem. Soc.* **2008**, *130*, 17913.

79 Park, S. R. *et al. Appl. Microbiol. Biotechnol.* **2010**, *85*, 1227.

80 (a) Seiple, I. B.; Zhang, Z. Y.; Jakubek, P.; Langlois-Mercier, A.; Wright, P. M.; Hog, D. T.; *et al. Nature* **2016**, *533*, 338; for a highlight, see: (b) Yan, M.; Baran, P. S. *Nature* **2016**, *533*, 326.

81 (a) Evans, D. A.; Wood, M. R.; Trotter, B. W.; Richardson, T. I.; Barrow, J. C.; Katz, J. L. *Angew. Chem., Int. Ed.* **1998**, *37*, 2700; (b) Evans, D. A.; Dinsmore, C. J.; Watson, P. S.; Wood, M. R.; Richardson, T. I.; Trotter, B. W.; *et al. Angew. Chem., Int. Ed.* **1998**, *37*, 2704.

82 (a) Nicolaou, K. C.; Nataranjan, S.; Li, H.; Jain, N. F.; Hughes, R.; Solomon, M. E.; *et al. Angew. Chem., Int. Ed.* **1998**, *37*, 2708; (b) Nicolaou, K. C.; Jain, N. F.; Nataranjan, S.; Hughes, R.; Solomon, M. E.; Li, H.; *et al. Angew. Chem., Int. Ed.*

1998, *37*, 2714; (c) Nicolaou, K. C.; Takayanagi, M.; Jain, N. F.; Nataranjan, S.; Koumbis, A. E.; Bando, T.; *et al. Angew. Chem., Int. Ed.* **1998**, *37*, 2717; (d) Nicolaou, K. C.; Li, H.; Boddy, C. N. C.; Ramanjulu, J. M.; Yue, T.-Y.; Natarajan, S.; Chu, X.-J.; *et al. Chem. Eur. J.* **1999**, *5*, 2584; (e) Nicolaou, K. C.; Boddy, C. N. C.; Li, H.; Koumbis, A. E.; Hughes, R.; Natarajan, S.; *et al. Chem. Eur. J.* **1999**, *5*, 2602; (f) Nicolaou, K. C.; Koumbis, A. E.; Takayanagi, M.; Natarajan, S.; Jain, N. F.; Bando, T.; *et al. Chem. Eur. J.* **1999**, *5*, 2622.

83 (a) Boger, D. L.; Miyazaki, S.; Kim, S. H.; Wu, J. H.; Loiseleur, O.; Castle, S. L. *J. Am. Chem. Soc.* **1999**, *121*, 3226; (b) Boger, D. L.; Miyazaki, S.; Kim, S. H.; Castle, S. L.; Wu, J. H.; Loiseleur, O.; *et al. J. Am. Chem. Soc.* **1999**, *121*, 10004.

84 (a) Boger, D. L.; Kim, S. H.; Miyazaki, S.; Strittmatter, H.; Weng, J.-H.; Mori, Y.; *et al. J. Am. Chem. Soc.* **2000**, *122*, 7416; (b) Boger, D. L.; Kim, S. H.; Mori, Y.; Weng, J.-H.; Rogel, O.; Castle, S. L.; *et al. J. Am. Chem. Soc.* **2001**, *123*, 1862. cEvans, D. A.; Katz, J. L.; Peterson, G. S.; Hintermann, T. J. *J. Am. Chem. Soc.* **2000**, *122*, 12411.

85 (a) Nicolaou, K. C.; Mitchell, H. J.; Jain, N. F.; Winssinger, N.; Hughes, R.; Bando, T. *Angew. Chem., Int. Ed.* **1999**, *38*, 240; (b) Nicolaou, K. C.; Mitchell, H. J.; Jain, N. F.; Bando, T.; Hughes, R.; Winssinger, N.; Natarajan, S.; *et al. Chem. Eur. J.* **1999**, *5*, 2648.

86 (a) Okano, A.; Isley, N. A.; Boger, D. L. *Proc. Natl. Acad. Sci. U.S.A.* **2017**, *114*, E5052; for a commentary, see: (b) *Proc. Natl. Acad. Sci. U.S.A.* **2017**, *114*, 6656.

87 The Boger Group, *Website*. Available at: www.scripps.edu/boger/ (accessed February 2018).

88 Rossiter, S. E.; Fletcher, M. H.; Wuest, W. M. *Chem. Rev.* **2017**, *117*, 12415.

89 (a) Okano, A.; Isley, N. A.; Boger, D. L. *Chem. Rev.* **2017**, *117*, 11952; see also: (b) Boger, D. L. *J. Org. Chem.* **2017**, *82*, 11961.

2

Selected Procedure-Economical Enantioselective Total Syntheses of Natural Products

Pei-Qiang Huang

Fujian Provincial Key Laboratory of Chemical Biology, Department of Chemistry and iChEM, Xiamen University, Xiamen, China

With the aim of illustrating the principles of efficiency in the total syntheses of natural products, selected concise enantioselective total syntheses are presented in this chapter. Because both "step-economy" and "pot-economy" are used in literature to describe concise total syntheses, we opt to use "procedure-economy"[1] in the chapter title to cover the two categories, and leave the terms "step-economy" and "pot-economy" for individual synthesis. With the intention to illustrate the current status of the total synthesis in terms of efficiency, only the shortest enantioselective total syntheses of natural products (to date) are selected, which are presented in order from 1 to 28 procedures. The step count refers to the longest linear sequence (LLS) whether it was indicated or not. Due to limited space, about half of them are presented with analysis, and others are listed in Table 2.2 later with key information provided. Considering that enantioselective synthesis is popular in contemporary organic chemistry, and that total syntheses should represent the state-of-the–art of organic synthesis, this chapter will not include total syntheses of unnatural enantiomers (with two exception) and of racemates, even in cases where the natural products have been isolated as racemates. Please note that, some of the shortest total syntheses gave lower overall yields compared with alternative ones with longer steps, reflecting the current states of organic synthesis in terms of selectivity control.

Efficiency in Natural Product Total Synthesis, First Edition. Edited by Pei-Qiang Huang, Zhu-Jun Yao, and Richard P. Hsung.

2.1 One-Step/One-Pot Enantioselective Total Synthesis of Natural Products/Drugs

2.1.1 Robinson's One-Step Synthesis of Tropinone

In 1917, R. Robinson reported the seminal one-step synthesis of achiral tropinone (**1**). The synthesis consisted of the condensation of succinaldehyde (**2**) with a calcium salt of acetone dicarboxylic acid and methylamine in aqueous solution,[2] acidification of the initial condensation product (calcium tropinone dicarboxylate **4**), and a spontaneous decarboxylation of *β*-keto diacid **5** to give, in one pot, tropinone (**1**) at a 42% yield (Scheme 2.1).

This landmark synthesis is significant on several aspects. First, compared with the previous 19-step synthesis of tropinone, reported in 1901 by another great chemist Willstätter, Robinson's one-pot synthesis was revolutionary. Second, this was not only a breakthrough at that time, but also set a high standard for modern organic synthesis, and is still admired by today's organic chemists. Third, what is amazing is that at this time, Robinson had already put forward the concept of developing "an economical process."[2] Moreover, as can be seen from the following sentence extracted from his paper, he presented to readers the concept and primary method of synthetic analysis:

> By imaginary hydrolysis at the points indicated by the dotted lines, the substance may be resolved into succinaldehyde (**2**), methylamine, and acetone, and this observation suggested a line of attack of the problem which has resulted in a direct synthesis.[2]

It is also important to note that behind these analyses was his idea of biogenetic synthesis and of imitating plant biogenetic synthesis in a chemical laboratory. As a result of this effort, he was able to advance the first rational theories of the biogenesis of alkaloids.[3] In today's organic chemistry, Robinson's synthesis of tropinone is taken as a model of biomimetic synthesis.

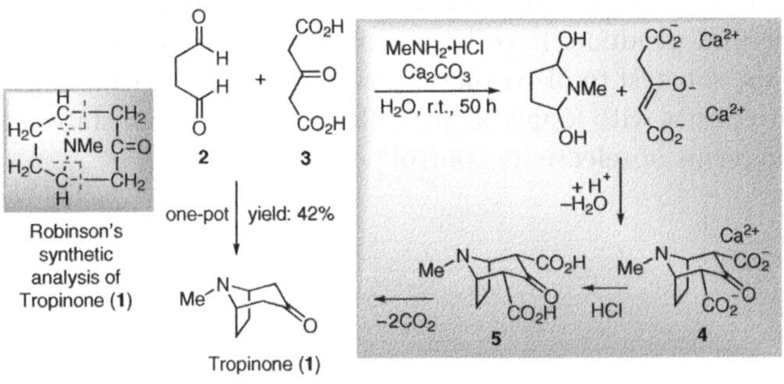

Scheme 2.1 Robinson's synthetic analysis and one-step synthesis of tropinone (**1**).

2.1.2 Hayashi's One-Pot Synthesis of (+)-ABT-341

ABT-341 (**6**) is a highly potent, selective, and orally bioavailable DPP4 inhibitor developed by Abbott Laboratories.[4] Previously, it was synthesized in 11 steps.[4] As a continuation of their endeavor to develop pot-economical total synthesis (*vide infra*), Y. Hayashi and colleagues reported in 2011 a high-yielding one-pot synthesis of ABT-341 (**6**) (Scheme 2.2).[5] The one-pot four-component reaction sequence consists of an diphenylprolinol silyl ether **7a** catalyzed enantioselective Michael reaction of acetaldehyde with nitroalkene **8**, a domino Michael reaction/Horner–Wadsworth–Emmons reaction of the nitro intermediate with vinyl phosphonate, epimerization, *O*-deprotection, amidation, and reduction. From this one-pot procedure, ABT-341 (**6**) was obtained at a 63% yield over six steps in a highly diastereoselective and enantioselective manner.

2.2 Two-Step/Two-Pot Enantioselective Total Synthesis of Natural Products

2.2.1 Hayashi's Two-Pot Synthesis of (−)-Oseltamivir

The re-emergence of influenza as a global major health problem in the last decade has attracted worldwide attention. After the global pandemic of AH1N1 influenza that resulted in many deaths, the high potential risk of a worldwide

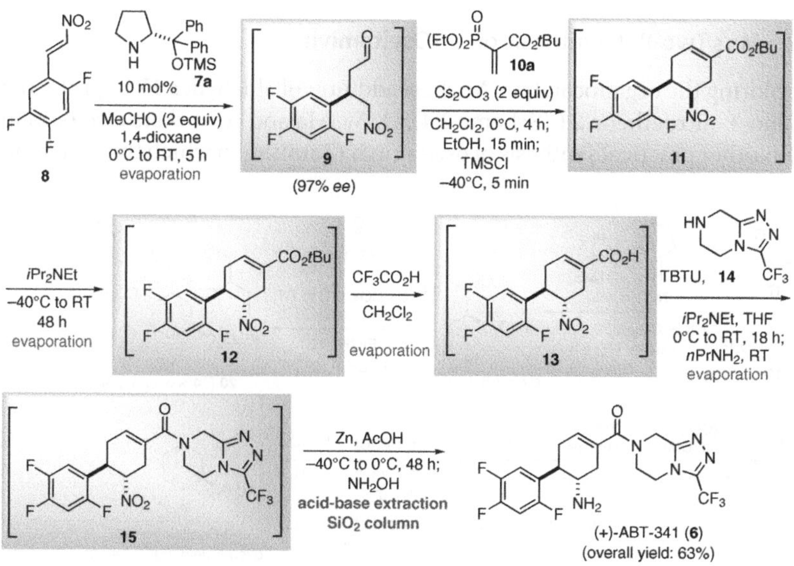

Scheme 2.2 Hayashi's one-pot synthesis of (+)-ABT-341 (**6**).

spread of the more deadly avian H5N1 influenza virus attracted serious concern. Neuraminidase inhibitors (−)-oseltamivir phosphate (Tamiflu®, **16**) and zanamivir (Relenza) turned out to be effective antiviral flu drugs that have found widespread clinical application for the treatment of influenza caused by the AH1N1 virus.[6] In 2009, the best yield that had achieved for the total synthesis of oseltamivir (**16**) was approximately 35%. Thus, great effort has been devoted to developing more efficient synthetic approaches. In this context, organocatalysis turned out to be a powerful tool toward that goal. As a result of their endeavor to develop pot-economical total synthesis, in 2009, Yujiro Hayashi and coworkers published a high-yielding synthesis of (−)-oseltamivir (**16**) by three "one-pot" procedures.[7a] The method was later improved to two "one-pot" reaction sequences with one column chromatography purification, which afforded (−)-oseltamivir (**16**) at a 60% overall yield (Scheme 2.3).[7b] The first one-pot reaction sequence consists of the diphenylprolinol silyl ether (**7a**) mediated asymmetric Michael reaction, a domino Michael addition reaction/Horner–Wadsworth–Emmons reaction (and retro-aldol reaction/Horner–Wadsworth–Emmons reaction and retro-Michael reactions to convert the side product to the desired product), and a thiol Michael reaction, as well as a base-catalyzed isomerization. It is remarkable that the first one-pot sequence not only built the cyclohexane ring bearing all the required functionalities, but also formed five stereogenic centers with correct stereochemistries.

Recently, the synthesis was further improved to a one-pot reaction,[7c] which could be completed in 1 hour.[7d]

2.2.2 Ma's Two-Pot Synthesis of (−)-Oseltamivir

By exploring the organocatalytic Michael addition of aldehydes **17** to protected 2-amino-1-nitroethene **21** (Scheme 2.4), D.-W. Ma and coworkers disclosed in 2010 another practical synthesis of oseltamivir (Tamiflu) in which the acylanino

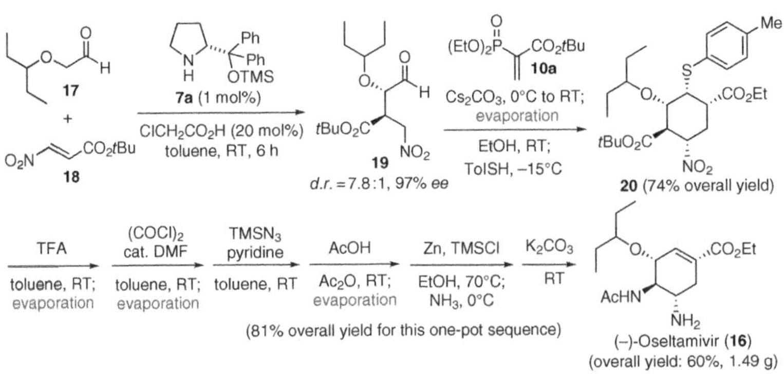

Scheme 2.3 Hayashi's two-pot synthesis of (−)-oseltamivir (**16**).

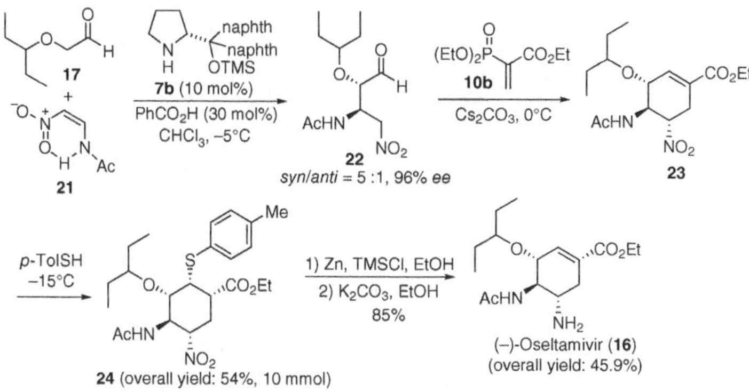

Scheme 2.4 Ma's two-pot enantioselective synthesis of (–)-oseltamivir (**16**).

group found in the target molecule was installed with correct stereochemistry in the first step.[8] The overall yield of this five-step two-pot synthesis was as high as 45.9%.

2.2.3 Li's Two-Step Chemoenzymatic Total Synthesis of Aszonalenin

Aszonalenin (**25a**) and its derivatives display remarkable activity as substance P inhibitors for the human neurokinin-1 receptor. S.-M. Li and coworkers[9] developed a two-step total synthesis of four aszonalenin stereoisomers **25a-25d** from (*R*)-benzodiazepinedione (**26**) and (*S*)-**26**, available from the condensation reaction of D- or L-tryptophan with isatoic anhydride (**27**)[10] (Scheme 2.5). The method features a chemoenzymatic synthesis using (*R*)-**26** and (*S*)-**26** as substrates and AnaPT and CdpNPT as catalysts. AnaPT from *Neosartorya fischeri* was used to catalyze the C3-prenylation of benzodiazepinedione (**26**), while CdpNPT from *Aspergillus fumigatus* catalyzed the N1-prenylation of tryptophan-containing cyclic dipeptides. The stereoselectivity of the one-step reaction was about 100% and the conversion rates reached 85–100%.

2.2.4 Ishikawa's Two-Step Total Syntheses of (+)-WIN 64821 and (+)-Naseseazine B

(+)-WIN 64821 (**28**), (–)-ditryptophenaline (**29**) and (+)-naseseazine B (**30**) belong to a large group of tryptophan-based dimeric diketopiperazine alkaloids isolated from fungi, which show fascinating biological activities. On the basis of their proposed biosynthetic pathway, and with the consideration that true biosynthesis should be carried out in aqueous media without special protective groups on the substrates, in 2013, H. Ishikawa and coworkers developed a two-step syntheses of (+)-WIN 64821 and (+)-naseseazine B (Scheme 2.6).[11a]

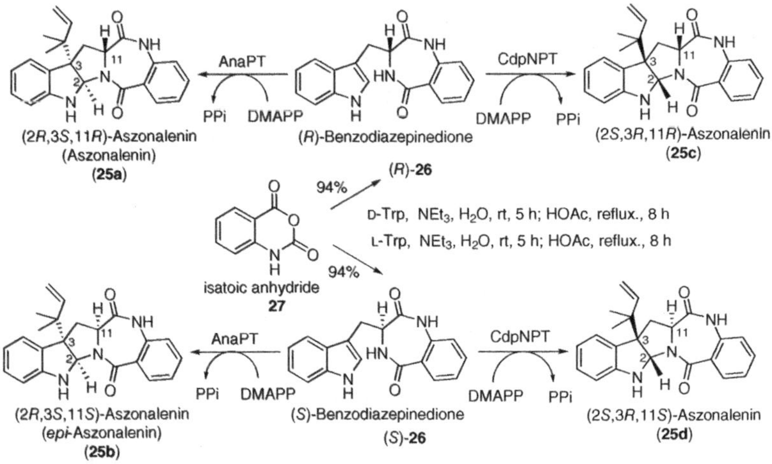

Scheme 2.5 Li's two-step chemoenzymatic total synthesis of four stereoisomers of aszonalenin.

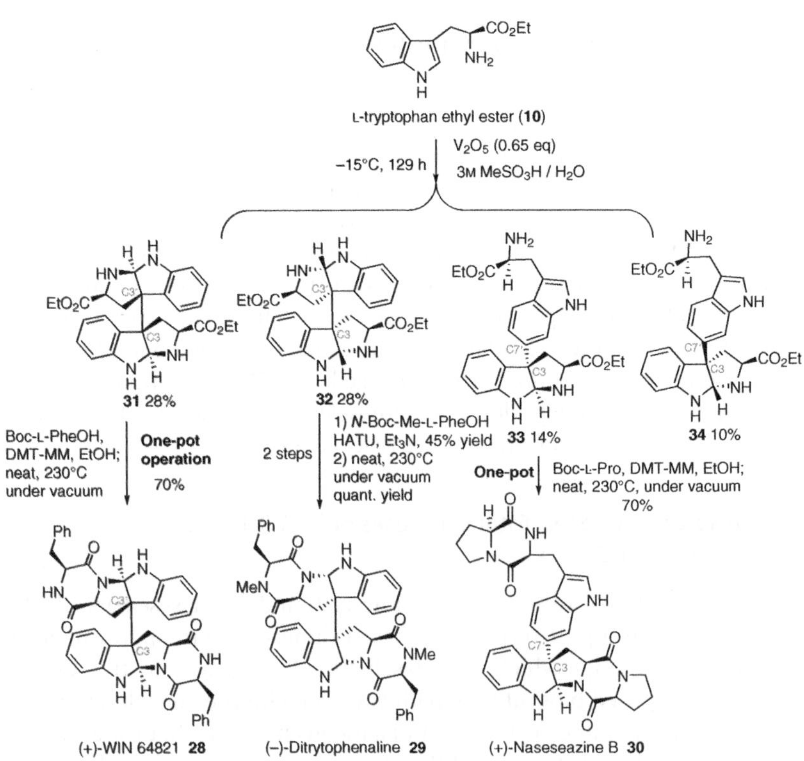

Scheme 2.6 Ishikawa's two-step enantioselective total synthesis of (+)-WIN 64821 (**28**) and (+)-naseseazine B (**30**).

The synthesis started with the V_2O_5-mediated dimerization of L-tryptophan ethyl ester in aqueous methanesulfonic acid, which gave C_2-symmetrical compounds **31** and **32** each at a 28% yield, along with **31** at a 14% yield (28% yield by using $Mn(OAc)_3$ in place of V_2O_5 as an oxidant). By a one-pot operation, compounds **31** and **33** were converted to (+)-WIN 64821 (**28**) and (+)-naseseazine B (**30**) (both 70% yield), respectively. On the other hand, compound **32** was transformed, at a 45% overall yield, to (−)-ditryptophenaline (**29**) via a two-step procedure. Thus, the concise syntheses of (+)-WIN 64821 (**28**), (−)-ditryptophenaline (**29**), and (+)-naseseazine B (**30**) were accomplished in overall yields of 20, 13, and 20%, respectively. It is worth mentioning that, in the same year, S. E. Reisman and coworkers reported a five-step syntheses affording (+)-naseseazines A and B in overall yields of 25 and 19%, respectively.[12] Recently, H. Ishikawa extended his methodology to the collective synthesis of tryptophan alkaloids eurocristatine, 15,15'-bis-epieurocristatine, 1'-(2-phenylethylene)-ditryptophenaline, WIN 64745, cristatumin C, asperdimin, naseseazine A, as well as ditryptoleucine A.[11b]

2.3 Three-Step/Three-Pot Enantioselective Total Synthesis of Natural Products

2.3.1 Carreira's Three-Step Asymmetric Total Syntheses of (+)-Aszonalenin and (−)-Brevicompanine B

In 2014, E. M. Carreira and coworkers developed a flexible catalytic method for the selective reverse prenylation of 3-substituted-1*H*-indoles at C3.[13] Modification of the reaction conditions by using 9-BBN-*n*C_6H_{13} and KHMDS as reagents allowing extension of the reaction to a D- or L-tryptophan methyl ester, which resulted, without racemization, in hexahydropyrrolo[2,3-b]indole (+)-*exo*-**37** or *ent*-**37** at a 58% yield and >20:1 d.r. (Scheme 2.7).[13b] (+)-*exo*-**37** was converted to (+)-aszonalenin (**25a**) by coupling with 2-aminobenzoic acid followed by $AlMe_3$-mediated lactamization. Similarly, coupling of (−)-*exo*-**37** to (*R*)-Fmoc-Leu, followed by deprotection and cyclization led to brevicompanine B (**35**) as single diastereomer. The latter is a plant growth regulator isolated from *Penicillium brevicompactum.* This three-step synthesis of **35** from the commercially available L-tryptophan methyl ester (39% overall yield) is far superior to the only synthesis reported so far (9 steps, 11% overall yield).[14]

2.3.2 Husson's Three-Step Asymmetric Total Synthesis of (−)-Sibirine

Although most biomimetic syntheses lead to racemic natural products, use of a chiral auxiliary may resulted in an enantioselective synthesis. This has been

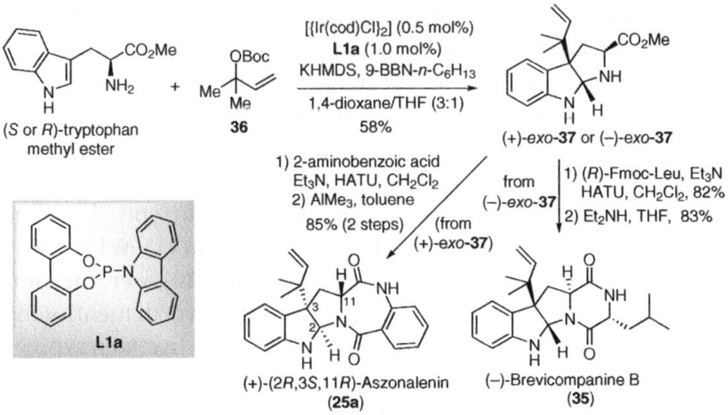

Scheme 2.7 Carreira's three-step asymmetric total syntheses of (+)-aszonalenin (**25a**) and (−)-brevicompanine B (**35**).

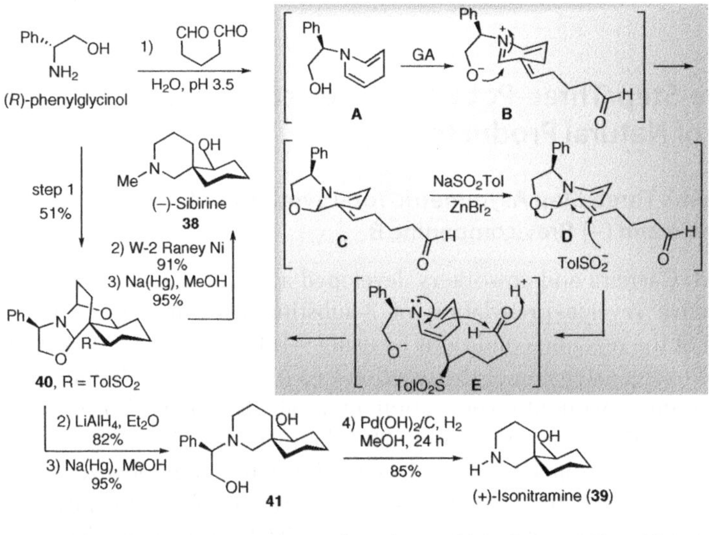

Scheme 2.8 Husson's three-step total synthesis of (−)-sibirine (**38**) and four-step access to (+)-isonitramine (**39**).

demonstrated by H.-P. Husson and coworkers in the enantioselective biomimetic synthesis of *Nitraria* alkaloids. Thus, by a combination of their (*R*)-phenylglycinol-based CN(*R*,*S*)-methodology[15a] with a biogenetic hypothesis, they develop an expeditious one-pot synthesis of tetracyclic spirocompound **40** (51% yield) by the condensation of (*R*)-phenylglycinol with glutaraldehyde (2.5 equiv) and sodium *p*-toluenesulfinate (Scheme 2.8). The method involves a series of cascade reactions enabling the rapid build-up of molecular complexity from relatively simple starting materials. Compound **40** is then converted in two steps to (–)-sibirine (**38**) (86% yield for two steps) and in three steps to (+)-isonitramine (**39**) (66% yield for three steps), respectively.[15b]

2.3.3 MacMillan's Three-Step Asymmetric Total Synthesis of (+)-Frondosin B

The frondosins are a family of marine sesquiterpenes that exhibit a range of biological activities such as anti-inflammatory properties and have potential application in anticancer and HIV therapy. Previous total syntheses were long and complicated requiring 10–20 steps with low overall yields (1–13%).[16] D. W. C. MacMillan and coworkers disclosed a highly concise three-step total synthesis of (+)-frondosin B (**42**) at a 50% overall yield (Scheme 2.9).[17a] The method relied on the organocatalytic enantioselective Friedel–Crafts alkylation method developed by the same group,[17b] with the modification of using boronic acid **43** as a precursor of trifluoroborate salt **43a**. The resulting aldehyde (*R*)-**45** (84% yield, 93% *ee*) was subjected to addition of a vinylic lithium reagent generated by the Shapiro reaction of the known hydrazone **46**. Finally, treatment of allylic alcohol **47** with BBr_3 resulted in the closure of the seven-membered ring, olefinic bond migration, and *O*-demethylation to give (+)-frondosin B (**42**) directly along with its double-bond isomer in 3.6:1 ratio with an isolated yield of 88%.

2.3.4 Hayashi's Three-Pot Total Synthesis of (–)-PGE_1 Methyl Ester

By extending their methodology of pot economy,[5,7] in 2013, Hayashi and colleagues accomplished a three-pot total synthesis of prostaglandin A_1 and E_1

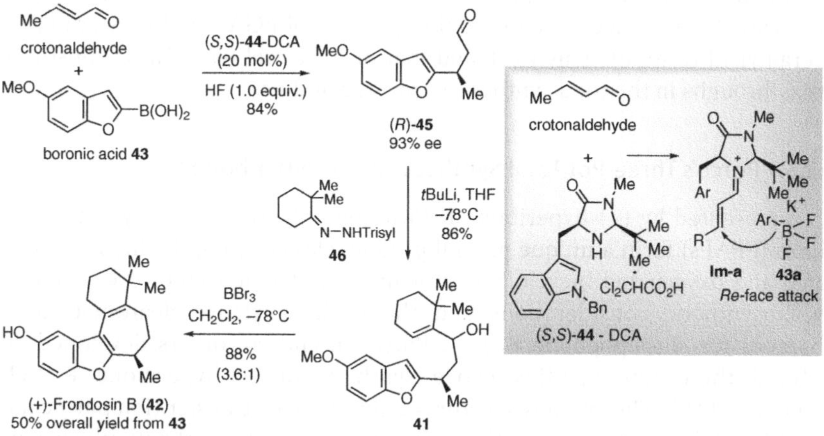

Scheme 2.9 MacMillan's three-step asymmetric total synthesis of (+)-frondosin B (**42**).

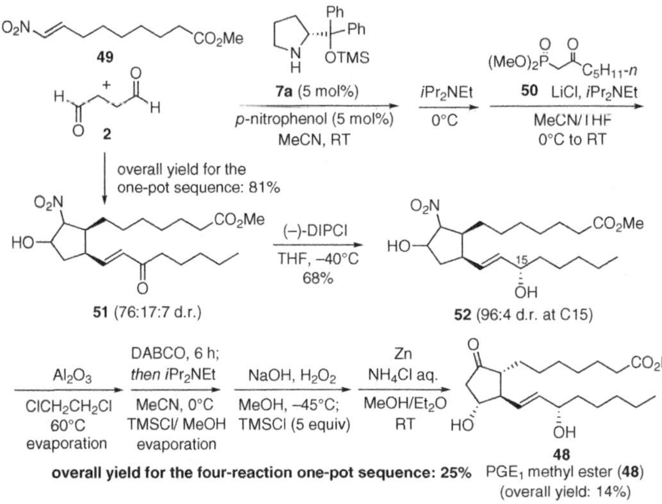

Scheme 2.10 Hayashi's three-pot enantioselective total synthesis of (–)-PGE_1 methyl ester (**48**).

methyl esters (Scheme 2.10).[18] The synthesis relied once again on the (*R*)-diphenylprolinol silyl ether (**7a**) catalyzed asymmetric Michael reaction of succinaldehyde (**2**) with nitroalkene **49**, which gave three diastereomers at the ratio of 76:17:7. Subsequent *i*-Pr_2EtN-promoted an intramolecular Henry reaction and Horner–Wadsworth–Emmons reaction with **50** affording, in one pot, the scaffold of prostaglandin **51** at a 81% yield. Stereoselective reduction of the ketone group with (–)-DIPCl followed by another one-pot reaction furnished PGE_1 methyl ester (**48**) at a 14% overall yield from **49**. Transformation of a synthetic intermediate allowed the synthesis of PGA_1 methyl ester in 25% overall yield from **49**. This work and that of Aggarwal (*vide infra*) constitute breakthroughs in the total synthesis of prostaglandins.

2.3.5 Porco's Three-Pot Total Synthesis of (–)-Hyperibone K

As represented by (+)-hyperibone K (**53**), the polyprenylated acylphloroglucinols (PPAPs) form a unique natural product class bearing the highly substituted and oxygenated bicyclo[3.3.1]nonane-1,3,5-trione framework. Inspired by biosynthetic considerations and the facile alkylative dearomatization observed for clusiaphenone B, J. A. Porco, Jr. and coworkers developed, in 2010, a three-step enantioselective synthesis of (+)-hyperibone K (**53**) (Scheme 2.11).[19] The synthesis features a chiral phase-transfer (ion pair) catalyst (**56**) catalyzed enantioselective dearomative annulation of naturally occurring clusiaphenone B (**54**) with **55** producing (–)-**57** at a 71% yield (90% *ee*).

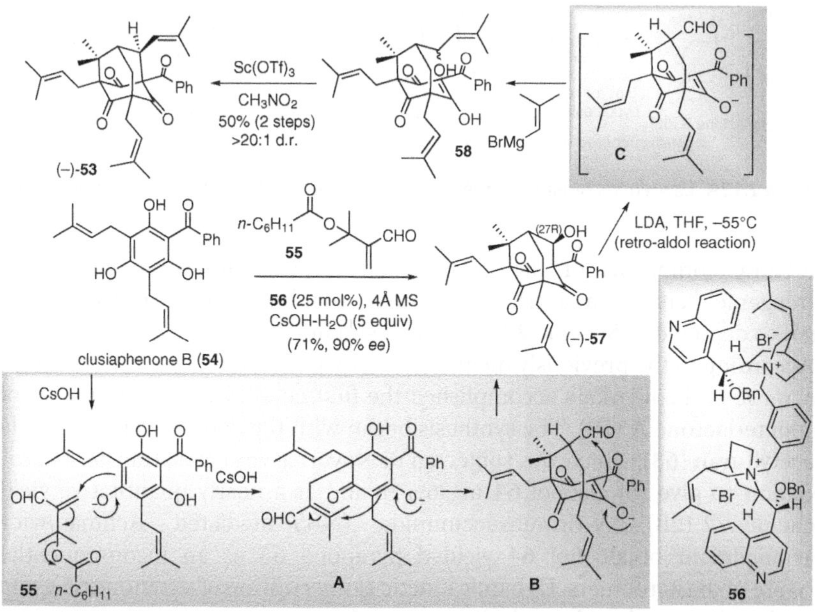

Scheme 2.11 Porco, Jr.'s three-step total synthesis of (–)-hyperibone K (**53**).

LDA-Mediated retro-aldol reaction yielded aldehyde intermediated **C**, which was subjected to a Grignard reagent addition to afford allylic alcohol **58**. $Sc(OTf)_3$-promoted intramolecular cationic cyclization of **58** proceeded with high (>20:1) diastereoselectivity to furnish (–)-**53** at a 50% yield over two steps.

2.4 Four-Step Enantioselective Total Synthesis of Natural Products

2.4.1 Lawrence's Four-Step Total Synthesis of (–)-Angiopterlactone A

Angiopterlactones A (**59**) and B (**60**) are two unique lactones isolated from rhizomes of *Angiopteris caudatiformis*.[20] The latter are used as ingredients in the treatment of "cough with lung heat," venomous snake bites, furuncles, and bleeding wounds in Tu folk medicine in China. This congener phenomenon led Zou to suggest that (+)-angiopterlactone B (**59**) may be derived from the coisolated (–)-angiopterlactone A (**60**) by an intramolecular Michael reaction (Scheme 12A).[20] Following this biogenetic hypothesis, A. L. Lawrence further suggested that the letter might be formed by an intermolecular oxa-1,4-conjugate addition reaction between γ-lactone **61** and δ-lactone **62**.[21a] Thus, despite the

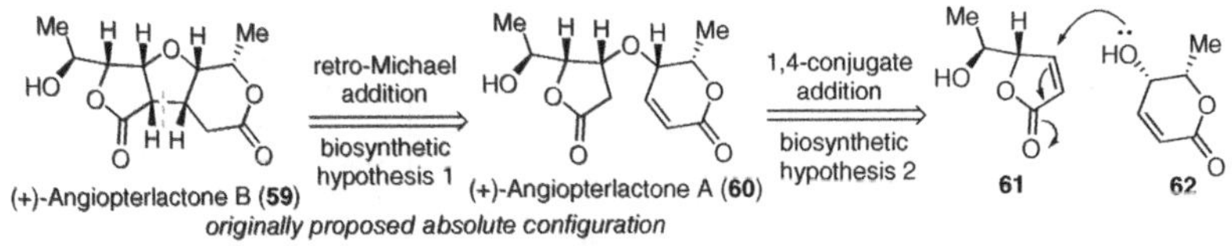

Scheme 2.12A Lawrence's bio-inspired retrosynthetic analysis of (–)-angiopterlactone A (**59**).

structural complexity of angiopterlactones A (**59**) with seven contiguous stereogenic centers in a tricyclic system, the bio-inspired retrosynthetic analysis depicted in Scheme 2.12A is quite simple.

According to the previously mentioned retrosynthetic analysis, in 2017 A. L. Lawrence and coworkers accomplished the first enantioselective synthesis of angiopterlactone A (**59**). The synthesis began with the commercially available 2-acetylfuran (**63**), which was subjected to Noyori asymmetric transfer hydrogenation to give (*S*)-alcohol **64** in 96% *ee* and at a nearly quantitative yield (Scheme 2.12B). *N*-Bromosuccinimide (NBS)-mediated Achmatowicz rearrangement of alcohol **64** yielded pyranone **65** as an inconsequential mixture of diastereomers. Dynamic kinetic isomerization of pyranone **65** using a tandem Brønsted acid and iridium catalysis gave δ-lactone **62** at a 62–71% yield. Under carefully defined conditions, δ-lactone **62** was converted into (–)-angiopterlactone B (**59**) at a 25% isolated yield. The yield of 25% is apparently quite low. However, the result is impressive taking into consideration that in a single pot (–)-angiopterlactone B (**59**) was isolated as the sole isolable diastereomer among 72 possible ones through a complex reaction sequence involving the interconversions between lactone **62**, **61**, and its epimer **66**, tandem Michael addition, and an oxa-1,4-conjugate addition reaction. The enantioselective total syntheses of both enantiomers of **59** by Bhattacharya allows revising the sign of optical rotation of natural angiopterlactone B as (–)-(**59**).[21b]

2.4.2 Maimone's Four-Step Synthesis of (+)-Cardamom Peroxide

Isolated from *Amomum krervanh* Pierre (Siam cardamom), cardamom peroxide (**67**) displays potent activity against *P. falciparum* (EC_{50} = 170 nM). The 1,2-dioxepane motif found in **67** is distinct from the trioxane found in artemisin (qinghaosu); moreover, its biosynthetic origins remain mysterious and the absolute configuration remains unknown. Guided by biosynthetic logic, in 2014 T. J. Maimone and X. Hu developed a four-step enantiospecific synthesis of cardamom peroxide (**67**) from (–)-myrtenal and molecular oxygen (Scheme 2.13).[22] Recognizing the target molecule as a dimeric monoterpene, the construction of the carbocyclic ring system was simplified. Thus, by McMurray coupling of the inexpensive (–)-myrtenal, C_2-symmetric triene **68** was obtained in one step. Triene **68** was subjected to a [4+2] cycloaddition

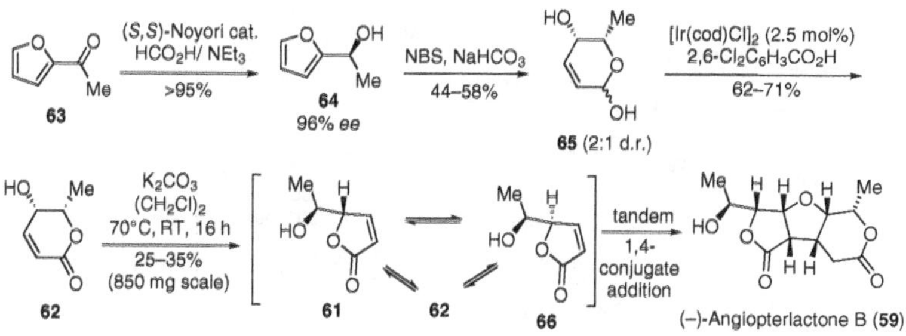

Scheme 2.12B Lawrence's four-step total synthesis of (–)-angiopterlactone A (**59**).

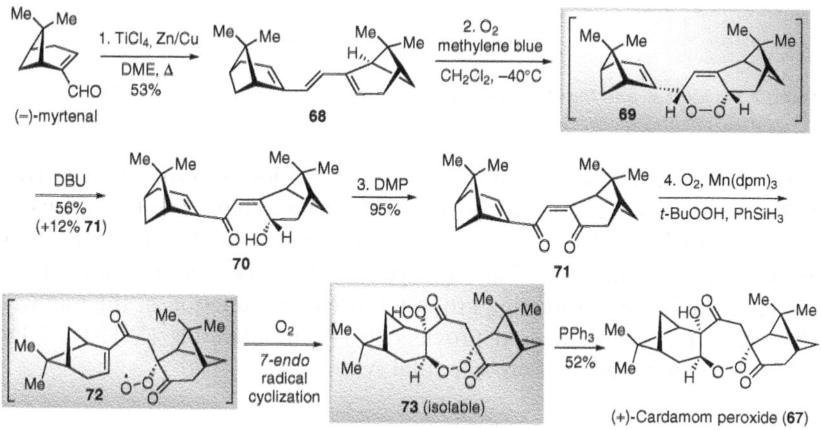

Scheme 2.13 Maimone's four-step total synthesis of (+)-cardamom peroxide (**67**).

reaction with singlet oxygen to generate an endoperoxide intermediate **69** that upon treatment with DBU led, via the Kornblum–DeLaMare rearrangement, to dienone **70**. The latter is convertible to **71** by both air and Dess–Martin periodinane oxidation. By slow addition of $PhSiH_3$, an unusual manganese-catalyzed tandem hydroperoxidation, cyclization reaction yielded **73** that when treated with PPh_3 furnished **67** at a 52% yield. The overall yield of this four-step total synthesis was 18%, and >0.5 g of **67** was prepared.

2.4.3 Xie, Lai, and Ma's Four-Step Total Synthesis of (–)-Chimonanthine

To develop an asymmetric synthesis of the cyclotryptamine alkaloid (–)-chimonanthine (**74**), Xie and coworkers first investigated a catalytic asymmetric bromocyclization of tryptamines.[23] They discovered that by employing readily available DABCO-derived bromine salt **B1** as a brominating reagent

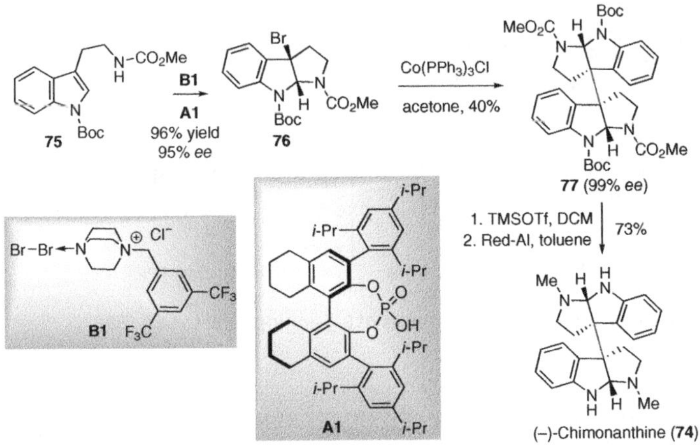

Scheme 2.14 Xie, Lai, and Ma's four-step total synthesis of (–)-chimonanthine (**74**).

and 8*H*-*R*-TRIP **A1** as a catalyst, the bromocyclization of tryptamine derivative **75**, running at a gram scale, produced pyrroloindoline **76** at a 96% yield with 95% *ee* (Scheme 2.14).[23] $Co(PPh_3)_3Cl$-catalyzed homodimerization of **76** afforded bispyrroloindoline 77 at a 40% yield and 99% *ee*. Removal of the *tert*-butoxycarbonyl (Boc) protecting group followed by reduction of the methylcarbamate moiety with Red-Al provided (–)-chimonanthine (**74**) at a 73% yield, which completed the four-step total synthesis.

2.4.4 Huang's Four-Step Total Synthesis of (–)-Chaetominine

(–)-Chaetominine (**78** in Scheme 2.15) is an alkaloid first isolated from the solid-substrate culture of *Chaetomium* sp. IFB-E015, an endophytic fungus on apparently healthy *Adenophora axilliflora* leaves.[24a] Subsequently, several homologs and diastereomers of (–)-chaetominine have been isolated from different fungi, which include (–)-pseudofischerine (**79**), (–)-aniquinazoline D (**80**), (–)-isochaetominines A–C (**81–83**), and (+)-14-*epi*-isochaetominine C (**84**).[24b] In 2009, P.-Q. Huang and coworkers reported a four-step enantioselective total synthesis of (–)-chaetominine (**78**) (Scheme 2.15).[25a,b] The method consists of two amide bond-forming reactions, a quinazolinone ring formation reaction, and a epoxidation-triggered double cyclization reaction. The latter reaction afforded (–)-chaetominine (**78**) at a 41% yield, along with the monocyclization diastereomer **88** at a 50% yield. Diastereomer **88** could be converted to a diastereomer of chaetominine (**89**). To date, this is the shortest and also the most efficient enantioselective total synthesis of (–)-chaetominine (**78**). Further extension of this stereodivergent method allowed access to all the reported homologs and diastereomers of (–)-chaetominine **79–84** in four to six steps.[25c–f] Considering the fact that each synthetic sequence produced two

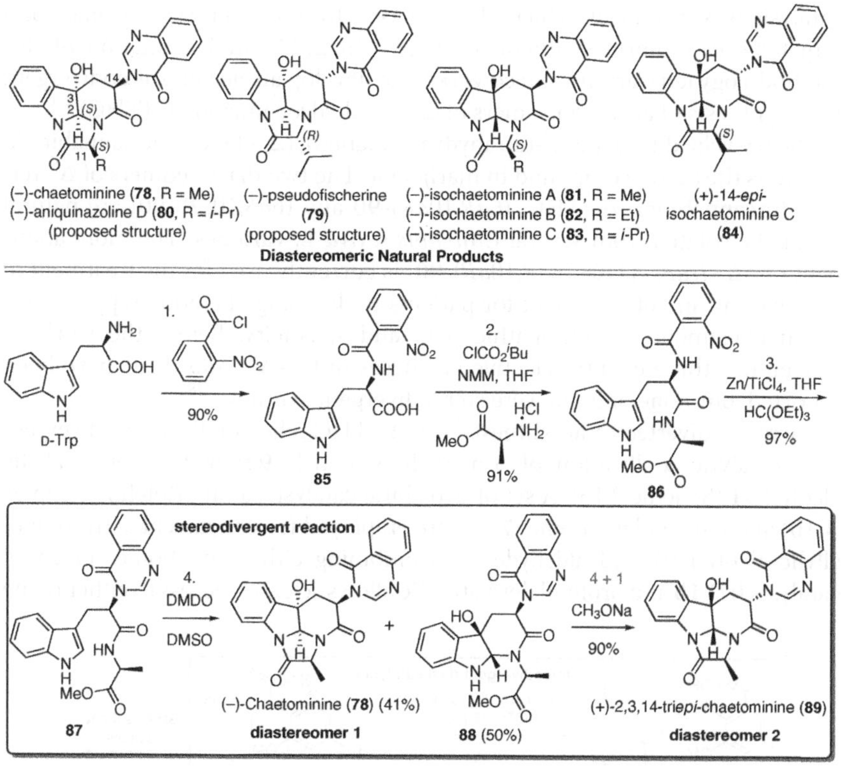

Scheme 2.15 Huang's stereodivergent and collective total synthesis of the chaetominine family of alkaloids.

or three diastereomers, the combined overall yield for two (or three) diastereomers is quite high. Thanks to this efficient method, stereochemistries of several reported alkaloids of this family have been revised.[25c,d,f]

2.5 Five-Step/Pot Enantioselective Total Synthesis of Natural Products

2.5.1 Carreira's Five-Step Total Syntheses of Δ^9-Tetrahydrocannabinols

The phenomenon that many natural products and their diastereomers are found in nature as congeners (see Section 2.4.4) means it is important to develop a general, stereodivergent total synthesis to fully probe the biology of natural products.[26] In recent years, E. M. Carreira and coworkers have developed a powerful iridium-catalyzed enantioselective transformation methodology that allows two adjacent chiral centers to establish in one reaction.[13] On the basis of this methodology, the concept of stereodivergent dual catalysis was advanced, which allows access to any given stereoisomer of a target by an uniform synthetic strategy.[27] In 2014, implementation of this methodology led Carreira and coworkers to develop an elegant stereodivergent total synthesis of all stereoisomers of Δ^9-tetrahydrocannabinols (THC).[27c]

The tricyclic diterpene Δ^9-tetrahydrocannabinol (Δ^9-THC), the cannabinoid family, is the active compound in marijuana. The two diastereomers of Δ^9-tetrahydrocannabinol [Δ^9-THC, (6*aR*,10*aR*)-**90** and (6*aS*,10*aR*)-**90**] are natural products isolated from several *Cannabis sativa* L. varieties. The more abundant (−)-Δ^9-*trans*-THC [(6*aR*,10*aR*)-**90**] is currently used for the treatment of anorexia, as an anti-nauseant for patients undergoing chemotherapy, and for the management of neuropathic pain and spasticity. Before the work by Carreira,[27c] three enantioselective syntheses of (−)-Δ^9-*trans*-THC have been reported, but none of them were stereodivergent synthesis.[28]

Carreira's diverted total synthesis[27c] of Δ^9-THC relied on the stereodivergent dual catalytic α-allylation of 5-methylhex-5-enal (**92**) with racemic allylic alcohol **91** (Scheme 2.16). A set of two chiral catalysts, an Ir/(P,olefin) complex (**L1b**) and a secondary amine (**7c**), were employed for concurrent activation of allylic alcohol **91** and aldehyde **92**. Combining either enantiomer of chiral catalysts Ir/**L1b** and prolinol derivative **7c** allows selective access to either of the

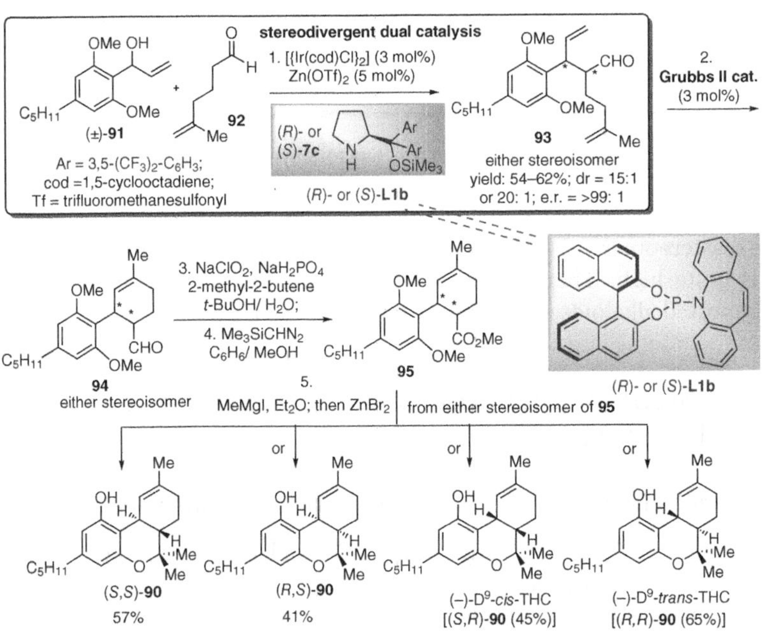

Scheme 2.16 Carreira's five-step stereodivergent dual catalysis-based total syntheses of all stereoisomers of Δ^9-THC (**90**).

four stereoisomers of **93**, and thus the four stereoisomers of Δ^9-THC (**90**) each in four steps via identical synthetic sequences and conditions (Scheme 2.16).

2.5.2 Studer's Five-Step Total Syntheses of (+)-Machaeriols B and D

The biological and medical importance of (−)-Δ^9-tetrahydrocannabinol has been discussed in Section 2.5.1. In addition to (−)-Δ^9-*trans*-THCs, its isomer (−)-Δ^8-*trans*-THC (**96**) is also a natural product. Machaeriols such as antimicrobial and antimalarial (+)-machaeriol B (**97**) and (+)-machaeriol D (**98**) are a group of natural products structurally in related with cannabinoids, but with different absolute configuration. In 2015, A. Studer and F. Klotter reported a five-step syntheses of (+)-machaeriol B, (+)-machaeriol D, and a four-step synthesis of (+)-*ent*-**96** starting from commercially available (*S*)-perillic acid (**99**, Scheme 2.17).[29] The divergent strategy features the stereospecific palladium-catalyzed decarboxylative arylation (**100** to **102**), a method developed by the same group,[30] and a one-pot stereocontrolled hydroboration-oxidation or reduction of the intermediary borane, as key steps.

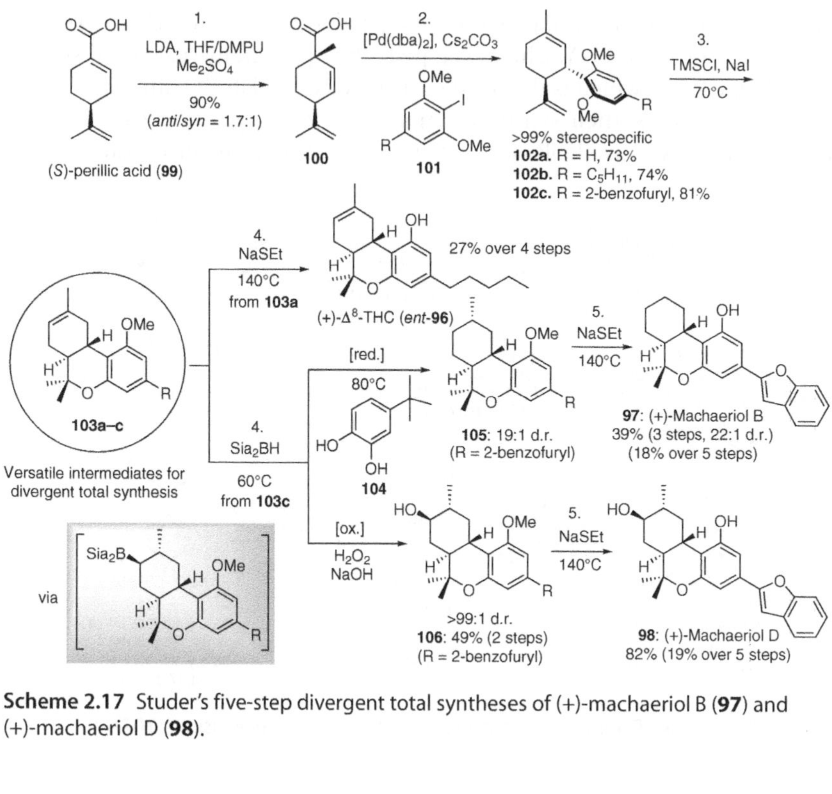

Scheme 2.17 Studer's five-step divergent total syntheses of (+)-machaeriol B (**97**) and (+)-machaeriol D (**98**).

Deprotonation of **99** with LDA and methylation of the resulting dienolate with dimethyl sulfate yielded carboxylic acid **100** along with complete α-selectivity and moderate *trans*-diastereoselectivity (Scheme 2.17). Decarboxylative γ-arylation of the diastereomeric mixture **100** [Pd(dba)$_2$] (10 mol%), Cs_2CO_3 (1.1 equiv.) with iodide **101a–c** proceeded with complete stereoselectivity affording **102a–c**. Treatment of **102a–c** with TMSCl/NaI resulted in *O*-demethylation and concomitant Δ^9 to a Δ^8 double-bond migration to give Δ^8-tetrahydrodibenzopyran derivatives **103a–c**. Further demethylation of **103a** with NaSEt in DMF at 140°C yielded (+)-Δ^8-*trans*-THC (*ent*-**96**). Next, Renaud's one-pot hydroboration – radical *C-B* reduction method,[31] was adopted for the transformation of **103c** to **97**. Thus hydroboration of **103c** with disiamylborane (Sia$_2$BH) and subsequent radical reduction with 4-*tert*-butylcatechol (**104**) yielded **105**, which was further demethylated to afford (+)-machaeriol B (**97**). Alternatively, hydroboration of **103c** followed by oxidative workup (H_2O_2, NaOH) afforded secondary alcohols **106** as a single diastereoisomer (>99:1 d.r.), which was deprotected to give (+)-machaeriol D (**98**).

2.5.3 Cook's Five-Pot Total Synthesis of (+)-Artemisinin (Qinghaosu)

In view of the need for reliable, low-cost production of artemisinin (**107**), in 2012, S. P. Cook and C.-Y. Zhu developed a five-pot, gram-scale total synthesis (Scheme 2.18), which is the most cost-effective to date.[32] Different from all previous syntheses, their synthesis employed achiral cyclohex-2-en-1-one as a cheap and easily available starting material, and an unconventional [4+2]

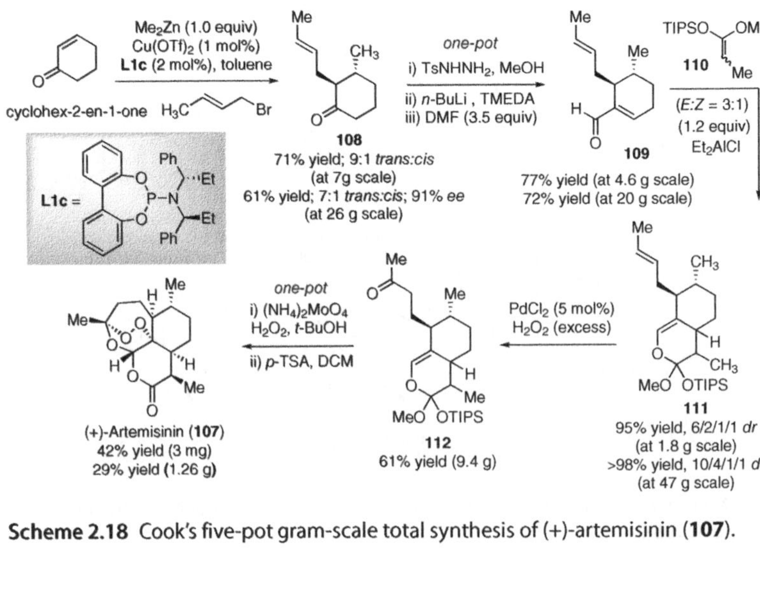

Scheme 2.18 Cook's five-pot gram-scale total synthesis of (+)-artemisinin (**107**).

reaction for the installation of the six-membered lactone of **107**. The synthesis relied on the one-pot asymmetric conjugate addition-trapping of zinc enolates[33] with crotyl bromide, which afforded ketone **108** at a 61% yield (7:1 *trans:cis*, 91% *ee*). In this reaction, it is advantageous to use Alexakis' ligand (**L1c**)[34] as the chiral ligand, and crotyl bromide as a latent methyl vinyl ketone (MVK). Cyclohexanone derivative **108** was converted to *α,β*-enal **109** by the two-step Shapiro process in one pot. In the presence of low-cost dimethyl- or diethylaluminum chloride, the [4+2] reaction between *α,β*-enal **109** and silyl ketene acetal **110** produced an orthoester **111** at a ≥95% yield. It is important to note that although **111** was formed as a mixture of four diastereomers, improving the selectivity or separation of the diastereomers were unnecessary because two of the three stereogenic centers proved to be irrelevant to the synthesis of **107**. Thus, under modified Wacker oxidation conditions ($PdCl_2$, H_2O_2), crude **111** was converted into the desired methyl ketone **112** at a 61% yield, along with ethyl ketone (not shown) at a 30% yield. The last step, namely the oxidative rearrangement of **112** to produce **107** turned out to be challenging. Finally, this was achieved by employing *in situ* formed singlet oxygen from the decomposition of H_2O_2 by ammonium molybdate to oxidize the enol olefin of **112** (diastereomeric mixture), and the products were treated with *p*-toluenesulfonic acid to produce **107** at a 29–42% yield.

Overall, this is an elegant synthesis with every step proceeding harmoniously. It is remarkable that only five pots are required for the construction of three rings containing a sensitive peroxy bridge. The *de novo* strategy and the careful consideration of each step forged not only a step-economical synthesis but also the most cost-effective synthesis. The brevity is due to its redoxy economy (only one oxidation step used) and being free of a protecting group. Not using expensive reagents renders the synthetic route cost-effective.

2.5.4 Corey's Five-Step Total Synthesis of Aflatoxin B_2

In 2005, E. J. Corey and G. Zhou described a short, enantioselective total synthesis of aflatoxin B_2 (**113**) (Scheme 2.19).[35] The method relied on the (*R*)-oxazaborolidinium triflimide (**cat. 1**)-catalyzed asymmetric [3+2]-cycloaddition between 2-methoxy-1,4-benzoquinone and 2,3-dihydrofuran, which afforded **114** as the major product at a 65% yield (91% *ee*). Compound **114** was transformed to aldehyde **115** by regioselective *ortho*-formylation and triflate ester formation. The resulting aldehyde was subjected to addition with MeMgBr and the product was oxidized to deliver methyl ketone **116** at a 85% yield. Baeyer–Villiger oxidation of methyl ketone **116**, reductive removal of the triflate group with Raney-Ni/H_2, and deacetylation gave the tricyclic phenol **117** at a yield of approximately 40% overall. In the presence of $ZnCO_3$, cyclo-condensation of tricyclic phenol **117** with *β*-bromo-*α,β*-enone **118** produced aflatoxin B_2 (**113**) at a 36% yield.

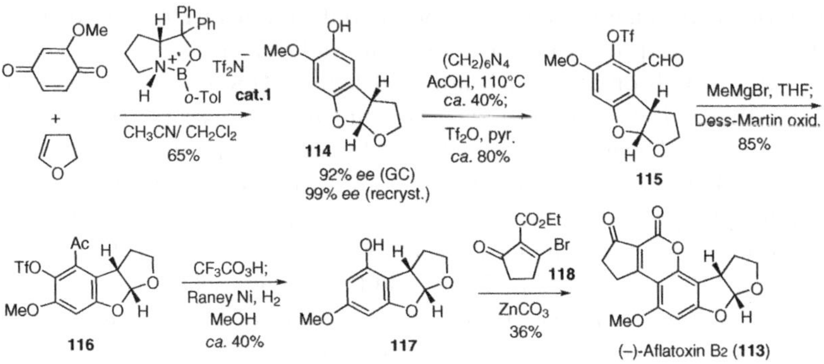

Scheme 2.19 Corey's five-step total synthesis of (−)-aflatoxin B_2 (**113**).

2.6 Six-Step Enantioselective Total Synthesis of Natural Products

2.6.1 Comins' Six-Step Total Synthesis of (*S*)-Camptothecin

Camptothecins are among the most important anticancer alkaloids of the twenty-first century.[36] The parent member, (*S*)-camptothecin (CPT, **119**, Scheme 2.20) is a pentacyclic quinoline alkaloid first discovered and isolated by M. E. Wall and coworkers in 1966 from the bark of *Camptotheca acuminate*, or Chinese Happy Tree ("Xi Shu"), a tree native to China used as a cancer

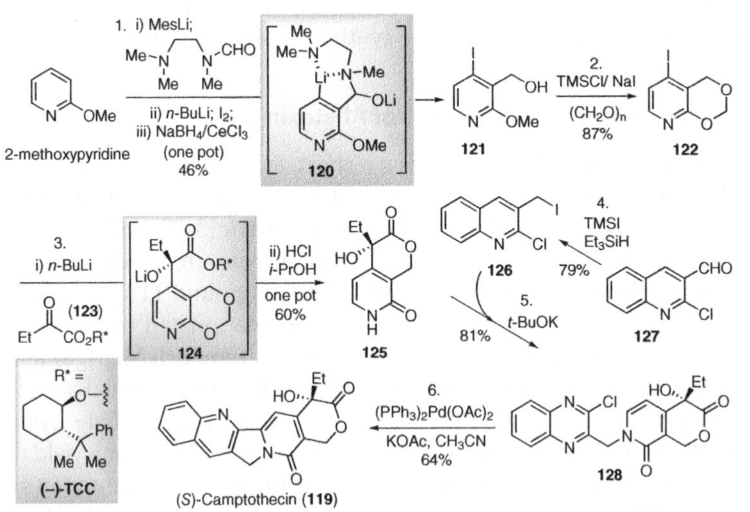

Scheme 2.20 Comins' six-step total synthesis of (+)-camptothecin (CPT, **119**).

treatment in the Traditional Chinese Medicine (TCM).[37] CPT showed remarkable anticancer activity in preliminary clinical trials. But its low solubility and high toxicity prevent its medicinal application. Nevertheless, this alkaloid has been proven to be an important lead compound, from which topotecan (Hycamtin), irinotecan (Camptosar), and belotecan, have been developed as anticancer drugs, and several CPT analogs are being evaluated as potential anticancer drug candidates.[36] Over the past few decades, the synthesis of CPT and analogs has attracted considerable attention, which has culminated in a number of synthetic approaches.[36,38,39] Among several elegant approaches thus developed, Comins and coworkers' six-step total synthesis, reported in 2001, remains the shortest asymmetric route to date.[39]

Comins' route to (*S*)-camptothecin employed commercially available 2-methoxypyridine and 2-chloroquinoline-3-carbaldehyde as the starting materials (Scheme 2.20). The synthesis commenced with a one-pot process: deprotonation with mesityllithium generated C3 lithiated intermediate, which upon addition onto *N*-formyl-*N,N',N'*-trimethylethylenediamine formed an *α*-amino alkoxide; addition of *n*-BuLi effected α-amino alkoxide directed lithiation at C-4 to give the dianion **120**, which was exposed to iodine, and workup with aqueous $NaBH_4/CeCl_3$ to give **121** at a 46% yield. The latter was converted directly to 1,3-dioxane **122** upon treating with NaI/ TMSCl/ paraformaldehyde. Lithium-halogen exchange of the iodopyridine derivative **122** followed by the addition onto ketoester **123** bearing a chiral auxiliary generated alkoxide **124**, which was converted to **125** via one-pot protonation, acetal hydrolysis, and lactonization. *t*-BuOK-mediated coupling of fragments **125** with **126**, available in one step from commercially available 2-chloro-3-quinolinecarboxaldehyde (**127**) afforded **128** (>99% *ee*), which was transformed to (*S*)-camptothecin (**119**) via an intramolecular Heck reaction.

At the end of their paper,[39] Comins concluded: "This 6-step synthesis is practical and should be amenable to the large-scale preparation of CPT analogs of medicinal importance." This turned out to be true. On the basis of the powerful Comins route and the work by the group from Institut Henri Beaufour, the Lavergnee–Comins route was forged for the efficient synthesis of homocamptothecins. The latter form an especially interesting class of anticancer agents, which retain the potent cytotoxicity of camptothecins while having very different pharmacodynamic properties.[40]

2.6.2 Krische's Six-Step Total Synthesis of (–)-Cyanolide A

How many steps will be required to synthesize highly functionalizd glycosidic 16-membered macrodiolide cyanolide A (**129**)? The total syntheses reported before 2013 used 12–17 steps (LLS; 18–22 steps).[41] In 2013, M. J. Krische and coworkers' recent work showed that its enantioselective total synthesis could be achieved in only six steps (LLS) from neopentyl glycol (**130**, Scheme 2.21).[42]

Recognizing that cyanolide A (**129**) is a molecule of C_2 symmetry, the synthesis is simplified. Moreover, it was further simplified by taking advantage of Krische's powerful redox-economical methodology, featuring the two-directional, enantioselective carbonyl allylation from the alcohol oxidation level.[43]

The synthesis begins with the double allylation of **130**, which was achieved by treating with [{Ir(cod)Cl}$_2$], allyl acetate, 4-chloro-3-nitrobenzoic acid, and (*S*)-Cl,MeO-biphep[42] (Scheme 2.21). Taking advantages of Fuwa's tandem cross-metathesis/oxa-Michael cyclization method,[44] the desired *cis*-2,6-disubstituted pyran **132** was produced at a 76% yield (10:1 d.r.) from **131** and ethyl vinyl ketone. Glycosylation of **132** and successive exposure of the resulting **133** to ozone and Li(*s*-Bu)$_3$BH yielded **134** as a 5:1 mixture of diastereomers at C9. Compound **134** was converted to hydroxy acid **135** via site-selective oxidation. Finally, subjecting of **135** to the macrodiolide formation conditions delivered cyanolide A (**129**).

The high efficiency of this enantioselective total synthesis relied firstly on Krische's powerful redox-economical methodology, which provided a C_2-symmetric chiral building block in one pot and in excellent diastereo- and enantioselectivities from an achiral starting material **130**. The two-directional reactions in the first and second step, and the tandem cross-metathesis/oxa-conjugate

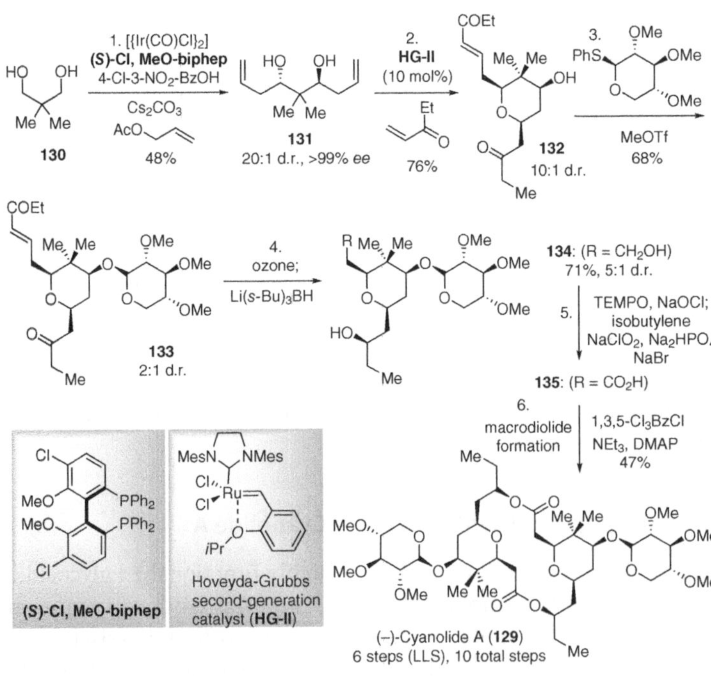

Scheme 2.21 Krische's six-step total synthesis of (−)-cyanolide A (**129**).

addition/cyclization in the second step allowed a rapid generation of molecular complexity (see **132**) in a highly stereoselective manner in just two steps! Moreover, the synthesis proceeded in the absence of protecting groups, chiral auxiliaries, or premetalated carbon nucleophiles.

It is worth mentioning that cyanolide A exhibits potent molluscicidal activity (LC_{50} = 1.2 μM) against the water snail *Biomphalaria glabrata*. The latter is a vector of the human parasitic disease schistosomiasis, which has infected more than 200 million people.[45]

2.7 Seven-Step Enantioselective Total Synthesis of Natural Products

2.7.1 Baran's 7–10-Step Total Syntheses of Hapalindole-Type Natural Products

While most of the syntheses selected in this chapter involve recently reported total syntheses, we present the seminal work of P. S. Baran and coworkers on hapalindole-type natural products that was published 10 years ago.[46a] The reason resides not only in the fact that they have pioneered the enantioselective total syntheses of this large family of alkaloids that exhibit potent and exciting biological activities, and possess intriguing and unprecedented molecular architectures, but more importantly, several general guidelines for achieving the efficacy in the total synthesis (Table 2.1) have been followed from the very beginning; namely, during the planning stage (retrosynthetic analysis) of their synthesis. Among the eight general guidelines they summarized, they have paid much attention to implementing the concepts of protecting group-free and redox-economy in the total synthesis. These efforts led them to develop very concise total syntheses (7–10 steps) of several members of the hapalindole family of natural products.

The synthesis of (−)-fischerindole I (**136**) [and (+)-welwitindolinone A (**137**)] from (*S*)-carvone oxide is depicted in Scheme 2.22. *In situ* protection of the ketone group in (*S*)-carvone oxide (**138**) by deprotonation with LHMDS generated enolate intermediate **139** that was subjected to addition with vinyl Grignard reagent to afford alcohol **140**. Chlorination of **140** using NCS/Ph_3P proceeded with inversion of configuration to yield chloroketone **141**. The crucial coupling of **141** with indole using the reaction that Baran and coworkers invented specifically for forming this type of C–C bond[46c,d] furnished indole **142** as a single diastereomer. Treatment of **142** with Montmorillonite K-10 (DCE, 120°C, microwave, 6 min) furnished the expected Friedel–Crafts cyclization product (no shown) in 40% yield (57% BRSM), which was further subjected to stereocontrolled reductive amination to yield amine **143**. *N*-Formylation followed by dehydration afforded

Table 2.1 General guidelines for the total syntheses adopted and summarized by Baran:[46a]

(1) Minimizing the use of redox reactions that do not form C–C bonds.
(2) Maximizing the percentage of C–C bond-forming reactions.
(3) Developing convergent strategy.
(4) Adopting tactics that allow a linear escalation of the overall oxidation level of intermediates during assembly of the molecular framework.
(5) Maximizing the use of cascade (tandem) reactions.
(6) Developing protecting-group-free synthesis or minimizing the use of protecting groups.
(7) Developing new methodologies to achieve synthetic efficiency.
(8) Developing biomimetic/bio-inspired synthesis or taking into consideration of possible biosynthetic pathway.

(In addition to atom- and step economy)

11-*epi*-fischerindole G (**144**). Oxidative dehydration of **144** was achieved by treating with DDQ in the presence of water, which might pass through α,β-unsaturated imine **145** to give (–)-fischerindole I (**136**). By this route, more than 2 g of fischerindole I (**136**) have been prepared. In addition, in one step, **136** was converted to (+)-welwitindolinone A (**137**). Moreover, an eight-step, enantiospecific, protecting-group-free total synthesis of (–)-hapalindole U (**150**), and the first total synthesis of an ambiguine alkaloid, (+)-ambiguine H (**151**, 10 steps) have been accomplished starting from commercially available *p*-menth-1-en-9-ol (**149**) (Scheme 2.22).[46a,b]

2.7.2 Aggarwal's Seven-Step Total Synthesis of (+)-PGF$_2\alpha$

Prior to Hayashi's three-pot enantioselective synthesis of PGE$_1$ methyl ester (**48**, Scheme 2.10),[18] in 2012, Aggarwal and coworkers reported a seven-step synthesis of PGF$_2\alpha$ (**152**, Scheme 2.23).[47a] This breakthrough further demonstrated the power of organocatalysis in the enantioselective total synthesis of bioactive natural products and important medicinal agents. As we have discussed in Chapter 1, since the late 1960s, the intensive investigation from both academic and industrial scientists has led to the successful development of several PG-based pharmaceuticals. Among them, latanoprost (**153**) and bimatoprost (**154**) are two antiglaucoma drugs developed from PGF$_2\alpha$ (**152**, Figure 2.1). Latanoprost is a "blockbuster" drug for Pfizer. Its sales in 2010 reached \$1.7 billion.[48] While lower in sales, bimatoprost (\$849 million in sales in 2011) is more efficient in intraocular pressure reduction than latanoprost, and is also an active ingredient in Latisse, a new prescription medicine to treat hypotrichosis of the eyelashes. The industrial processes for the manufacture

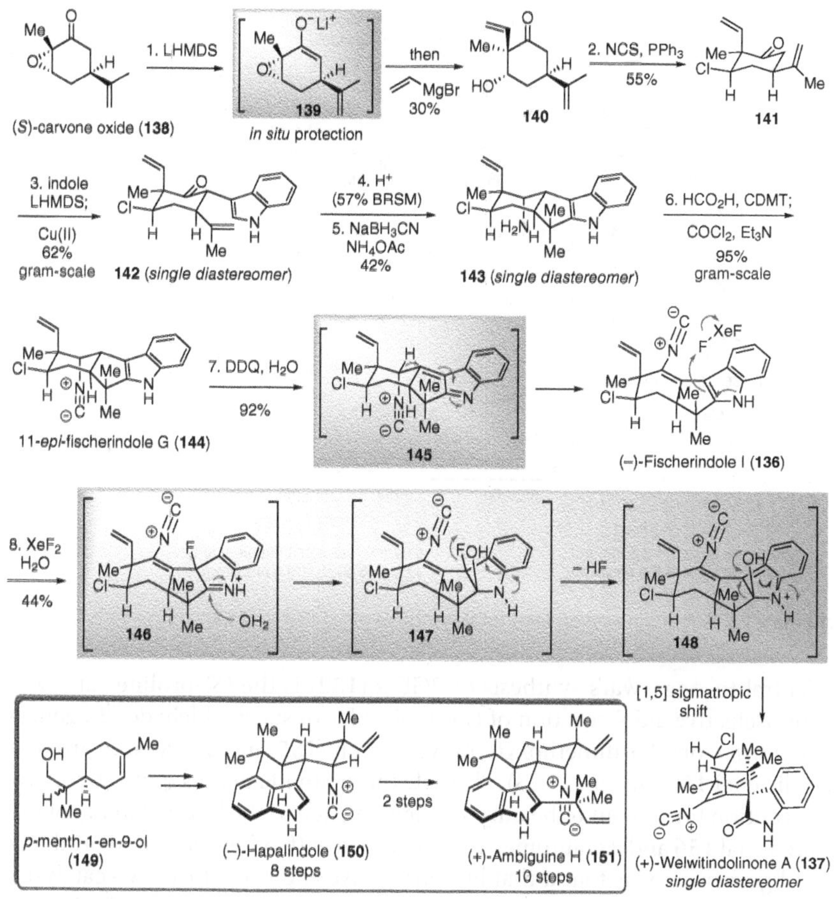

Scheme 2.22 Baran's 7–10-step total syntheses of hapalindole-type natural products.

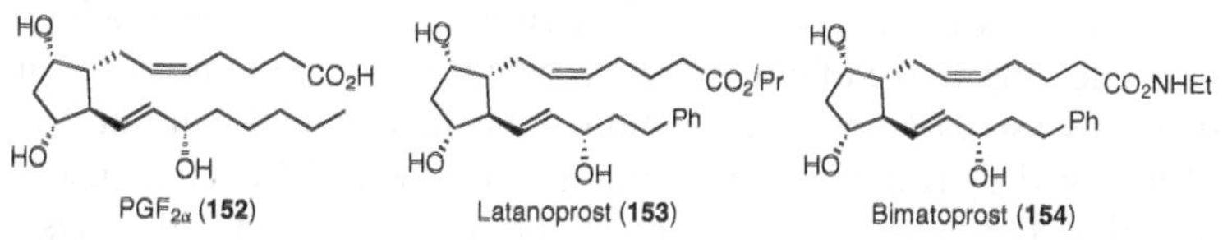

Figure 2.1 Structures of $PGF_2\alpha$ and antiglaucoma drugs Latanoprost (**153**) and Bimatoprost (**154**).

of latanoprost and bimatoprost are based on the strategy originally developed by E. J. Corey and coworkers in 1969 for the synthesis of the parent prostaglandin $PGF_2\alpha$,[49] which requires 19 and 17 steps, respectively.[50] The *de novo* organocatalysis-based synthetic route is of great value academically and commercially.

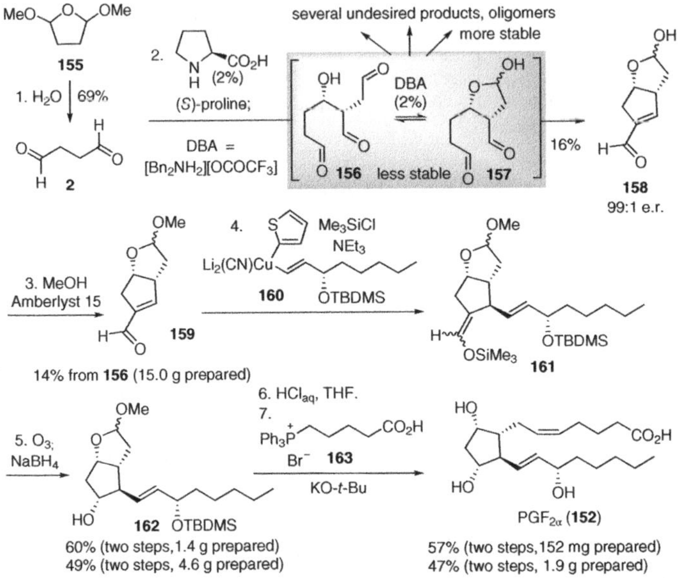

Scheme 2.23 Aggarwal's seven-step total synthesis of (+)-$PGF_2\alpha$ (**152**).

Central to Aggarwal's synthesis of $PGF_2\alpha$ (**152**) is the (*S*)-proline catalyzed enantioselective aldol reaction of two molecules of succinaldehyde (**2**), generated *in situ* from 2,5-dimethoxytetrahydrofuran (**155**) in hot water, to give the bicyclic enal **158** (Scheme 2.23).[47a] While the synthetic plan is quite attractive, its execution turned out to be quite challenging due to the lower stability of the desired trial **156** and its tautomer **157** compared with other undesired products and oligomers. It was found that by consecutive addition of the two catalysts, (*S*)-proline and $[Bn_2NH_2][OCOCF_3]$, the desired enal **158** could be obtained at a 16% yield with an enantiomeric ratio of 99:1 and on a multigram scale. The poor yield of **158** is partially recompensed by running the reaction at a high concentration allowing 15.0 g of pure methoxy acetal **159** to be synthesized at a 14% yield over two steps. Conjugate addition of the mixed vinylic cuprate **160** to enal **159** followed by trapping the resulting adduct with TMSCl furnishes the silyl enol ether **161**. The later was subjected to chemoselective ozonolysis of the electron-rich silyl enol ether and reduction of the resulting ketone with $NaBH_4$ to give the cyclopentanol derivative **162**. Both the conjugate addition and reduction reactions proceeded with complete stereoselectivity at the newly created stereogenic centers. Treatment of **162** with aqueous HCl resulted in the cleavage of the silyl ether (TBDMS) and the transformation of the acetal to hemi-acetal (not shown), which was subjected to Wittig reaction with an ylide generated *in situ* from the phosphonium salt **163** and KO*t*-Bu to give $PGF_2\alpha$ (**152**). It is worth mentioning that although the (*S*)-proline-catalyzed one-pot

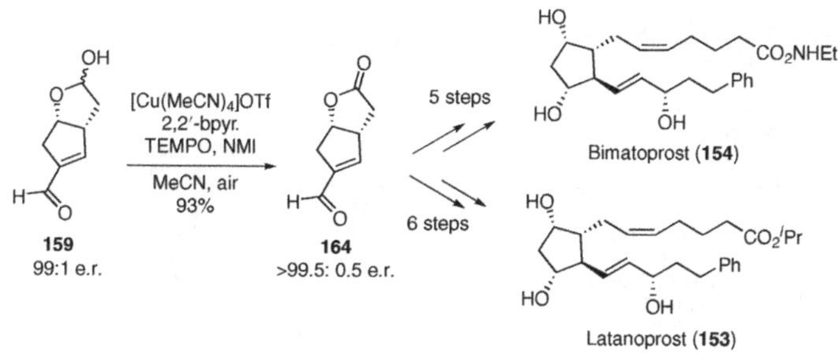

Scheme 2.24 Aggarwal's short syntheses of antiglaucoma drugs Latanoprost (**153**) and Bimatoprost (**154**) via a bicyclic enal lactone intermediate (**164**).

reaction produced compound **158** at a low yield, the one-pot construction of an advanced $PGF_2\alpha$ precursor with concurrent formation of two chiral centers in excellent enantioselectivity, two C-C bonds, and a bicyclic ring system is remarkable. Moreover, the brevity and elegance of the method allowed a gram scale synthesis of $PGF_2\alpha$ (**152**), which has attracted a great deal of attention of chemists from both academia and the pharmaceutical industry.

Recently, this methodology has been applied to the concise synthesis of latanoprost (**153**) and bimatoprost (**154**), in which a crystalline lactone with higher enantiopurity was used as the key intermediate (Scheme 2.24).[47b]

2.7.3 Echavarren's Seven-step Total Syntheses of Aromadendrane Sesquiterpenes

Echavarren's seven-step syntheses of aromadendrane sesquiterpenes illustrates the power of a gold(I)-catalyzed cascade reaction in the efficient and stereodivergent synthesis of natural products.[51] Belonging to aromadendrane sesquiterpenes, (–)-epiglobulol (**165**), (–)-4*α*,7*α*-aromadendranediol (**166**), and (–)-4*β*,7*α*-aromadendranediol (**167**) are found in many essential oils from a variety of plant species and display diverse bioactivities. These natural products feature a tricyclic ring system containing six stereogenic centers. In 2014, A. M. Echavarren and coworkers disclosed a seven-step approach allowing the syntheses of (–)-epiglobulol (**165**), (–)-4*α*,7*α*-aromadendranediol (**166**), and (–)-4*β*,7*α*-aromadendranediol (**167**) at 12, 15, and 17% overall yields, respectively, from (*E*,*E*)-farnesol (**168**), which are the shortest total syntheses of these natural compounds (Scheme 2.25). The syntheses relied on a gold(I)-catalyzed stereodivergent cascade reaction to form the tricyclic aromadendrane core in a single step (via **171A** in Scheme 2.25), a methodology developed from the same group.[52] The dienyne (*S*,*E*)-**170** was prepared in four steps (Sharpless asymmetric epoxidation; conversion of allylic alcohol to the corresponding chloride; *n*-BuLi-mediated rearrangement to give propargylic alcohol;

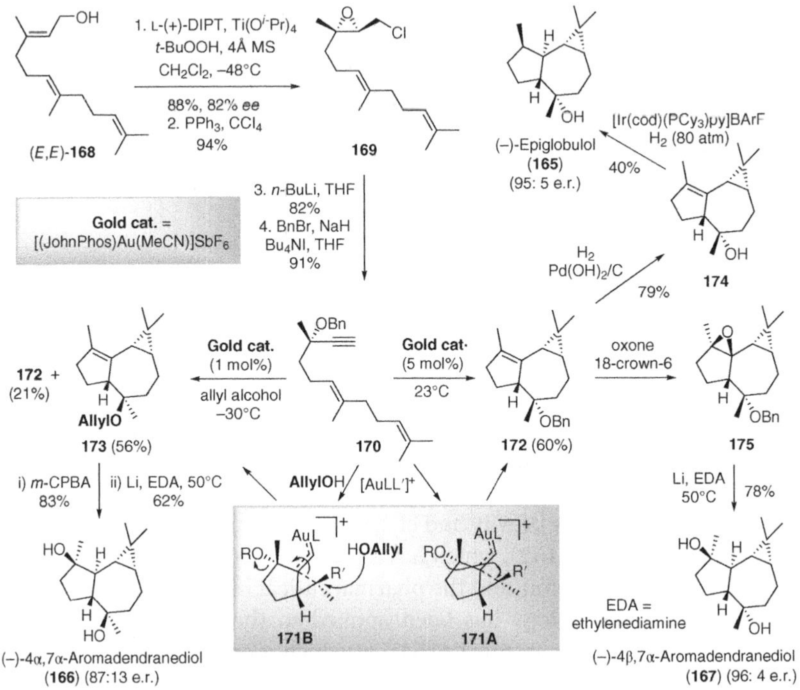

Scheme 2.25 Echavarren's seven-step total synthesis of (–)-sesquiterpenes **165–167**.

benzylation); and a 62% overall yield from (*E*,*E*)-farnesol (**168**). Exposing (*S*,*E*)-**170** to the cationic gold(I) complex [(JohnPhos)Au(MeCN)]SbF_6 gave **172** at a 60% yield.

Interestingly, when the gold-catalyzed reaction was performed in the presence of allyl alcohol as an external nucleophile and at a lower gold catalyst loading (1 mol%), the allyl ether **173**, a C4 epimeric analog of **172** was obtained at a 56% yield along with **172** at a 21% yield. Debenzylation and hydrogenation of **172** gave **165** in a 95:5 e.r. (via **171B** in Scheme 2.25). Alternatively, epoxidation of **172** with dimethyldioxirane (DMDO, generated *in situ* from oxone®) yielded **175** stereoselectively. Reductive epoxide opening and ether cleavage with Li in ethylenediamine yielded **167** at 78% (96:4 e.r.). On the other hand, **173** was transformed to **166** by selective epoxidation with *m*CPBA and reductive opening of the epoxide/allyl cleavage.

2.7.4 Zhu's Seven-Step Total Synthesis of Peganumine A

Peganumine A (**176**), a dimeric tetrahydro-*β*-carboline alkaloid isolated in 2014 from the seeds of *Peganum harmala* L., was reported to display

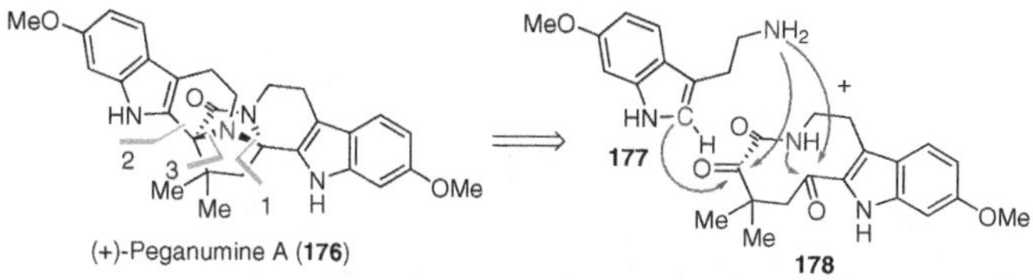

Scheme 2.26 Key retrosynthetic analysis of peganumine A via strategic bond disconnections.

significant cytotoxic activity against MCF-7, PC-3, HepG2 cells, and a selective effect on HL-60 cells with an IC_{50} value of 5.8 μM. The first enantioselective disclosed recently by J. P. Zhu and coworkers[53] relies on the strategic bond disconnections on the central 2,7-diazabicyclo[2.2.1]heptan-3-one core (Scheme 2.26), which implies, in a forward sense, the asymmetric one-pot condensation of **177** and **178** to give **176**. Another key feature of the approach resides in building the macrolactam **178** by either the intramolecular Passerini reaction[54] or the Ugi reaction of ω-isocyano-γ-oxoaldehyde.[55]

The synthesis started from 6-methoxytryptamine (**177**), which was converted to organostannane **179** in three steps (Scheme 2.27). The Liebeskind–Srogl cross-coupling[56] between a thiol ester **180b** (derived in two steps from 3,3-dimethylpent-4-enoic acid **180a**) and organostannane **179**, afforded ketone **181** at a 95% yield. Dehydration of keto-formamide **181** produced ω-isocyano-γ-oxoaldehyde **182**. The Ugi four-center three-component reaction between isocyano-aldehyde **182**, *N*-methylhydroxylamine, and acetic acid in MeOH afforded **183** at a 75% yield. The reaction may involve the one-pot Ugi reaction to deliver **178a**, β-elimination to afford the α-iminolactam **178b**, and hydrolysis of the latter. Finally, condensation of **183** with **177** in the presence of Jacobsen's chiral thiourea catalyst[57] (*S*)-**185** (0.2 equiv.) and benzoic acid, followed by treatment of the resulting imine with TFA produced (+)-peganumine A (**176**) (69%, 96:4 e.r.). The scenario involves one-pot imine formation, an enantioselective Pictet–Spengler reaction, a transannulation,[58] and aminal formation.

The efficiency of this total synthesis deserves comment. First, with a 33% overall yield, this seven-step enantioselective total synthesis is highly efficient in terms of both steps and overall yield. The efficiency gained firstly from the strategic disconnection-based assembly of the complex ring system via multicomponent Ugi reaction-triggered cascade reaction (**182–183**). Second, this synthesis features a singular strategy of late-stage installation of stereogenic centers: All the starting materials and intermediates are achiral, which alleviates the need for any stereochemical control. The enantioselective control in the last step was achieved by an unprecedented thiourea-catalyzed enantioselective Pictet–Spengler reaction. It is worth mentioning that this ingenious

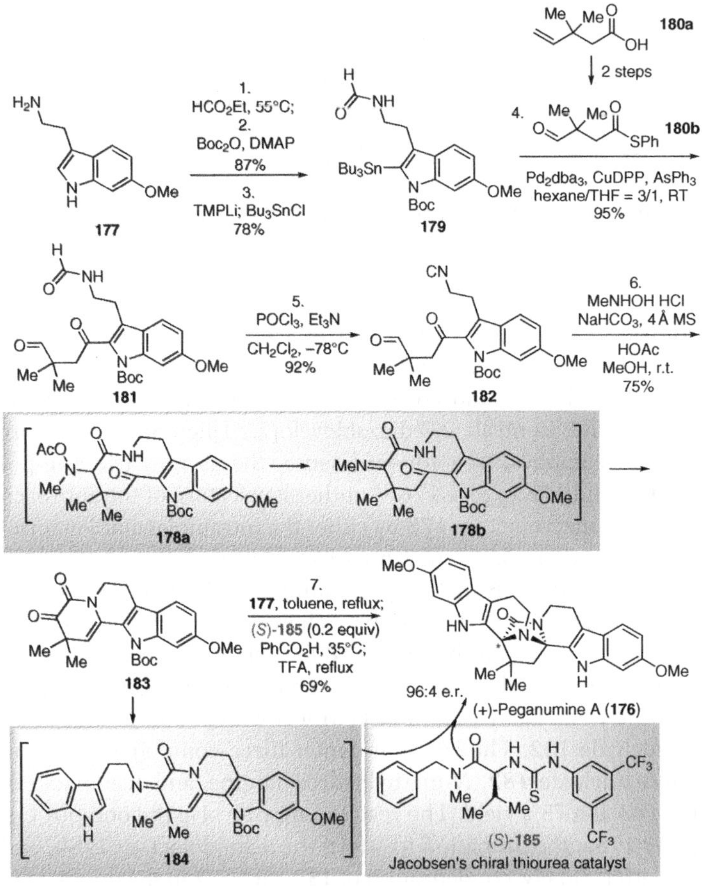

Scheme 2.27 Zhu's seven-step total synthesis of (+)-peganumine A (**176**).

total synthesis is not without luck, as the author noted: "Since the thiourea of type **185** has not been applied to the Pictet–Spengler reaction of ketone, (*S*)-**185** was chosen arbitrarily, as no empirical model could be followed to predict the stereochemical outcome."[53]

2.7.5 Rychnovsky's Seven-Step Synthesis of *Lycopodium* Alkaloid (+)-Fastigiatine

Lycopodium alkaloids are a huge family that includes nearly 300 natural products with impressive and challenging polycyclic structures and interesting biological activities.[59] First known in the TCM as "*Qian Ceng Ta*" ("thousand-laid pagodas"), the herb was used to treat a variety of diseases including contusions, schizophrenia, fever, inflammation, and so on. It is also called "Jin Bu Huan" (more valuable than gold) referring to its medicinal efficacy and availability in minuscule amounts.[59,60]

In recent years, the synthesis of complex *Lycopodium* alkaloids has attracted considerable attention.[59,61] In 2016, S. D. Rychnovsky and coworkers reported a concise six-step synthesis of (+)-fastigiatine (**186**) from (*R*)-3-methylcyclohexanone (**191**).[62] Structurally, (+)-fastigiatine contains five rings, six stereogenic centers including one aza- and one all-carbon quaternary centers. Rychnovsky's retrosynthetic analysis (Scheme 2.28) is based on a plausible biomimetic transannular Mannich reaction proceeding through enamino-iminium intermediate **187**. In this manner, the structural complexity is dramatic reduced. Further simplification is envisaged by comparing the *cis*-benzo[7]annulene **189** with (+)-fastigiatine (**186**): It contains 12 of the carbon atoms and three of the stereogenic centers in the correct absolute configurations. Compound **189** should be accessible from the known (*R*)-5-methylcyclohex-2-en-1-one (**190**) by a Diels–Alder reaction followed by a ring expansion reaction.

The synthesis commenced with commercially available (*R*)-3-methylcyclohexanone (**191**), which was converted in one pot to (*R*)-5-methylcyclohex-2-en-1-one (**190**)[63] (Scheme 2.29). The Diels–Alder reaction of **190** with diene **192** gave decalinone derivative **193** as a 14:1 mixture of diastereomers favoring the *cis* version. Cyclopropanation of **193** with dibromocarbene provided **194**, which was subsequently treated with $AgClO_4$ to give the ring expansion product α-bromoenone **189** as a diastereomeric mixture (*cis:trans* = ~3:1). The Suzuki coupling[64] between the borane derived from **195** and bromoenone **189** proceeded with concomitant epimerization leading to a thermodynamic diastereomeric mixture of **196** (*cis:trans* = 2:3). Conjugate addition of the latter with cuprate **197c**, generated *in situ* from phenylthiocarbamate **197a** via reductive lithiation and Li–Cu exchange of **197b**, afforded **188** as a very complex mixture of products containing epimers possible at C4, C10, and C12. When worked up the reaction with K_2CO_3/methanol, the desired intramolecular aldol reaction occurred to produce tricyclic product **198** and its

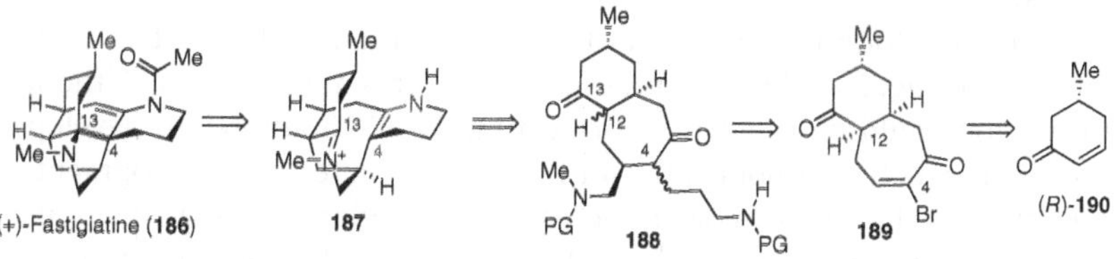

Scheme 2.28 Rychnovsky's retrosynthetic analysis of (+)-fastigiatine (**186**).

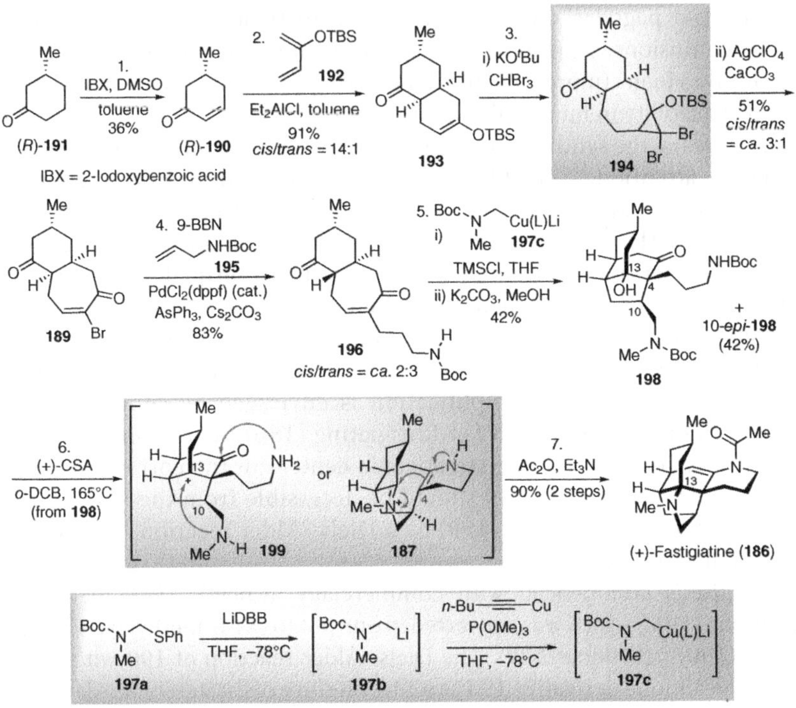

Scheme 2.29 Rychnovsky's seven-step total synthesis of (+)-fastigiatine (**186**).

C-10 epimer each at a 42% yield. Treatment of **198** with (+)-10-camphorsulfonic acid (CSA) in 1,2-dichlorobenzene (*o*-DCB) at elevated temperatures cleaved the two Boc protecting groups. Under acidic reaction conditions, two scenarios may occur: the hydroxyl group at C-13 left to generate a tertiary cation **199**, which was captured by the *N*-Me group, and cyclic enamine formation; or a retro-aldol occurred that permitted the formation of intermediate **187** en route to a transannular Mannich reaction.[58] Acylation of the crude reaction mixture produced (+)-fastigiatine (**186**) at a 90% yield over two steps.

It is worth mentioning that due to the configurational lability at C-12, the first three intermediates in the synthesis are mixtures at this center. Although epimerization could be avoided by protecting the C-13 ketone, to develop a concise approach, the authors opted for the current route. The development of cuprate **197c** as the "methylene amine" synthon avoided late-stage manipulations, which constitutes another valuable tactic for achieving brevity of the synthetic route.

2.8 Eight-Step Enantioselective Total Synthesis of Natural Products

2.8.1 Overman's Eight-Step Synthesis of (+)-*Trans*-Clerodane Iterpenoid

In 2015, Overman and coworkers put forward a convergent fragment-coupling strategy in which a structurally complex target molecule is disconnected at a quaternary carbon stereocenter to fragments of comparable complexity (Scheme 2.30).[65a] The value of this tactic was demonstrated by the eight-step enantioselective total synthesis of *trans*-clerodane **200** (Scheme 2.31).[65b,c] As represented by compound **200**, the *trans*-clerodane subset belongs to the clerodane family of diterpenoid natural products composed of more than 650 secondary metabolites isolated from various plant sources.

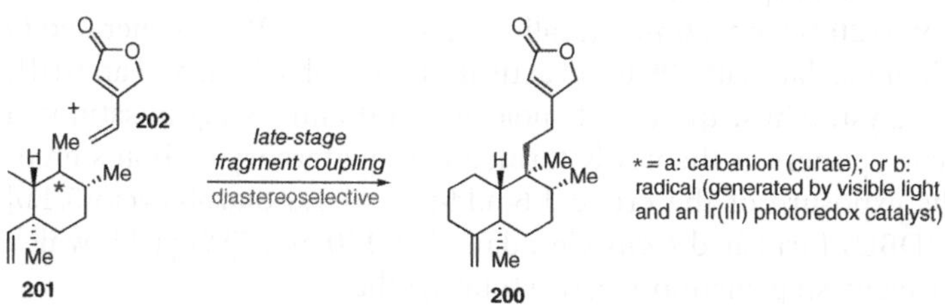

Scheme 2.30 Overman's fragment-coupling strategy for the synthesis of (+)-*trans*-clerodane diterpenoid.

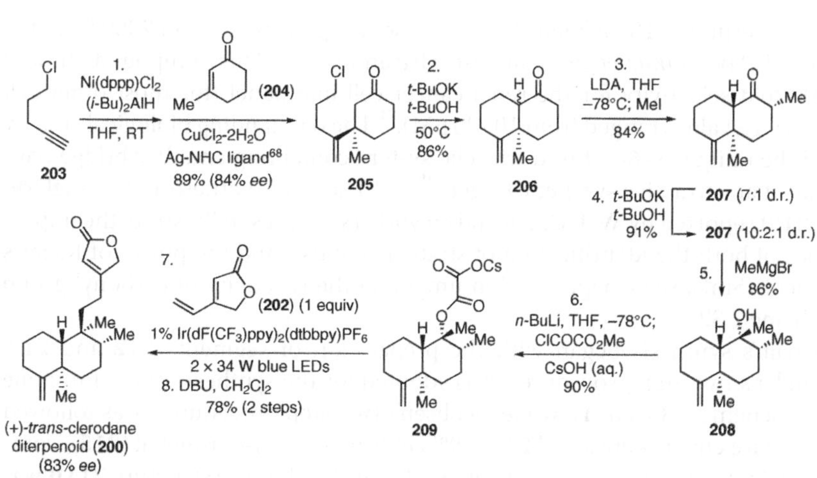

Scheme 2.31 Overman's eight-step enantioselective total synthesis of (+)-*trans*-clerodane diterpenoid (**200**).

The challenge for the total synthesis of *trans*-clerodane diterpenoids resides in the construction of two all-carbon quaternary centers.[66] Central to the synthesis of cyclohexanone **205** is the enantioselective building of the quaternary carbon stereocenter. This was achieved by utilizing Hoveyda's Ni-catalyzed regioselective hydroalumination[67] and catalytic enantioselective conjugate addition[68] methods to Piers' racemic synthetic route (Scheme 2.31).[69] In this manner, cyclohexanone **205** was prepared at a 89% yield and 84% *ee*. *t*-BuOK-mediated cyclization of **205** afforded decalone **206** as a 2.8:1 mixture of *trans/cis* diastereomers. Methylation of decalone **206** with LDA/ MeI, followed by *t*-BuOK-catalyzed equilibration provided **207** as a diastereomeric mixture at a 10:2:1 ratio. Addition of methylmagnesium bromide to decalone **207** delivered decalinol **208**, which was isolated as a single diastereomer at a 66% yield. Next, by the one-pot acylation/saponification procedure,[70] alcohol **208** was converted to its activated form, oxalate salt **209**, suitable for the visible-light photoredox catalysis.[70] The second quaternary center was installed by 1,6-addition of 4-vinylfuran-2-one (**202**) with a *trans*-decalin tertiary radical **201b**, generated *in situ* from the hemioxalate salt **209** upon activation by visible light and an Ir(III) photoredox catalyst. Thus, the target molecule **200** and its *β,γ*-unsaturated isomer as single epimers at the newly formed C-9 quaternary carbon stereocenter. Finally, exposure of the crude 1,6-adducts to 1,8-diazabicyclo[5.4.0]undec-7-ene (DBU) furnished *trans*-clerodane (+)-**200** at a 78% yield, which completed the eight-step enantioselective total synthesis.

2.8.2 Chain's Eight-Step Synthesis of (−)-Englerin A

(−)-Englerin A (**210**, Scheme 2.32) is a sesquiterpene isolated in 2009 from the bark of *Phyllanthus engleri*, an east African plant.[71] This compound showed 1000-fold selectivity for the renal cancer cell line panel (for most other cell lines, GI_{50} values ranged from 10–20 μM).[71] The outstanding biological activity and the unique 5–6–5 fused tricyclic system containing an ether bridge have made it an attractive synthetic target.[71b,72,73] The brief synthesis of (−)-englerin A (**210**) reported by W. J. Chain and coworkers[73] serves to illustrate the importance of both the identification of strategic bonds[74] and the power of Kagan's reagent (SmI_2, see Chapter 6)[75] in umpoling the reactivity of carbonyl group (Scheme 2.32).

Chain's synthesis began with the preparation of segments **212** and **213**. Winkler's two-step procedure[76] was adopted for the synthesis of the furanone **212** (Scheme 2.33, eq. **1**), while Jacobsen's two-step procedure[77] was followed to prepare enantio-enriched **213** (79% *ee*) from (*R*)-(+)-citronellal (**215**), commercially available in 77% *ee* (Scheme 2.33, eq. 2). The core structure **211B** was constructed in two steps. The addition of lithium enolate generated from the 3-furanone **212** to 5-methylcyclopentenecarboxaldehyde (**213**) gave preferentially **217** out of eight possible diastereomers at a good yield (75% combined

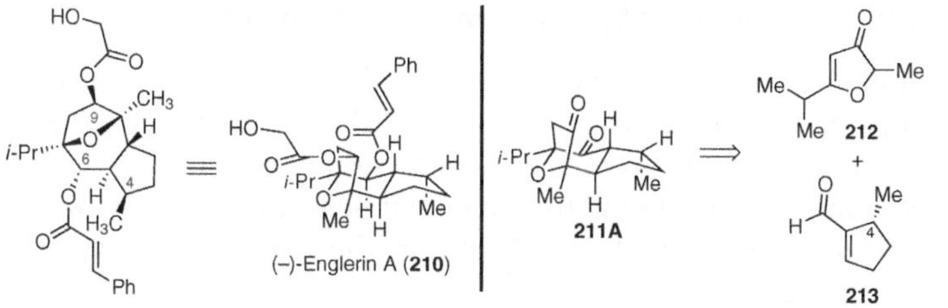

Scheme 2.32 Structure of (–)-englerin A (**210**) and Chain's idea for its synthesis.

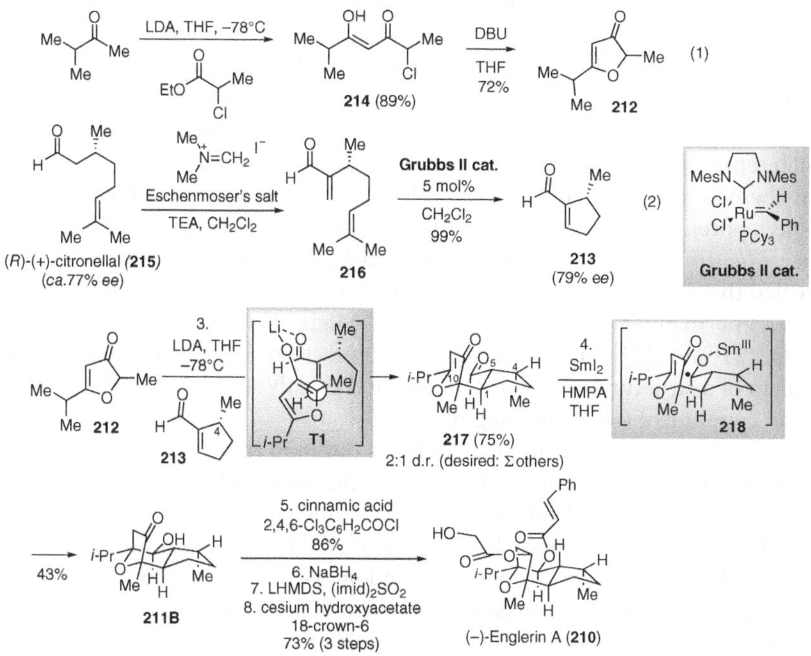

Scheme 2.33 Chain's eight-step total synthesis of (–)-englerin A (**210**).

yield, 2:1 d.r. desired: Σ others). A possible scenario is that the C4 methyl-bearing stereogenic center serves as a key controlling element (see **T1**). For the second key reaction, namely the reductive carbonyl-alkene cyclization, Kagan's reagent (SmI_2) turned out to be superiors compared with other single-electron reducing agents. Thus, treatment of the Michael adduct **217** with SmI_2 in the presence of HMPA in THF produced, via **218**, keto alcohol **211B** at a 43% yield. In this key transformation, not only was the second critical carbon–carbon bond formed to afford the tricyclic core structure, but also the correct oxidation level and stereochemistry at C6 were established. Compound **211B** was converted into the target molecule **210** in four conventional steps.

2.8.3 Shenvi's Eight-Step Total Synthesis of (−)-Jiadifenolide

Neurodegeneration diseases such as Alzheimer's disease are affecting more and more older people. Isolated in 2009 from the pericarps of *Illicium jiadifengpi* collected in southwestern China,[79a] the sesquiterpenoid (−)-jiadifenolide (**219**, Scheme 2.34) was found to be a potent neurotrophic agent, which represents a valuable small-molecule lead for the potential therapeutic treatment of neurodegenerative diseases.[79]

Structurally, (−)-jiadifenolide (**219**) possesses a densely functionalized caged pentacyclic ring system featuring four rings emanating from the central, highly substituted B-ring cyclohexane, and seven contiguous stereocenters. The important medicinal potential, scarcity from natural source, and challenging structure rend (−)-jiadifenolide an attractive synthetic target. Among several elegant enantioselective syntheses so far reported,[80,81] Shenvi's eight-step, and gram scale total synthesis of (−)-jiadifenolide (**219**) is remarkably concise,[81] which represents another model of strategic bond disconnection.[74]

In Shenvi's retrosynthetic analysis (Scheme 2.34), compound **220** was identified as a potentially versatile intermediate for the synthesis of jiadifenolide and other *Illicium* terpenes. A two-bond retro-Michael addition disconnection revealed the possibility to form the tetracyclic compound **220** in a one-pot and convergent manner by a tandem double Michael addition between fragments **221** and **222**. Thus, after this disconnection, a challenging synthetic problem of a structurally complex molecule is greatly simplified to the synthesis and coupling of two simple segments.

The synthesis of (−)-jiadifenolide (**219**) is outlined in Scheme 2.35.[81] Chiral butenolide **221** was synthesized from (+)-citronellal (**215**) by a three-step procedure consisting of the dehydration of citronellal, ozonolysis of the alkene, and a hetero-Pauson–Khand (hPK) reaction.[82] The synthesis of acetylbutenolide **222** involves a known one-pot tandem protocol: thermolysis of dioxinone **224**, trapping the acylketene intermediate **225** with an α-hydroxy acetone, and silica gel mediated intramolecular Knoevenagel condensation of the keto *β*-ketoester **226**. The key double Michael addition proceeded smoothly to deliver the desired adduct **220** at 70% and in excellent diastereoselectivity (20:1 d.r.). Hydroxylation of the acidic *α*-position of *β*-ketolactone **220** via oxidation of the enol tautomer with *m*CPBA proceeded stereoselectively from

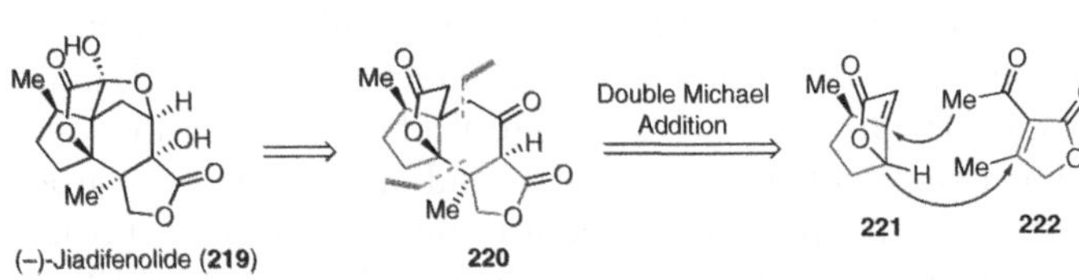

(−)-Jiadifenolide (**219**) **220** **221** **222**

Scheme 2.34 Retrosynthetic analysis of (−)-jiadifenolide (**219**).

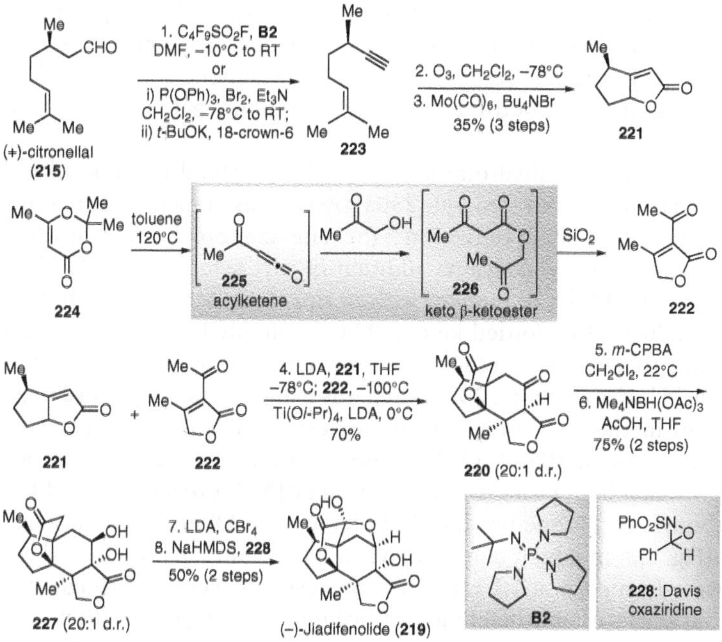

Scheme 2.35 Shenvi's eight-step gram-scale synthesis of (−)-jiadifenolide (**219**).

the *α*-face to produce the more stable *cis*-fused ring system. The newly introduced hydroxyl group served in turn as a directing group in the stereoselective reduction of the proximal ketone, which yielded diol **227** in excellent diastereoselectivity (20:1 d.r.). Finally, without using any protecting group, the other lactone in **227** was subjected to deprotonation-*α*-bromination and deprotonation-oxidation with Davis' racemic oxaziridine (**228**), which directly afforded (−)-jiadifenolide (**219**).

2.8.4 Maimone's Eight-Step Total Synthesis of (+)-Chatancin

(+)-Chatancin (**229**) is a diterpene of cembrane biosynthetic origin isolated from a soft coral (*Sarcophyton* sp.) off the coast of Okinawa, Japan,[83] and was reported to inhibit both platelet-activating factor (PAF) induced platelet aggregation (IC_{50} = 2.2 mm) and PAF receptor binding (IC_{50} = 0.32 mm). The chemical synthesis of chatancin[84,85] presents formidable challenges due both to the structural complexity and the inherent instability. Structurally, chatancin possesses a complex carbon skeleton featuring two *cis*-decalin motifs folded into a unique polycyclic arrangement by a hemiketal bridge, and containing seven stereocenters (six of which are contiguous). Moreover, it is extremely acid sensitive, it rapidly dehydrates to anhydrochatancin under even mildly acidic conditions.[85]

The general strategy for cembranoids syntheses features linear precursor assembly followed by macrocyclization.[86] However, such an approach generally leads to lengthy synthetic routes. In 2015, T. J. Maimone and Y.-M. Zhao reported an alternative bond-forming orchestration that is shown in Scheme 2.36.[85]

The synthesis used (*S*)-dihydrofarnesal (**230b**) as a chiral building block, which is available either from farnesal (**230a**) by List's asymmetric counteranion (**cat. 2**)-directed catalytic reduction[87] (Scheme 2.37) or from L-dihydrofarnesol. The vinylogous Mukaiyama addition of silyl ketene acetal **232**[88] to (*S*)-dihydrofarnesal (**230b**), in tandem with *in situ* Dess–Martin oxidation of the resulting adduct **233** afforded ketone **234**. By simply heating compound **234** in refluxing toluene, the acetone extrusion with concomitant cyclization occurred to yield hydroxypyrone **235**, which was triflated *in situ* to afford vinyl triflate **236**. The methoxycarbonylation of the electron-deficient **236** called for Fürstner's catalyst system ($Pd(OAc)_2$/DPEPhos),[89] producing the desired product (not shown) at a high yield. An intramolecular [4+2] cycloaddition of the methoxycarbonylated pyrone produced diastereomers **237a** and its diastereomer **237b** (Fig. 2.2) in a high combined yield (90%). Although the two diastereomers were formed in a 1:1 ratio, this process, proceeding through two boat transition states, established four stereocenters in a single operation. The final task in constructing the chatancin cembrane ring system was the formation of the C1–C11 bond. This was achieved via allylic chlorination with SO_2Cl_2, and subsequent *in situ* formation and addition of allylic zinc reagent, which afforded the desired **239** as a single diastereomer at a 80% yield from **237a**. Hydrogenation of **239** furnished (+)-chatancin (**229**) at a 93% yield.

This protecting-group-free total synthesis is quite concise. In the synthesis, the first stereogenic center was established by List's asymmetric counteranion-directed catalytic reduction, which demonstrated that organocatalysis is useful not only for enantioselective C-C bond formation, but also for the asymmetric reduction. This allowed employing naturally occurring achiral building block farnesal (**230a**) as a pro-template of the target molecule: with only one chain elongation needed (**230b** to **233**). Moreover, the strategic utilization of this building block also provided another handle for the stereospecific formation of last ring (**237a**→**239** via **238**). The core structure (**237a**) of the target molecule was

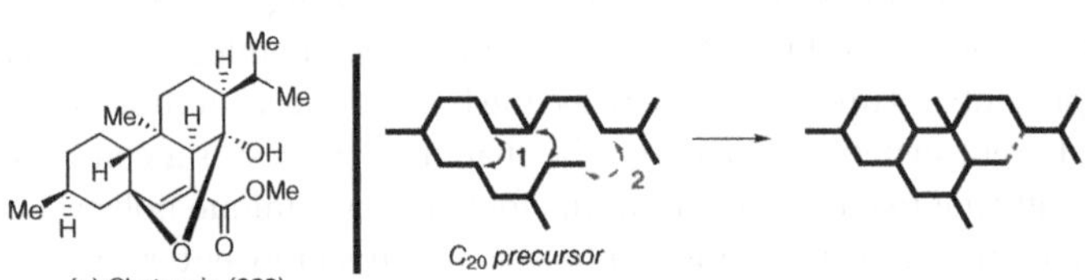

(+)-Chatancin (**229**)

Scheme 2.36 Maimone's abiotic synthetic strategy to diterpene (+)-chatancin (**229**).

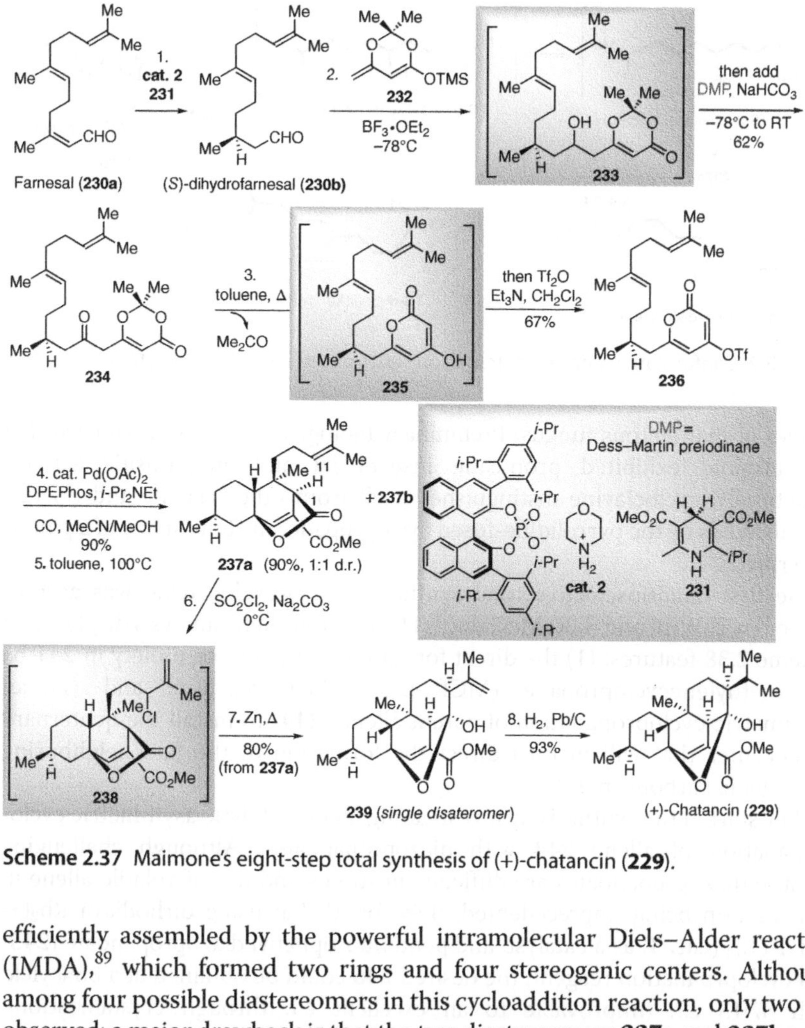

Scheme 2.37 Maimone's eight-step total synthesis of (+)-chatancin (**229**).

efficiently assembled by the powerful intramolecular Diels–Alder reaction (IMDA),[89] which formed two rings and four stereogenic centers. Although among four possible diastereomers in this cycloaddition reaction, only two are observed; a major drawback is that the two diastereomers **237a** and **237b** were formed in a 1:1 ratio. Conformation analysis (Figure 2.2) showed that cycloadducts **237a** and **237b** were likely to arise from favorable chairlike transition states. However, due to a lack of control elements such as allylic strain, there is no preference for **TS-a** or **TS-b**. Nevertheless, this eight-step synthesis has been achieved with a high overall yield (13% from **230b**).

2.8.5 Wipf's Eight-Step Total Synthesis of (−)-Cycloclavine

(−)-Cycloclavine (**240**) is an ergot alkaloid isolated from the seeds of the African morning glory shrub, *Ipomoea hildebrandtii*, and *Aspergillus japonicus*, a

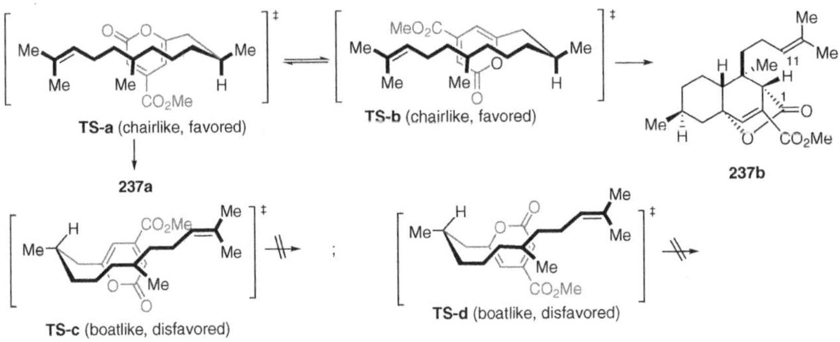

Figure 2.2 Conformation analysis of transition states of the Diels–Alder reaction.

species of filamentous fungus. Preliminary biological evaluations showed that this alkaloid exhibited promising insecticidal and antiparasitic activity. Structurally, cycloclavine distinguishes itself from other clavine and lysergic acid alkaloids by the pyrrolidine-fused cyclopropane sub-unit in lieu of a piperidine ring.

The first enantioselective total synthesis of (−)-cycloclavine was accomplished by P. Wipf and S. R. McCabe.[90] The retrosynthetic analysis displayed in Scheme 2.38 features: (1) the direct formation of the enone moiety in **241** by the methylenecyclopropane (MCP)-Diels–Alder reaction; and (2) an asymmetric cyclopropanation of parent allene (**244**) to install the quaternary stereocenter that will in turn direct the formation of the two neighboring stereogenic carbons in **240**.

They started the synthesis by investigating metal-catalytic asymmetric cyclopropanation of allene **244** with diazopropanoates. Although challenging because diazopropanoates are difficult substrates and use of volatile allene in this reaction being unprecedented, they found that using dirhodium Rh_2(*S*-TBPTTL)$_4$ (**cat. 3**) as a catalyst, and pentafluorophenyl diazopropanoate (**245**) as a cycloprpanation reagent, the desired **246** could be obtained at a 86% yield with 87:13 e.r. (improvable to an excellent e.r. through crystallization). Transamidation with anion generated from vinylogous amide **243** yielded amide **247**, which was deprotonated and *O*-silylated to give the highly reactive

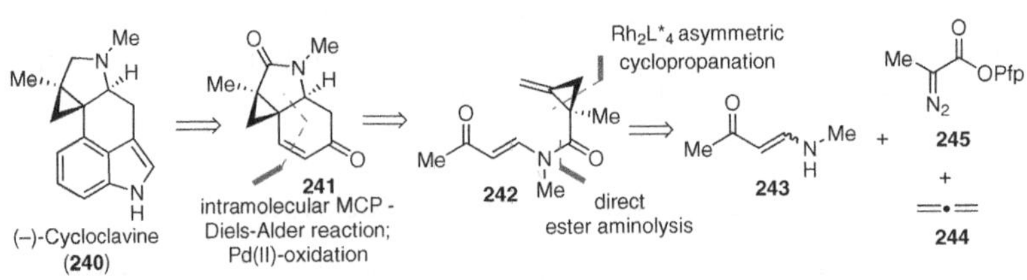

Scheme 2.38 Wipf's retrosynthetic strategy to (−)-cycloclavine (**240**).

diene **248**. This diene underwent an *anti*-selective intramolecular [4+2] cycloaddition with the pendant methylenecyclopropane upon microwave irradiation. Desilylation of the crude enol ether adducts afforded separable diastereomers of ketone **249** in 4.8:1 d.r. at C5. A $Pd(DMSO)_2(TFA)_2$-catalyzed direct dehydrogenation of ketone **249** provided **241** in >99% *ee* after recrystallization. As the second spotlight of the synthesis, a new 2,2,6,6-tetramethyl-1-piperidinyloxy (TEMPO) carbamate capable to generate a more thermodynamically stable and chemoselective furyl lithium species **250**, was developed, which ensured the introduction of the indole moiety via the key intramolecular strain-promoted Diels–Alder methylenecyclopropane (IMDAMC) reaction. Thus, addition of **250** to **241** afforded **251** as a 1.5:1 mixture of diastereomers at C3. Heating a solution of compound **251** in toluene at 135°C resulted in the expected tandem IMDAF reaction, aromatization, and thermolysis of the TEMPO carbamate protecting group to deliver directly the desired, protecting-group-free indole **252**. Finally, reduction of lactam **252** completed the enantioselective total synthesis of (–)-cycloclavine (**240**) in eight steps and a 7.1% overall yield (Scheme 2.39).

This total synthesis features the extensive use of cycloaddition reactions: Three of eight steps are cycloaddition reactions, which include one cyclopropanation (**244** to **246**) and two intramolecular Diels–Alder reactions[89] (**248–249**) and (**251–252**). The efficiency also relies on the development of new chemistries. In particular, the unprecedented asymmetric cyclopropanation of allene (**244**) with **246** not only forges the substituted cyclopropane with

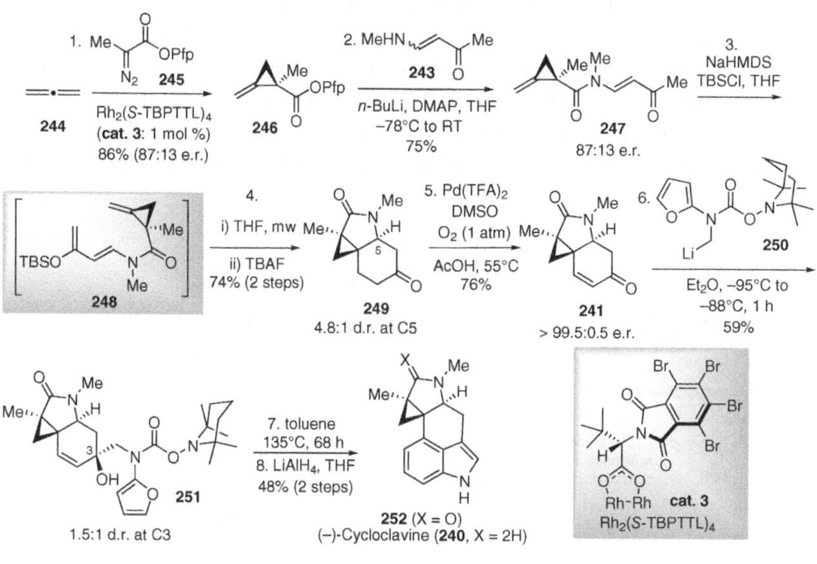

Scheme 2.39 Wipf's eight-step total synthesis of (–)-cycloclavine (**240**).

correct stereochemistry found in the target molecule, but also provides handles for the direct amidation and subsequent Diels–Alder reaction. A drawback of the synthesis is that the addition of furyl lithium **250** to enone **241** proceeds with low diastereoselectivity (1.5:1).

2.8.6 Shenvi's Eight-Step Total Synthesis of (−)-Neothiobinupharidine

The nuphar dimers (e.g., **253**, Scheme 2.40) comprise a family of 11 topologically complex alkaloids isolated from the fresh water plant *Nuphar lutea*. Recent studies disclosed that these compounds exhibited significant anticancer activities.[91] In 2013, R. A. Shenvi and D. J. Jansen reported the first total synthesis of a thiaspirane nuphar dimer (−)-neothiobinupharidine (**253**),[92] which also constituted the shortest synthesis to date of the monomeric nuphar quinolizidine series in an enantiomerically pure form.

In LaLonde's proposal for the biogenesis of the thiaspirane alkaloids' gross structure,[93] iminium ion **254** was postulated as the monomeric intermediate for the dimerization. In light of this hypothesis, the retrosynthesis of Shenvi focused on *N-epi*-dehydronupharidine **255**, which was expected to serve as a precursor of iminium **254** via a Polonovski–Potier reaction[94] (Scheme 2.40). The stereotetrad of **255** was envisioned to arisen through substrate controlled reactions onto lactam **256**, which in turn could be accessible from cyclopentanone **257** through Beckman rearrangement. Enantioenriched cyclopentanone derivative **257** could be available from 3-methyl-2-cyclopentenone (**258**).

The synthesis commenced with the asymmetric reductive α-allylation of 2-cyclopentenone (**258**). A new procedure merging Buchwald's conjugate reduction reaction[95] with a modified Tsuji–Trost allylation[96] was developed for this purpose, which led to ketone **257** in one pot with high diastereoselectivity (>10:1 d.r.) and high enantioselectivity (97.5:2.5 e.r., Scheme 2.41). Cyclopentanone **257** was converted into the corresponding oxime and the

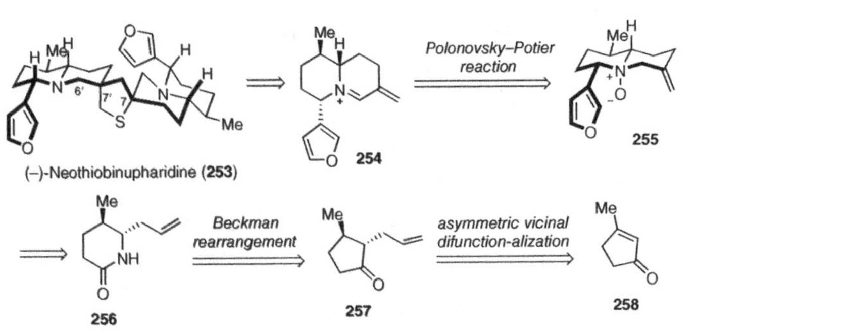

Scheme 2.40 Shenvi's retrosynthetic analysis of nuphar dimer (−)-neothiobinupharidine.

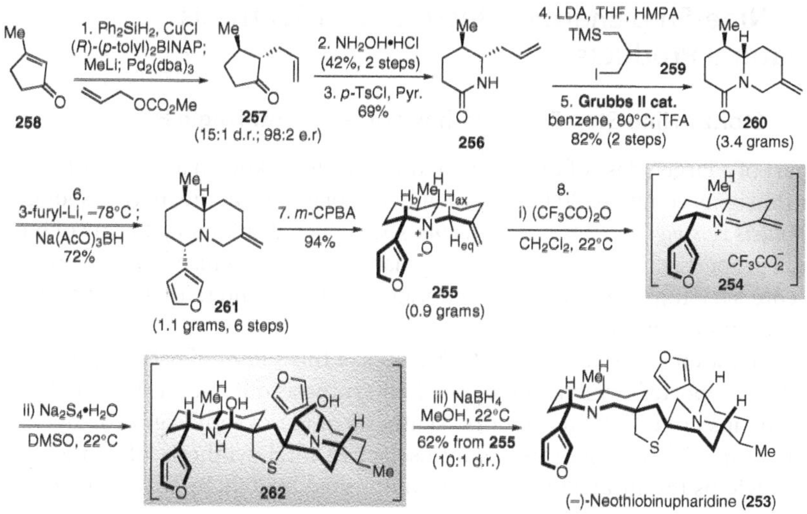

Scheme 2.41 Shenvi's eight-step total synthesis of (–)-neothiobinupharidine (**253**).

latter subjected to a Beckman rearrangement to give lactam **256**. Next, 2-piperidinone **256** was converted to quinolizidinone **260** by Vanderwal's allylsilane-ring-closing metathesis (RCM) methodology[97] and *in situ* protodesilylation with trifluoroacetic acid. Reductive furan-3-ylation of the quinolizidone by a variation of Fowler's procedure[98] utilizing sodium triacetoxyborohydride[99] as a reducing agent yielded quinolizidine derivative **261**. Exposure of **261** to *m*CPBA resulted in a clean conversion to *trans*-quinolizidine *N*-oxide **255**. Treatment of **255** with trifluoroacetic anhydride resulted in the desired Polonovski–Potier elimination to give *α,β*-unsaturated iminium ion **254**, which implied a selective elimination of the equatorial proton (H_{eq}) *syn-* to the *N*-oxide. The key dimerization reaction was achieved by treating **254** with an excess of Na_2S_4 in DMSO, which produced **262** in 90% conversion and at high diastereoselectivity (10:1). Reduction of the crude mixture of **262** with $NaBH_4$ furnished neothiobinupharidine (**253**) at a 62% yield from *N*-oxide **255**.

This eight-step asymmetric total synthesis is quite concise. This should be attributed to the bio-inspired strategy, which centered on the iminium ion **254**, its generation and reaction, including the new method for the reductive allylation of cyclopentenones, Beckman rearrangement, Vanderwal's allylsilane-RCM methodology using the bifunctional reagent **259**, Polonovski–Potier reaction, and so on. All the functional groups were well positioned, and all reactions were organized in a streamlined manner without detour, and no protecting group.

2.9 Nine-Step Enantioselective Total Synthesis of Natural Products

2.9.1 Stoltz's Nine-Step Total Synthesis of (–)-Cyanthiwigin F

As part of a larger class of diterpene natural products known as the cyathanes, cyanthiwigins comprise 30 known members isolated from the marine sponges *Epipolasis reiswigi* and *Myrmekioderma styx*, which display a vast array of biological properties. These molecules are characterized by a fused [5–6–7] tricyclic core skeleton containing four contiguous stereocenters that include two quaternary stereocenters.

Only a few enantioselective total syntheses of this subset of natural products have been reported, which include the first synthesis of (+)-cyanthiwigin U by A. J. Phillips *et al.*,[100a,c] and the first synthesis of (+)-cyanthiwigin AC by T. J. Reddy *et al.* [100b] The shortest enantioselective total synthesis of a member cyanthiwigins (cyanthiwigin F) was accomplished in 2008 by B. M. Stoltz and coworkers.[101a]

In view of providing a general approach to cyanthiwigin family of diterpenes, Stoltz's retrosynthetic analysis of (–)-cyanthiwigin F (**263**, Scheme 2.42) focused on a quick construction of the tricyclic cyathane skeleton. Thus, a disconnection of the five-membered A-ring of cyanthiwigin F (**263**) lead back to bicyclic precursor **264**. A retro-RCM disconnection deduced tetraolefin **265** as a further simplified precursor, which was expected to be accessible via enantioenriched diketone **266**. The diketone **266** provided an excellent platform for implementing Stoltz's powerful enantioselective alkylation technology.[66e,g,102]

Since their first report in 2008,[101a] the details of the method and the second-generation synthesis of the cyanthiwigin natural product core were disclosed in 2011,[101b] and 2016,[101c] respectively. As such, the synthetic route presented here represents the endeavor of Stoltz's group that continued for almost 10 years. The synthesis commenced with the preparation of the known bis(β-ketoester) **268**. A new two-step synthesis from succinic acid was developed that gave **268** as a 1:1 mixture of racemic and *meso* diastereomers (Scheme 2.43). Tremendous effort has been made in implementing the key double catalytic enantioselective allylic alkylation for the

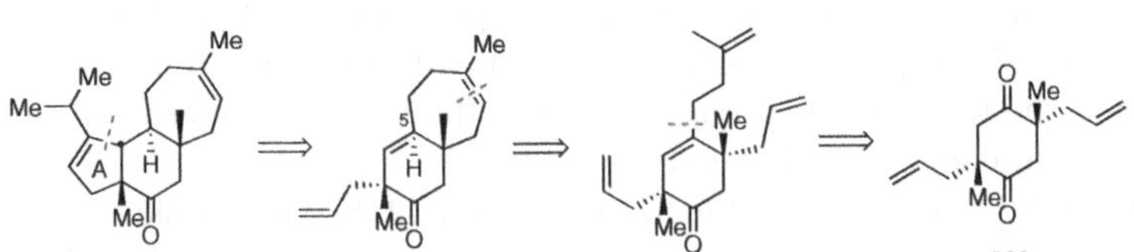

(–)-Cyanthiwigin F (**263**) **264** **265** **266**

Scheme 2.42 Stoltz's retrosynthetic analysis of (–)-cyanthiwigin F.

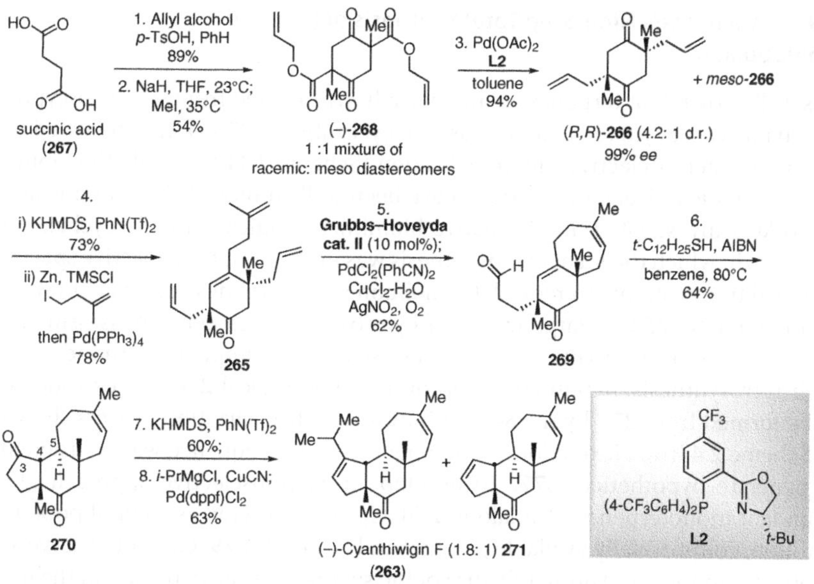

Scheme 2.43 Stoltz's nine-step total synthesis of (−)-cyanthiwigin F.

direct transformation of bis(β-keto ester) **268** to enantio-enriched diketone **266**. The recently reoptimized conditions shown in Scheme 2.43 allowed the running of the reaction at a 10-g scale and in an economical manner. This exceptional transformation demonstrates the power of stereoablative enantioselective alkylation methodology.[66e,g,102] Indeed, the reaction enabled concurrent formation of two independent C-C bonds and the establishing of two all-carbon quaternary stereocenters corresponding to positions C6 and C9 of the cyanthiwigin core with the requisite stereochemistries in good yield, diastereoselectivity, and excellent enantioselectivity from a complex mixture of racemic and *meso*-diastereomers of the starting material. Transformation of **266**, via a vinyl triflate, to tetraene **265** with direct sp^2–sp^3 bond formation was achieved with Negishi cross-coupling. Compound **265** was converted to bicyclic aldehyde **269** by RCM using a modified version of Grubbs–Hoveyda catalyst II[103] followed by anti-Markovnikov Tsuji–Wacker oxidation using Grubbs' conditions.[104] Radical-induced cyclization[105] generated the tricyclic natural product core (**270**) as a single diastereomer that was further converted into cyanthiwigin F (**263**) via selective enol triflate formation and Pd-catalyzed coupling reaction with an *i*-Pr-organocuprate reagent. However, from the last reaction a substantial amount of reduced side product **271** was isolated (**263**: **271** = 1.8: 1), which is a drawback of this synthesis. On the other hand, key intermediate **270** was converted to cyanthiwigins B (4 steps) and G (7 steps).[101b]

2.9.2 Maimone's Nine-Step Total Synthesis of (–)-6-*Epi*-Ophiobolin N

It is well known that terpenes are biologically synthesized from simple polyprenyl chains via enzymatic cationic cascade cyclizations (**273** → **275**) followed by regio- and stereoselective functionalization (Scheme 2.44A). While the biomimetic cationic polyene cyclizations have been well studied, T. J. Maimone and coworkers investigated an alternative biomimetic radical cascade cyclization reaction, in which the form of biosynthesis was imitated but the reactive species changed from cationic to radical. Ophiobolin sesterterpenes such as (–)-6-*epi*-Ophiobolin N (**272**), a large family of cytotoxic sesterterpenes possessing an intricate 5–8–5 fused ring systems, were selected to implement the idea. In their retrosynthesis, hypothetical carbon-centered radical **276** was considered to be formed from **277** by the 8-*endo*/5-*exo*-cascade cyclization process shown in Scheme 2.44(B). Next, a four-component coupling sequence was envisioned to generate hypothetical **277**. To develop an enantioselective approach, the abundant monoterpene (–)-linalool (**278**, C_{10}) was selected as a chiral pool. In addition, commercially available geraniol and farnesol (**279**, C_{15}) could serve as another component. Another feature of the synthetic strategy resides in the use the naturally occurring building blocks (**278–280**) but to forge the bonds in an abiotic radical fashion. They thus developed a rapid (four or five steps)

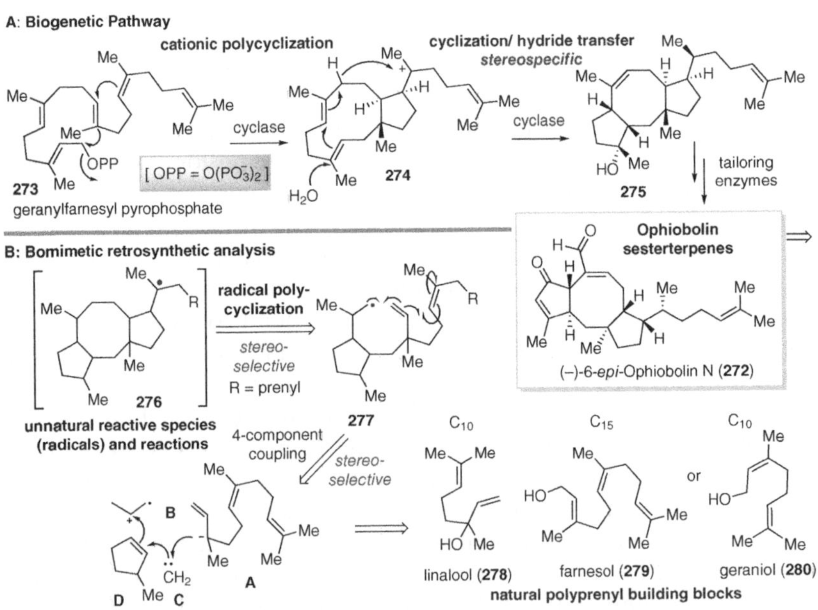

Scheme 2.44 Maimone's bio-inspired synthetic strategy for ophiobolin ring systems.

enantioselective synthetic entry into complex 5–8–5 fused tricycles with requisite handles, convertible into diverse terpenoids, and applied this to the concise enantioselective total synthesis of (–)-6-*epi*-ophiobolin N (**272**).[106]

The synthesis of (–)-6-*epi*-ophiobolin N (**272**) started from the preparation of cyclopentenone **282** from monoterpene (–)-linalool (**278**, Scheme 2.45).[106] This was achieved by a two-step sequence consisting of RCM followed by *in situ* silylation, and Ru-catalyzed allylic oxidation. On the other hand, alkyl iodide **283**, a precursor of another component (**284**) was prepared from farnesol (**279**) via Charette's asymmetric cyclopropanation[107] and iodation. Next, **283** was subjected to lithium-halogen exchange, which resulted in an anionic cyclopropane fragmentation to generate organocopper reagent **284**. The three-component reaction with enone **282** and 2,2,2-trichloroacetyl chloride yielded cyclopentanone **285** (3:1 d.r.), which was subjected to chemoselective reduction of ketone and acetylation to deliver the key intermediate **286**. In the presence of benzothiophene-based TADDOL monothiol **cat. 4**,

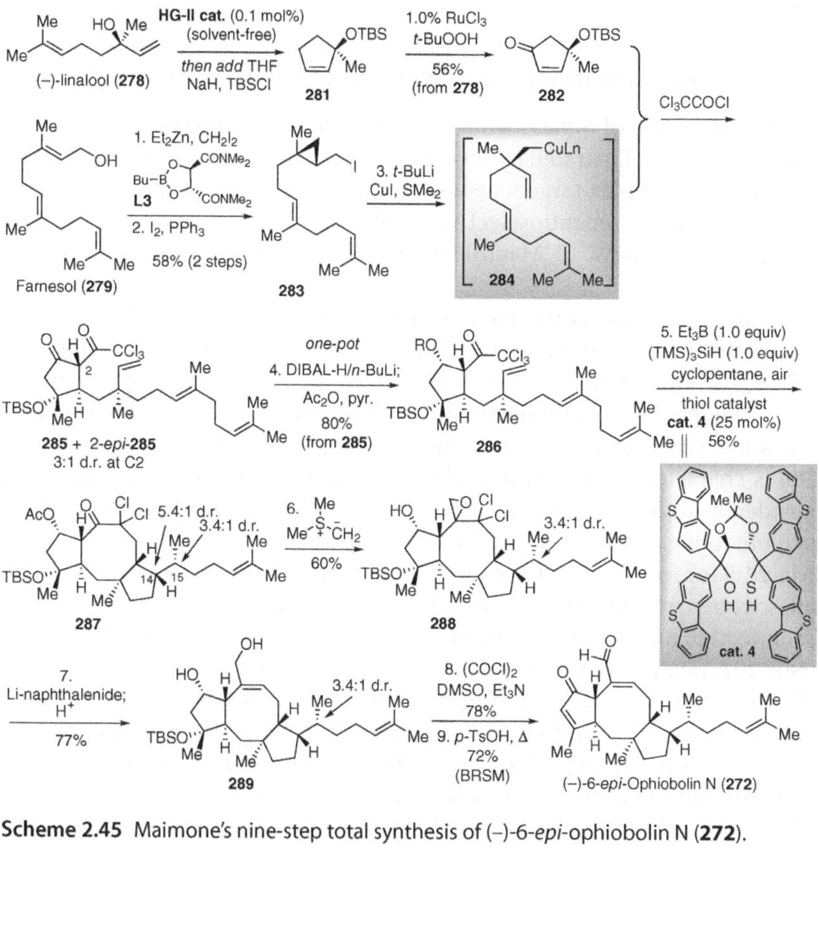

Scheme 2.45 Maimone's nine-step total synthesis of (–)-6-*epi*-ophiobolin N (**272**).

triethylborane/air-initiated reductive radical cyclization of trichloroketone **286** afforded the desired cyclization product **287**, along with its C14 epimer (5.4:1 d.r.) as well as C15 epimer (3.4:1 d.r.) in a combined yield of 56%. This key abiotic radical polycyclization formed two ring and three stereogenic centers albeit in modest diastereoselectivities. The next task was one carbon homologation onto the ketone carbonyl group. This was accomplished via a Corey–Chaykovsky epoxidation reaction,[108] which proceeded with concomitant cleavage of the acetate group to yield **288**. The reductive cleavage of the epoxide in **288** with lithium naphthalenide proceeded with concomitant dechlorination to afford **289**. Finally, Swern oxidation and treatment of the resulting keto-enal with *p*-TsOH afforded (–)-6-*epi*-ophiobolin N (**272**).

It is worth mentioning that the nine-step enantioselective total synthesis of (–)-6-*epi*-ophiobolin N (**272**) is remarkably concise, which also lays the foundation for the efficient synthesis of other complex 5–8-n fused terpenes. However, due to a lack of complete diastereocontrol in both the three-component reaction and the radical polycyclization reaction, the overall yield was only 2%.

2.9.3 MacMillan's Nine-Step Total Synthesis of (–)-Vincorine

Although the value of organocatalysis in the efficient total synhesis of natural products has been illustrated by several elegant examples, its power in the conquest of more complex targets needs to be demonstrated. In this context, the concise and highly enantioselective total synthesis of (–)-vincorine (**290**) accomplished by D. W. C. MacMillan and B. D. Horning[109] provides an outstanding example.

(–)-Vincorine (**290**) is the parent compound of the akuammiline alkaloid subclass that belongs to the *Vinca* alkaloid family. Many members of this class of alkaloids have been successfully developed as medicinal agents for use in cancer therapy and as immunosuppressive drugs, and so on, which include vinblastine, vincristine, vindesine, and vinorelbine. Structurally, (–)-vincorine (**290**) is characterized by a tetracyclic core that incorporates two challenging appended bridged rings: the strained seven-membered azepanyl ring, and the pyrroloindoline motif, as well as four chiral centers, and an olefin motif.

MacMillan's concise synthesis of (–)-vincorine (**290**) originates from his concise retrosynthesis outlined in Scheme 2.46. An unconventional C–C bond disconnection on the strained seven-membered azepanyl system deduced precursor **291**, implied a radical cyclization. A triple disconnection on compound **291** suggested an ambitious organocatalytic enantioselective Diels–Alder reaction–iminium cyclization cascade to build the tetracyclic ring system from simple achiral tryptamine derivative **292** and methyl (*E*)-4-oxobut-2-enoate (**293**). For the build-up of the seven-membered azepanyl ring, a 7-exo-dig radical cyclization utilizing an acyl telluride as an effective alkyl radical

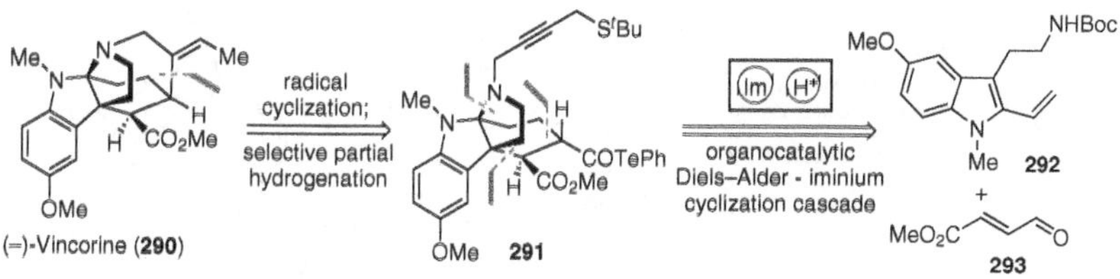

Scheme 2.46 MacMillan's retrosynthetic analysis of (+)-vincorine (**290**).

precursor was conceived. The tactical use of propargylic sulfide as a radical acceptor affords the possibility to control the stereochemistry in the formation of the ethylidene in a stepwise manner.

The requisite 2-vinyltryptamine derivative **292** was prepared in two steps from commercially available 5-methoxy-*N′*-Boc-tryptamine (**294**) via *N*-methylation and metalation/Negishi coupling (Scheme 2.47). Employing the first-generation Diels–Alder imidazolidinone catalyst **5**-HBF_4 combination, the key Diels–Alder reaction/cyclization cascade between diene **292** and methyl (*E*)-4-oxobut-2-enoate **293** proceeded in an efficient and stereoselective manner to afford the tetracyclic adduct **297** at a 70% yield and 95% *ee*. Remarkably, through this cascade transformation two rings, three bonds, and three stereocenters, including the all-carbon quaternary center C8, with the correct relative and absolute stereochemistry were formed in one pot and in a highly enantioselective and highly diastereoselective manner. The scenario may involve the condensation of catalyst **5** onto enal **293** to generate an iminium ion-activated dienophile **Im-b**. The Diels–Alder reaction of **Im-b** with diene **292** adopted transition state **TS-A** with orientation of the reactive π-system away from the catalyst *gem*-dimethyl group to facilitate effective shielding of one π-face by the benzyl group, which resulted in a highly enantioselective and *endo*-selective cycloaddition to deliver cycloadduct **295**. Brønsted acid-catalyzed tautomerization of enamine **295** to iminium ion **296** triggered the intramolecular 5-*exo*-trig cyclization to yield the tetracyclic product **297**. The key intermediate **297** was converted in four steps (oxidation; acyl telluride formation; removal of Boc; *N*-reductive propargylation with aldehyde **300**) to the radical cyclization precursor **291**. The desired radical cyclization was undertaken by heating a solution of acyl telluride **291** in 1,2-dichlorobenzene at 200°C to provide the allene product **300**. Finally, catalytic hydrogenation at −15°C allowed chemo- and stereoselective saturation of the terminal C=C bond from the less hindered face to yield (−)-vincorine (**290**) at a 80% yield as a single olefin isomer. Thus, the enantioselective total synthesis of (−)-vincorine (**290**) was accomplished in nine steps and a 9% overall yield from commercially available starting materials.

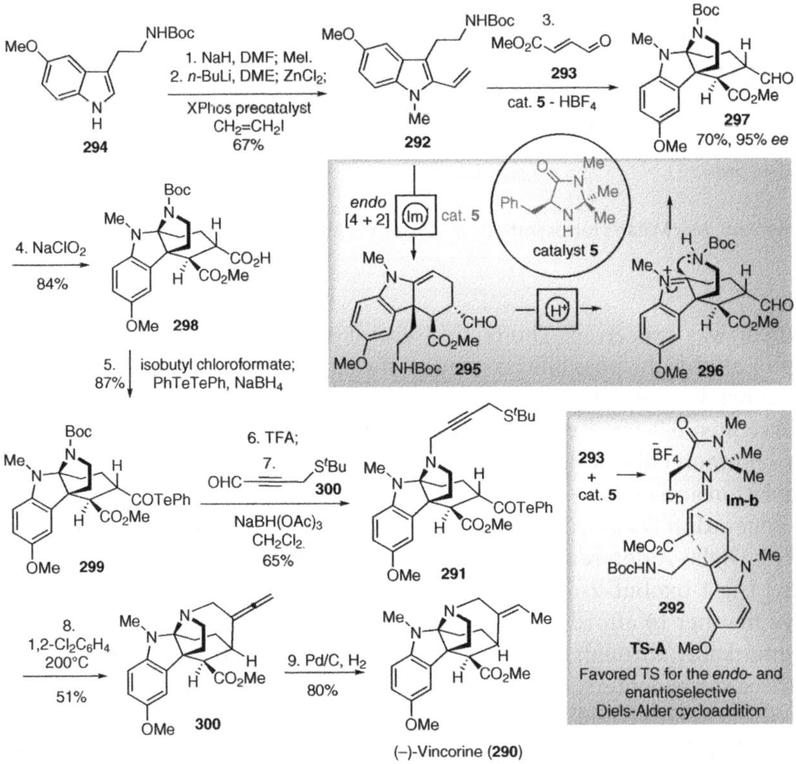

Scheme 2.47 MacMillan's nine-step total synthesis of (−)-vincorine (**290**).

2.9.4 Ramharter's Nine-Step Total Synthesis of (+)-Lycoflexine

Lycoflexine (lycobergine, **301**) belongs to the fawcettimine-related alkaloids, which constitute a class of the *Lycopodium* alkaloids family.[59,61] Surprisingly, although the isolation of lycoflexine (**301**) dated the early 1970s,[110] no total synthesis was reported before 2010. Its structure features a complex tetracyclic ring system including a spiro-annulated six-membered ring and four stereogenic centers including two adjacent all-carbon quaternary carbon centers.

The first enantioselective total synthesis of (+)-lycoflexine (**301**) was accomplished by J. Ramharter and coworkers in 2010.[111] The retrosynthetic analysis is shown in Scheme 2.48, which features a bio-inspired transannular Mannich reaction[110] to assembly the spiro-ring from **302**. The latter was envisioned to be generated from substrate **303**, which can be accessed from **304** via an enynene RCM reaction. Cyclohexanones **304** and **305** could be synthesized from optically active enone **190**.

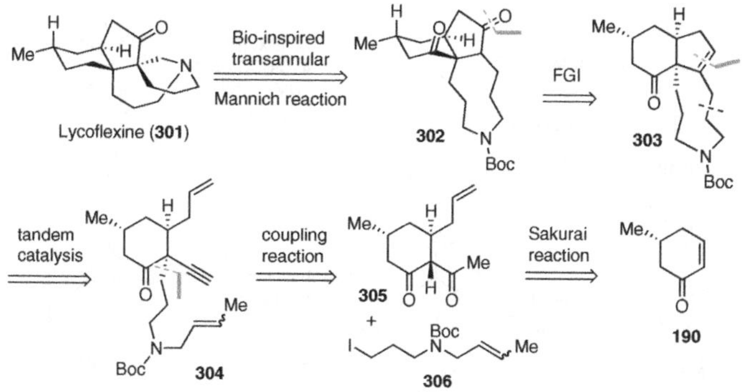

Scheme 2.48 Ramharter's retrosynthetic analysis of (+)-lycoflexine (**301**).

The chiral building block (*R*)-**190**[63] could have been prepared in one pot from (*E*)-pent-3-en-2-one (**307**) and acetaldehyde by organocatalysis (80% *ee*;[112a] 91% *ee*[112b]). Subjection of enone (*R*)-**190** to tandem Sakurai allylation/aldol addition led to hydroxyl-ketone **308** (Scheme 2.49). Oxidation with IBX gave *β*–diketone **305**, which was alkylated with iodocarbamate **306** to yield compound **309**. Triflation of the less hindered carbonyl group with Comins' reagent,[113] and treatment of the resulting vinyl triflate **310** with pyridine yielded the dienyne **304**. In the light of Grubbs' findings that metathesis catalysts can be converted into active hydrogenation catalysts by treatment with dihydrogen,[114] a tandem enynene RCM and selectively hydrogenation the less-substituted double bond in **311** was achieved to give the desired tricyclic carbamate **303** at a 52% yield from **304**. Tandem hydroboration, *in situ* oxidation of the organoborane with IBX, delivered **302** as an inconsequential diastereomeric mixture. Finally, treatment of **302** with dilute aqueous HCl and an excess of formaldehyde led to the tandem *N*-deprotection, iminium ion generation, and biomimetic transannular Mannich reaction to furnish (+)-lycoflexine (**301**) at a 64% yield from **303**.

This concise nine-step synthesis of lycoflexine (**301**) from **307** (with an overall yield of 13% from **190**) features the extensive use of tandem and one-pot reactions (Sakurai/aldol, enynene RCM/hydrogenation tandem catalysis, hydroboration/oxidation, *N*-Boc deprotection/transannular Mannich cyclization[58]).

It is noteworthy that both diastereomers of **302** can be used to synthesize another *Lycopodium* alkaloid fawcettimine (**314**) (Scheme 2.49) as demonstrated by C. H. Heathcock in 1986.[115] As R. Sarpong pointed out in his recent concept article,[61c] the landmark synthesis by Heathcock had a lasting influence on other strategies employed in the synthesis of fawcettimine-type *Lycopodium* alkaloids (the Heathcock-inspired approach to fawcettimine).

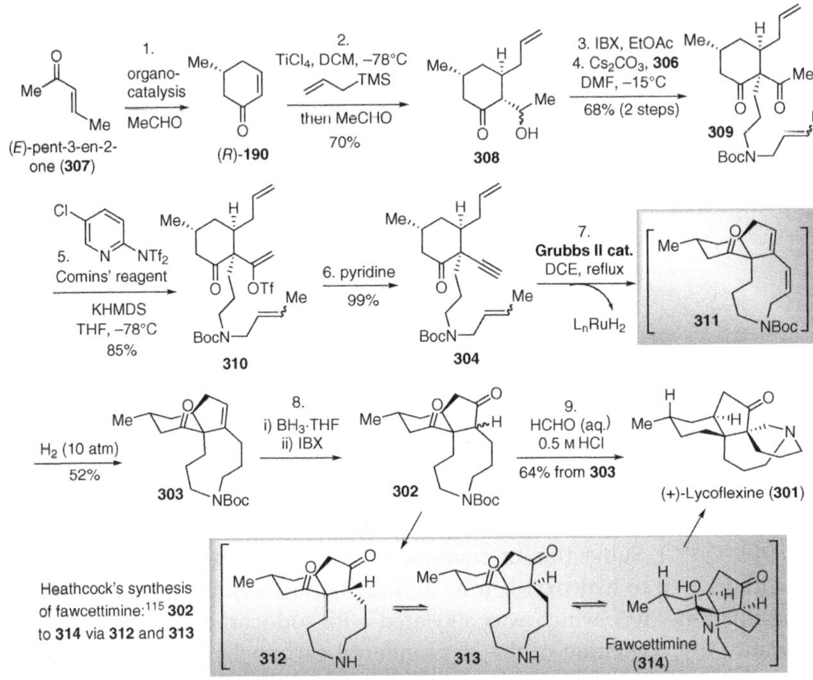

Scheme 2.49 Ramharter's nine-step total synthesis of (+)-lycoflexine (**301**).

2.9.5 Gao's and Theodorakis' Nine-Step Total Syntheses of (+)-Fusarisetin A

Fusarisetin A (**315**, Scheme 2.50) is a tetramic acid-based natural product isolated from the soil fungus *Fusarium* sp. FN080326.[116] Structurally, this molecule is characterized by a challenging 6,6,5,5,5-fused pentacyclic ring system bearing 10 stereogenic centers. Biologically, this compound displayed significant activity in inhibiting acinar morphogenesis, cell migration, and invasion in MDAMB-231 cells without significant cytotoxicity.[116] The complex molecular architecture and remarkable bioactivity makes this molecule an attractive synthetic target.[117–120] Thus, soon after the report of its isolation in May 2011 by Ahn *et al.*,[116] in December 2011, A. Li and coworkers disclosed the first total synthesis that was achieved in only 13 steps from (−)-citronellal (*ent*-**215**).[117] The efficiency of synthesis of a natural product of such complexity is really remarkable. Although the final synthetic product turned out to be the unnatural enantiomer (−)-fusarisetin A, in addition to the novel chemistry developed, their efforts allowed a revision of the absolute configuration as (+)-fusarisetin A. In 2012, E. A. Theodorakis and colleagues reported a nature-inspired total synthesis of (−)-fusarisetin A, which was achieved in only nine steps from

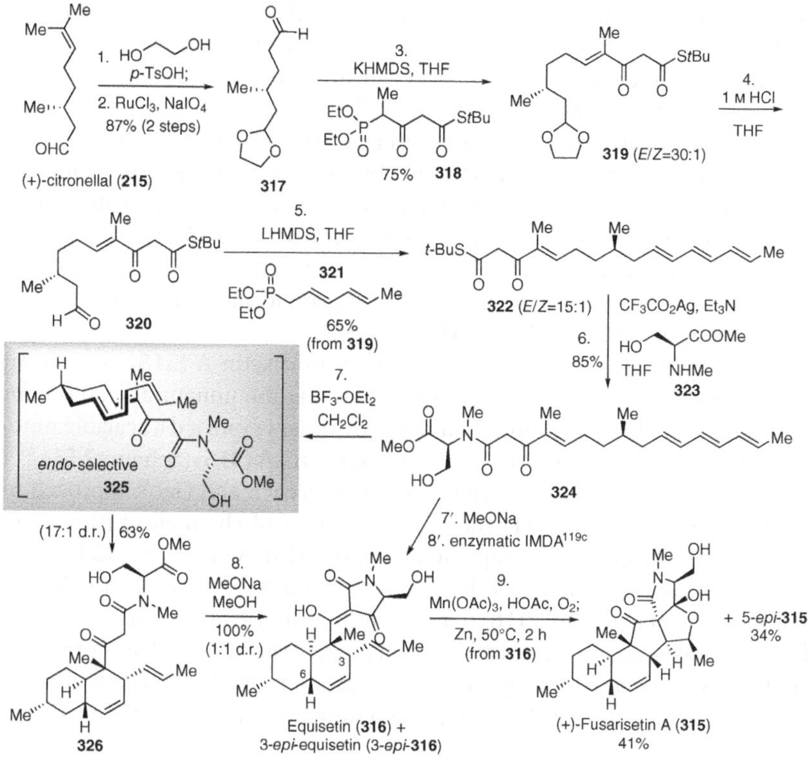

Scheme 2.50 Gao's nine-step total synthesis of (+)-fusarisetin A (**315**).

(−)-citronellal (*ent*-**215**).[118a] Although this total synthesis was also misled by the initial erroneous assignment of the absolute configuration of the natural product,[116] the first protecting-group-free nine-step total synthesis is very concise and highly venerable. Moreover, they proposed a plausible biosynthetic pathway that fusarisetin derives biogenetically from equisetin, another secondary metabolite from *Fusarium* sp., via a sequence involving a stabilized radical.

The first enantioselective and bio-inspired total synthesis of (+)-fusarisetin A (**315**) was achieved in early 2012 by S. H. Gao and coworkers.[119a,b] The efficient strategy relied on their own biosynthetic considerations. One of the key points of their biosynthetic hypothesis is that polyenoylamino acid corresponds to intermediate **324**, which has all the required elements of equisetin (**316**) and fusarisetin A (**315**), and is a key intermediate to both **316** and **315** so the former could be transformed into the latter through a cyclization sequence (Scheme 2.50).[119a,b] They thus developed a six-step synthesis of the key intermediate **324** from (+)-citronellal (**215**). $BF_3{\cdot}Et_2O$ promoted IMDA of **324** proceeded, via the chairlike transition state **325** and in an *endo*-selective mode, to deliver the desired *trans*-decalin **326** with excellent diastereoselectivity (17:1 d.r.). Base-mediated Dieckmann cyclization of *trans*-decalin **326** produced equisetin (**316**) and 3-*epi*-equisetin (3-*epi*-**316**) (1:1 d.r.) at a quantitative yield. The bio-inspired transformation of **316** to **315** was realized in one pot through an aerobic radical reaction (Mn^{3+}/O_2) or a visible-light promoted reactive oxygen species (ROS) process followed by a reduction to afford (+)-fusarisetin A (**315**) at a 41% yield along with 5-*epi*-**315** at a 34% yield. Recently, they have shown that the Diels–Alderase Fsa2 can catalyze the IMDA reaction to deliver the *trans*-decalin ring system of the molecule in a stereoselective manner (see Scheme 2.50, **324** to **316**).

Attracted by the remarkable bioactivities of fusarisetin A (**315**) as a potent cancer migration inhibitor, soon after synthesizing the unnatural enantiomer of fusarisetin A (**315**),[118a] Theodorakis and coworkers reported a scalable nine-step enantioselective total synthesis of (+)-fusarisetin A (**315**) (Scheme 2.51).[118b] The major differences between Theodorakis' approach and Gao's approach to (+)-fusarisetin A (**315**) are: (1) the order/direction of chain elongation from (*R*)-citronellal (**215**); (2) the stage at which the IMDA reaction (**329** versus **324**) is performed; and (3) the oxidants and reductants used for the biomimetic transformation of equisetin (**316**) to fusarisetin A (**315**). It is worth mentioning that Theodorakis' approach is scalable with most of the intermediates prepared in multigram scale and 201 mg of the final product obtained, which allowed them to undertake *in vitro* biological screening.[118b,c] It has been shown that

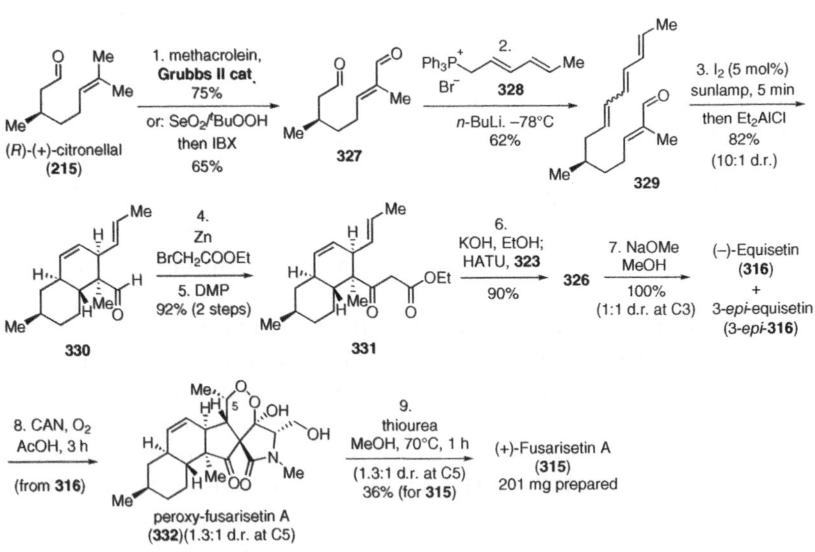

Scheme 2.51 Theodorakis' nine-step total synthesis of (+)-fusarisetin A (**315**).

(+)-fusarisetin A (**315**) and 5-*epi*-fusarisetin A are potent cancer migration inhibitors, while (−)-fusarisetin A and (−)-equisetin (**316**) are significantly less potent.[118b]

It is noteworthy that all the total syntheses of fusarisetin A reported so far[117–120] are based on the naturally occurring enantiopure chiral building blocks (*R*)-citronellal (**215**)[118b,119] (or (*S*)-citronellal for the synthesis of (−)-fusarisetin A[117,118a]) and (*S*)-carvone.[120a,b] In addition, in all but one approach, IMDA reaction has been employed as the key reaction to build the *trans*-decalin ring system, demonstrating once again the power of this type of reaction in the efficient total synthesis.[89]

2.10 Ten/Eleven-Step Enantioselective Total Syntheses of Natural Products

2.10.1 Lin's 10-Step Total Synthesis of (−)-Huperzine A

Huperzine A (**333**) is a *Lycopodium* alkaloid isolated in minuscule amounts (200 ppm from the dry plant) from the clubmoss *Huperzia serrata* (Thunb.) (*Lycopodium serratum*), which is used in various formulas in traditional Chinese medicine to alleviate problems with memory loss, promote circulation, and for fever and inflammation. Huperzine A was found to be a potent, reversible inhibitor of acetylcholinesterase (AChE). This inhibition results in increased levels of the neurotransmitter acetylcholine and a corresponding improvement in cognitive ability.[59a,61c] Huperzine A is also neurotrophic and neuroprotective, is used clinically to treat AD symptoms in China, and is marketed as powdered *H. serrata* in the USA for memory impairment.[121,122a] Structurally, huperzine A (**333**) belongs to the lycodine subgroup, which features a bicyclo[3.3.1]nonane core flanked by a piperidine ring.

Since the seminal work of A. P. Kozikowski,[122] the synthesis of huperzine A (**333**) has attracted considerable attention.[123] Among the reported enantioselective total syntheses of (−)-huperzine A (**333**), that developed by G.-Q. Lin and coworkers in 2012 remains the most efficient.[124a]

Lin's synthesis used enantiopure (*R*)-pulegone [(*R*)-**334**] as the chiral building block, which was converted to triflate **335** by treatment with LDA/$PhNTf_2$ followed by ozonolysis (Scheme 2.52). Triflate **335** was subjected to Buchwald–Hartwig coupling reaction[125] that delivered **336** at a 91% yield with complete conservation of the optical purity. The α-alkylation of **336** with **337** was realized via enolization with 2.5 equiv of LDA at −70°C, furnishing **339** in excellent stereoselectivity (96% *ee*, >20/1 d.r.). The latter was quantitatively reduced to give allylic alcohol **340** as an inconsequential diastereomeric mixture, which was subjected to intramolecular Heck reaction, Ley–Griffith oxidation,[126] and Grignard reagent addition to deliver compound **342** at a 47% yield over four steps.

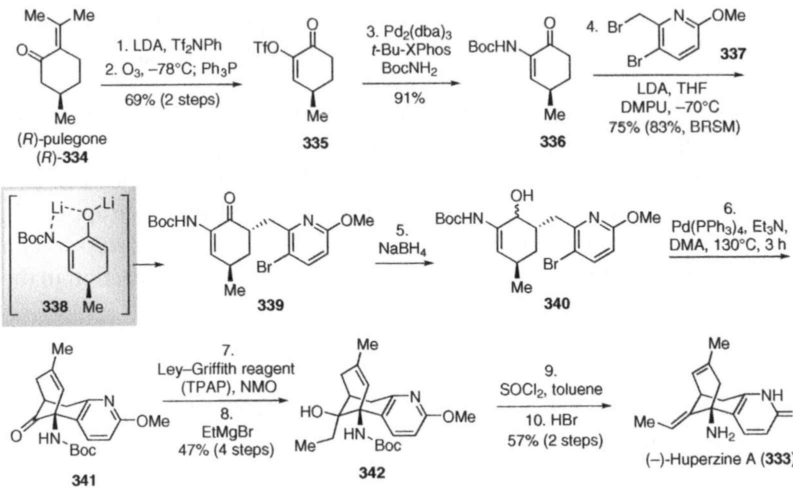

Scheme 2.52 Lin's 10-step total synthesis of (–)-huperzine A (**333**).

Finally, dehydration and deprotection produced huperzine A (**333**). Thus, the synthesis of huperzine A was completed in 10 steps from (*R*)-pulegone with a 17% overall yield. By the same approach, the first asymmetric total syntheses of huperzine B and huperzine U have been achieved in overall yields of 10 and 9%, respectively.[124b]

2.10.2 Trauner's 10-Step Total Synthesis of (+)-Loline

Loline alkaloids are found in a few genera of important forage grasses (*Poaceae*) infected with an endophytic fungus such as *Acremonium coenophialum*. The first loline alkaloid temuline (**343**) was isolated as early as 1892. Loline alkaloids exhibit several biological activities including resistance to a broad spectrum of insect pests, inhibition of seed germination, and seed growth of alfalfa annual ryegrass. Despite their low molecular weight, the intriguing semi-cage skeleton bearing four contiguous stereogenic centers, including two vicinal amino alcohol moieties, and the 2,7-oxygen (ether) bridge, makes the total synthesis of this group of alkaloids particularly challenging.[127,128] Before the work by Trauner, only one racemic[127a] and one asymmetric total synthesis[127b,c] of loline (**344**) appeared in 1986 and 2000, respectively. The latter took 20 steps to complete the total synthesis from (–)-malic acid. The breakthrough was made by Trauner and coworkers, who reported in 2011 an elegant, concise, and versatile approach to (+)-norloline (**343**, 9 or 8 steps), (+)-loline (**344**, 10 or 9 steps), and (+)-*N*-formylloline (**345**, 11 or 10 steps).[128]

Trauner's synthesis (Scheme 2.53)[128] employed an achiral divinyl carbinol **346** as the starting material, which was desymmetrized to give chiral epoxide **347** (99% *ee*) by a known procedure utilizing Sharpless asymmetric epoxidation reaction.[129] Epoxide **347** was converted, in one pot, to diene **348** by nucleophilic addition and *N*-protection. An RCM of diene **348** with Grubbs generation II catalyst produced diol **349**, which was converted to cyclic sulfite (**350**). Nucleophilic addition of sulfite **350** with lithium azide proceeded regioselectively to give azido alcohol **351**. The key transannular aminobromination of olefinic carbamate **351** occurred regio- and stereoselectively and with concomitant *N*-deprotection to yield bromopyrrolizidine **354**. Next, to ensure the subsequent formation of the emblematic ether bridge of the loline alkaloids via a S_N2 reaction, it was necessary to inverse the configuration at the bromine-bearing carbon (C7). Thus, bromopyrrolizidine **354** was subjected to a reaction with LiCl to give chloropyrrolizidine **355**. Exposure of **355** to potassium carbonate in a microwave apparatus resulted in the formation of an ether bridge to give azide **356** at an excellent yield. Under these conditions, only 5-*exo*-tet nucleophilic substitution occurred and no elimination product was observed. These two steps could also be performed in one pot. Hydrogenation of azide **356** in the presence of di-*tert*-butyl pyrocarbonate yielded *N*-Boc temuline **357**, which was reduced to afford (+)-loline (**344**), and thus completed the total synthesis. Moreover, hydrogenation of azide **356** gave temuline (norloline) (**343**), formylation of loline (**344**) yielded *N*-formylloline (**345**).

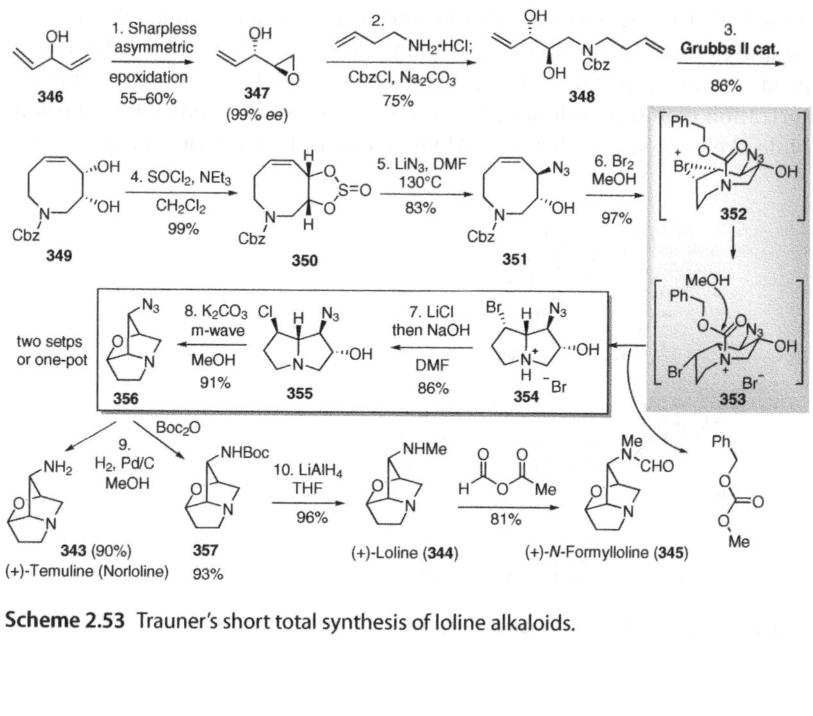

Scheme 2.53 Trauner's short total synthesis of loline alkaloids.

The elegance of this synthesis arises from its brevity and high chemo-, regio-, and stereoselectivity. The former benefits from their effort to maximize the synthetic efficiency at the stage of synthetic planning according to the principles of redox economy and protecting-group economy. Indeed, although two protecting groups (Cbz and Boc) have been used, they did not result in any supplementary step.

2.10.3 Zhai's 10-Step Total Synthesis of (+)-Absinthin

(+)-Absinthin (**358**, Scheme 2.54) is a triterpene isolated from wormwood, *Atemisia absinthium* L. in 1953. In TCM, the bitter herb absinthium has been used as stomachic tonic and anthelminthic, and used to treat acute bacillary dysentery. Structurally, (+)-absinthin (**358**) is a heptacyclic compound containing 14 chiral centers that include three chiral quaternary ones. Surprisingly, despite the challenging structure and physiological activity exhibit by absinthin, no reports on the synthesis of this terpene appeared before 2005.

How many steps does one need to complete an enantioselective total synthesis of such a challenging target? In 2005, by accomplishing the first enantioselective total synthesis of (+)-absinthin (**358**), H.-B. Zhai and coworkers gave their answer: 10 steps and a 7.18% overall yield from commercially available materials![130]

To address the synthesis of a complex target such as absinthin, getting information about the molecule from literature is imperative. This is particularly true for absinthin. It is known that absinthin is a dimeric guaianolide, which decomposes to artabsin (**359**) on heating[131] (Scheme 2.54). This information allowed the assumption that the triterpene might be generated from the dimerization of **360**, a cyclopentadiene-containing ring system within the guaianolide skeleton via a self-Diels–Alder reaction. In addition, thermolysis of

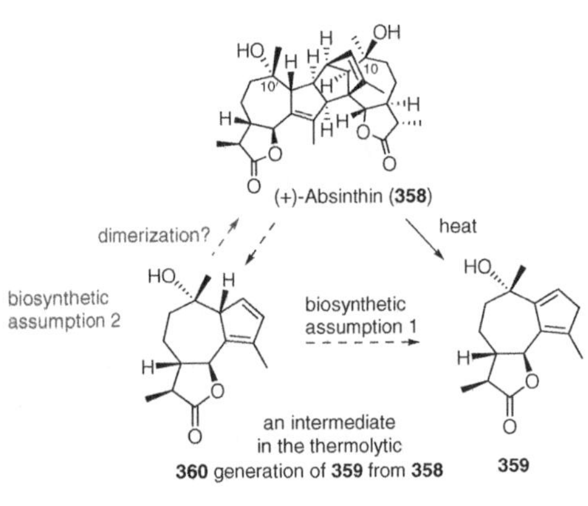

Scheme 2.54 Reported transformations related to (+)-absinthin (**358**).

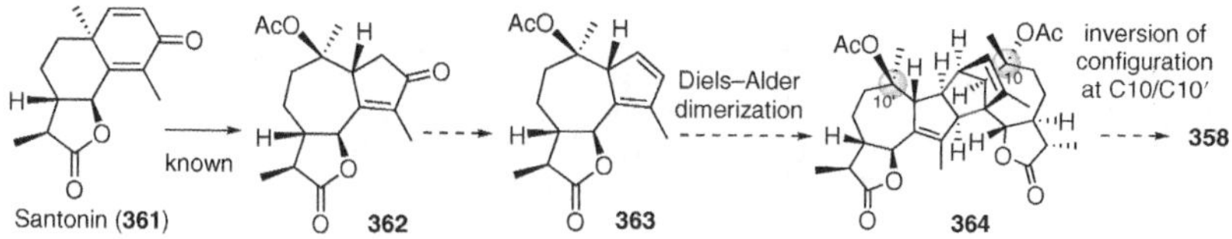

Scheme 2.55 Zhai's bio-inspired synthetic plan to (+)-absinthin (**358**).

358 to generate **359** might pass through **360**. Moreover, it has been reported that photolysis of santonin (**361**), a commercially available compound,[133] yielded *O*-acetylisophotosantonic lactone **362**.[132] The structure of **362** is reminiscent of artabsin (**359**).

Thus a bio-inspired building-block-based dimerization approach was conceived for the total synthesis of **358** (Scheme 2.55). After the Diels–Alder-type dimerization, all that remains to be done is the inversion of configuration at C10/C10′.

The synthesis started from the photolysis of the commercially available santonin (**361**) to give *O*-acetylisophotosantonic lactone (**362**, Scheme 2.56). The ketone was reduced and the resulting hydroxyl group was subjected to Mitsunobu arylselenylation to give selenides **366**. Oxidation of the latter with $NaIO_4$ resulted in tandem elimination to yield cyclopentadiene (**363**). Remarkably, the key biomimetic dimerization of diene **363** proceeded smoothly to give heptacyclic product **364** at a 72% yield simply by standing neat diene **363** in a nitrogenated atmosphere at room temperature for 10 days. Moreover, the Diels–Alder reaction proceeded regio- and stereospecifically, which implies that two identical Diels–Alder partners **363** approach each other from the less hindered face of the diene moieties, and in a head-to-head orientation (with regard to the lactone carbonyl groups) to minimize steric interactions. Bis-deacetylation of **364** yielded 10-*epi*-10′-*epi*-absinthin (**367**). Finally, the inversion of configuration at C10/C10′ was achieved in four steps consisting of: (1) kinetic dehydration (to form two terminal alkenes); (2) selective dihydroxylation (of the newly formed olefinic bonds); (3) oxidative cleavage of the resulting vicinal diols; and (4) addition of methyl lithium to the two newly formed ketone groups in **368**. Remarkably, the addition of MeLi proceeded chemoselectively (at the keto carbonyls, C10 and C10′) and stereoselectively (from the less hindered β-face of the carbonyl at C10, and the α-face of that at C10′) to give (+)-absinthin (**358**) at an excellent yield.

This work demonstrated the value and power of strategies employing renewable resource-based building blocks/chirons, and bio-inspired reactions for the total synthesis of natural products.

2.10.4 Baran's 11-Step Total Synthesis of (−)-Maoecrystal V

P S. Baran and coworkers illustrated, through their recent conquest of (−)-maoecrystal V (**369**), how long a journey may need to forge a singular, concise,

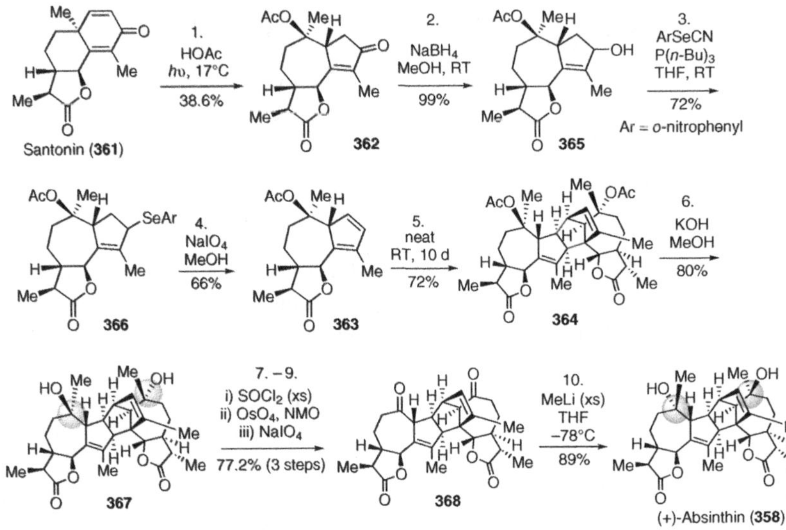

Scheme 2.56 Zhai's 10-step total synthesis of (+)-absinthin (**358**).

scalable, and enantioselective total synthesis.[134] The *ent*-kaurane diterpene maoecrystal V (**369**)[135] possesses a densely interlocked array of ring systems containing six stereogenic centers including adjacent vicinal quaternary stereocenters. The intriguing structure combined with the reported exciting bioactivity renders it an attractive synthetic target.[136] All previous successful total syntheses have relied on the Diels–Alder cycloaddition reaction that requires 18–35 steps (0.03–1% overall yield). Among them, only two are enantioselective total syntheses.[135e,f] To develop a practical enantioselective total synthesis, Baran envisioned an alternative convergent approach. In the light of Xu's biosynthetic hypothesis[137] featuring an unusual pinacol-type shift of cation **370** to build the ring system (Scheme 2.57), a bio-inspired retrosynthetic analysis relied on a pinacol shift was performed.

The first chiral center was forged by the enantioselective conjugate addition of an allyl silane to 2-cyclohexenone to deliver **375** (Scheme 2.58). Under specially defined conditions employing TADDOL-derived phosphine-phosphite **L4** as the chiral ligand, the reaction proceeded at a high yield and excellent enantioselectivity (99% *ee*) on a 20-g scale.[138] Chiral cyclohexanone **375** was converted to bicyclic ketone **373** via a one-pot regioselective α-acetoxylation (LiTMP; Davis oxaziridine **228**; Ac_2O in THF/DMPU), intramolecular Sakurai reaction[139] ($EtAlCl_2$), one-pot *O*-methylation-deacylation, and oxidation. It is worth noting that only after an extensive evaluation of more than 50 Lewis acids has $EtAlCl_2$ been discovered to be the Lewis acid to catalyze the disfavored (according to

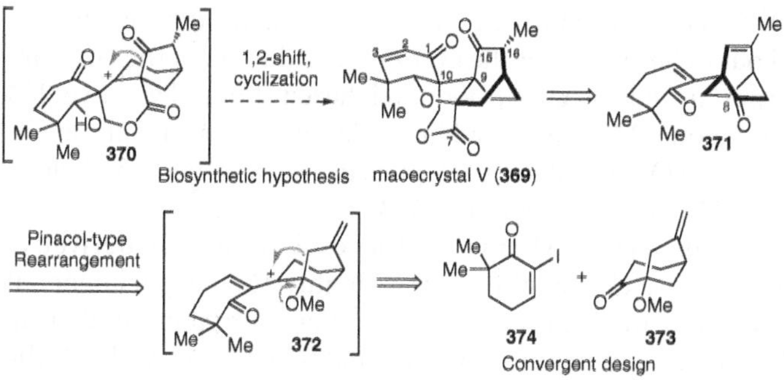

Scheme 2.57 Baran's bio-inspired retrosynthetic analysis of (−)-maoecrystal V (**369**).

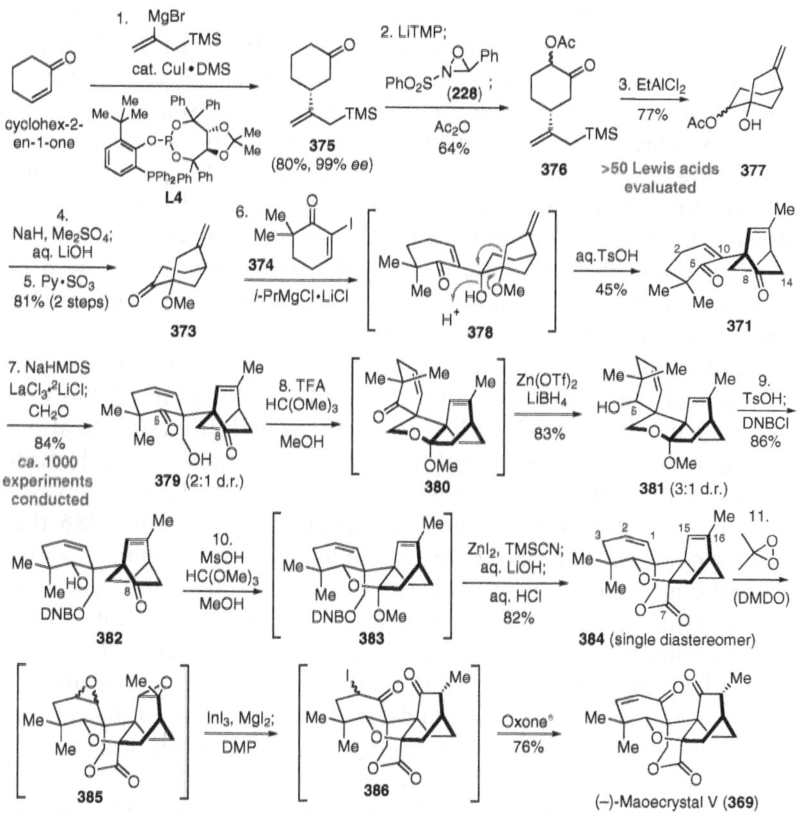

Scheme 2.58 Baran's 11-step total synthesis of (−)-maoecrystal V (**369**).

the Baldwin rules)[140] 5-*enolendo-exo-trig* cyclization of **376**. In addition, the acetoxy group turned out to be crucial for the cyclization. Addition of the Grignard reagent derived from an unprotected α-iodocyclohexen-2-one derivative **374** with [3.2.1]bicyclic ketone **373** yielded adduct **378**, which on heating in the presence of aqueous *p*-TsOH resulted in key pinacol rearrangement and olefin isomerization to give **371** at a 45% yield from **373**.

Deconjugative α-hydroxymethylation of the cyclohexenone derivative **371** turned out to be quite challenging. After roughly 1000 experiments, it was found that addition of $LaCl_3 \cdot 2LiCl$ to the dienolate generated from enone **371**, followed by quenching with formaldehyde gas yielded the desired adduct **379** and its diastereomer at a 2:1 diastereomeric ratio. The next task was the selective reduction of the more sterically hindered ketone at C5. This was achieved via internal protection in the form of acetal **380**. After screening more than 100 reducing conditions, $LiBH_4/Zn(OTf)_2$ combination was selected for the reduction of the ketone at C5, which gave **381** in 3:1 d.r. Acetal in **381** was cleaved and the liberated primary hydroxyl group was protected as dinitrobenzoate (DNB) **382**. The DNB group, selected to render intermediates crystalline for stereochemical assignment, turned out once again to be a fortuitous choice for the subsequent transformations. Treatment of γ-hydroxyl-ketone **382** with a 6:1 ratio of $CH(OMe)_3$:MeOH and methanesulfonic acid generated the strained acetal **383**, which not only formed the penultimate ring of the target molecule, but also set the stage for introducing C7 carbon and the last ring. A ZnI_2-catalyzed reaction of acetal intermediate **383** with TMSCN followed by saponification and acidification led to the direct formation of lactone **384** as a single diastereomer. All that remained was the transformation of olefin at C1-C2 to enone and to unmask another alkene (C15-C16) to a ketone with the generation of the last chiral center. Bis-epoxidation of **384** with DMDO yielded diepoxide **385**, which was treated with InI_3/MgI_2 to afford the corresponding iodohydrin, and promote a stereospecific 1,2-hydride shift to install the C16 stereochemistry. The C-1 hydroxyl group was oxidized with Dess–Martin periodinane to produce α-iodoketone **386** that was further oxidized with oxone® to give an α-iodoso intermediate, which was spontaneously eliminated to deliver (–)-maoecrystal V (**369**). All these reaction sequences have been performed in one pot, thus completed the 11-step enantioselective total synthesis of (–)-maoecrystal V at a 2.7% overall yield. With >80 mg of (–)-**369** synthesized, a screening against 32 different cancer cell lines has been undertaken, which allowed the conclusion that despite the original report, maoecrystal V (**369**) exhibits virtually no cytotoxicity in any cancer cell line tested.

It is worth noting that this remarkably concise total synthesis relied first on an ingenious synthetic strategy design. However, the successful realization of the synthetic plan was highly dependent on experimentation.

2.11 Fourteen/Fifteen-Step Enantioselective Total Synthesis of Natural Products

2.11.1 Baran's 14-Step Total Synthesis of (−)-Ingenol

Classified as being in the phorboid family, diterpene ingenol (**387**, Scheme 2.59) was isolated from *Euphorbia ingens* in 1968.[141] Various ester derivatives of ingenol have shown remarkable biological properties to mimic diacylglycerol and function as endogenous activators of protein kinase (PKC), to exhibit anti-tumor or tumor promoting, antileukemic, and anti-HIV properties. Among them, ingenol 3-angelate (ingenol mebutate, **388**) was found to be a *Euphorbia peplus* metabolite. Under the trade name "Picato," this compound was approved by the FDA in early 2012 as a first-in-class topical treatment for actinic keratosis, a pre-cancerous skin condition. In addition, **388** has undergone clinical testing for the topical treatment of both basal cell carcinoma and squamous cell carcinoma.[142] Hence, a large supply of ingenol and its analogs through chemical synthesis became imperative for the development of new therapeutic agents. Moreover, the complex structure of ingenol also renders it a challenging yet attractive target for synthetic organic chemists. The key feature of ingenol resides in the incorporation of an unusual *trans*-intrabridgehead bicyclo[4.4.1]undecane motif (known as in/out isomerism or inside/outside framework)[143] within the highly oxygenated tertracyclic ring system, which brings about great strain to the molecule.

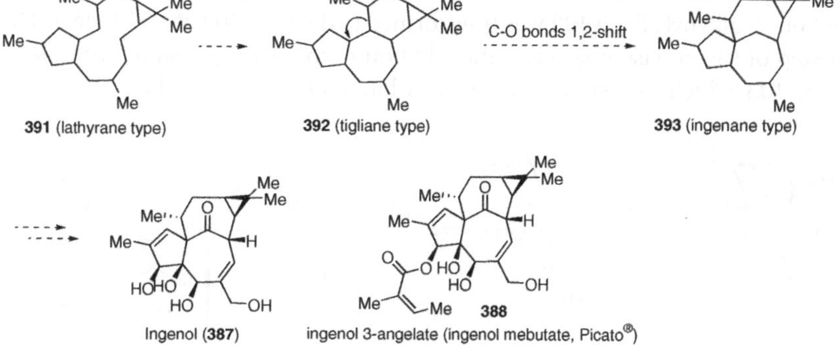

Scheme 2.59 Plausible biosynthetic pathway for *Euphorbia* diterpenes.

Over the past 35 years, much effort has been devoted to the total synthesis of ingenol,[144] however, to date, only four total syntheses including two enantioselective ones have been accomplished.[145,146] Among them, Baran's 14-step enantioselective total synthesis of (−)-ingenol, first reported in 2013,[146] remains the most concise until now. Baran's synthesis is inspired by the suggested biosynthesis of *Euphorbia* diterpenes (Scheme 2.59),[147] which features a two-phase strategy,[148] namely building of skeleton by cyclases and functionalization of the built skeleton by oxidases (see Section 2.92, Scheme 2.44).

Accordingly, the retrosynthetic analysis is depicted in Scheme 2.60. In the oxidase phase, the hydroxyl groups at C3 and C20 were first removed, which could be introduced from the synthetic perspective by allylic oxidations. The remaining two hydroxyl groups at C4 and C5 and the *in/out*-bicyclic system in **394** could be installed by a stereoselective dihydroxylation of the corresponding olefin (**395**) and a vinylogous pinacol rearrangement, respectively. In the cyclase phase, intermediate diene **395** was envisaged to be available from (+)-3-carene (**396**), ethynyl magnesium bromide, and aldehyde **397**. The enantioselective synthesis was envisioned to rely on the directing effect of dimethyl cyclopropanyl group in (+)-3-carene (**396**).

As outlined in Scheme 2.61, Baran's synthesis commenced with the elaboration of (+)-3-carene (**396**) into an intermediate (**398**) allowing the introduction of the methyl group at C11 and subsequent coupling with **397**. For this purpose, **396** was subjected to chlorination with *N*-chlorosuccinimide (NCS) and ozonolysis, which gave *α*-chloroketone **399** at a 48% yield over two steps. Next, a one-pot protocol consisting of reductive enolate formation-methylation, and deprotonation with lithium bis(trimethylsilyl)amide (LiHMDS) followed by addition of the resulting enolate with **397** was developed, which afforded an *α*-methylated aldol product **400** as a single diastereomer at a 44% yield. Through this three-step sequence, **396** was elaborated into **400** with two C–C bonds formed in a highly diastereoselective manner. Addition of ethynyl magnesium bromide followed by one-pot protection of the two hydroxyl groups provided **402** in 71% yield. Under the conditions established by Brummond and coworkers for the allenic Pauson–Khand-type reaction,[149] alkynyl allene **402** was transformed to dienone **403** at a 72% yield. The last step of the cyclase phase was the addition of methylmagnesium bromide to ketone **403**, which gave stereoselectively carbinol **395** at a 80% yield.

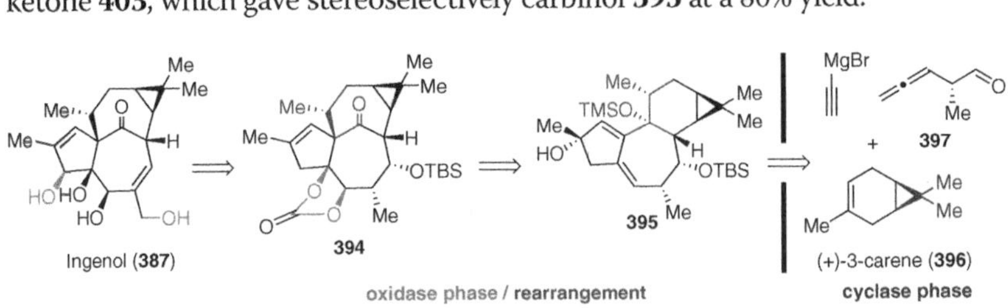

Scheme 2.60 Baran's bio-inspired retrosynthetic analysis of (−)-ingenol (**387**).

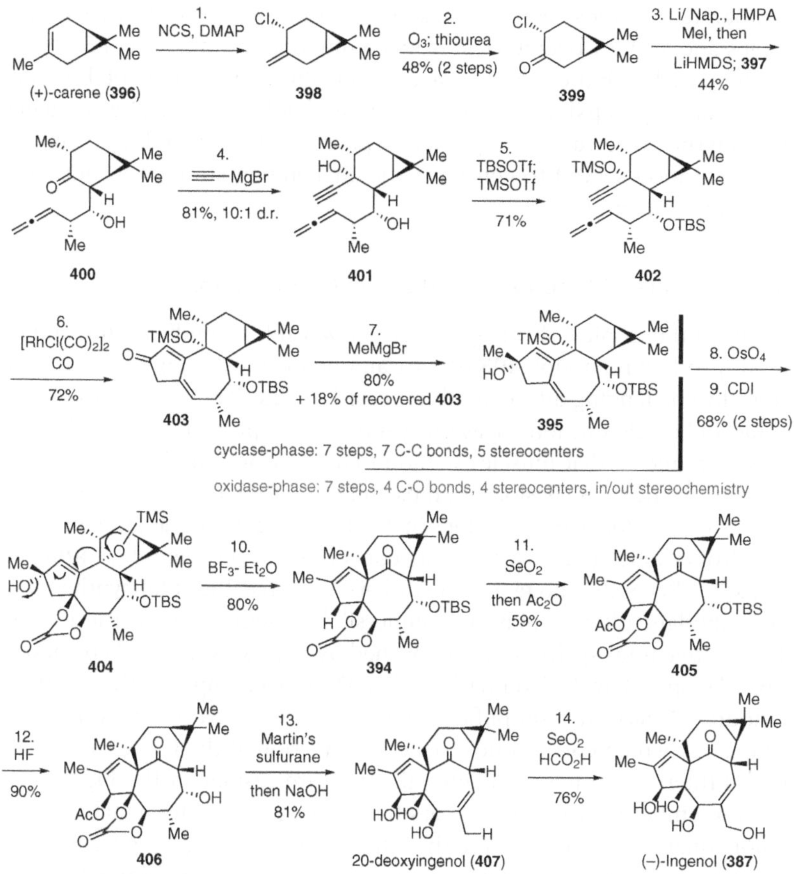

Scheme 2.61 Baran's 14-step total synthesis of (–)-ingenol (**387**).

Dihydroxylation of **395** with OsO_4 followed by protecting the resulting diol with *N,N*-carbonyldiimidazole (CDI) afforded carbonate **404** as a single stereoisomer at a 68% yield over two steps. The key pinacol rearrangement was realized by treating **404** with $BF_3{\cdot}Et_2O$ in CH_2Cl_2 at a low temperature, followed by quenching with Et_3N and MeOH, which established the strained *in/out* stereochemistry and delivered ingenane **394** at a 80% yield. Allylic oxidation of **394** with SeO_2, and *in situ* protection of the resulting hydroxyl group provided acetate **405** as a single diastereomer. Cleavage of the TBS group with HF gave alcohol **406** at a 90% yield. Successive treatment of **406** with Martin's sulfurane and NaOH produced the dehydration and hydrolytic product 20-deoxyingenol (**407**) at a 81% yield. Finally, exposure of diene **407** to SeO_2 resulted in regioselective hydroxylation[150] yielding **387** at 76%.

Thus, Baran's enantioselective total synthesis of ingenol (**387**) was achieved in 14 steps and an 1.2% overall yield from commercially available (+)-3-carene (**396**). Compared with the previous racemic and enantioselective total syntheses (45, 46, and 37 steps), this approach is remarkably concise. The efficiency gained from the powerful bio-inspired two-phase strategy including a pinacol-type 1,2-migration served to construct the challenging strained in/out bicyclo[4.4.1]undecane motif.

2.11.2 Reisman's 15-Step Total Synthesis of (+)-Ryanodol

Isolated from the tropical shrub *Ryania speciosa* Vahl, ryanodine (**408**) and its hydrolysis product ryanodol (**408a**)[151] are among the most highly oxidized and synthetically challenging diterpenoids reported to date. These terpenes modulate intracellular calcium-ion release at ryanodine receptors, ion channels critical for skeletal and cardiac muscle excitation-contraction coupling and synaptic transmission.[152] In 1979, Deslongchamps and coworkers disclosed a total synthesis of (+)-ryanodol (**408a**) (37 steps in its LLS)[153] via its degradation product (+)-anhydroryanodol (**418**), which was converted to (+)-ryanodol (**408a**) in two steps.[153d] Recently, Inoue and coworkers reported the total syntheses of several ryanodane diterpenoids and epimers.[154] Presented here is the 15-step enantioselective total synthesis accomplished by S. E. Reisman and coworkers in 2016.[155a] Inspired by the landmark work of Deslongchamps,[153d] the synthesis of ryanodol (**408a**) was simplified to that of (+)-anhydroryanodol (**409**) (Scheme 2.62), which corresponds to a disconnection at the C1–C15 bond of ryanodol (**408a**). The most difficult sub-structure resides in the highly oxidized five-membered A-ring that bears five stereogenic centers, including an all-carbon quaternary center and the other four bonding to oxygen. It is challenging to develop a concise synthesis of **408a** with minimal redox, protecting group, and functional group transformations. Central to the retrosynthetic analysis of (+)-anhydroryanodol (**409**) is the use of the

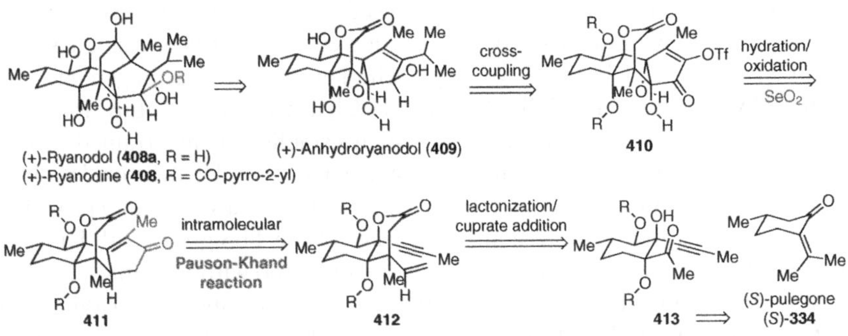

Scheme 2.62 Reisman's retrosynthetic analysis of (+)-ryanodol (**408a**).

Pauson–Khand reaction[156] to build the carbon framework and a SeO_2-mediated oxidation to install three oxygen atoms in a single step (Scheme 2.62).

The synthesis commenced with the dihydroxylation (KHMDS; Davis oxaziridine **228**) of (*S*)-pulegone [(*S*)-**334**], which gave **414** as a single diastereomer at a 42–50% yield (Scheme 2.63). After bis-*O*-protection, the addition of propynylmagnesium bromide afforded an adduct with an equatorially disposed propynyl group as the major diastereomer (5:1 d.r.). Chemoselective ozonolysis of the olefinic group in the adduct yielded methyl ketone **413**, which was subjected to addition with ethoxyethynyl magnesium bromide and subsequent Ag-catalyzed cascade cyclization and elimination to give *α*,*β*-unsaturated lactone **416**. Conjugate addition made the olefinic alkyne **412** a single diastereomer, which established the all-carbon quaternary center and set the stage for the key Pauson–Khand reaction. For the key Pauson–Khand cyclization, because the standard conditions using stoichiometric amounts of $Co_2(CO)_8$[156a] failed to deliver **411** at both high yield and diastereoselectivity,

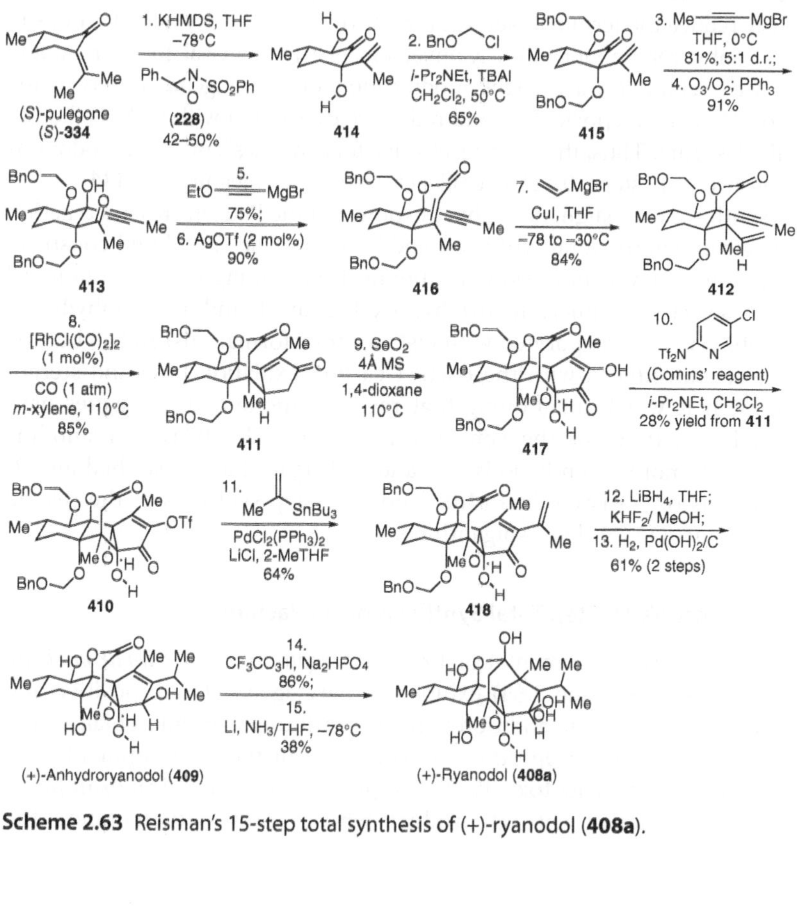

Scheme 2.63 Reisman's 15-step total synthesis of (+)-ryanodol (**408a**).

an alternative catalytic procedure was adopted. Thus, in the presence of 1 mol % of $[RhCl(CO)_2]_2$, exposure of **412** to an atmosphere of carbon monoxide yielded the desired cyclization product **411** as a single diastereomer at a 85% yield. This reaction not only forged the basic skeleton of (+)-ryanodol (**408a**), but also allowed running at a multigram scale to provide enough material for further transformations, which include chemo- and stereoselective hydroxylation at C3, C4, and C12, and incorporation of an isopropyl unit at C2.

Treatment of **411** with 10 equiv. of SeO_2 in the presence of freshly activated 4 Å MS at 110°C afforded **417**, which was exposed to the Comins' reagent[113] to give vinyl triflate **410** in 28% isolated yield over two steps. Considering that through this reaction sequence three hydroxyl groups have been introduced in a stereospecific manner, the overall yield of accounts for an average efficiency of ~70% yield per transformation. Next, **410** was converted to (+)-anhydroryanodol (**409**) via a three-step sequence. Palladium-catalyzed cross-coupling allowed installation of the final three carbons to deliver **418** at a 64% yield. 1,2-Reduction of the enone moiety with $LiBH_4$ proceeded stereoselectively to furnish the corresponding alcohol, which was subjected to catalytic hydrogenolysis conditions to simultaneously reduce the terminal olefin and remove the benzyloxymethyl groups to yield (+)-anhydroryanodol (**409**). Conversion of this material to (+)-ryanodol (**408a**) was achieved via a slight modification of Deslongchamps's two-step protocol,[153d] namely, epoxidation and reductive cyclization with Li/NH_3 (liq.) to build the last ring. Thus, the enantioselective total synthesis of (+)-ryanodol was accomplished in 15 steps (0.42% overall yield) from (*S*)-pulegone [(*S*)-**334**].

This synthetic campaign highlights the power of the Pauson–Khand reaction as strategic one in the total synthesis, and how strategic C–O bond constructions can be achieved in a redox-economical, protecting-group economical and stereoselective manner: the five hydroxyl groups found in (+)-anhydroryanodol (**409**) were incorporated with just two transformations, and all but the C3–OH are introduced with the correct carbon oxidation level and stereochemistry. It is worth mentioning that recently, Inoue and colleagues have developed a method for the conversion of ryanodol (**408a**) to ryanodine (**408**)[154e] and, more recently, Reisman and colleagues have described an 18-step synthesis of (+)-ryanodine (**408**) and a 19-step synthesis of the related natural product (+)-20-deoxyspiganthine.[155b]

2.11.3 Johnson's 15-Step Total Synthesis of (+)-Pactamycin

Isolated in 1961 from a fermentation broth of *Streptomyces pactum ar pactum*, Pactamycin (**419**, Scheme 2.64) exhibits activity against Gram-positive and Gram-negative bacteria, in addition to potent antitumor, antimicrobial, antiviral, and antiprotozoal properties. However, its further development as a drug was hampered by its toxicity. This represents a prototypical example of a promising bioactive natural product whose complexity hampers the

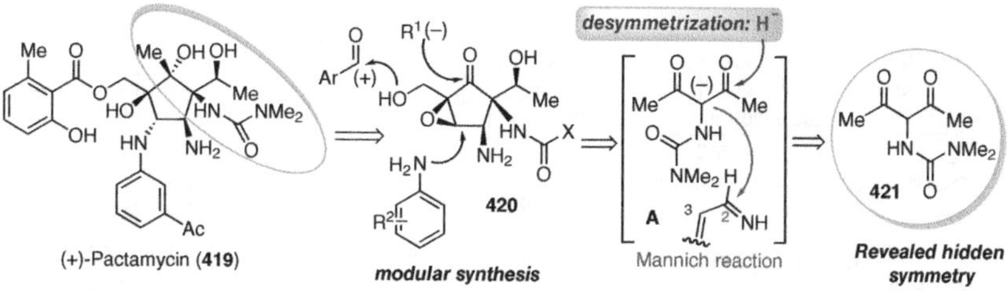

Scheme 2.64 Johnson's retrosynthetic analysis of (+)-pactamycin (**419**).

investigation of structure-activity relationships (SARs). It has been suggested that the structural complexity of pactamycin renders these and related structural modifications are "inaccessible by synthetic organic chemistry."[157] The synthetic challenge of pactamycin (**419**) stems from the fully functionalized cyclopentane core harboring six contiguous stereogenic centers including three quaternary ones and several polar substituents. The landmark first enantioselective total synthesis of pactamycin (**419**) was accomplished in 2011 by Hanessian and coworkers.[158] The 32-step total synthesis reflects the challenge of the molecule. Two years later, J. S. Johnson and coworkers reported an elegant 15-step enantioselective total synthesis.[159] In view of paving a concise route enabling not only the synthesis of the natural product itself, but also analogs for future SAR studies, in planning their synthetic route, Johnson *et al.* emphasized both modular construction and introduction of functionality in the final desired form of pactamycin, as well as late-stage introduction of the aniline- and salicylate-binding elements to **420** (Scheme 2.64). In the strategy aspect, the recognition of a hidden symmetry in the C4–C8 moiety (see α-ureido-2,4-pentanedione **421**) of pactamycin (**419**) found the base of their retrosynthetic analysis. This simplifies the formation of the fully substituted C1 center via a Mannich reaction (**A**, Scheme 2.64), since only the diastereoselective C2–amino incorporation needs to be taken into consideration. It is expected that once the chiral center at C2 is established, it could be used to direct a site- and diastereoselective diketone monoreduction (desymmetrization) to set the C2/C1/C7 stereotriad. Thus the entire functionalized carbon-core skeleton of pactamycin could be assembled in a rapid, modular manner.

The requisite pronucleophile **421** was synthesized in two steps from commercially available chemical acetylacetone (Scheme 2.65). The first challenge was the stereoselective Mannich reaction with an appropriately configured imine electrophile. An adaptation of Schaus conditions[160] to the new nucleophile-electrophile pair turned out to be fruitful. It was found that cinchonidine (**cat. 6**) catalyzed a Mannich reaction of **421** with cinnamaldehyde-derived imine **422** that afforded adduct **423** at 70% and in a 98:2 enantiomeric ratio (e.r.). It is worth noting that although the Mannich reaction established an incorrect configuration at C2, fortunately, the advancement of this stereochemical mistake turned

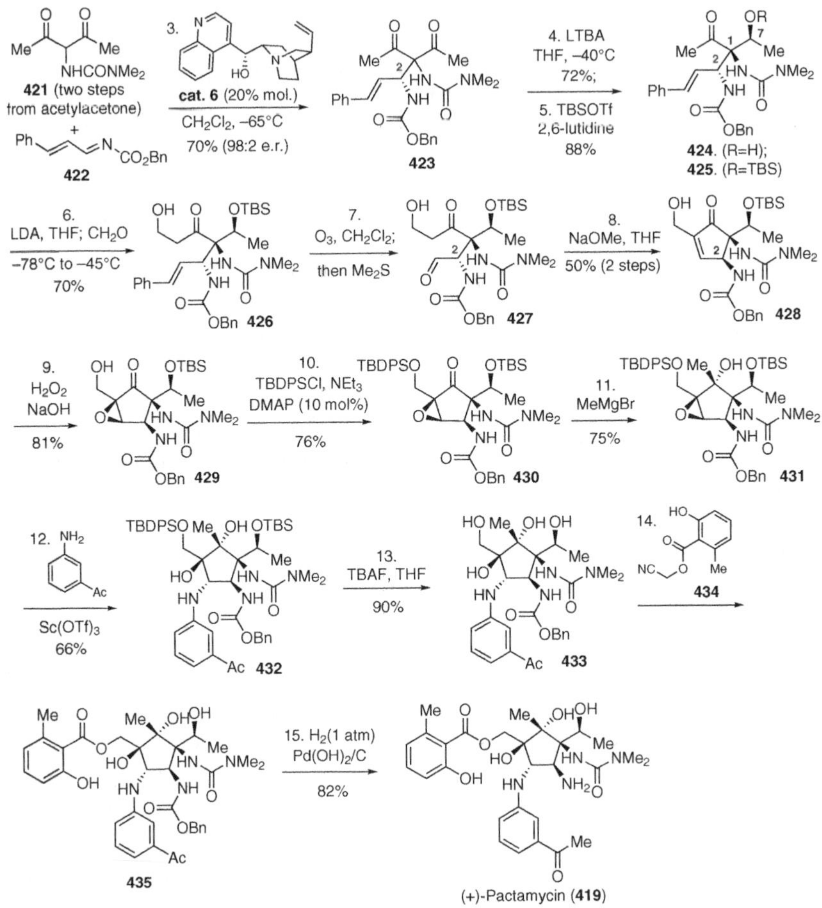

Scheme 2.65 Johnson's 15-step total synthesis of (+)-pactamycin (**419**).

out to be critical to orchestrate downstream stereochemical outcomes that allowed efficient completion of the synthesis. Moreover, the strategic use of cinnamyl imine **422** as a partner for the Mannich reaction translated to the installation all five carbons of the pactamycin core, with appropriate functional handles, in this very first C–C bond-forming reaction. The desymmetrization of **423** was achieved by reducing with lithium tri(*tert*-butoxy)aluminum hydride (LTBA), which delivered hydroxyketone **424** in a high diastereoselectivity (>10:1 ratio of 5: Σ [other diastereomers]). The stereochemistry of **424** turned to be (1*R*,2*R*,7*S*), which implies that the incorrect C2 isomer in **423** was parlayed into the correct C1/C7 configurations in **424**.

The next task was the installation of the C4 side chain and cyclization to assemble the cyclopentenone core. Thus, the C7 hydroxyl group was protected, and the resulting **425** was subjected to deprotection-addition with formaldehyde to give hydroxylmethylated adduct **426**. Alkene ozonolysis followed by MeONa-mediated intramolecular aldol condensation provided **428** with the five-membered pactamycin core built. Remarkably, a spectacular scenario appeared: the aldol condensation reaction proceeded with concomitant complete epimerization at C2, thus forging the correct stereochemistry at C2!

Inspired by the work of Hanessian,[158] the enone **428** was epoxidized and the hydroxyl group protected as TBDPS ether. Addition of methyl Grignard bromide to ketone **430** afforded carbinol **431** with correct stereochemistry implying that addition took place from the concave face. This unusual stereochemical outcome might benefit from the directing effect of urea's functionality, judging the correctness of the strategic choice to incorporate this functionality in its native form from the outset. $Sc(OTf)_3$-promoted ring opening of the epoxide in **431** with 3-acetylaniline proceeded regioselectively to deliver **432** at a 66% yield, which competed functionalization of the pactamycin core. Finally, *O*-bis-de-silylation, selective aroylation of the primary hydroxyl group using **434** as a ketene precursor,[161] and *N*-deprotection afforded (+)-pactamycin (**419**), thus completing the 15-step total synthesis at a 1.9% overall yield. It is worth mentioning that the epoxide **431**, corresponding to **420** in Scheme 2.64, being a versatile intermediate for analog synthesis, has been prepared in gram quantities on a single pass.

In summary, although this campaign of total synthesis of pactamycin (**419**) is not without thrilling scenarios, with the several correct strategic decisions from the outset, and some luck and serendipity, overall it has been a pleasant journey almost free of frustration. It is nice to end this chapter with this completely synthetic legend.

2.12 Other Procedure-Economical Enantioselective Total Syntheses of Natural Products

Due to limited space, we regret not being able to enumerate all the shortest enantioselective total syntheses. In addition to those presented in Sections 2.1.2–2.11.3, some other elegant ones (50 synthetic routes) are compiled in Table 2.2 with the key information provided.

2.13 Conclusion

In this chapter, besides the Robinson's synthesis of achiral tropinone, we selected the 94 shortest enantioselective total syntheses of individual natural products. An inspection of the 44 enantioselective total syntheses presented in Sections 2.1.2–2.11.3 (see Table 2.3) allows us to draw the following conclusions:

Table 2.2 Another 50 examples of the shortest enantioselective total syntheses of natural products.

3-step enantioselective total synthesis	
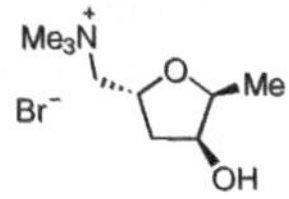	**(+)-*epiallo*-Muscarine** 3 steps, 20% overall yield, 90% *ee* J. Hartung, P. Kunz, *et al. Synlett* **2003**, 51
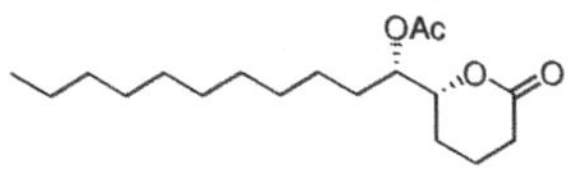	**(−)-(5*R*,6*S*)-6-Acetoxyhexadecan-5-olide** 3 steps, 57.8% overall yield, 100% *ee* Y. Li, B. Sun, *et al. Tetrahedron: Asymmetry* **2005**, *16*, 1305
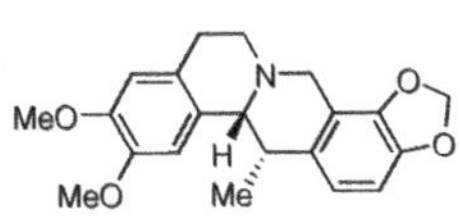	**(+)-Cavidine** and 11 other tetrahydroprotoberberine alkaloids 3 steps, 47–64% overall yields, 91–96% *ee* R. B. Tong and S. Q. Zhou *Org. Lett.* **2017**, *19*, 1594
5-step enantioselective total synthesis	
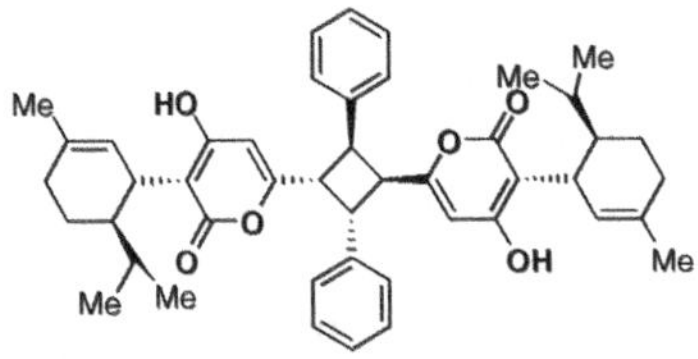	**(+)-Katsumadain C** 5 steps, 11.4% overall yield, 89% *ee* Z. Yang, Y.-F. Tang, P.-T. Zhang, *et al. Org. Lett.* **2012**, *14*, 162
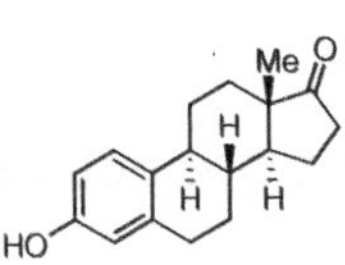	**(+)-Estrone** 5 steps, 18.4% overall yield, 99.8% *ee* B. List, S. Prévost, *et al.*, *Angew. Chem. Int. Ed.* **2014**, *53*, 8770
6- and 7-step enantioselective total synthesis	
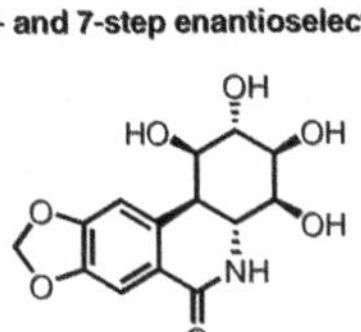 (+)-7-Deoxypancratistatin 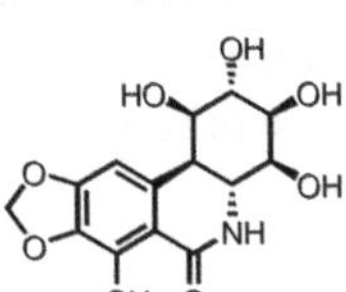 (+)-Pancratistatin	**(+)-7-Deoxypancratistatin** 6 steps, 12% overall yield, 96% *ee* **(+)-Pancratistatin** 7 steps, 19% overall yield, 96% *ee* D. Sarlah, L. W. Hernandez, *et al. J. Am. Chem. Soc.* **2017**, *139*, 15656
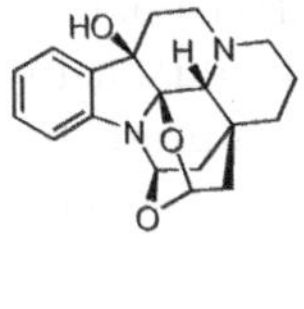	**(−)-Terengganensine A** 7 steps, 23% overall yield, 90% *ee* J. Zhu, C. Piemontesi, *et al. Angew. Chem. Int. Ed.* **2016**, *55*, 6656

Table 2.2 (Continued)

8-step enantioselective total synthesis	
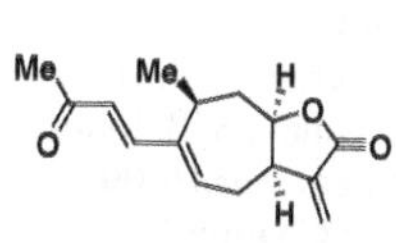 (+)-8-*epi*-Xanthatin 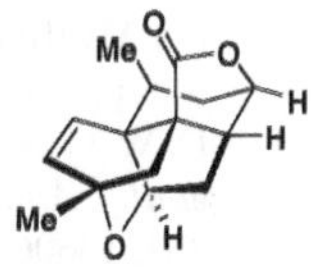 (–)-Xanthipungolide	**(+)-8-*epi*-Xanthatin** 7 steps, 15% overall yield, 95% *ee* **(–)-Xanthipungolide** 8 steps, 6% overall yield, 95% *ee*
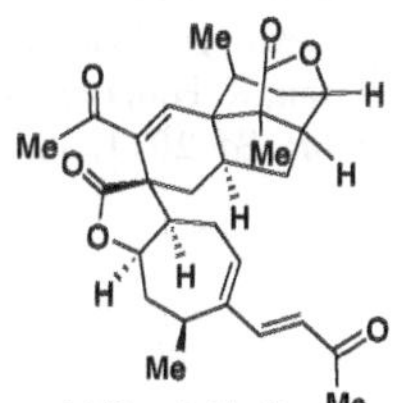 (–)-Pungiolide D	**(–)-Pungiolide D** 8 steps, 6% overall yield, 95% *ee* Y.-F. Tang, J. Feng, *et al. Angew. Chem. Int. Ed.* **2017**, *56*, 16323.
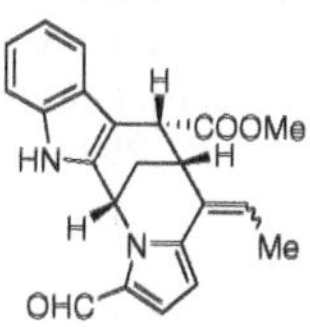	**(–)-(*E*)- and (*Z*)-Alstoscholarines** 8 steps, 14% overall yield, 93% *ee* J. Zhu, T. Gerfaud, *et al. Angew. Chem. Int. Ed.* **2011**, *50*, 3954
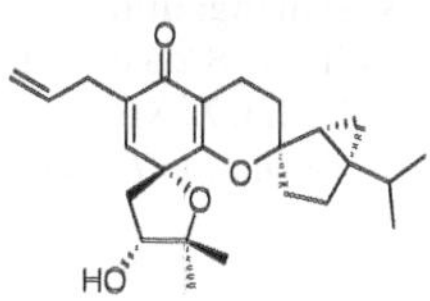	**(–)-Spirooliganones A** and **B** (not shown) 8 steps, 1.9%, 5.2% overall yield Z.-X. Xie, L. Wei, *et al. Org. Lett.* **2014**, *16*, 2784
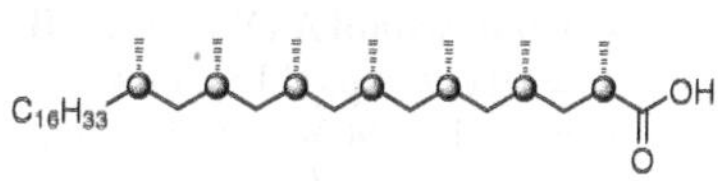	**(+)-Phthioceranic Acid** 8 steps (LLS) E. Negishi, S. Xu, *et al. Angew. Chem. Int. Ed.* **2015**, *54*, 9319
9-step enantioselective total synthesis 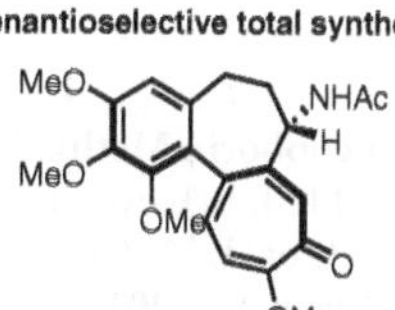	**(–)-Colchicine** 9 steps, 9.2% overall yield, >99% *ee* C.-C. Li, B. Chen, *et al. Chem. Sci.* **2017**, *8*, 4961–4966

(*Continued*)

Table 2.2 (Continued)

9-step enantioselective total synthesis	
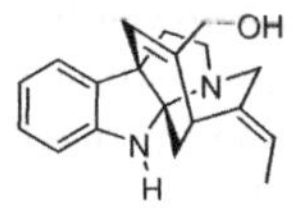	**(+)-Minfiensine** 9 steps, 21% overall yield, 96% *ee* D. W. C. MacMillan, S. B. Jones, *et al. J. Am. Chem. Soc.* **2009,** *131*, 13606 (see Chapter 7)
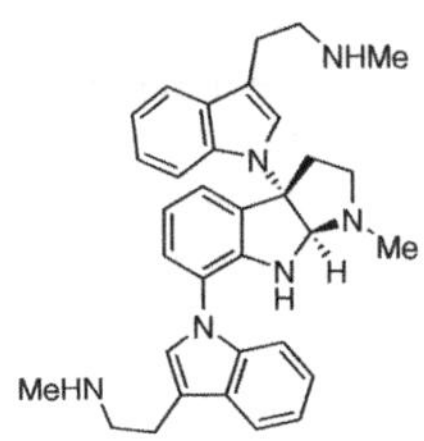	**(+)-Psychotrimine** 9 steps, 7% overall yield, 100% *ee* P. S. Baran, K. Foo, *et al. Angew. Chem. Int. Ed.* **2011**, *50*, 2716
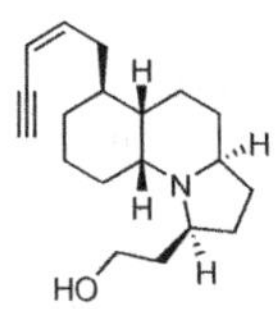	**(−)-Gephyrotoxin** 9 steps, 14% overall yield M. D. Smith, S. Chu, *et al. Angew. Chem. Int. Ed.* **2014**, *53*, 13826
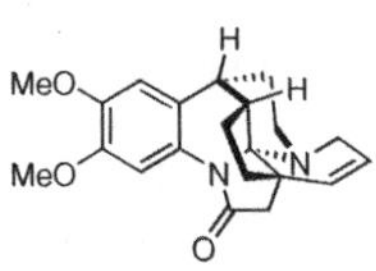	**(−)-Isoschizogamine** 9 steps (LLS), 8.2% overall yield, 83% *ee* J. Zhu, Z. Xu, *et al. Angew. Chem. Int. Ed.* **2015**, *54*, 14937
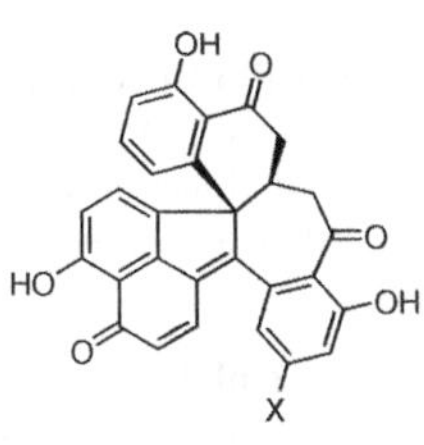	**(+)-Dalesconol A** (X = H) and **B** (X = OH) 9 steps (LLS), 11% overall yield, 96% *ee* W.-J. Tang, G.-Q. Zhao, *et al. J. Am. Chem. Soc.* **2017**, *139*, 3360
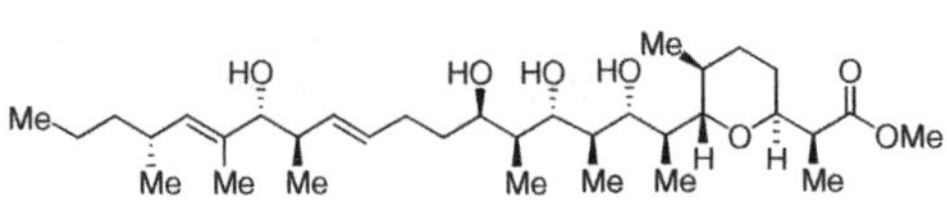	**(+)-Zincophorin Methyl Ester** 9 steps (LLS), 10% overall yield J. L. Leighton, L.-A. Chen, *et al. J. Am. Chem. Soc.* **2017**, *139*, 4568

Table 2.2 (Continued)

10-step enantioselective total synthesis	
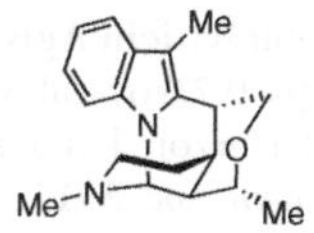	**(–)-Alstoscholarisine A** 10 steps, 5.3% overall yield, 99% *ee* Y. R. Yang, X. Liang, *et al.* *J. Am. Chem. Soc.* **2016**, *138*, 2560
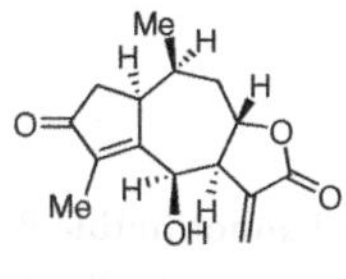	**(+)-Mikanokryptin** 10 steps, 6% overall yield T. J. Maimone, X. R. Hu, *et al. Angew. Chem. Int. Ed.* **2017**, *56*, 1624
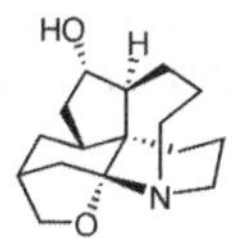 (–)-Huperzine Q 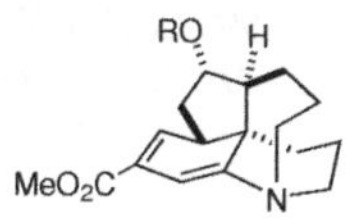 (+)-Lycopladine B (R = H) (+)-Lycopladine C (R = Ac)	**(–)-Huperzine Q** 10 steps, 1.7% overall yield, 93% *ee* **(+)-Lycopladine B**, 12 steps **(+)-Lycopladine C**, 13 steps X. G. Lei, B. K. Hong, *et al. Angew. Chem. Int. Ed.* **2015**, *54*, 1011 *Chem. Asian J.* **2017**, *12*, 1557
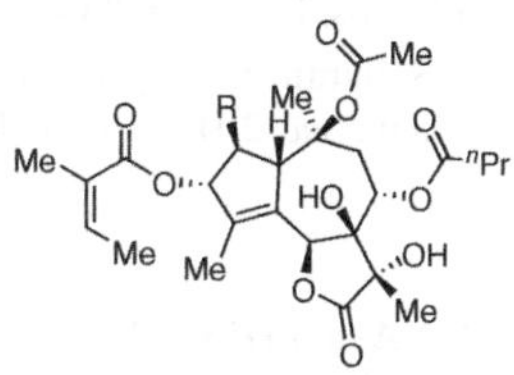	**(–)-Nortrilobolide** (R = H), 10 steps (LLS), 13.3%, overall yield **(–)-Thapsigargin** (R = $OCOC_7H_{15}n$) 12 steps (LLS),. 5.8% overall yield P. A. Evans and D. Z. Chen *J. Am. Chem. Soc.* **2017**, *139*, 6046
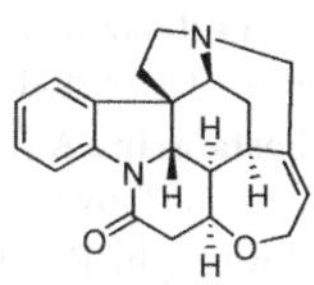	**Strychnine** 10 steps, 3.0% overall yield D. Y.-K. Chen, G. S. Lee, *et al. Chem. Eur. J.* **2017**, *23*, 16189
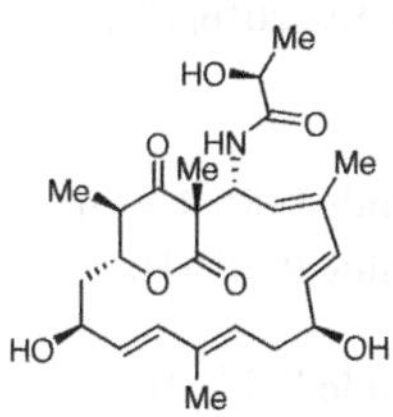 **(–)-Lankacidinol** 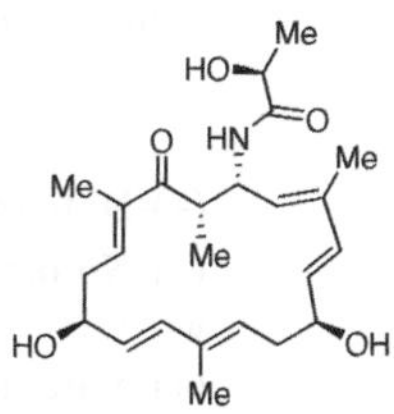 **(–)-Lankacyclinol**	**(–)-Lankacidinol** 10 steps, 2.34% overall yield **(–)-Lankacyclinol** 10 steps, 3.16% overall yield Zheng, K.; Shen, D.; Hong, R. *J. Am. Chem. Soc.* **2017**, *139*, 12939

(*Continued*)

Table 2.2 (Continued)

11-step enantioselective total synthesis 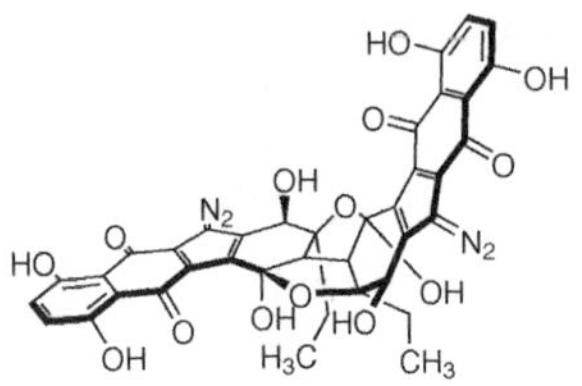	**(−)-Lomaiviticin Aglycon** 11 **steps**, 0.7% overall yield, 91% *ee* S. B. Herzon, L. Lu, *et al. J. Am. Chem. Soc.* **2011**, *133*, 7260
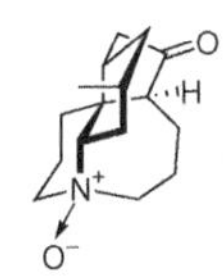	**(−)-Lannotinidine B** 11 steps, 15.9% overall yield, > 99% *ee* Z.-J. Yao, H.-M. Ge, *et al. J. Am. Chem. Soc.* **2012**, *134*, 12323
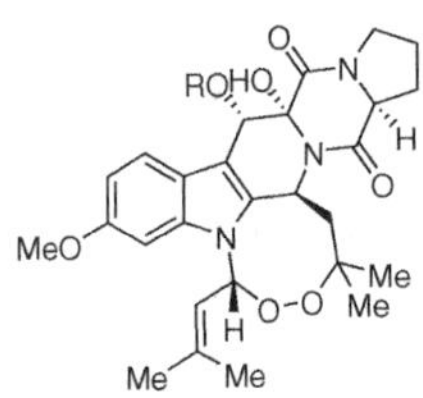	**(−)-Verruculogen** (R = H) 11 steps, 2.0% overall yield **(−)-Fumitremorgin A** (R = prenyl) 12 steps, 1.9% overall yield P. S. Baran, Y. Feng, *et al. J. Am. Chem. Soc.* **2015**, *137*, 10160
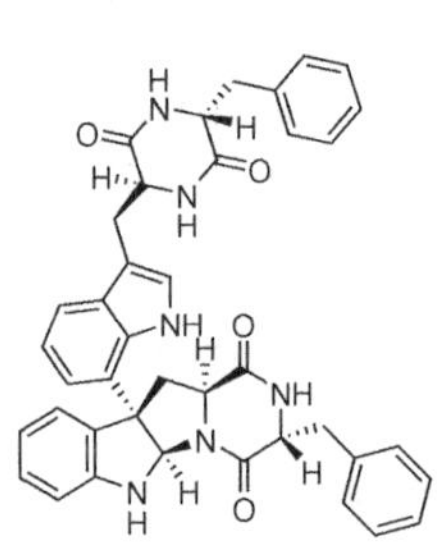	**(+)-Asperazine** 11 steps, 6.8%, overall yield **(+)-*iso*-Pestalazine A** (not shown) 11 steps, 8.0% overall yield **(+)-Pestalazine A** (not shown) 11 steps, 4.8% overall yield M. Movassaghi, R. P. Loach, *et al. J. Am. Chem. Soc.* **2016**, *138*, 1057
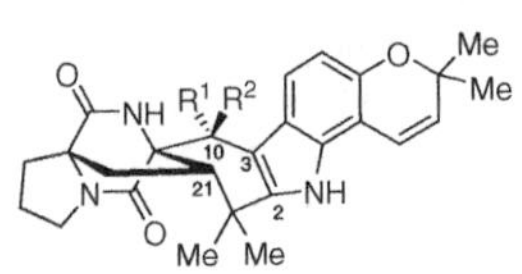	**(+)-Notoamide I** (R^1, R^2 = O) **(+)-Notoamide R** (R^1 = H, R^2 = OH) **(+)-Notoamide F** (R^1 = H, R^2 = OMe) 11, 12, and 13 steps, 6.6, 5.6, 5.3% overall yield C.-Z. Li, D.-Q. Sun, *et al. Angew. Chem. Int. Ed.* **2016**, *55*, 10435
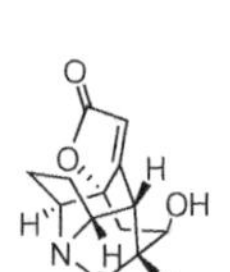	**(−)-Virosaine A** 11 steps, 9.7% overall yield, >99% *ee* J. L. Gleason and J. M. E. Hughes *Angew. Chem. Int. Ed.* **2017**, *56*, 10830

Table 2.2 (Continued)

12-step enantioselective total synthesis	
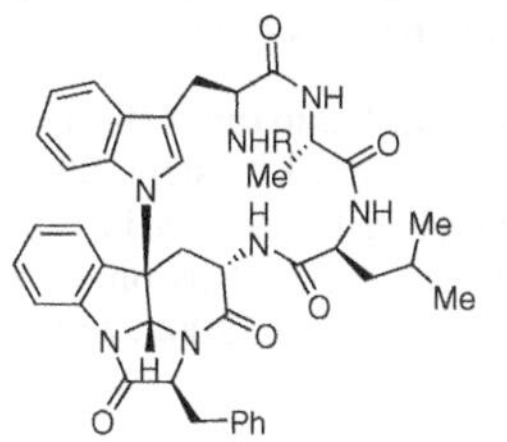	**(–)-Kapakahines F** (R = H) and **B** (R = L-Phe) 12, 14 steps, 4.5%, 3.7% overall yield P. S. Baran, T. Newhouse, *et al. J. Am. Chem. Soc.* **2009**, *131*, 6360
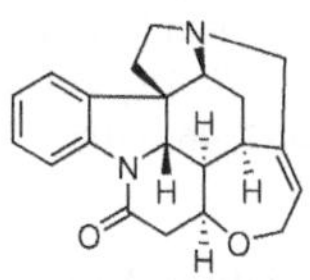	**(–)-Strychnine** 12 steps, 6.3% overall yield, 97% *ee* D. W. C. MacMillan, S. B. Jones, *et al. Nature* **2011**, *475*, 183 (Collective synthesis, 34 steps for 6 natural products) (cf. chapter 7)
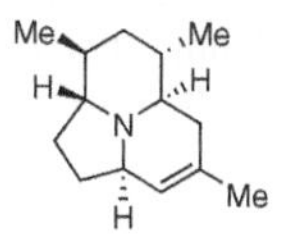	**Alkaloid (–)-205B** 12 steps, 8% overall yield D. L. Comins and S. V. Tsukanov *Angew. Chem. Int. Ed.* **2011**, *50*, 8626*J. Org. Chem.* **2014**, *79*, 9074; J. K. Cha and N. N. Rao *J. Am. Chem. Soc.* **2015**, *137*, 2243
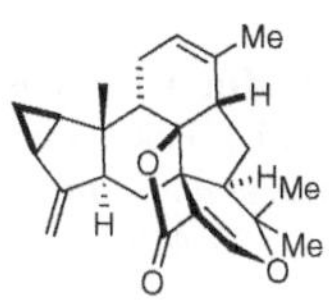	**(–)-Bolivianine** 12 steps, 2.3% overall yield B. Liu, C. C. Yuan, *et al. J. Am. Chem. Soc.* **2013**, *135*, 9291 *Chem. Commun.* **2015**, *51*, 6179 *Chin. Chem. Lett.* **2017**, *28*, 113
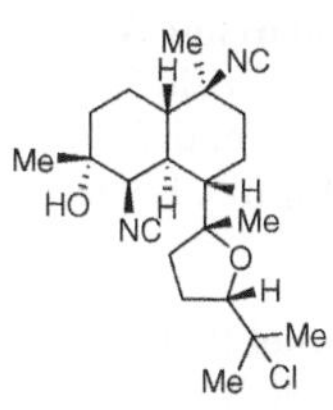	**(+)-Kalihinol B** 12 steps, 0.15% overall yield, 78% *ee* C. D. Vanderwal, M. E. Daub, *et al. J. Am. Chem. Soc.* **2015**, *137*, 4912
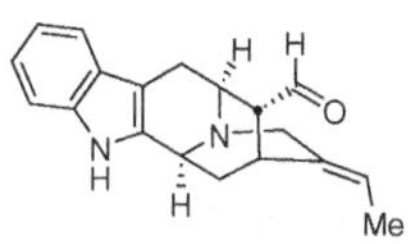	**(+)-Vellosimine, (+)-*N*-methylvellosimine, (+)-10-methoxyvellosimine** 12 steps, 5.5–6.7% overall yield, 93% *ee* T. Gaich and S. Krüger *Angew. Chem. Int. Ed.* **2015**, *54*, 315
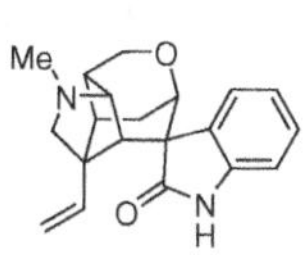	**(+)-Gelsemine** 12 steps, 4.8% overall yield, >99% *ee* H. B. Zhai, F. G. Qiu, *et al. Nat. Commun.* **2015**, 6: 7204

(*Continued*)

Table 2.2 (Continued)

13-step enantioselective total synthesis	
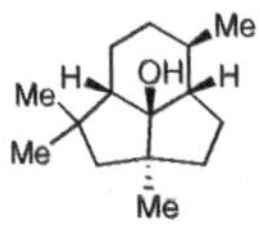	**(−)-Presilphiperfolan-8-ol** 13 steps, 4.4% overall yield S. A. Snyder and P. F. Hu *J. Am. Chem. Soc.* **2017**, *139*, 5007
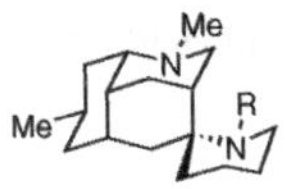	**(+)-Nankakurine A** (R = H) and **B** (R = Me) 13, 14 steps, 20, 16% overall yield, >99% *ee* L. E. Overman, B. L. Nilsson, *et al. J. Am. Chem. Soc.* **2008**, *130*, 11297
14-step enantioselective total synthesis	
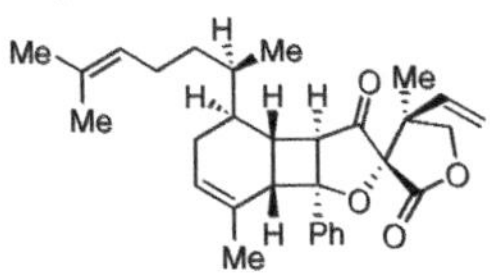	**(−)-Biyouyanagin A** 14 steps, 9.3% overall yield K. C. Nicolaou, D. Sarlah, *et al. Angew. Chem. Int. Ed.* **2007**, *46*, 4708
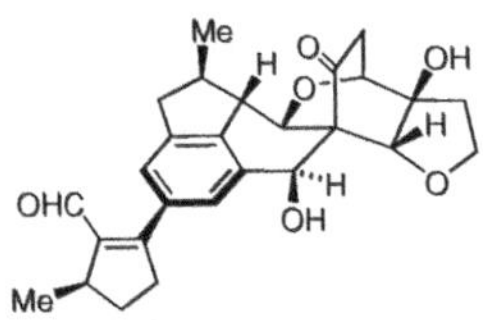	**(−)-Incarviatone A** 14 steps, 0.38% overall yield, 99% *ee* X. G. Lei, B. K. Hong, *et al. J. Am. Chem. Soc.* **2015**, *137*, 11946
15-step enantioselective total synthesis	
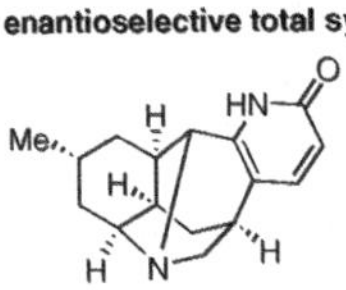	**(+)-Lyconadin A** 15 steps, 4.3% overall yield T. Fukuyama, T. Nishimura, *et al. J. Am. Chem. Soc.* **2011**, *133*, 418
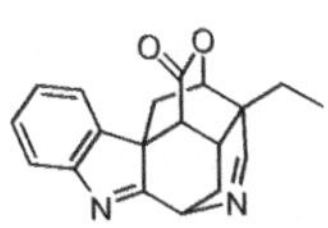	**(+)-Scholarisine A** 15 steps, 1.2% overall yield S. A. Snyder and M. W. Smith *J. Am. Chem. Soc.* **2013**, *135*, 12964
19-step enantioselective total synthesis	
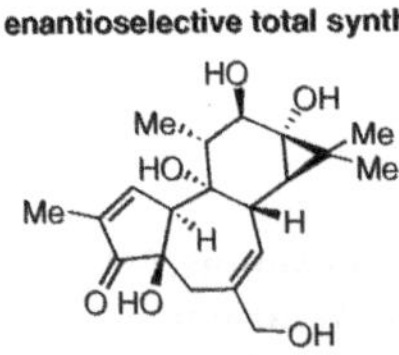	**(+)-Phorbol** 19 steps, 0.05% overall yield P. S. Baran, S. Kawamura, *et al. Nature* **2016**, *532*, 90
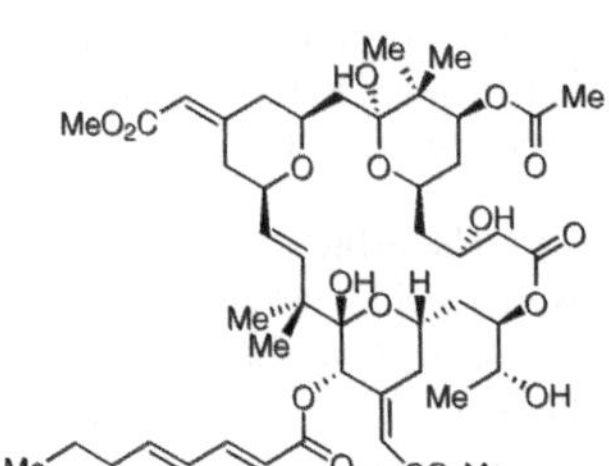	**(−)-Bryostatin 1** 19 steps (LLS, 29 total steps) 4.8% overall yield, > 2 g produced P. A. Wender, C. T. Hardman, *et al. Science* **2017**, *358*, 218

Table 2.2 (Continued)

20-step enantioselective total synthesis 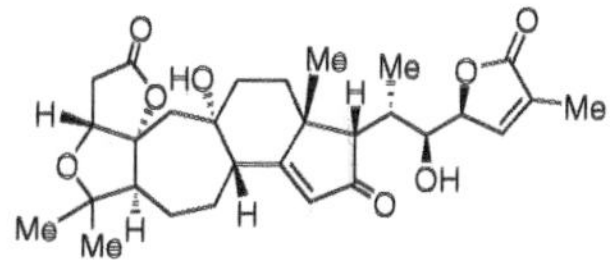	**(+)-Propindilactone G** 20 steps (LLS) Z. Yang, L. You, *et al. J. Am. Chem. Soc.* **2015**, *137*, 10120
25-step enantioselective total synthesis 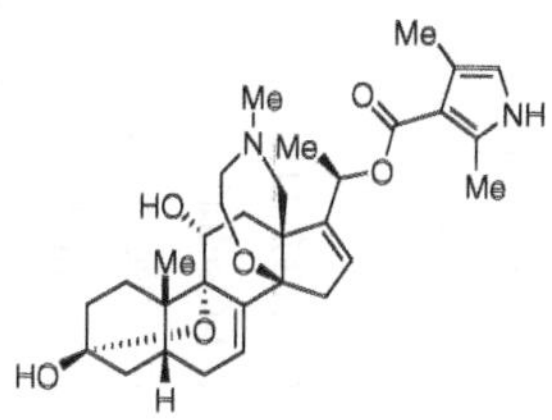	**(−)-Batrachotoxin [(−)-BTX]** 24 steps, 0.25% overall yield J. Du Bois, M. M. Logan, *et al. Science* **2016**, *354*, 865
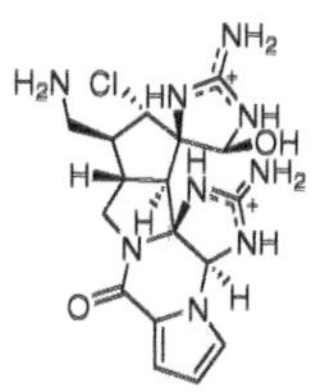	**(−)-Palau'amine** 25 steps, 0.015% overall yield P. S. Baran, I. B. Seiple, *et al. Angew. Chem. Int. Ed.* **2010**, *49*, 1095*J. Am. Chem. Soc.* **2011**, *133*, 14710
28-step enantioselective total synthesis 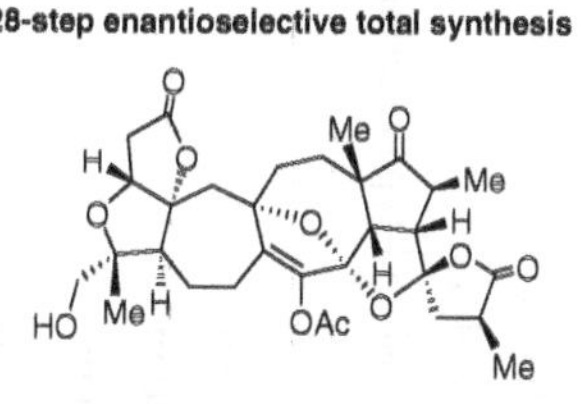	**(+)-Lancifodilactone G Acetate** 28 steps (LLS) Z. Yang, D. D. Liu, *et al. J. Am. Chem. Soc.* **2017**, *139*, 5732

Table 2.3 Key strategies used in the 39 enantioselective total syntheses presented in Sections 2.1.2–2.11.3.

Main Strategy	Key Feature	Natural Products	Section
Bio-inspired synthesis (cascade reaction)	PGF; CBB	(+)-WIN 64821	2.2.4
	CBB (chiral auxiliary)	(−)-Sibirine	2.3.2
	chiral phase-transfer (ion pair) catalysis	(−)-Hyperibone K	2.3.5
	asymmetric catalysis	(−)-Angiopterlactone A	2.4.1
	revealing the hidden symmetry; CBB	(+)-Cardamom peroxide	2.4.2
	PGF; CBB	(−)-Chaetominine	2.4.4
	CBB; Diels–Alder reaction, transannular reaction	(+)-Fastigiatine	2.7.5
	asymmetric catalysis; PGF	(−)-Neothiobinupharidine	2.8.6
	asymmetric catalysis; CBB; radical reaction	(−)-6-*epi*-Ophiobolin N	2.9.2
	PGF, RCM, CBB, transannular reaction	(+)-Lycoflexine	2.9.4
	IMDA, CBB	(+)-Fusarisetin A (Gao; Theodorakis)	2.9.5
	Diels–Alder reaction; CBB	(+)-Absinthin	2.10.3
	asymmetric catalysis	(−)-Maoecrystal V	2.10.4
	CBB; Pauson–Khand-type reaction	(−)-Ingenol	2.11.1
Organocatalysis	one-pot reaction	(+)-ABT-341	2.1.2
	one-pot reaction	(−)-Oseltamivir	2.2.1
	one-pot reaction	(−)-Oseltamivir	2.2.2
	redox economy	(+)-Frondosin B	2.3.3
	one-pot reaction	(−)-PGE_1 methyl ester	2.3.4
	chiral anionic phase-transfer catalysts	(−)-Chimonanthine	2.4.3

	one-pot reaction	(+)-$PGF_2\alpha$	2.7.2
	cascade reaction; Diels–Alder reaction; radical reaction	(−)-Vincorine	2.9.3
Organometallic asymmetric catalysis	Ir; CBB;	(+)-Aszonalenin	2.3.1
	dual catalysis, RCM	Δ^9-Tetrahydrocannabinols	2.5.1
	complexity-building reaction cascades	(+)-Artemisinin	2.5.3
	[3+2]-cycloaddition	(−)-Aflatoxin B_2	2.5.4
	asymmetric double allylation; cascade reaction, PGF	(−)-Cyanolide A	2.6.2
	cascade reaction	(−)-Aromadendrane Sesquiterpenes	2.7.3
	cyclopropanation, IMDA	(−)-Cycloclavine	2.8.5
	asymmetric alkylation of *in situ*-generated cyclic ketone enolates; RCM	(−)-Cyanthiwigin F	2.9.1
	RCM, one-pot reaction; transannular reaction	(+)-Loline	2.10.2
Strategic bond disconnections (convergent fragment-coupling Strategy)	organocatalysis; Ugi 3-component reaction; transannular	(+)-Peganumine A	2.7.4
	reaction asymmetric catalysis; photoredox reaction	*trans*-Clerodane diterpenoid	2.8.1

(Continued)

Table 2.3 (Continued)

Main Strategy	Key Feature	Natural Products	Section
	Kagan's reagent (SmI_2); CBB	(−)-Englerin A	2.8.2
	double Michael addition, CBB	(−)-Jiadifenolide	2.8.3
	PGF; IMDA; organocatalysis	(+)-Chatancin	2.8.4
	CBB; coupling reactions	(−)-Huperzine A	2.10.1
	Pauson–Khand reaction; CBB; one-pot reaction	(+)-Ryanodol	2.11.2
	revealing the hidden symmetry; organocatalysis	(+)-Pactamycin	2.11.3
Chemoenzymatic catalysis		Aszonalenin	2.2.3
Coupling reaction	CBB; one-pot reaction	(+)-Machaeriol B and (+)-Machaeriol D	2.5.2
One-pot process	chiral auxiliary	(+)-Camptothecin	2.6.1
Synthetic economy	several general guidelines followed, CBB	hapalindole-type natural products	2.7.1

Notes: CBB = chiral building block; PGF = protecting group-free; RCM = ring-closing metathesis; IMDA = intramolecular Diels–Alder reaction.

1) To achieve a highly procedure-economical total synthesis of a natural product at least two principles of synthetic economy should be applied.
2) Bio-inspired synthesis, organocatalysis, organometallic asymmetric catalysis, and strategic bond disconnections constitute four major strategies to achieve high procedure economy.
3) The Diels–Alder reaction, conjugate addition, RCM, Pauson–Khand reaction, and transannular reactions are the most commonly used strategic reactions for achieving high procedure economy.
4) The chiral building block-based approach is still the most commonly used (18/39) and a reliable strategy for enantioselective total synthesis of natural products.

Acknowledgements

Financial support from the National Key R&D Program of China (grant No. 2017YFA0207302), the National Natural Science Foundation of China (21332007, 21472153, and 21672176) and the Program for Changjiang Scholars and Innovative Research Team in University of Ministry of Education, China gratefully acknowledged.

We thank Ms. Yan-Jiao Gao and Ms. Ling-Ling Li for their assistance in the preparation of this manuscript, and thank graduate students Feng Han, Jiang-Feng Wu, Ting-Ting Chen, Ya-Nan Zhu, and Yu-Cheng Liu for their help in drawing schemes/figures.

References

1 (a) The term "procedure-economy/procedure-economical" was coined previously, see: Huang, P.-Q.; Wang, Y.; Luo, S.-P.; Geng, H.; Ruan, Y.-P.; Wang, A.-E *Tetrahedron Lett.* **2015**, *56*, 1255; (b) we define a reaction procedure as one in which a reactant is transformed to a product with only one workup (one extraction) or purification irrespective of the number of steps and transformations. The number of steps is counted from commercially available materials instead of from known compounds. Consequently, the steps counts here may differ from those in the original papers.
2 (a) Robinson, R. *J. Chem. Soc., Trans.* **1917**, *111*, 762; for a recent viewpoint on Robinson's landmark synthesis of tropinone, see: (b) Medley, J. W.; Movassaghi, M. *Chem. Commun.* **2013**, *49*, 10775.
3 Robinson R. *J. Chem. Soc., Trans.* **1917**, *111*, 876.
4 Pei, Z.; Li, X.; von Geldern, T. W.; Madar, D. J.; Longenecker, K.; Yong, H.; *et al.* *J. Med. Chem.* **2006**, *49*, 6439.
5 Ishikawa, H.; Honma M.; Hayashi, Y. *Angew. Chem. Int. Ed.* **2011**, *50*, 2824.

6 (a) Magano, J. *Chem. Rev.* **2009**, *109*, 4398; (b) Magano, J. *Tetrahedron* **2011**, *67*, 7875; (c) Shibasaki, M.; Kanai, M. *Eur. J. Org. Chem.* **2008**, 1839.

7 (a) Ishikawa, H.; Suzuki, T.; Hayashi, Y. *Angew. Chem. Int. Ed.* **2009**, *48*, 1304; (b) Ishikawa, H.; Suzuki, T.; Orita, H.; Uchimaru, T.; Hayashi, Y. *Chem. Eur. J.* **2010**, *16*, 12616; (c) Mukaiyama, T.; Ishikawa, H.; Koshino, H.; Hayashi, Y. *Chem. Eur. J.* **2013**, *19*, 17789; (d) Hayashi, Y.; Ogasawara, S. *Org. Lett.* **2016**, *18*, 3426.

8 Zhu, S. L.; Yu, S. Y.; Wang, Y.; Ma, D. W. *Angew. Chem. Int. Ed.* **2010**, *49*, 4656.

9 Yin, W.-B.; Cheng, J.; Li, S.-M. *Org. Biomol. Chem.* **2009**, *7*, 2202.

10 Barrow C. J.; Sun, H. H. *J. Nat. Prod.* **1994**, *57*, 471.

11 (a) Tadano, S.; Mukaeda Y.; Ishikawa, H. *Angew. Chem. Int. Ed.* **2013**, *52*, 7990; (b) Tadano, S.; Sugimachi, Y.; Sumimoto, M.; Tsukamoto, S.; Ishikawa, H. *Chem.–Eur. J.* **2016**, *22*, 1277.

12 Kieffer, M. E.; Chuang, K. V.; Reisman, S. E. *J. Am. Chem. Soc.* **2013**, *135*, 5557.

13 (a) Schafroth, M. A.; Sarlah, D.; Krautwald, S.; Carreira, E. M. *J. Am. Chem. Soc.* **2012**, *134*, 20276; (b) Ruchti J.; Carreira E. M. *J. Am. Chem. Soc.* **2014**, *136*, 16756.

14 Matsumura, K.; Kitahara, T. *Heterocycles* **2001**, *54*, 727.

15 (a) Husson H.-P.; Royer, J. *Chem. Soc. Rev.* **1999**, *28*, 383; (b) François, D.; Lallemand, M. C.; Selkti, M.; Tomas, A.; Kunesch N.; Husson, H.-P. *Angew. Chem. Int. Ed.* **1998**, *37*, 104.

16 (a) Inoue, M.; Frontier A. J.; Danishefsky, S. *J. Angew. Chem. Int. Ed.* **2000**, *39*, 761; (b) Inoue, M.; Carson, M. W.; Frontier A. J.; Danishefsky, S. J. *J. Am. Chem. Soc.* **2001**, *123*, 1878; (c) Hughes C. C.; Trauner, D. *Angew. Chem. Int. Ed.* **2002**, *41*, 1569; Correction: Hughes C. C.; Trauner, D. *Angew. Chem. Int. Ed.* **2002**, *41*, 2227; (d) Hughes C. C.; Trauner, D. *Tetrahedron* **2004**, *60*, 9675; (e) Kerr, D. J.; Willis A. C.; Flynn, B. L. *Org. Lett.* **2004**, *6*, 457; (f) Ovaska, T. V.; Sullivan, J. A.; Ovaska, S. I.; Winegrad J. E.; Fair, J. D. *Org. Lett.* **2009**, *11*, 2715; (g) Olson J. P.; Davies, H. M. L. *Org. Lett.* **2008**, *10*, 573; correction: Olson J. P.; Davies, H. M. L. *Org. Lett.* **2010**, *12*, 1144; (h) Oblak, E. Z.; Van Heyst, M. D.; Li, J.; Wiemer, A. J.; Wright, D. L. *J. Am. Chem. Soc.* **2014**, *136*, 4309.

17 (a) Reiter, M.; Torssell, S.; Lee S.; MacMillan, D. W. C. *Chem. Sci.* **2010**, *1*, 37; (b) Lee, S.; MacMillan, D. W. C. *J. Am. Chem. Soc.* **2007**, *129*, 15438.

18 Hayashi, Y.; Umemiya, S. *Angew. Chem. Int. Ed.* **2013**, *52*, 3450.

19 Qi, J.; Beeler, A. B.; Zhang, Q.; Porco, J. A. Jr. *J. Am. Chem. Soc.* **2010**, *132*, 13642.

20 Yu, Y.-M.; Yang, J.-S.; Peng, C.-Z.; Caer, V.; Cong, P.-Z.; Zou, Z.-M.; *et al.* *J. Nat. Prod.* **2009**, *72*, 921.

21 (a) Thomson, M. I.; Nichol, G. S.; Lawrence A. L. *Org. Lett.* **2017**, *19*, 2199; (b) Kotammagari, T. K.; Gonnade, R. G.; Bhattacharya, A. K. *Org. Lett.* **2017**, *19*, 3564.

22 Hu, X. R.; Maimone, T. J. *J. Am. Chem. Soc.* **2014**, *136*, 5287.

23 Xie, W.-Q.; Jiang, G.-D.; Liu, H.; Hu, J.-D.; Pan, X.-X.; Zhang, H.; *et al.* *Angew. Chem. Int. Ed.* **2013**, *52*, 12924.

24 (a) Jiao, R.-H.; Xu, S.; Liu, J.-Y.; Ge, H.-M.; Ding, H.; *et al. Org. Lett.* **2006**, *8*, 5709; (b) Liao, L.; You, M.; Chung, B. K.; Oh, D.-C.; Oh, K.-B.; Shin, J. *J. Nat. Prod.* **2015**, *78*, 349.

25 (a) Huang, P.-Q.; Liu, L.; Peng, Q. Method for synthesizing cytotoxic alkaloid chaetominine and its epimer. Faming Zhuanli Shenqing Gongkai Shuomingshu. Chinese Patent CN 101463074 A, June 24, **2009**; (b) Peng, Q.-L.; Luo, S.-P.; Xia, X.-E.; Liu, L.-X.; Huang, P.-Q. *Chem. Commun.* **2014**, *50*, 1986; (c) Xu, C.-P.; Luo, S.-P.; Wang, A.-E; Huang, P.-Q. *Org. Biomol. Chem.* **2014**, *12*, 2859; (d) Luo, S.-P.; Peng, Q.-L.; Xu, C.-P.; Wang, A.-E; Huang, P.-Q. *Chin. J. Chem.* **2014**, *32*, 757; (e) Mao, Z.-Y.; Geng, H.; Zhang, T.-T.; Ruan, Y.-P.; Ye, J.-L.; Huang, P.-Q. *Org. Chem. Front.* **2016**, *3*, 24; (f) Huang, P.-Q.; Mao, Z.-Y.; Geng, H. *Chin. J. Org. Chem.* **2016**, *36*, 315.

26 Szpilman, A. M.; Carreira, E. M. *Angew. Chem. Int. Ed.* **2010**, *49*, 9592.

27 (a) Krautwald, S.; Sarlah, D.; Schafroth, M. A.; Carreira, E. M. *Science* **2013**, *340*, 1065; (b) Krautwald, S.; Schafroth, D.; Sarlah, D.; Carreira, E. M. *J. Am. Chem. Soc.* **2014**, *136*, 3020; (c) Schafroth, M. A.; Zuccarello, G.; Krautwald, S.; Sarlah, D.; Carreira, E. M. *Angew. Chem. Int. Ed.* **2014**, *53*, 13898; (d) Krautwald, S.; Carreira, E. M. *J. Am. Chem. Soc.* **2017**, *139*, 5627.

28 (a) Evans, D. A.; Barnes, D. M.; Johnson, J. S.; Lectka, T.; Matt, P. V.; Miller, S. J.; *et al. J. Am. Chem. Soc.* **1999**, *121*, 7582; (b) Trost, B. M.; Dogra, K. *Org. Lett.* **2007**, *9*, 861; (c) Cheng, L.-J.; Xie, J.-H.; Chen, Y.; Wang, L.-X.; Zhou, Q.-L. *Org. Lett.* **2013**, *15*, 764.

29 Klotter, F.; Studer, A. *Angew. Chem. Int. Ed.* **2015**, *54*, 8547.

30 Chou, C.-M.; Chatterjee, I.; Studer, A. *Angew. Chem. Int. Ed.* **2011**, *50*, 8614.

31 Villa, G.; Povie, G.; Renaud, P. *J. Am. Chem. Soc.* **2011**, *133*, 5913.

32 Zhu, C.-Y.; Cook, S. P. *J. Am. Chem. Soc.* **2012**, *134*, 13577.

33 (a) Alexakis, A.; Bäckvall, J. E.; Krause, N.; Pàmies, O.; Diéguez, M. *Chem. Rev.* **2008**, *108*, 2796; (b) Feringa, B. L. *Acc. Chem. Res.* **2000**, *33*, 346; (c) Feringa, B. L.; Pineschi, M.; Arnold, L. A.; Imbos, R.; de Vries, A. H. M. *Angew. Chem. Int. Ed. Engl.* **1997**, *36*, 2620.

34 Alexakis, A.; Rosset, S.; Allamand, J.; March, S.; Guillen, F.; Benhaim, C. *Synlett* **2001**, *1375.*

35 Zhou, G.; Corey, E. J. *J. Am. Chem. Soc.* **2005**, *127*, 11958.

36 (a) Martino, E.; Volpe, S. D.; Terribile, E.; Benetti, E.; Sakaj, M.; Centamore, A.; *et al. Bioorg. Med. Chem. Lett.* **2017**, *27*, 701; (b) Wall, M. E.; Wani, M. C. *Alkaloids* **1998**, *50*, 509; (c) Pantaziz, P.; Giovanella, B. C.; *Eds. Ann. N. Y. Acad. Sci.* **1996**, *803*; (d) Potmesil, M.; Pinedo, H., Eds. *Camptothecins: New Anticancer Agents*; Boca Raton, FL: CRC Press, **1995**.

37 Wall, M. E.; Wani, M. C.; Cook, C. E.; Palmer, K. H.; McPhail, A. T.; Sim, G. A. *J. Am. Chem. Soc.* **1966**, *88*, 3888.

38 (a) Chen, L.; Chen, F.-E. *Synlett* **2017**, *28*, 1134; (b) Li, K.; Ou, J.-J.; Gao, S.-H. *Angew. Chem. Int. Ed.* **2016**, *55*, 14778; (c) Xu, P.; Chen, D.-S.; Xi, J.; Yao, Z.-J. *Chem. Asian J.* **2015**, *10*, 976; (d) Yu, S.-B.; Huang, Q.-Q.; Luo, Y.; Lu, W. *J. Org.*

Chem. **2012**, *77*, 713; (e) Yao, Y.-S.; Liu, J.-L.; Xi, J.; Miu, B.; Liu, G.-S.; Wang, S.; *et al. Chem. Eur. J.* **2011**, *17*, 10462; (f) Zhao, L.; Xiong, F.-J.; Chen, W.-X.; Chen, F.-E. *Synthesis* **2011**, 4045; (g) Liu, G.-S.; Dong, Q.-L.; Yao, Y.-S.; Yao, Z.-J. *Org. Lett.* **2008**, *10*, 5393; (h) Zhang, L.-P.; Bao, Y.; Kuang, Y.-Y.; Chen, F.-E. *Helv. Chim. Acta* **2008**, *91*, 2057; (i) Zhou, H.-B.; Liu, G.-S.; Yao, Z.-J. *Org. Lett.* **2007**, *9*, 2003.

39 Comins, D. L.; Nolan, J. M. *Org. Lett.* **2001**, *3*, 4255.

40 Curran, D. P. *Compt. Rend. Chim.* **2008**, *11*, 1574.

41 Lee, K.; Lanier, M. L.; Kwak, J.-H.; Kim, H.; Hong, J. *Nat. Prod. Rep.* **2016**, *33*, 1393.

42 Waldeck, A. R.; Krische, M. *J. Angew. Chem. Int. Ed.* **2013**, *52*, 4470.

43 (a) Dechert-Schmitt, A.-M. R.; Schmitt, D. C.; Gao, X.; Itoh, T.; Krische, M. *J. Nat. Prod. Rep.* **2014**, *31*, 504; (b) Lu, Y.; Kim, I. S.; Hassan, A.; Del Valle, D. J.; Krische, M. J. *Angew. Chem. Int. Ed.* **2009**, *48*, 5018; (c) Gao, X.; Han, H.; Krische, M. J. *J. Am. Chem. Soc.* **2011**, *133*, 12795.

44 Fuwa, H.; Noto, K.; Sasaki, M. *Org. Lett.* **2010**, *12*, 1636.

45 Steinmann, P.; Keiser, J.; Bos, R.; Tanner, M.; Utzinger, J. *Lancet Infect. Dis.* **2006**, *6*, 411.

46 (a) Baran, P. S.; Maimone, T. J.; Richter, J. M. *Nature* **2007**, *446*, 404; (b) Maimone, T. J.; Ishihara, Y.; Baran, P. S. *Tetrahedron* **2015**, *71*, 3652; (c) Baran, P. S.; Richter, J. M. *J. Am. Chem. Soc.* **2004**, *126*, 7450; (d) Baran, P. S.; Richter, J. M. *J. Am. Chem. Soc.* **2005**, *127*, 15394.; (e) Richter, J. M.; Ishihara, Y.; Masuda, T.; Whitefield, B. W.; Llamas, T.; Pohjakallio, A.; Baran, P. S. *J. Am. Chem. Soc.* **2008**, *130*, 17938.

47 (a) Coulthard, G.; Erb, W.; Aggarwal, V. K. *Nature* **2012**, *489*, 278; (b) Prévost, S.; Thai, K.; Schützenmeister, N.; Coulthard, G.; Erb, W.; Aggarwal, V. K. *Org. Lett.* **2015**, *17*, 504.

48 EPSRC Research Performance and Economic Impact Report 2012–2013. Available at: www.epsrc.ac.uk/newsevents/pubs/research-performance-and-economic-impact-report-2012-2013/ (accessed February 2018), **2013**.

49 Corey, E. J.; Winshenker, N. M.; Schaaf, T. K.; Huber, W. *J. Am. Chem. Soc.* **1969**, *91*, 5675.

50 Nair, S. K.; Henegar, K. E. In *Modern Drug Synthesis*; Li, J. J., Johnson, D. S. (eds); Hoboken, NJ: John Wiley & Sons, Inc., **2010**, Ch. 21, pp. 329–338.

51 Carreras, J.; Livendahl, M.; McGonigal, P. R.; Echavarren, A. M. *Angew. Chem. Int. Ed.* **2014**, *53*, 4896.

52 Jiménez-NuÇez, E.; Raducan, M.; Lauterbach, T.; Molawi, K.; Solorio, C. R.; Echavarren, A. M. *Angew. Chem. Int. Ed.* **2009**, *48*, 6152.

53 Piemontesi, C.; Wang, Q.; Zhu, J. *J. Am. Chem. Soc.* **2016**, *138*, 11148.

54 Banfi, L.; Riva, R. *Org. React.* **2005**, *65*, 1.

55 (a) Zhu, J. *Eur. J. Org. Chem.* **2003**, 1133; (b) Dömling, A. *Chem. Rev.* **2006**, *106*, 17.

56 (a) Wittenberg, R.; Srogl, J.; Egi, M.; Liebeskind, L. S. *Org. Lett.* **2003**, *5*, 3033; (b) Prokopcová, H.; Kappe, C. O. *Angew. Chem. Int. Ed.* **2009**, *48*, 2276.

57 (a) Taylor, M. S.; Jacobsen, E. N. *J. Am. Chem. Soc.* **2004**, *126*, 10558. (b) Raheem, I. T.; Thiara, P. S.; Peterson, E. A.; Jacobsen, E. N. *J. Am. Chem. Soc.* **2007**, *129*, 13404.

58 Reyes, E.; Uria, U.; Carrillo, L.; Vicario, J. L. *Tetrahedron* **2014**, *70*, 9461.

59 (a) Ma, X.; Gang, D. R. *Nat. Prod. Rep.* **2004**, *21*, 752; (b) Kobayashi, J.; Morita, H. In *The Alkaloids: Chemistry and Biology*, Cordell, G. A. (ed.); New York: Academic Press, **2005**; vol. 61, Chapter 1, pp 1–57; (c) Siengalewicz, P.; Mulzer, J.; Rinner, U., In *The Alkaloids: Chemistry and Biology*, Knölker, H.-J. Ed.; New York: Academic Press, **2013**, vol. 72, Ch. 1, pp. 1–151.

60 Zangara, A. *Pharmacol. Biochem. Behav.* **2003**, *75*, 675.

61 (a) Saha, M.; Carter, R. G. *Synlett* **2017**, *28*, 2212; (b) Zhong, L.-R.; Yao, Z.-J. *Sci. China: Chem.* **2016**, *59*, 1079; (c) Murphy, R. A.; Sarpong, R. *Chem.–Eur. J.* **2014**, *20*, 42; (d) Wang, X.; Li, H.; Lei, X. *Synlett* **2013**, *1032*. (e) Nakayama, A.; Kitajima, M.; Takayama, H. *Synlett* **2012**, *23*, 2014.

62 Samame, R. A.; Owens, C. M.; Rychnovsky, S. D. *Chem. Sci.* **2016**, *7*, 188.

63 (a) Nicolaou, K. C.; Zhong, Y. L.; Baran, P. S. *J. Am. Chem. Soc.* **2000**, *122*, 7596; (b) Yamamoto, K.; Hentemann, M. F.; Allen, J. G.; Danishefsky, S. J. *Chem.–Eur. J.* **2003**, *9*, 3242; (c) Caine, D.; Procter, K.; Cassell, R. A. *J. Org. Chem.* **1984**, *49*, 2647.

64 Miyaura, N.; Ishiyama, T.; Sasaki, H.; Ishikawa, M.; Satoh, M.; Suzuki, M. *J. Am. Chem. Soc.* **1989**, *111*, 314.

65 (a) Müller, D. S.; Untiedt, N. L.; Dieskau, A. P.; Lackner, G. L.; Overman, L. E. *J. Am. Chem. Soc.* **2015**, *137*, 660. Correction: *J. Am. Chem. Soc.*, **2015**, *137*, 4874. (b) Slutskyy, Y.; Jamison, C. R.; Lackner, G. L.; Müller, D. S.; Dieskau, A. P.; Untiedt, N. L.; *et al. J. Org. Chem.* **2016**, *81*, 7029; (c) Jamison, C. R.; Overman, L. E. *Acc. Chem. Res.* **2016**, *49*, 1578.

66 (a) Ling, T.; Rivas, F. *Tetrahedron* **2016**, *72*, 6729; (b) Tian, L.; Luo, Y.-C.; Hu, X.-Q.; Xu, P.-F. *Asian J. Org. Chem.* **2016**, *5*, 580; (c) Büschleb, M.; Dorich, S.; Hanessian, S.; Tao, D.; Schenthal, K. B.; Overman, L. E. *Angew. Chem. Int. Ed.* **2016**, *55*, 4156; (d) Long, R.; Huang, J.; Gong, J.-X.; Yang, Z. *Nat. Prod. Rep.* **2015**, *32*, 1584; (e) Liu, Y.-Y.; Han, S.-J.; Liu, W.-B.; Stoltz, B. M. *Acc. Chem. Res.* **2015**, *48*, 740; (f) Quasdorf, K. W.; Overman, L. E. *Nature* **2014**, *516*, 181; (g) Hong, A. Y.; Stoltz, B. M. *Eur. J. Org. Chem.* **2013**, 2745.

67 Gao, F.; Hoveyda, A. H. *J. Am. Chem. Soc.* **2010**, *132*, 10961.

68 May, T. L.; Dabrowski, J. A.; Hoveyda, A. H. *J. Am. Chem. Soc.* **2011**, *133*, 736.

69 (a) Piers, E.; Wai, J. S. M. *Can. J. Chem.* **1994**, *72*, 146; (b) Piers, E.; Yeung, B. W. A. *J. Org. Chem.* **1984**, *49*, 4567.

70 Nawrat, C. C.; Jamison, C. R.; Slutskyy, Y.; MacMillan, D. W. C.; Overman, L. E. *J. Am. Chem. Soc.* **2015**, *137*, 11270.

71 (a) Ratnayake, R.; Covell, D. G.; Ransom, T. T.; Gustafson, K. R.; Beutler, J. A. *Org. Lett.* **2009**, *11*, 57; (b) Wu, Z.-H.; Zhao, S.-Z.; Fash, D. M.; Li, Z.-W.; Chain, W. J.; Beutler, J. A. *J. Nat. Prod.* **2017**, *80*, 771.

72 (a) Nelson, R.; Gulias, M.; Mascarenas, J. L.; Lopez, F. *Angew. Chem. Int. Ed.* **2016**, *55*, 14359; (b) Hanari, T.; Shimada, N.; Kurosaki, Y.;

Thrimurtulu, N.; Nambu, H.; Anada, M.; Hashimoto, S. *Chem. - Eur. J.* **2015**, *21*, 11671; (c) Zhang, J.; Zheng, S.; Peng, W.; Shen, Z. *Tetrahedron Lett.* **2014**, *55*, 1339; (d) Zahel, M.; Kessberg, A.; Metz, P. *Angew. Chem., Int. Ed.* **2013**, *52*, 5390; (e) Wang, J.; Chen, S. G.; Sun, B. F.; Lin, G. Q.; Shang, Y. J. *Chem. - Eur. J.* **2013**, *19*, 2539; (f) Radtke, L.; Willot, M.; Sun, H.; Ziegler, S.; Sauerland, S.; Strohmann, C.; *et al. Angew. Chem. Int. Ed.* **2011**, *50*, 3998; (g) Nicolaou, K. C.; Kang, Q.; Ng, S. Y.; Chen, D. Y.-K. *J. Am. Chem. Soc.* **2010**, *132*, 8219; (h) Zhou, Q.; Chen, X.; Ma, D. *Angew. Chem. Int. Ed.* **2010**, *49*, 3513; (i) Molawi, K.; Delpont, N.; Echavarren, A. M. *Angew. Chem. Int. Ed.* **2010**, *49*, 3517; (j) Willot, M.; Radtke, L.; Könning, D.; Fröhlich, R.; Gessner, V. H.; Strohmann, C.; Christmann, M. *Angew. Chem. Int. Ed.* **2009**, *48*, 9105.

73 (a) Li, Z.; Nakashige, M.; Chain, W. J. *J. Am. Chem. Soc.* **2011**, *133*, 6553; (b) Chain, W. J. *Synlett* **2011**, 2605.

74 (a) Corey, E. J., Cheng, X.-M., Eds.; *The Logic of Chemical Synthesis*; John Wiley & Sons, Inc.: New York, **1989**; (b) Hill, N.; Paruch, K.; Svenda, J. *Tetrahedron* **2016**, *72*, 3345.

75 Kagan, H. B. *Tetrahedron* **2003**, *59*, 10351.

76 Winkler, J. D.; Oh, K.; Asselin, S. M. *Org. Lett.* **2005**, *7*, 387.

77 Chavez, D. E.; Jacobsen, E. N. *Org. Lett.* **2003**, *5*, 2563.

78 Takano, S.; Inomata, K.; Samizu, K.; Tomita, S.; Yanase, M.; Suzuki, M.; *et al. Chem. Lett.* **1989**, *1283*.

79 (a) Kubo, M.; Okada, C.; Huang, J.-M.; Harada, K.; Hioki, H.; Fukuyama, Y. *Org. Lett.* **2009**, *11*, 5190; (b) Shenvi, R. A. *Nat. Prod. Rep.* **2016**, *33*, 535.

80 (a) Li, L.-B.; Shen, Y.; Zhang, Y.-D. Chinese *J. Org. Chem.* **2016**, *36*, 439; (b) Cheng, X.-Y.; Micalizio, G. C. *J. Am. Chem. Soc.* **2016**, *138*, 1150; (c) Shen, Y.; Li, L.; Pan, Z.; Wang, Y.; Li, J.; Wang, K.; Wang, X.; *et al. Org. Lett.* **2015**, *17*, 5480; (d) Paterson, I.; Xuan, M.; Dalby, S. M. *Angew. Chem. Int. Ed.* **2014**, *53*, 7286. (e)Siler, D. A.; Mighion, J. D.; Sorensen, E. J. *Angew. Chem. Int. Ed.* **2014**, *53*, 5332. (f) Xu, J.; Trzoss, L.; Chang, W. K.; Theodorakis, E. A. *Angew. Chem. Int. Ed.* **2011**, *50*, 3672.

81 Lu, H.-H.; Martinez, M. D.; Shenvi, R. A. *Nat. Chem.* **2015**, *7*, 604.

82 (a) Crowe, W. E.; Vu, A. T. *J. Am. Chem. Soc.* **1996**, *118*, 1557; (b) Kablaoui, N. M., Hicks, H. A.; Buchwald, S. L. *J. Am. Chem. Soc.* **1996**, *118*, 5818.

83 Sugano, M.; Shindo, T.; Sato, A.; Iijima, Y.; Oshima, T.; Kuwano, H.; Hata, T. *J. Org. Chem.* **1990**, *55*, 5803.

84 Soucy, P; L'Heureux, A; Toró, A.; Deslongchamps, P. *J. Org. Chem.* **2003**, *68*, 9983.

85 Zhao, Y.-M.; Maimone, T. J. *Angew. Chem. Int. Ed.* **2015**, *54*, 1223.

86 Tius, M. A. *Chem. Rev.* **1988**, *88*, 719.

87 Mayer, S.; List, B. *Angew. Chem. Int. Ed.* **2006**, *45*, 4193.

88 Fettes, A.; Carreira, E. M. *J. Org. Chem.* **2003**, *68*, 9274.

89 (a) Heravi, M. M.; Vavsari, V. F. *RSC Adv.* **2015**, *5*, 50890; (b) Takao, K.; Munakata, R.; Tadano, K. *Chem. Rev.* **2005**, *105*, 4779.

90 McCabe, S. R.; Wipf, P. *Angew. Chem. Int. Ed.* **2017**, *56*, 324.

91 (a) Matsuda, H.; Morikawa, T.; Oda, M.; Asao, Y.; Yoshikawa, M. *Bioorg. Med. Chem. Lett.* **2003**, *13*, 4445. (b) Matsuda, H.; Yoshida, K.; Miyagawa, K.; Nemoto, Y.; Asao, Y.; Yoshikawa, M. *Bioorg. Med. Chem. Lett.* **2006**, *16*, 1567. (c) Ozer, J.; Eisner, N.; Ostrozhenkova, E.; Bacher, A.; Eisenreich, W.; Benharroch, D.; Golan-Goldhirsh, A.; Gopas, J. *Cancer Biol. Ther.* **2009**, *8*, 1860.

92 (a) Jansen, D. J.; Shenvi, R. A. *J. Am. Chem. Soc.* **2013**, *135*, 1209; (b) Tada, N.; Jansen, D. J.; Mower, M. P.; Blewett, M. M.; Umotoy, J. C.; *et al. ACS Cent. Sci.* **2016**, *2*, 401.

93 Wong, C. F.; LaLonde, R. T. *Experientia* **1975**, *31*, 15.

94 (a) Polonovski, M.; Polonovski, M. *Bull. Soc. Chim. Fr.* **1927**, *1190*; (b) Cave, A.; Kan-Fan, C.; Potier, P.; Le Men, J. *Tetrahedron* **1967**, *23*, 4681.

95 Chae, J.; Yun, J.; Buchwald, S. L. *Org. Lett.* **2004**, *6*, 4809.

96 (a) Tsuji, J.; Takahashi, H.; Moriwaka, M. *Tetrahedron Lett.* **1965**, *6*, 4387; (b) Trost, B. M.; Fullerton, T. J. *J. Am. Chem. Soc.* **1973**, *95*, 292; (c) Lu, Z.; Ma, S. *Angew Chem. Int. Ed.* **2007**, *47*, 258.

97 Dowling, M. S.; Vanderwal, C. D. *J. Org. Chem.* **2010**, *75*, 6908.

98 Fowler, F. W.; Hwang, Y. C. *J. Org. Chem.* **1985**, *50*, 2719.

99 Abdel-Magid, A. F.; Mehrman, S. J. *Org. Proc. Res. Dev.* **2006**, *10*, 971.

100 (a) Pfeiffer, M. W. B.; Phillips, A. J. *J. Am. Chem. Soc.* **2005**, *127*, 5334; (b) Reddy, T. J.; Bordeau, G.; Trimble, L. *Org. Lett.* **2006**, *8*, 5585; (c) Pfeiffer, M. W. B.; Phillips, A. *J. Tetrahedron Lett.* **2008**, *49*, 6860.

101 (a) Enquist, J. A., Jr.; Stoltz, B. M. *Nature* **2008**, *453*, 1228; (b) Enquist, J. A., Jr.; Virgil, S. C.; Stoltz, *B. M. Chem. - Eur. J.* **2011**, *17*, 9957; (c) Kim, K. E.; Stoltz, B. M. *Org. Lett.* **2016**, *18*, 5720.

102 Bhat, V.; Welin, E. R.; Guo, X.-L.; Stoltz, B. M. *Chem. Rev.* **2017**, *117*, 4528.

103 Stewart, I. C.; Ung, T.; Pletnev, A. A.; Berlin, J. M.; Grubbs, R. H.; Schrodi, Y. *Org. Lett.* **2007**, *9*, 1589.

104 Kim, K. E.; Li, J.; Grubbs, R. H.; Stoltz, B. M. *J. Am. Chem. Soc.* **2016**, *138*, 13179.

105 Yoshikai, K.; Hayama, T.; Nishimura, K.; Yamada, K.; Tomioka, K. *J. Org. Chem.* **2005**, *70*, 681.

106 Brill, Z. G.; Grover, H. K.; Maimone, T. J. *Science* **2016**, *352*, 1078, correction: *Science* **2016**, *353*, 1507.

107 Charette, A. B.; Juteau, H.; Lebel, H.; Molinaro, C. *J. Am. Chem. Soc.* **1998**, *120*, 11943.

108 Corey, E. J.; Chaykovsky, M. *J. Am. Chem. Soc.* **1965**, *87*, 1353.

109 Horning B. D.; MacMillan, D. W. C. *J. Am. Chem. Soc.* **2013**, *135*, 6442.

110 Ayer, W. A.; Fukazawa, Y.; Singer, P. P. *Tetrahedron Lett.* **1973**, *14*, 5045.

111 Ramharter, J.; Weinstabl, H.; Mulzer, J. *J. Am. Chem. Soc.* **2010**, *132*, 14338.

112 (a) Carlone, A.; Marigo, M.; North, C.; Landa, A.; Jørgensen, K. A. *Chem. Commun.* **2006**, 4928; (b) Carpenter, R. D.; Fettinger, J. C.; Lam, K. S.; Kurth, M. J. *Angew. Chem. Int. Ed.* **2008**, *47*, 6407.

113 Comins, D. L.; Dehghani, A. *Tetrahedron Lett.* **1992**, *33*, 6299.

114 (a) Louie, J.; Bielawski, C. W.; Grubbs, R. H. *J. Am. Chem. Soc.* **2001**, *123*, 11312; (b) Furstner, A.; Leitner, A. *Angew. Chem. Int. Ed.* **2003**, *42*, 308; (c) Wasilke, J.-C.; Obrey, S. J.; Baker, R. T.; Bazan, G. C. *Chem. Rev.* **2005**, *105*, 1001.

115 (a) Heathcock, C. H.; Smith, K. M.; Blumenkopf, T. A. *J. Am. Chem. Soc.* **1986**, *108*, 5022; (b) Heathcock, C. H.; Blumenkopf, T. A.; Smith, K. M. *J. Org. Chem.* **1989**, *54*, 1548.

116 Jang, J.-H.; Asami, Y.; Jang, J.-P.; Kim, S.-O.; Moon, D. O.; Shin, K.-S.; *et al. J. Am. Chem. Soc.* **2011**, *133*, 6865, correction: *J. Am. Chem. Soc.* **2012**, 134, 7194.

117 Deng, J.; Zhu, B.; Lu, Z.-Y.; Yu, H.-X.; Li, A. *J. Am. Chem. Soc.* **2012**, *134*, 920.

118 (a) Xu, J.; Caro-Diaz, E. J. E.; Trzoss, L.; Theodorakis, E. A. *J. Am. Chem. Soc.* **2012**, *134*, 5072; (b) Xu, J.; Caro-Diaz, E. J. E.; Lacoske, M. H.; Hung, C.-I.; Jamora, C.; Theodorakis, E. A. *Chem. Sci.* **2012**, *3*, 3378; (c) Caro-Diaz, E. J. E.; Aung, A.; Xu, J.; Varghese, S.; Theodorakis, E. A. *Org. Chem. Front.* **2014**, *1*, 135.

119 (a) Yin, J.; Wang, C.; Kong, L.; Cai, S.; Gao, S. *Angew. Chem., Int. Ed.* **2012**, *51*, 7786; (b) Yin, J.; Kong, L.; Wang, C.; Shi, Y.; Cai, S.; Gao, S. *Chem. - Eur. J.* **2013**, *19*, 13040; (c) Li, X.-J.; Zheng, Q.-F.; Yin, J.; Liu, W.; Gao, S. *Chem. Commun.* **2017**, *53*, 4695.

120 (a) Huang, J.; Fang, L.; Long, R.; Shi, L.-L.; Shen, H.-J.; Li, C.-C.; *et al. Org. Lett.* **2013**, *15*, 4018; (b) Huang, J.; Fang, L.; Gong, J.; Li, C.; Yang, Z. *Tetrahedron* **2015**, *71*, 3720; (c) Liu, C.-G.; Zeng, Z.-X.; Chen, R.-Z.; Jiang, X.-J.; Wang, Y.-L.; Zhang, Y.-D. *Org. Lett.* **2016**, *18*, 624.

121 (a) Olsen, E. K.; Hansen, E.; Moodie, L. W. K.; Isaksson, J.; Sepčić, K.; Cergolj, M.; *et al. Org. Biomol. Chem.* **2016**, *14*, 1629; (b) Williams, P.; Sorribas, A.; Howes, M.-J. R. *Nat. Prod. Rep.* **2011**, *28*, 48; (c) Houghton, P. J.; Ren, Y.-H.; Howes, M.-J. *Nat. Prod. Rep.* **2006**, *23*, 181.

122 (a) Kozikowski, A. P.; Tuckmantel, W. *Acc. Chem. Res.* **1999**, *32*, 641; (b) Xia, Y.; Kozikowski, A. P. *J. Am. Chem. Soc.* **1989**, *111*, 4116.

123 Zheng, S.-Y.; Yu, C.-H.; Shen, Z.-W. *Chin. J. Org. Chem.* **2013**, *33*, 2261.

124 (a) Ding, R.; Sun, B.-F.; Lin, G.-Q. *Org. Lett.* **2012**, *14*, 4446; (b) Ding, R.; Fu, J.-G.; Xu, G.-Q.; Sun, B.-F.; Lin, G.-Q. *J. Org. Chem.* **2014**, *79*, 240.

125 (a) Guram, A. S.; Rennels, R. A.; Buchwald, S. L. *Angew Chem. Int. Ed. Engl.* **1995**, *34*, 1348; (b)Louie, J.; Hartwig, J. F. *Tetrahedron Lett.* **1995**, *36*, 3609.

126 (a) Griffith, W. P.; Ley, S. V.; Whitcombe, G. P.; White, A. D. *J. Chem. Soc., Chem. Commun.* **1987**, *1625*; (b) Ley, S. V.; Norman, J.; Griffith, W. P.; Marsden, S. P. *Synthesis* **1994**, *639*.

127 (a) Tufariello, J. J.; Meckler, H.; Winzenberg, K. *J. Org. Chem.* **1986**, *51*, 3556; (b) Blakemore, P. R.; Schulze, V. K.; White, J. D. *Chem. Commun.* **2000**, 1263;

(c) Blakemore, P. R.; Kim, S. K.; Schulze, V. K.; White, J. D.; Yokochi, A. F. T. *J. Chem. Soc., Perkin Trans. 1* **2001**, 1831; (d) Miller, K. E.; Wright, A. J.; Olesen, M. K.; Hovey, M. T.; Scheerer, J. R. *J. Org. Chem.* **2015**, *80*, 1569; (e) Ye, J.-L.; Liu, Y.; Yang, Z.-P.; Huang, P.-Q. *Chem. Commun.* **2016**, *52*, 561 (*N*-acetyl norloline).

128 Cakmak, M.; Mayer, P.; Trauner, D. *Nat. Chem.* **2011**, *3*, 543.

129 Smith, D. B.; Zhaoyin, W.; Schreiber, S. L. *Tetrahedron* **1990**, *46*, 4793.

130 Zhang, W.-H.; Luo, S.-J.; Fang, F.; Chen, Q.-S.; Hu, H.-W.; Jia, X.-S.; *et al.* *J. Am. Chem. Soc.* **2005**, *127*, 18.

131 Vokac, K.; Samek, Z.; Herout, V.; Sorm, F. *Tetrahedron Lett.* **1968**, *3855*.

132 (a) Barton, D. H. R.; De Mayo, P.; Shafiq, M. *J. Chem. Soc.* **1957**, *929*; (b) Blay, G.; Cardona, L.; Garcia, B.; Lahoz, L.; Pedro, J. R. *J. Org. Chem.* **2001**, *66*, 7700.

133 α-Santonin is commercially available from Aldrich Company ($17.40/5 g; $28.60/10 g) according to a 2003–2004 product catalog.

134 Cernijenko, A.; Risgaard, R.; Baran, P. S. *J. Am. Chem. Soc.* **2016**, *138*, 9425.

135 Li, S.-H.; Wang, J.; Niu, X.-M.; Shen, Y.-H.; Zhang, H.-J.; Sun, H.-D.; *et al.* *Org. Lett.* **2004**, *6*, 4327.

136 (a) Gong, J.; Lin, G.; Sun, W.; Li, C.-C.; Yang, Z. *J. Am. Chem. Soc.* **2010**, *132*, 16745; (b) Peng, F.; Danishefsky, S. J. *J. Am. Chem. Soc.* **2012**, *134*, 18860; (c) Lu, P.; Gu, Z.; Zakarian, A. *J. Am. Chem. Soc.* **2013**, *135*, 14552; (d) Lu, P.; Mailyan, A.; Gu, Z.; Guptill, D. M.; Wang, H.; Davies, H. M. L.; Zakarian, A. *J. Am. Chem. Soc.* **2014**, *136*, 17738; (e) Zheng, C.; Dubovyk, I.; Lazarski, K. E.; Thomson, R. J. *J. Am. Chem. Soc.* **2014**, *136*, 17750; (f) Zhang, W.-B.; Shao, W.-B.; Li, F.-Z.; Gong, J.-X.; Yang, Z. *Chem. - Asian J.* **2015**, *10*, 1874.

137 Han, Q.-B.; Cheung, S.; Tai, J.; Qiao, C.-F.; Song, J.-Z.; Tso, T.-F.; *et al.* *Org. Lett.* **2006**, *8*, 4727.

138 Robert, T.; Velder, J.; Schmalz, H.-G. *Angew. Chem., Int. Ed.* **2008**, *47*, 7718.

139 Hosomi, A.; Sakurai, H. *Tetrahedron Lett.* **1976**, *17*, 1295.

140 Baldwin, J. E.; Lusch, M. J. *Tetrahedron* **1982**, *38*, 2939.

141 Hecker, E. *Cancer Res.* **1968**, *28*, 2338.

142 Vasas, A.; Rédei, D.; Csupor, D.; Molnár, J.; Hohmann, J. *Eur. J. Org. Chem.* **2012**, 5115.

143 Alder, R. W.; East, S. P. *Chem. Rev.* **1996**, *96*, 2097.

144 (a) Kim, S.; Winkler, J. D. *Chem. Soc. Rev.* **1997**, *26*, 387; (b) Kuwajima, I.; Tanino, K. *Chem. Rev.* **2005**, *105*, 4661; (c) Cha, J. K.; Epstein, O. L. *Tetrahedron* **2006**, *62*, 1329.

145 (a) Winkler, J. D.; Rouse, M. B.; Greaney, M. F.; Harrison, S. J.; Jeon, Y. T. *J. Am. Chem. Soc.* **2002**, *124*, 9726; (b) Tanino, K.; Onuki, K.; Asano, K.; Miyashita, M.; Nakamura, T.; Takahashi, Y.; Kuwajima, I. *J. Am. Chem. Soc.* **2003**, *125*, 1498; (c) Nickel, A.; Maruyama, T.; Tang, H.; Murphy, P. D.; Greene, B.; Yusuff, N.; Wood, J. L. *J. Am. Chem. Soc.* **2004**, *126*, 16300.

146 (a) Jørgensen, L.; McKerrall, S. J.; Kuttruff, C. A.; Ungeheuer, F.; Felding, J.; Baran, P. S. *Science* **2013**, *341*, 878, correction: *Science* **2016**, *351*, 133; (b) McKerrall, S. J.; Jørgensen, L.; Kuttruff, C. A.; Ungeheuer, F.; Baran, P. S. *J. Am. Chem. Soc.* **2014**, *136*, 5799, correction: *J. Am. Chem. Soc.* **2015**, *137*, 14545.
147 (a) Hecker, E. *Pure Appl. Chem.* **1977**, *49*, 1423; (b) Adolf, W.; Hecker, E. *Isr. J. Chem.* **1977**, *16*, 75; (c) Schmidt, R. J. *Bot. J. Linn. Soc.* **1987**, *94*, 221.
148 (a) Ishihara, Y.; Baran, P. S. *Synlett* **2010**, *1733*; (b) Chen, K.; Baran, P. S. *Nature* **2009**, *459*, 824; (c) Mendoza, A.; Ishihara, Y.; Baran, P. S. *Nat. Chem.* **2012**, *4*, 21; (d) Cherney, E. C.; Green, J. C.; Baran, P. S. *Angew. Chem. Int. Ed.* **2013**, *52*, 9019.
149 Brummond, K. M.; Chen, H.-F.; Fisher, K. D.; Kerekes, A. D.; Rickards, B.; Sill, P. C.; *et al. Org. Lett.* **2002**, *4*, 1931.
150 Shibuya, K. *Synth. Commun.* **1994**, *24*, 2923.
151 (a) Wiesner, K.; Valenta, Z.; Findlay, J. A. *Tetrahedron Lett.* **1967**, *8*, 221; (b) Srivastava, S. N.; Przybylska, M. *Can. J. Chem.* **1968**, *46*, 795.
152 Sutko, J. L.; Airey, J. A.; Welch, W.; Ruest, L. *Pharmacol. Rev.* **1997**, *49*, 53.
153 (a) Bélanger, A.; Berney, D. J. F.; Borschberg, H. J.; *et al. Can. J. Chem.* **1979**, *57*, 3348; (b) Deslongchamps, P.; Bélanger, A.; Berney, D. J. F.; *et al. Can. J. Chem.* **1990**, *68*, 115; (c) Deslongchamps, P.; Bélanger, A.; Berney, D. J. F.; *et al. Can. J. Chem.* **1990**, *68*, 127; (d) Deslongchamps, P.; Bélanger, A.; Berney, D. J. F.; *et al. Can. J. Chem.* **1990**, *68*, 153; (e) Deslongchamps, P.; Bélanger, A.; Berney, D. J. F.; *et al. Can. J. Chem.* **1990**, *68*, 186.
154 (a) Nagatomo, M.; Koshimizu, M.; Masuda, K.; Tabuchi, T.; Urabe, D.; Inoue, M. *J. Am. Chem. Soc.* **2014**, *136*, 5916; (b) Nagatomo, M.; Hagiwara, K.; Masuda, K.; Koshimizu, M.; Kawamata, T.; Matsui, Y.; *et al. Chem.–Eur. J.* **2016**, *22*, 222; (c) Masuda, K.; Koshimizu, M.; Nagatomo, M.; Inoue, M., *Chem.–Eur. J.* **2016**, *22*, 230; (d) Koshimizu, M.; Nagatomo, M.; Inoue, M. *Angew. Chem. Int. Ed.* **2016**, *55*, 2493; (e) Masuda, K.; Nagatomo, M.; Inoue, M. *Chem. Pharm. Bull.* **2016**, *64*, 874; (f) Inoue, M. *Acc. Chem. Res.* **2017**, *50*, 460.
155 (a) Chuang, K. V.; Xu, C.; Reisman, S. E. *Science* **2016**, *353*, 912; (b) Xu, C.; Han, A.; Virgil, S. C.; Reisman, S. E. *ACS Cent. Sci.* **2017**, *3*, 278.
156 (a) Khand, I. U.; Knox, G. R.; Pauson, P. L.; Watts, W. E.; Foreman, M. I. *J. Chem. Soc. Perkin Trans. 1* **1973**, 977; (b) Boñaga, L. V. R.; Krafft, M. E. *Tetrahedron* **2004**, *60*, 9795; (c) Torres, R. R. The *Pauson–Khand Reaction: Scope, Variations and Applications*. Hoboken, NJ: John Wiley & Sons, Inc., **2012**.
157 Lu, W.; Roongsawang, N.; Mahmud, T. *Chem. Biol.* **2011**, *18*, 425.
158 Hanessian, S.; Vakiti, R. R.; Dorich, S.; Banerjee, S.; Lecomte, F.; Del Valle, J. R.; *et al. Angew. Chem. Int. Ed.* **2011**, *50*, 3497.
159 Malinowski, J. T.; Sharpe, R. J.; Johnson, J. S. *Science* **2013**, *340*, 180.
160 Lou, S.; Taoka, B. M.; Ting, A.; Schaus, S. E. *J. Am. Chem. Soc.* **2005**, *127*, 11256.
161 Shen, R.; Lin, C. T.; Bowman, E. J.; Bowman, B. J.; Porco, J. A., Jr. *Org. Lett.* **2002**, *4*, 3103.

3

Diels–Alder Cascades in Natural Product Total Synthesis

Richard P. Hsung[1], Zhi-Xiong Ma[1], Lichao Fang[2], and John B. Feltenberger[3]

[1] *Division of Pharmaceutical Sciences, School of Pharmacy, University of Wisconsin-Madison, Madison, WI, USA*
[2] *China Novartis Institutes for BioMedical Research, Shanghai, China*
[3] *Medicinal Chemistry Center, School of Pharmacy, University of Wisconsin-Madison, Madison, WI, USA*

3.1 Introduction

Natural product total synthesis continues to be one of the most challenging endeavors for synthetic chemists. Innovative approaches to complex natural products featuring design efficiency and practicality are particularly coveted, and heavily pursued. Two powerful concepts for rapidly accessing molecular complexity, the Diels–Alder reaction and cascade sequences, have long been applied in total synthesis work. Both reaction manifolds in isolation have been extensively reviewed in the literature,[1,2] yet a unified account encompassing all variety of Diels–Alder cascades remains unreported.[3] This chapter will present here a contemporary review of the Diels–Alder cascades as a transformative concept in natural product syntheses as reported over the last 20 years.

First discovered over 80 years ago,[4] the Diels–Alder reaction is the prototypical cycloaddition reaction, uniting a diene with a dienophile to form a cyclohexene ring, generally with high regio- and diastereoselectivity.[1] Key pioneering efforts in the early 1950s lead by Stork, Gates, and Tschudi, as well as Woodward, established the Diels–Alder reaction as a pre-eminent strategy for construction of complex frameworks in natural products.[5] Later advances in catalysis,[6] as well as asymmetric variations,[7] have led to significant improvements in the Diels–Alder reaction as applied to total synthesis efforts.

Cascade reactions also enjoy a rich history in total synthesis, owing to the tremendous potential to rapidly build complex scaffolds that comes from combining several synthetic steps into a single operation. In the nearly 100 years

Efficiency in Natural Product Total Synthesis, First Edition. Edited by Pei-Qiang Huang, Zhu-Jun Yao, and Richard P. Hsung.

since Robinson's one-pot synthesis of tropinone in 1917,[8] chemists have pursued a wide array of cascade reaction strategies in natural product synthesis. Many of the benefits of a well-designed cascade are just as important today as in the nascent years of total synthesis, including superior step and atom economies, time and resource efficiency, and reduced labor and waste streams.[9] Indeed, many of these attributes associated with cascade reactions, while providing the chemist with the practical benefits from reduced reaction time, effort, and material costs, can also be classified as "green chemistry" due to the reduction in waste produced.[10]

The cascade reaction is nebulously defined due to the lack of a singular reaction framework for bond making (or breaking). Even the terminology in this field is somewhat unsettled, with the terms "cascade," "domino," "tandem," "consecutive," "serial," "multistep," and others representing generally the same concept.[2,11] For our subjective purposes, we will use these terms interchangeably. We consider cascade reactions to be any process in which two or more bond-forming transformations take place under the same reaction conditions, and which the subsequent reactions result as a consequence of the functionality formed in the previous step. Hence, two or more isolated reactions occurring on the same substrate, also known as multidirectional reactions, have been excluded from consideration in this chapter.[12]

Elegant examples of Diels–Alder cascades featured as the key step in total syntheses reported over the last two decades are presented here. While there may be some overlap of reaction categories, we have made an attempt to organize the chapter into several sections based on general cascade themes. In Section 3.2, we will describe cascades involving the coupling of the diene to the dienophile, followed by Diels–Alder cycloaddition. Presented in Section 3.3 are examples where the diene or dienophile are generated *in situ* and subsequently undergo Diels–Alder cycloaddition. Section 3.4 highlights examples of rearrangement-initiated Diels–Alder cascades, and likewise Section 3.5 showcases the cyclization-initiated Diels–Alder cascades. Finally, in Section 3.6 are examples of cascades initiated by Diels–Alder reaction, followed by additional transformations.

Given the breadth of fantastic chemistry over this time period, we have decided against an exhaustive review of Diels–Alder cascades in favor of a more focused book chapter. First, we have decided to focus exclusively on Diels–Alder cycloadditions yielding carbocycles, which naturally excludes examples of hetero-Diels–Alder cascades, although there have been a number of excellent syntheses reported.[13] Likewise, we have excluded the Diels–Alder/aromatization cascade sequence;[14] although this represents an excellent approach to complex aromatic frameworks, we felt that the overall loss of stereochemical information initially built through the cycloaddition reaction via the ensuing aromatization was not appropriate in the context of this review. And lastly, while ultimately validating the significance of these tandem

Diels–Alder approaches to natural product assembly, we have excluded examples utilizing the same cascade strategy to similar natural products in an effort to avoid redundancy.

3.2 Cascades Initiated by Coupling of a Pre-Formed Diene and Dienophile

In 2010, Cho disclosed a total synthesis of (±)-galanthamine **5** employing a tandem Stille coupling/Diels–Alder reaction strategy.[15] In the synthesis shown in Scheme 3.1, a C3-selective Stille coupling of **1** and **2** was performed to afford intermediate **3** that would undergo intramolecular Diels–Alder reaction under the thermal reaction conditions to give a mixture of products. The best result (45% yield, *endo*-**4**: *exo*-**4** = 2.5:1) was obtained with a combination of $Pd(PPh_3)_4$/CuI/DMF. The *endo*-**4** product was finally converted to the target natural product in several additional steps.

In 1999, Overman and coworkers completed the first enantioselective total syntheses of (+)-aloperine **16a**, (+)-*N*-methylaloperine **16b**, and (+)-*N*-allylaloperine **16c** utilizing an *in situ* generated silyl-tethered intramolecular Diels–Alder cycloaddition cascade strategy (Scheme 3.2).[16] Chiral diene **8** was assembled via an efficient Heck coupling of triflate **7** and (*R*)-2-vinyl piperidine **6**. At this stage, the authors noted that care was required to unveil the free piperidine. Cleavage of **8** using TMSI/2,6-lutidine and methanol afforded the free secondary amine **9**, without erosion of enantiopurity. Treatment of **9** with silyl triflate **10a** at a low temperature provided a 5:1 mixture of tetracyclic Diels–Alder products **12** and **13** directly, which were partially characterized by ^{1}H NMR, but ultimately unstable toward aqueous workup. It was noted that the intramolecularity of the Diels–Alder reaction was critical to the success of the

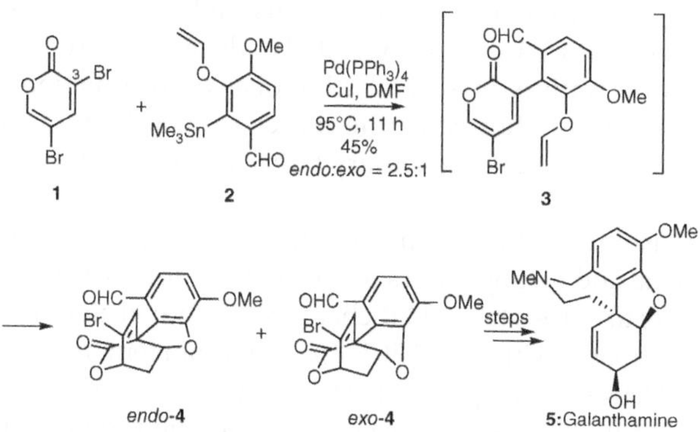

Scheme 3.1 Cho's total synthesis of (±)-galanthamine via a Stille coupling/IMDA cascade.

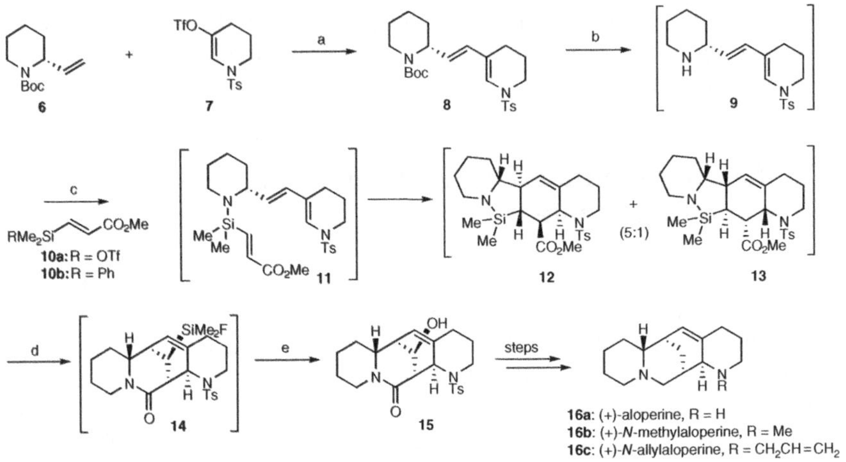

Reagents and conditions: (a) Pd(OAc)$_2$ (0.14 equiv), K_2CO_3, *n*-Bu$_4$NOTf, DMF, H_2O, 55°C, 83%; (b) TMSI (2.5 equiv), 2,6-lutidine (4.0 equiv), CH_2Cl_2, rt; then MeOH; NaOH; (c) **10a** (1.9 equiv), Et_3N (3.75 equiv), CH_2Cl_2, 0°C to rt, 3.5 h; (d) Pyridine·HF, rt, concentrate, then mesitylene, reflux, concentrate; (e) H_2O_2 (15 equiv), KF (6.5 equiv), $KHCO_3$ (2.8 equiv), MeOH/THF (1:1), reflux 1.5 h, 63% over 4 steps.

Scheme 3.2 Overman's total synthesis of (+)-aloperine featuring an *in situ* silyl-tethered IMDA cascade.

cycloaddition, given that the intermolecular cycloaddition of **9** with vinylsilane **10b** failed to afford cycloaddition products at temperatures as high as 165 °C.

The crude products **12** and **13** were treated directly with anhydrous HF•pyridine, which cleaved the N–Si bond while also placing a fluoride substituent on silicon, which served to activate the C–Si bond toward oxidative cleavage. After a solvent change to mesitylene, the crude mixture was heated to 165 °C to induce the intramolecular lactamization, affording tetracycle **14**. After another solvent exchange, silane **14** was subjected to Tamao–Fleming oxidation, affording the desired tetracyclic alcohol **15** at a 63% overall yield as the first isolated and fully characterized product in this multistep, one-pot sequence. The tetracyclic alcohol **15** was then converted to the natural product (+)-aloperine **16a** and its derivatives in several additional transformations. This concise total synthesis effort showcases the power of *N-Si* bond as a readily introduced and easily removable temporary tether for intramolecular Diels–Alder reactions.

The Nozaki–Hiyama–Kishi (NHK) reaction was employed by Uenishi and coworkers to unite the diene and dienophile fragments in their total synthesis of (−)-ircinianin **22** *en route* to (+)-wistarin **23** (Scheme 3.3).[17] When aldehyde **18** and vinyliodide **19** were treated with 0.1% $NiCl_2$-$CrCl_2$ at room temperature and reacted over 18 h, the NHK reaction proceeded to afford the coupled product **20β** with β-OH isolated in 20% yield, as well as the tricyclic product **21** produced from intramolecular Diels–Alder reaction in 60% yield as a single

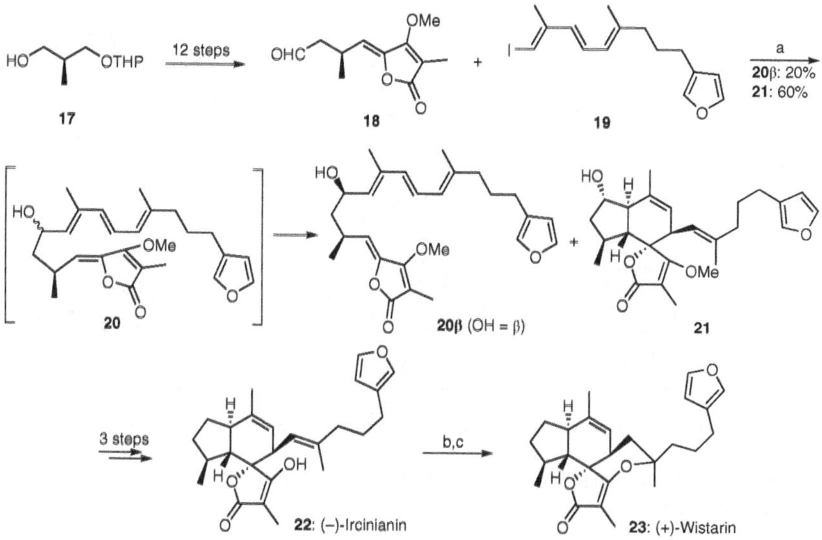

Reagents and conditions: (a): 0.1% $NiCl_2$-$CrCl_2$, DMSO, rt, **20**β: 20% and **21**: 60%; (b): I_2, K_2CO_3, $CHCl_3$, 0°C, 5 min, 71% yield; (c) Bu_3SnH, cat. Et_3B, benzene 7°C, 77%.

Scheme 3.3 Uenishi's total syntheses of (–)-ircinianin and (+)-wistarin featuring the NHK reaction/IMDA cascade.

cyclization product among four possible stereoisomers. Notably, this was the first successful case of a $NiCl_2$–$CrCl_2$-promoted coupling reaction between an iodotriene and an aldehyde. When the reaction was stopped after 1 h, it was observed that aldehyde **18** was completely consumed, and a mixture of tricycle **21**, **20β**, and **20α** (α-OH) in a total yield of 89% at a 1:2:2 ratio, indicating the diastereoselectivity of the NHK reaction to be 3:2. Interestingly, a Diels–Alder reaction of **20α** did not occur spontaneously at room temperature in the absence of $CrCl_2$, indicating its role as a Lewis acid catalyst for the cycloaddition. Tricycle **21** was further elaborated in three steps to afford (–)-ircinianin **22**, which was then converted to (+)-wistarin **23** by iodination/cyclization and further hydro-deiodination under radical conditions.

3.3 Simple Transformations to Diene/Dienophiles Followed by the Diels–Alder Cascade

In 1994, Martin Kuehne and coworkers described the synthesis of the *strychnos* family member (±)-echitamidine **30** utilizing a tandem retro-Michael/condensation/intramolecular Diels–Alder reaction cascade.[18] As shown in Scheme 3.4, treatment of azepine **24** with aldehyde **25** refluxing in toluene

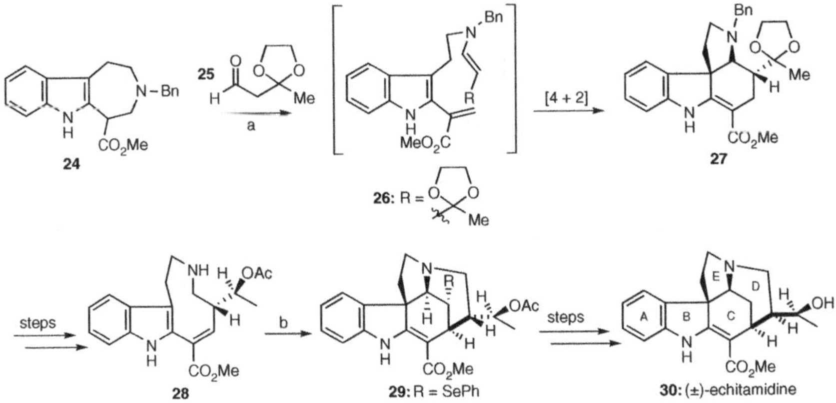

Reagents and conditions: (a) **25**, toluene, 4Å MS, reflux 36 h, 92%; (b) $CHOCH_2SePh$, toluene, reflux, 81%.

Scheme 3.4 Kuehne's total synthesis of (±)-echitamidine via tandem retro-Michael/condensation/IMDA.

afforded 92% of the IMDA product **27**, by way of enamine **26**. Reductive cleavage of the C/E rings could not be achieved in the presence of the ketal, so **27** was hydrolyzed and after several additional steps, indoloazonine **28** was formed. Treatment of **28** with phenylselenylacetaldehyde afforded another intramolecular Diels–Alder adduct **29**, which was elaborated to the natural product (±)-echitamidine **30**.

Martin and coworkers employed a novel domino Stille coupling/Diels–Alder reaction strategy to rapidly assemble the ABC tricyclic core en route to the enantioselective synthesis of Ircinal A **37** and related manzamine alkaloids (Scheme 3.5).[19] The dienophile precursor **34** was assembled in four steps from secondary amine **33** and chiral *N*-boc pyrrolidinone **31**. Treatment of **34** with vinyl tributylstannane in the presence of Pd(0) afforded triene **35** that spontaneously underwent intramolecular Diels–Alder cycloaddition to produce **36** as the sole product in a 68% yield. It should be noted that the single stereocenter of **35** controls the absolute and relative stereochemistry at the remaining stereocenters comprising the ABC tricylic subunit of **36**. Further elaboration of **36**, including sequential ring-closing metathesis (RCM) reactions produced the requisite 13- and 8-membered D and E rings of Ircinal A **37**, respectively. To further establish the identity of their synthetic Ircinal A **37**, Martin and coworkers converted a small portion to manzamine A following Kobayashi's protocol.[20]

In 2005, Crimmins reported the enantioselective total synthesis of 11-acetoxy-4-deoxyasbestinin D **44** utilizing a tandem olefination/IMDA strategy.[21] As shown in Scheme 3.6, synthesis of the oxonene core of 11-acetoxy-4-deoxyasbestinin D **44** was approached using the asymmetric

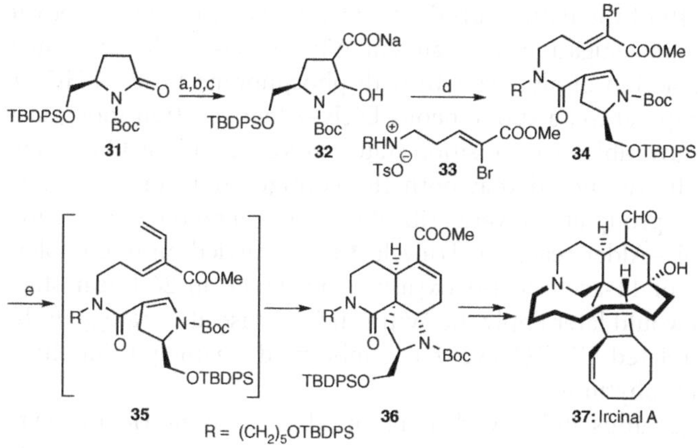

Reagents and conditions: a) LHMDS, CO_2, –78°C; b) $NaBH_4$, 0°C then H^+; c) $NaCO_3$, 95% over 3 steps; d) $(COCl)_2$, then **33**, Et_3N, 79%; e) vinyltributyltin, $Pd(PPh_3)_4$, toluene, heating, 68%.

Scheme 3.5 Martin's total synthesis of Ircinal A featuring a Stille coupling/IMDA cascade.

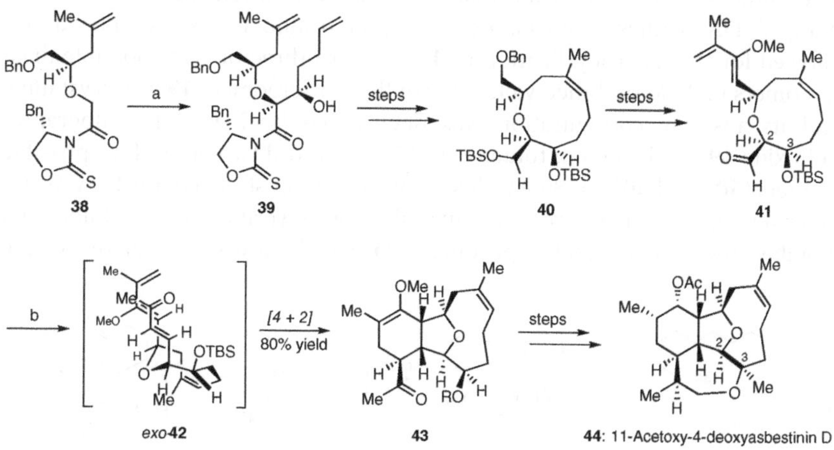

Reagents and conditions: (a) $TiCl_4$, *i*-Pr_2NEt, NMP, 4-pentenal, CH_2Cl_2, –78°C, 70%, 95:5 *dr*; (b) $Ph_3P{=}CHC(O)Me$, Tol. 110°C, 80%.

Scheme 3.6 Crimmins' total synthesis of 11-acetoxy-4-deoxyasbestinin D via a Wittig/IMDA cascade.

glycolate aldol reaction of 4-pentanal with the titanium enolate of thioimide **38** in the presence of NMP to give the *syn*-aldol adduct **39** in good yield and diastereoselectivity (70%, >95:5 *dr*). Subsequent steps including reductive removal of the chiral auxiliary, protection of the diol, and RCM established the oxonene core **40** of the natural product. Further elaboration to aldehyde **41** allowed for the investigation of a tandem Wittig/Diels–Alder sequence. After the Wittig reaction of aldehyde **40** with phosphorane $Ph_3P=CHC(O)Me$ at elevated temperature, a spontaneous Diels–Alder reaction took place through the more favorable *exo*-transition state *exo*-**42**, providing **43** as a single diastereomer. It was noted that both the configuration and choice of hydroxyl protecting group at C3 were critical for controlling the diastereoselectivity of the Diels–Alder reaction. Tricycle **43** was carried on to complete the total synthesis of 11-acetoxy-4-deoxyasbestinin D **44**, in 26 linear steps overall. Crimmins would later apply the Wittig/IMDA cascade strategy in the total synthesis of related C2-C11 cyclized cembranoids, namely Briarellin J, Vigulariol and Sclerophytin A.[22]

In 2003, Deslongchamps and coworkers reported the asymmetric total synthesis of (+)-chatancin **52** utilizing a tandem pyrolysis/TADA sequence.[23] As shown in Scheme 3.7, assembly of the terpenoid skeleton by alkylation of the lithium salt of sulfone **45** with allyl bromide **46** (derived from (*S*)-citronellol) afforded pivolate **47** at a 92% yield. The pivolate **47** was elaborated in several steps to dione **48**, which was poised for an intramolecular Michael reaction. The authors noted that while macrocyclization is often the bottleneck for many TADA syntheses efforts, the use of β-keto sulfoxide as a Michael donor allowed for the macrocyclization of **48** to afford three of eight possible diastereomers in 80% combined yield. One of the diastereomers **49** was crystalline and its absolute configuration was determined by X-ray crystallography. Sulfoxide **49** was heated in toluene at 115 °C to afford two out of four possible isomers (**50a** and **50b**) at 80% yields. While the exact stereochemistry was not determined, their structures were unambiguously verified by NMR and MS. Notably the major kinetic product **50a** could slowly isomerize to the

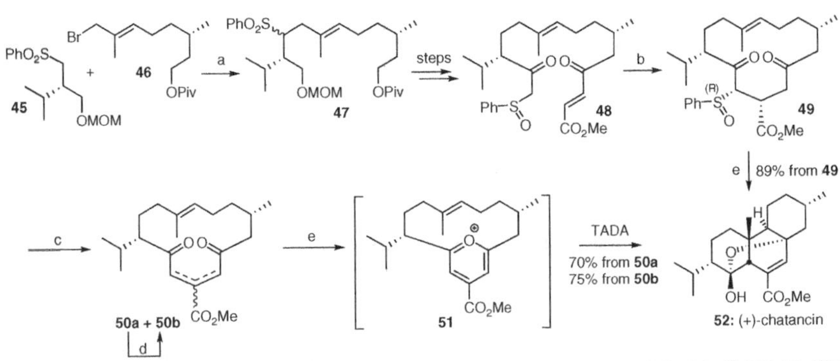

Reagents and conditions: (a) **45**, *n*-BuLi, THF, –78°C to –25°C, then **46**, 92%; (b) Cs_2CO_3, acetone, [**48**] = 6.4 mM, 15°C, 7 h, 80%, *ca* 12:59:29 mixture of diastereoisomers; (c) toluene, 110–115°C, 15 min, 80%, **50a/50b** *ca* 5:3; (d) HBF_4, NaOAc, acetone, 48 h, 80%; (e) DMSO/H_2O, 105°110°C, 3–18 h, up to 89%.

Scheme 3.7 Deslongchamps' total synthesis of (+)-chatancin via a pyrophane TADA cascade.

thermodynamic product **50b** under buffered acid conditions. Heating **50a** and **50b** in aqueous DMSO at 110 °C produced (+)-chatancin **52** at a 70 and 75% yield, respectively. It was noticed that the presence of water in the medium was critical to the TADA reaction, as the aprotic pyrolysis conditions in the conversion of **49** to **50a/50b** resulted in no TADA product being observed. The critical role of water for the TADA reaction suggests a central pyrylium intermediate **51**. Notably, directly heating any one of the three diastereomers **49** under the same aqueous conditions provided chatancin **52** at an up to 89% yield, in a tandem elimination/pyrolysis/TADA process. These findings validated the possibility of TADA in the biosynthesis of chatancin **52**.

Chen and coworkers recently unveiled an interesting total synthesis of Echinopines A **62** and B **63**.[24] Central to their synthetic strategy was the employment of a cycloisomerization/Diels–Alder cascade to construct a [5/6/7] tricycle (**Scheme 3.8**), which was ultimately converted to a [5/7] bicycle en route to the total synthesis of Echinopines A **62** and B **63**.

Construction of the key ene-yne **57** commenced with the $TiCl_4$-mediated Hosomi–Sakurai reaction of allylsilane **53** with aldehyde **54** to furnish alcohol **55** at a 74% yield as a mixture of C10 epimers. Further protection/deprotection of **55** followed by Swern oxidation led to aldehyde **56**, which underwent a two-step Henry reaction and dehydration to produce the targeted nitroalkene **57**. The desired palladium-catalyzed ene-yne cycloisomerization produced the transient diene **58**, which underwent a spontaneous intramolecular Diels–Alder

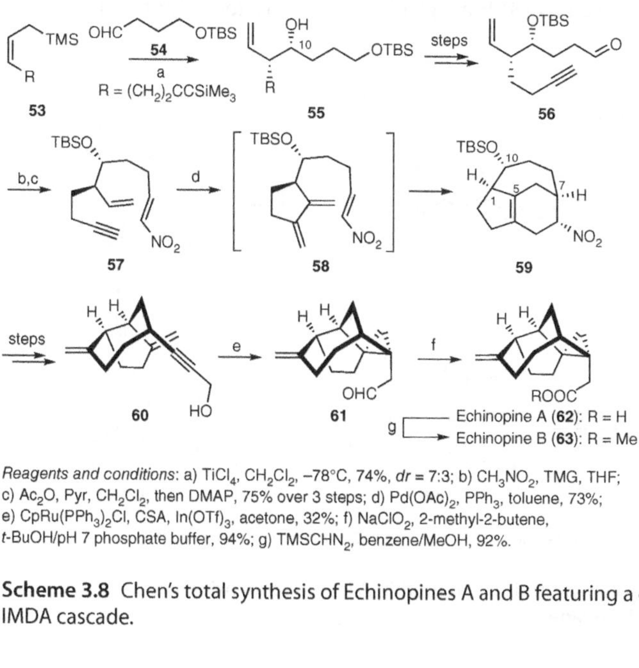

Reagents and conditions: a) $TiCl_4$, CH_2Cl_2, −78 °C, 74%, *dr* = 7:3; b) CH_3NO_2, TMG, THF; c) Ac_2O, Pyr, CH_2Cl_2, then DMAP, 75% over 3 steps; d) $Pd(OAc)_2$, PPh_3, toluene, 73%; e) $CpRu(PPh_3)_2Cl$, CSA, $In(OTf)_3$, acetone, 32%; f) $NaClO_2$, 2-methyl-2-butene, *t*-BuOH/pH 7 phosphate buffer, 94%; g) $TMSCHN_2$, benzene/MeOH, 92%.

Scheme 3.8 Chen's total synthesis of Echinopines A and B featuring a cycloisomerization/IMDA cascade.

cycloaddition under the reaction conditions to afford nitro-[5/6/7] tricycle **59** at a 73% yield as an inconsequential mixture of C10 epimers. Additional steps included a facially selective reduction of the double bond of **59**, as well as cleavage of the six-membered ring, eventually leading to diene-yne **60**. Treatment of **60** with optimized cycloisomerization conditions reported by Trost using an organoruthenium catalyst with $In(OTf)_3$ afforded the desired aldehyde **61**.[25] Pinnick oxidation of aldehyde **61** afforded natural product Echinopine A **62**, followed by methylation to yield Echinopine B **63**. While the [5/6/7] → [5/7] → [3/5/5/7] ring forming sequence may appear counterintuitive at first, the power of the key cycloisomerization/Diels–Alder cascade, paired with a facially selective hydrogenation allowed Chen *et al.* to control the C1, C5, and C7 stereochemistry, which would prove to be a challenging feat via direct [5/7] bicycle construction.

In 2003, Mulzer accomplished the enantioselective total synthesis of elisabethin A **69**, featuring a tandem oxidative dearomatization/intramolecular Diels–Alder reaction sequence.[26] A Diels–Alder precursor **66** was furnished via a diastereoselective alkylation of oxazolidinone **64** with alkyl iodide **65**, followed by reduction of the chiral auxiliary, oxidation, and Wittig olefination. As shown in Scheme 3.9, deprotection of bis-silyl ether **66** with TBAF, followed by oxidation using aqueous $FeCl_3$ afforded quinone **67**, which was detectable by TLC and NMR, but cyclized *in situ* to afford **68** at a 91% overall yield. The authors noted the significance of this tandem IMDA reaction sequence was threefold: (1) the use of a terminal (*Z*)-olefin was unprecedented at the time of publication, (2) the mild and biomimetic conditions (aqueous medium, ambient temperature), as well as (3) high yield and stereoselectivity for the overall

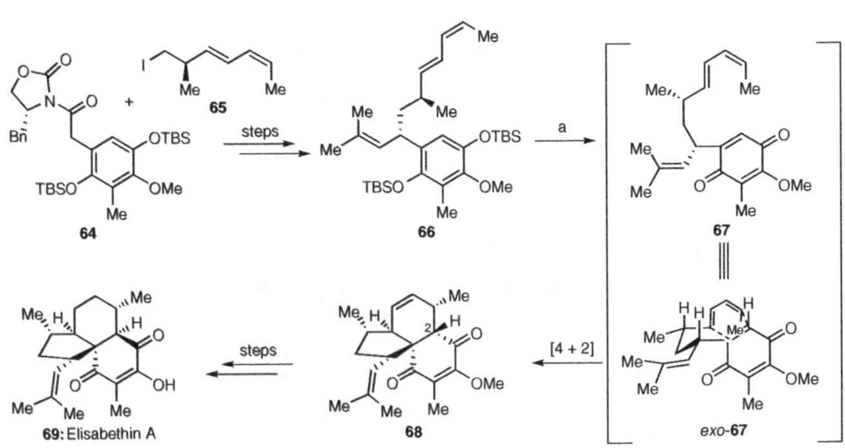

Reagents and conditions: (a) TBAF (2.4 equiv), THF, 1 h, rt; then $FeCl_3$ (10 equiv), H_2O, 6 h, rt, 91%.

Scheme 3.9 Mulzer's total synthesis of elisabethin A featuring an oxidative dearomatization/IMDA cascade.

transformation. The observed stereochemistry could be rationalized by means of an *exo*-transition state, *exo*-**67**. The total synthesis was completed in a further three-step sequence comprising of selective hydrogenation of the disubstituted olefin, base-catalyzed epimerization at C2, and cleavage of the methyl ether to afford elisabethin A **69**.

In 2004, Shair and coworkers published an elegant tandem oxidative dearomatization/transannular Diels–Alder sequence to build the skeletal A, C, and D rings *en route* to the biomimetic synthesis of (–)-longithorone A **78**.[27] Outlined in Scheme 3.10, Shair's divergent strategy called for the generation of two [12]-paracyclophanes **73** and **74** *via* intramolecular ene-yne metathesis reactions, both of which were synthesized from a common intermediate **72**, which was conveniently assembled in two steps from vinyl iodide **70** and aryl bromide **71**. The key to atropisomerism control in the [12]-paracyclophane syntheses was the utilization of a strategically placed, removable stereogenic benzylic group that disfavored the undesirable rotamers due to $A_{1,3}$ strain (not shown). The intermolecular Diels–Alder reaction of **73** and **74** was accomplished using Me_2AlCl at –20 °C, affording a mixture of **75** and another

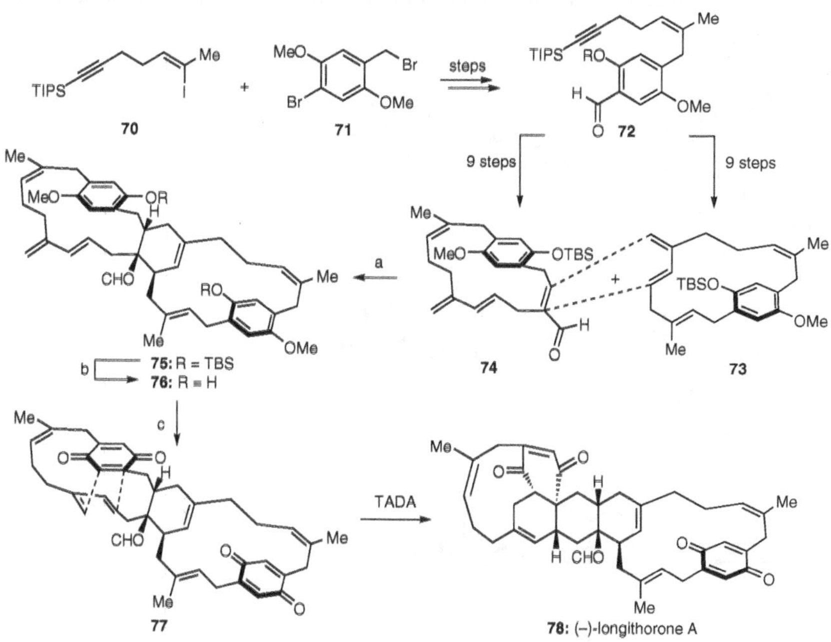

Reagents and conditions: (a) Me_2AlCl, CH_2Cl_2, –20°C, 70%, 1:1.4 mix of diastereomers; (b) TBAF, THF, 0°C; (c) PhIO, $MeCN\text{-}H_2O$, 0 to 25°C, 90% over two steps.

Scheme 3.10 Shair's total synthesis of (–)-longithorone A via an oxidative dearomatization/TADA cascade.

diastereomer in 70% yield and 1:1.4 ratio disfavoring **75**, with the cycloaddition being completely *endo*-selective, but lacking facial selectivity. It was conjectured that the necessity of a Lewis acid catalyst, as well as poor substrate-based diastereoselectivity in the intermolecular Diels–Alder reaction would implicate that the biosynthesis of (−)-longithorone A **78** may involve a Diels–Alderase to produce a similar cycloaddition event.

Desilylation of **75** followed by oxidative dearomatization using iodosylbenzene afforded bisquinone **77**, which was observable by NMR and TLC; however, it underwent transannular Diels–Alder cycloaddition at room temperature to generate (−)-longithorone A **78** at a 90% yield from **75**.

3.4 Rearrangement-Initiated Diels–Alder Cascades

Burke and coworkers disclosed a convergent enantioselective synthesis of indanomycin **83** in 1994 (Scheme 3.11).[28] Burke's strategy for the synthesis of the "right wing" tetrahydroindan core **82** consisted of a planned Ireland–Claisen rearrangement of silyl ketene acetal **79**. Heating **79** at 135 °C in toluene, followed by acidic workup and treatment with diazomethane resulted in the

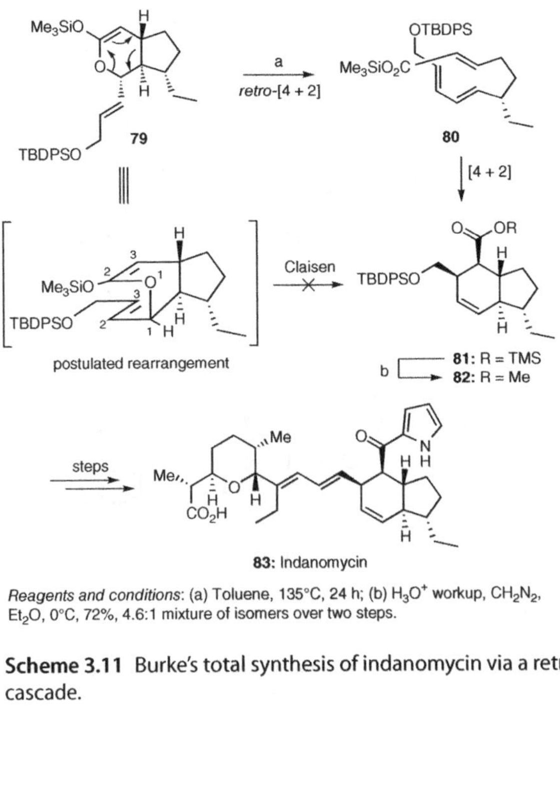

Reagents and conditions: (a) Toluene, 135°C, 24 h; (b) H_3O^+ workup, CH_2N_2, Et_2O, 0°C, 72%, 4.6:1 mixture of isomers over two steps.

Scheme 3.11 Burke's total synthesis of indanomycin via a retro hetero Diels–Alder/IMDA cascade.

production of **82** at a 72% yield as a 4.6:1 mixture of separable diastereomers. The lack of stereospecificity was inconsistent with the rigid transition state required for the Claisen rearrangement. They concluded that the ring strain in the desired Claisen transition state **79** for the [3,3] sigmatropic migration was energetically inaccessible, and instead a retro-hetero-Diels–Alder took place to give an intermediate trienoate **80**, which spontaneously underwent an intramolecular Diels–Alder reaction via the *endo* transition state to afford the major hydridene product **81**. By means of lowering the reaction temperature to 105 °C, the intermediate trienoate **80** was isolated and characterized. Since the resultant product **81** of the tandem retro-hetero-Diels–Alder/intramolecular Diels–Alder reaction gave the desired Ireland–Claisen product, Burke termed this a "mock Claisen" rearrangement. Hydridene **81** was elaborated further to complete the total synthesis of indanomycin **83**.

Nicolaou and coworkers developed a photo-induced generation and trapping of hydroxyl-*o*-quinonedimethanes as the key transformation for the total synthesis of several of the hamigerans **88a,b** and **89a,b** (Scheme 3.12).[29] One challenge for the synthesis of this class of natural products was the establishment of the correct relative stereochemistry at the four contiguous stereocenters situated at C5, C6, C9, and C10. Irradiation of benzaldehyde **84a** in deoxygenated benzene afforded tricycle **86a** at a 91% yield, presumably via the *in situ* generated diene **85**. It was found that the relative configuration of the substituents was controlled by the single C6 stereocenter of the benzaldehyde **84a**. Subsequent attempts to elaborate the scaffold of **86a** and invert the

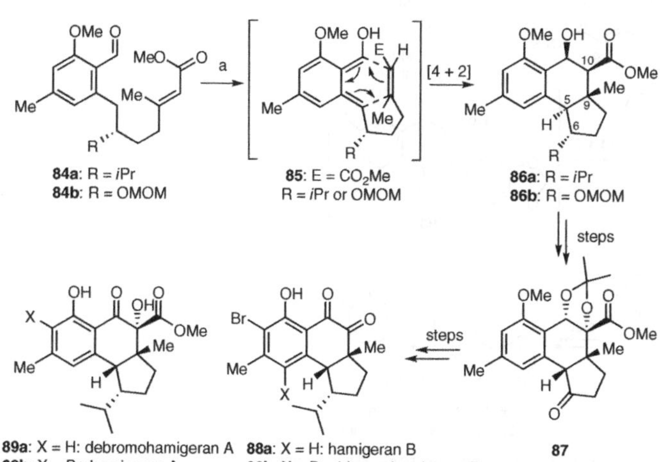

Reagents and conditions: a) benzene, *hν* (450-W Hanovia lamp, pyrex filter), 25°C, 20 min, **86a**: 91% or **86b**: 92%

Scheme 3.12 Nicolaou's total synthesis of hamigerans featuring photoenolization/IMDA cascades.

stereochemistry at C5 failed under a variety of conditions. An alternative route with benzaldehyde **84b** incorporating an oxygen functionality at C6 was found to undergo photoenolization/IMDA cascade efficiently with tricycle **86b** isolated at a 92% yield. This oxygen handle proved effective at allowing C5 epimerization, and subsequent access to hamigerans **88a,b** and **89a,b**, highlighting the synthetic power of intramolecular trapping of photochemically generated hydroxyl-*o*-quinondimethanes in total synthesis.

In 2009, Hsung and coworkers developed a *de novo* cascade of pericyclic ring-openings of amidobenzocyclobutanes and Oppolzer-type *N*-tethered intramolecular Diels–Alder cycloadditions.[30] This tandem process was initiated by linking together an increasingly versatile enamide functional group with the powerful benzyne chemistry in a thermally driven [2 + 2] cycloaddition. An application of this cascade methodology was later described in the total syntheses of benzophenanthridine alkaloids (±)-chelidonine **96** and (±)-norchelidonine **97** (Scheme 3.13).[31]

Enamide **90** was rapidly prepared in three steps from commercially available 5-bromo-1,3-benzodioxole-4-carboxaldehyde, including Sonogashira coupling, reductive amination, and Cu(I)-catalyzed amidation of vinyl bromide. The benzyne precursor **91** was conveniently synthesized from sesamol in two

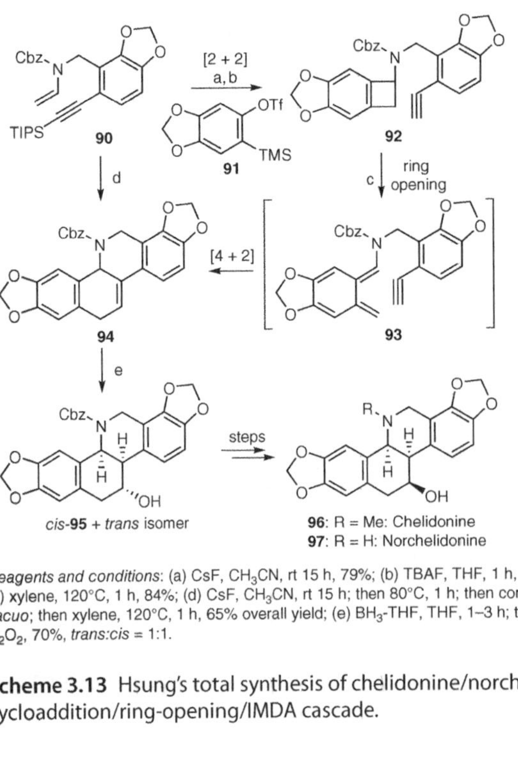

Reagents and conditions: (a) CsF, CH_3CN, rt 15 h, 79%; (b) TBAF, THF, 1 h, 95%; (c) xylene, 120°C, 1 h, 84%; (d) CsF, CH_3CN, rt 15 h; then 80°C, 1 h; then concentrate *in vacuo*; then xylene, 120°C, 1 h, 65% overall yield; (e) BH_3-THF, THF, 1–3 h; then NaOH, H_2O_2, 70%, *trans:cis* = 1:1.

Scheme 3.13 Hsung's total synthesis of chelidonine/norchelidonine via a [2 + 2] cycloaddition/ring-opening/IMDA cascade.

steps. As shown in Scheme 3.13, a [2 + 2] cycloaddition reaction of enamide **90** with **91** by treatment with CsF at room temperature in CH_3CN, afforded amidobenzocyclobutane **92** after removal of the terminal TIPS protection with TBAF. Heating of **92** at 120 °C initiated the cascade sequence via ring-opening of an amidobenzocyclobutane ring to form intermediate **93**, followed by an intramolecular Diels–Alder reaction, giving tetracycle **94** at a good yield. More importantly, the authors further found that this three-step sequence could be performed in a one-pot operation, that is, heating the [2 + 2] reaction mixture at 85 °C to achieve deprotection, removing CsF by filtration, and then subjecting the crude product to cascade thermal conditions. This one-pot operation gave **94** with a yield similar to the three-step sequence, while saving two column purifications, thus tremendously improved the overall efficiency. Tetracycle **94** was carried on for several more transformations, including hydroboration to alcohol *cis*-**95,** en route to the expeditious synthesis of (±)-chelidonine **96** and (±)-norchelidonine **97**.

Stoltz's group reported a concise total synthesis of basiliolide B **103** by using an Ireland–Claisen rearrangement/IMDA cascade reaction as the key step (Scheme 3.14).[32] The precursor **100** was synthesized by esterification of readily available allylic alcohol **98** and iodopyrone acid **99**. When **100** was treated with *N,O*-bis(trimethylsilyl)acetamide while heating in toluene, the Ireland–Claisen rearrangement/IMDA cascade occurred smoothly to give tricyclic acid **102** in 67% yield as a 2:1 mixture of diastereomers at C8. Both diastereomers were finally converted to basiliolide B **103** and *epi*-8-basiliolide B *epi*-**103**. Later on, Stoltz also achieved the total syntheses of the transtaganolides, basiliolide C, and *epi*-8-basiliolide C with a similar cascade strategy.[33]

Nicolaou and coworkers reported an elegant synthesis employing a well-designed tandem double Claisen rearrangement/Diels–Alder reaction sequence to build the 4-oxatricyclo[4.3.1.0]decan-2-one ring system of 1-*O*-methylforbesione **109**.[34] As shown in Scheme 3.15, the tandem sequence

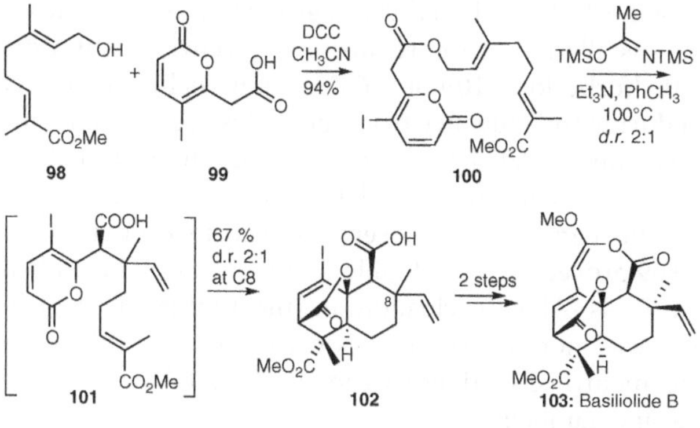

Scheme 3.14 Stoltz's total synthesis of basiliolide B via an Ireland–Claisen/IMDA cascade.

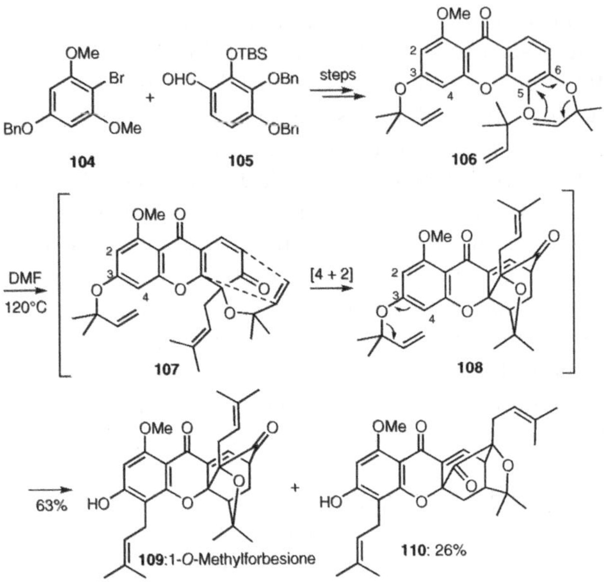

Scheme 3.15 Nicolaou's total synthesis of 1-*O*-methylforbesione featuring a double Claisen rearrangement/Diels–Alder cascade.

commenced with xanthone intermediate **106** possessing the three requisite *O*-*α*,*α*-dimethylallyl groups. Xanthone **106** was accessed from simple starting materials **104** and **105**. Upon heating the key intermediate **106** in DMF at 120 °C, the Claisen rearrangement transferred the 6-*O*-α,α-dimethylallyl group to the xanthone C5 position, which was followed by the spontaneous intramolecular Diels–Alder reaction of the newly formed diene with the terminal alkene of the remaining 5-*O*-*α*,*α*-dimethylallyl group to form the desired complex 4-oxatricyclo[4.3.1.0]decan-2-one skeleton. An additional Claisen rearrangement of the 3-*O*-*α*,*α*-dimethylallyl group to the xanthone C4 position resulted in the installation of the requisite C4 prenyl substitution, thus affording 1-*O*-methylforbesione **109** at a 63% overall yield. Byproduct **110** was also generated during the sequence from the undesired Claisen rearrangement of the 5-*O*-*α*,*α*-dimethylallyl group to xanthone C6 position. Two minor byproducts involved the regioisomeric Claisen products at C2, resulting in a less than 3% overall yield. While the overall transformation of **106** to 1-*O*-methylforbesione **109** proceeded via a double Claisen reaction in tandem with an intramolecular Diels–Alder cycloaddition, the timing of the two Claisen rearrangements was not determined. Nicolaou later employed a similar Claisen rearrangement/Diels–Alder cascade in the total syntheses of 1-O-methyllateriflorone and gambogin.[35]

3.5 Cyclization-Initiated Diels–Alder Cascades

Trauner recently disclosed a biomimetic synthesis of a series of antimalarial naphthoquinones.[36] Although pinnatal **116** was originally synthesized in a multiple-step sequence, it was discovered that the target could be obtained by heating dialdehyde **112** and 1,4-naphthylquinone **113** in the presence of ethylenediaminediacetate via a condensation/6π-cyclization/intramolecular Diels–Alder reaction cascade, albeit with a low overall yield (Scheme 3.16).

The intramolecular ionic Diels–Alder reaction was introduced by Gassman's group in the 1980s, which utilized a vinyl oxocarbenium ion as an activated dienophile.[37] Later on, Sammakia's group developed an asymmetric version of the ionic Diels–Alder reaction in 1990s.[38] Here, in their total synthesis of (+)-dihydrocompactin **123**, Sammakia applied the remote stereocontrolled intramolecular ionic Diels–Alder reaction (Scheme 3.17).[39] Starting from the D-A precursor **117**, when treated with $Al(OTf)_3$ and triflic acid, an oxocarbenium ion intermediate **118** was presumably formed followed by an intramolecular ionic Diels–Alder reaction to give cyclized products **120** and **121** in a 1:1 ratio. The newly formed seven-membered ring enhanced the induction effect of chiral environment introduced by TMS-protected alcohol, further showing the superiority of the ionic D-A strategy in this special case. Both **120** and **121** could be converted to *trans*-decalin compound **122** through hydrolysis and base-catalyzed epimerization of *cis*- and *trans*-octalones *en route* to (+)-dihydrocompactin **123**. The ionic Diels–Alder strategy was also utilized by other groups in the synthesis of lycopodine and (+)-himbacine.[40,41]

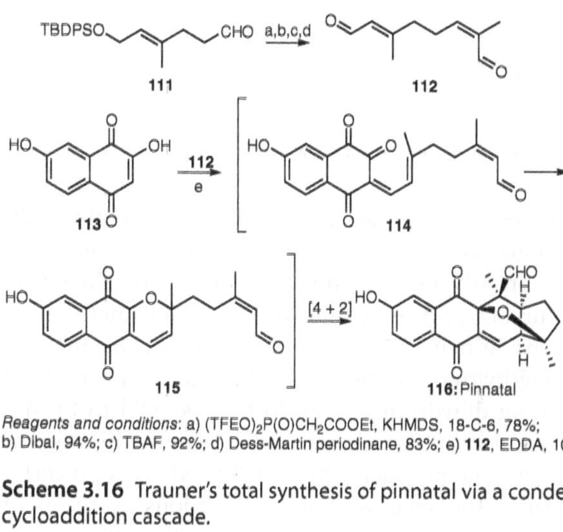

Reagents and conditions: a) $(TFEO)_2P(O)CH_2COOEt$, KHMDS, 18-C-6, 78%; b) Dibal, 94%; c) TBAF, 92%; d) Dess-Martin periodinane, 83%; e) **112**, EDDA, 10%.

Scheme 3.16 Trauner's total synthesis of pinnatal via a condensation/electrocyclization/cycloaddition cascade.

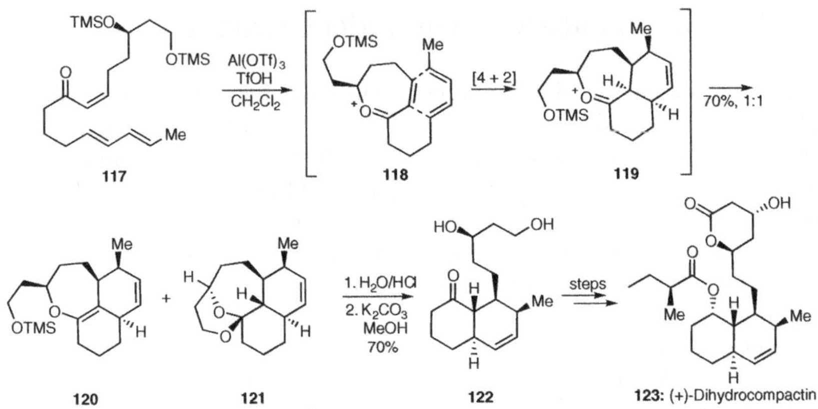

Scheme 3.17 Sammakia's total synthesis of (+)-dihydrocompactin using an ionic IMDA cascade.

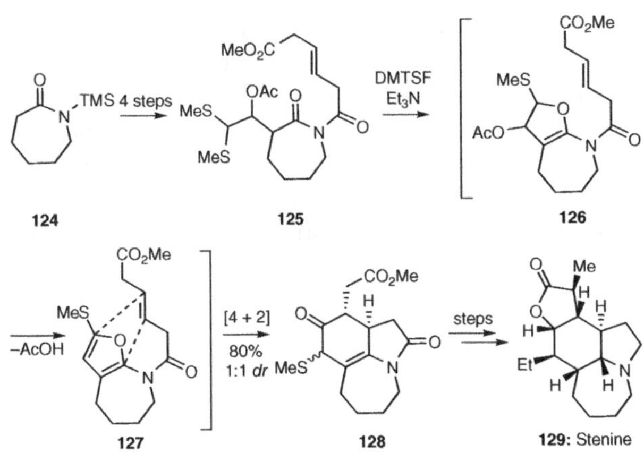

Scheme 3.18 Padwa's total synthesis of (±)-stenine via an IMDAF cascade.

In 2002, Padwa's group described a racemic synthesis of stenine **129** highlighting their intramolecular Diels–Alder Furan (IMDAF) reaction methodology for the construction of the azepinoindolecore.[42] As shown in Scheme 3.18, the D-A precursor **125** was synthesized from *N*-trimethylsilylcaprolactam **124** in four steps. When **125** was treated with DMTSF, methylsulfenylation of one of the methylthio groups triggered a thionium-promoted cyclization. Then, the resulting dihydrofuran **126** lost acetic acid to form a furan **127**, which readily reacted with the tethered alkene followed by oxygen-bridge opening to produce tricyclic product **128** in 80% yield. This powerful IMDAF strategy was critical to the successful and expeditious synthesis of stenine **129**, and was also employed by the Padwa group in the syntheses of strychnos skeleton,[43] lycoricidine,[44] and minfiensine.[45]

In 2009, Nicolaou's group reported an elegant total synthesis of hirsutellone B **135** featuring an epoxide opening/intramolecular Diels–Alder cascade (Scheme 3.19).[46] Starting from (R)-(+)-citronellal **130**, a five-step sequence delivered cyclization precursor **131**. When **131** was treated with $AlEt_2Cl$, it was assumed that the activation of epoxide and cleavage of TMS triggered elimination and epoxide opening via a chair-like transition state **132**, followed by an IMDA reaction through an *endo* transition state **133**. This tandem process afforded the tricyclic compound **134** as a single diastereoisomer in 50% yield. Elaboration of the tricyclic core and installation of the *p*-cyclophane over 17 additional steps provided hirsutellone B **135**.

In the total synthesis of momilactone A **140** reported by Deslongchamps' group, a transannular Diels–Alder (TADA) strategy was developed to construct the [6/6/6] tricyclic core with a *trans-syn-trans* (TST) ring junction stereochemistry.[47] The cyclization precursor **137** was achieved from commercially available neryl acetate **136** (Scheme 3.20). When **137** was treated with cesium carbonate in refluxing acetonitrile, the desired tandem process successfully delivered the tricyclic product **139** at a 60% yield. The reaction pathway was believed to proceed via an intramolecular allylic substitution followed by a TADA reaction through a sterically favored *endo* transition state, through a chair-boat-chair conformation **138**. Tricycle **139** was carried on to afford (±)-momilactone A **140** in the longest linear sequence of 29 steps.

Roush's group utilized a tandem olefination/transannular Diels–Alder cyclization sequence and a vinylogous Morita–Baylis–Hillman strategy to construct the [5/6/5] core skeleton of spinosyn A **146**.[48] As shown in Scheme 3.21, the cyclization precursor **142** was synthesized from readily available chiral alcohol **141**. When **142** was treated with DIPEA and LiCl, the intramolecular Horner–Wadsworth–Emmons olefination occurred, setting the stage for the transannular Diels–Alder reaction. This remarkable tandem sequence occurred with

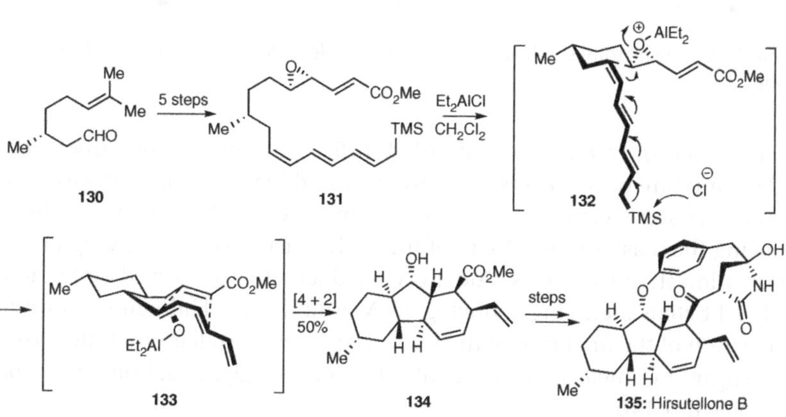

Scheme 3.19 Nicolaou's total synthesis of hirsutellone B featuring an epoxide opening/IMDA cascade.

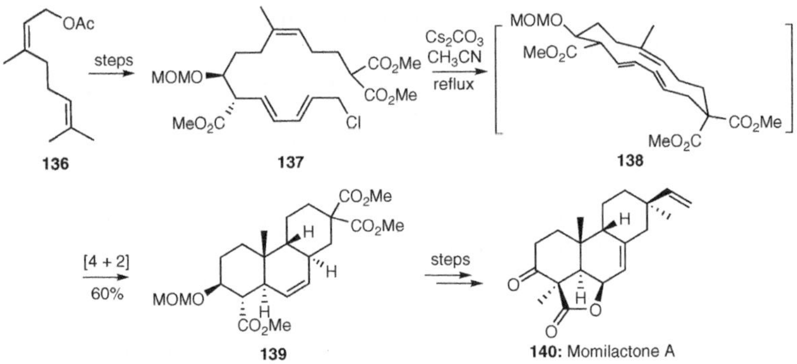

Scheme 3.20 Deslongchamps' total synthesis of (±)-momilactone A featuring a macrocyclization/TADA cascade.

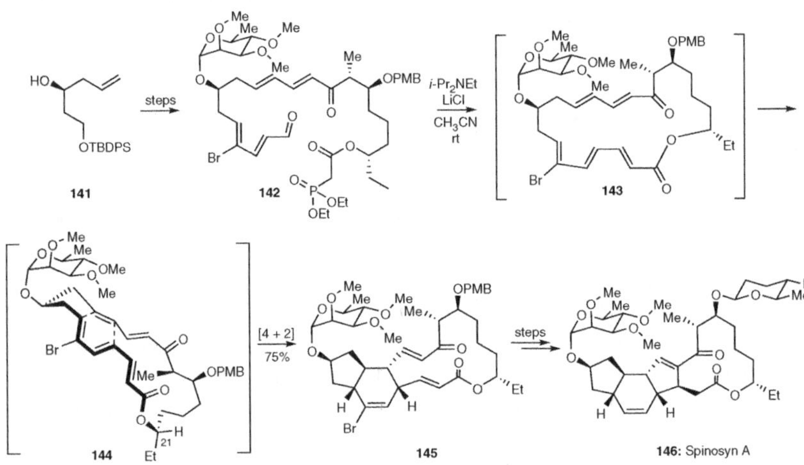

Scheme 3.21 Roush's total synthesis of (−)-spinosyn A featuring an olefination/TADA cascade.

high *E/Z* control (≥95:5), and afforded **145** as the major component of a 73:12:9:6 mixture of two *trans-* and two *cis-*fused diastereomers. It was noted that the C21 stereocenter was a stereochemical control element, capable of enhancing the diastereoselectivity of the TADA reaction by enforcing favorable transition state **144**. Diels–Alder cycloadduct **145** was carried on for additional eight steps including a vinylogous Morita–Baylis–Hillman reaction for construction of the final ring of the [5/6/5] core, and installation of the forosamine sugar by means of a highly β-selective glycosidation to afford (−)-spinosyn A **146**.

Porco and coworkers accomplished the asymmetric total synthesis of the quinone epoxide dimer (+)-torreyanic acid **152** employing a biomimetic tandem oxidation/6π-electrocyclization/Diels–Alder dimerization process (Scheme 3.22).[49] Synthesis of the key chiral quinone epoxide intermediate **148** was accomplished via a hydroxy group-directed epoxidation of **147**, followed by several additional transformations. Alcohol **148** was oxidized using Dess–Martin periodinane conditions, and the aldehyde **149** subsequently underwent 6π-electron electrocyclic ring closure to afford diastereomeric *syn*-**150** and *anti*-**150**. A spontaneous Diels–Alder reaction between *syn*-**150** and *anti*-**150** cleanly afforded **151** at an 80% overall yield. Interestingly, the facial selectivity for the Diels–Alder reaction is revealed by the conformational analysis of the *syn*-**150** and *anti*-**150** 2*H*-pyrans. For the *anti*-**150** 2*H*-pyran, the pentyl group adopts an equatorial orientation, whereas the *syn*-**150** conformation has the pentyl group in an axial orientation, which leaves the other face accessible to the dienophile, which suggests only *syn*-**150** can serve as the diene. Removal of the *tert*-butyl ester of **151** using TFA afforded (+)-torreyanic acid **152**. Realization of the tandem oxidation/6π-electrocyclization/Diels–Alder dimerization process has helped to shed light on the possible biosynthesis of (+)-torreyanic acid **152**, and has also been used in the synthesis of related epoxyquinoid dimers (+)-epoxyquinols A and B.[50]

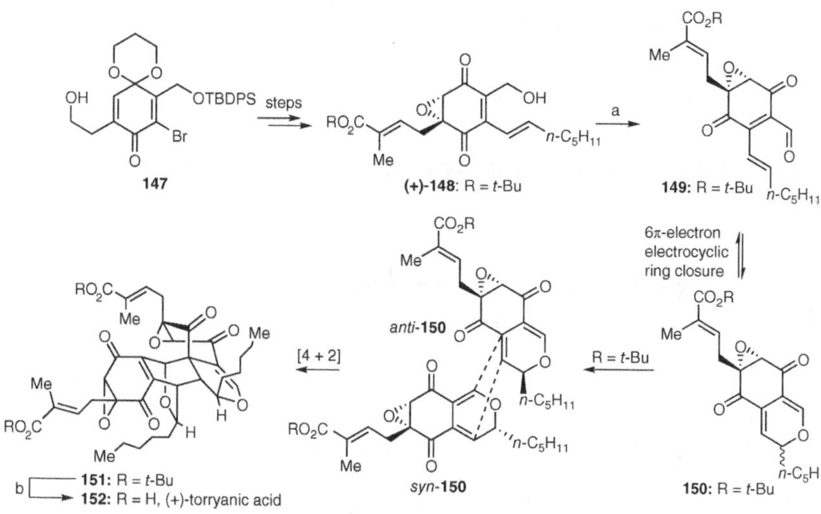

Reagents and conditions: (a) Dess-Martin periodinane, CH_2Cl_2, 1.5 h, 80%; (b) TFA/CH_2Cl_2, 0°C, 2 h 100%

Scheme 3.22 Porco's total synthesis of (+)-torreyanicacid via an oxidation/electrocyclization/Diels–Alder cascade.

3.6 Diels–Alder Initiated Cascades

Aubé reported an expeditious nine-step total synthesis of *rac*-stenine employing an intermolecular Diels–Alder reaction/Schmidt reaction strategy.[51] As shown in Scheme 3.23, a Diels–Alder reaction of diene **154** with cyclohexenone in the presence of $SnCl_4$ followed by a intramolecular Schmidt reaction afforded **156** and **157** (around 3:1) at a 70% yield. The major product **156** from the *exo*-selective Diels–Alder reaction could be easily transformed to the final natural product stenine **129**.

Inspired by the original biogenetic hypothesis by Jullian,[52] Liu's group developed a Diels–Alder/intramolecular hetero-Diels–Alder (IMHDA) cascade reaction in their total synthesis of bolivianine **163** (Scheme 3.24).[53] Starting from the commercially available (+)-verbenone **158**, onoseriolide **159** was

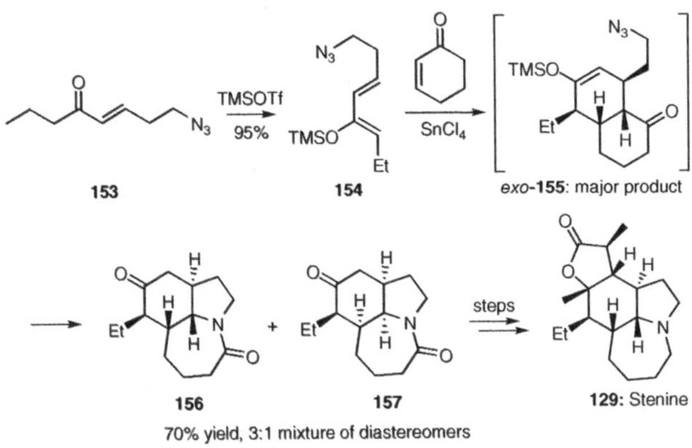

Scheme 3.23 Aubé's total synthesis of stenine via an intramolecular Diels–Alder/Schmidt cascade.

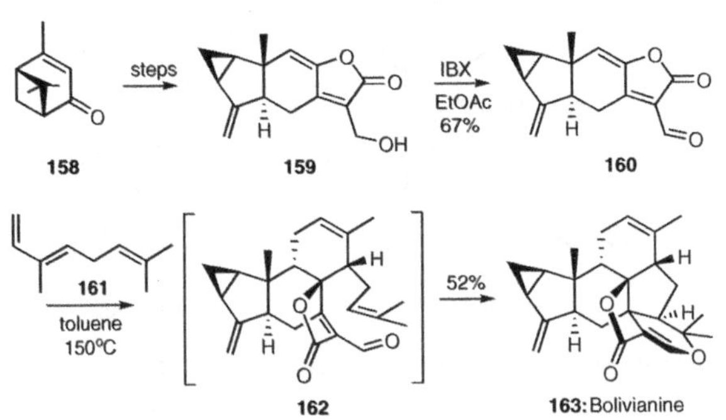

Scheme 3.24 Liu's total synthesis of bolivianine featuring a Diels–Alder/intramolecular hetero-Diels–Alder cascade.

synthesized in 12 steps featuring a dexterous assembly of a [3/5/6] skeleton via intramolecular cyclopropantion of an allylic metal carbene intermediate. Probing the biosynthetic pathway of bolivianine by treatment onoseriolide **159** with *β-E*-ocimene **161** in toluene at 150 °C, it was found that no Diels–Alder reaction was formed. Fortuitously, it was found that lowering the LUMO of the dienophile by oxidation of the primary alcohol to aldehyde **160** using IBX, and subsequent treatment with *β-E*-ocimene **161** under thermal conditions produced the natural product bolivianine **163** at a 52% yield in a highly efficient tandem Diels–Alder/IMHDA sequence to furnish three rings, four C-C bonds, and five stereogenic centers with excellent selectivity. These observations suggest the biogenetic route to bolivianine **163** proceeds via the oxidation of onoseriolide **159** to the aldehyde **160**, then a subsequent Diels–Alder reaction with *β-E*-ocimene **161**, presumably by Diels–Alderase catalysis, is followed by a spontaneous IMHDA reaction.

MacMillan recently published an elegant nine-step enantioselective total synthesis of (+)-minfiensine **173** employing an organocatalytic Diels–Alder reaction/amine cyclization cascade sequence.[54] In this synthesis shown in Scheme 3.25, an efficient Diels–Alder reaction of 2-vinylindole **164** with propynal **166** was developed employing chiral amine organocatalyst **165**, affording intermediate **167** that would cyclize spontaneously under the reaction conditions to give **170** at a 87% yield and 97% *ee* after subsequent reduction, thus building the tetracyclic carbazole framework in one single step. Tetracycle **170** could be elaborated to the final product with another five steps, representing an efficient synthesis of (+)-minfiensine **173**. MacMillan would later use similar organocatalytic Diels–Alder/amine cyclization cascade sequences in the enantioselective total syntheses of minovincine and vincorine.[55]

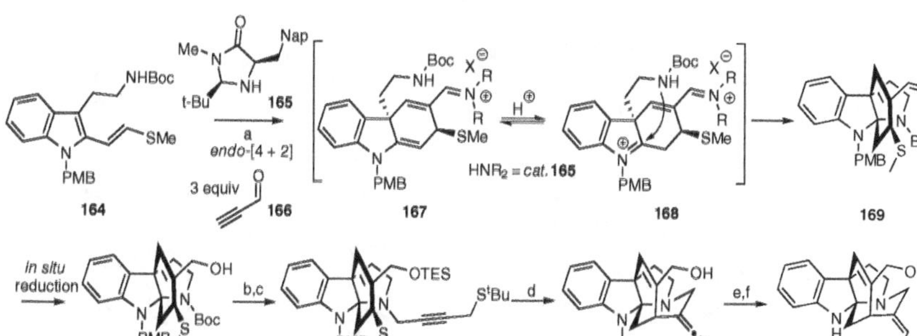

Reagents and conditions: a) **165**-TBA, Et_2O, –50°C, then $NaBH_4$, $CeCl_3$, MeOH, 87%, 96% *ee*; b) TESOTf, MeCN, 0°C, 84%; c) 4-(*tert*-Butylthio)but-2-ynal, $NaBH(OAc)_3$, CH_2Cl_2, rt, 96%; d) *t*-Bu_3SnH, AIBN, toluene, 110°C, 61%; e) Pd/C, H_2, THF, –15°C, ≥20:1 *E/Z*; f) PhSH, TFA, rt, 90%.

Scheme 3.25 MacMillan's total synthesis of (+)-minfiensine employing an organocatalytic Diels–Alder/cyclization cascade.

Sorensen and coworkers unveiled an asymmetric synthesis of cytotoxic natural product (+)-FR182877 **182**,[56] featuring a transannular Diels–Alder cascade inspired by a proposed biosynthesis.[57] Construction of the key transannular Diels–Alder precursor **176** commenced via a multiple-step sequence involving the Stille coupling of fragments **174** and **175** (Scheme 3.26). Elaboration of the resultant Weinreb amide **176** to β-keto ester **177**, followed by a palladium-catalyzed intramolecular allylation afforded **178** as a mixture of diastereomers. Selenylation of the stabilized enolate of **178** with PhSeBr followed by oxidative elimination (*m*-CPBA, CH_2Cl_2, −78 °C, then $CHCl_3$, 40 °C) afforded the 19-membered macrocyclic enone **179**, which spontaneously underwent intramolecular Diels–Alder reaction followed by an inverse demand *oxo*-Diels–Alder reaction to give the pentacyclic compound **181** in a 40% yield from **178**. This yield is actually very good considering that the elimination afforded a pair of 1:1 E/Z isomers at the newly formed C1–C19 olefin. Trace amounts of other isomers were also found. Finally global deprotection and lactone formation delivered the desired natural product (+)-FR182877 **182**. Evans and coworkers reported a similar cascade sequence in their total synthesis of (−)-FR182877 *ent*-**182**.[58]

An alternative approach to (−)-FR182877 *ent*-**182** reported by Nakada's group in 2009 combined the synthetic efficiency of the intramolecular Diels–Alder (IMDA) reaction and the intramolecular hetero-Diels–Alder (IMHDA) reaction, developing a one-pot tandem IMDA–IMHDA sequence (Scheme 3.27).[59] When the allylic alcohol **184** was subjected to slow oxidation using MnO_2 and

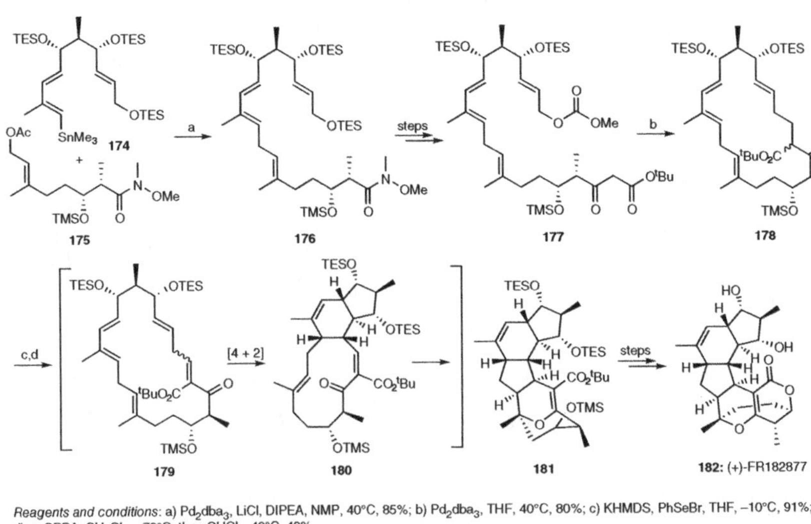

Reagents and conditions: a) Pd_2dba_3, LiCl, DIPEA, NMP, 40°C, 85%; b) Pd_2dba_3, THF, 40°C, 80%; c) KHMDS, PhSeBr, THF, −10°C, 91%; d) *m*-CPBA, CH_2Cl_2, −78°C; then $CHCl_3$, 40°C, 40%.

Scheme 3.26 Sorensen's total synthesis of (+)-FR182877 featuring an elimination/TADA/hetero-TADA cascade.

heating in toluene at 80 °C, the oxidation followed by tandem IMDA–IMHDA occurred to produce the desired tetracyclic intermediate **187** at a 28% yield. Analysis of other products revealed only undesired bicyclic side products with fused five- and six-membered rings and possessing an aldehyde, suggesting that the first IMDA reaction produced a mixture of diastereomers. Despite the moderate overall yield of **187**, this tandem IMDA/IMHDA process forming four rings and seven correct stereogenic centers from acyclic substrate is unique, and **187** was carried on to complete (−)-FR182877 *ent*-**182** in nine additional steps.

In 2003, Porco's group reported a remarkable intermolecular Diels–Alder dimerization/hemiketalization cascade in their total synthesis of (+)-panepophenanthrin **194**,[60] an ubiquitin-activating enzyme inhibitor.[61] As shown in Scheme 3.28, the key intermediate **189** was easily produced from readily available monomethoxy hydroquinone **188** in seven steps. When **189** was subjected to the silyl group deprotection conditions by using aqueous HF, natural product panepophenanthrin **194** was obtained at a 40% yield. After careful mechanistic investigation, Porco proposed that after the removal of TBS group, the intermediate **190** underwent Diels–Alder dimerization through a hydrogen bonding transition state **191** followed by an intramolecular hemiketalization reaction. A similar tandem strategy was also employed by Baldwin and Moses in their total synthesis of panepophenanthrin **194**,[62] as well as Liu's synthesis of Teucvin.[63]

Nicolaou disclosed an efficient synthesis of bisanthraquinone natural product BE-43472B **202**, in which an interesting cascade reaction initiated by Diels–Alder reaction was employed to construct the polycyclic system.[64] As shown in Scheme 3.29, a Diels–Alder reaction of diene **195** with dienophile **196** selectively generated intermediate **197**, which further underwent hemiketalization followed by intramolecular substitution afford **201** as a mixture of

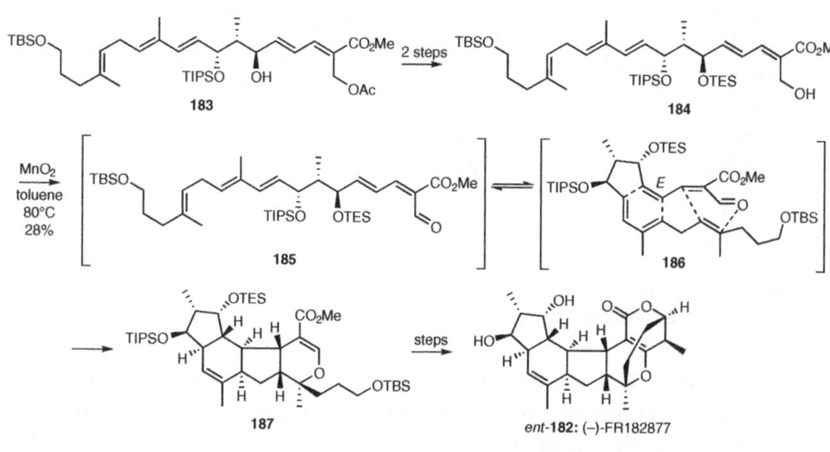

Scheme 3.27 Nakada's total synthesis of (−)-FR182877 featuring an IMDA/IMHDA cascade.

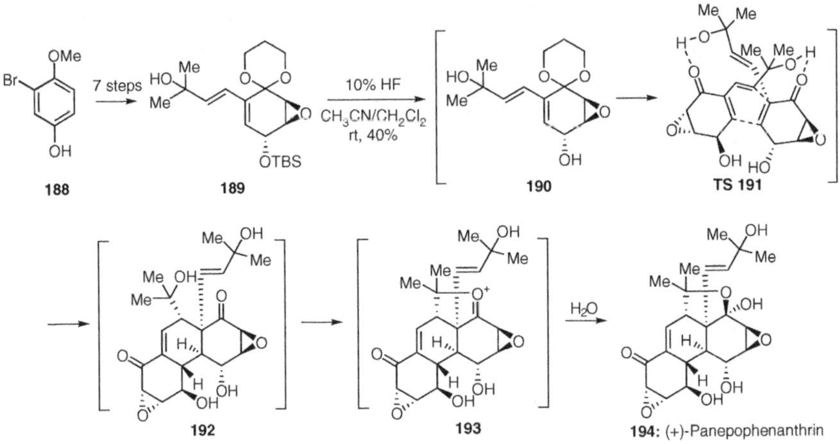

Scheme 3.28 Porco's total synthesis of panepophenanthrin via an intermolecular Diels–Alder dimerization/hemiketalization cascade.

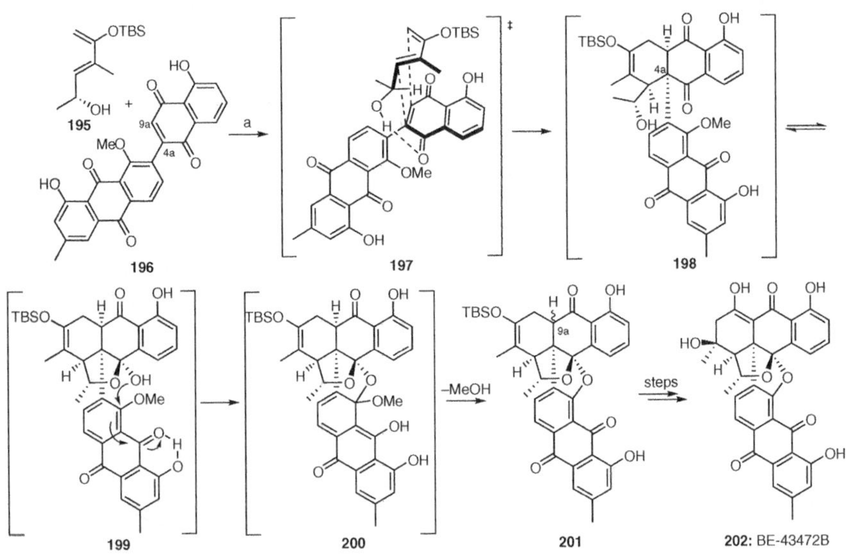

Reagents and conditions: a) CH_2Cl_2, 85°C, sealed tube, 2 days then toluene, 135°C, 1 day, 98%, *dr* = 1:2 at C9a

Scheme 3.29 Nicolaou's total synthesis of BE-43472B employing a Diels–Alder/hemiketalization/nucleophilic aromatic substitution cascade.

C9a isomers (α:β = 1:2) resulting from epimerization under elevated temperature. Ketal **201** could be further elaborated to the final natural product BE-43472B **202** with several additional steps. The success of this fascinating sequence was based on the highly selective Diels–Alder reaction. The authors proposed that an intermolecular hydrogen bonding between the free OH in **195** and the more Lewis basic carbonyl oxygen of the quinone moiety in **196** (see transition state **197**) was essential for the high stereoselectivity in forming the challenging quaternary stereocenter at C4a.

3.7 Concluding Remarks

We have presented here a collection of impressive Diels–Alder cascades reported over the past two decades demonstrating their awesome power in natural product syntheses. As we have shown, the design of Diels–Alder cascades to provide requisite architectures of substantial structural and stereochemical intricacy found in nature represents a formidable intellectual challenge for synthetic chemists. These cascades represent not only extraordinary feats of synthetic efficiency, but also elevate the art and science of total synthesis due to their aesthetic appeal. It is our belief that these beautiful Diels–Alder cascades will continue to captivate and inspire current and future generations of synthetic chemists in their own total synthesis efforts.

Acknowledgements

RPH would like to thank Laura and Edward Kremers Family Foundation for a generously endowed chair in natural products chemistry and also Vilas Trust and the University of Wisconsin for a Distinguished Achievement Professorship.

References

1 For an excellent review of the Diels–Alder reaction, see (a) Oppolzer, W. in *Comprehensive Organic Synthesis*; Trost, B. M., Fleming, I. (Eds); Pergamon: Oxford, **1991**, vol. 5, Chapter 4.1; pp. 315–399. For excellent reviews of Diels–Alder reactions in natural product syntheses, see (b) Bear, B. R.; Sparks, S. M.; Shea, K. J. *Angew. Chem. Int. Ed.* **2001**, *40*, 820–849. (c) Nicolaou, K. C.; Snyder, S. A.; Montagnon, T.; Vassilikogiannakis, G. *Angew. Chem. Int. Ed.* **2002**, *41*, 1668–1698.

2 For excellent reviews on cascade reactions, see: (a) Tietze, L. F. *Chem. Rev.* **1996**, *96*, 115–136. (b) Nicolaou, K. C.; Edmonds, D. J.; Bulger, P. G. *Angew. Chem. Int. Ed.* **2006**, *45*, 7134–7186. (c) Padwa, A.; Bur, S. K. *Tetrahedron*, **2007**, *63*, 5341–5378. (d) Jones, A. C.; May, J. A.; Sarpong, R.; Stoltz, B. M. *Angew. Chem. Int. Ed.* **2014**, *53*, 2556–2591.

3 For an excellent review on tandem sequences combining Diels–Alder reactions with sigmatropic rearrangements, see: Neuschütz, K.; Velker, J.; Neier, R. *Synthesis* **1998**, 227–255.

4 Diels, O.; Alder, K. *Justus Liebigs Ann. Chem.* **1928**, *460*, 98–122.

5 For early examples of the Diels–Alder reaction in natural product syntheses, see: (a) Stork, G.; Van Tamalen, E. E.; Friedman, L. J.; Burgstahler, A. W. *J. Am. Chem. Soc.* **1951**, *73*, 4501. (b) Stork, G.; Van Tamalen, E. E.; Friedman, L. J.; Burgstahler, A. W. *J. Am. Chem. Soc.* **1953**, *75*, 384–392. (c) Gates, M.; Tschudi, G. *J. Am. Chem. Soc.* **1952**, *74*, 1109–1110. (d) Gates, M.; Tschudi, G. *J. Am. Chem. Soc.* **1956**, *78*, 1380–1393. (e) Gates, M. *J. Am. Chem. Soc.* **1950**, *72*, 228–234. (f) Woodward, R. B.; Sondheimer, F.; Taub, D.; Heusler, K.; McLamore, W. M. *J. Am. Chem. Soc.* **1952**, *74*, 4223–4251.

6 For the first example of a Lewis-acid catalyzed Diels–Alder reaction, see: Yates, P.; Eaton, P. *J. Am. Chem. Soc.* **1960**, *82*, 4436–4437. For an excellent introduction to Lewis acid catalysis of the Diels–Alder reaction, see: Oppolzer, W. in *Comprehensive Organic Synthesis*, Trost, B. M., Fleming, I., Eds.; Pergamon: Oxford, **1991**, vol. 5, Chapter 4.1.4; pp. 339–345.

7 For an introduction to the use of chiral auxiliaries in Diels–Alder reactions, see: Oppolzer, W. in *Comprehensive Organic Synthesis*, Trost, B. M., Fleming, I. (Eds); Pergamon: Oxford, **1991**, vol. 5, Chapter 4.1.6; pp. 352–377.

8 Robinson, R. *J. Chem. Soc. Trans.* **1917**, 762–768.

9 (a) Trost, B. M. *Science* **1991**, *254*, 1471–1477. (b) Trost, B. M. *Angew. Chem.* **1995**, *107*, 285–307. *Angew. Chem. Int. Ed. Engl.* **1995**, *34*, 259–281.

10 (a) Anastas, P. T.; Warner, J. C. *Green Chemistry: Theory and Practice*, Oxford: Oxford University Press, **2000**, p. 135; (b) Matlack, A. S. *Introduction to Green Chemistry*, New York: Marcel Dekker, **2001**, p. 570.

11 (a) Corey, E. J.; Cheng, X.-M. *The Logic of Chemical Synthesis*; New York: John Wiley & Sons, Inc., **1989**. (b) Tietze, L. F.; Beifuss, U. *Angew. Chem., Int. Ed. Engl.* **1993**, *32*, 131–163. (c) Denmark, S. E.; Thorarensen, A. *Chem. Rev.* **1996**, *96*, 137–165. (d) Posner, G. H. *Chem. Rev.* **1986**, *86*, 831–844. (e) Evans, D. A.; Bryan, C. A.; Sims, C. L. *J. Am. Chem. Soc.* **1972**, *94*, 2891–2892.

12 For an excellent example of a multidirectional, double tandem Diels–Alder reaction, see: (a) Roush, W. R.; Sciotti, R. J. *J. Am. Chem. Soc.* **1994**, *116*, 6457–6458. (b) Roush, W. R.; Sciotti, R. J. *J. Am. Chem. Soc.* **1998**, *120*, 7411–7419.

13 For several elegant examples of hetero-Diels–Alder cascade in natural product synthesis, see: (a) Lumb, J.-P.; Trauner, D. *J. Am. Chem. Soc.* **2005**, *127*, 2870–2871. (b) Elliot, G. I.; Velcicky, J.; Ishikawa, H.; Li, Y.; Boger, D. L. *Angew. Chem.* **2006**, *118*, 636–638; *Angew. Chem. Int. Ed.* **2006**, *45*, 620–622. (c) Kuttruff, C. A.; Zipse, H.; Trauner, D. *Angew. Chem. Int. Ed.* **2011**, *50*, 1402–1405.

14 For selected recent examples of Diels–Alder/aromatization cascades, see: (a) Pottie, I. R.; Nandaluru, P. R.; Bodwell, G. *J. Synlett*, **2011**, *15*, 2245–2247. (b) Petronijevic, F. R.; Wipf, P. *J. Am. Chem. Soc.* **2011**, *133*, 7704–7707. (c) Nandaluru, P. R.; Bodwell, G. J. *Org. Lett.* **2012**, *14*, 310–313. (d) Nandaluru, P. R.; Bodwell, G. J. *J. Org. Chem.* **2012**, *77*, 8028–8037. (e) Nawrat, C. C.; Moody, C. J. *Org. Lett.* **2012**, *14*, 1484–1487. (f) Zhao, P.; Beaudry, C. M. *Org. Lett.* **2013**, *15*, 402–405.

15 Chang, J. H.; Kang, H.-U.; Jung, I.-H.; Cho, C.-G. *Org. Lett.* **2010**, *12*, 2016–2018

16 Brosius, A. D.; Overman, L. E.; Schwink, L. *J. Am. Chem. Soc.* **1999**, *121*, 700–709.

17 (a) Uenishi, J.; Kawahama, R.; Yonemitsu, O. *J. Org. Chem.* **1997**, *62*, 1691–1701. (b) Uenishi, J.; Kawahama, R.; Imakoga, T.; Yonemitsu, O. *Chem. Pharm. Bull.* **1998**, *46*, 1090–1093.

18 Kuehne, M. E.; Brook, C. S.; Frasier, D. A.: Xu, F. *J. Org. Chem.* **1994**, *59*, 5977–5982.

19 (a) Martin, S. F.; Humphrey, J. M.; Ali, A.; Hillier, M. C. *J. Am. Chem. Soc.* **1999**, *121*, 866–867. (b) Humphrey, J. M.; Liao, Y.; Ali, A.; Rein, T.; Wong, Y.-L.; Chen, H.-J.; *et al. J. Am. Chem. Soc.* **2002**, *124*, 8584–8592.

20 Kondo, K; Shigemori, H.; Kikuchi, Y.; Ishibashi, M.; Sasaki, T.; Kobayashi, J. *Org. Chem.* **1992**, *57*, 2480–2483.

21 (a) Crimmins, M. T.; Ellis, J. M. *J. Am. Chem. Soc.* **2005**, *127*, 17200–17201. (b) Crimmins, M. T.; Ellis, J. M. *J. Org. Chem.* **2008**, *73*, 1649–1660.

22 (a) Crimmins, M. T.; Mans, M. C.; Rodríguez, A. D. *Org. Lett.* **2010**, *12*, 5028–5031. (b) Crimmins, M. T.; Stauffer, C. S.; Mans, M. C. *Org. Lett.* **2011**, *13*, 4890–4893.

23 (a) Toró, A.; L'Heureux, A.; Deslongchamps, P. *Org. Lett.* **2000**, *2*, 2737–2740. (b) Soucy, P.; L'Heureux, A.; Toró, A.; Deslongchamps, P. *J. Org. Chem.* **2003**, *68*, 9983–9987.

24 (a) Peixoto, P. A.; Severin, R.; Tseng, C.-C.; Chen, D. Y.-K. *Angew. Chem. Int. Ed.* **2011**, *50*, 3013–3016. (b) Peixoto, P. A.; Richard, J.-A.; Severin, R.; Chen, D. Y.-K. *Org. Lett.* **2011**, *13*, 5724–5727

25 Trost, B. M.; Breder, A.; O'Keefe, B. M.; Rao, M.; Franz, A. W. *J. Am. Chem. Soc.* **2011**, *133*, 4766–4769.

26 Heckrodt, T. J.; Mulzer, J. *J. Am. Chem. Soc.* **2003**, *125*, 4680–4681.

27 (a) Layton, M. E.; Morales, C. A.; Shair, M. D. *J. Am. Chem. Soc.* **2002**, *124*, 773–775. (b) Morales, C.; A.; Layton, M. E.; Shair, M. D. *Proc. Natl. Acad. Sci. USA* **2004**, *101*, 12036–12041.

28 Burke, S. D.; Piscopio, A. D.; Kort, M. E.; Matulenko, M. A.; Parker, M. H.; Armistead, D. M.; *et al. J. Org. Chem.* **1994**, *59*, 332–347.

29 (a) Nicolaou, K. C.; Gray, D.; Tae, J. *Angew. Chem. Int. Ed.* **2001**, *40*, 3675–3678. (b) Nicolaou, K. C.; Gray, D.; Tae, J. *Angew. Chem. Int. Ed.* **2001**, *40*, 3679–3683. (c) Nicolaou, K. C.; Gray, D. L. F.; Tae, J. *J. Am. Chem. Soc.* **2004**, *126*, 613–627.

30 Feltenberger, J. B.; Hayashi, R.; Tang, Y.; Babiash, E. S. C.; Hsung, R. P. *Org. Lett.* **2009**, *11*, 3666–3669.

31 Ma, Z.-X.; Feltenberger, J. B.; Hsung, R. P. *Org. Lett.* **2012**, *14*, 2742–2745.

32 (a) Nelson, H. M.; Murakami, K.; Virgil, S. C.; Stoltz, B. M. *Angew. Chem., Int. Ed.* **2011**, *50*, 3688–3691. (b) Nelson, H. M.; Stoltz, B. M. *Tetrahedron Lett.* **2009**, *50*, 1699–1701. (c) Nelson, H. M.; Stoltz, B. M. *Org. Lett.* **2008**, *10*, 25–28.

33 (a) Nelson, H. M.; Gordon, J. R.; Virgil, S. C.; Stoltz, B. M. *Angew. Chem., Int. Ed.* **2013**, *52*, 6699–6703. (b) Gordon, J. R.; Nelson, H. M.; Virgil, S. C.; Stoltz, B. M. *J. Org. Chem.* **2014**, *79*, 9740–9747.

34 Nicolaou, K. C.; Li, J. *Angew. Chem. Int. Ed.* **2001**, *40*, 4264–4268.

35 (a) Nicolaou, K. C.; Sasmal, P. K.; Xu, H.; Namoto, K.; Ritzen, A. *Angew. Chem. Int. Ed.* **2003**, *42*, 4225–4229. (b) Nicolaou, K. C.; Sasmal, P. K.; Xu, H. *J. Am. Chem. Soc.* **2004**, *126*, 5493–5501. (c) Nicolaou, K. C.; Xu, H.; Wartmann, M. *Angew. Chem., Int. Ed.* **2005**, *44*, 756–761. (d) Hayden, A. E.; Xu, H.; Nicolaou, K. C.; Houk, K. N. *Org. Lett.* **2006**, *8*, 2989–2992.

36 Malerich, J. P.; Maimone, T. J.; Elliott, G. I.; Trauner, D. *J. Am. Chem. Soc.* **2005**, *127*, 6276–6283.

37 (a) Gassmann, P. G.; Singleton, D.; Wilwerding, J. J.; Chavan, S. *J. Am. Chem. Soc.* **1987**, *109*, 2182–2184. (b) Gassmann, P. G.; Chavan, S. P. *J. Org. Chem.* **1988**, *53*, 2392–2394. (c) Gassmann, P. G.; Chavan, S. P. *Tetrahedron Lett.* **1988**, *29*, 3407–3410. (d) Gassmann, P. G.; Chavan, S. P. *J. Chem. Soc., Chem. Commun.* **1989**, 837–839.

38 (a) Sammakia, T.; Berliner, M. A. *J. Org. Chem.* **1995**, *60*, 6652–6653. (b) Sammakia, T.; Berliner, M. A. *J. Org. Chem.* **1994**, *59*, 6890–6891.

39 Sammakia, T.; Johns, D. M.; Kim, G.; Berliner, M. A. *J. Am. Chem. Soc.*, **2005**, *127*, 6504–6505.

40 Grieco, P. A.; Dai, Y.; *J. Am. Chem. Soc.* **1998**, *120*, 5128–5129.

41 Tchabanenko, K.; Adlington, R. M.; Cowley, A. R.; Baldwin, J. E. *Org. Lett.* **2005**, *7*, 585–588.

42 (a) Ginn, J. D.; Padwa, A. *Org. Lett.* **2002**, *4*, 1515–1517. (b) Padwa, A.; Ginn, J. D. *J. Org. Chem.* **2005**, *70*, 5197–5206.

43 (a) Zhang, H.; Boonsombat, J.; Padwa, A. *Org. Lett.* **2007**, *9*, 279–282. (b) Boonsombat, J.; Zhang, H.; Chughtai, M. J.; Hartung, J.; Padwa, A. *J. Org. Chem.* **2008**, *73*, 3539–3550.

44 Padwa, A.; Zhang, H. *J. Org. Chem.* **2007**, *72*, 2570–2582.

45 Li, G.; Padwa, A. *Org. Lett.* **2011**, *13*, 3767–3769.

46 Nicolaou, K. C.; Sarlah, D.; Wu, T. R.; Zhan, W. *Angew. Chem., Int. Ed.* **2009**, *48*, 6870–6874.

47 (a) Germain, J.; Deslongchamps, P. *J. Org. Chem.* **2002**, *67*, 5269–5278. (b) Deslongchamps, P. *Aldrichim. Acta* **1991**, *24*, 43–56. (c) Deslongchamps, P. *Pure Appl. Chem.* **1992**, *64*, 1831–1847. (d) Marsault, E.; Toro, A.; Nowak, P.; Deslongchamps, P. *Tetrahedron* **2001**, *57*, 4243–4260. (e) Germain, J.; Deslongchamps, P. *Tetrahedron Lett.* **1999**, *40*, 4051–4054.

48 (a) Mergott, D. J.; Frank, S. A.; Roush, W. R. *Proc. Natl. Acad. Sci. USA* **2004**, *101*, 11955–11959. (b) Winbush, S. M.; Mergott, D. J.; Roush, W. R. *J. Org. Chem.* **2008**, *73*, 1818–1829.

49 (a) Li, C.; Lobkovsky, E.; Porco, J. A., Jr. *J. Am. Chem. Soc.* **2000**, *122*, 10484–10485. (b) Li, C.; Johnson, R. P.; Porco, J. A., Jr. *J. Am. Chem. Soc.* **2003**, *125*, 5095–5106.

50 (a) Li, C.; Bardhan, S.; Pace, E. A.; Liang, M.-C.; Gilmore, T. D.; Porco, J. A., Jr. *Org. Lett.* **2002**, *4*, 3267–3270. (b) Li, C.; Porco, J. A., Jr. *J. Org. Chem.* **2005**, *70*, 6053–6065.

51 (a) Golden, J. E.; Aubé, J. *Angew. Chem., Int. Ed.* **2002**, *41*, 4316–4318. (b) Zeng, Y.; Aubé, J. *J. Am. Chem. Soc.* **2005**, *127*, 15712–15713. (c) Frankowski, K. J.; Golden, J. E.; Zeng, Y.; Lei, Y.; Aubé, J. *J. Am. Chem. Soc.* **2008**, *130*, 6018–6024.

52 Acebey, L.; Sauvain, M.; Beck, S.; Moulis, C.; Gimenez, A.; Jullian, V. *Org. Lett.* **2007**, *9*, 4693–4696.

53 (a) Yuan, C.; Du, B.; Yang, L.; Liu, B. *J. Am. Chem. Soc.* **2013**, *135*, 9291–9294. (b) Du, B.; Yuan, C.; Yu, T.; Yang, L.; Yang, Y.; Liu, B.; Qin, S. *Chem. Eur. J.* **2014**, *20*, 2613–2622.

54 Jones, S. B.; Simmons, B.; MacMillan, D. W. C. *J. Am. Chem. Soc.* **2009**, *131*, 13606–13607.

55 (a) Laforteza, B. N.; Pickworth, M.; MacMillan, D. W. C. *Angew. Chem., Int. Ed.* **2013**, *52*, 11269–11272. (b) Horning, B. D.; MacMillan, D. W. C. *J. Am. Chem. Soc.* **2013**, *135*, 6442–6445.

56 (a) Vosburg, D. A.; Vanderwal, C. D.; Sorensen, E. J. *J. Am. Chem. Soc.* **2002**, *124*, 4552–4553. (b) Vanderwal, C. D.; Vosburg, D. A.; Weiler, S.; Sorensen, E. J. *J. Am. Chem. Soc.* **2003**, *125*, 5393–5407.

57 Vanderwal, C. D.; Vosburg, D. A.; Sorensen, E. *J. Org. Lett.* **1999**, *1*, 645–648.

58 (a) Evans, D. A.; Starr, J. T. *Angew. Chem. Int. Ed.* **2002**, *41*, 1787–1790. (b) Evans, D. A.; Starr, J. T. *J. Am. Chem. Soc.* **2003**, *125*, 13531–13540

59 (a) Suzuki, T.; Nakada, M.; *Tetrahedron Lett.* **2002**, *43*, 3263–3267. (b) Suzuki, T.; Tanaka, N.; Matsumura, T.; Hosoya, Y.; Nakada, M.; *Tetrahedron Lett.* **2006**, *47*, 1593–1598. (c) Suzuki, T.; Tanaka,N.; Matsumura, T.; Hosoya, Y.; Nakada, M.; *Tetrahedron Lett.* **2007**, *48*, 6483–6487. (d) Tanaka, N.; Suzuki, T.; Hosoya, Y.; Nakada, M.; *Tetrahedron Lett.* **2007**, *48*, 6488–6492. (e) Tanaka, N.; Suzuki, T.; Matsumura, T.; Hosoya, Y.; Nakada, M. *Angew. Chem., Int. Ed.* **2009**, *48*, 2580–2583.

60 Lei, X.; Johnson, R. P.; Porco, Jr. J. A. *Angew. Chem., Int. Ed.* **2003**, *42*, 3913–3917.

61 Sekizawa, R.; Ikeno, S.; Nakamura, H.; Naganawa, H.; Matsui, S.; Iinuma, H.; Takeuchi, T. *J. Nat. Prod.* **2002**, *65*, 1491–1493.

62 (a) Moses, J. E.; Commeiras, L.; Baldwin, J. E.; Adlington, R. M. *Org. Lett.* **2003**, *5*, 2987–3988. (b) Commeiras, L.; Moses, J. E.; Adlington, R. M.; Baldwin, J. E.; Cowey, A. R.; Baker, C. M.; Albrecht, B.; Grant, G. H. *Tetrahedron* **2006**, *62*, 9892–9901.

63 Liu, H.-J.; Zhu, J.-L.; Chen, I.-C.; Jankowska, R.; Han, Y.; Shia, K.-S. *Angew. Chem., Int. Ed.* **2003**, *42*, 1851–1853.

64 (a) Nicolaou, K. C.; Lim, Y. H.; Becher, J. *Angew. Chem. Int. Ed.* **2009**, *48*, 3444–3448. (b) Nicolaou, K. C.; Becker, J.; Lim, Y. H.; Lemire, A.; Neubauer, T.; Montero, A. *J. Am. Chem. Soc.* **2009**, *131*, 14812–14826.

4

Organometallics-Based Catalytic (Asymmetric) Synthesis of Natural Products

Hongbin Zhai[1], *Yun Li*[2], *Bin Cheng*[2], *Zhiqiang Ma*[3], *Peng Gao*[2], *Xin Chen*[4], *Weihe Zhang*[2], *Hanwei Hu*[2], *and Fang Fang*[2]

[1] *School of Chemical Biology and Biotechnology, Shenzhen Graduate School of Peking University, Shenzhen, China*
[2] *Department of Chemistry, Lanzhou University, Lanzhou, China*
[3] *Department of Chemistry, South China University of Technology, Guangzhou, China*
[4] *School of Pharmacy, Presbyterian College, Clinton, SC, USA*

4.1 Introduction

Organometallic chemistry plays an important role in natural product synthesis as it provides unique techniques for carbon–carbon and carbon-heteroatom bond formations that may not be easily accessible by other conventional means. In addition to the fundamental bond formation, its use in domino reactions can easily access structurally complex key intermediates in natural products synthesis; as a result, it allows chemists to access the targeted natural products in an efficient and stereoselective fashion. Moreover, natural product syntheses will benefit from the discovery of unprecedented organometallic transformations. In this chapter, we will cover a variety of metal complexes including Au, Ag, Pt, Co, Cu, Cr, Fe, Mn, Ni, Pd, Rh, and Ru with their applications to natural product syntheses.

4.2 Au-Catalyzed Reactions in Total Synthesis

Gold-catalyzed reactions are very synthetically attractive because they have several unique characteristics: They significantly increase molecular complexity; they can work on a diverse set of reactions; the reaction conditions are mild, often room temperature; Au(I) generally is oxygen-tolerant; minimal additives are required; they have excellent chemoselectivity; asymmetric control is possible; work-ups are usually straightforward; the pre-catalysts are

Efficiency in Natural Product Total Synthesis, First Edition. Edited by Pei-Qiang Huang, Zhu-Jun Yao, and Richard P. Hsung.

robust and readily accessible; they have orthogonal reactivity to many other transition-metal-catalyzed processes; and Au(I) and Au(III) do not readily cycle between oxidation states. Gold catalysts range from simple Au(I) and Au(III) halides and cationic gold complexes employing spectator ligands such as phosphines and *N*-heterocyclic carbenes. Gold can catalyze a variety of reactions including X–H addition across π–systems, indirect X-H addition, C-H addition across π–systems, C–X addition across π–systems, cycloadditions, enyne cycloisomerizations, and so forth.

Gold-catalyzed bis-spiroketalization was employed to construct the A and B rings of the marine toxin azaspiracid, setting up one of the outstanding examples of gold-catalyzed intramolecular O-H addition to alkynes. Gold-mediated hydroalkoxylation of the alkyne **4.1** afforded an enol ether that further cyclized under protic conditions to the desired structure (Scheme 4.1).[1]

Another absolutely stunning example of gold-catalyzed intramolecular O-H addition to alkynes comes from the total synthesis of bryostatin **16** using atom-economical and chemoselective approaches.[2] Bryostatin **16** is a parent structure from which almost all other bryostatins could in principle be accessed. Gold-catalyzed intramolecular O-H addition to alkyne was one of the two key steps in the concise total synthesis. Formation of the C ring of bryostatin via a *6-endo-dig* cyclization represented one of the biggest challenges, where a cationic gold complex, $[Au(PPh_3)]SbF_6$, was found to be superior to a number of other metal complexes. Inclusion of $NaHCO_3$ (10 equiv.) as a buffer for the reaction system in DCM-CH_3CN (10:1) led to the acid-sensitive 6-*endo* product **4.4** in 73% yield. The cyclization occurred in the presence of acetals, activated alkenes and unprotected alcohols (Scheme 4.2).

Gold complexes, particularly $[(Ph_3PAu)_3O]BF_4$, can catalyze the cyclization of an enol ether with an alkyne, resulting in the formation of dihydropyran derivatives. In the absence of an external nucleophile, the same reaction affords allenols after reduction that can be substrates for further gold-catalyzed cyclization. This propargyl Claisen rearrangement can be applied to highly substituted and labile species such as azadirachtin.[3,4]

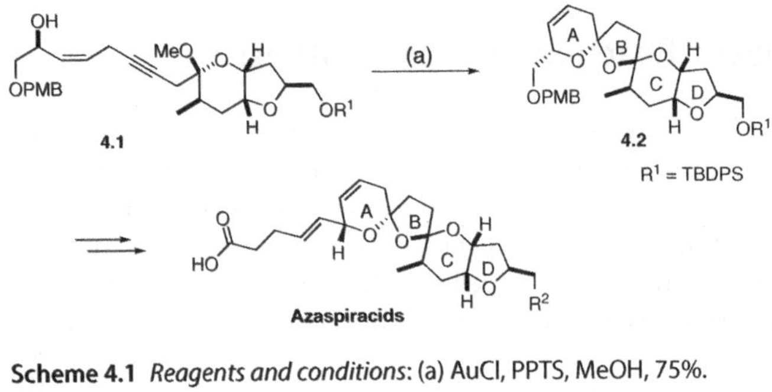

Scheme 4.1 *Reagents and conditions*: (a) AuCl, PPTS, MeOH, 75%.

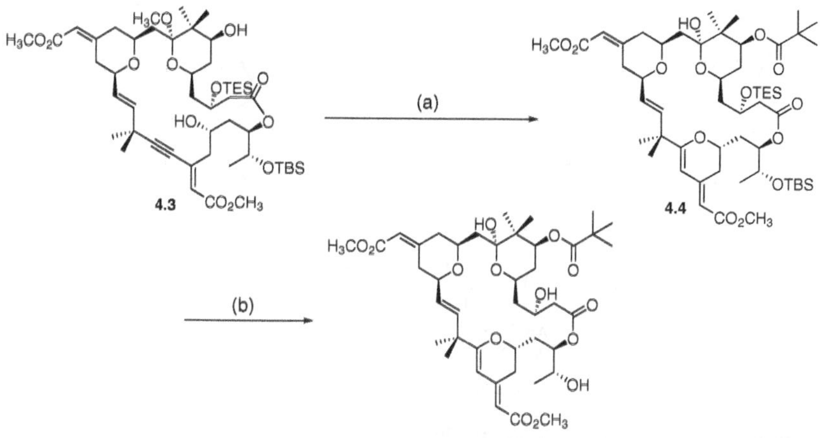

Scheme 4.2 *Reagents and conditions*: (a) (i) $AuCl(PPh_3)$ (20 mol%), $AgSbF_6$ (20 mol%), $NaHCO_3$, DCM/MeCN, 0°C RT, 73%; (ii) Piv_2O, DMAP, 50°C, 62%; (b) TBAF, THF, 52%.

Azadirachtin is a complex natural product that has been at the center of structural, biological, and synthetic studies ever since its isolation from the Indian neem tree *Azadirachtaindica* in 1968.[5] It has very complex molecular architecture including 16 contiguous stereogenic centers, seven of which are tetrasubstituted carbon atoms. Azadirachtin possesses a diverse array of oxygenated functionalities in addition to a rigid conformation imposed by intramolecular hydrogen-bonding. Moreover, its sensitivity to acids and bases together with its photo-instability make it particularly prone to rearrangement, thereby frustrating many synthesis plans. Steven Ley's group reported a successful approach that included a gold-catalyzed reaction as a key step to the final compound. Claisen rearrangement of propargylicenol ether **4.5** under gold(I)-catalyzed conditions was employed to construct the key C8-C14 bond while simultaneously installing the requisite allene for the ensuing radical cyclization leading to **4.6**, which is six steps away from azadirachtin. It is worthwhile noting that this step could also proceed under thermal conditions (microwave, 180 °C); however, with the gold catalyst, the reaction could be performed at room temperature in 80% yield (Scheme 4.3).

This gold-catalyzed cyclization reaction can also be applied to silylenol ethers. The cyclization of silylenol ethers onto alkynes is a powerful methodology for the formation of C–C bonds, with O–Si cleavage in the presence of alcohol/water additives. This process has been applied in the synthesis of natural products such as lycopladine,[6] platencin,[7] and fawcettimine[8] to name a few. The fawcettimine class of *Lycopodium* alkaloids consists of over 60 natural products that are typically tetracyclic compounds containing a single quaternary carbon center and are derived biosynthetically from the lycopodane core

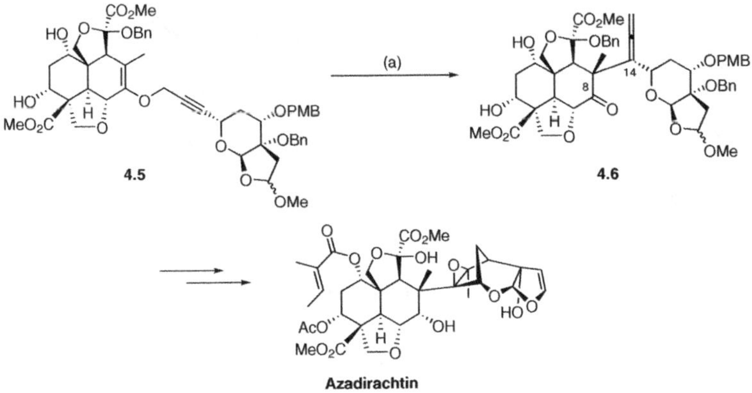

Scheme 4.3 *Reagents and conditions*: (a) $[(Ph_3PAu)_3O]BF_4$, CH_2Cl_2, RT, 80%.

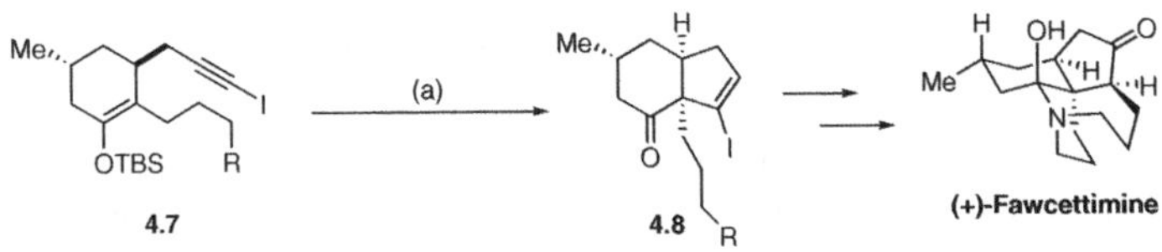

Scheme 4.4 *Reagents and conditions*: (a) $(Ph_3PAu)Cl$ (10%), $AgBF_4$ (10%). CH_2Cl_2/MeOH, 40°C.

through an oxidative rearrangement reaction. The first successful asymmetric total synthesis of (+)-fawcettimine from the Toste group was accomplished in 13 steps from crotonaldehyde. Ultimately, this work demonstrates that organocatalytic annulation and gold(I)-catalyzed cyclization reactions are an effective combination for the synthesis of complex molecules. Catalyzed by (triphenylphosphine)gold(I) tetrafluoroborate, *5-endo-dig* carbocyclization of **4.7** proceeded smoothly to furnish the hydrindanone core (**4.8**) of fawcettimine (Scheme 4.4). Notably, both the [4.3.0] bicyclic structure and quaternary carbon center were installed in a single step.

Gold compounds, especially $AuCl_3$, are also widely applied to benzannulation reactions. Oxonium (or iminium) species generated from the nucleophilic attack of a carbonyl (imine) onto a gold activated alkyne may be employed in cycloaddition reactions as 1,3- or 1,4-dipoles. Benzannulated products are formed after further rearrangement. Formal [4 + 2] benzannulation of o-alkynylbenzaldehydes was employed as the key step in the synthesis of (+)-rubiginone and (+)-ochromycinone[9] (Scheme 4.5).

Gold can also catalyze the intramolecular coupling of aldehydes/ketones and enynes through which unusual tricyclic structures can be accessed by intermolecular capture of enyne intermediates acting as all-carbon 1,4-dipoles.

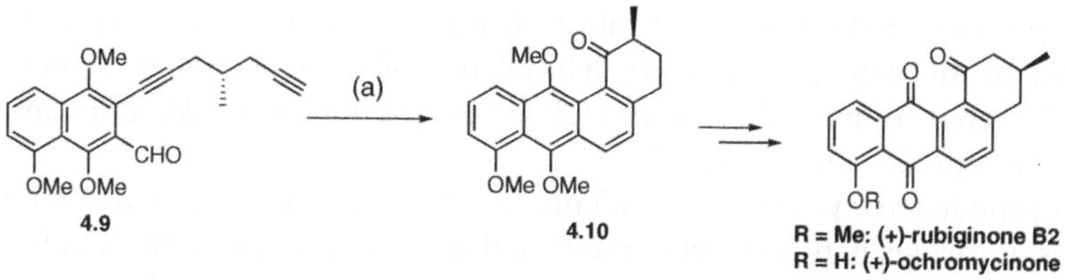

Scheme 4.5 *Reagents and conditions*: (a) $AuCl_3$, DCE, 50°C, 1 h, 84%.

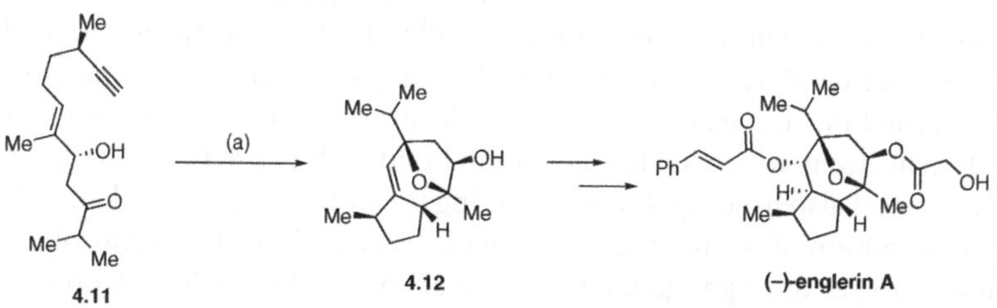

Scheme 4.6 *Reagents and conditions*: (a) AuCl, 20 min, CH_2Cl_2, 48%.

Collapse of the intermediate results in the formation of the cyclopropane ring and the release of the catalytic gold species. This process was widely applied in the synthesis of complex natural products such as (+)-orientalol F[10] and (−)-englerins A and B (Scheme 4.6).[11, 12]

4.3 Ag-Catalyzed Reactions in Total Synthesis

Silver, mainly used as a precious metal, has outstanding electrical conductivity and unique redox chemistry. The use of silver in organic chemistry today can be classified into two well-defined areas: Heterogeneous oxidation processes and homogeneous silver-mediated or catalyzed reactions.

Homogenous silver-mediated reactions, either stoichiometrically or catalytically, have had great development in the past two decades, especially within the last 10 years. Probably due to its moderate Lewis acidity, silver has been neglected in organic synthesis for a long time. Ag(I) can form complexes with π-donors (such as alkenes and aromatics) and n-donors (ethers and amines). Silver cation activation of olefin, propargyl, and allenyl substrates is involved in Ag(I)-mediated nucleophilic addition reactions. These reactions include cyclization of alkyl alcohols, acids, amines, and imines, cyclization of allenyl ketones and aldehydes, and nucleophilic cyclization of allenes. In addition to these reactions, nucleophilic intermolecular addition to alkynes and nucleophilic additions of hydroxyl and carboxylic groups to unactivated alkenes have been reported recently.[13]

While promoting the heterocyclization through C-O bond formation, the substrates for Ag-mediated reaction can be allenes (allenols, allenones, allenoic acids and related compounds), alkynes (alkynols, alkynoic acid, alkynones and related compounds), and alkenes.[14]

The Ag-mediated cyclization of allenic alcohols has been applied to the syntheses of various natural products and analogs. Verrucosidin, citreoviridin, and their metabolite citreal have been prepared via an Ag-promoted stereoselective cyclization of dihydroxyallene **4.13**[15] (Scheme 4.7).

The cyclization of α-allenones was also successfully applied to total synthesis of natural products. The furanocembranes kallolide A and B were obtained through two Ag-catalyzed cyclizations as key steps.[16, 17] In the synthesis of kallolide A, the first step was an allenone cyclization using a catalytic amount of $AgNO_3$ in acetone, which led to a key furan building block **4.16** in very good yield. The second also used $AgNO_3$ for building the butenolide part of kallolide A. An isomerization of the intermediate allenic ester with triphenylphosphine was required to get the right geometry of the allenic moiety, before deprotection of the acid and cyclization to the corresponding butenolide. This step required an excess of sliver reagent (Scheme 4.8).

$AgNO_3$-catalyzed cyclization of an appropriate acetylenic acid has been the key step for the synthesis of dechloro analog of the antibiotic and algicidal cyanobacterin isolated from the freshwater cyanobacterium *Syctonemahofmanni*. This reaction allowed the benzylidene lactone structure of this antibiotic analog to be assembled efficiently and stereoselectively (Scheme 4.9).[18, 19]

Ag-catalyzed cyclization has been employed elegantly in two total syntheses of lissoclinolide with highly conjugated structure. In one synthesis, protected (*E*)-pent-2-en-4-ynol was coupled with methyl 2,3-dibromoprop-2-enoate and,

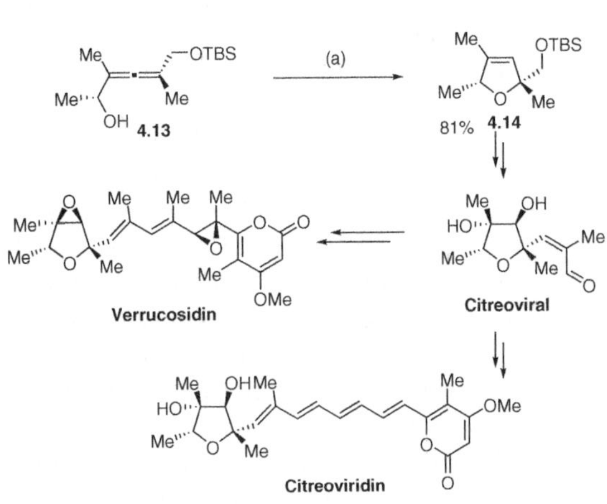

Scheme 4.7 *Reagents and conditions*: (a) $AgNO_3$ (0.8 equiv.), $CaCO_3$, H_2O/acetone.

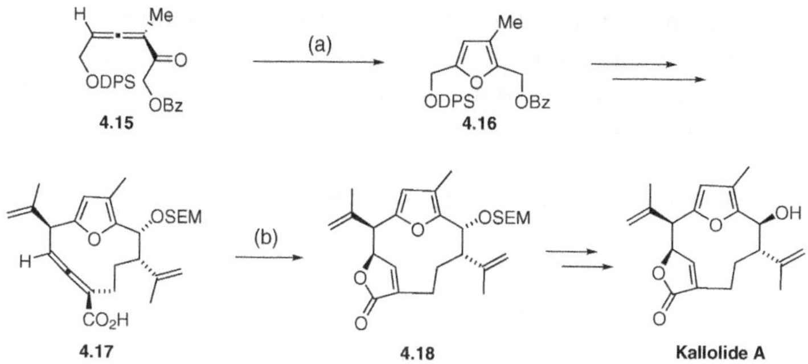

Scheme 4.8 Reagents and conditions: (a) $AgNO_3$ (0.1 equiv.), 88%; (b) $AgNO_3$, MeOH, 60%.

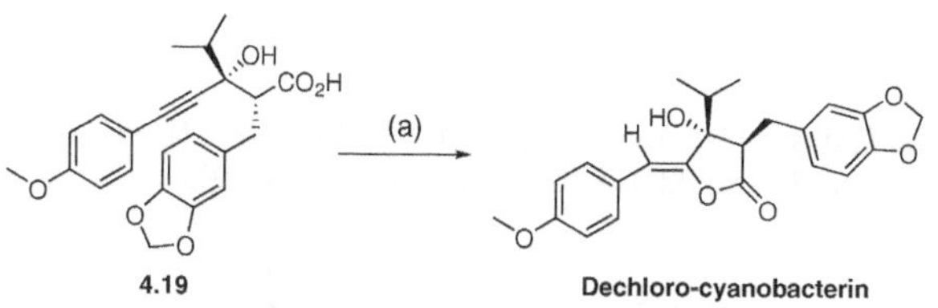

Scheme 4.9 *Reagents and conditions*: (a) $AgNO_3$, MeOH, 82%.

the resulted bromodienyneester was converted into the corresponding acid **4.22**, which cyclized to form **4.23** with silver nitrate as a catalyst in acetone. The alkylidenebromolactone **4.23** so formed was then homologated by coupling again, yielding the natural product in only seven steps (Scheme 4.10).[20] In the other synthesis, the monoprotected hex-2-en-4-yn-1,6-diol was homologated by the Linstrumelle's coupling, then further coupled with a vinylzirconocene, and homologated again to produce a trienyne acid **4.20**. The latter was then quantitatively cyclized upon treatment by methanolic silver nitrate (Scheme 4.10). Deprotection furnished the natural product in good overall yield (38% over seven steps).[21]

Ag can also mediate the heterocyclization through C-N bond formation. The substrates include allenes (allenamines, allenimines, and oximes, and allenamides and related compounds), alkynes, and alkenes. The total synthesis of clavepictines, antibiotic, and antitumor quinolizidine alkaloids isolated from a tunicate, was cleverly designed and realized with Ag-catalyzed cyclization of the piperidinoallene **4.24** as the key step (Scheme 4.11). Control of the configuration of allenes imposed a favored transition state, which afforded the desired *cis*-quinolizidine derivative in 48% yield. Only 7% of vinyl isomer was isolated.[22]

Ag salt is also well-known for catalyzing the [3,3]-sigmatropic rearrangement, through which the heterocyclization reaction can be accomplished.

Quantitative chiral transfer was achieved starting from an enantio-enriched propargyl ester. This was applied to the synthesis of (−)-ascofuranone, and the chirality was introduced by the asymmetric reduction of the alkynone with (S)-alpine borane (Scheme 4.12).[23]

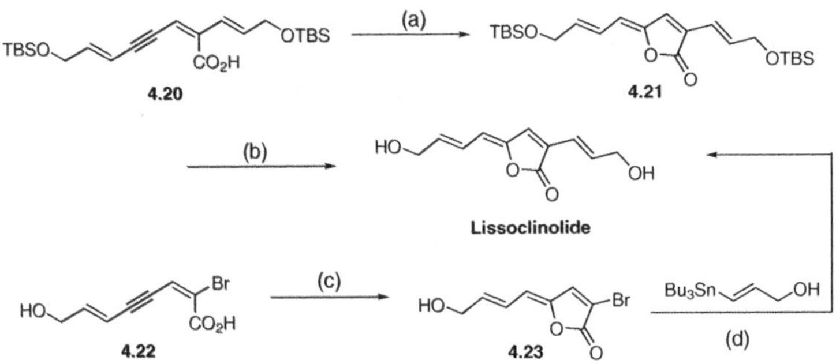

Scheme 4.10 *Reagents and conditions*: (a) 5% $AgNO_3$, MeOH, 100%; (b) HF, MeCN, 0°C, 85%; (c) 20% $AgNO_3$, acetone, 20°C, 41 h, 77%; (d) $PdCl_2(PhCN)_2$, CuI, $AsPh_3$, NMP, 59%.

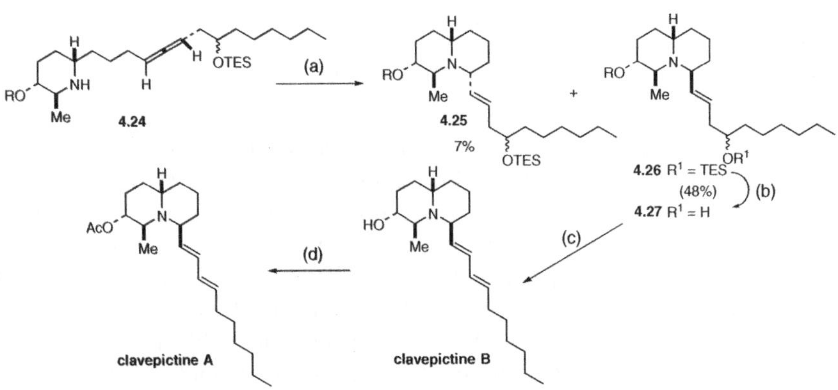

Scheme 4.11 *Reagents and conditions*: (a) $AgNO_3$ (7% for **4.25**, 48% for **4.26**); (b) *n*-Bu_4NF, 81%; (c) (i) Martin sulfurane; (ii) *n*-Bu_4NF, 84% (2 steps); (d) Ac_2O, pyr., 74%.

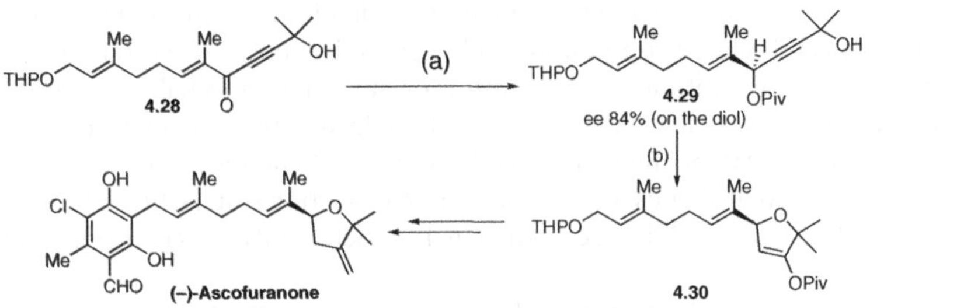

Scheme 4.12 *Reagents and conditions*: (a) (i) (S)-Alpine borane; (ii) PivCl, DMAP, *ee* 84% (on the diol); (b) $AgBF_4$, 65%.

4.4 Pt-Catalyzed Reactions in Total Synthesis

As a member of noble metals, Pt itself is not active and exhibits a remarkable resistance to corrosion, even at high temperatures. Pt compounds/complexes, on the other hand, have been used as catalysts in organic chemistry for a long time because they, like Au, are convenient tools for generating molecular complexity and thus are widely applied in total synthesis of natural products.

4.4.1 Pt-Catalyzed Enyne Cycloisomerization Reactions

The transition-metal-catalyzed enyne cycloisomerization is among the most important strategies for the synthesis of functionalized cyclic structures [24] and is therefore the first noble metal-catalyzed process to be implemented into total synthesis of natural products.[25] Through this process, the structural complexity can be rapidly increased starting with relatively simple acyclic subunits containing ene and yne fragments. Due to the highly catalytic capability of gold and platinum complexes in these reactions, a diverse array of cyclic products have been delivered under mild conditions with excellent chemoselectivity and high synthetic efficiency. A notable example comes from the synthesis of cortistatin core.[26] The desired cycloheptadiene (**4.32**, Scheme 4.13) could be readily obtained by treating **4.31** with a catalytic amount of $PtCl_2$. This cycloisomerization occurred without significant amount of isomeric by-products generated.

The Pt together with other noble metals, catalyzed propargyl ester chemistry are always well behaved in a stereochemical regard. They allow the configuration of the substrate to be translated with high fidelity into the resulted cyclopropane product. This feature was very essential in the total synthesis of (−)-α-cubebene and (−)-cubebol (Scheme 4.14).[27] Activation of the triple bond in propargylic esters **4.33** by Pt might stimulate either anchimeric participation of the adjacent carboxylate as the primary step or being attacked by a nearby double bond if enyne reactivity prevails (Ohloff–Rautenstrauch rearrangement, Scheme 4.14). In any case, the system furnished a cyclopropylenol ester, which upon hydrolysis afforded the cyclopropyl carbonyl derivative (**4.34**) as the final product. Consequently, propargylic esters may be regarded as simple, safe, and convenient alternatives to α-diazocarbonyl

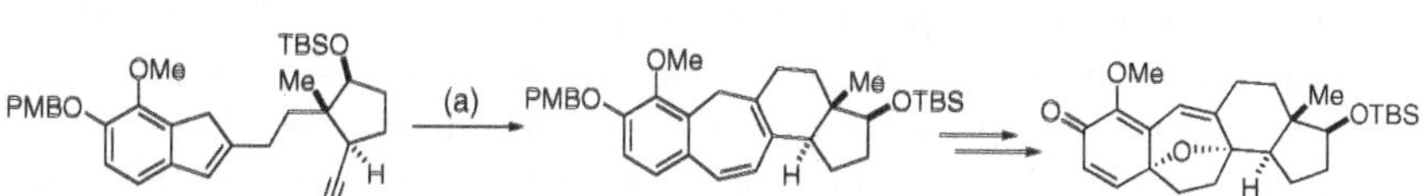

4.31 4.32 Cortistatin core

Scheme 4.13 *Reagents and conditions*: (a) $PtCl_2$ (10 mol%), RT −40°C, 61%.

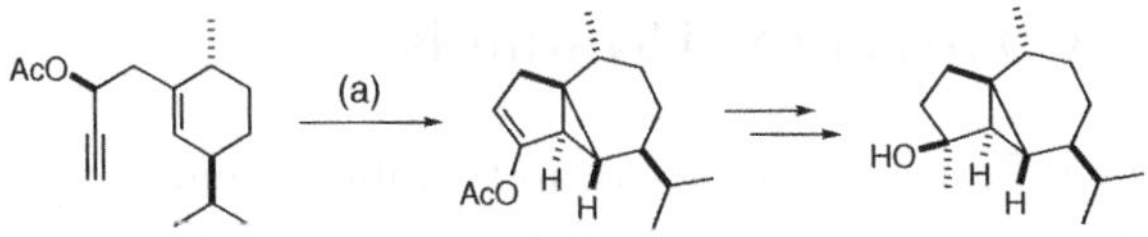

4.33 4.34 (−)-cubebol

Scheme 4.14 *Reagents and conditions*: (a) $PtCl_2$, tol. 80°C, 92%.

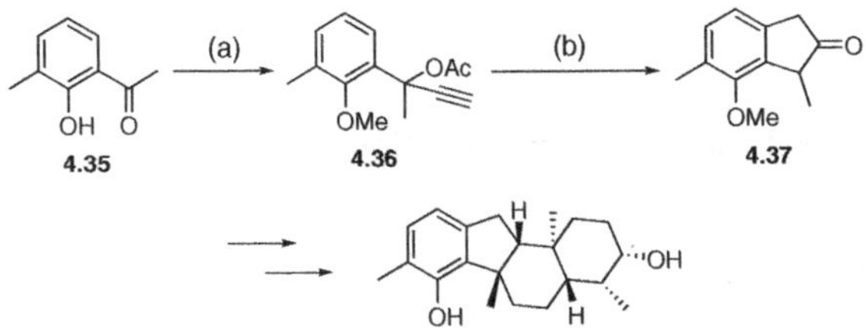

4.35 4.36 4.37

(±)-Dasyscyphin D

Scheme 4.15 *Reagents and conditions*: (a) (i) K_2CO_3, MeI, acetone; (ii) ethynylmagnesium bromide; (iii) Ac_2O, Py, DCM, 95% (3 steps); (b) $PtCl_2$ (5 mol%), tol. 60°C, then HCl, 87%.

derivatives that are commonly used for the preparation of such products. It is interesting to note that in addition to $PtCl_2$, $[Cu(MeCN)_4]BF_4$[28] and gold complexes[29] have also been reported to catalyze the key transformation forging the tricyclic skeleton of the targets in good yields.

Pt can also catalyze pentannulation of propargyl acetates and this was elegantly employed in the efficient total synthesis of (±)-dasyscyphin D.[30] One of the key steps involved a $PtCl_2$-catalyzed pentanulation reaction and the target compound was accomplished in nine steps with a 22.6% overall yield (Scheme 4.15).

4.4.1.1 Unconventional Nucleophiles

While enynes are the first substrates for platinum catalysis, mechanistic analysis suggested that only the alkyne unit needs to be activated by the π-acid, and the alkene merely serves as a nucleophile,[31] therefore, it can be replaced by other suitable reaction partners such as amines, alcohols, ethers, epoxides, acetals, oxime derivatives, aziridines, triazenes, sulfoxides, hydroxylamine derivatives, and many others. A representative case is the total synthesis of the antibiotic erypoegin H and congeners. A remarkably clean cycloisomerization of compound **4.38**, on exposure to catalytic amounts of $PtCl_2$ in toluene under a CO atmosphere, formed the desired benzofuran derivative **4.39**. In the presence of powdered molecular sieves this reaction proceeded exceedingly well and afforded **4.39** in 84% yield on a multigram scale (Scheme 4.16).[32]

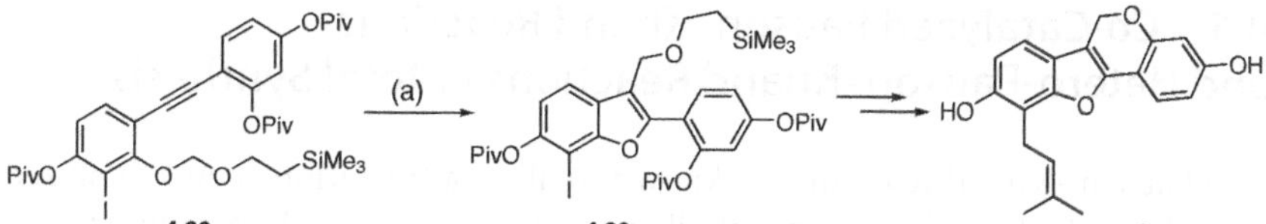

4.38 4.39 erypoegin H

Scheme 4.16 *Reagents and conditions*: (a) $PtCl_2$ (10 mol%), CO (1 atm), tol. 80 °C, 84%.

4.4.1.2 Cascade Reactions

Because of the charged character of the putative reactive intermediates, Pt-catalyzed reactions are predisposed to amalgamation with cationic processes of, for example, the Friedel–Crafts-, Wagner–Meerwein-, pinacol-, Nazarov-, or Prins-types. There are some excellent examples of these Pt-catalyzed cascade reactions in the literature including the synthesis of nakadomarin A[33] and the highly cytotoxic tylophora alkaloids.[34, 35] In the synthesis of nakadomarin A by Hongbin Zhai's group, π-acid catalysis was successfully linked to prototype Friedel–Crafts type process. An *N*-acyliminium ion, generated in situ upon attack of the enamide on the proximal alkyne, triggered a cyclization leading to the tetracyclic core of nakadomarin A (Scheme 4.17).

4.4.1.3 Formation of Quaternary Chiral Centers

Various Pt-catalyzed reactions can help the construction of (all-carbon) quaternary centers, which would be difficult to achieve by other means. The total synthesis of (+)-fawcettidine, for example, nicely adapted a platinum catalyzed addition of the enamide to alkyne for this very purpose (Scheme 4.18).[36]

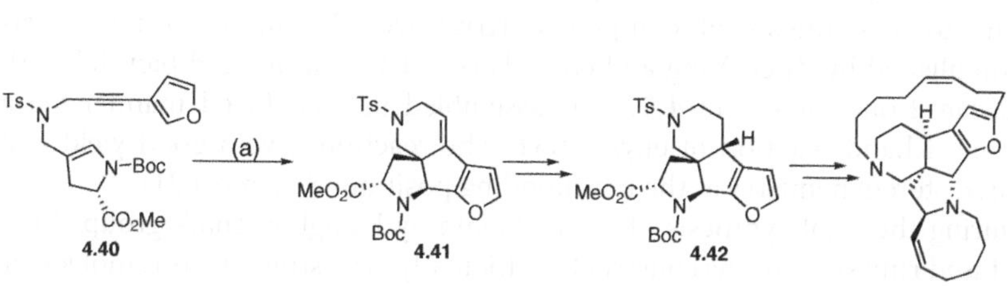

4.40 4.41 4.42 (−)-nakadomarin A

Scheme 4.17 *Reagents and conditions*: (a) $PtCl_2$, tol. heat, 50%.

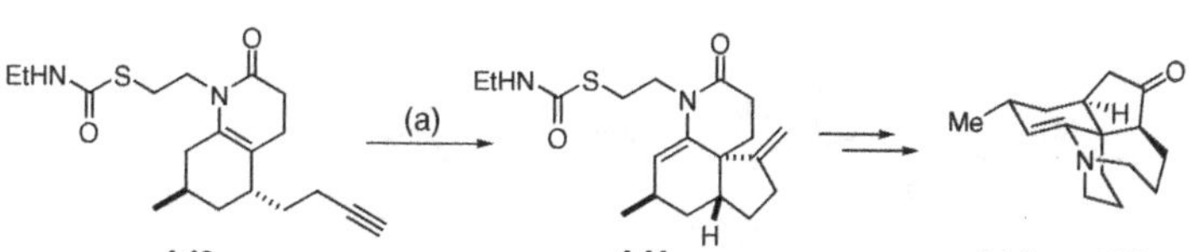

4.43 4.44 (+)-fawcettidine

Scheme 4.18 *Reagents and conditions*: (a) $PtCl_2$ (10 mol%), tol. 90 °C, 87%.

4.5 Co-Catalyzed Pauson–Khand Reactions and Hetero-Pauson–Khand Reactions in Total Synthesis

The Pauson–Khand reaction[37] is known to all as a transition-metal catalyzed formal [2+2+1] cycloaddition of alkynes, alkenes, and carbon-monoxide to form cyclopentenones. The first Pauson–Khand reaction was reported by I. U. Khand and P. L. Pauson using a cobalt complex as catalyst to give cyclopentenone at a good yield.[38] Since then, this efficient reaction has become one of the most important methods for constructing cyclopentenones and bicyclic enones. Here are some excellent examples and applications in total synthesis.

During the enantioselective total synthesis of (−)-pentalenene by Joseph Fox's Group,[39] the biosynthetic precursor to the pentalenolactone family of antibiotics,[40] the cyclopropenyne **4.45**, was used as the intramolecular Pauson–Khand reaction precursor to establish the quaternary center in **4.46** with good selectivity. Compound **4.46** was transferred to (−)-pentalenene in steps to complete the first enantioselective synthesis of the target molecular (Scheme 4.19).

In the laboratory of Chisato Mukai, the total syntheses of (−)-magellanine, (+)-magellaninone and (+)-paniculatine were accomplished through Pauson–Khand reactions.[41] Enyne **4.47** was subjected to Pauson–Khand reaction conditions based on Sugihara's procedure[42] to construct the bicycle[4.3.0] carbon framework in compound **4.48** with good stereoselectivity. Then **4.48** was transferred to enyne **4.49** in a few steps, the second Pauson–Khand reaction was carried out under similar conditions to produce the bicycle[3.3.0] skeleton, which could be converted into 6,5,5,6-tetracyclic frameworks of (−)magellanine, (+)-magellaninone, and (+)-paniculatine in a few steps (Scheme 4.20).

The total synthesis of complex nortriterpenoid schindilactone A was accomplished by Zhen Yang and coworkers.[43] The oxa-bridged bicycle[6.3.0] undecan-2-one moiety in **4.52** was assembled in a *cis*-fused manner via a Pauson–Khand reaction of enyne **4.51**. The reaction gave a good yield and desired stereochemistry at the new bonding position (Scheme 4.21).

During the total synthesis of (−)-jiadifenin by Hongbin Zhai's group,[44] the 5,6-fused ring system was constructed efficiently via a strategic intramolecular Pauson–Khand reaction. Enyne **4.53** (**4.53α** was major, inseparable) subjected to Pauson–Khand reaction conditions furnished cyclopentenones **4.54**, which was the desired core structure of jiadifenin (Scheme 4.22).

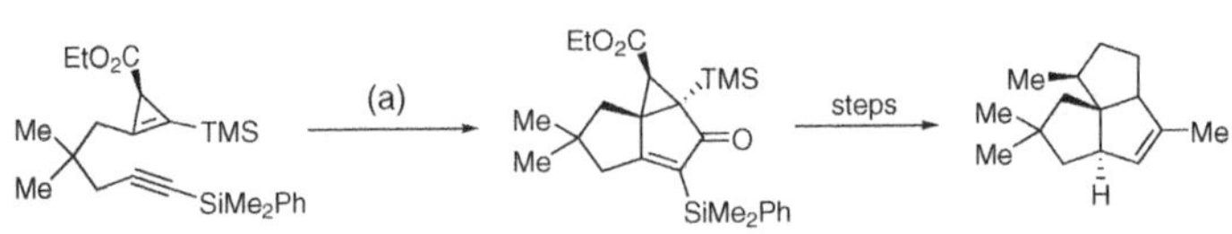

4.45 **4.46** **(−)-pentalenene**

Scheme 4.19 *Reagents and conditions*: (a) $Co_2(CO)_8$, tol., 36 °C, then TMTU, CO (1 atm), 45 °C, 14 h, 64%.

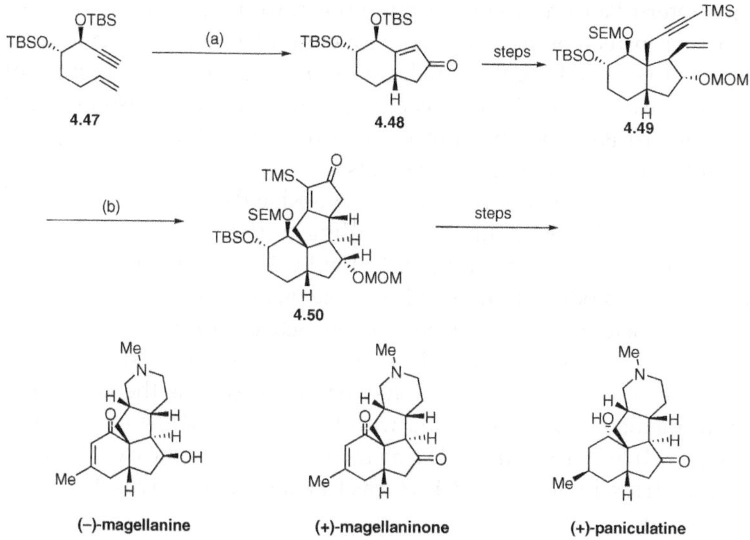

Scheme 4.20 *Reagents and conditions*: (a) $Co_2(CO)_8$, Me_2S, $(CH_2Cl)_2$, 45 °C, 89%; (b) $Co_2(CO)_8$, Me_2S, $(CH_2Cl)_2$, 18 °C, 79%.

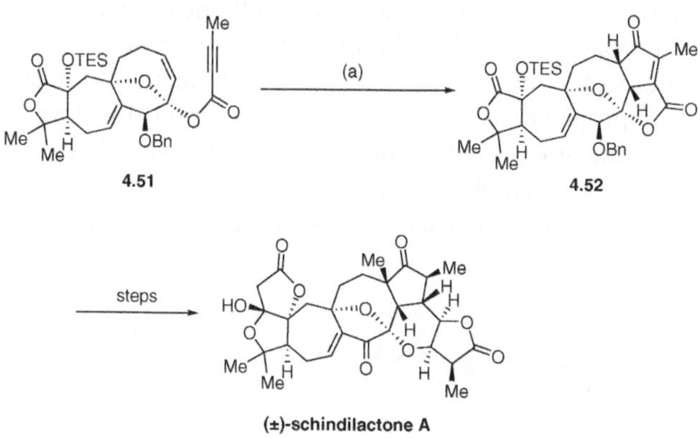

Scheme 4.21 *Reagents and conditions*: (a) $Co_2(CO)_8$, TMTU, CO, PhH, 70 °C, 4 h, 74%.

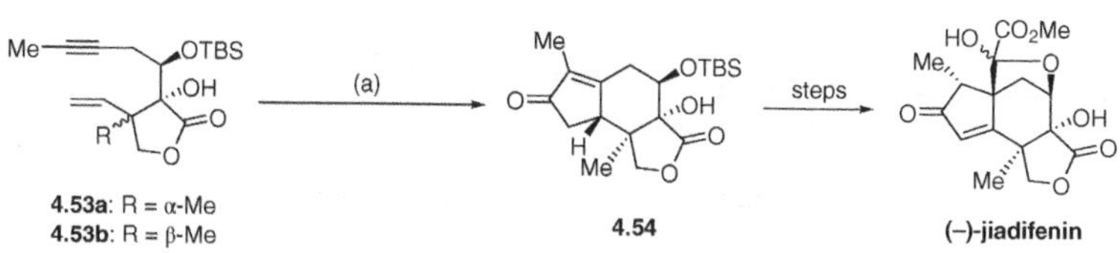

Scheme 4.22 *Reagents and conditions*: (a) $Co_2(CO)_8$, Bu_3PS, toluene, RT –75 °C, 67%.

The term Hetero-Pauson–Khand reaction refers to the use of the imine or carbonyl partner instead of the alkene component under Pauson–Khand reaction conditions to give α,β-unsaturated lactams (aza-Pauson–Khand reaction) or butenolides (oxa-Pauson–Khand reaction).[45] As with the Pauson–Khand reaction, the hetero-Pauson–Khand reaction also has very important applications in the total synthesis of natural products.

In the laboratory of Chisato Mukai, an aza-Pauson–Khand reaction was employed for synthesis of (±)-physostigmine.[46] According to the $Co_2(CO)_8$-catalyzed aza-Pauson–Khand reaction, the pyrrolo[2, 3-*b*]-indol-2-one **4.56** was prepared in 55% yield from the carbodiimide **4.55**. This was the first demonstration of the use of $Co_2(CO)_8$ in the hetero-Pauson–Khand reaction (Scheme 4.23).

Hongbin Zhai's group accomplished an efficient total synthesis of (±)-merrilactone A with a hetero-Pauson–Khand reaction as the key step.[47] The ynal intermediate **4.57** was treated with $[Mo(CO)_3(DMF)_3]$ under a CO atmosphere in THF to furnish the B/D ring in one step. After the oxa-Pauson–Khand reaction, tricyclic compound **4.58** was transferred to merrilactone A in a few steps (Scheme 4.24).

4.6 Cu-Catalyzed Reactions in Total Synthesis

Copper is one of the most useful transition metals in synthetic organic chemistry. Since 1960s, organocopper reagents have become the most popular synthetic tools in the total synthesis of natural products.[48] Copper-mediated conjugate addition and arene cyclopropanation are two types of common transformations, these reactions found great application in total syntheses of natural products. Here we have some examples in which copper-mediated reactions have been used as the key steps to construct target molecules efficiently.

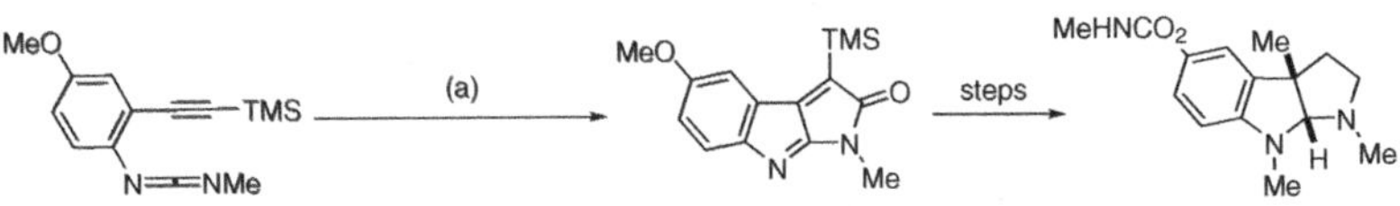

Scheme 4.23 *Reagents and conditions*: (a) $Co_2(CO)_8$, TMTU, CO, PhH, 70 °C, 55%.

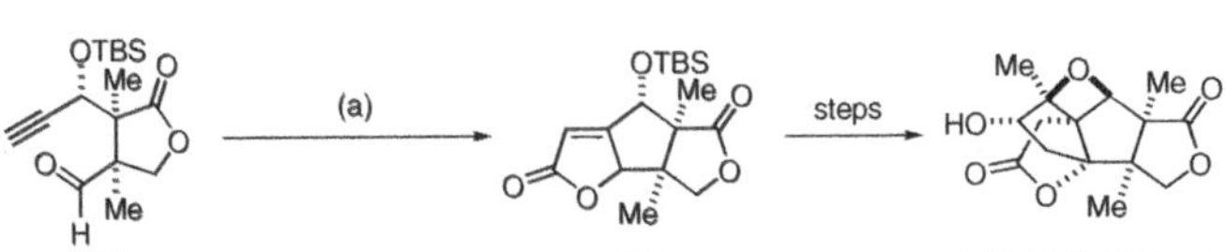

Scheme 4.24 *Reagents and conditions*: (a) $Mo(CO)_3(DMF)_3$, CO, THF, RT, 69%.

4.6.1 Asymmetric Conjugate Addition

Prostaglandins are unsaturated carboxylic acids, consisting of a 20-carbon skeleton that also contains a five-membered ring, and are based upon the fatty acid, arachidonic acid. There are a variety of structures containing one, two, or three double bonds in the side chains. On the five-membered ring there may also be double bonds, a ketone, or hydroxyl groups.

One of the most efficient strategies to construct whole frameworks of the prostaglandins is the three-component coupling process, which involves the organometallic mediated conjugate addition of the side-chain unit to 4-oxygenated 2-cyclopentenones followed by trapping of the enolate species by organic halides with α side-chain structures (Scheme 4.25).

An earlier extensive study on this possibility showed extreme difficulty in achieving the direct alkylation.[49] In 1985, R. Noyori and coworkers finally realized this earnestly desired convergent synthesis.[50] The success relied on the lithium (or copper) to tin transmetalation in the enolate stage. PGE_2 can be obtained from **4.64** by removal of the protecting groups. The acetylenic compounds of the **4.65** type served as common intermediates for the general synthesis of the PG family (Scheme 4.26).

FK-506 has interesting biological properties for immunosuppressive activity, in addition, FK-506 presents a significant challenge to the synthetic organic chemists. Robert Ireland and coworkers reported a convergent route for the total synthesis of FK-506, which involved an efficient copper-mediated conjugate addition as the key step.[51]

With the completion of synthesis of the two fragments **4.67** and **4.68**, the use of hexynylcopper as its *bis*-HMPT complex gave a reproducible 70% yield of the ketone **4.69**. The steroselectivity was high and no diasteromeric conjugate addition product was observed or isolated. The success of this coupling procedure provides an ideal alternative to the problem of trisubstituted olefin synthesis encountered in previous syntheses[52] (Scheme 4.27).

Natural product iso[7]-levuglandin D2, with an α,β-disubstituted ketone as the core structure was synthesized with a convergent synthetic strategy via copper-mediated conjugate addition.[53] The higher order cyanocuprate **4.72** was generated by transmetalation of vinylstannane **4.71** with $Li_2Me_2Cu(CN)$. The highly functionalized vinylcyanocuprate **4.72** added to the enone **4.70** to produce a mixture of diastereomeric conjugate addition products **4.73** in 65% total yield (Scheme 4.28).

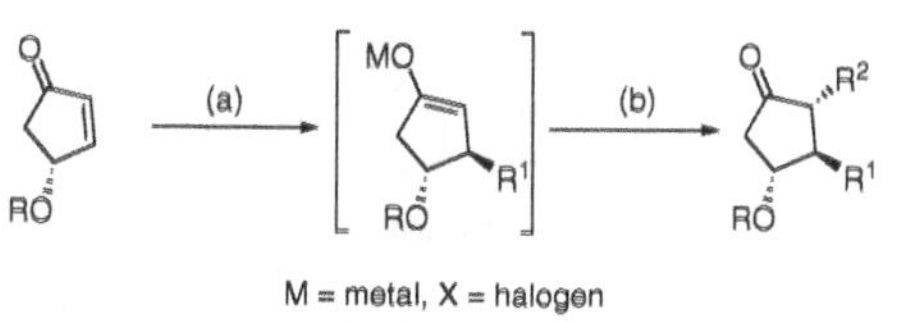

Scheme 4.25 *Reagents and conditions*: (a) $[R^1M]$; (b) $[R^2M]$.

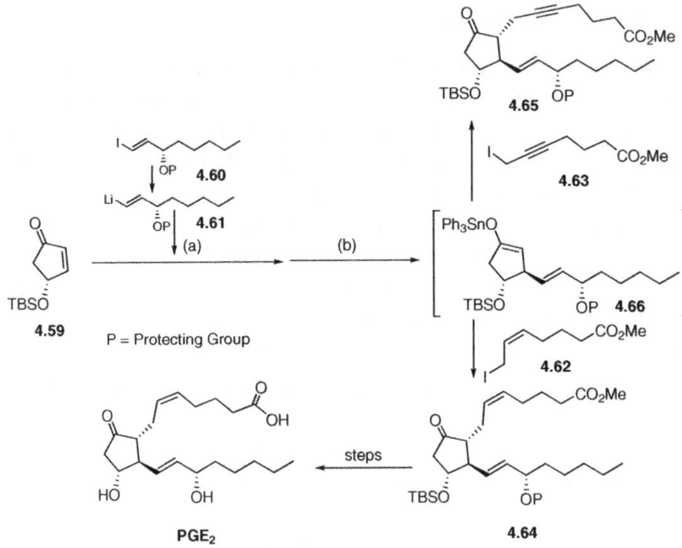

Scheme 4.26 *Reagents and conditions*: (a) CuI (1 eq), Bu_3P (2.6 eq), 1 h, −78 °C, 69%; (b) Ph_3SnCl, HMPA (11 equiv.), 30 min, −78 °C.

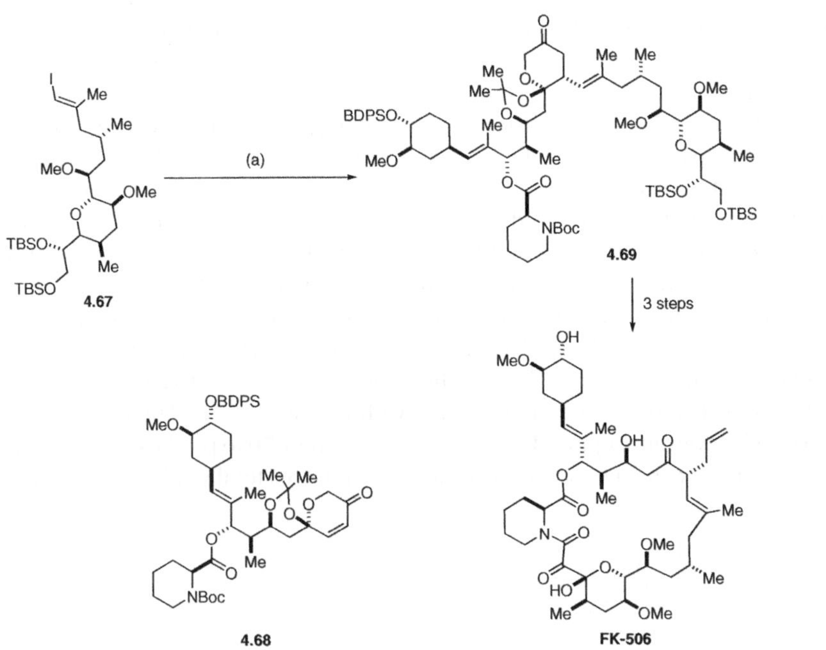

Scheme 4.27 *Reagents and conditions*: (a) (i) *t*-BuLi, −78 °C; (ii) BuCCCu•2HMPT, −23 °C; (iii) **4.68**, TMSCl, 70% (3 steps).

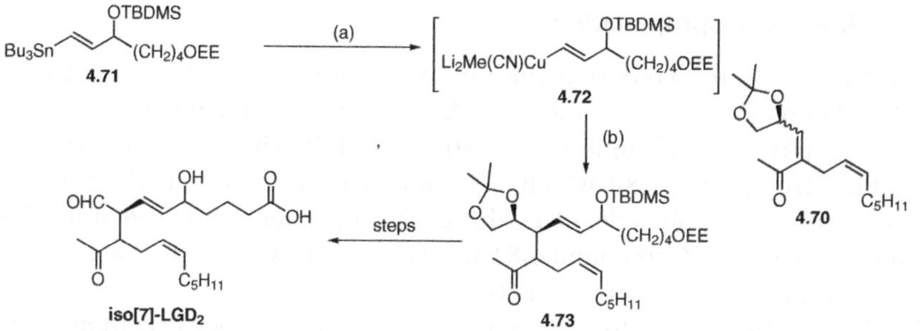

Scheme 4.28 *Reagents and conditions*: (a) CuCN, LiMe, THF, RT; (b) **4.70**, THF, 65%.

Phthioceranic acid is a heptamethyl-branched fatty acid isolated from mycobacterium tuberculosis. Its orderly structure facilitated an iterative method for constructing its core framework. Ben Feringa and coworkers adopted the copper-catalyzed asymmetric conjugate addition to construct the chiral center iteratively, which showed high efficiency.[54]

Initially, thioester **4.74** was submitted to a **4.78**·CuBr complex catalyzed enantioselective 1,4-addition with MeMgBr to afford the product **4.75** in 95% yield and 98% *ee*. Following the asymmetric conjugate addition reaction was a two-step transformation of **4.75** into **4.76** by reduction of the thioester to its corresponding aldehyde and subsequent Wittig olefination. All seven chiral centers with methyl groups were introduced efficiently by repeating the sequence of 1,4-addition, reduction, and Wittig olefination (Scheme 4.29).

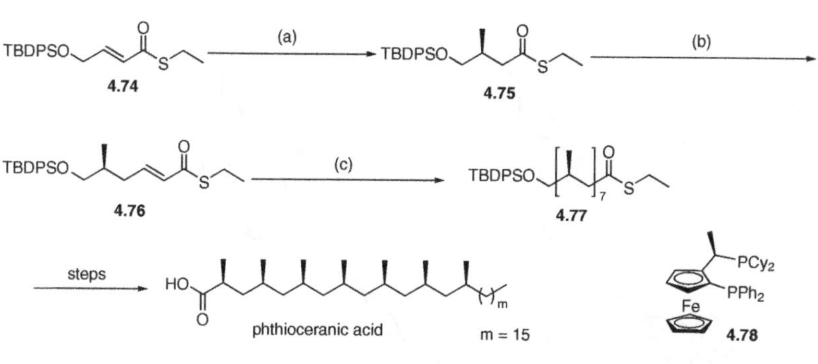

Scheme 4.29 *Reagents and conditions*: (a) MeMgBr (1.2 equiv.), **4.78**·CuBr (1 mol%), t-BuOMe, −78°C, 95%, 98% *ee*; (b) (i) DABALH (1.2 equiv.), DCM, −78°C, 3 h; (ii) $Ph_3PCHCOSEt$ (1.2 equiv.), DCM, reflux, 16 h, 70% (2 steps); (c) iterative steps A and B, 6 times (without step B in the last sequence), 8% from **4.74**.

4.6.2 Arene Cyclopropanation

Lewis Mander's Group reported efficient total syntheses of polycyclic diterpenoids: gibberellins (±)-GA_{103} and (±)-GA_{73}.[55] The cyclopropane in GA_{103} was synthesized from diaza ketone **4.79**, treatment of **4.79** with $Cu(acac)_2$ provided unstable norcaradiene **4.80**, which was trapped in situ by 3-methylfuran-2,5-dione to construct polycycle **4.81**. This efficient sequence proceeded in 75% yield over two steps. Compound **4.81** was subsequently elaborated in steps to (±)-GA_{103} and (±)-GA_{73} (Scheme 4.30).

In Yong Qin's laboratory, a copper-catalyzed cyclopropanation/ring-opening/iminium cyclization reaction (CRI reaction) was designed to synthesize complex indole alkaloids.[56] Diazo compound **4.82** was treated with CuOTf to form the stable cyclopropane intermediate **4.83** (diastereomeric isomers in a 1.6:1 ratio). Reduction of the azide group in **4.83** resulted in a two-step cascade reaction of cyclopropane ring collapse and ring closure with an *in situ* generated aniline, to give the pentacyclic compound **4.84**, which was the key intermediate in the total synthesis of (±)-communesin F (Scheme 4.31).[57]

Similar to (±)-communesin F synthesis, the CRI reaction strategy was applied in the total synthesis of indole alkaloids (±)-minfiensine[58] and (±)-vincorine[59] by Yong Qin and coworkers. The key intermediates **4.86** and **4.88** were synthesized efficiently via a three-step, one-pot cascade reaction including cyclopropanation, ring opening, and ring closure from **4.85** and **4.87**, respectively (Scheme 4.32).

During the enantioselective total synthesis of (+)-salvileucalin B, Sarah Reisman's Group used an intramolecular copper-catalyzed arene cyclopropanation reaction to construct the unusual norcaradiene core **4.90** efficiently.[60] The key step resulted in the development of a copper-catalyzed cyclopropanation to access a norcaradiene bearing a fully substituted cyclopropane ring. Compound **4.90** was transferred to the complex diterpenoid natural product (+)-salvileucalin B in a few steps (Scheme 4.33).

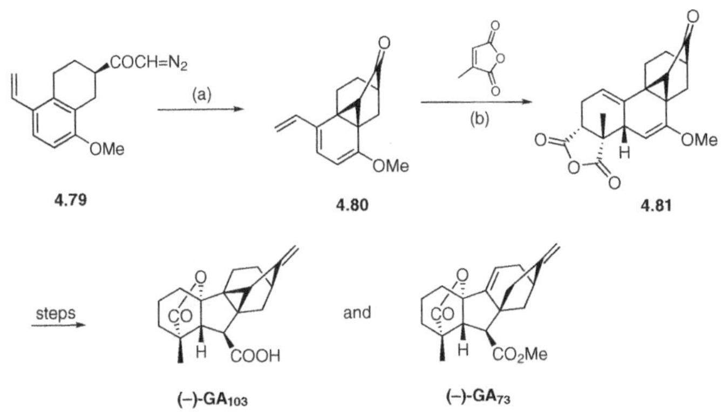

Scheme 4.30 *Reagents and conditions*: (a) $Cu(acac)_2$, $ClCH_2CH_2Cl$, D; (b) 18 h, 22 °C, 75% (2 steps).

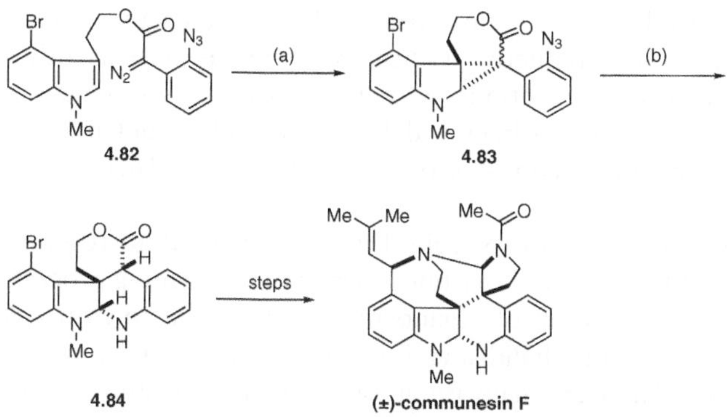

Scheme 4.31 *Reagents and conditions*: (a) CuOTf, CH_2Cl_2, RT, 88% (1.6:1); (b) PBu_3, aq. THF, 0 °C.

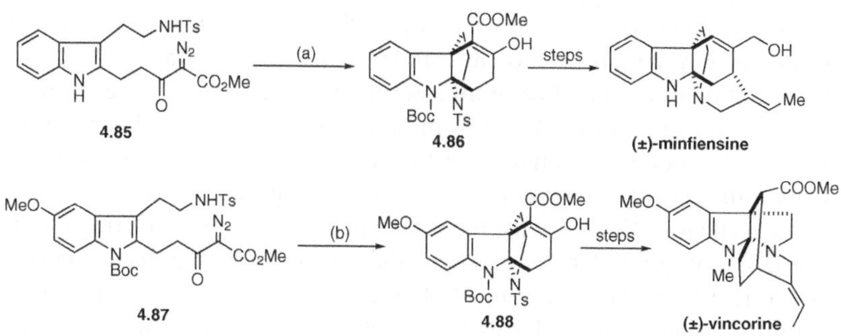

Scheme 4.32 *Reagents and conditions*: (a) CuOTf, CH_2Cl_2, RT, 50%; (b) CuOTf, CH_2Cl_2, RT, 52%.

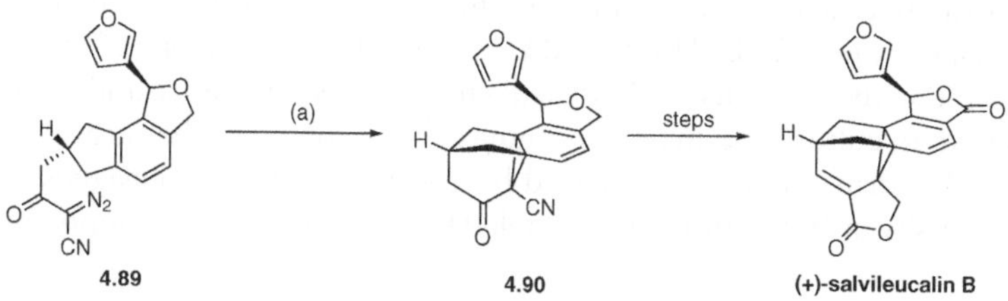

Scheme 4.33 *Reagents and conditions*: (a) $Cu(hfacac)_2$, CH_2Cl_2, 120 °C, uwave, 1 min, 65%.

4.7 Chromium-Catalyzed Reactions in Total Synthesis

Chromium-mediated reactions are frequently applied in a wide range of important reactions in organic chemistry. Tremendous examples show the efficiency of high-valent chromium compounds in promoting a number of oxidation reactions. As for the low-valent chromium compounds, the most frequently applied catalyst precursor is $CrCl_2$,[61, 62, 63] which is applied stoichiometrically or catalytically[64, 65] in Nozaki–Kishi–Hiyama reactions,[66] Reformatsky reactions,[67] and reductive radical reactions as well as other carbon–carbon bond-forming reactions. Other frequently used catalysts include Cr(III) compounds that have been widely applied in olefin polymerization, asymmetric ring-opening of epoxides, or asymmetric hetero-Diels–Alder reactions.

First reported in the late 1970s, the Nozaki–Kishi–Hiyama (NHK) reaction has become an important and versatile carbon–carbon bond-forming process, involving the nucleophilic addition of organochromium reagent to carbonyl compounds.[68] It couples organic halides and aldehydes and is based on the Cr(II)/Cr(III) redox pair. There are many unique and important features about the NHK reactions that render them particularly well suited to application in total synthesis. First, they have pronounced chemoselectivity of the organochromium(III) reagents for aldehydes, even in the presence of ketones. Second, a wide range of substrates including allyl, propargyl, alkenyl, alkynyl, aryl halides, alkenyl triflates, sulfonates, and phophates, are suitable precursors for the formation of organochromium intermediates. Third, they are compatible with an array of functional groups in both reaction partners. Fourth, they have unique stereochemical characteristics, for instance, γ-mono-substituted allylchromium reagents in general lead to the corresponding homoallyl alcohols with excellent *anti*-selectivity, independent of whether the starting halide is (*E*)- or (*Z*)-configured. Finally, alkenyl halides react with complete retention of their double bond geometry in NHK reactions. As a result, chromium-induced inter- or intramolecular carbon–carbon bond formations have been used as key steps in the synthesis of many complex targets with examples including halichondrin B.

Halichondrin B is a natural product originally isolated from the marine sponge *Halichondriaokadai* by Hirata and Uemura in 1986 (Figure 4.1).[69] It has shown extreme activity against L-1210 leukemia, P-388 leukemia, and B-16 melanoma both in culture and in *in vivo* studies.[69] The complete chemical synthesis of halichondrin B that was achieved by Yoshito Kishi and colleagues at Harvard University in 1992[70] ultimately enabled the discovery and development

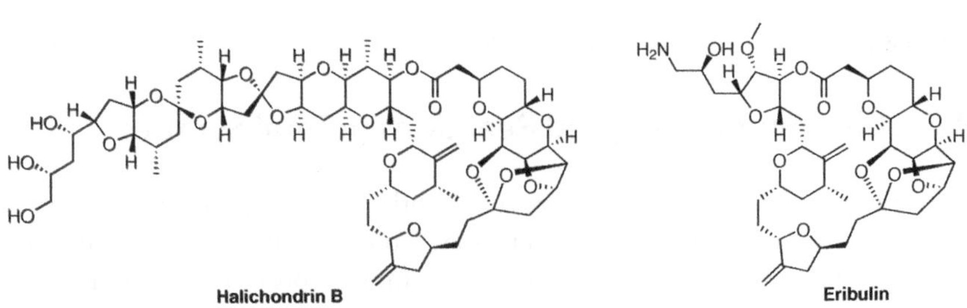

Figure 4.1 The structure of halichondrin B and eribulin.

of the structurally simplified and pharmaceutically optimized analog eribulin (Figure 4.1),[71] which was approved by the U.S. Food and Drug Administration on November 15, 2010 to treat patients with metastatic breast cancer who have received at least two prior chemotherapy regimens for late-stage disease, including both anthracycline- and taxane-based chemotherapies.

The total synthesis of halichondrin B may represent the most sophisticated application of chromium catalysts to facilitate the construction of very complex natural products. The NHK reaction was utilized four times throughout the whole synthesis, from the simple substrates **4.92** with (*E*)-methyl 3-iodoacrylate to the very complex substrates **4.100** and **4.101** to build the complete structure of the target compound. Each of the Cr mediated reactions went nicely under mild reaction conditions and with good stereoselectivity. $NiCl_2$ is used as a co-catalyst in all these four coupling reactions and was found to be necessary. Very recently, a similar synthetic approach was employed to successfully lead to the total synthesis of halichondrin C by the same group (Scheme 4.34).[72]

The well-known Reformatsky reaction was introduced for the first time in 1887. Its classic form consists of the zinc-induced formation of β-hydroxyalkanoates from ethyl α-haloacetates in a reaction with aldehydes or ketones. Later on, the reaction was ultimately found to be applicable to alkyl 2-haloalkanoates in general as well as to "remote" (3-, 4-, 5-, or higher) haloalkanoates. And the scope of this reaction was also extended beyond the use of aldehydes or ketones as electrophiles. Additionally, a number of other metals and catalysts were found to promote the role analogous to that of zinc.[73] Therefore, "Reformatsky reactions are defined as those resulting from metal insertions into carbon-halogen bonds activated by carbonyl-, carbonyl-derived, or carbonyl-related groups in vicinal or vinylogous positions with practically all kinds of electrophiles."[67] Generally, the Reformatsky reaction can be taken as subsuming all enolate formations by oxidative addition of a metal or a low-valent metal salt or complex into a carbon-halogen bond (or carbon–leaving group bond) activated by a vicinal carbonyl-derived group, followed by a reaction of the enolates thus formed with an appropriate electrophile. Reformatsky reactions are one of the most useful methods for the formations of carbon–carbon bonds, presenting a valuable tool in modern organic synthesis with a broad applicability and great versatility in numerous inter- and intramolecular reactions involving a great variety of electrophiles.[67]

Reformatsky substrates such as α-halo ketones, esters, and nitriles readily react with chromium(II) salts, in most cases chromium dichloride, in the presence of carbonyl electrophiles to give the corresponding aldol compounds. Most reactions proceed smoothly within a few minutes or hours between room temperature and 60 °C, usually in THF, DMF, or acetonitrile. Two most reactive Reformatsky substrates are two-halo ketones and vinylogous compounds. Esters, especially acetates, are much less reactive and may require either

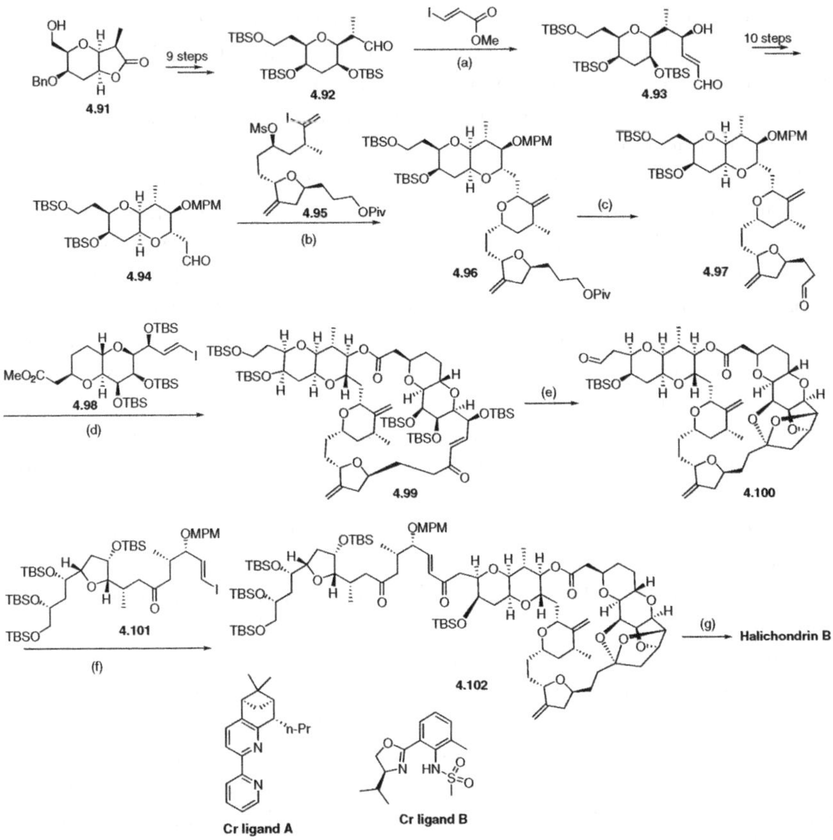

Scheme 4.34 *Reagents and conditions*: (a) $NiCl_2$ (1%)-$CrCl_2$, without ligand A 1.3:1, with ligand A 8.5:1; (b) $NiCl_2$ (0.5%)–$CrCl_2$, without ligand B 3.5:1, with ligand B 20:1, 50–60%; (c) (i) $LiAlH_4$, (ii) DMP; (d) (i) $NiCl_2$ (0.1%)–$CrCl_2$; 2) DMP, 7% (2 steps); (iii) DDQ; (iv) LiOH; (v) Yamaguchi, 63% (3 steps); (e) (i) TBAF; (ii) PPTS, 64% (2 steps); (iii) p-$NO_2PhCOCl$, pyr.; (iv) TBSOTf; (v) K_2CO_3, CH_3OH; (vi) DMP; (f) (i) $NiCl_2$ (0.1%)-$CrCl_2$; (ii) DMP, 60% (2 steps); (g) (i) TBAF; (ii) DDQ (pH 7 buffer); (iii) CSA, 50–60% (3 steps).

lithium iodide addition and/or slightly elevated temperatures. Most chromium-Reformatsky reactions have been performed with equal amounts of reagents or excess of the usually cheaper electrophiles. However, as with other chromium(II) reactions, the inverse ratio of reagents with excess halide/Cr(II) gives improved yields.

Tonantzitlolone is a new diterpenoid isolated from the endemic Mexican plant *Stillingiasanguinolenta* with unknown absolute stereochemistry. This molecule became a new target and a synthetic challenge for organic chemists because of its complex structure and high activity and selectivity against human breast and kidney cancer cell lines. A. Kirschning *et al.* explored a convergent synthesis of tonantzitlolone and prepared a key fragment of the molecule (Scheme 4.35).[74] They employed a $CrCl_2$-promoted asymmetric Reformatsky reaction, rarely used for the synthesis of natural products. The desired chiral aldehyde was obtained starting from geraniol in two steps. The $CrCl_2$-mediated Reformatsky reaction between the enantiopure α-bromoacyloxazolidinone and the enantiopure carbonyl moiety occurred at room temperature, as described by L. A. Wessjohann *et al.* with a moderate yield giving the tetrahydrofuran as a mixture of diastereomers (*syn*/*anti* = 3.5:1).[75] *O*-Silylation with TESCl eased the separation of the two stereoadducts. Another three steps gave the key fragment **4.106**. Although this synthetic approach did not eventually lead to the completion of the target molecule, it showed the Cr catalyst's ability in facilitating the coupling of the labile aldehyde **4.103** with **4.104** in mild conditions to give **4.105** at a moderate yield with moderate stereoselectivity (*syn*/*anti* = 3.5:1).

Pinacol coupling reaction, the reductive carbonyl coupling reaction leading to 1,2-diols, is conceptually the most direct approach that leads to this structural pattern. It is also one of the oldest and best studied carbon–carbon bond-forming reactions. Since the 1990s, exciting developments have resulted in efficient protocols to the coupling of two different carbonyl compounds mixed in approximately equimolar amounts. Virtually all of them involve the use of metals in a low oxidation state acting in single electron transfer (SET) processes.

Low-valent chromium is one of the metals frequently used in pinacol coupling reactions that have received much attention. In the last 20 years, a number of studies on its use in pinacol cross coupling reactions have been reported. Takai was the first to show that Cr(II), in the presence of iodotrimethylsilane, is effective in coupling acroleinacetals with aldehydes in a crosswise fashion for the generation of 1,2-diol derivatives.[76] More recently, Takai reported pinacol cross-coupling reactions between vinyl ketones and aldehydes

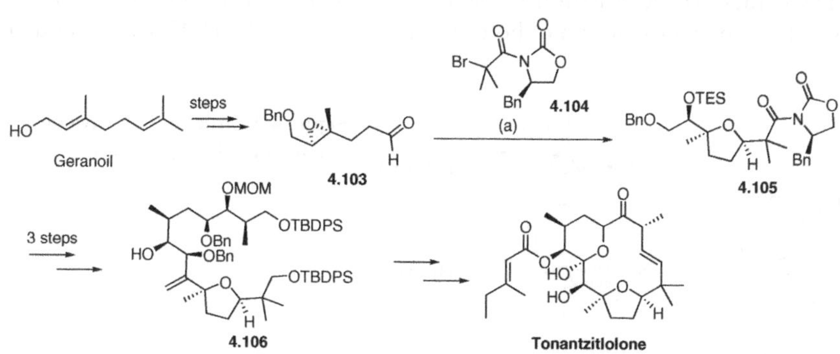

Scheme 4.35 *Reagents and conditions*: (a) (i) $CrCl_2$, **4.104**, THF, RT; (ii) TESCl, imidazole, DMAP.

by the use of an excess of chromium chloride and chlorotriethylsilane as a scavenger.[77] Based on the Takai chromium-mediated pinacol cross coupling and on the methodology developed by Fürstner for a catalytic Nozaki-Hiyama reaction, Boeckmann developed the first catalyzed reductive carbonyl cross-coupling protocol. In this protocol, acetals of acrolein or methacrolein and aldehydes were efficiently coupled in the presence of chlorotrimethylsilane and stoichiometric amounts of manganese as a reducing agent.[78] Sodium iodide was incorporated into the mixture under optimized conditions. The alcohols were obtained generally at excellent yields and good selectivities in favor of the anti-isomer. Groth and coworkers extended the method to the pinacol cross coupling reaction of α,β-unsaturated carbonyl compounds and aldehydes to form 1,2-diols using 10 mol% of chromium. Vinyl ketones and acroleins were coupled with aldehydes in moderate to good yields and with variable diastereoselectivity.[79, 80, 81] Their method is a remarkable tool for the synthesis of complex natural products due to the high tolerance of steric demanding and functionalized substrates. The procedure will have a key function in the intended totally synthesis of the natural product cotylenol. Cotylenol induces the differentiation of acute myeloid of leukemia cells in a new way, which opens up an alternative approach for the therapy of leukemia. Groth's group is working on the synthesis of the eight-membered ring of cotylenol by employing the Cr-catalyzed pinacol cross-coupling strategy (Figure 4.2).

The Diels–Alder reaction, first documented in 1928 by Nobel Prize laureates Otto Paul Hermann Diels and Kurt Alder, is a cycloaddition reaction between a conjugated diene and a substituted alkene, termed the dienophile, to form a substituted cyclohexene system. This reaction can be applied to the formation of hetero six-membered ring too, which is called a hetero-Diels–Alder (HAD) reaction.[82] The Diels–Alder reaction is widely known as a powerful reaction, especially in its asymmetric variants, for the construction of six-membered rings with excellent regio- and stereocontrol. The hetero-Diels–Alder (HDA) reaction is somewhat less well known but is extremely useful for the construction of heterocycles.[82] In 1982 the HDA was extended

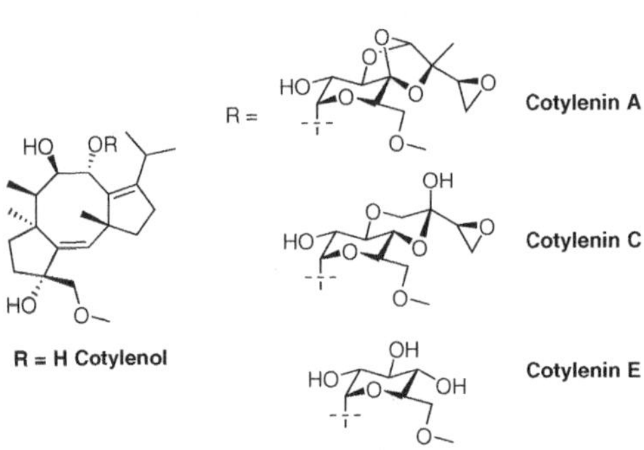

Figure 4.2 Structures of cotylenin sub-family natural products.

to unactivated aldehyde heterodienophiles by Danishefsky and coworkers through the use of Lewis acid catalysis.[83] Since that time, the reaction has received much attention because of the ease with which the dihydropyranone products of the reaction of an aldehyde with Danishefsky's diene can be elaborated to numerous pyran containing natural products.

A wide range of Lewis acids based on Al, B, Ti, Zr, Cu, and Co were developed as catalysts for the enantioselective HDA reaction. In 1998, Jacobsen and coworkers investigated the Lewis acid catalyzed HDA reaction of benzaldehyde with Danishefsky's diene and found that chromium(III) salen complexes (**4.107a** and **4.107b**) catalyze the addition of Danishefsky's diene to selected aliphatic, vinylic, and aromatic aldehydes in good yields (65–98%) and good enantioselectivities (85:15–96:4 er) (Figure 4.3).[84] Although the Cr-salen complexes are effective catalysts for the HDA of numerous aldehydes, enantiomeric ratios greater than 92:8 are seldom observed. Exploration of tridentate Schiff base complexes led to the discovery of complexes **4.108a** and **4.108b** that are highly selective catalysts for the HDA of numerous aliphatic and aromatic aldehydes with both activated and unactivated dienes.[85] Complex **4.108a** is also a competent catalyst for the inverse-electron-demand HDA of α,β-unsaturated aldehydes with ethyl vinyl ether to afford substituted pyrans.[86] The Jacobsen HDA catalysts also effect catalyst-controlled diastereoselective reactions in both the normal (81–99% yield, 99:1 er, 1.4:1–33:1 dr),[87] and inverse-electron-demand HDA ((85% yield, 99:1 er, 7:1–97:3 dr).[88] The utility of the Jacobsen HDA method has been demonstrated through extensive application in natural product synthesis, including such targets as fostriecin, (+)-ambruticin, FR901464, several iridoid natural products,[89] (–)-dactylolide,[90] and gambierol.[91]

Dactylolide was isolated from a marine sponge belonging to the genus *Dactylospongia* found off the coast of Vanuatu. It exhibited cytotoxicity against L1210 and SK-OV-3 tumor cell lines, with 63 and 40% inhibition, respectively, at 3.2 μg/mL.[92] Structurally, dactylolide possesses an unsaturated 18-membered lactone ring containing a 2,6-*cis*-substituted tetrahydropyran and an aldehyde

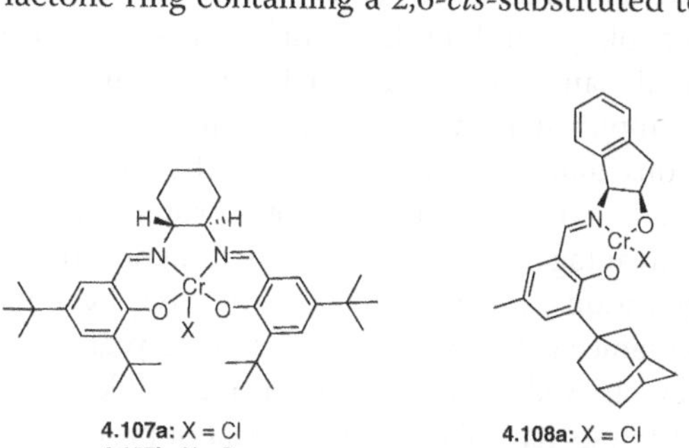

4.107a: X = Cl
4.107b: X = Br

4.108a: X = Cl
4.108b: X = SbF_6

Figure 4.3 Highly selective catalysts for the HDA of Danishefsky's diene with aldehydes.

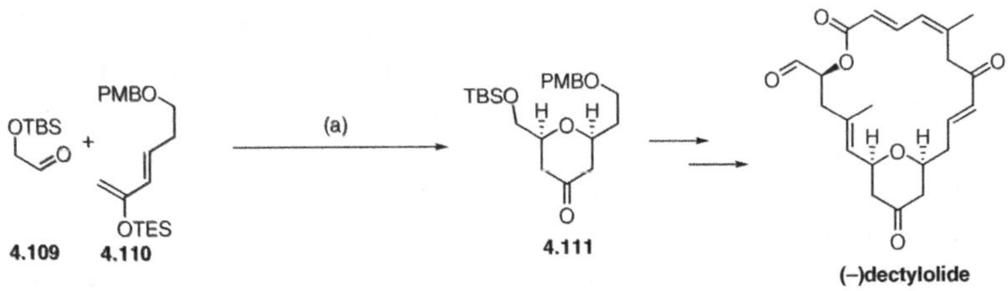

Scheme 4.36 *Reagent and conditions*: (a) 4 Å molecular sieve ligand **4.108a**, then TBAF, AcOH, THF, 82%.

side chain. However, the relative stereochemistry at the aldehyde-bearing C19-stereo-center was not reported on isolation. The enantioselective total synthesis of (–)-dacetylolide was reported by McLeod's group in 2006.[90] The absolute stereochemistry of the tetrahydropoyran was established by a Cr-catalyzed asymmetric Jacobsen HDA reaction (Scheme 4.36).

4.8 Fe-Mediated Coupling Reactions in Total Synthesis

Compared to other metal complexes in the tool box of organic chemists that are derived from heavy or rare metals with toxicity and prohibitive prices preventing their large-scale applications, iron, as one of the most abundant metals on Earth, is extremely cheap, in abundant supply, is nontoxic, and is environmentally benign. Moreover, many iron salts and complexes are commercially available, or described in the literature. Therefore, it is not surprising that iron is gaining importance in modern synthetic organic chemistry and some very efficient processes able to compete with other metal-catalyzed ones have emerged, also in the field of asymmetric catalysis.

The field of iron-catalyzed C–C bond formation reactions has developed rapidly in the past few decades. Many major accomplishments and efficient processes for addition, cross-coupling, and cycloaddition reactions have emerged over the last 10 years. Significant progress has also been made in enantioselective transformations as exemplified by the achievements in Diels–Alder reactions, 1,3-dipolar cycloadditions, and sulfoxidations. While the number of reactions is still limited, these results constitute the basis of a promising new area of research. The numerous advantages of this metal make it highly attractive, especially for large-scale applications, and iron catalysts will surely become an even more powerful tool for organic synthesis in the forthcoming years.

In the 1970s, Kochi and coworkers showed that cross-coupling is possible under iron catalysis.[93] Since then, iron catalysts have shown a great deal of promise in the cross-coupling of alkyl electrophiles and are rapidly becoming attractive alternatives for transition-metal catalysis. Substantial advances have been made to increase the range of nucleophilic and electrophilic partners amenable to such transformations. Important examples include acid chlorides; alkenyl elctrophiles such as alkenly halides, alkenylsufones, sulfides, phosphates and triflates; aryl electrophiles such as aryl halides, triflates, and tosylates; and alkyl halides.[94]

4.8.1 Reaction with Acid Chlorides

Back in 1953, it was reported by Cook *et al.* that catalytic $FeCl_3$ was highly effective in improving the addition of *n*-butylmagnisium chloride to acetyl chloride, furnishing 2-hexanone at a >70% yield compared to 31% for the uncatalyzed process.[95] However, only a few decades later this iron-catalyzed ketone formation process became generalized to allow a variety of functionalized acid chlorides, acid cyanides, and thioesters to react with primary, secondary, and tertiary alkyl and aryl Grignard reagents or diorganozinc derivatives in good to excellent yields under notably mild conditions.[96, 97, 98] The preparation of ketone **4.113**, which served as a key building block for a concise total synthesis of the actin-binding macrolides of the latrunculin family, is representative (Scheme 4.37).[99] In this particular case, the iron-catalyzed process proved far superior to the uncatalyzed and even to the copper-catalyzed addition reactions. It was reported that by using [Fe(acac)$_3$] as an efficient, cheap, and nontoxic catalyst for the cross-coupling of Grignard reagents with functionalized acid chlorides at low temperatures, the ketone **4.113** was formed at a well reproducible 80% yield. Although the known predisposition of amino acid chlorides for racemization is responsible for a slight decrease in optical purity, a single recrystallization of the crude material (*ee* 87%) from hexane conveniently solved this problem and increased the *ee* value to 99% by HPLC.

4.8.2 Reaction with Alkenyl Electophiles

As one of the most effective metal catalysts for the promotion of the reactions between Grignard reagents and organic halides, iron, in the form of $FeCl_3$, was reported by Kochi *et al.* in 1971 to catalyze the reaction between a variety of Grignard reagents and organic halides at a concentration less than 10^{-5} M.[93]

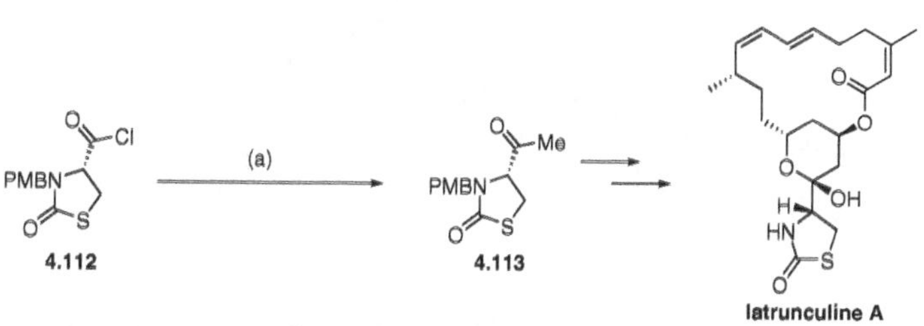

Scheme 4.37 *Reagent and conditions*: (a) $PdCl_2(PPh_3)_2$ (10 mol%), CO (65 atm), EtOH, 60°C, 81%.

$Fe(acac)_3$ was later found by the same group to be a superior precatalyst because it is easier to handle and usually results in higher overall yields.[100] Importantly, this reaction has been found to be stereospecific. To improve Kochi's original protocol, which required 3 equiv. of the alkenyl halide to reach full conversion of the Grignard reagent, several groups have found that by choice of the appropriate medium, it is possible to obtain high yields while employing nearly equimolar amounts of the reaction partners.[101, 102] Besides alkenyl halides, related electrophiles such as alkenylsulfones, sulfides, phosphates, and triflates are also suitable coupling partners.[97, 102, 103, 104] The iron-catalyzed bond formation usually outperforms the uncatalyzed addition of the Grignard reagent to other electrophilic sites in the substrate, thus rendering functional groups such as esters, ketones, enones, carbamates, or acetals compatible. Although the majority of iron-catalyzed processes utilized Grignard reagents, other nucleophiles can be employed. Most notable are organomanganese chlorides, which rival their magnesium analogs in terms of yield and reactivity. One of the chemical advantages of iron catalysts over other better known noble metal systems is that iron catalysts can provide the cross-coupling compound in high yield, avoiding the Grignard homocoupling that plagues its nickel- and palladium based relatives.[105] Consequently, this methodology has been successfully implemented in target-oriented synthesis, among which the total synthesis of (−)-cubebene and the late-stage application en route to ciguatoxin (**4.114**–**4.115**) are particularly striking (Scheme 4.38).[106, 107]

4.8.3 Reaction with Aryl Halides

While studying the iron-catalyzed cross-coupling reactions using aryl halides as the electrophilic partners, a significant and unexpected reactivity trend was

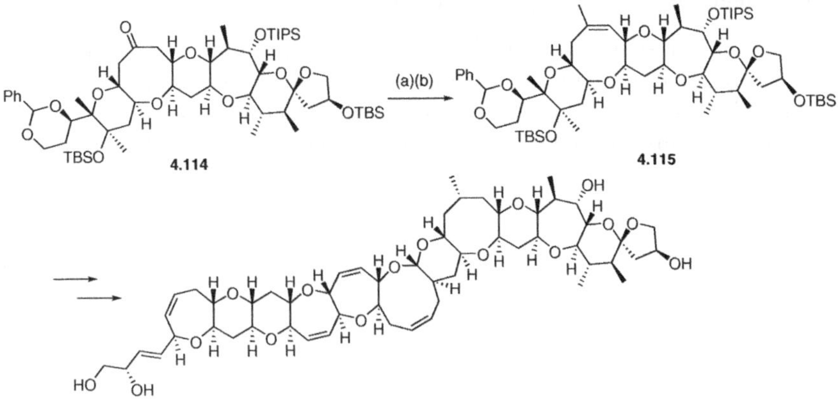

Scheme 4.38 *Reagent and conditions*: (a) KHMDS, $PhNTf_2$; (b) $Fe(acac)_3$, MeMgBr, 89% for two steps.

reported in 2002 by Fürstner and coworkers.[108, 109] Their study showed that aryl iodides and bromides, preferred substrates for palladium- and nickel-catalyzed processes, led to substantial formation of the reduced product, the corresponding aryl chlorides, triflates, or tosylates resulted in a rapid and selective cross-coupling on treatment with alkyl Grignard reagents in the presence of FeX_3 (X = Cl, acac). A wide variety of aryl and heteroaryl substrates take part in this process, although electron-rich arenes will need the more reactive triflates. In the meantime, the scope of the nucleophilic partner is also reasonably broad, as primary and secondary alkyl Grignard reagents function well, provided that [Fe(salen)Cl] is used as the precatalyst in the latter case. Furthermore, alkyl zincates and organomanganese reagents were found to be appropriate, whereas organolithium species usually failed to deliver the desired products. Interestingly, in reactions performed with FeX_3 (X = Cl, acac) in THF/NMP, allyl, vinyl, and aryl Grignard reagents uniformly led to poor yields. However, it was later found that by using arylcopper reagents[110] or a FeF_3-based catalyst system,[111] an aryl–aryl bond with electron-poor aryl iodides or aryl chlorides can be formed efficiently. In addition to the precatalyst being inexpensive, stable, and benign, $Fe(acac)_3$-catalyzed alkyl-aryl cross-coupling processes are distinguished by exceptionally high reaction rates and are readily amenable to scale-up. Note, however, that the reaction is sensitive to steric hindrance, as aryl chlorides bearing *ortho*-substituents generally afford lower yields than their *para*-substituted counterparts. As an additional asset, (hetero) arenes with more than one halide or pseudohalide are amenable to either selective monosubstitution or sequential one-pot cross-coupling. These features are exemplified by largely catalysis-based olfactory macrocycle (+)-muscopyridine (Scheme 4.39).[112]

As mentioned earlier, the success of iron-catalyzed cross-coupling did not immediately translate into an effective solution for aryl-aryl bond formation. Homocoupling of the Grignard reagent remains the major concern and the reactions were essentially limited to electron-deficient haloarenes. Recent reports, however, suggest that by modification of the precatalyst composition,

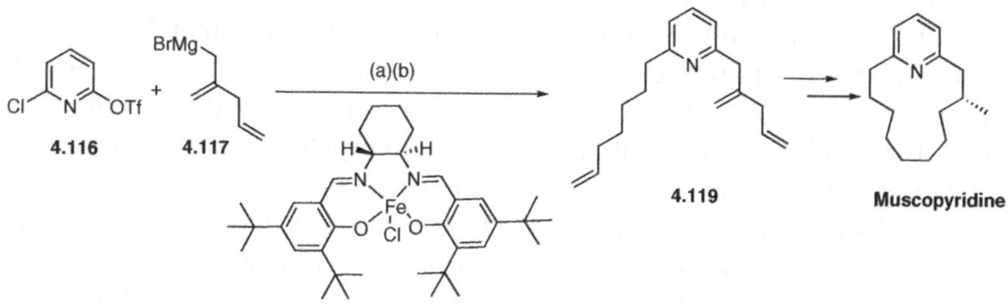

Scheme 4.39 *Reagent and conditions*: (a) Fe complex (5 mol%), NMP, THF, 0°C; (b) Fe complex (5 mol%), 6-heptenylmagnesium bromide, 80% for two steps.

the formation of a variety of biaryl frameworks can be realized. Homocoupling of the Grignard reagent can be effectively reduced when FeF_3 is employed in combination with an *N*-heterocyclic carbene ligand.[111] The specific effect of fluoride is demonstrated by the addition of KF to a $FeCl_3$ catalyst precursor, which otherwise provides predominately homocoupling products.

4.8.4 Reaction with Alkyl Halides

Due to the high barrier to oxidative addition and the proclivity of the once-formed metal–alkyl complexes toward hydride elimination, alkyl halides have been tough substrates for transition-metal catalyzed cross-coupling. Even with the aid of more active palladium or nickel catalyst complexes, it took several rounds of careful optimization of the ligand sets until appropriate conditions were found to allow such substrates to undergo cross-coupling.[113] Therefore, it is really surprising to see that relatively unsophisticated iron compounds proved competent for the cross-coupling of a variety of primary and secondary alkyl halides. Three distinct solutions were presented in 2004.[114, 115, 116] Nakamura discovered that tetramethylethylenediamine (TMEDA) largely suppressed elimination in the reaction of cycloheptyl bromide with PhMgBr.[114] Immediately following that, Hayashi presented an alternative protocol devoid of any additive but requiring Et_2O as the solvent.[116] In the same year, Fürstner's group reported that the ferrate complex $[Fe(C_2H_4)_4][Li(tmeda)]_2$ induced reactions of aromatic Grignard reagents with a variety of alkyl, allyl, and propargyl halides, even in the presence of esters, ketones, nitriles, isocyanates, or *tert*-amines.[117] More recently, the iron-catalyzed cross-coupling of alkylhalides was extended to alkenyl and even alkyl Grignard reagents as the nucleophiles.[118, 119]

4.8.5 Related Iron-Catalyzed C–C Bond Formations

Although not cross-coupling in the strictest sense, some related iron-catalyzed carbon–carbon bond-forming reactions also have a similar low-valent iron generated by the procedures similar to those listed previously and this is capable of inducing a variety of other preparatively useful C–C bond formations. Particularly relevant in the present context are the *syn*-carbometalation reactions of alkynes, strained cycloalkenes, or electron-deficient olefins, which supposedly generate transient alkenyl or alkyl iron intermediates similar to those obtained by direct oxidative insertion into organic halides. After transmetalation to magnesium, these reactive intermediates can be quenched with different electrophiles. In an alternative format, the presence of a propargylic leaving group in the substrate allows allenes to be accessed in a regio- and stereoselective manner. This concept was particularly successful with propargyl epoxides and has already met the rigors of natural product total synthesis, which is exemplified by the total synthesis of amphidinolide X and

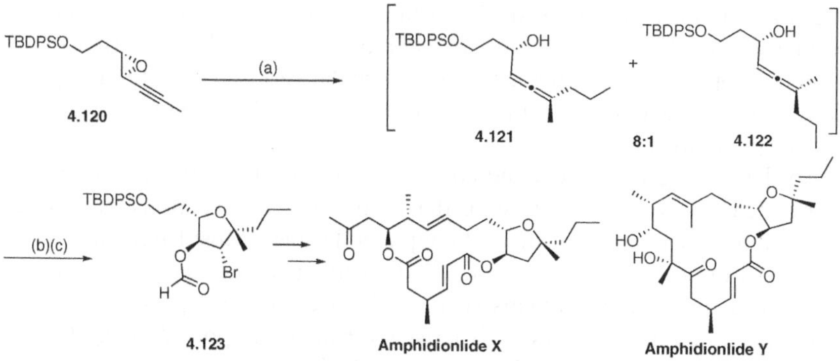

Scheme 4.40 *Reagent and conditions*: (a) $Fe(acac)_3$ (cat), PrMgBr, Tol. 62%, *syn:anti* = 8:1; (b) $AgNO_3$, $CaCO_3$, acetone; (c) NBS, DMF/H_2O (15:1), 59% for two steps.

amphidinolide Y[120] (Scheme 4.40). The reducing capability of low-valent iron is further demonstrated by cycloisomerization reactions of the alder-ene or [5 + 2]-type, as well as by higher order cycloadditions, all of which are likely triggered by oxidative cyclization of suitable unsaturated substrates.[121] Examples are known for both *in situ* generated catalysts and structurally defined ferrate complexes, which corroborates the supposedly close relationship between these different forms of low-valent iron.[121]

4.8.6 Iron-Catalyzed C–O, C–S, and C–N Cross-Coupling

Very recent reports show that certain iron salt/ligand combinations can affect C–N,[122] C–O,[123] and C–S-bond[124] formation. Even though fairly high reaction temperatures are required, these reports constitute an interesting step toward benign alternatives for C–X bond formation.

4.9 Mn-Mediated Coupling Reactions in Total Synthesis

Manganese-mediated coupling reactions are radical coupling reactions between enolizable carbonyl compounds and unsaturated compounds initiated by a manganese(III) salt, typically $Mn(OAc)_3$, $Mn(acac)_3$, or $Mn(pic)_3$. Copper(II) acetate ($Cu(OAc)_2$) is sometimes used as a cooxidant to assist in the oxidation of intermediate radicals to carbocations. From both economic and environmental points of view, the use of manganese salts for coupling reactions is attractive because they are readily available, cheap, and more environmentally benign in comparison to other transition metals. Trivalent manganese occupies a rather unusual place among metal oxidants in higher oxidations states and is particularly useful in carbon–carbon bond-forming reactions. Numerous novel region-, chemo-, and stereoselective synthetic methods have been developed in both inter- and intramolecular reactions and their applications to the construction of complex natural and biologically active compounds have been demonstrated.

Trivalent manganese-mediated coupling reactions all go through oxidative free-radical mechanism. Those oxidative free-radical reactions have been extensively developed over the past couple of decades. Most frequently, trivalent manganese compounds are applied to mediate the oxidative coupling reaction of unsaturated systems such as aldehyde, ketone, β-diketoacids, β-diketones, and β-diketoesters, with active methylene compounds (e.g., alkenes, alkynes, 1,3-alkadienes, 1,3-alkadiynes, and 1-alken-3-ynes).[125, 126] This reaction can happen inter- or intramolecularly to furnish complex cyclic compounds that are difficult to access by other means.

Intramolecular cyclization of olefinic β-diketoester (**4.124**) has been used as a key step in the synthesis of dihydropallescensin D, the natural furano sesquiterpene found in the marine sponge *Disideapallescens*.[127] The cyclization proceeds with high regio- and stereoselectivity and affords bicycle[3,4,1] decane (X) as a single stereoisomer with a *cis* relationship between bridgehead hydrogens and the carbomethoxy group. Copper(II) acetate ($Cu(OAc)_2$) is used as a cooxidant to promote an oxidative deprotonation of the intermediate tertiary radical adduct. The regioselectivity of the elimination step is determined by the higher thermodynamic stability of regioisomer (**4.125**) with the exocyclic double bond (Scheme 4.41).[128]

Trivalent manganese compounds, typically $Mn(OAc)_3$, can mediate the cycloperoxidation reactions carried out in the presence of oxygen to afford the cyclic peroxides such as 1,2-dioxanols. This methodology was successfully extended to synthesize a great variety of azabicyclic peroxides.[129] Another extension of this method was applied to substituted quinolinones.[130] A number of quinolone alkaloids exist in plants and are well known for their wide range of biological activities such as antimalarial, antitumor, antiparasitic, anthelmintic, cytotoxic, local anesthetic, and insect pheromone-like activities.

The third application of trivalent manganese in coupling reactions is its mediation of lactonization reactions. Many studies have been carried out in this field, particularly by Snider, regarding intramolecular reactions.[126] The intermolecular reactions with acetic acid derivatives, such as cyanoacetic acids, chloroacetic

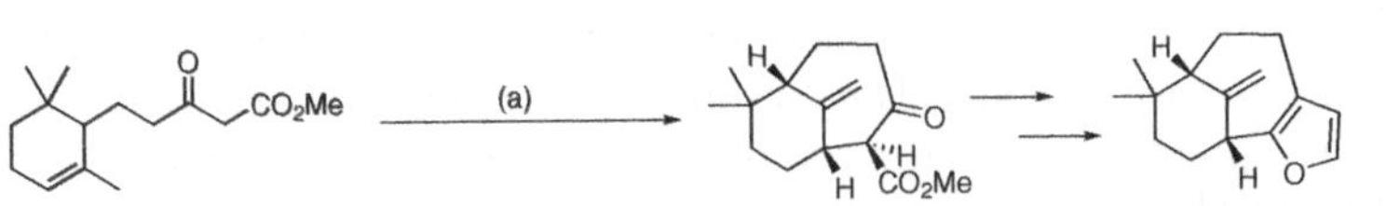

4.124 **4.125** **Dihydropallescensin D**

Scheme 4.41 *Reagent and conditions*: (a) $Mn(OAc)_3$, $Cu(OAc)_2$, HOAc, 25 °C, 3 h, 61%.

acids, 3-chloropropionic acids, monomethylmalonates, and malonic acids, were extensively investigated.[131] Reactions of malonic acids and derivatives with alkenes,[132] cycloalkenes, and cinnamic acid[133] were reported to afford variously substituted γ-lactones. This reaction was intelligently applied in the first step toward the total synthesis of (±)-paeoniflorigenin by Corey's research group.[134] Paeoniflorigenin is the nonglucoside part of paeoniflorin, an active ingredient of the essential oils from *Paeonialactiflora* that have diverse biological activity.[135] The first step of the total synthesis consists of the lactonization of unsymmetrically 1,4-cyclohexadiene (**4.126**) with cyanoacetic acid. Lactone (**4.127**) is formed chemo-, regio-, and stereoselectively; that is, the electron-rich double bond is exclusively attached by an electrophilic educt radical and the bicyclic[4,3,0] system with *trans*-configuration is formed. It is noteworthy that lactonization of the unsubstituted cyclohexene with cyanoacetic acid at elevated temperatures produces a mixture of four stereoisomers. The observed *trans*-stereoslectivity may be produced by the bulky substituent at the double bond (Scheme 4.42).

Mediation of the oxidative cyclization of α-(methylthio)acetamides, enamides, and β-ketocarboxamides is another application for Mn(III) compounds in forming carbon–carbon bonds. α-(Methylthio)acetamides, enamides, and β-ketocarboxamides have been found to undergo $Mn(OAc)_3$ oxidation to afford cyclic products. β-Amido esters, for example, were shown to cyclize in a 5- or 4-exo manner to give substituted pyrrolidines or β-lactams, respectively. Mn (III) can also mediate the oxidative cyclization of allyl α-methyl-β-ketoesters. The mechanistic studies by DFT calculations on the intramolecular free-radical cyclization of substituted allyl α-methyl-β-ketoester radicals were studied by Sung and Wang,[136] and it was concluded that the free-radical cyclization of allyl α-methyl-β-ketoester with $Mn(OAc)_3$ gave a *cis* product as a major product, while the same oxidative free-radical cyclization of substituted allyl α-methyl-β-ketoester with $Mn(OAc)_3$ produced *trans*-products as the major products. This approach is also widely applied in the total synthesis of natural products, especially the cyclic asymmetric natural products. An outstanding example is the key step of constructing the bicyclic skeleton of podolactons, which are nor- or *bis*-norditerpenic compounds isolated from different plants of the genus *Podocarpus* and from filamentous fungi.[137] These compounds present a wide range of biological activities including antitumor, ansecticidal, antifeedant, allelopathic, and fungicide activities.[138] One of the two key steps in

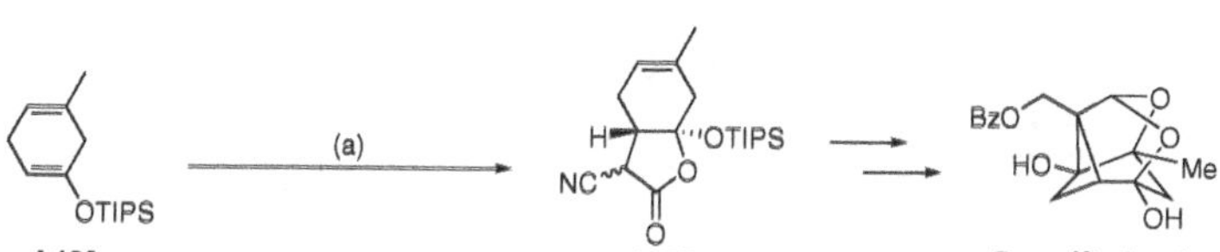

4.126 4.127 Paeoniflorigenin

Scheme 4.42 *Reagent and conditions*: (a) $NCCH_2CO_2H$ (20 eq.), $Mn_3O(OAc)_7$ (4 equiv.), KOAc, MeCN, rt, 15 h, 48%.

the total synthesis is the formation of the bicyclic intermediate **4.129** from allyl α-methyl β-ketoesters acyclic precursor **4.128** through the oxidative free-radical cyclization process using a 2:1 molar ratio of $Mn(OAc)_3$ and $Cu(OAc)_2$ in de-aerated AcOH in 68% yield[139] (Scheme 4.43).

Finally, manganese(III) can mediate alkylation reactions as well. A general example of $Mn(OAc)_3$ mediated alkylation is the oxidative addition of an aldehyde or ketone to alkenes. In this approach, $Mn(OAc)_3$ generated α-carbon radical of the carbonyl compound which is added to the double bond of the alkene with subsequent proton abstraction resulting in the formation of a corresponding alkylated carbonyl compound. Enol acetates and alkynes were also reported to undergo this type of addition reaction. This type of reaction has made a solid contribution in the synthesis of pheromones that constitute a group of naturally occurring compounds that effect chemical communication among insects and animals. Another very elegant application of this reaction exists in the straightforward synthesis of pyrenophorin, a natural fungicidal. By using $Cu(OAc)_2$ to effect oxidative deprotonation of adduct radicals, the sulfonyl group in the substrate **4.130** will allegedly direct regioselective deprontonation in adduct radicals, producing exo-methylene derivative **4.131**. γ,δ–Enone **4.131**, isolated in moderate yield (27%) was converted into the known precursor **4.132** in three steps. This method constitutes a facile access to functionalize γ–keto acrylates, and a formal synthesis of natural fungicide pyrenophorin (Scheme 4.44).[140]

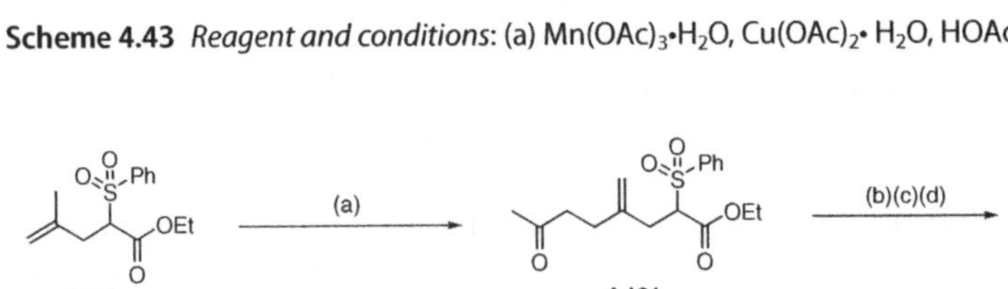

Scheme 4.43 *Reagent and conditions*: (a) $Mn(OAc)_3{\cdot}H_2O$, $Cu(OAc)_2{\cdot}H_2O$, HOAc, 68%.

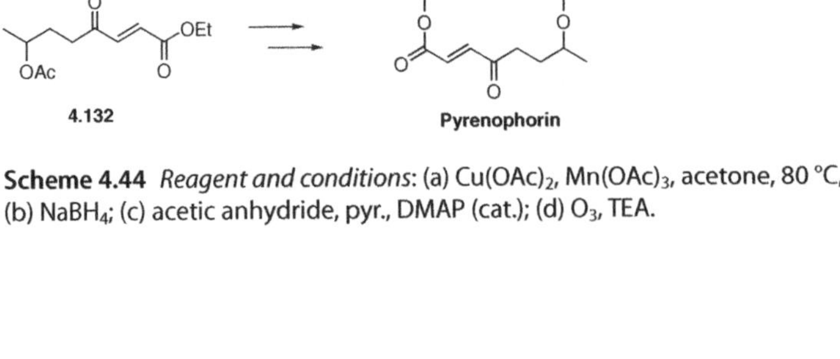

Scheme 4.44 *Reagent and conditions*: (a) $Cu(OAc)_2$, $Mn(OAc)_3$, acetone, 80 °C, 20 h, 27%; (b) $NaBH_4$; (c) acetic anhydride, pyr., DMAP (cat.); (d) O_3, TEA.

4.10 Ni-Catalyzed Reactions in Total Synthesis

Nickel-catalyzed reactions have emerged as powerful methods for the preparation of functionally and stereochemically enriched compounds. Among the large number of reactions that can be catalyzed by nickel, the two major and most important reaction types are Ni-catalyzed cycloaddition reactions and Ni-catalyzed coupling reactions.

4.10.1 Ni-Catalyzed Cycloadditions

Ni compounds, in most cases, $Ni(COD)_2$, have long been found to be able to promote cycloaddition reactions as they provide new opportunities for highly selective cycloaddition reactions[141] since the complexation of nickel to a reactant significantly modifies its reactivity, thus allowing for improved reactivity and novel chemistry. Since the mid-1900s, a huge number of cycloaddition reactions using Ni catalysts have been developed including [2+1] cycloadditions,[142] [2+2] cycloadditions,[143, 144] $[3_2\sigma+2_2\pi]$ cycloadditions,[145] $[2_2\pi+2_2\pi+2_2\pi]$ cycloadditions,[146] [2+2+2] cycloadditions,[147] tandem [3+2] and [2+1] cycloadditions,[148] $[2_2\pi+2_2\pi+2_2\pi]$ homo-Diels–Alder cycloadditions,[143] $[4_2\pi+2_2\pi]$ Diels–Alder cycloadditions,[149] $[2_2\pi+2_2\pi+2_2\pi+2_2\pi]$ cycloadditions,[150] and $[4_2\pi+4_2\pi]$ cycloadditions.[151] While searching for an efficient synthetic approach to eight-membered carbocycles fused with other rings that commonly exist in natural products, Wender and coworkers sought to explore an intramolecular [(4+4)] reaction that could have application in natural product synthesis. They were very successful in constructing the 5/8 and 6/8 bicyclic key intermediates through this approach and accomplished the total synthesis of a few natural products containing the fused eight-membered carbocycles such as sesquiterpene asteriscanolide (Scheme 4.45).[152]

4.10.2 Ni-Catalyzed Coupling Reactions

The other extensively explored reaction utilizing Ni catalysts is the Ni-catalyzed coupling reaction including the coupling of carbonyls with unsaturated compounds, the coupling between unsaturated compounds, and the cross-coupling

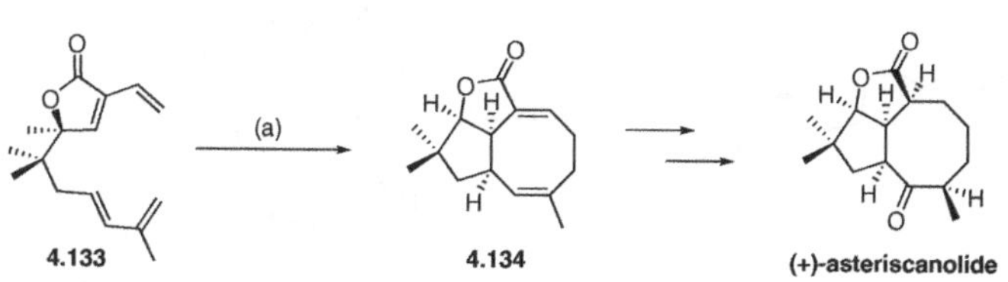

Scheme 4.45 *Reagent and conditions*: (a) $Ni(COD)_2$, Ph_3P, 90 °C, 67%.

of halides. This class of reactions provides a facile access to functionally enriched complex molecules bearing multiple stereocenters and is especially useful in the total synthesis of natural products.

Catalyzing the coupling reactions between carbonyls and unsaturated compounds is found to be one of the most important roles that Ni catalysts plan in synthetic organic chemistry.[153] A wide range of coupling substrates has been found to be effective in this process. Aldehyde, ketones, enones, imines, and isocyanoates have been used as carbonyl electrophiles; alkynes, allenes, enynes, alkenes, or dienes can be employed as unsaturated donors. Depending on the nature of the organometallic reagents employed, the mode of the coupling can be reductive,[153] alkylative,[154] or borylative.[155]

Among various transformations in the Ni-catalyzed coupling reaction between carbonyls and unsaturated compounds, the reductive aldehyde-alkyne coupling represents one of the most intensively investigated categories.[156] The mechanism suggests that the reaction is initiated by the oxidative cyclization of two π components with nickel to afford nickelacyclopentene, which is the rate-determining step and controls regio- and enatioselectivity.[157] Organozincs are involved as reducing agents in the first examples of nickel-catalyzed reductive couplings of aldehydes and alkynes. Subsequent studies illustrated that Et_3SiH and Et_3B are more effective reducing agents than Et_2Zn in promoting the reductive cyclization pathway. And the Et_3SiH variant was utilized in the total synthesis of three members of the allopumiliotoxin family. The Et_3B variant was involved in the elegant total syntheses of amphidinolides T1 and T4 (Scheme 4.46).[158] Cyclization of ynal **4.135** with $Ni(COD)_2/PBu_3$ and Et_3B as a reducing agent allowed the efficient preparation of allylic alcohol **4.136**, which was converted to amphidinolide T1 and amphidinolides T4. A key

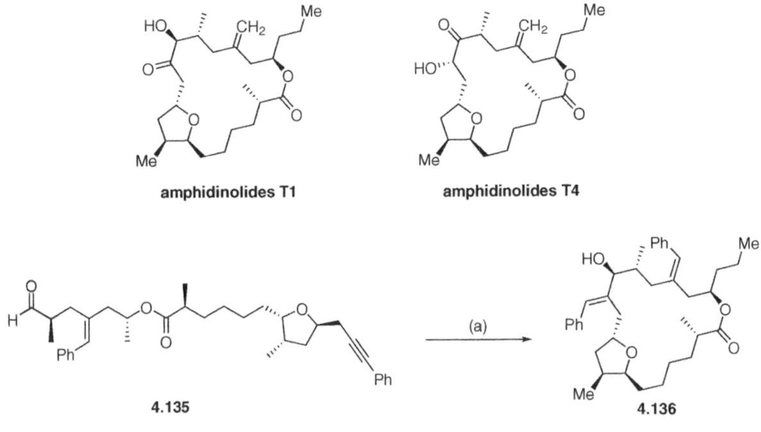

Scheme 4.46 *Reagent and conditions*: (a) $Ni(cod)_2$, 20 mol%, PBu_3, 40 mol%, Et_3B, tol, 60 °C, 44%, > 10:1 d.r.

feature of the approach was the use of an aromatic alkyne, which directed the regiochemistry to favor the desired exocyclization process.

Another very valuable coupling component, an epoxide moiety, can also participate in this type of coupling process. Recently, the reductive epoxide-alkyne coupling reactions have been applied to natural product synthesis. Considerable efforts were made to utilize this method in the preparation of a key intermediate for the total synthesis of (–)-gloeosporone, where epoxides and alkynes are assembled to furnish large rings with high region- and enantioselectivity (Scheme 4.47).[159]

Catalyzing the coupling reactions between two unsaturated compounds is another very important application for Ni catalysts. The two unsaturated compounds can be alkenes with alkenes, alkenes with alkynes, and alkynes with alkynes, all of which have long been documented as powerful ways to construct carbon–carbon bonds. The nickel-catalyzed coupling of an alkene, an alkyne, and a main-group organometallic reagent provides an excellent way to control the stereochemistry of challenging tri- and tetrasubstituted alkenes. An apparent requirement for the coupling or cyclization to proceed effectively is that the alkene unit must be electron deficient. Both internal and terminal alkynes are cleanly tolerated as the acetylenic components, and enones, alkylidene malonates, nitroalkenes, and unsaturated imides are tolerated as the electron-deficient alkene components. The scope of this reaction is very broad. Several total synthesis applications of this reaction class have been demonstrated, including the total synthesis of isodomoic acid G (Scheme 4.48).[160] Isodomoic

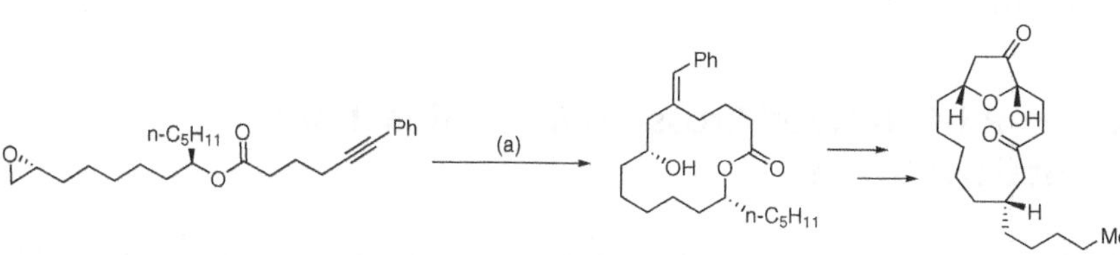

4.137 4.138 (–)-Gloeosporone

Scheme 4.47 *Reagent and conditions*: (a) $Ni(cod)_2$, PBu_3, Et_3B, THF.

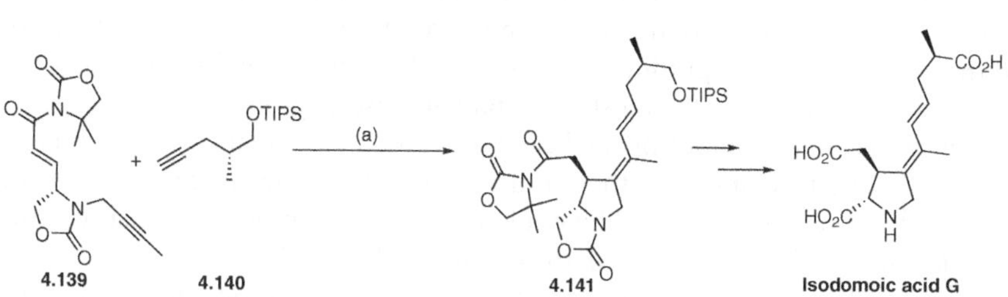

4.139 4.140 4.141 Isodomoic acid G

Scheme 4.48 *Reagent and conditions*: (a) Cp_2ZrHCl, $Ni(COD)_2$, $ZnCl_2$, THF, 74%.

G contains an exocyclic alkene at C4, which is more structurally complex than kainoid amino acids. The nickel-catalyzed cyclization of alkyne **4.139** with the vinylzirconium reagent **4.140** was used to prepare directly the isodomoic acid G core structure. Notably, the pyrrolidine unit, the C2/C3 relative stereochemistry, and the complete densely functionalized 1,3-diene were assembled in a single operation in a completely selective fashion to allow an efficient total synthesis of this natural product.

Nickel is by far the most versatile metal for the cross-coupling of halides. It is able to couple organic electrophiles like alkyl halides, aryl halides, alkenyl halides, and phenol derivatives with a variety of organometallic reagents, including zinc, boron, silicon, tin, and indium compounds. Asymmetric processes with racemic starting materials have also been developed with Ni catalysts. These reactions show a broad scope with respect to both coupling partners and Ni catalysts have found very powerful in mediating a number of coupling reactions including Nigishi-type coupling, Suzuki-type coupling, Hiyama-type coupling, Stille-type coupling, and Sonogashira-type coupling. Due to the excellent functional group tolerance, high chemoselectivity, and excellent stereoselectivity of organozinc reagents in many reactions, organozinc compounds constitute ideal organometallic reagents for the construction of complex polyfunctional molecules. In this respect, the possibility of building alkyl–alkyl bonds in the presence of functional groups has opened up the way to new retrosynthetic disconnections. In addition, Ni complexes can switch among several oxidation states, resulting in a rich reactivity with surprising catalytic cycles and promising new possibilities that will probably be discovered in the near future.

4.11 Pd-Catalyzed Cross-Coupling Reactions in Total Synthesis

Palladium-catalyzed cross-coupling reactions, currently predominating in the field of transition-metal catalyzed reactions, are highly essential to organic chemistry. Allowing for efficient and convenient building up of complex molecules through C–C and C–hetero bond formation, these reactions offer a new paradigm for disconnection analysis and revolutionarily change in the way of thinking about natural product synthesis.[161] There are several reasons that the Pd-catalyzed reaction is the most important and useful protocol used in organic synthesis: (1) most importantly, Pd catalysts offer an abundance of possibilities for C–C bond formation; (2) there is a wide functional group compatibility of Pd-catalyzed reactions; and (3) by not being so oxygen-moisture-sensitive, this enables them to be widely used in the ordinary laboratories.

In this chapter, we will review the recent applications involving Pd-catalyzed reactions in natural product synthesis, some representative and exciting examples of which will be carefully discussed where their efficiencies and advantages in building up complex molecular architectures will be highlighted.

4.11.1 Heck Reactions in Total Synthesis

The palladium-catalyzed C-C coupling between aryl halides or vinyl halides and alkenes in the presence of a base is referred as the "Heck reaction." This reaction was the first example of a carbon–carbon bond-forming reaction that followed a Pd(0)/Pd(II) catalytic cycle. Recent developments in the catalysts and reaction conditions have resulted in a much broader range of applications including the preparation of new advanced materials, total syntheses of natural products, biologically active pharmaceutical agents, and so on. The Heck reaction can bring about unprecedented structural changes, particularly in its intramolecular version. Nowadays, the catalytic asymmetric variant of the Heck reaction has emerged as a reliable method for enantioselective C–C bond formation, which has made this powerful reaction much flourish more in natural product synthesis.

In the first total synthesis of scopadulcic acid B, Overman's group presented the innovative and pioneering approach toward the formidable tetracyclic structure of the target molecular by using an intramolecular Heck reaction (Scheme 4.49).[162] In this strategy, the key *bis*-cyclization of **4.131** was accomplished in the presence of a 10 mol% coordinatively unsaturated catalyst prepared from $Pd(OAc)_2$ and Ph_3P. The initial oxidative addition of the carbon–iodine bond followed by the first intramolecular cyclization to generate alkylpalladium intermediate **4.144**, without undergoing β-hydride elimination

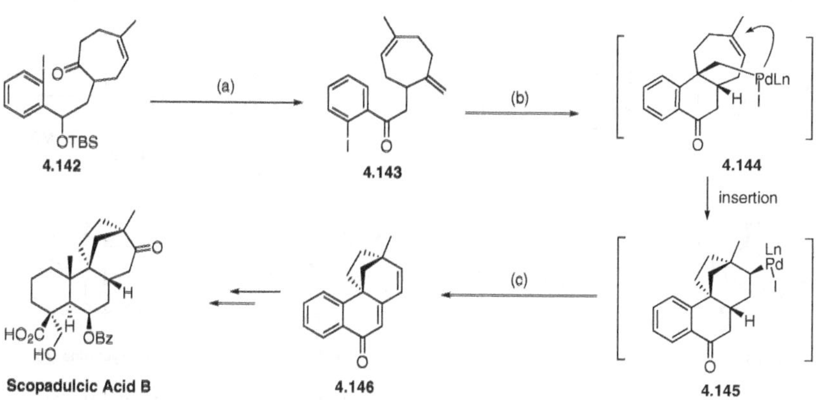

Scheme 4.49 *Reagent and conditions:* (a) (i) $Ph_3P{=}CH_2$, THF; (ii) TBAF, THF, 23°C; (iii) PCC; (b) 5 mol% $Pd(OAc)_2$, Ph_3P (1:2) MeCN, 80°C; (c) 80–85% from **4.143** (10–14 g scale), then DDQ, PhCl, reflux, 63%.

owing to the absence of any suitable hydrogen atoms, which was further under through the second 1,2-insertion reaction leading to the desired bridged polycyclic scaffold and with two quaternary centers installed. The cyclization could be conducted on a 10–14-g scale and provide the mixture of enones in a combined yield of 80–85%. This mixture was further oxidized by DDQ in methanol, which led to the clean dienone **4.146** in 63% yield. This advanced intermediate could then be elaborated in a number of steps to complete the total synthesis of scopadulcic acid B. This synthesis shows the remarkable efficiency of the Heck reaction in the construction of sterically crowded ring systems and quaternary stereocenters too.

(−)-Strychnine, the most famous member of the strychnos alkaloids that has seven rings and six stereogenic centers, is definitely as one of the most complex natural products of its size. Miwako Mori and coworkers reported a novel method for synthesizing indole derivative **4.149** by means of the Heck cyclization of 2-bromoaniline intermediate **4.148**, which was prepared from the easily accessible starting material **4.147** by using the asymmetric allylic alkylation reaction and relative transformations (Scheme 4.50).[163] The resulting compound **4.149** was treated with five steps of routine transformations to form tetracylic compound **4.150**, which was subjected to the second round of intramolecular Heck reaction to construct the six-membered lactam **4.151** in 46% yield. Isomerization of the double bond of **4.151** by treatment with *i*PrONa in

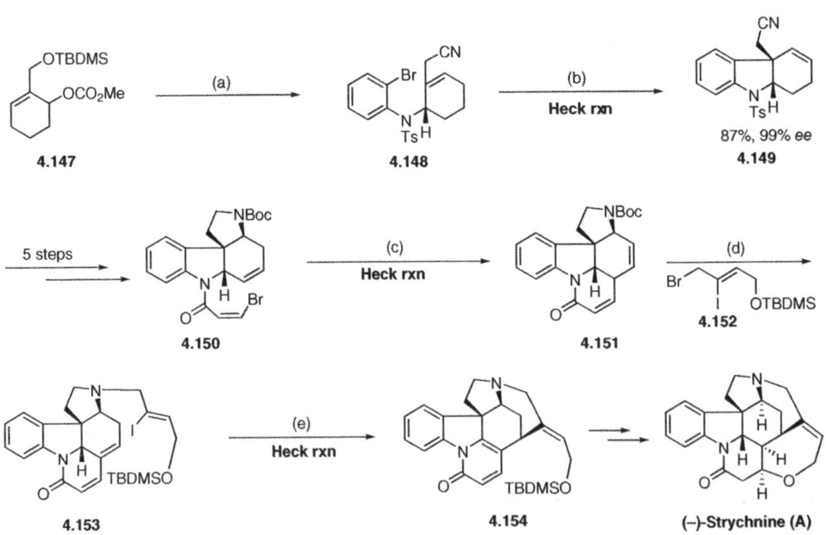

Scheme 4.50 *Reagent and conditions*: (a) (i) Pd(0), (*s*)-BINAPO; (ii) HCl; (iii) PBr_3; (iv) NaCN; 69% for four steps; (b) 2 mol% $Pd(OAc)_2$, $PPhMe_2$, Ag_2CO_3, DMSO, 90°C, 17 h; (c) 10 mol% $Pd(OAc)_2$, 20 mol% PPh_3, *i*Pr_2Net, DMSO, 80°C, 1.5 h; (d) (i) NaO*i*-Pr; (ii) TFA; (iii) Li_2CO_3, DMF; (e) $Pd(OAc)_2$, Bu_4NCl, K_2CO_3, DMF, 70°C, 0.5 h.

2-propanol followed by removal of the Boc group and alkylation with **4.152** afforded compound **4.153**, which was then converted into pentacyclic compound **4.154** by using the third round of Heck cyclization. Further synthetic transformations including LAH reduction, desilyl protection, and base mediate cyclization successfully delivered the desired product (−)-strychnine. In this synthetic example, three ring systems together with three stereocenters were constructed via intramolecular Heck cyclization, which demonstrated the importance of palladium catalysis in the modern synthetic organic chemistry.

4.11.2 Suzuki Reactions in Total Synthesis

The palladium-catalyzed cross-coupling reaction between organoboron compounds and organic halides or triflates provides a powerful and general method for the formation of C–C bonds known as Suzuki cross coupling.[164] More and more attention has been focused on the Suzuki reaction within the laboratory and in industry due to its mild reaction conditions; commercial availabilities and stabilities of many boronic acids; compatibility with water and oxygen; the ease of by-product removal; the wide function group tolerance; and so on. In recent years, tremendous progress has been made in the development of Suzuki coupling reactions. A large number of ligands have been successfully applied in these reactions that meant this reaction could be performed under very mild conditions, even with relative inactivate substrates such as chloride. Indeed, it has become one of the most reliable and widely used cross-coupling reactions in total synthesis and the pharmaceutical industry as well.

The stereocontrolled total synthesis of 4-hydroxydictyolactone was described by David R. Williams and coworkers in 2009 (Scheme 4.51).[165] The synthesis started from the known compound ester **4.155** and was later steadily converted to the ester **4.156**, which was then employed to the Ireland–Claisen rearrangement condition and effectively established the backbone and three of the contiguous stereocenters. After several straightforward transformations, compound **4.158** was successfully converted to trialkene **4.159**. A comprehensive condition investigation has been made into the sequential intramolecular B-alkyl Suzuki cross-coupling reaction. It was found that the yield of ring closure product was largely dependent on the substituent on the hemiketal group and the palladium catalyst as well. When the methoxy hemiketal was applied as substrate and using $Pd(PPh_3)_4$ as catalyst under a much diluted concentration ($C = 0.005\,M$), the desired product **4.160** could be isolated in 66% yield. The completion of total synthesis of 4-hydroxydictyolactone was achieved by two steps of additional oxidation and TBS cleavage procedure. This synthesis has documented the development of the B-alkyl Suzuki cross-coupling reaction as a useful strategy for cyclizations to directly afford complex (*E*)-cyclononene systems.

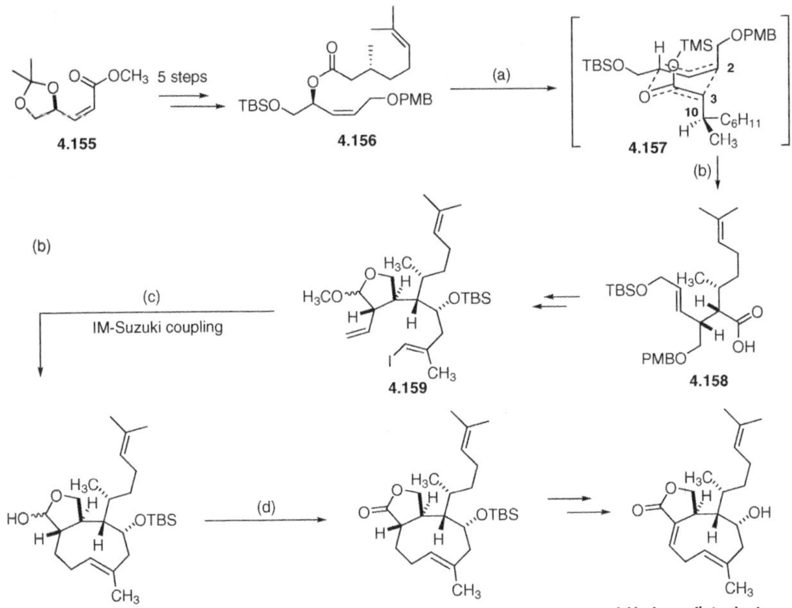

Scheme 4.51 *Reagent and conditions*: (a) TMSCl, LDA, 85%, d.r = 94:6; (b) then reflux; (c) 9-BBN (1.5 equiv.), THF, then $Pd(PPh_3)_4$, NaOH, CH_3CN/H_2O, then aq. AcOH, THF, 85°C, 66%; (d) TPAP, NMO,CH_2Cl_2, 79%.

In 2002, Brian M. Stoltz's group reported the first total synthesis of the biologically significant *bis*-indole alkaloid dragmacidin D,[166] which was retrosynthetically divided into two indole fragments and one pyrazine fragment (Scheme 4.52). The pyrazine fragment **4.162** and commercially available indole fragment **4.163** were firstly coupled at room temperature Suzuki reaction that afforded the coupled indolopyrazine **4.164.** The second crucial Suzuki coupling between indolopyrazine **4.164** and bronic-indole **4.165** was conducted under the carefully controlled conditions (50 °C, 72 h), and the desired *bis*-indole alkoxy pyrazine **4.167** was formed in a good yield, with complete selectivity for coupling of the pyrazinyl bromide in the presence of the indolyl bromide. Further synthetic transformations involving oxidation state adjustments and protecting group manipulations afforded advanced aminoketone **4.167**. Final installation of the aminoimidazolium unit with cyanamide followed by treatment with trifluoroacetic acid provided dragmacidin D in 86% yield. In this synthesis, thermal and electronic modulation provide the key to a series of palladium-catalyzed Suzuki cross-coupling reactions that furnished the core structure of dragmacidins, and meanwhile demonstrated the power of Suzuki reaction in the construction of biaryl structures.

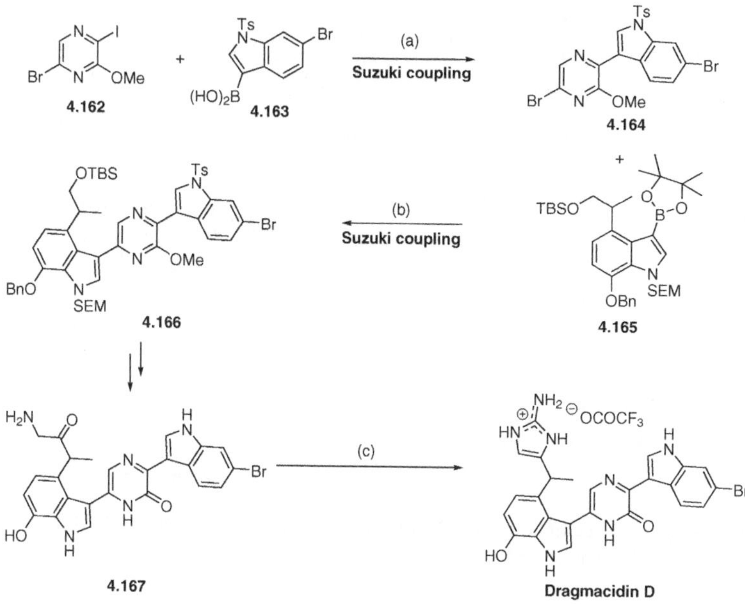

Scheme 4.52 *Reagent and conditions*: (a) $Pd(PPh_3)_4$, MeOH, PhH, Na_2CO_3, H_2O 23°C, 72 h, 71%; (b) $Pd(PPh_3)_4$, MeOH, PhH, Na_2CO_3, H_2O, 50°C, 72 h, 82%; (c) H_2NCN, EtOH, 60°C, then TFA, H_2O, CH_3CN, 86%.

4.11.3 Stille Reactions in Total Synthesis

The Pd(0)-catalyzed coupling reaction between an organostannane and organic halides or triflates is known as Stille cross-coupling. The main advantages of organo-tins over other elements derivatives are the ease of their preparation and stability, as well as the wide functional group tolerance. The main disadvantages are their toxicities and the difficulties in removing traces of tin by-products from the reaction mixture. In the past two decades, the Stille reaction has become one of the most powerful synthetic tools in organic chemistry. The success of Stille coupling is largely attributed to the mild conditions. Nowadays, Stille reactions are frequently being used in natural product synthesis both in its intermolecular or more attractive intramolecular fashion. An elegant example applying the Stille coupling reaction as the key strategy was reported by Chuo Chen and coworkers in the total synthesis of nakiterpiosin (Scheme 4.53).[167] The electrophile of a triflate fragment **4.168** was synthesized from commercially available **4.170** via a furanyl intramolecular Diels–Alder approach. The original chirality was introduced by an asymmetric Noyori reduction and the related hydoxy group was converted to the bromide, which is the required functionality in the natural product. The aryl tin fragment **4.169** was prepared from **4.171**, during which a Sharpless asymmetric epoxidation reaction,

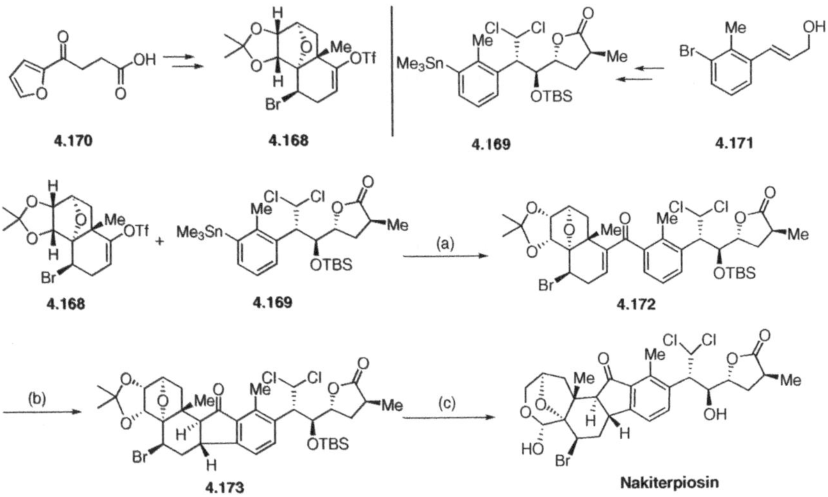

Scheme 4.53 *Reagent and conditions*: (a) CO, CuI, $Pd(PPh_3)_4$; (b) (i) hν; (ii) iPr_2NH; (c) (i) TFA; (ii) $NaIO_4$; (iii) $BF_3{\cdot}OEt_2$,Et_3SiH; (iv) TBAF.

Yamamoto epoxide rearrangement, and a vinylogous Mukayama aldol reaction were involved. The carbonylative coupling of **4.168** and **4.169** was achieved with a modified Stille's protocol using $Pd(PPh_3)_4$/CuCl in DMSO under 1 atm CO. The CuCl additive and DMSO solvent provided dramatic rate enhancement and were crucial to the success of this reaction. The photo-Nazarov cyclization reaction of aryl vinyl ketones was employed to complete the construction of the cyclopentenones skeleton of nakiterpiosin. After removal of protecting group and diol motif break down, the completion of total synthesis of nakiterpiosin was successfully achieved. This synthesis shows the efficiency of late-stage carbonylative Stille cross-coupling reaction in the natural product synthesis.

Another synthetic example utilizing Stille coupling in natural product total synthesis was reported by Jack E. Baldwin and colleagues in 2003 (Scheme 4.54).[168] This elegant synthetic strategy was inspired by the biosynthetic considerations that constitute a Stille coupling together with an intramolecular Diels–Alder reaction cascade. The author was first set to attempt the Stille coupling of fragments **4.174** and **4.176;** however, it was subsequently found that protection of the free hydroxyl group of **4.174** was required to permit efficient Stille cross-coupling. Thus, TES was selected as the protecting group of this secondary alcohol and subsequently subjected to the Stille coupling reaction condition. Interestingly, the coupled product was found to be unstable and easily dimerized on standing overnight to give **4.176** as a single diastereoisomer in 75% overall yield from **4.175**. The completion of the total synthesis of panepophenanthrin was achieved after the removal of TES. This

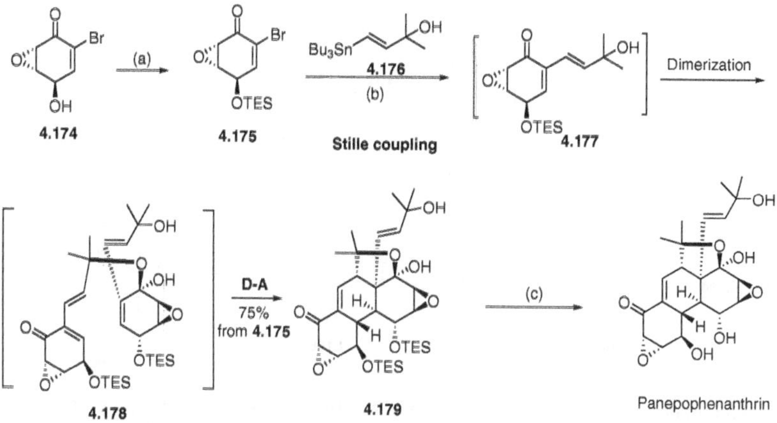

Scheme 4.54 *Reagent and conditions*: (a) TESCl, 2,6-lutidine, 94%; (b) $Pd_2(dba)_3$, $AsPh_3$, 110°C, toluene; (c) NH_4F, MeOH, 85%.

synthetic example undoubtedly shows the efficiency of the Stille coupling reaction in the synthesis of a complex natural product, particularly when combined with other classic transformations such as a Diels–Alder reaction.

4.11.4 Tsuji–Trost Reactions in Total Synthesis

The Pd-catalyzed allylation of carbon nucleophiles with allylic compounds via π-allylpalladium complexes is called the Tsuji–Trost reaction. It is one of the most synthetically valuable C–C and C–hetero (nitrogen, oxygen, and sulfur-based soft nucleophiles can also be used in Tsuji–Trost allylation reactions) bond-forming reactions that have emerged since the last quarter of the twentieth century. Different from most transition-metal-catalyzed processes, asymmetric allylic akylations involve the reaction at sp^3 instead of sp^2 centers. The ability to transform achiral, prochiral, or chiral racemic material to enantiopure material using the catalytic amount of Pd and ligand have made this transformation a very powerful tool in organic synthesis.

The most impressive synthetic example employing asymmetric allylic alkylation was illustrated in the total synthesis of (−)-cyanthiwigin that was described by Stoltz and coworkers in 2008 (Scheme 4.55).[169] The synthesis began with a Claisen–Dieckmann sequence that converted diallyl succinate **4.180** to *bis*(β-ketoester) **4.181** as a 1:1 mixture of racemic and *meso*-diastereoisomers. This mixture was exposed to the condition of allylic alkylation and was readily transformed to a *bis*(allylated) ketone **4.183** with a 75% yield and 99% enantiomeric excess as a 4.4:1 mixture of diastereoisomers. Selective monoenolization of diketone **4.183** with potassium bis(trimethylsilyl) amide (KHMDS) and trapping with $PhN(Tf)_2$ led to a corresponding

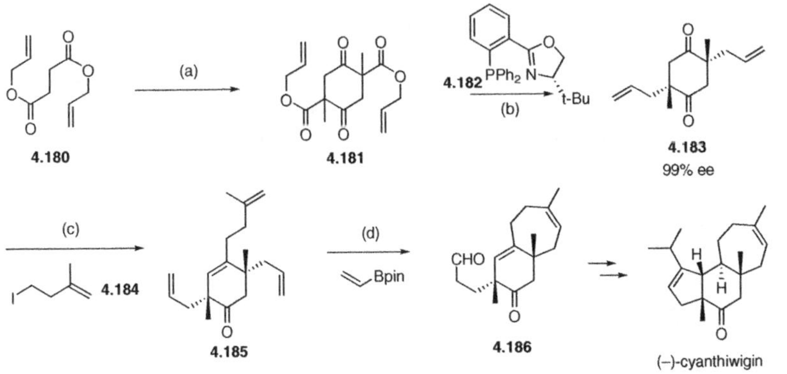

Scheme 4.55 Reagents and conditions: (a) (i) allyl alcohol, NaH, PhMe, reflux; (ii) K_2CO_3, MeI acetone, reflux, 51%. (b) $Pd(dmdba)_2$, Et_2O, 25°C, 78% yield of 4.182. (c) (i) KHMDS, $PhN(Tf)_2$, THF, 73%; (ii) Zn, TMSCl, then, $Pd(PPh_3)_4$, 78%. (d) 10% Grubbs Cat, PhH, 60°C, $NaBO_3$, THF, 53%.

trifluoromethanesulfonate that was further employed to Pd-catalyzed Negishi cross-coupling to give the alkene side chain installed intermediate **4.185** with a good yield. A tandem ring-closing metathesis–cross-metathesis sequence with Grubbs' ruthenium catalyst was carried out to give the bicyclic aldehyde **4.186**. Elaboration of this advanced intermediate to the natural product cyanthiwigin was achieved in a further three steps. This short synthetic route highlights the utility of Pd-catalyzed double enantioselective allylation in the construction of complex molecules with high levels of stereocenters.

As illustrated by the recent work of Armen Zakarian, synthetic strategy that exploit intramolecular allylic dearomatization reaction in the building of complex molecular skeletons represents one of the most powerful and efficient methods in the total synthesis of spiro natural products (Scheme 4.56).[170] In this total synthesis endeavor, the author started from the readily available material **4.187** and **4.188** using a serials of efficient transformation to synthesize ayllic carbonate **4.189**. Intramolecular allylic dearomatization of complex substrate **4.189** was investigated concentrating on palladium(0) and iridium(I) catalysis, only palladium gave a good yield and stereoselectivity. The desired cyclization could be achieved cleanly upon treatment with $Pd(dba)_2$ and $P(OPh)_3$, and afforded an inseparable mixture of atropisomeric products **4.190** and **4.191** together with diastereomer **4.192** in a 84% combined yield (92% based on recovered starting material, d.r. 6:3:2, respectively). The synthesis of tatanans B and C was readily accomplished in two direct steps from **4.190** and **4.191**, respectively. The atropisomeric tatanans B and C were separated by preparative reverse phase HPLC.

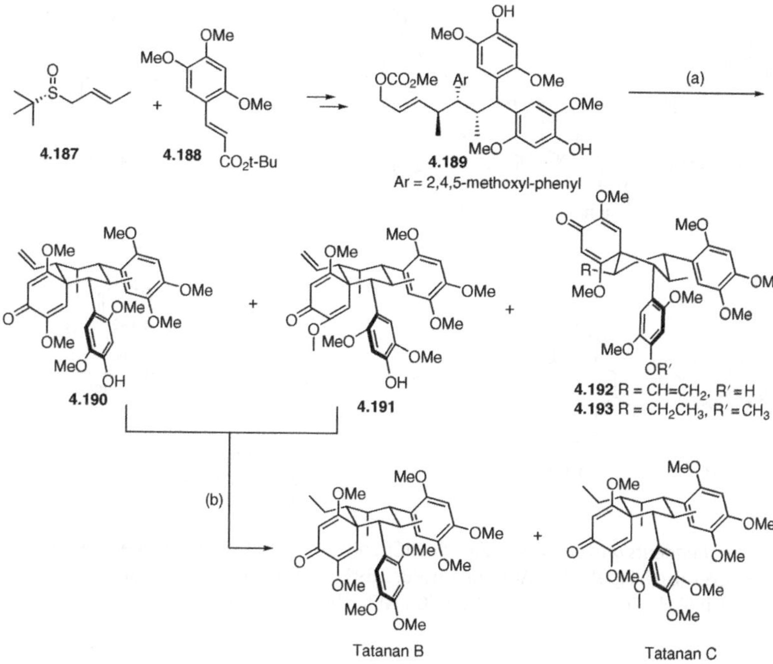

Scheme 4.56 *Reagents and conditions*: (a) $Pd(dba)_2$, $P(OPh)_3$, CH_2Cl_2, 25°C, 2 h, 84% (92% brsm) d.r. 6:3:2. (b) (i) NaH, CH_3I, THF; (ii) H_2, Lindlar cat, EtOAc, 78%.

4.11.5 Negishi Reactions in Total Synthesis

The Pd- or Ni-catalyzed stereoselective cross-coupling of organozincs and aryl-, alkenyl-, or alkynyl halides is known as the Negishi cross-coupling. These versatile coupling reactions using organozinc reagents are usually the most reactive in Pd-catalyzed cross-coupling reactions and do not require the use of additives (e.g., bases as in Suzuki cross couplings) to boost reactivity; therefore, this allows for a much greater functional group tolerance. The reaction is mostly used for the coupling of two C(sp2) carbons but C(sp2)-C(sp) as well as C(sp2)–C(sp3) couplings. The applications of organozinc compounds range from asymmetric synthesis of biologically relevant molecules, and new materials, as well as combinatorial libraries.

In 2007, Gregory Fu and coworkers provided a very efficient method for the remote stereochemical control by using a nickel (II)-catalyzed asymmetric Neigish-coupling reaction that was described in the formal synthesis of fluvirucinine A1 (Scheme 4.57).[171] The synthetic campaign was started from racemic allylic chloride **4.195** and alkylzinc reagent **4.196** in the presence of (*S*)-BnCH$_2$-Pybox and NiCl$_2$·glyme to generate an unsaturated ester **4.197** at

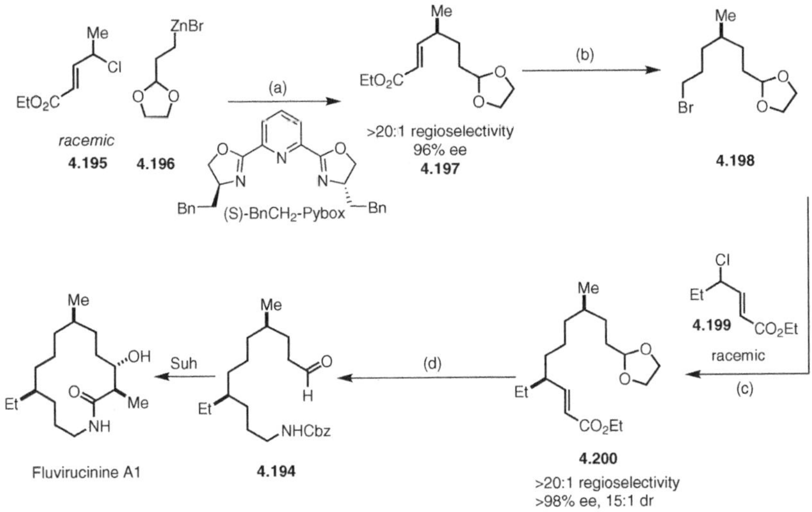

Scheme 4.57 *Reagents and conditions:*(a) 5% $NiCl_2$, glyme, (S)-$BnCH_2$-Pybox 4 eq., NaCl, DMA-DME, 93%. (b) (i) H_2, Pd/C; $LiAlH_4$; (ii) Ph_3PBr_2, 89%. (c) 5% $NiCl_2$, glyme, (S)-$BnCH_2$-Pybox, 82%. (d) (i) H_2, Pd/C, $LiAlH_4$; (ii) CbzN-SO_2-NEt_3, then HCl, 58%.

a 93% yield and 96% *ee*. Elaboration over two steps to a bromide **4.198**, followed by the second sequence of Ni-catalyzed asymmetric Negishi coupling with racemic secondary chloride **4.199** provided the ester **4.200** in a good yield and with >98% *ee* and 15:1 *dr* ratio. A reduction/amination sequence delivered the desired aldehyde **4.194** that could be further converted to fluvirucinine A1 using Suh's synthetic route.[172] This method exemplifies the efficiency of Negishi coupling in the construction of $C(sp^3)$–$C(sp^3)$ bond asymmetrically.

4.11.6 Pd-Catalyzed Domino Reactions in Total Synthesis

Domino (cascade) reactions are of great appeal to organic chemists, which offer an efficient protocol to construct structurally complicate molecules with high atom economy and environmental benefits. As cascade reactions allow the formation of multiple covalent bonds in a single step, complex products could be more efficiently synthesized from simple starting materials with no need of isolation or purification of intermediates involved. The chemistry of palladium is generally quite well understood and the Pd catalysts tolerate various functional groups. Therefore, there is a huge potential to make it an ideal catalyst for domino reactions. In this section, we reviewed the recent progress of domino reactions based on Pd-catalyzed reactions, especially their application in the synthesis of natural products.

In the laboratory of Nobutaka Fujii, a novel synthetic approach to lysergic acid was developed utilizing a Pd-catalyzed cascade pathway (Scheme 4.58).[173] The key allene precursor was prepared from the chiral propargyl alcohol **4.201** by using nosyl hydrazine under Mitsunobu conditions, which was originally reported by Myer's Group. Subsequent cleavage of the benzylidene group of **4.202** with PTSA gave the allenic amide **4.203** (dr = 94:6) in 85% yield. Treatment of **4.203** with 5 mol% of $Pd(PPh_3)_4$ and K_2CO_3 in DMF at 100 °C provided the tetracyclic product **4.204** in 76% yield with good diastereoselectivity (dr = 92:8). This reaction allowed the formation of two rings in a single step and realized the quick access to the carbon framework of lysergic acid. Further synthetic transformations included several oxidative state adjustments and protecting group manipulations, which led to lysergic acid in an additional six steps.

The cascade reaction completely meets the requirements of modern synthetic chemistry and fulfills the demand for ecology and economy. Another impressive example of a Pd-catalyzed domino process in natural product synthesis was reported by Lutz F. Tietze and coworkers.[174] They accomplished the first total synthesis of linoxepin in 10 synthetic steps and a 30% overall yield in a protecting group-free manner. In this work, the key intermediate **4.209** was prepared from aryl iodide **4.207** via a Sonongashira coupling with propargyl alcohol. Intermediate **4.209** was sequentially treated with $Pd(OAc)_2$ and Xphos as te ligand, the cascade reaction that initiated from a 7-exo Heck cyclization and followed by a silyl-terminated 6-exo Heck reaction smoothly gave the

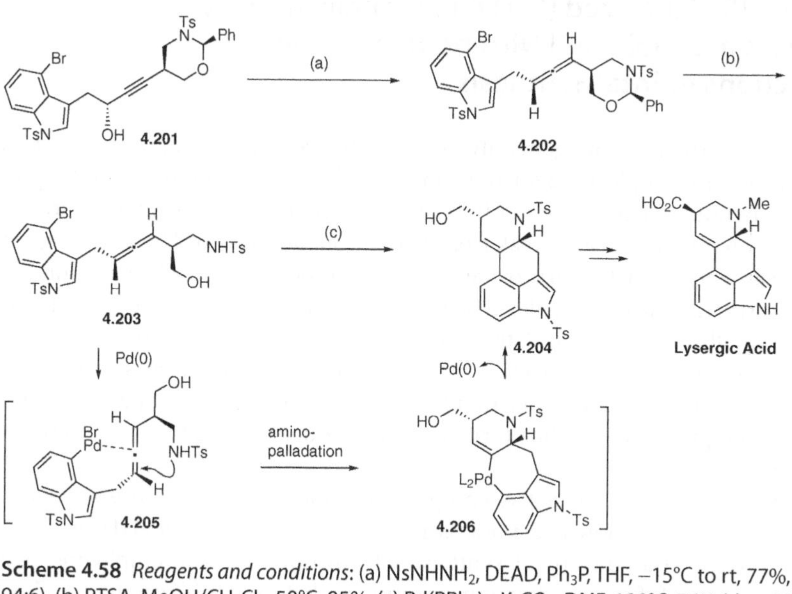

Scheme 4.58 *Reagents and conditions*: (a) $NsNHNH_2$, DEAD, Ph_3P, THF, −15°C to rt, 77%, (dr = 94:6). (b) PTSA, $MeOH/CH_2Cl_2$, 50°C, 85%. (c) $Pd(PPh_3)_4$, K_2CO_3, DMF, 100°C, 76% (dr = 92:8).

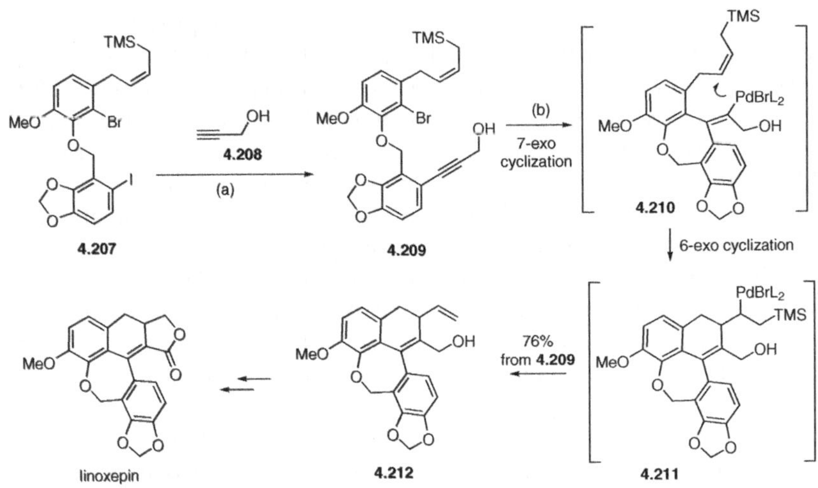

Scheme 4.59 *Reagents and conditions*: (a) $Pd(PPh_3)_4$ (7 mol%), CuI (10 mol%), dioxane, 60°C, 98%. (b) $Pd(OAc)_2$, XPhos, DME, 80°C.

double cyclic product **4.212** in 76% yield. After four steps of ordinary transformation, the linoxepin was successfully synthesized in a very efficient fashion (Scheme 4.59).

4.12 Rh-Catalyzed (C–H Functionalization by Metal Carbenoid and Nitrenoid Insertion) Reactions in Total Synthesis

Rhodium carbenoids are generally prepared by treatment of diazo precursors with rhodium complexes to form the metal carbene bond and release nitrogen. In a carbenoid, the metal center binds to the carbene carbon center via strong δ donation interaction and weak π back donation interaction. The advantage of metal carbenoid is that it can provide practical method for C–C and heteroatom–C bond formation, which enables efficient synthesis of natural products and bioactive small molecules. To date, rhodium carbenoids have been successfully applied to several stereoselective reactions with chiral auxiliaries or ligands, including lactone formation, lactam formation, CvH insertion/cope rearrangement, cyclopropanation, ylide formation, subsequent cycloaddition, and so on (Scheme 4.60).

Lithospermic acid is an active component in medicinal herb *Danshen* and it has been recently shown to possess potent and nontoxic anti-HIV activity. Yu[175] developed a convergent synthesis of lithospermic acid employing two successive C–H functionalization reactions; that is, Rh(I)-catalyzed C–H

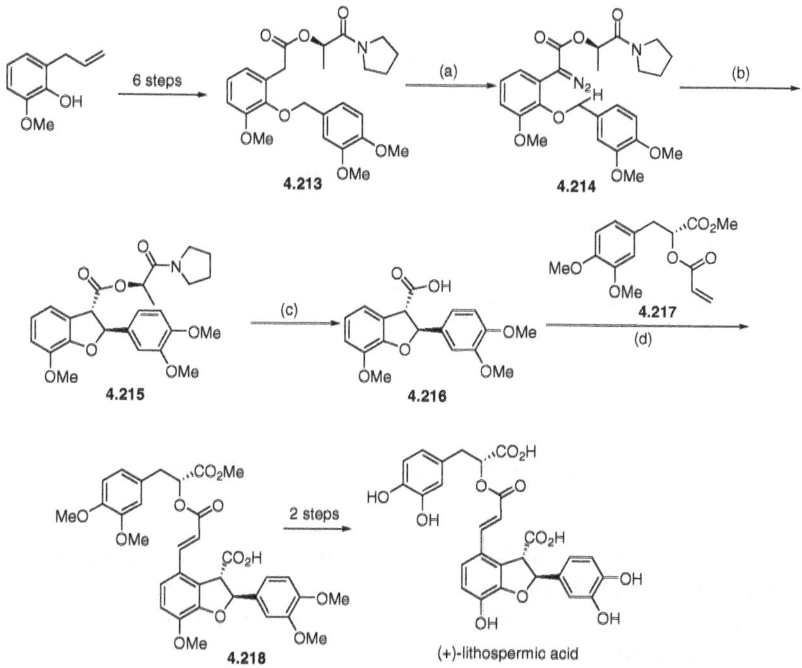

Scheme 4.60 *Reagents and conditions:* (a) *p*-ABSA, DBU, 82%. (b) Rh_2(S-DOSP)$_4$, 85%. (c) $Ba(OH)_2$, 86%. (d) $Pd(OAc)_2$/ligand, O_2 (1 atm), $KHCO_3$, *tert*-Amyl-OH, 85°C.

insertion to forge a dihydrobenzofuran core and a late-stage olefination of dihydrobenzofuran core catalyzed by a $Pd(OAc)_2$/ligand. As illustrated in Scheme 4.60, the cyclization precursor **4.213** was prepared in six steps from *o*-eugenol. Treatment of **4.213** with a diazo transfer reagent in the presence of DBU at room temperature afforded a diazo compound **4.214** in 82% yield. When **4.214** was treated with Davies's catalyst (Rh_2(*S*-DOSP)$_4$), the desired trans-dihydrobenzofuran core **4.215** was readily obtained through C–H functionalization in 85% yield with good d.r. (8:1). Hydrolysis of the auxiliary and subsequent intermolecular C–H olefination of compound **4.216** furnished the backbone **4.218**, which was elaborated to lithospermic acid after two routine steps of modification.

In 1999, Davies reported an unusual combined C–H insertion/cope rearrangement in a formal asymmetric synthesis of (+)-sertraline.[176] Subsequently, the synthetic utility of this novel transformation was displayed by the asymmetric total synthesis of diterpenes (–)-colombiasin A and (–)-elisapterosin B (Scheme 4.61).[177] Treatment of racemic dihydronaphthalene **4.219** with diazoester **4.220** in the presence of the chiral rhodium catalyst Rh_2(R-DOSP)$_4$ afforded compound **4.221** with all three stereocenters formed

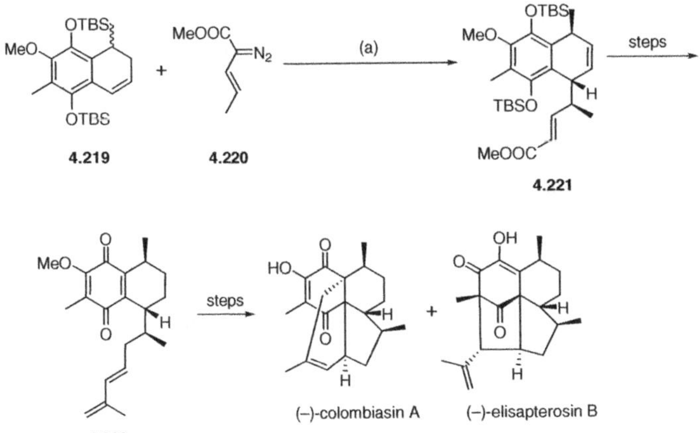

Scheme 4.61 *Reagents and conditions*: (a) 2 mol% Rh_2(R-DOSP)$_4$, 2,2-DMB.

in a single operation in good enantiomeric excess. Several routine steps converted **4.221** to **4.222**, which could be elaborated to both colombiasin and elisapterosin by an intramolecular [4+2] cycloaddition and Lewis acid catalyzed [5+2] cycloaddition, respectively.

Since nitrogen atoms are ubiquitous in natural products and pharmaceuticals, and C–N formation plays a fundamental role in organic synthesis, transition-metal nitrenoids (especially rhodium nitrenoids) have drawn considerable attention of synthetic community due to their powerful amination capability. Du Bois used tethered substrates (carbamates, sulfamates, ureas, guanidines, sulfonamides, etc.) for intramolecular amination with good regiocontrol and also developed an intermolecular amination system that employs $TcesNH_2$ and $Rh_2(esp)_2$. Both amination methods have been successfully developed into corresponding asymmetric versions by means of substrate control or by ligand control.[178] The building blocks accessed with Du Bois's methods are widely occurring in natural products and Du Bois has successfully applied his own methods in the synthesis of alkaloids.

(+)-Gonyautoxin 3, a representative paralytic shellfish poison, contained *bis*-guanidinium motifs. The challenge for its chemical synthesis stems from the heteroatom-rich tricyclic structure and the dicationic nature of it further complicates its final purification. In Du Bois's synthetic work (Scheme 4.62),[179] *L*-serine methyl ester was selected as the starting material, ester **4.223** could be easily prepared through amidation and protection of hydroxyl with a silyl group. Reduction of **4.223** to aldehyde and then condensation with allylamine in the presence of $BF_3 \cdot Et_2O$ gave the *trans*-substituded urea **4.224** with >20:1 diastereoselectivity, which could be elaborated to the requisite cyclization precursor **4.225** in a further four steps. Amination of **4.225** delivered compound **4.226** via a possible aziridine intermediate resulted from the insertion

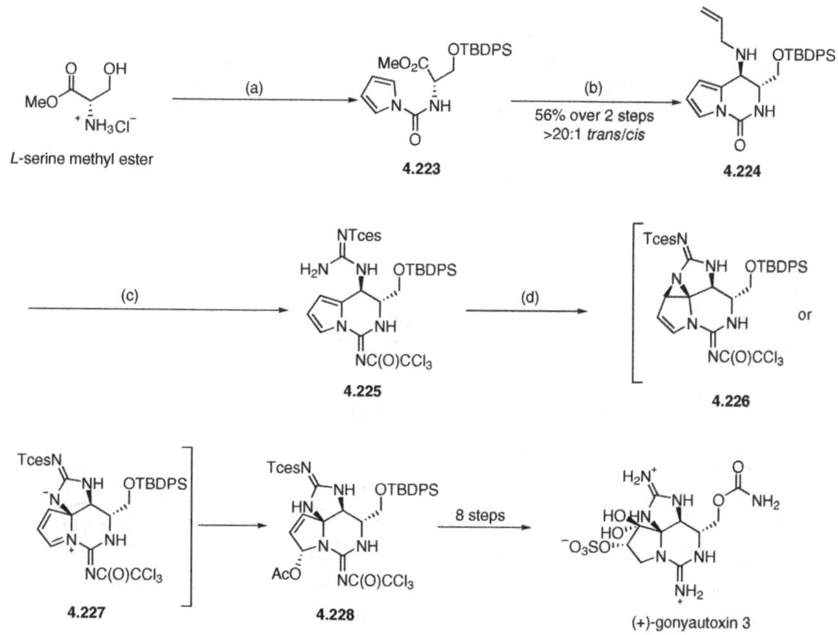

Scheme 4.62 Reagents and conditions: (a) (i) pyrrole-1-carboxylic, acid, DCC, Et_3N, DCM, 65%. (ii) $tBuPh_2SiCl$, imidazole, DMF, 97%. (b) (i) DIBAL-H, DCM, −90°C; (ii) allyl amine, $BF_3{\bullet}Et_2O$, (c) (i) $Pd(PPh_3)_4$, DCM, 1,3-dimethylbarbituric acid, then Na_2CO_3, TcesN=(SMe)Cl, 94%; (ii) EtOTf, 2,4,6-tri-*tert*-butylpyrimidine, DCM, 47°C, 78%; (iii) NH_3, NH_4OAc, MeOH, 60°C, 82%. (d) 5 mol% $Rh_2(esp)_2$, MgO, $PhI(AcO)_2$, 61%.

of pyrrole by the rhodium nitrenoid of the tethered guanidine and subsequent ring opening by acetic acid. A further eight steps decoration of the tricyclic backbone finished the synthesis of (+)-gonyautoxin 3.

Tetrodotoxin (TTX), another paralytic shellfish poison, featured with a densely oxygenated cyclohexane framework and affixed unique *ortho*-acid and guanidine aminal functionalities. Bois developed an approach involving stereospecific metal-mediated nitrene and carbene C-H insertion reactions as the key transformations.[180] Carboxylic acid **4.229** was prepared in seven steps from *D*-isoascorbic acid. Treatment of **4.229** with acyl chloride and then diazomethane delivered diazo compound **4.230**, cyclization of which proceeded smoothly with 1.5 mol% $Rh_2(HNCOCPh_3)_4$ exclusively affording cyclic ketone **4.231**. After multiple steps of manipulation, carbamate **4.233** was obtained, which is the precursor for installing the tetrasubstituted carbinolamine at C8a through stereospecific C–H amination. Treatment **4.233** With 10 mol% $Rh_2(HNCOCF_3)_4$ afforded the desired product **4.234** in the yield of 77%. With the C8a nitrogen center in place, straightforward assembly of the target molecule tetrodotoxin was accomplished by a sequence of routine manipulations (Scheme 4.63).

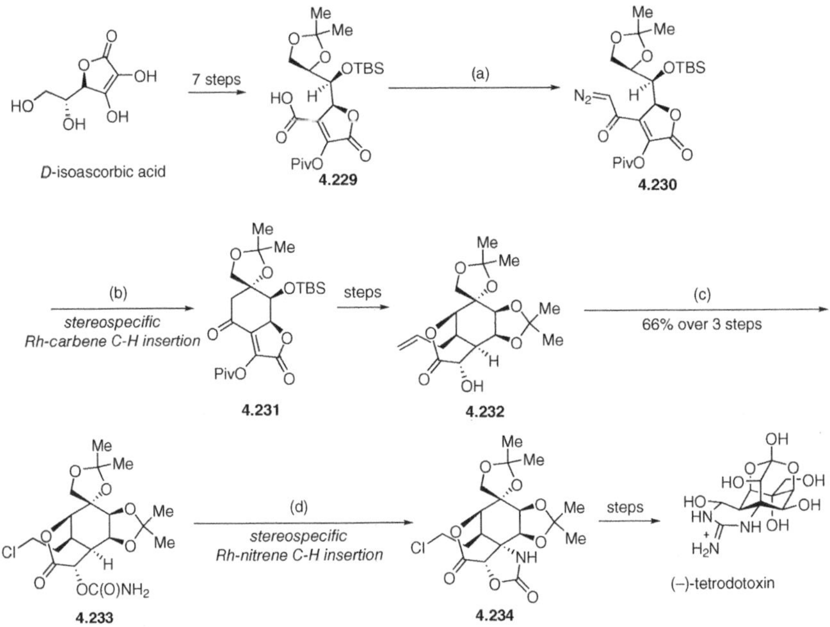

Scheme 4.63 *Reagents and conditions*: (a) $(COCl)_2$, cat. DMF, THF, then CH_2N_2, CH_2Cl_2, 63–70%. (b) $Rh_2(HNCOCPh_3)_4$ (1.5 mol%), CCl_4. (c) (i) $Cl_3CC(O)NCO$, CH_2Cl_2; Zn, MeOH (ii) O_3, then $NaBH_4$, CH_2Cl_2/MeOH (iii) $MeSO_2Cl$, C_5H_5N, DCE. (d) $Rh_2(HNCOCF_3)_4$ (10 mol%), $PhI(OAc)_2$, MgO, C_6H_6, 65°C, 77%.

4.13 Ru-Catalyzed RCM and RCAM in Total Synthesis

Alkene units widely appear in a large class of naturally occurring bioactive compounds and pharmaceutical molecules. Among the reactions that form the double bond, the classical carbonyl alkenylation reactions were represented by the Wittig reaction and its variants such as the Horner–Wadsworth–Emmons reaction, Julia olefination, Peterson reaction, and so on. The C–C cross-coupling reactions such as Stille coupling, Heck reaction, Negishi coupling, and so on are also frequently used methods to introduce alkene units. During the last decade, olefine metathesis has come to the fore as a very powerful tool for the syntheses of alkenes, providing an expedient route to unsaturated molecules that are often challenging or impossible to prepare by any other means. The term "olefin metathesis" was introduced in 1967, which means the metal-catalyzed redistribution of carbon–carbon double bonds, and also can be viewed as a "change-your-partners" dance as Chauvins' mechanism described. The metal-catalyzed redistribution of carbon–carbon triple bonds is called alkyne metathesis. With the advance in evolution of new catalysts, understanding of the mechanism and catalyst–substrate interactions, more and more researchers employed olefin and alkyne metathesis reactions in the syntheses of natural products and polymer chemistry. The Nobel Prize for Chemistry in 2005 was awarded jointly to Yves Chauvin, Robert H. Grubbs, and Richard R. Schrock for the development of the metathesis method in organic synthesis.[181] In this section we will introduce some classical applications of alkene metathesis and alkyne metathesis and demonstrate their power in complex molecule synthesis.

Macrocyclization by ring-closing metathesis is often used as the key step in the synthesis of natural product, but the *E*/*Z* selectivity of the newly formed double bond is rarely predictable, depending on many factors. Nowadays, there are two major approaches to address the stereoselectivity of ring-closing metathesis (RCM) reactions. One approach is a two-step procedure involving ring-closing alkyne metathesis (RCAM) followed by stereoselective reduction of the macrocyclic alkyne unit to the corresponding *E* or *Z* double bond. Another approach is that RCM reaction undergoes with a silyl group at one of the internal olefin positions in diolefinic substrate followed by protodesilylation.[182]

(−)-Nakadomarin A, a complex marine alkaloid, contained a fused 6/5/5/5 tetracyclic core system flanked with fused 8-membered and bridging 15-membered rings. Total synthesis of (−)-nakadomarin A was first accomplished by Nishida and coworkers,[183] and olefin RCM was used to assemble both of the eight- and 15-membered azacycles. With diene **4.235** in hand, it set the stage for the formation of the eight-membered ring by an RCM reaction. A problem that arose at this stage was the high reactivity of the terminal alkyne under RCM conditions. When alkyne **4.235** was exposed to Grubbs^{2nd} catalyst, no cyclization product was obtained. So, the terminal alkyne **4.235** was protected as a dicobalt hexacarbonyl complex, which was then treated with Grubbs^{2nd} catalyst, and a facile RCM ensued to furnish azocine lactam as a single *Z*-isomer in 83% yield. **4.237**, a precursor for the second RCM reaction, was obtained via reductive decomplexation, deprotection of the Boc and *N*-acylation. When **4.237** was treated with Grubbs^{1st} catalyst, ring F was formed to give a mixture of geometrical isomers, from which *Z*-isomer could only be obtained by column chromatography on silica gel in 26% yield. Finally, reduction of bislactam of *Z*-isomer with Red-Al resulted in the first asymmetric total synthesis of (−)-nakadomarin A (Scheme 4.64).

In Kerr's synthesis of nakadomarin A,[184] on treatment of **4.238**, a diamine substrate, with Grubbs^{1st} catalyst gave a 5:3 mixture of *trans*- to desired *cis*-cycloalkene **4.238a**, which was not separable by column chromatography on silica gel, or HPLC on standard or reverse phase columns. Finally, they followed Nishida's route and introduced a same amide side chain into the cyclization precursor to make the subsequent separation easier, and accomplished total synthesis of (+)-nakadomarin A.

Recently, in Dixon's[185] and Zhai's[186] synthesis of nakadomarin A, the (+)-CSA-assisted *Z*-selective olefin RCM involving a more polar diamine and a

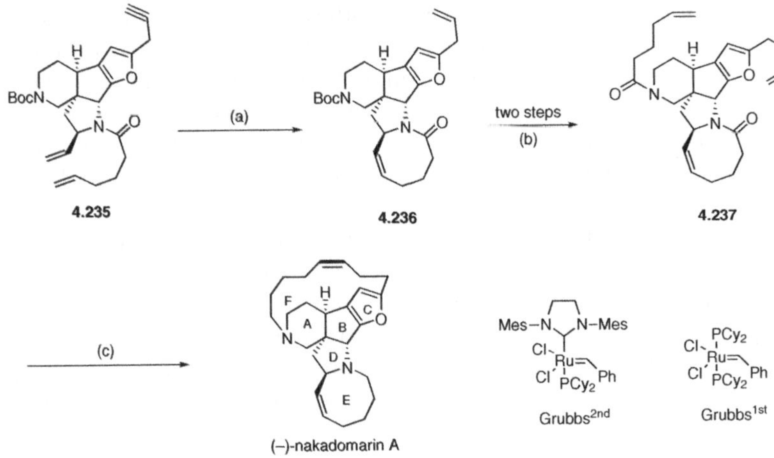

Scheme 4.64 *Reagents and conditions*: (a) (i) $Co_2(CO)_8$, DCM, 91%; (ii) Grubbs[1st] (25 mol%), DCM (1.0 mM), reflux, 1.5 h, 83%; (iii) nBu_3SnH, benzene, 75%. (b) (i) TFA, DCM; (ii) 5-hexenoyl chloride, TEA, DCM, 92%. (c) (i) Grubbs[2nd] (20 mol%), DCM (0.5 mM), reflux, 24 h, *Z* isomer 26%, *E* isomer 46%; (ii) red-Al, toluene, reflux, 92%.

monoamine precursor substrate for F ring generation proceeded smoothly, while the *E*/*Z* selectivity was reversed compared to the amine-free lactam/amide substrates utilized by Nishida and Kerr; Moreover, the product **4.238a** and (24*E*)-**4.238a** could easily be separated by semi-preparative SFC-HPLC. The geometrical selectivity culminated in Hoveyda's recent report[187] that RCM of the strained **4.239** with tungsten complex afforded the natural product in 63% yield (69% based on recovered substrate) and 94% *Z*-selectivity compared to previous attempts. Table 4.1 summarizes the *Z*/*E* selectivity and separability of the isomeric products from the RCM reactions.

On the other hand, In Funk's synthesis of nakadomarin A,[190] the *E*/*Z* selectivity problem in the construction of the macrocycle was circumvented by utilizing a RCAM/semi-hydrogenation strategy with *Z*-cycloalkene as a single geometrical isomer. The diyne precursor **4.240** was subjected to the Schrock carbyne catalyst in chlorobenzene allowing cyclization to afford **4.241** in 77% yield on a gram scale. The air-stable molybdenum nitride complex [(pyridine)$(Ph_3SiO)_3MoN$] recently developed by Fürstner was also effective, giving comparable results with slightly lower catalyst loading. The *cis* double bond of the 15-membered ring was then introduced by straightforward Lindlar reduction of cycloalkyne **4.241,** unaccompanied with the *E*-olefin isomer (Scheme 4.65).

Amphidinolide V, featured with densely arrayed alkene units, was isolated from symbiotic dinoflagellates *Amphidinium* sp., which exhibited cytotoxicity against murine lymphoma L1210 and human epidermoid carcinoma KB *in vitro*. Fürstner devised a flexible approach to construct the

Table 4.1 The *Z/E* selectivity and separability of the isomeric products from the RCM reactions of **4.237**, **4.238**, and **4.239**.

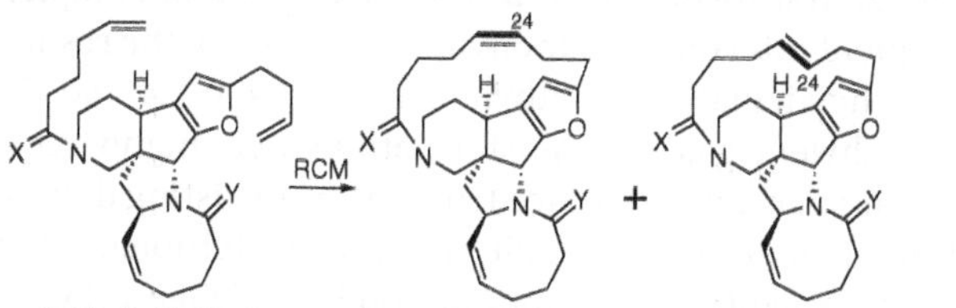

4.238: X = H_2, Y = O
4.237: X = Y = O
4.239: X = Y = H_2

4.238a: X = H_2, Y = O
4.237a: X = Y = O
4.239a: X = Y = H_2

(24*E*)-**4.238a**: X = H_2, Y = O
(24*E*)-**4.237a**: X = Y = O
(24*E*)-**4.239a**: X = Y = H_2

RCM Products	Data Source	RCM Type (Z/E Ratio)	Z/E Isomer Separability
4.238a/(24*E*)-**4.238a**	Zhai[186]	*Z*-selective (*Z*/*E*, 2:1)	separable (SFC-HPLC)
4.238a/(24*E*)-**4.238a**	Nishida[188]	regular (*Z*/*E*, 1:2)	*inseparable*
ent-**4.238a**/*ent*-(24*E*)-**4.238a**	Kerr[184]	regular (*Z*/*E*, 3:5)	*inseparable*
4.237a/(24*E*)-**4.237a**	Nishida[183]	regular (*Z*/*E*, 1:1.8)	Separable (column chromatography)
ent-**4.237a**/*ent*-(24*E*)-**4.237a**	Nishida[189]	regular (*Z*/*E*, 2:3)	Separable (column chromatography)
ent-**4.237a**/*ent*-(24*E*)-**4.237a**	Kerr[184]	regular (*Z*/*E* unavailable)	separable (column chromatography)
4.239a/(24*E*)-**4.239a**	Dixon[185]	*Z*-selective (*Z*/*E*, 63:37)	separable(NP-HPLC)
4.239a/(24*E*)-**4.239a**	Nishida[188]	obtained by lactam reduction	*inseparable*
ent-**4.239a**/*ent*-(24*E*)-**4.239a**	Kerr[184]	obtained by lactam reduction	*inseparable*

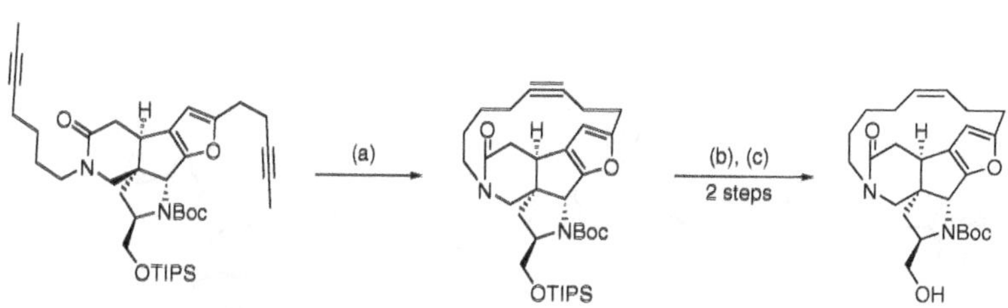

4.240 **4.241** **4.242**

Scheme 4.65 *Reagent and conditions*: (a) 25 mol%, $(tBuO)_3WCtBu$, PhCl, 80°C, 3 h, 77% or 20 mol% $[(pyridine)(Ph_3SiO)_3MoN]$, PhMe, 80°C, 16 h, 80%; (b) H_2, Lindlar; (c) TBAF, 80% for two steps.

distinctives-*trans-diene* unit of amphidinolide V.[191] Diyne **4.243** was selected as the substrate for ring-closing reaction, which proceeded smoothly with the molybdenum-based catalyst formed *in situ* upon activation of complex **4.246** with DCM. Subsequent intermolecular enyne metathesis of the resulting cycloalkyne installed the vicinal exo-methylene branches at C4 and C5 of amphidinolide V with ethylene gas. All transformations were highly regioselective without any appreciable interference from the pre-existing double bonds. Lastly, installation of an unsaturated side chain with additional three operations furnished the total synthesis of amphidinolide V (Scheme 4.66).

Flueggine A and virosaine B, two securinega alkaloids, were isolated from the twigs and leaves of *Flueggea virosa* by Ye *et al.* in 2011 and 2012, respectively. Recently, Yang reported the first total syntheses of (−)-flueggine A and (+)-virosaine B with tandem RRCMs and 1,3-dipolar cycloadditions as the key steps.[192] When enyne **4.247** was subjected to several commercially available metathesis catalysts, including Grubbs^{2nd}, Hoveyda–Grubbs, and Zhan-1b catalysts, Zhan-1b catalyst provided the best results to afford **4.249** as the sole product in 64% yield. The author speculated the whole process is that the less sterically encumbered terminal olefins in enyne **4.247** reacted with the ruthenium catalyst first, and then underwent relay metathesis via the active ester–carbene complexes **4.248**, the subsequent enyne cyclizaton proceeded smoothly to give **4.249**. The subsequent investigation involving removal of the Boc group with TFA, followed by a K_2CO_3-mediated cyclization, afforded (−)-norsecurinine in 64% yield. To complete the synthesis of (−)-norsecurinine, **4.249** was initially converted to an allylic bromide by treatment with NBS in the presence of AIBN, which could directly annulate in the presence of $AgSbF_6$ in acetone to give (−)-norsecurinine in 85% yield. Further, chemoselectively oxidization and a 1,3-dipolar addition effectively constructed the complex

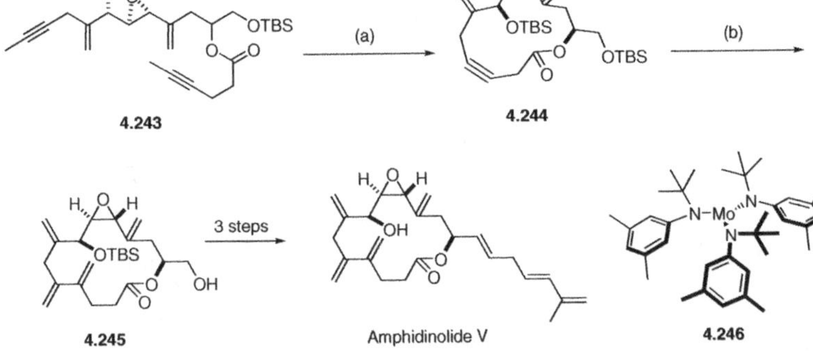

Scheme 4.66 Synthesis of amphidinolide V. *Reagent and conditions*: (a) Mo (20%), 85°C, DCM/PhMe, 66%; (b) Grubbs^{2nd} (2 mol%), C_2H_4 (1.8 atm), toluene, 45°C, 90%.

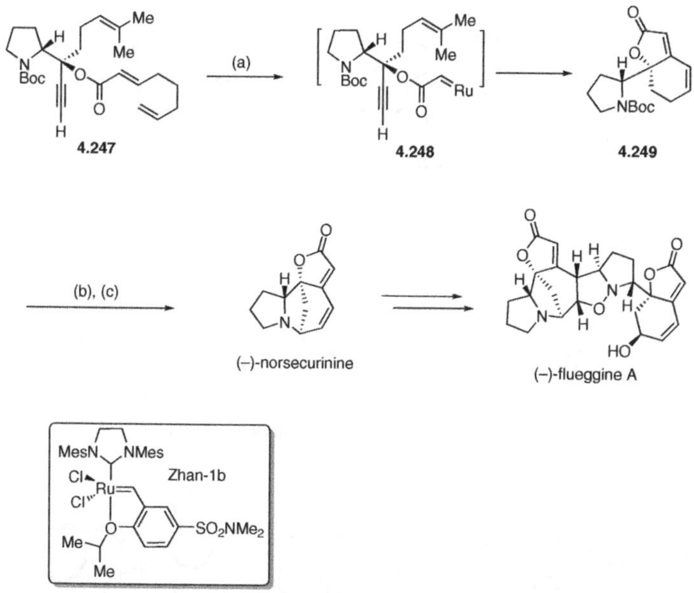

Scheme 4.67 Syntheses of flueggine A and virosaine B. *Reagent and conditions:* (a) 5 mol% Zhan-1b, toluene, reflux, 64%; (b) NBS, AIBN, CCl_4, 80°C, 60%; (c) TFA, CH_2Cl_2, 0°C, then K_2CO_3, $n\mathrm{Bu}_4\mathrm{NI}$, THF, RT, 64% or $AgSbF_6$, acetone, RT, 85%.

molecular scaffold of (–)-flueggine A in four steps with 57% overall yield (Scheme 4.67).

The formation of conjugated dienes or polyenes by RCM remains a formidable challenge, as not only it requires rigorous control over the configuration of the newly formed double bond, but also it remains difficult to discriminate between the two olefinic sites of a given 1,3-diene substrate. Because activation of the internal double bond could result in ring contraction, it may become the dominant or even exclusive pathway. Fürstner has developed a strategy that positioning of a silyl group on the diene unit allowed *E,Z*-configured macrocyclic 1,3-diene to be formed in good to excellent yields with unprecedented levels of regio- and stereocontrol, which was successfully applied to the total synthesis of lactimidomycin (Scheme 4.68).[193] Lactimidomycin, a potent translation and cell migration inhibitor, featured with *E, Z*-configured 1,3-diene, was isolated from *Streptomyces amphibisporus* in 1992. When **4.250** was subjected to complex **4.254**, the desired product **4.251** was obtained in a moderate yield and contaminated with significant amounts of the 11-membered homolog **4.252**, which was the result of alkene isomerization and could not be removed by flash chromatography. With complex **4.255**, designed for metathesis reactions of sterically hindered substrates, a lower yield was obtained. Reaction with the molybdenum alkylidene **4.256** only gave dimeric products. Gratifyingly,

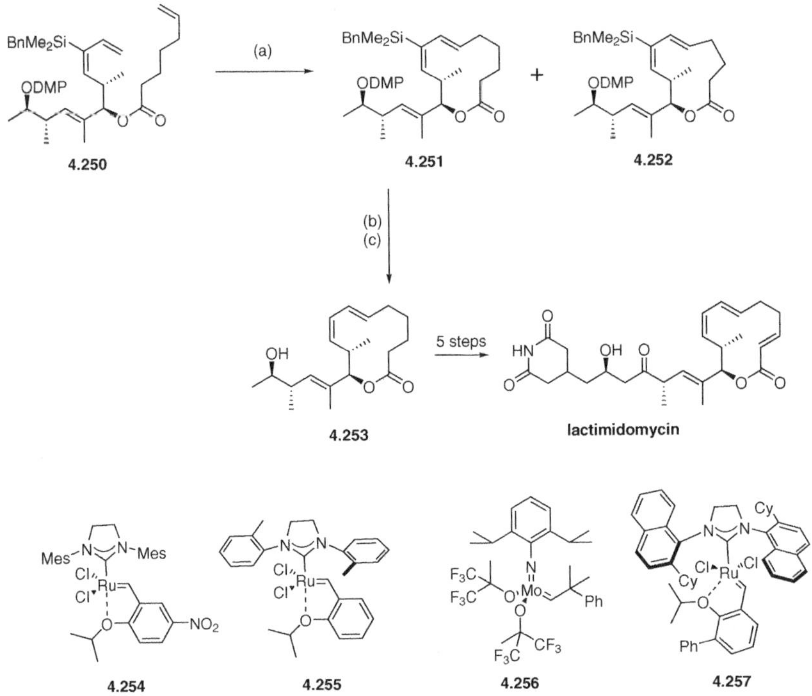

Scheme 4.68 Synthesis of lactimidomycin. *Reagent and conditions*: (a) conditions see Table 4.2; (b) DDQ, DCM, pH = 7 buffer 0°C, 87%; (c) TBAF, THF, 60°C, 85%.

using the more encumbered ruthenium carbine **4.257** largely suppressed the competing isomerization, which could result in losing of one methylene unit from the tether and give the desired product **4.251** in appreciable yield, while only very small amounts of the 11-membered congener **4.252** were detected. The resulting product **4.253** could be elaborated into lactimidomycin in another five steps. Table 4.2 summarizes the results of catalyst optimization for the RCM of substrate **4.250**.

Table 4.2 Catalyst optimization for the RCM of substrate **4.250**.

Entry	Catalyst (mol%)	T (°C)	Yield (%)	**4.251:4.252**
1	**4.254** (20)	80	56	63:37
2	**4.254** (20)	120	54	79:21
3	**4.255** (2*20)	80	35	nd
4	**4.256**	60	dimer	
5	**4.257**	120	76–78	95:5

Spirofungins A and B, initially isolated as a mixture from *Streptomyces violaceusniger* Tü4113, were reported to inhibit growth of *Candida albicans.* Kozmin developed a stereocontrolled and efficient strategy enabling the exclusive formation of bistramide A, using a cyclic silane as a temporary connection to hold the two arms with the aim to avoid epimerization of the spiroketal subunit.[194] The synthesis began with cyclopropenone acetal metathesis by subjecting alkene **4.258** to cyclopropenone acetal **4.259** in the presence of Grubbs^{2nd} catalyst. The ring-opening metathesis proceeded smoothly to give diene **4.260** upon subsequent desilylation. After the requisite dialkoxysilane connector was introduced, chemoselective removal of the 1,3-dioxane was efficiently achieved using oxalic acid to give ketone **4.262**. Exposure of **4.262** to the Grubbs^{2nd} catalyst resulted in the cyclization to afford the 15-membered dienone **4.263** in 85% yield, with complete chemoselectivity only at the two terminal alkenes of trienone **4.262**. Subsequently, a seven-step transformation afforded triene **4.264**, which was further elaborated via cross metathesis with methyl acrylate to afford **4.265**. Installation of another diene subunit allowed for the completion of the synthesis of spirofungin A (Scheme 4.69).

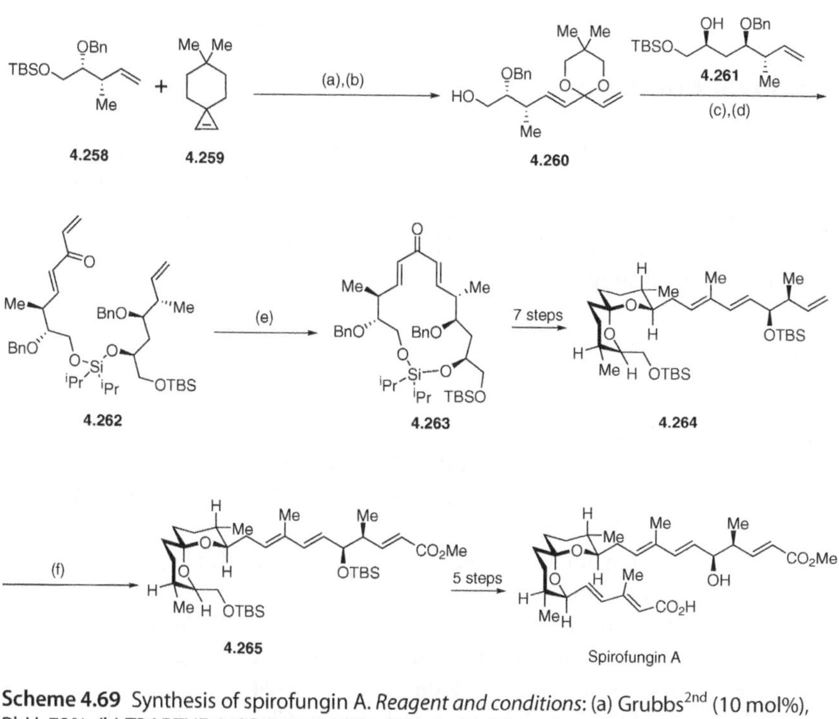

Scheme 4.69 Synthesis of spirofungin A. *Reagent and conditions*: (a) Grubbs^{2nd} (10 mol%), PhH, 72%; (b) TBAF,THF, 20°C, 84%; (c) $(iPr)_2SiCl_2$, imid.; (d) oxalic acid, silica gel, DCM, 65% for two steps; (e) Grubbs^{2nd} (30 mol%), PhH, 85%; (f) CH_2=$CHCO_2Me$, Grubbs^{2nd} (10 mol%), 65%.

4.14 Conclusion

In this chapter, several synthetic syntheses of natural products that were achieved by utilizing organometallics-based methodologies were reviewed. The scope and the limitation of some methodologies were well explored considering the different reactivities of each metal complex. These methodologies have been well applied by rapid construction of advanced intermediates and the core skeleton of the targeted natural products starting from simple and acyclic precursors. Examples can be seen in Au- and Pt-catalyzed cycloisomerization reactions for englerins A; Co-mediated Pauson–Khand reactions for syntheses of jiadifenin, merrilactone A; and Cu-promoted cyclopropanation/ring-opening/iminium cyclization reactions for (±)-minfiensine and (±)-vincorine. Although a great deal of organometallics-based natural product syntheses have been exemplified, there are many natural products that still remain synthetically challenging; therefore, more applications and development of new organometallics are anticipated. For example, for Fe-mediated chemical reactions, a better understanding of the mechanism and broader applications in natural product synthesis are needed given the limited examples and the abundance of Fe.

Acknowledgements

We thank the NSFC (21732001, 21672017, 21472072, 21290183), Shenzhen Science and Technology Innovation Committee (JCYJ20150529153646078, JSGG20160229150510483), Program for Changjiang Scholars and Innovative Research Team in University (PCSIRT: IRT_15R28), and "111" Program of MOE for financial support.

References

1 Li, Y. F.; Zhou, F.; Forsyth, C. J. *Angew. Chem. Int. Ed.* **2007**, *46*, 279–282.
2 Trost, B. M.; Dong, G. B. *Nature* **2008**, *456*, 485–488.
3 Veitch, G. E.; Beckmann, E.; Burke, B. J.; Boyer, A.; Maslen, S. L.; Ley, S. V. *Angew. Chem. Int. Ed.* **2007**, *46*, 7629–7632.
4 Veitch, G. E.; Beckmann, E.; Burke, B. J.; Boyer, A.; Ayats, C.; Ley, S. V. *Angew. Chem. Int. Ed.* **2007**, *46*, 7633–7635.
5 Butterworth, J.; Morgan, E. D. *Chem. Commun. (London).* **1968**, 23–24.
6 Staben, S. T.; Kennedy-Smith, J. J.; Huang, D.; Corkey, B. K.; LaLonde, R. L.; Toste, F. D. *Angew. Chem. Int. Ed.* **2006**, *45*, 5991–5994.
7 Nicolaou, K. C.; Tria, G. S.; Edmonds, D. J. *Angew. Chem. Int. Ed.* **2008**, *47*, 1780–1783.

8 Linghu, X.; Kennedy-Smith, J. J.; Toste, F. D. *Angew. Chem. Int. Ed.* **2007**, *46*, 7671–7673.
9 Sato, K.; Asao, N.; Yamamoto, Y. *J. Org. Chem.* **2005**, *70*, 8977–8981.
10 Jimenez-Nunez, E.; Molawi, K.; Echavarren, A. M. *Chem. Commun.* **2009**, 7327–7329.
11 Zhou, Q. H.; Chen, X. F.; Ma, D. W. *Angew. Chem. Int. Ed.* **2010**, *49*, 3513–3516.
12 Molawi, K.; Delpont, N.; Echavarren, A. M. *Angew. Chem. Int. Ed.* **2010**, *49*, 3517–3519.
13 Naodovic, M.; Yamamoto, H. *Chem. Rev.* **2008**, *108*, 3132–3148.
14 Weibel, J. M.; Blanc, A.; Pale, P. *Chem. Rev.* **2008**, *108*, 3149–3173.
15 Marshall, J. A.; Pinney, K. G. *J. Org. Chem.* **1993**, *58*, 7180–7184.
16 Marshall, J. A.; Liao, J. L. *J. Org. Chem.* **1998**, *63*, 5962–5970.
17 Marshall, J. A.; Wallace, E. M.; Coan, P. S. *J. Org. Chem.* **1995**, *60*, 796–797.
18 Haga, Y.; Okazaki, M.; Shuto, Y. *Biosci. Biotechnol. Biochem.* **2003**, *67*, 2183–2193.
19 Haga, Y.; Okazaki, M.; Shuto, Y. *Biosci. Biotechnol. Biochem* **2003**, *67*, 2215–2223.
20 Rossi, R.; Bellina, F.; Biagetti, M.; Mannina, L. *Tetrahedron Lett.* **1998**, *39*, 7799–7802.
21 Xu, C. D.; Negishi, E. *Tetrahedron Lett.* **1999**, *40*, 431–434.
22 Du Ha, J.; Cha, J. K. *J. Am. Chem. Soc.* **1999**, *121*, 10012–10020.
23 Saimoto, H.; Yasui, M.; Ohrai, S.; Oikawa, H.; Yokoyama, K.; Shigemasa, Y. *Bull. Chem. Soc. Jpn.* **1999**, *72*, 279–284.
24 Zhang, L. M.; Sun, J. W.; Kozmin, S. A. *Adv. Synth. Catal.* **2006**, *348*, 2271–2296.
25 Furstner, A.; Szillat, H.; Gabor, B.; Mynott, R. *J. Am. Chem. Soc.* **1998**, *120*, 8305–8314.
26 Simmons, E. M.; Hardin, A. R.; Guo, X.; Sarpong, R. *Angew. Chem. Int. Ed.* **2008**, *47*, 6650–6653.
27 Furstner, A.; Hannen, P. *Chem. Eur. J.* **2006**, *12*, 3006–3019.
28 Fehr, C.; Galindo, J. *Angew. Chem. Int. Ed.* **2006**, *45*, 2901–2904.
29 Furstner, A.; Hannen, P. *Chem. Commun.* **2004**, 2546–2547.
30 Zhang, L.; Xie, X. G.; Liu, J.; Qi, J.; Ma, D. H.; She, X. G. *Org. Lett.* **2011**, *13*, 2956–2958.
31 Furstner, A.; Stelzer, F.; Szillat, H. *J. Am. Chem. Soc.* **2001**, *123*, 11863–11869.
32 Furstner, A.; Heilmann, E. K.; Davies, P. W. *Angew. Chem. Int. Ed.* **2007**, *46*, 4760–4763.
33 Deng, H. B.; Yang, X. B.; Tong, Z. L.; Li, Z.; Zhai, H. B. *Org. Lett.* **2008**, *10*, 1791–1793.
34 Furstner, A.; Kennedy, J. W. J. *Chem. Eur. J.* **2006**, *12*, 7398–7410.
35 Obika, S.; Yasui, Y.; Yanada, R.; Takemoto, Y. *J. Org. Chem.* **2008**, *73*, 5206–5209.
36 Kozak, J. A.; Dake, G. R. *Angew. Chem. Int. Ed.* **2008**, *47*, 4221–4223.

37 (a) Torres, R. R. (Eds), *The Pauson-Khand Reaction: Scope, Variations and Applications*, John Wiley & Sons, Ltd: Chichester, UK, **2012**. (b) Lee, H.-W.; Kwong, F.-Y. *Eur. J. Org. Chem.* **2010**, 789–811.

38 (a) Khand, I. U.; Knox, G. R.; Pauson, P. L.; Watts, W. E. *J. Chem. Soc. Chem. Commun.* **1971**, *36*. (b) Khand, I. U.; Knox, G. R.; Pauson, P. L.; Watts, W. E. *J. Chem. Soc. Perkin Trans.* **1973**, *1*, 975–977. (c) Khand, I. U.; Knox, G. R.; Pauson, P. L.; Watts, W. E.; Foreman, M. L. *J. Chem. Soc. Perkin Trans.* **1973**, *1*, 977–981.

39 Pallerla, M. K.; Fox, J. M. *Org. Lett.* **2007**, *9*, 5625–5628.

40 Seto, H.; Yonehara, H. J. *J. Antibiot.* **1980**, *33*, 92–93.

41 Kozaka, T.; Miyakoshi, N.; Mukai, C. *J. Org. Chem.* **2007**, *72*, 10147–10154.

42 Sugihara, T.; Yamada, M.; Yamaguchi, M.; Nishizawa, M. *Synlett*, **1999**, *6*, 771–773.

43 Xiao, Q.; Ren, W.-W.; Chen, Z.-X.; Sun, T.-W.; Li, Y.; Ye, Q.-D. *et al. Angew. Chem. Int. Ed.* **2011**, *50*, 7373–7377.

44 Yang, Y.; Fu, X.; Chen, J.; Zhai, H. *Angew. Chem. Int. Ed.* **2012**, *51*, 9825–9828.

45 (a) Buchwald, S. L.; Hicks, F. A. In *Comprehensive Asymmetric Catalysis*; Jacobsen, E. N., Pfaltz, A., Yamamoto, H. (Eds); Springer: Heidelberg, **1999**, *2*, p. 491. (b) Adrio, J.; Carretero, J. C. *J. Am. Chem. Soc.* **2007**, *129*, 778–779 and references therein.

46 Mukai, C.; Yoshida, T.; Sorimachi, M.; Odani, A. *Org. Lett.* **2006**, *8*, 83–86.

47 Chen J.; Gao, P.; Yu, F.; Yang, Y.; Zhu, S.; Zhai, H. *Angew. Chem. Int. Ed.* **2012**, *51*, 5897–5899.

48 Krause, N.; *Modern Organocopper Chemistry*, Wiley-VCH: Weinheim, Germany, 1st Edn, **2002**.

49 (a) Noyori, R.; Suzuki, M. *Angew. Chem. Int. Ed.* **1984**, *23*, 847–876. (b) Patterson, J. W., Jr.; Fried, J. H. *J. Org. Chem.* **1974**, *39*, 2506–2509. (c) Davis, R.; Untch, K. G. *J. Org. Chem.* **1979**, *44*, 3755–3759. (d) Suzuki, M.; Kawagishi, T.; Suzuki, T.; Noyori, R. *Tetrahedron. Lett.* **1982**, *23*, 4057–4060. (e) Suzuki, M.; Kawagishi, T.; Noyori, R. *Tetrahedron. Lett.* **1982**, *23*, 5563–5566. (f) Tanaka, T.; Toru, T.; Okamura, N.; Hazato, A,; Sugiura, S.; Manabe, K.; *et al. Tetrahedron. Lett.* **1983**, *24*, 4103–4104. (g) Suzuki, M.; Yanagisawa, A.; Noyori, R. *Tetrahedron. Lett.* **1984**, *25*, 1383–1386.

50 Suzuki, M.; Yanagisawa, A.; Noyori, R. *J. Am. Chem. Soc.* **1985**, *107*, 3348–3349.

51 Ireland, R. E.; Gleason, J. L.; Gegnas, L. D.; Highsmith, T. K. *J. Org. Chem.* **1996**, *61*, 6856–6872.

52 (a) Jones, T. K.; Mills, S. G.; Reamer, R. A.; Askin, D.; Desmond, R.; Volante, R. P, *et al. J. Am. Chem. Soc.* **1989**, *111*, 1157–1159. (b) Jones, T. K.; Reamer, R. A.; Desmond, R.; Mills, S. G. *J. Am. Chem. Soc.* **1990**, *112*, 2998–3017. (c) Nakatsuka, M.; Ragan, J. A.; Sammakia, T.; Smith, D. B.; Uehling, D. E.; Schreiber, S. L. *J. Am. Chem. Soc.* **1990**, *112*, 5583–5601. (d) Jones, A. B.; Villalobos, A.; Linde, R. G.; Danishefsky, S. J. *J. Org. Chem.* **1990**, *55*, 2786–2797.

(e) Gu, R. L.; Sih, C. J. *Tetrahedron. Lett.* **1990**, *31*, 3287–3290. (f) Smith, A. B., III; Chen, K.; Robinson, D. J.; Laakso, L. M.; Hale, K. J. *Tetrahedron. Lett.* **1994**, *35*, 4271–4274.
53 Roy, S. C.; Nagarajan, L.; Salomon, R. G. *J. Org. Chem.* **1999**, *64*, 1218–1224.
54 Horst, B. T.; Feringa, B. L.; Minnaard, A. *Org. Lett.* **2007**, *9*, 3013–3015.
55 King, G. K.; Mander, L. N.; Monck, N. J. T.; Morris, J. C.; Zhang, H. *J. Am. Chem. Soc.* **1997**, *119*, 3828–3829.
56 Zhang, D.; Song, H.; Qin, Y. *Acc. Chem. Res.* **2011**, *44*, 447–457.
57 Yang, J.; Wu, H.; Shen, L.; Qin, Y. *J. Am. Chem. Soc.* **2007**, *129*, 13794–13795.
58 Shen, L.; Zhang, M.; Wu, Y.; Qin, Y. *Angew. Chem. Int. Ed.* **2008**, *47*, 3618–3621.
59 Zhang, M.; Huang, X.; Shen, L. Qin, Y. *J. Am. Chem. Soc.* **2009**, *131*, 6013–6020.
60 Levin, S. L.; Nani, R. R.; Reisman, S. E. *J. Am. Chem. Soc.* **2011**, *133*, 774–776.
61 Avalos, M.; Babiano, R.; Cintas, P.; Jimenez, J. L.; Palacios, J. C. *Chem. Soc. Rev.* **1999**, *28*, 169–177.
62 Furstner, A. *Chem. Rev.* **1999**, *99*, 991–1045.
63 Smith, K. M. *Coordin. Chem. Rev.* **2006**, *250*, 1023–1031.
64 Furstner, A.; Shi, N. Y. *J. Am. Chem. Soc.* **1996**, *118*, 12349–12357.
65 Kuroboshi, M.; Tanaka, M.; Kishimoto, S.; Goto, K.; Mochizuki, M.; Tanaka, H. *Tetrahedron. Lett.* **2000**, *41*, 81–84.
66 Hargaden, G. C.; Guiry, P. J. *Adv. Synth. Catal.* **2007**, *349*, 2407–2424.
67 Ocampo, R.; Dolbier, W. R. *Tetrahedron.* **2004**, *60*, 9325–9374.
68 Gibson, V. C.; Spitzmesser, S. K. *Chem. Rev.* **2003**, *103*, 283–315.
69 Hirata, Y.; Uemura, D. *Pure. Appl. Chem.* **1986**, *58*, 701–710.
70 Aicher, T. D.; Buszek, K. R.; Fang, F. G.; Forsyth, C. J.; Jung, S. H.; Kishi, Y.; *et al. J. Am. Chem. Soc.* **1992**, *114*, 3162–3164.
71 Towle, M. J.; Salvato, K. A.; Budrow, J.; Wels, B. F.; Kuznetsov, G.; Aalfs, K. K.; *et al. A. Cancer. Res.* **2001**, *61*, 1013–1021.
72 Yamamoto, A.; Ueda, A.; Bremond, P.; Tiseni, P. S.; Kishi, Y. *J. Am. Chem. Soc.* **2012**, *134*, 893–896.
73 Orsini, F.; Sello, G. *Curr. Org. Synth.* **2004**, *1*, 111–135.
74 Wittenberg, R.; Beier, C.; Drager, G.; Jas, G.; Jasper, C.; Monenschein, H. et al. *Tetrahedron. Lett.* **2004**, *45*, 4457–4460.
75 Wessjohann, L. A.; Scheid, G.; *Synthesis*, **1999**, *1*, 1–36.
76 Takai, K.; Nitta, K.; Utimoto, K. *Tetrahedron Lett.* **1988**, *29*, 5263–5266.
77 Takai, K.; Morita, R.; Toratsu, C. *Angew. Chem. Int. Ed.* **2001**, *40*, 1116–1119.
78 Boeckman, R. K.; Hudack, R. A. *J. Org. Chem.* **1998**, *63*, 3524–3525.
79 Jung, M.; Groth, U. *Synlett.* **2002**, 2015–2018.
80 Groth, U.; Jung, M.; Vogel, T. *Chem.Eur.* **2005**, *11*, 3127–3135.
81 Fischer, S.; Groth, U.; Jung, M.; Lindenmaier, M.; Vogel, T. *Tetrahedron Lett.* **2005**, *46*, 6679–6682.
82 Danishefsky, S. J.; Deninno, M. P. *Angew. Chem. Int. Ed.* **1987**, *26*, 15–23.
83 Danishefsky, S.; Kerwin, J. F.; Kobayashi, S. *J. Am. Chem. Soc.* **1982**, *104*, 358–360.

84 Schaus, S. E.; Branalt, J.; Jacobsen, E. N. *J. Org. Chem.* **1998**, *63*, 403–405.
85 Dossetter, A. G.; Jamison, T. F.; Jacobsen, E. N. *Angew. Chem. Int. Ed.* **1999**, *38*, 2398–2400.
86 Gademann, K.; Chavez, D. E.; Jacobsen, E. N. *Angew. Chem. Int. Ed.* **2002**, *41*, 3059–3061.
87 Joly, G. D.; Jacobsen, E. N. *Org. Lett.* **2002**, *4*, 1795–1798.
88 Chavez, D. E.; Jacobsen, E. N. *Org. Lett.* **2003**, *5*, 2563–2565.
89 Taylor, M. S.; Jacobsen, E. N. *J. Am. Chem. Soc.* **2003**,*125*, 11204–11205.
90 Louis, I.; Hungerford, N. L.; Humphries, E. J.; McLeod, M. D. *Org. Lett.* **2006**, *8*, 1117–1120.
91 Majumder, U.; Cox, J. M.; Johnson, H. W. B.; Rainier, J. D. *Chem. Eur.* **2006**, *12*, 1736–1746.
92 Cutignano, A.; Bruno, I.; Bifulco, G.; Casapullo, A.; Debitus, C.; Gomez-Paloma, L. *et al. J. Org. Chem.* **2001**, 775–778.
93 Krusic, P. J.; Kochi, J. K. *J. Am. Chem. Soc.* **1971**, *93*, 846–860.
94 Sherry, B. D.; Furstner, A. *Accounts of Chemical Research*, **2008**, *41*, 1500–1511.
95 Percival, W. C.; Wagner, R. B.; Cook, N. C. *J. Am. Chem. Soc.* **1953**, *75*, 3731–3734.
96 Cardellicchio, C.; Viandanese, V.; Marchese, G.; Ronzini, L. *Tetrahedron Lett.* **1985**, *26*, 3595–3598.
97 Scheiper, B.; Bonnekessel, M.; Krause, H.; Furstner, A. *J. Org. Chem.* **2004**, *69*, 3943–3949.
98 Scheiper, B.; Bonnekessel, M.; Krause, H.; Furstner, A. *J. Org. Chem.* **2004**, *69*, 3943–3949.
99 Furstner, A.; De Souza, D.; Turet, L.; Fenster, M. D. B.; Parra-Rapado, L.; Wirtz, C. *et al. Chem. Eur,.* **2007**, *13*, 115–134.
100 Neumann, S. M.; Kochi, J. K. *J. Org. Chem.* **1975**, *40*, 599–606.
101 Molander, G. A.; Rahn, B. J.; Shubert, D. C.; Bonde, S. E. *Tetrahedron Lett.* **1983**, *24*, 5449–5452.
102 Cahiez, G.; Avedissian, H. *Synthesis*, **1998**, 1199–1205.
103 Furstner, A.; Martin, R. *Chem. Lett.* **2005**, *34*, 624–629.
104 Itami, K.; Higashi, S.; Mineno, M.; Yoshida, J. *Organic Lett.* **2005**, *7*, 1219–1222.
105 Berthon-Gelloz, G.; Hayashi, T. *J. Org. Chem.* **2006**, *71*, 8957–8960.
106 Hamajima, A.; Isobe, M. *Organic Lett.* **2006**, *8*, 1205–1208.
107 Hamajima, A.; Isobe, M. *Angew. Chem. Int. Ed.* **2009**, *48*, 2941–2945.
108 Furstner, A.; Leitner, A.; Mendez, M.; Krause, H. *J. Am. Chem. Soc.* **2002**, *124*, 13856–13863.
109 Furstner, A.; Leitner, A. *Angew. Chem. Int. Ed.* **2002**, *41*, 609–612.
110 Sapountzis, I.; Lin, W. W.; Kofink, C. C.; Despotopoulou, C.; Knochel, P. *Angew. Chem. Int. Ed.* **2005**, *44*, 1654–1657.
111 Hatakeyama, T.; Nakamura, M. *J. Am. Chem. Soc.* **2007**, *129*, 9844–9845.

112 Furstner, A.; Leitner, A. *Angew. Chem. Int. Ed.* **2003**, *42*, 308–311.
113 Frisch, A. C.; Beller, M. *Angew. Chem. Int. Ed.* **2005**, *44*, 674–688.
114 Nakamura, M.; Matsuo, K.; Ito, S.; Nakamura, B. *J. Am. Chem. Soc.* **2004**, *126*, 3686–3687.
115 Nakamura, M.; Ito, S.; Matsuo, K.; Nakamura, E. *Synlett.* **2005**, 1794–1798.
116 Nagano, T.; Hayashi, T. *Org. Lett.* **2004**, *6*, 1297–1299.
117 Martin, R.; Furstner, A. *Angew. Chem. Int. Ed.* **2004**,*43*, 3955–3957.
118 Guerinot, A.; Reymond, S.; Cossy, J. *Angew. Chem. Int. Ed.* **2007**, *46*, 6521–6524.
119 Dongol, K. G.; Koh, H.; Sau, M.; Chai, C. L. L. *Adv. Synth. Catal.* **2007**, *349*, 1015–1018.
120 Furstner, A.; Kattnig, E.; Lepage, O. *J. Am. Chem. Soc.* **2006**, *128*, 9194–9204.
121 Furstner, A.; Majima, K.; Martin, R.; Krause, H.; Kattnig, E.; Goddard, R.; Lehmann, C. W. *J. Am. Chem. Soc.* **2008**, *130*, 1992–2004.
122 Correa, A.; Bolm, C. *Angew. Chem. Int. Ed.* **2007**, *46*, 8862–8865.
123 Bistri, O.; Correa, A.; Bolm, C. *Angew. Chem. Int. Ed.* **2008**, *47*, 586–588.
124 Correa, A.; Carril, M.; Bolm, C. *Angew. Chem. Int. Ed.* **2008**, *47*, 2880–2883.
125 Linker, T. *J. Organomet. Chem.* **2002**, *661*, 159–167.
126 Snider, B. B. *Chem. Rev.* **1996**, *96*, 339–363.
127 Cimino, G.; Destefano, S.; Guerriero, A.; Minale, L. *Tetrahedron Lett.* **1975**, *16*, 1425–1428.
128 White, J. D.; Somers, T. C.; Yager, K. M. *Tetrahedron Lett.* **1990**, *31*, 59–62.
129 Kumabe, R.; Nishino, H.; Yasutake, M.; Nguyen, V. H.; Kurosawa, K. *Tetrahedron Lett.* **2001**, *42*, 69–72.
130 Kumabe, R.; Nishino, H. *Tetrahedron Lett.* **2004**, *45*, 703–706.
131 Snider, B. B.; Che, Q. L. *Heterocycles*, **2004**, *62*, 325–332.
132 Lamarque, L.; Meou, A.; Brun, P. *Tetrahedron*, **1998**, *54*, 6497–6506.
133 Meou, A.; Lamarque, L.; Brun, P. *Tetrahedron Lett.* **2002**, *43*, 5301–5304.
134 Corey, E. J.; Wu, Y. J. *J. Am. Chem. Soc.* **1993**. *115*, 8871–8872.
135 Aimi, N.; Inaba, M.; Watanabe, M.; Shibata, S. *Tetrahedron*, **1969**, *25*, 1825–1838.
136 Sung, K. S.; Wang, Y. Y. *J. Org. Chem.* **2003**, *68*, 2771–2778.
137 Andersen, N. R.; Rasmussen, P. R.. *Tetrahedron Lett.* **1984**, *25*, 465–468.
138 Hosoe, T.; Nozawa, K.; Lumley, T. C.; Currah, R. S.; Fukushima, K.; Takizawa, K.; *et al. Chem. & Pharma. Bull.* **1999**, *47*, 1591–1597.
139 Barrero, A. F.; Herrador, M. M.; del Moral, J. F. Q.; Valdivia, M. V. *Organic Lett.* **2002**, *4*, 1379–1382.
140 Breuilles, P.; Uguen, D.. *Tetrahedron Lett.* **1984**, *25*, 5759–5762.
141 Lautens, M.; Klute, W.; Tam, W. *Chem. Rev.* **1996**, *96*, 49–92.
142 Kanai, H.; Hiraki, N. *Chem. Lett.* **1979**, 761–762.
143 Lautens, M.; Edwards, L. G.; Tam, W.; Lough, A. *J. Am. Chem. Soc.* **1995**, *117*, 10276–10291.
144 Saito, S.; Hirayama, K.; Kabuto, C.; Yamamoto, Y. *J. Am. Chem. Soc.* **2000**, *122*, 10776–10780.

145 Bapuji, S. A.; Motherwell, W. B.; Shipman, M. *Tetrahedron Lett.* **1989**, *30*, 7107–7110.
146 Sato, Y.; Nishimata, T.; Mori, M. *J. Org. Chem.* **1994**, *59*, 6133–6135.
147 Seo, J.; Chui, H. M. P.; Heeg, M. J.; Montgomery, J. *J. Am. Chem. Soc.* **1999**, *121*, 476–477.
148 Tang, X. Q.; Montgomery, J. *J. Am. Chem. Soc.* **2000**, *122*, 6950–6954.
149 Wender, P. A.; Jenkins, T. E. *J. Am. Chem. Soc.* **1989**, *111*, 6432–6434.
150 Brun, P.; Tenaglia, A.; Waegell, B. *Tetrahedron Lett.* **1983**, *24*, 385–388.
151 Wender, P. A.; Ihle, N. C. *J. Am. Chem. Soc.* **1986**, *108*, 4678–4679.
152 Wender, P. A.; Ihle, N. C.; Correia, C. R. D. *J. Am. Chem. Soc.* **1988**, *110*, 5904–5906.
153 Kimura, M.; Tamaru, Y. *Top. Curr. Chem.* **2007**, *279*, 173–207.
154 Amarasinghe, K. K. D.; Montgomery, J. *J. Am. Chem. Soc.* **2002**, *124*, 9366–9367.
155 Ogoshi, S.; Tonomori, K.; Oka, M.; Kurosawa, H. *J. Am. Chem. Soc.* **2006**, *128*, 7077–7086.
156 Montgomery, J.; Sormunen, G. J. *Top. Curr. Chem.* **2007**, *79*, 1–23.
157 McCarren, P. R.; Liu, P.; Cheong, P. H. Y.; Jamison, T. F.; Houk, K. N. *J. Am. Chem. Soc.* **2009**, *31*, 6654–6655.
158 Colby, E. A.; O'Brien, K. C.; Jamison, T. F. *J. Am. Chem. Soc.* **2005**, 4297–4307.
159 Trenkle, J. D.; Jamison, T. F. *Angew. Chem. Int. Ed.* **2009**, *48*, 5366–5368.
160 Ni, Y.; Amarasinghe, K. K. D.; Ksebati, B.; Montgomery, J. *Org. Lett.* **2003**, *5*, 3771–3773.
161 Nicolaou, K. C.; Bulger, P. G.; Sarlah, D. *Angew. Chem. Int. Ed. Engl.* **2005**, *44*, 4442–89.
162 Overman, L. E.; Ricca, D. J.; Tran, V. D. *J. Am. Chem. Soc.* **1992**, *115*, 2042–2044.
163 Nakanishi, M.; Mori, M. *Angew. Chem. Int. Ed. Engl.* **2002**, *41*, 1934–1936.
164 Heravi, M. M.; Hashemi, E. *Tetrahedron.* **2012**, *68*, 9145–9178.
165 Williams, D. R.; Walsh, M. J.; Miller, N. A. *J. Am. Chem. Soc.* **2009**, *131*, 9038–45.
166 Garg, N. K.; Sarpong, R.; Stoltz, B. M. *J. Am. Chem. Soc.* **2002**, *124*, 13179–13184.
167 Gao, S.; Wang, Q.; Chen, C. *J. Am. Chem. Soc.* **2009**, *131*, 1410–1412.
168 Moses, J. E.; Commeiras, L.; Baldwin, J. E.; Adlington, R. M. *Org. Lett.* **2003**, *5*, 2987–2988.
169 Enquist, J. A., Jr.; Stoltz, B. M. *Nature.* **2008**. *453*, 1228–1231.
170 Xiao, Q.; Jackson, J. J.; Basak, A.; Bowler, J. M.; Miller, B. G.; Zakarian, A. *Nat. Chem.* **2013**, *5*, 410–416.
171 Son, S.; Fu, G. C. *J. Am. Chem. Soc.* **2008**, *130*, 2756–2757.
172 Suh, Y.-G.; Kim, S.-A.; Jung, J.-K.; Shin, D.-Y.; Min, K.-H.; Koo, B.-A. *et al.* *Angew. Chem. Int. Ed.* **1999**, *38*, 3545–3547.

173 Inuki, S.; Iwata, A.; Oishi, S.; Fujii, N.; Ohno, H. *J. Org. Chem.* **2011**, *76*, 2072–2083.
174 Tietze, L. F.; Duefert, S. C.; Clerc, J.; Bischoff, M.; Maass, C.; Stalke, D. *Angew. Chem. Int. Ed. Engl.* **2013**, *52*, 3191–3194.
175 Wang, D.-H.; Yu, J.-Q. *J. Am. Chem. Soc.* **2011**, *133*, 5767–5769.
176 Davies, H. M. L.; Stafford, D. G.; Hansen, T. *Org. Lett.* **1999**, *1*, 233–236.
177 Davies, H. M. L.; Dai, X.; Long, M. S. *J. Am. Chem. Soc.* **2006**, *128*, 2485–2490.
178 Roizen, J. L.; Harvey, M. E.; Bois, J. D. *ACH - Models Chem.* **2012**, *45*, 911–922.
179 Mulcahy, J. V.; Bois, J. D. *J. Am. Chem. Soc.* **2008**, *130*, 12630–12631
180 Hinman, A.; Bois, J. D. *J. Am. Chem. Soc.* **2003**, *125*, 11510–11511.
181 Schrock, R. R.; Chauvin, Y.; Grubbs, R. H. *Angew. Chem. Int. Ed.* **2006**, *45*, 3741–3765.
182 Wang, Y.; Jimenez, M.; Hansen, A. S.; Raiber, E.-A.; Schreiber, S. L.; Young, D. W. *J. Am. Chem. Soc.* **2011**, *133*, 9196–9199.
183 Ono, K.; Nakagawa, M.; Nishida, A. *Angew. Chem. Int. Ed.* **2004**, *43*, 2020–2023.
184 Young, I. S.; Kerr, M. A. *J. Am. Chem. Soc.* **2007**, *129*, 1465–1469.
185 Jakubec, P.; Cockfield, D. M.; Dixon, D. J. *J. Am. Chem. Soc.* **2009**, *131*, 16632–16633.
186 Cheng, B.; Wu, F.; Yang, X.; Zhou, Y.; Wan, X.; Zhai, H. *Chem-Eur. J.* **2007**, *17*, 12569–12572.
187 Yu, M.; Wang, C.; Kyle, A. F.; Jakubec, P., Dixon, D. J.; Schrock, R. R.; Hoveyda, A. H. *Nature.* **2011**, *479*, 88–93.
188 Nakagawa, M.; Nagata, T.; Ono, K.; Nishida, A. *J. Syn. Org. Chem. Jpn.* **2005**, *63*, 200–210.
189 Nagata, T.; Nakagawa, M.; Nishida, A. *J. Am. Chem. Soc.* **2003**, *125*, 7484–7485.
190 Nilson, M. G.; Funk, R. L. *Org. Lett.* **2010**, *12*, 4912–4915.
191 Fürstner, A.; Larionov, O.; Flügge, S. *Angew. Chem. Int. Ed.* **2007**,*46*, 5545–5548.
192 Wei, H.; Qiao, C.; Liu, G.; Yang, Z.; Li, C. *Angew. Chem. Int. Ed.* **2013**, *52*, 620–624.
193 Gallenkamp, D.; Fürstner, A. *J. Am. Chem. Soc.* **2011**, *133*, 9232–9235.
194 Marjanovic, J.; Kozmin, S. A. *Angew. Chem. Int. Ed.* **2007**, *46*, 8854–8857.

5

C–H Activation-Based Strategy for Natural Product Synthesis

Hongbin Zhai[1], Yun Li[2], and Fang Fang[2]

[1] *School of Chemical Biology and Biotechnology, Shenzhen Graduate School of Peking University, Shenzhen, China*
[2] *Department of Chemistry, Lanzhou University, Lanzhou, China*

5.1 Introduction

The introduction of functional groups via C–H bond activation has significant synthetic potential owing to the ubiquitous nature of such bonds in organic substances as well as a considerable atom-economical feature due to there being no need for pre-functionalization of the substrates. It offers a new paradigm for organic synthesis and has been revolutionary in changing the way of synthesis. To date, the C–H activation discipline is still in its infancy, since chemists have not fully succeeded in providing generally applicable and practical guidelines for organic synthesis, particularly in complex natural product synthesis. However, more and more examples of total synthesis involving the C–H bond activation concept have been reported during recent years.[1, 2, 3] In this chapter, some of the latest and representative examples will be discussed and we hope this attracts readers' attention to this intriguing field.

5.2 Recently Completed Total Syntheses of Natural Products via a C–H Activation Approach

In 2002, Sames and coworkers reported the total synthesis of (–)-rhazinilam via a platinum-mediated asymmetric C–H bond activation (Scheme 5.1).[4] This demonstrated one of the earliest applications of C–H bond activation in the context of total synthesis. Sames' pioneering work applied a platinum-promoted intramolecular C–H dehydrogenation reaction to discriminate

Efficiency in Natural Product Total Synthesis, First Edition. Edited by Pei-Qiang Huang, Zhu-Jun Yao, and Richard P. Hsung.

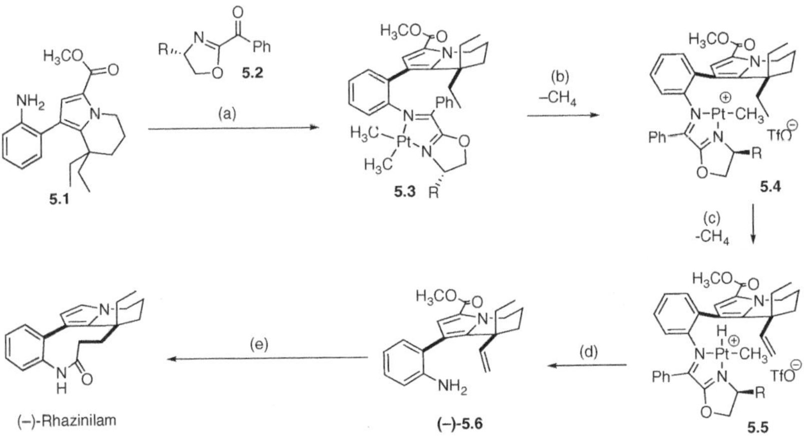

Scheme 5.1 *Reagent and conditions*: (a) (i) TsOH, toluene; (ii) $[Me_2Pt(m\text{-}SMe_2)]_2$, 29–36%; (b) TfOH, CH_2Cl_2; (c) CF_3CH_2OH 90% (NMR); (d) (i) KCN; (ii) NH_2OH, 60% from **5.3**; (e) (i) Pd–C, dppb, HCOOH, CO; (ii) NaOH (aq), MeOH, 52% over two steps.

between the two enantiotopic ethyl groups adjacent to the pyrrole motif. After liberating the platinum and removing the auxiliary of the Schiff base, the chiral aniline (−)-**5.6** was afforded in modest yield and good diastereoselectivity. (−)-Rhazinilam was successfully synthesized in two additional transformations. Despite the stoichiometric amount of platinum used in this synthesis, the success of this early example of C–H activation in total synthesis is still remarkable and proves to be highly innovative.

In 2010, Chen's group reported the total synthesis of celogentin C[5] in which the direct regio- and stereoselective activation of the C–H bond of a Leu moiety together with subsequent coupling with the iodotryptophan compound **5.9** was featured as the key transformation (Scheme 5.2). This methodology was first reported by Corey *et al.* in 2006.[6] Chen's group first applied this quinoline carboxamide directed C–H functionalization protocol in the synthesis of the β-tryptophan-substituted leucine domain of celogentin. The C–H activation reaction worked very well and could possibly be performed at a multi-gram scale. The desired β-tryptophan-substituted leucine **5.10** was obtained in 85% yield as a single stereoisomer.

Two years later, Baran's group reported very similar C–H activation logic in the total synthesis of piperaborenine B.[7] The imide motifs found in these natural products were treated as latent carboxylate-based directing groups for C–H functionalization. In this work, a photo-induced 4π-electrocyclization of methyl coumalate and sequential amide condensation wereadopted to prepare the key precursor of C–H arylation (Scheme 5.3).

It was found that the addition of hexafluoroisopropanol (HFIP) and pivalic acid was rather critical to the success of the direct arylation at the position of

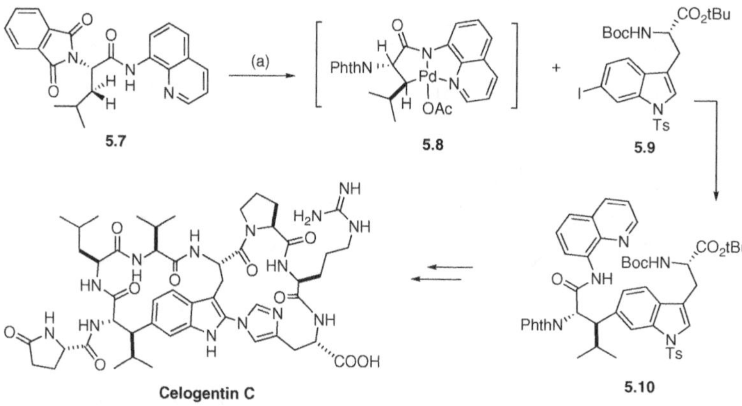

Scheme 5.2 *Reagent and conditions*: (a) $Pd(OAc)_2$, AgOAc, *t*BuOH, 110 °C, 36 h.

the cyclobutane backbone. The desired cyclobutane **5.16** could be isolated in 52% yield. The second round of arylation took place after the configuration inversion of the directing group under basic conditions. After the hydrolization of ester and amide functionalities and subsequential coupling with the dihydropyridone, the author completed the total synthesis of piperaborenine B in an additional three steps (Scheme 5.3).

Undirected catalytic sp^3 C–H activation was definitely regarded as the most challenging work in organic synthesis. In 2008, Baudoin and coworkers discovered that the sp^3 C–H bond could be directly activated by palladium(0) to trigger cyclopalladation and to form benzocyclobutenes when reductive elimination dominated.[8] One year later, the same group reported the total synthesis of coralydine, using benzocyclobutene as the key intermediate.[9] The desired benzocyclobutene **5.23** was successfully synthesized from the corresponding aryl bromide **5.21** via a Pd-catalyzed C–H activation/intramolecular C–C bond coupling reaction in 72–78% yield. The carboxylate group was latterly converted to an amino group by using a Cutius rearrangement and afforded **5.24** in 69% overall yield. A thermal tandem electrocyclic ring-opening/6π-electrocyclization reaction was successfully carried out and the desired dihydroisoquinoline **5.26** was isolated in 52% yield. A further reduction and cyclization of pyridine afforded coralydine (Scheme 5.4).

In 2011, Yu's group illustrated a highly convergent total synthesis of (+)-lithospermic acid by exploiting a carboxylic acid directed C–H activation/Heck reaction cascade to combine the two fragments of the targeted molecule.[10] In this work, the optically active trans-dihydrobenzofuran **5.30** was prepared through an asymmetric intramolecular carbene insertion reaction of diazo compound **5.29**. The critical C–H olefination reaction between **5.30** and acrylate **5.31** was carried out under the previously reported condition,

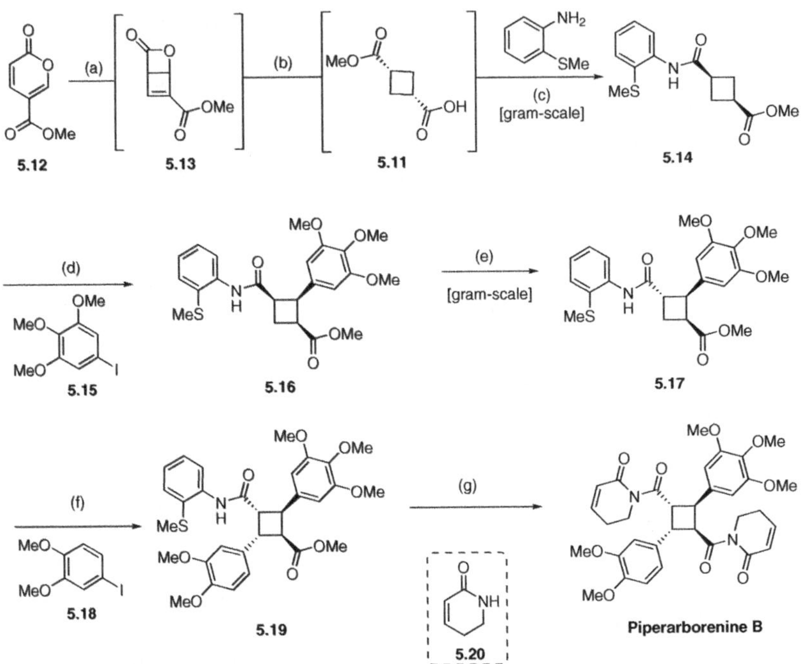

Scheme 5.3 *Reagent and conditions*: (a) hν; (b) Pt/C, H_2; (c) EDC, 61% from **5.12**; (d) cat. $Pd(OAc)_2$, Ag_2CO_3, PivOH, HFIP, **5.15**, 52%; (e) LiOtBu, >10:1 d.r., 79%; (f) cat. $Pd(OAc)_2$, Ag_2CO_3, **5.18**, tBuOH, 75 °C, 46%; (g) (i) Boc_2O (x-ray); (ii) LiOH, H_2O_2; (iii) $(COCl)_2$, **5.20**, 57% over three steps.

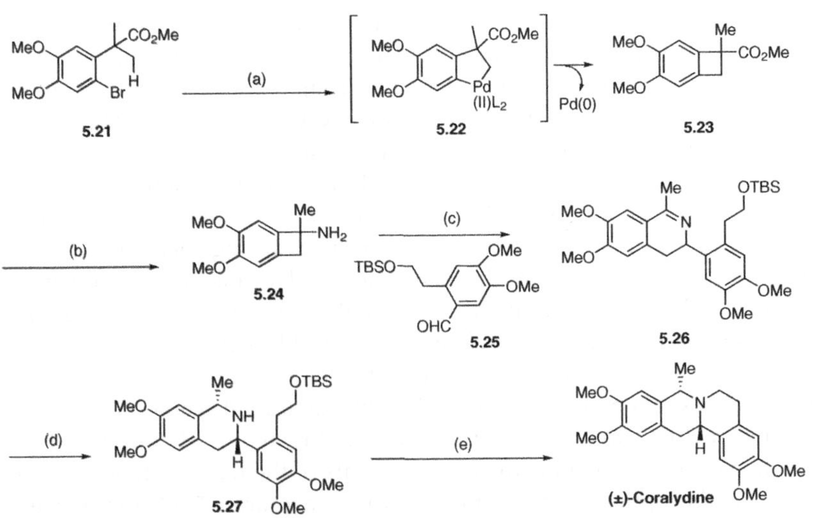

Scheme 5.4 *Reagent and conditions*: (a) $Pd(OAc)_2$ (10 mol%), $P(tBu)_3$ (20 mol%), K_2CO_3, DMF, 140 °C, 72–80%; (b) (i) NaOH, MeOH/H_2O, reflux, 90%; (ii) DPPA, Et_3N, toluene, reflux, then aq. HCl, 80 °C, 69%; (c) **5.25**, CH_2Cl_2, 20 °C, then DMF, 160 °C, 52%; (d) $NaBH_4$, MeOH, 68%, d.r. = 6:1; (e) (i) $n$$Bu_4NF$, THF, 20 °C; (ii) HBF_4, then PPh_3, DIAD, THF, reflux, 54%, two steps.

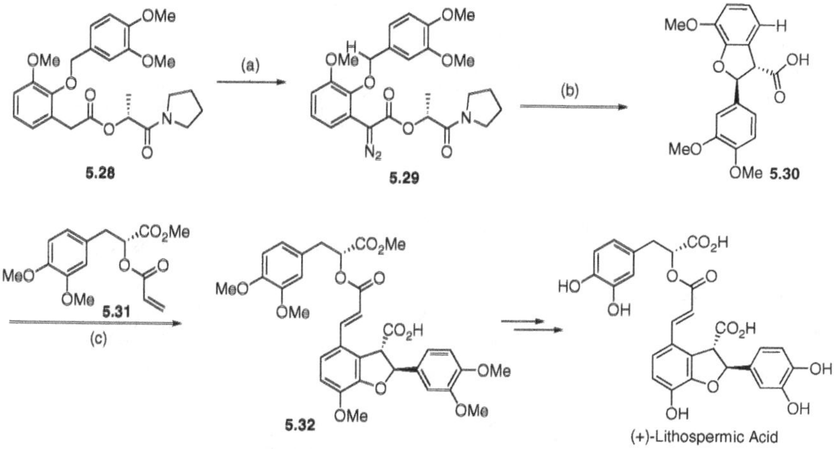

Scheme 5.5 *Reagent and conditions*: (a) *p*-ABSA, 82%; (b) (i) $Rh_2(S\text{-}DOSP)_4$; (ii) $Ba(OH)_2$, MeOH/THF = 1:1, 73%, two steps, d.r. = 8:1; (c) $Pd(OAc)_2$/Ac-Ile-OH, O_2 (1 atm), $KHCO_3$ (2 equiv), *tert*-Amyl-OH, 85 °C, 93%.

developed by their own group, and the desired product **5.32** was isolated at a decent yield of 93%. With an additional two-step demethylation procedure, the total synthesis of (+)-lithospermic acid was successfully achieved (Scheme 5.5).

Without a doubt, C–H functionalization offers a more atom economic and efficient way to build up molecular architectures. The total synthesis of dragmacidin D, which was accomplished by Itami and coworkers,[11] fully illustrated the advantage of C–H functionization in the construction of C–C bonds in total synthesis. The synthesis began with the directly coupling reaction of iodoindole derivative **5.33** and thiophene **5.34**. It was found that the triisopropylsilyloxy moiety within thiophene **5.34** played a very important role in site-selectivity. The thiophene moiety in the coupled product **5.35** was sequentially reduced by Raney Ni to afford the methyl ketone **5.36**. This molecule was then primed for the Pd-catalyzed C–H/C–H coupling reaction with pyrazine *N*-oxide **5.37** efficiently delivered **5.38** in 50% yield. Treatment of **5.38** with $(CF_3CO)_2O$ then produced pyrazinone **5.39**, which was then subjected to an oxidative C–H/C–H coupling reaction of with 6-bromoindole in the presence of CF_3SO_3H and air to afford the corresponding coupling product **5.41** with simultaneous removal of the two MOM groups. Finally, introduction of the guanidinium functionality completed the synthesis of dragmacidin D (Scheme 5.6).

The palladium-catalyzed Catellani reaction, which allows the formation of multiple chemical bonds in a single step, has become an attractive way to construct highly complex frameworks. Norbornene in this transformation is used to facilitate an *ortho* C–H functionalization between an aryl iodide

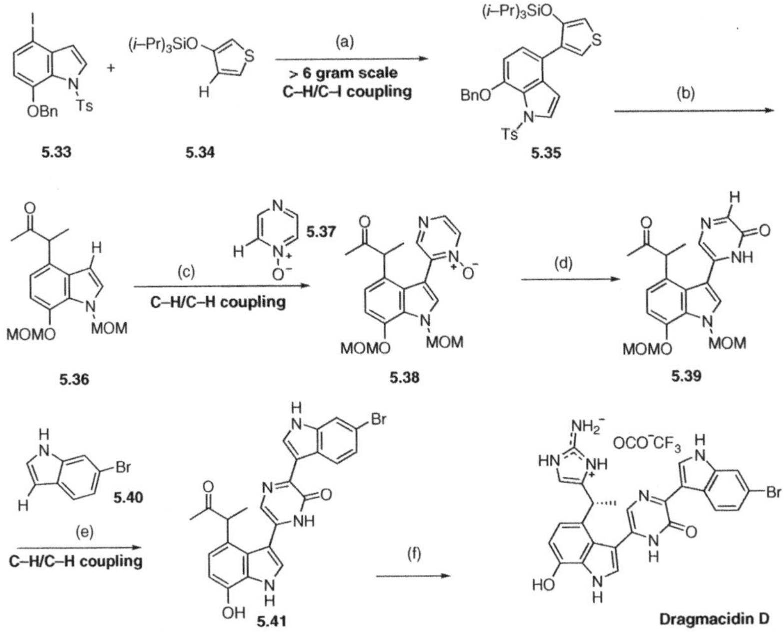

Scheme 5.6 *Reagent and conditions*: (a) $Pd(OAc)_2$, $P[OCH(CF_3)_2]_3$, Ag_2CO_3, 60%; (b) (i) Bu_4NF, AcOH, then Raney Ni; (ii) $Mg(OMe)_2$; (iii) MOM-Cl, 70%, three steps; (c) $Pd(OAc)_2$, AgOAc, 50%; (d) $(CF_3CO)_2O$; (e) CF_3SO_3H, air, 57%, two steps; (f) (i) *i*-Pr_2NEt, Me_3SiOTf, then NBS; (ii) (Boc)guanidine; then CF_3CO_2H, 37%, two steps.

and an alkyl halide. An elegant example of this kind of C–H activation in the total synthesis of (+)-linoxepin was reported by Mark Lautens and coworkers in 2013 (Scheme 5.7).[12] Leveraging their experience with the Catellani reaction, the readily prepared ether **5.42**, iodolactone **5.43** and acrylate **5.44** were unified through a palladium-catalyzed Catellani reaction to afford the desired tricycle **5.45** smoothly in 92% yield. The precise mechanism of this transformation is still unclear, but it is known that this catalytic cycle is initiated by Pd(0) oxidatively adding into the Ar-I bond and followed by an insertion into norbornene so then base-mediated C–H activation occurs to form palladacycle **5.46**. This intermediate is sequentially oxidized by iodide **5.43** to produce the Pd(IV) intermediate **5.48**. Expulsion of norbornene followed by a Mizoroki–Heck type reaction completes the catalytic cycle to afford **5.45** in a single step. The oxidative cleavage of double bond of intermediate **5.45** and a $TiCl_4$ mediated intramolecular aldol reaction were carried out to produce tetracyclic lactone **5.50**. The resultant lactone **5.50** was subjected to Mizoroki–Heck reaction conditions and successfully furnished the total synthesis of (+)-linoxepin.

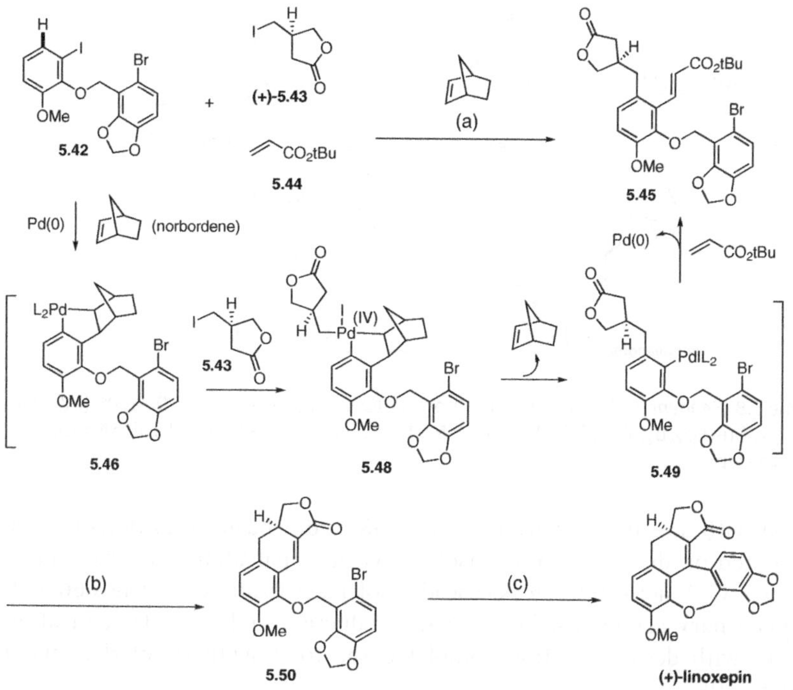

Scheme 5.7 *Reagent and conditions*: (a) $Pd(OAc)_2$ (10 mol%), PPh_3(22 mol%), DMF, Cs_2CO_3, 92%; (b) (i) OsO_4, $NaIO_4$, TEBAC, THF/H_2O 5:3, RT, 98%; (ii) $TiCl_4$, NEt_3, CH_2Cl_2, −78–25 °C, −RT, 3 h, 49%; (c) $PdCl_2$ (20 mol%), PPh_3 (44 mol%), CsOAc (10.0 equiv), DMF, 75 °C, 4 h, 76%.

Iridium-catalyzed borylation of arenes, which was developed by Hartwig and coworkers,[13] became a very popular methodology for direct functionalization of aromatic ring systems. Sarpong's total synthesis of complanadine fully showcased the power of this reaction.[14] This group demonstrated a highly divergent solution to this pseudodimeric alkaloid by starting with the same intermediate triflate **5.51.** The triflate group in this material was then removed in the palladium-catalyzed reducing condition to give pyridine **5.52**, then an iridium-catalyzed regioselective C–H borylation realized the desired functionalization at the C3 position of the pyridine motif. The resultant boronic ester **5.53** was directly coupled with the original triflate **5.51** under Suzuki reaction conditions to give complanadine after deprotection of the Boc group (Scheme 5.8).

Another total synthesis involving iridium-catalyzed C–H functionalization was reported by Gaunt and coworkers in 2008.[15] A site-selective iridium-catalyzed C–H broylation of C3 ester pyrrole and a subsequent Suzuki–Miyaura coupling reaction were carried out for the synthesis of pyrrole intermediate **5.56** in 63% overall yield as a one-pot process. This compound was then

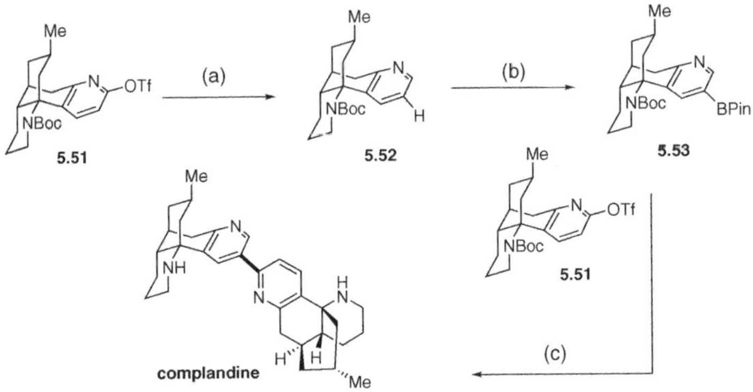

Scheme 5.8 *Reagent and conditions*: (a) 5 mol% $Pd(OAc)_2$, dppf, NH_4O_2CH, 90%; (b) [Ir(COD)(OMe)]$_2$/dtBu-dipy, $B_2(pin)_2$, THF, 75%; (c) (i) **5.51**, $PdCl_2$(dppf), K_3PO_4, Et_3SiH; (ii) 6 N HCl, 42%, two steps.

reacted with steadily prepared iodide **5.58** under basic conditions to give *N*-alkylated product **5.59** in 85% isolated yield. A palladium-catalyzed intramolecular C–H activation/Heck cascade took place to afford pyrrole **5.60** with the quaternary center installed. Further reduction and macrolactamization together with decarboxylation completed the total synthesis of rhazinilam (Scheme 5.9).

Rhodium-catalyzed C–H functionalization, especially in Rh-catalyzed carbenoid C–H insertion, has been extensively studied to date. Here, we will select one example with a non-carbene C–H activation mode, because synthetic rhodium carbene examples have been detailed in Section 4.11. In 2012, Cheng and coworkers reported an effective [RhCp*]-catalyzed three-component reaction of aryl aldehyde, methylamine, and alkyne to prepare isoquinolinium salts through C–H activation and applied this methodology to the total synthesis of oxychelerythrine (Scheme 5.10).[16] The treatment of benzaldehyde **5.61** with alkyne **5.62** and methylamine **5.63** in the presence of $(RhCp^*Cl_2)_2$, $AgBF_4$, and $Cu(OAc)_2$ gave the isoquinolinium salt **5.64** in 82% isolated yield. The possible mechanism for this catalytic C–H activation process is likely initiated by the coordination of the *in situ* generated imine nitrogen atom to the rhodium species, and subsequent *ortho* C–H bond activation to form the five-membered rhodacycle **5.66**. Regioselective insertion of the alkyne **5.62** into the rhodium–carbon bond and the following sequential elimination of **5.68** afforded the final isoquinolinium salt **5.64**. The resultant compound was further oxidized by $K_3[Fe(CN)_6]$ in aqueous methanol to produce lactam **5.65**. After a successful acid-catalyzed ring closure and a dehydration reaction, oxychelerythrine was obtained in 95% yield.

Direct C–H oxidation is another important topic in the C–H bond activation field. It allows the direct installation of oxygen into sluggish alkanes, which has proved to be one of the most useful methods in organic synthesis. In 2009,

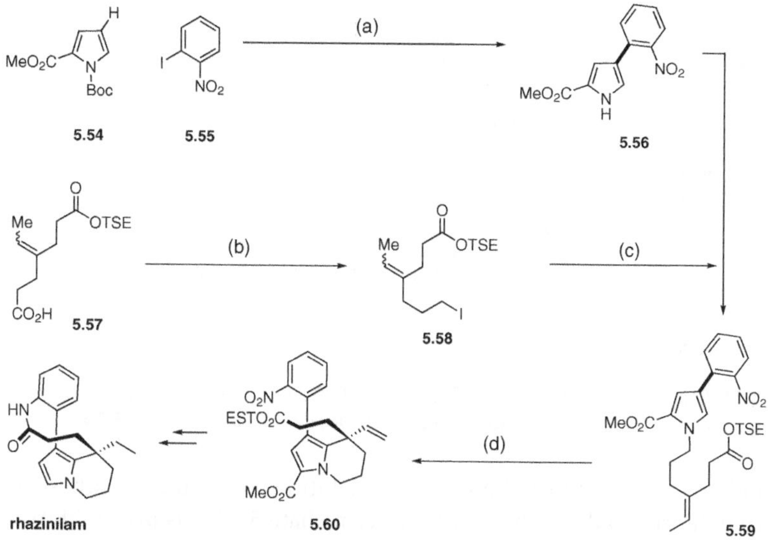

Scheme 5.9 *Reagent and conditions*: (a) 2 mol% [{IrCl(cod)}$_2$], B_2pin_2, nC_6H_{14}, mwave, 100 °C, then 2 mol% $Pd(OAc)_2$, 4 mol%, S-Phos, K_3PO_4, **5.55**, *n*BuOH, 100 °C, then DMF, 110 °C 63% yield; (b) (i) $ClCO_2Me$, Et_3N, $NaBH_4$, Et_3SiH; (ii) I_2, PPh_3, 50% over two steps; (c) **5.56**, NaH, DMF, 0 °C to RT then **5.58**, 85%; (d) 10 mol% $Pd(OAc)_2$, PivOH, 20 mol% NaOtBu, O_2, 10 mol% DMF, 110 °C, 60%.

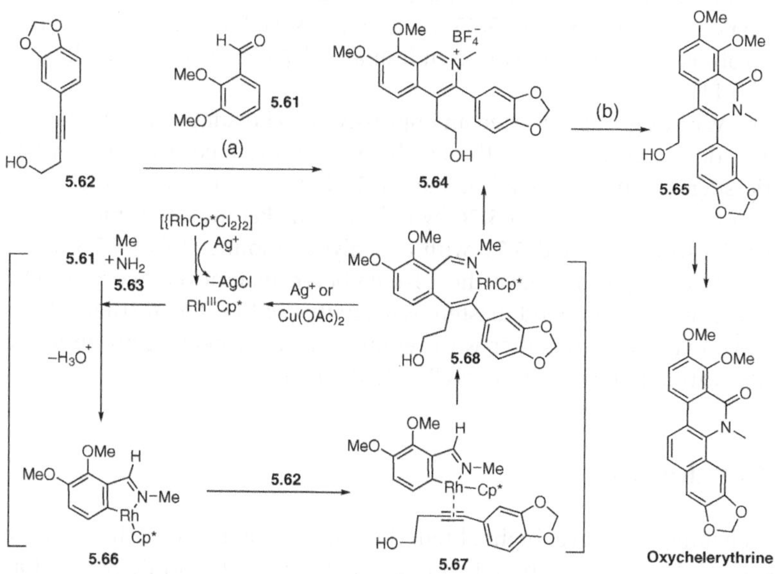

Scheme 5.10 *Reagent and conditions*: (a) $MeNH_2$, [{RhCp*Cl$_2$}$_2$], $AgBF_4$, $Cu(OAc)_2$, *tert*-amyl alcohol, 110 °C, 3 h, 82%; (b) $K_3[Fe(CN)_6]$, CsOH, $MeOH/H_2O$, 90%.

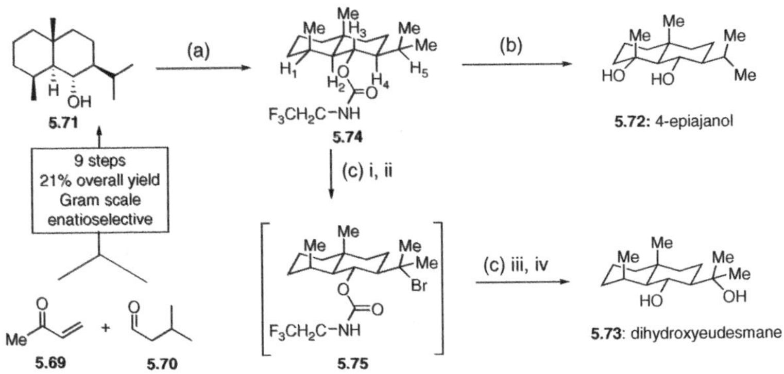

Scheme 5.11 *Reagent and conditions*: (a) CF_3CH_2NCO; 99%; (b) (i) TFDO, (ii) NaOMe, 78%, two steps; (c) (i) CH_3CO_2Br, (ii) Sunlamp, (iii) $AgCO_3$, AcOH; (iv) LiOH, 43% + 39% **5.74**.

Baran and coworkers reported an efficient solution to the total synthesis of eudesmane diterpenoids.[17] The bicyclic intermediate **5.71** was prepared from readily available compound **5.69** and **5.70** in nine steps with a 21% overall yield. The trifluoroethyl carbamate directing group was then installed in a quantitative yield and delivered **5.74** (Scheme 5.11).

A site-selective oxidation was carried out and delivered the corresponding alcohol with TFDO (1 eq.). The origin of this selectivity may be interpreted by strain-release effects in the transition state during oxidation. Basic hydrolysis of this resultant carbamate led to 4-epiajanol **5.72** in 78% overall yield. The site-specific halogenation of **5.74** to **5.75** was accomplished followed by cyclization and hydrolysis to afford dihydroxyeudesmane **5.73** in a 43% overall yield (Scheme 5.11).

In 2009, White's group reported an impressive C–H oxidative macrolactonization reaction to efficiently synthesize the erythromycin core at the very late synthetic stage.[18] The advanced precursor **5.77** of C–H oxidative macrolactonization was converted from **5.76** by reducing the ketone and removing the auxiliary. Treatment of acid **5.77** with a catalytic amount of Cat **5.78** in the presence of benzoquinone gave the 14-membered macrocyclic core of **5.80** with high regio-, chemo-, and diastereoselectivity (>40:1). The further saturation of terminal alkene together with removal of the protecting group led to 6-deoxyerythronolide B in 79% overall yield (Scheme 5.12).

5.3 Conclusion

In this chapter, we have highlighted the most recent synthetic research that uses a C–H activation-based strategy. Apparently, this strategy puts forward a complementary bond-disconnection option compared to conventional

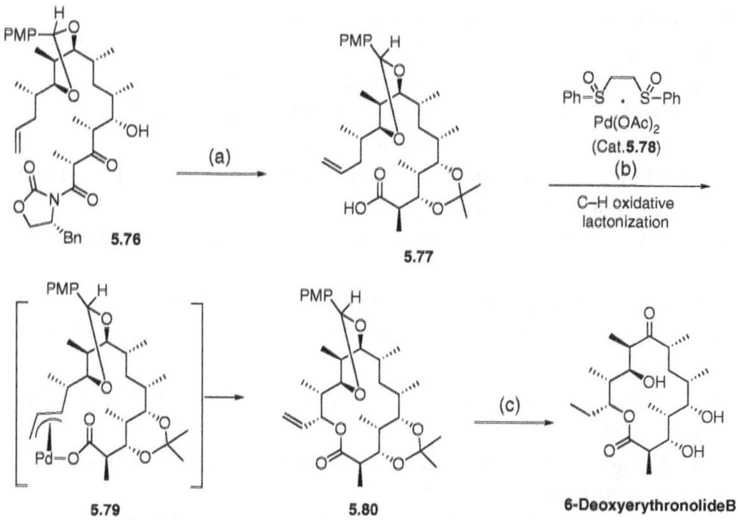

Scheme 5.12 *Reagent and conditions*: (a) (i) $Zn(BH_4)_2$ (1.6 equiv.), (ii) CSA (catalyst), 2,2-dimethoxypropane; (iii) LiOH (aq, 2.0 equiv.), 72% over three steps; (b) 1,4-benzoquinone, 56%, (>40:1 d.r.); (c) (i) $Pd(OH)_2$/C, H_2, iPrOH, (ii) TPAP, NMO, DCM, 0 °C, (iii) 1 M HCl (aq, 11 equiv.), 79% over three steps.

methodologies, which has been demonstrated by the synthetic efficiency of selected molecules. However, the field is far from maturity. The current applications to natural product synthesis, specifically for late-stage C–H activation of complex molecules, are few. Therefore, discovery of more new reactions and milder reaction conditions will broaden the substrate's scope and enable its application to more structurally challenging natural products.

Acknowledgements

We thank the NSFC (21732001, 21672017, 21472072, 21290183), Shenzhen Science and Technology Innovation Committee (JCYJ20150529153646078, JSGG20160229150510483), Program for Changjiang Scholars and Innovative Research Team in University (PCSIRT: IRT_15R28), and "111" Program of MOE for financial support.

References

1 Gutekunst, W. R.; Baran, P. S. *Chem. Soc. Rev.* **2011**, *40*, 1976–1991.
2 Giri, R.; Shi, B. F.; Engle, K. M.; Maugel, N.; Yu, J. Q. *Chem. Soc. Rev.* **2009**, *38*, 3242–3272.

3 Chen, D. Y.; Youn, S. W. *Chem. Eur. J.* **2012**, *18*, 9452–9474.
4 Johnson, J. A.; Ning, L.; Sames, D. *J. Am. Chem. Soc.* **2002**, *124*, 6900–6903.
5 Feng, Y. Q.; Chen, G. *Angew. Chem. Int. Ed.* **2010**, *49*, 958–961.
6 Reddy, B. V.; Reddy, L. R.; Corey, E. J. *Org. Lett.* **2006**, *8*, 3391–3394.
7 Gutekunst, W. R.; Baran, P. S. *J. Am. Chem. Soc.* **2011**, *133*, 19076–19079.
8 Chaumontet, M.; Piccardi, R.; Audic, N.; Hitce, J.; Peglion, J. L.; Clot, E. *et al.* *J. Am. Chem. Soc.* **2008**, *130*, 15157–15166.
9 Chaumontet, M.; Piccardi, R.; Baudoin, O. *Angew. Chem. Int. Ed.* **2009**, *48*, 179–182.
10 Wang, D. H.; Yu, J. Q. *J. Am. Chem. Soc.* **2011**, *133*, 5767–5769.
11 Mandal, D.; Yamaguchi, A. D.; Yamaguchi, J.; Itami, K. *J. Am. Chem. Soc.* **2011**, *133*, 19660–19663.
12 Weinstabl, H.; Suhartono, M.; Qureshi, Z.; Lautens, M. *Angew. Chem. Int. Ed.* **2013**, *52*, 5305–5308.
13 Ishiyama, T.; Takagi, J.; Ishida, K.; Miyaura, N.; Anastasi, N. R.; Hartwig, J. F. *J. Am. Chem. Soc.* **2002**, *124*, 390–391.
14 Fischer, D. F.; Sarpong, R. *J. Am. Chem. Soc.* **2010**, *132*, 5926–5927.
15 Beck, E. M.; Hatley, R.; Gaunt, M. *J. Angew. Chem. Int. Ed.* **2008**, *120*, 3004–3007.
16 Jayakumar, J.; Parthasarathy, K.; Cheng, C. H. *Angew. Chem. Int. Ed.* **2012**, *51*, 197–200.
17 Chen, K.; Baran, P. S. *Nature*, **2009**, *459*, 824–828.
18 Stang, E. M.; Christina White, M. *Nat. Chem.* **2009**, *1*, 547–551.

6

Recent Applications of Kagan's Reagent (SmI_2) in Natural Product Synthesis

Erica Benedetti[1], Cyril Bressy[2], Michael Smietana[3], and Stellios Arseniyadis[4]

[1] *Université Paris Descartes, Paris, France*
[2] *Aix Marseille Université, Marseille, France*
[3] *Université de Montpellier, Montpellier, France*
[4] *Queen Mary University of London, London, UK*

6.1 Background

First introduced by Kagan and coworkers in 1977,[1] samarium(II) iodide (SmI_2) has since become a popular reagent for the formation of carbon-carbon bonds with many useful applications in the total synthesis of natural products and other biologically active molecules.[2] Due to its large reduction potential (up to −2.05 V in the presence of HMPA),[3] SmI_2 can activate a variety of functional groups ranging from alkyl halides to carbonyl compounds, as well as conjugated double bonds.[4] Notably, the SmI_2-mediated reduction of alkyl halides can generate either radical species via a single-electron transfer or organosamarium reactive intermediates through two successive single-electron transfers (Scheme 6.1a). In a similar way, carbonyl compounds can generate either ketyl radicals or reactive carbanions upon treatment with SmI_2 under appropriate conditions (Scheme 6.1b).

Thus, careful optimization of the reaction conditions and use of additives such as Lewis bases[5] and proton sources[6] are usually required in order to control the rate of reductions as well as the chemo- or stereoselectivity of the SmI_2-mediated transformations.

Figure 6.1 showcases the various types of transformations that can be promoted by SmI_2. They span from Reformatsky type reactions (Type A), carbonyl/alkene reductive reactions (Type B), Pinacol-type couplings (Type C), to fragmentation reactions (Type D).

Efficiency in Natural Product Total Synthesis, First Edition. Edited by Pei-Qiang Huang, Zhu-Jun Yao, and Richard P. Hsung.

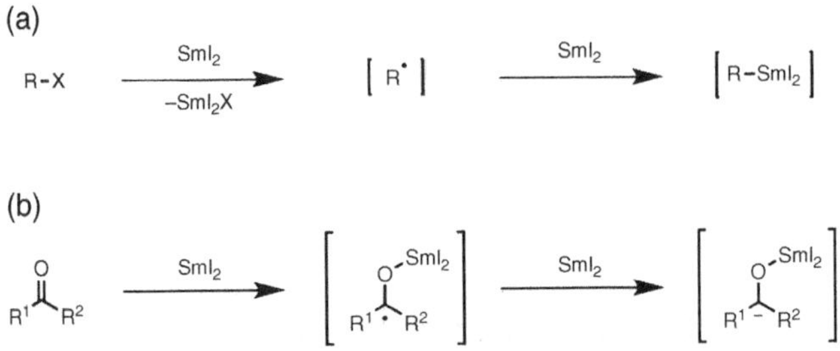

Scheme 6.1 SmI_2-mediated reduction of alkyl halides and carbonyl compounds.

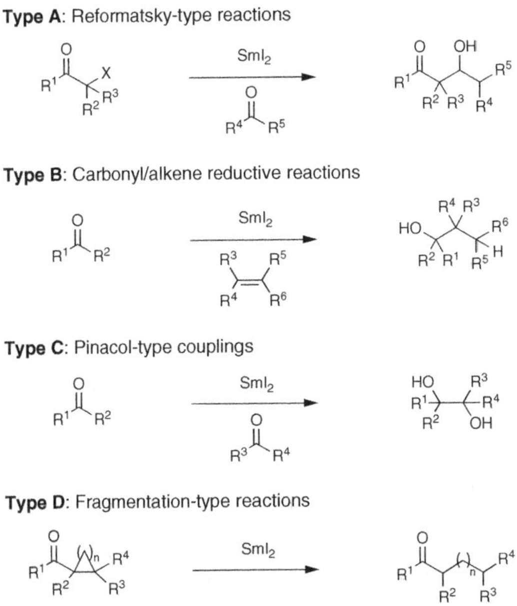

Figure 6.1 Types of transformations promoted by SmI_2.

The first part of this chapter will underline the versatility of SmI_2 by describing different types of SmI_2-mediated reactions together with some of the most elegant applications in natural product synthesis. Due to the large amount of work that has been devoted to this field, the selection of examples described in the upcoming sections will by no means cover every single SmI_2-mediated transformation, but is meant to illustrate to the reader the useful reactivity and tunable selectivity of this valuable reagent.[7]

6.1.1 The Reformatsky Reaction

The Reformatsky reaction is a reductive condensation of an α-halo ester with an aldehyde (or a ketone) in the presence of different metal salts, to form the

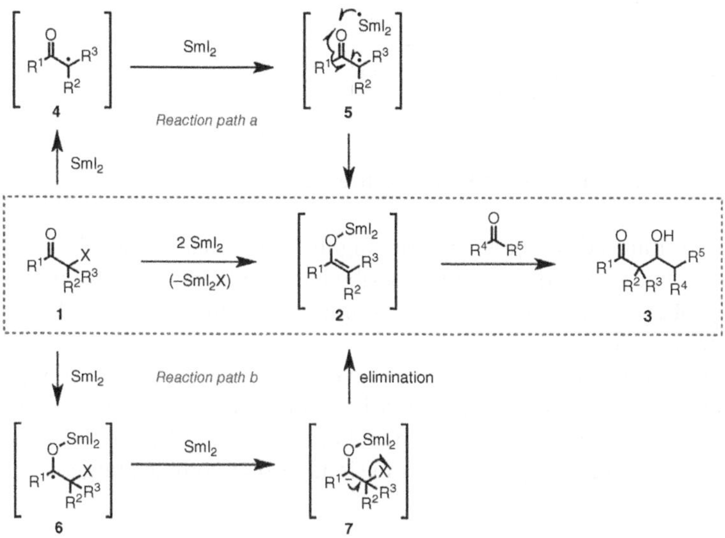

Scheme 6.2 SmI_2-mediated Reformatsky reaction.

corresponding β-hydroxy-esters.[8] First described by Kagan *et al.* in 1977.[1] the SmI_2-mediated version of this reaction proceeds through a reductive cleavage of the α-halo ester to form a Sm^{III} enolate, which then reacts with carbonyl groups in an aldol-like manner (Scheme 6.2).[9] Two different mechanisms have been proposed for the formation of the Sm^{III} enolate **2**, depending on the nature of the starting material **1**.[10] Hence, when the C–X bond of **1** is reductively labile, treatment with SmI_2 may induce a single-electron reduction to form intermediate **4**, followed by a second single-electron reduction to produce enolate **2** (Scheme 6.2, reaction path a).

On the other hand, if the C–X bond of the starting material **1** is not prone to undergo a direct reductive cleavage, exposure to SmI_2 will allow the formation of the ketyl radical **6** and subsequent generation of the carbanion **7** via two successive single-electron reductions. A spontaneous elimination reaction eventually generates the key intermediate **2** (Scheme 6.2, reaction path b).

Due to its high functional group tolerance and mild conditions, the SmI_2-mediated Reformatsky reaction has nowadays become a popular and widely used alternative to the original zinc-promoted variant, as well as to traditional aldol reactions. As such, this process recently found several interesting applications in the field of natural product synthesis.

6.1.2 Carbonyl/Alkene Reductive Reactions

In SmI_2-mediated carbonyl/alkene reductive reactions, a single-electron transfer allows the conversion of carbonyl compounds such as **8** to the corresponding ketyl radicals **9**. The latter can then subsequently react with both activated and

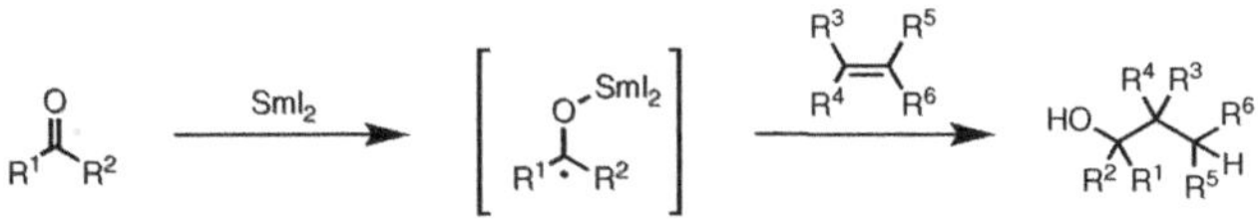

8 9 10

Scheme 6.3 SmI_2-mediated carbonyl/alkene reductive cyclizations.

inactivated unsaturated systems to form highly functionalized alcohols **10** (Scheme 6.3). These reactions, first described by Molander and Kenny in 1987,[11] are probably considered as the most important SmI_2-promoted transformations in total synthesis.

Although both intermolecular and intramolecular variants of the carbonyl/alkene reductive reactions are widely described in the literature, intramolecular cyclizations to form four- to eight-membered rings are by far the most popular processes in natural product synthesis and other complex targets.[12]

6.1.3 Pinacol-Type Couplings

In 1983, Kagan and coworkers described the first example of a SmI_2-promoted pinacol reaction[13]; a process that involves a reductive coupling between two carbonyl groups to generate vicinal diols or other related species.[14] Five years later, Molander and Kenny reported the first intramolecular version of this reaction,[15] which later proved to be a particularly useful process for the formation of cyclic compounds. The intramolecular pinacol coupling proceeds through a single-electron reduction of a dicarbonyl compound **11** to form the ketyl radical **12**. This intermediate then attacks the second carbonyl group to produce the oxygen radical **13**. Reduction of the latter, followed by quenching of the resulting alkoxide, finally generates the expected pinacol product **14** (Scheme 6.4).

Due to chelation of the ketyl radical **12** to SmI_2 (Scheme 6.4), intramolecular pinacol couplings generally occur in a highly stereocontrolled fashion to form

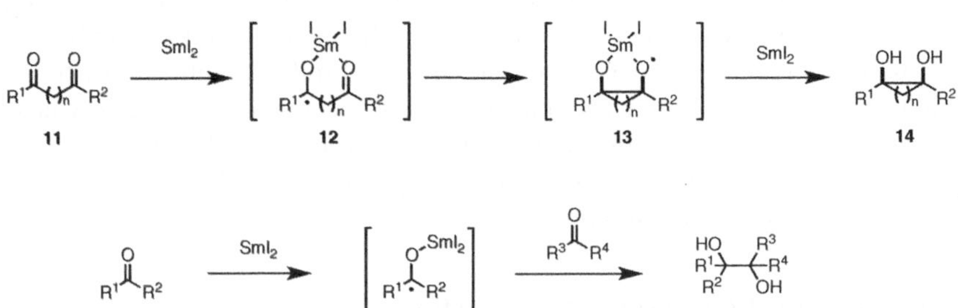

11 12 13 14

8 9 15

Scheme 6.4 Intra- and intermolecular SmI_2-mediated pinacol-type couplings.

the *cis*-diol product preferentially. On the contrary, intermolecular variants of these reactions usually suffer from scarce efficiency and poor stereoselectivity.

6.1.4 Fragmentation Reactions

SmI_2 can also promote different classes of fragmentation reactions, which usually involve strained C–C bonds that can be found in cyclopropane and cyclobutane systems as well as in heterocycles characterized by weak heteroatom-heteroatom bonds. The fragmentation process is mechanistically related to the reduction of α-halogenated carbonyl compounds and usually proceeds via the formation of ketyl radicals (Scheme 6.5).

SmI_2-promoted fragmentations have been extensively exploited in natural product synthesis as particularly useful means to introduce complex ring systems or sterically hindered substituents onto the target compound.[16]

6.2 SmI_2-Mediated Reactions in Natural Product Synthesis

6.2.1 Synthesis of (+)-Acutiphycin

The convergent total synthesis of (+)-acutiphycin (**19**), a complex macrolide possessing potent *in vivo* antineoplastic activities, was reported by Moslin and Jamison in 2006 (Scheme 6.6).[17, 18] The construction of this macrocycle was achieved via the assembly of four simple fragments and ethoxyethyne (Scheme 6.6) using an intermolecular SmI_2-mediated Reformatsky coupling reaction as a key step.

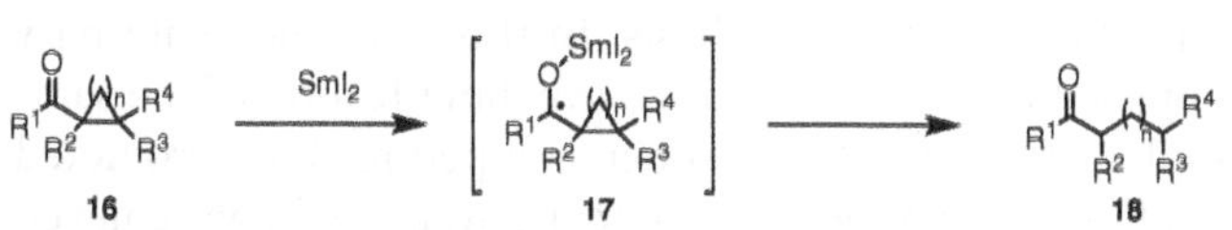

Scheme 6.5 SmI_2-mediated fragmentation reactions.

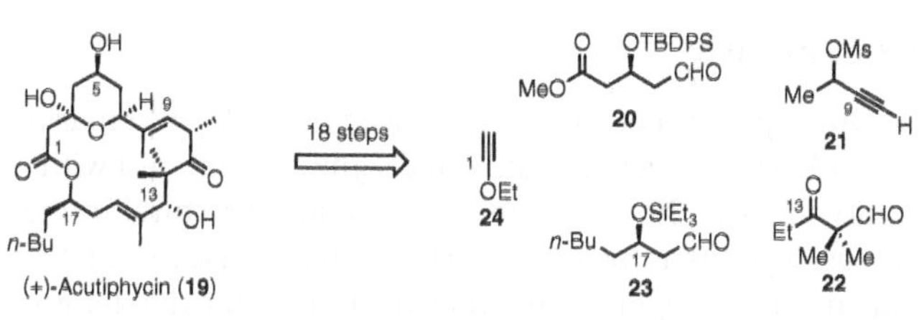

Scheme 6.6 Retrosynthetic analysis of acutiphycin by Moslin and Jamison.

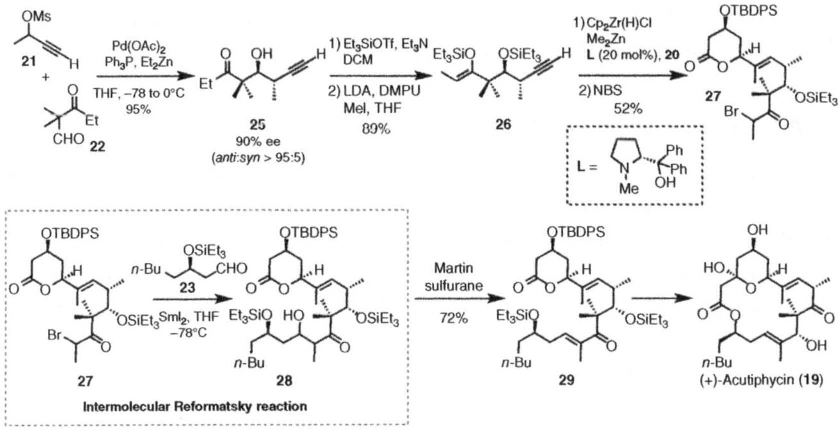

Scheme 6.7 Synthesis of acutiphycin by Moslin and Jamison.

Moslin and Jamison's synthesis began with the preparation of α-bromoketone **27**, which was obtained as a single regioisomer in five steps starting from alkyne **21** and aldehyde **22** (Scheme 6.7).

The α-bromoketone was then successfully coupled with aldehyde **23** in the presence of SmI_2 to form the expected product **28**, which was subsequently dehydrated using Martin sulfurane affording the target enone **29** at a 72% overall yield for the two-step procedure (Scheme 6.7). A short sequence of manipulations finally allowed the conversion of compound **29** to (+)-acutiphycin (**19**). Usually rarely employed in total synthesis because of the many possible side reactions that can occur, this example of intermolecular SmI_2-mediated Reformatsky reaction[19] proved to be a valuable alternative to other commonly used coupling strategies such as the Horner–Wadsworth–Emmons olefination, the Mukaiyama aldol, or cross-metathesis. In this case, the quaternary center of **27**, was supposed to restrain the oxidative dimerization of the samarium enolate as well as to prevent any other competing SmI_2-promoted processes. As a result, the total synthesis of (+)-acutiphycin could be achieved in 18 steps (longest linear sequence) starting from commercially available materials.

6.2.2 Synthesis of Brevetoxin B

Brevetoxin B (BTX-B, **30**), a potent neurotoxin produced naturally by a marine microorganism known as *Karenia brevis*, has attracted significant interest within the synthetic community for over 30 years due to its unique structural framework containing *trans*-fused polytetrahydropyran ring systems (Figure 6.2).[20]

In 1999, Nakata and coworkers reported an efficient strategy for the iterative synthesis of *trans*-fused polycyclic ethers based on SmI_2-induced carbonyl/alkene

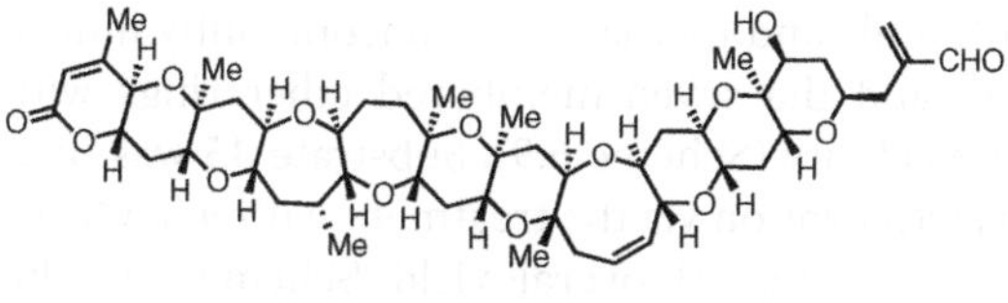

Brevetoxin-B (**30**)

Figure 6.2 Structure of brevetoxin B (BTX-B).

reductive intramolecular cyclizations.[21] The authors later successfully employed this useful method for the formation of several ether rings en route to BTX-B.

Nakata's synthesis of brevetoxin B relied on a convergent strategy involving the assembly of segments **31** and **32** (Scheme 6.8).[22]

Segment **31** was prepared following an innovative two-directional approach from bis(β-alkoxy acrylate) **34**, which, in turn, could be obtained in 18 steps from the commercially available tri-*O*-acetyl-D-glucal **33** (Scheme 6.9).

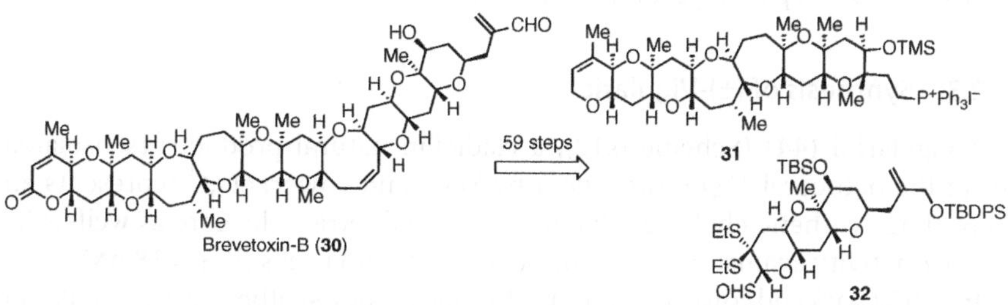

Scheme 6.8 Retrosynthetic analysis of brevetoxin B by Nakata and coworkers.

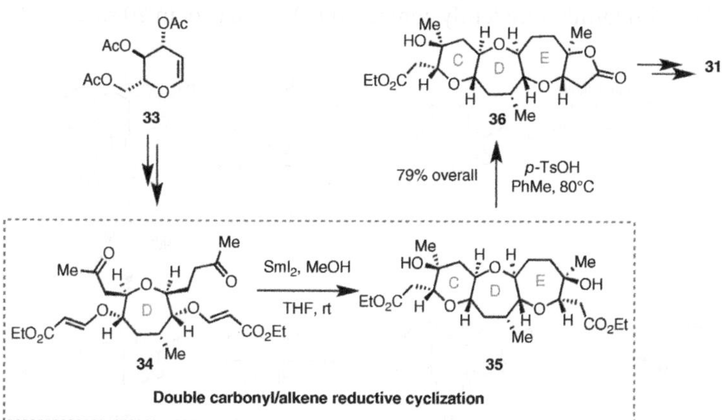

Scheme 6.9 SmI_2-mediated double carbonyl/alkene reductive cyclization.

Compound **34** was treated with SmI_2 and methanol to concomitantly induce the formation of both the six- and the seven-membered ether rings with complete control of the stereoselectivity (Scheme 6.9). Substrate **35** was then submitted to a chemoselective lactonization via the treatment with *p*-TsOH to form the corresponding tetracycle **36** at a 79% overall yield (Scheme 6.9). The latter was further elaborated to the desired segment **31**.

The synthesis of **32** started with the preparation of aldehyde **40** starting from commercially available 2-deoxy-D-ribose **37** through thioacetalization, benzylidation, hetero-Michael reaction, and deprotection of the thioacetal (Scheme 6.10).

A SmI_2-induced carbonyl/alkene reductive intramolecular cyclization was then used as the key step to generate the desired *trans*-fused bicycle **41** at an excellent yield as a single diastereoisomer (Scheme 6.10). The latter was eventually converted to the desired segment **32**.

The two key fragments **31** and **32** were finally coupled together following a procedure previously reported by Nicolaou and coworkers[23] (Scheme 6.11). The total synthesis of brevetoxin B was eventually achieved with high stereoselectivity in 59 steps (longest linear sequence).

6.2.3 Synthesis of (±)-Vigulariol

(±)-Vigulariol (**44**) (Scheme 6.12), a cladiellin natural product first isolated from the octocoral *Vigularia juncea* by Sheu and coworkers,[24] represents an appealing synthetic challenge due to its unusual tetracyclic core as well as its *in vitro* activity against human-lung adenocarcinoma cells ($IC_{50} = 18$ nM).

In 2007, Clark and coworkers reported a concise total synthesis of (±)-vigulariol (**44**) using a SmI_2-induced carbonyl/alkene reductive intramolecular cyclization as a key synthetic step.[25] Clark's approach, which can also lead to other members of the cladiellin family, efficiently generated (±)-vigulariol in 20 steps and a

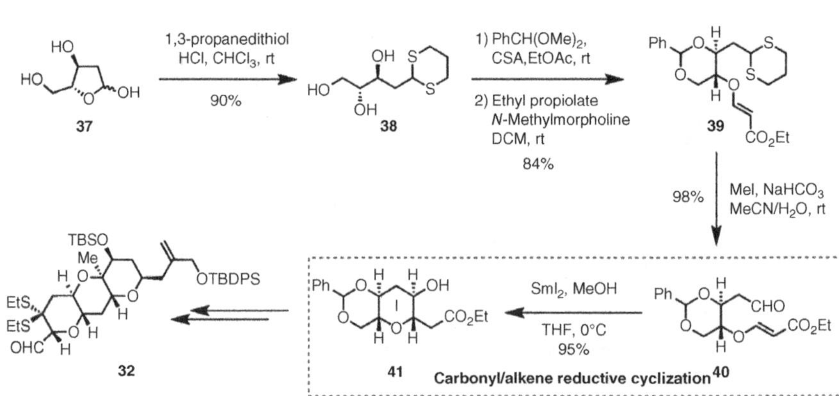

Scheme 6.10 SmI_2-mediated carbonyl/alkene reductive cyclization.

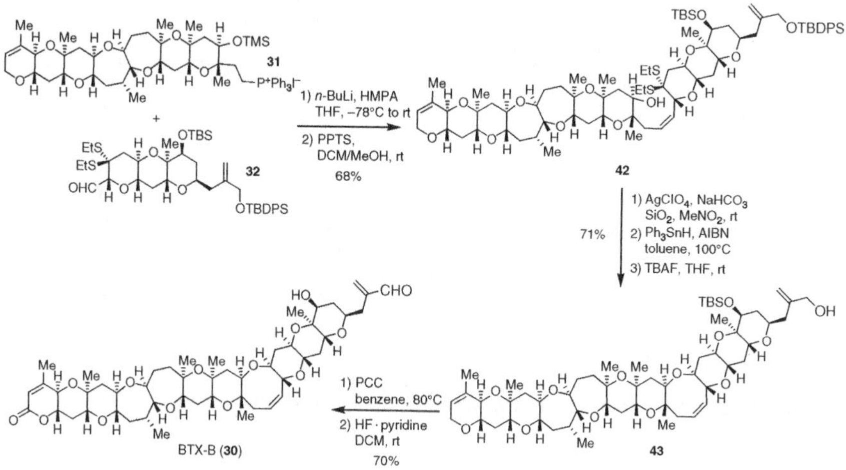

Scheme 6.11 End game.

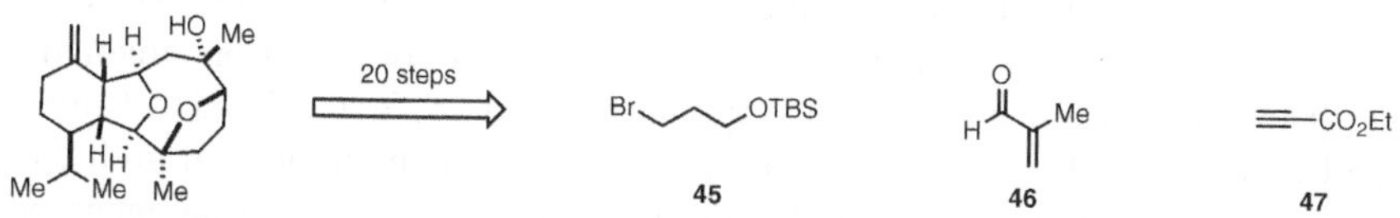

Scheme 6.12 Retrosynthetic analysis of vigulariol by Clark and coworkers.

4.0% overall yield starting from inexpensive and commercially available starting materials (Scheme 6.12).

The cyclization precursor **49** was easily prepared from bromide **45** using a four-step sequence, which included an initial Grignard reaction, alkylation of the allylic alcohol **48**, removal of the TBS protecting group, and Swern oxidation of the resulting primary alcohol (Scheme 6.13). The carbonyl/alkene reductive cyclization of compound **49** was then induced using SmI_2 in the presence of methanol to afford tetrahydropyranol **50**, which was isolated as a single diastereoisomer at a 76% yield (Scheme 6.13).

Protection of the hydroxyl group and cleavage of the ethyl ester, followed by conversion of the resulting carboxylic acid into an anhydride and treatment with diazomethane afforded the diazo ketone **51** at a good yield. This compound successfully underwent a reaction with copper(II) hexafluoroacetylacetonate to generate intermediate **52** via the formation of an oxonium ylide and subsequent rearrangement with a three-carbon ring expansion (Scheme 6.13). Alkene **52** was finally converted into (±)-vigulariol (**44**) through a short sequence of manipulations.

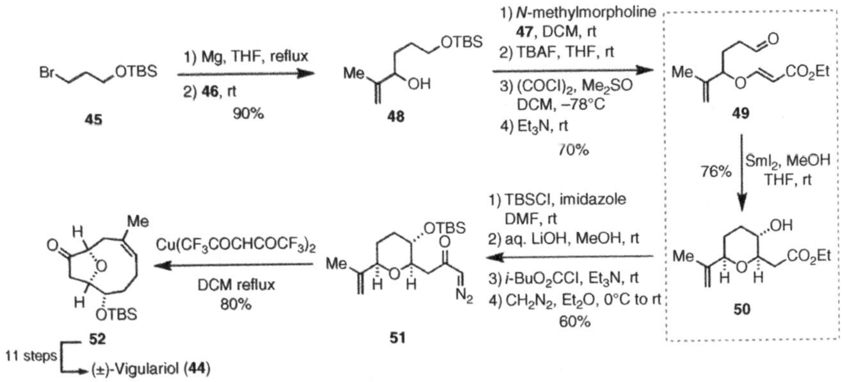

Scheme 6.13 SmI_2-mediated carbonyl/alkene reductive cyclization in the synthesis of vigulariol.

6.2.4 Synthesis of Diazonamide A

In the course of their second total synthesis of the marine-derived antitumor agent diazonamide A **53** (Scheme 6.14),[26, 27] Nicolaou *et al.* described an unprecedented SmI_2-mediated hetero-pinacol macrocyclization involving both an aldehyde and an oxime moiety. Diazonamide A, first isolated by Fenical and Clardy in 1991,[28] possesses a unique and challenging molecular architecture, which includes a cyclic polypeptide backbone, a strained halogenated heteroaromatic core and a quaternary center connecting the two major macrocyclic subunits. Following Nicolaou's strategy, this molecule was synthetized in 32 steps (longest linear sequence) starting from commercially available substrates (Scheme 6.14).

Intermediates **57** and **58** were easily obtained starting from 4-bromoindole **54**, tyrosine methyl ester **55** and 7-bromoisatin **56** in eight and 12 steps, respectively (Scheme 6.15).

These intermediates smoothly underwent a Suzuki coupling in the presence of $Pd(dppf)Cl_2$, a deprotection/oxidation sequence using TBAF and SO_3•py,

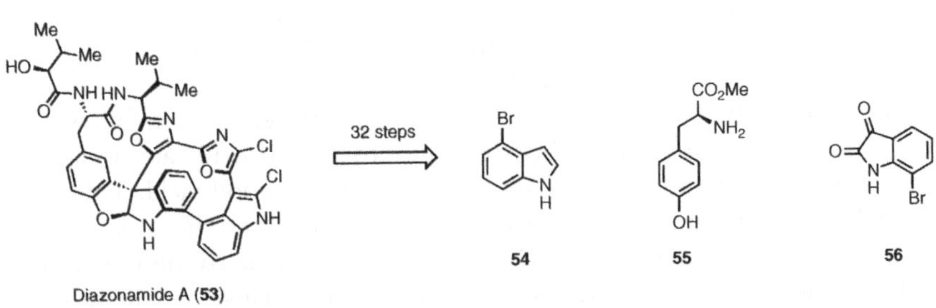

Scheme 6.14 Retrosynthetic analysis of diazonamide A by Nicolaou and coworkers.

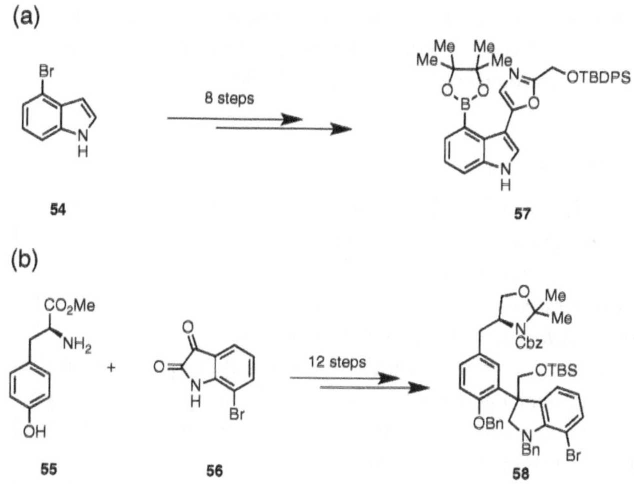

Scheme 6.15 Initial steps in the synthesis of diazonamide A: (a) Eight steps and (b) 12 steps.

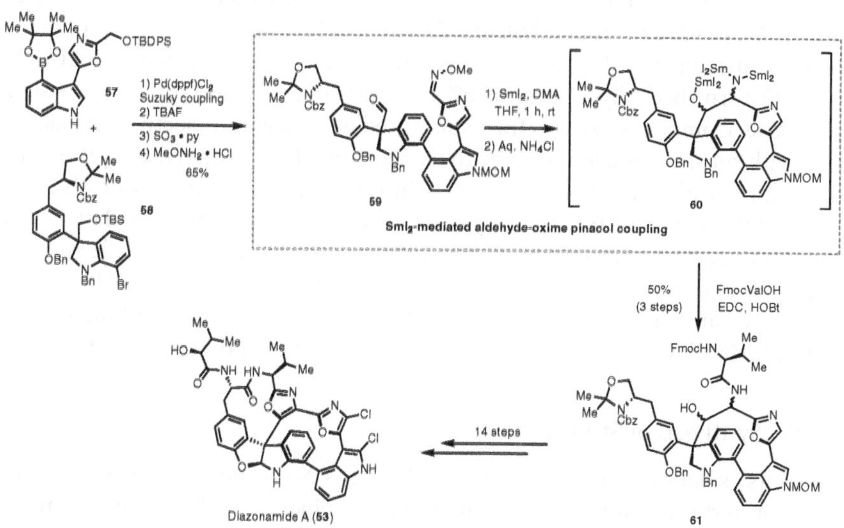

Scheme 6.16 SmI_2-mediated aldehyde-oxime pinacol coupling in the synthesis of diazonamide A.

and a selective capture of the more reactive aldehyde with $MeONH_2$•HCl to furnish the hetero-pinacol macrocyclization precursor **59** at a 65% overall yield for the three-step sequence (Scheme 6.16). The latter was then submitted to the key hetero-pinacol cyclization cascade in the presence of a large excess of SmI_2 and DMA (9 equiv. and 36 equiv., respectively) to generate the desired macrocyclic product **60**. A subsequent peptide coupling with Fmoc-protected L-valine, EDC, and HOBt in DMF eventually furnished compound **61** at a 50% yield, which was converted to diazonamide A after 14 additional steps.

6.2.5 Synthesis of Epothilone A

In 2001, Carreira and Bode described a concise and fully stereocontrolled synthesis of epothilone A **62** (Scheme 6.17) using an intriguing SmI_2-mediated isoxazoline reductive fragmentation reaction.[29]

The epothilones were first isolated by Höfle and coworkers in 1978 from cultures of *Soraginium cellulosum* collected in South Africa.[30] These molecules represent a new class of anticancer chemotherapeutics that share the ability with taxanes to prevent cancer cell proliferation by inhibiting tubulin polymerization. Due to their potent biological activity, epothilones have attracted considerable interest from the synthetic community.

Carreira's total synthesis of epothilone A (**62**) started with the preparation of the key isoxazoline precursor **63** from aldehyde **64** and 3-methylbutyn-3-ol **65** (Scheme 6.18). Indeed, addition of **64** to **65** followed by a treatment with BzCl and Et_3N afforded compound **66** at a 72% yield. Next, acetylene deprotection of **66** and subsequent reduction of the resulting propargylic benzoate with $LiAlH_4$ in Et_2O furnished allylic alcohol **67** at a 85% overall yield for the two-step sequence. Compound **67** was then submitted to a diastereoselective cycloaddition with substrate **68**, followed by the protection of the hydroxyl group, and an olefination with 2-methylthiazole-4-carboxaldehyde (**70**), to provide the key isoxazoline precursor **63** at a 70% overall yield (Scheme 6.18). The latter was then subjected to a smooth N–O bond cleavage upon treatment with SmI_2 (4 equiv.) in THF at 0 °C,[31] while a subsequent imine hydrolysis with $B(OH)_3$ generated the desired intermediate **71** at a 75% yield. A short sequence of manipulations finally allowed the conversion of compound **71** to epothilone A **62**.

6.2.6 Synthesis of Strychnine

Since its isolation from the seeds of the trees *Strychnos ignatii Bergius* by Pelletier and Caventou in 1818,[32] its structure elucidation by Robinson more

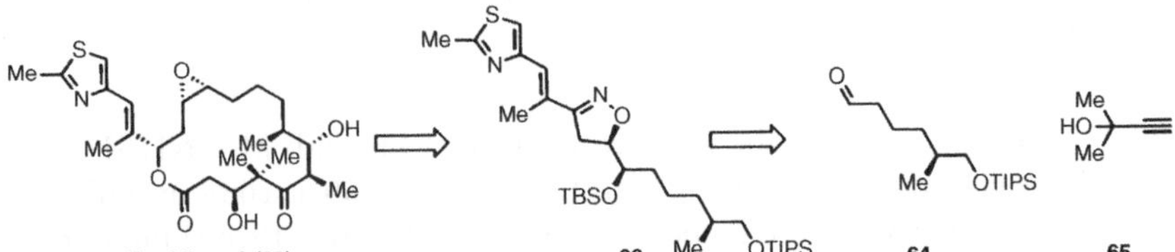

Scheme 6.17 Retrosynthetic analysis of epothilone A by Carreira and coworkers.

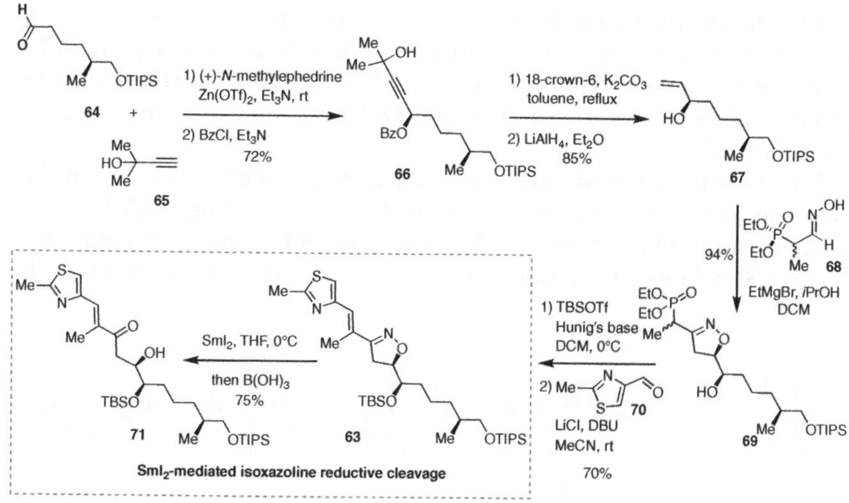

Scheme 6.18 SmI_2-mediated isoxazoline reductive cleavage in the synthesis of epothilone A.

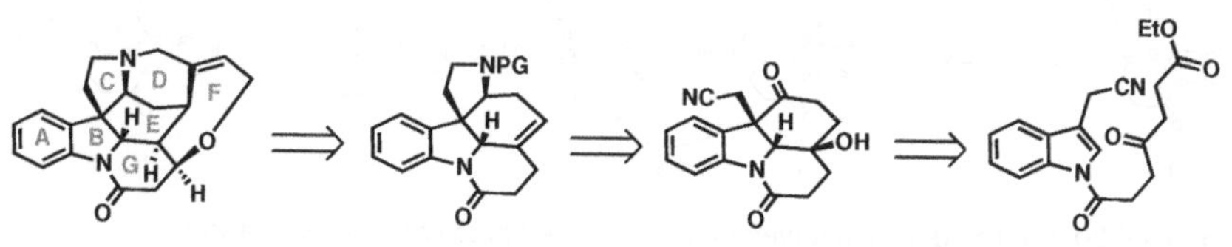

Scheme 6.19 Retrosynthetic analysis of strychnine by Beemelmanns and Reissig.

than a century later,[33] and its first total synthesis by Woodward in 1954,[34] strychnine has been the focus of tremendous synthetic efforts,[35] which were recently highlighted in an instructive review by Overman and Cannon.[36]

The originality of Beemelmanns' and Reissig' synthesis[37] relied on a particularly effective SmI_2-induced reductive dearomatization cascade reaction, which allowed them to rapidly construct the ABEG tetracyclic core of the molecule bearing three stereogenic centers, including one quaternary carbon (Scheme 6.19). Their retrosynthetic analysis also included a regioselective elimination of a tertiary hydroxyl group, a Heck-type cyclization, and a diastereoselective reductive amination for the construction of the five-membered C ring.

The synthesis started by the preparation of the cyclization precursor **75**, which was obtained at a multigram scale through the acylation of commercially available 3-indolylacetonitrile with 4-oxopimelic acid monoester. With this compound in hand, a thorough investigation of the key SmI_2-induced cyclization was then performed. Interestingly, after screening various reaction conditions, treatment with SmI_2/HMPA in the presence of bromo-acetonitrile afforded the best results with the formation of the tetracyclic compound **74** at a 77% yield along with 5% of the over-reduced product **76**, <5% of the dealkylated product 77 and 4% of the spiro-lactone **78** bearing an exo-methylene group (Scheme 6.20).

The rest of the synthesis involved a reduction of the nitrile moiety with concomitant formation of the pentacyclic amine **79** using Raney nickel under a hydrogen atmosphere (Scheme 6.21). Alkylation of the secondary amine with tosylate **80** followed by a Martin sulfurane-mediated dehydration led to the

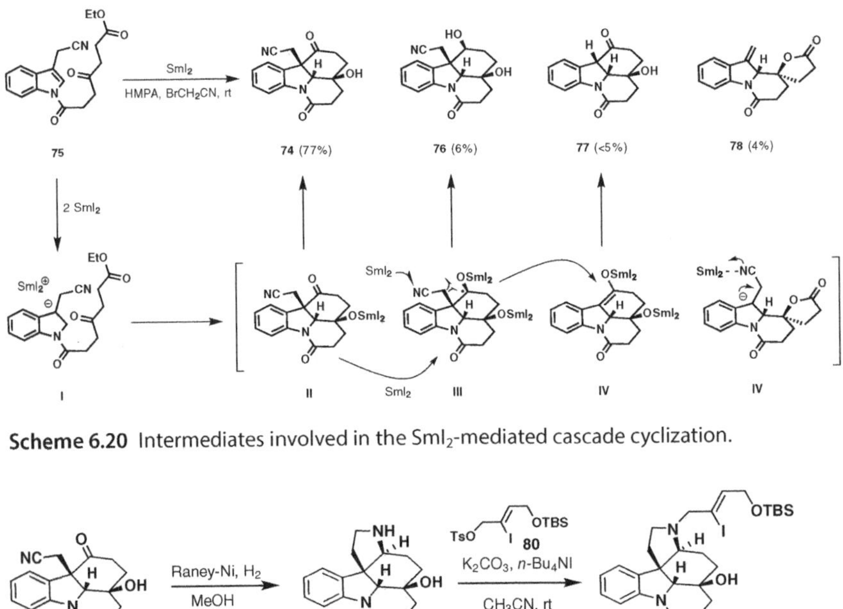

Scheme 6.20 Intermediates involved in the SmI_2-mediated cascade cyclization.

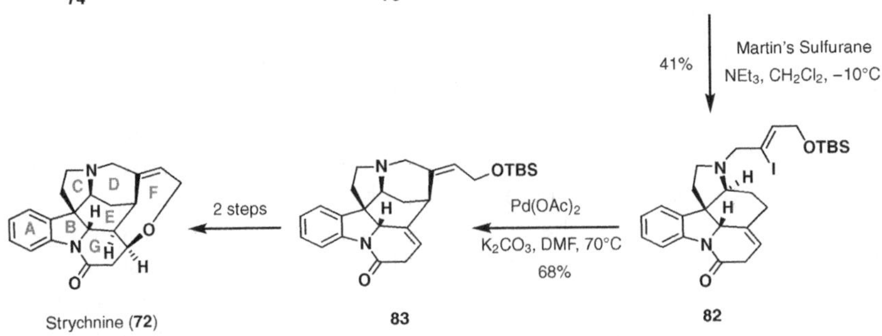

Scheme 6.21 End game.

pentacyclic core bearing a pendant vinyl iodide moiety. The latter was eventually subjected to Heck coupling conditions, which gave the hexacyclic TBS-protected isostrychnine **83** at a 68% yield, while two additional steps completed the synthesis.

6.2.7 Synthesis of the ABC Ring of Paclitaxel

Paclitaxel, also referred to as Taxol, was discovered as part of a National Cancer Institute program aimed at unveiling natural products with anticancer activities. Isolated from a crude extract from the bark of the Pacific yew tree *Taxus brevifolia*,[38] this diterpene was shown to be effective against many tumor cell lines by inhibiting cell division and chromosome segregation through tubulin targeting and microtubule stabilization.[39]

Several total syntheses of paclitaxel have been reported during the past two decades, notably by Holton, Nicolaou, Danishefsky, Wender Kuwajima, Mukaiyama, and Takahashi.[40] In 2015, Chida and coworkers reported an elegant approach toward the synthesis of the ABC ring of paclitaxel via a SmI_2-mediated reductive cyclization of an allylic benzoate possessing a pendant aldehyde moiety (Scheme 6.22).[41]

The synthesis of the cyclization precursor **91** started by a Shapiro coupling between the A-ring hydrazone **87** and the C-ring aldehyde **88**, which afforded the corresponding coupled product **89** at a 92% yield (Scheme 6.23). The latter was then converted to the key cyclization precursor at a roughly 12% overall yield after several additional steps. The cyclization was then performed using SmI_2 in a HMPA/THF mixture at 40 °C, to afford the desired eight-membered carbocycle at a 66% yield as a 1.5:1 epimeric mixture (**92**:**93**). The use of HMPA

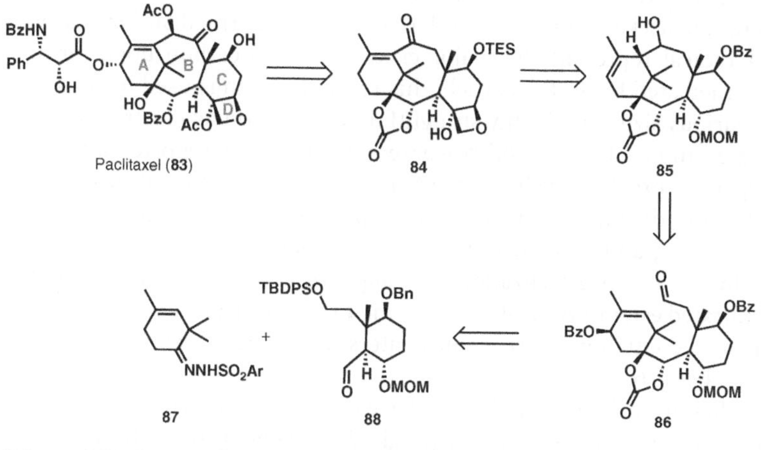

Scheme 6.22 Retrosynthetic analysis of paclitaxel by Chida and coworkers.

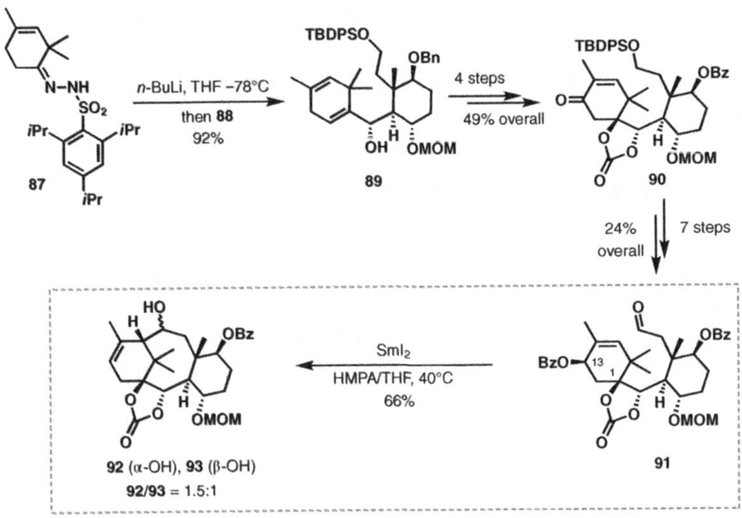

Scheme 6.23 SmI_2-mediated cyclization in the synthesis of paclitaxel.

as a cosolvent appeared to be crucial as the reaction performed in the absence of HMPA gave no cyclized product but the corresponding primary alcohol instead. Moreover, the *syn* relationship between the benzoate group at C13 and the oxygen at C1 appeared to be crucial, which implied a possible chelation of the Sm.

6.2.8 Miscellaneous

As illustrated by the number of examples reported so far, cross-couplings via the generation of a ketyl- or an imine-derived radical intermediate, or a carbon-centered radical represent the major class of carbon-carbon bond-forming reactions mediated by SmI_2. As such, this reagent was thus used to construct rings of various sizes in a highly straightforward fashion as illustrated by the following examples. Procter and coworkers, for instance, reported a particularly elegant 4-exo-trig cyclization of an aldehyde on an α,β-unsaturated lactone to form the four-membered core of pestalotiopsin A (Scheme 6.24).[42]

Likewise, Yang and coworkers reported the total synthesis of pseudolaric acid A using a 5-exo-trig cyclization of an aliphatic ketone onto a trisubstituted α,β-unsaturated ester to generate the five-membered ring and control the three newly formed contiguous stereogenic centers with a high level of diastereoselectivity (Scheme 6.25).[43]

Takahashi's group[44] employed a similar carbon-carbon forming strategy to access the tetrahydropyran core of aspergillide A (Scheme 6.26). In their case, the radical generated from aldehyde **100** underwent a 6-exo-trig cyclization to

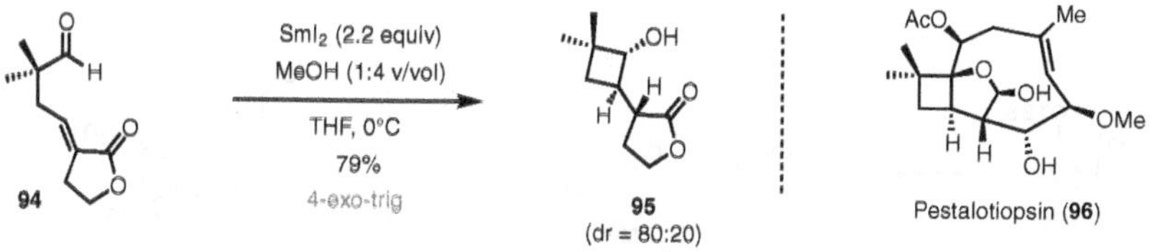

Scheme 6.24 Synthesis of pestalotiopsin A using a 4-exo-trig cyclization by Procter and coworkers.

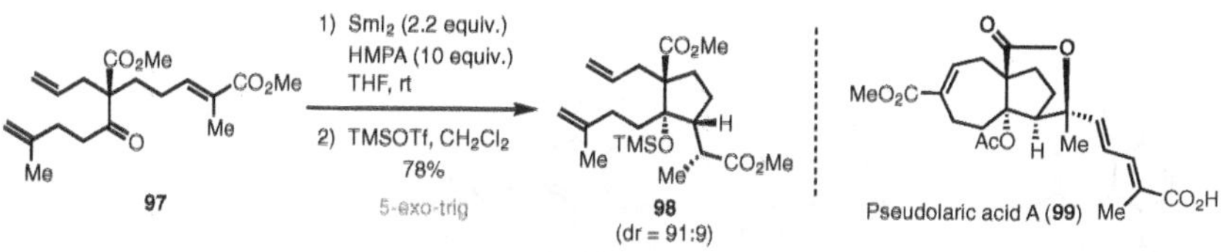

Scheme 6.25 Synthesis of pseudolaric acid A using a 5-exo-trig cyclization by Yang and coworkers.

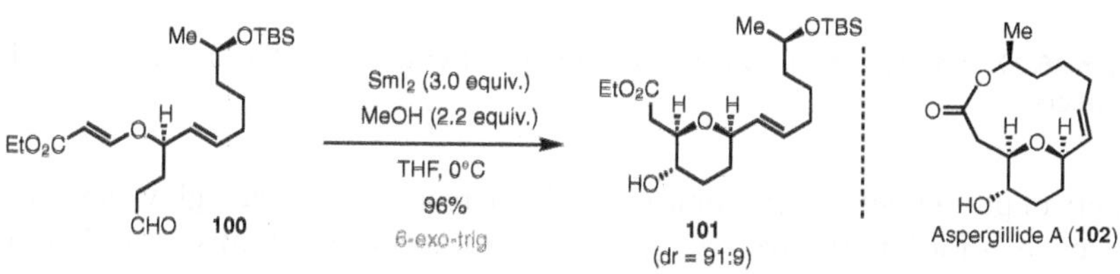

Scheme 6.26 Synthesis of aspergillide A using a 6-exo-trig cyclization by Takahashi and coworkers.

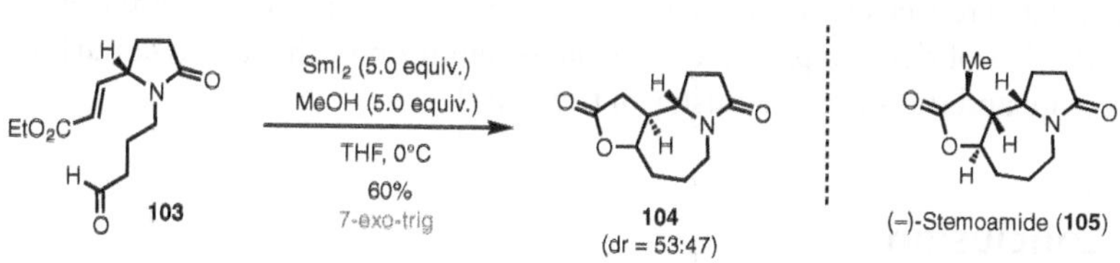

Scheme 6.27 Synthesis of (−)-stemoamide using a 7-exo-trig cyclization by Honda and coworkers.

afford the macrolactone precursor **101** at both an excellent yield and diastereoselectivity.

More recently, Honda and coworkers used SmI_2 to promote cross-coupling between an aldehyde and an α,β-unsaturated ester to build the seven-membered ring present in (−)-stemoamide through a 7-exo-trig cyclization (Scheme 6.27).[45]

In a similar fashion, Yao and coworkers used a SmI_2-mediated aldehyde/enone conjugate addition to construct the highly strained cyclobutanol

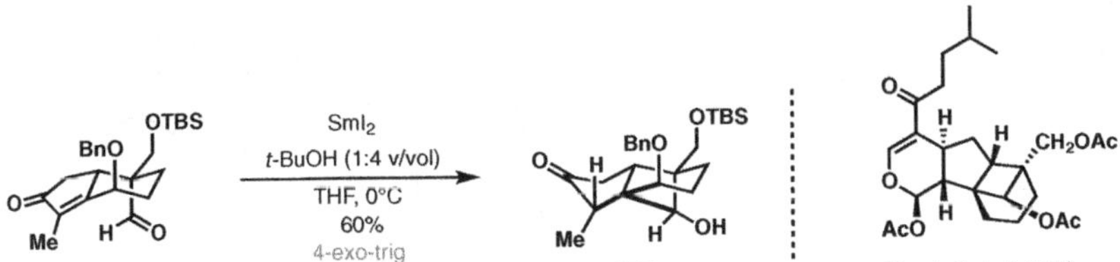

106

107
(dr = 80:20)

Plumisclerin A (108)

Scheme 6.28 Synthesis of the tricyclo[4,3,1,0^{1,5}]decane core of plumisclerin A by Yao and coworkers.

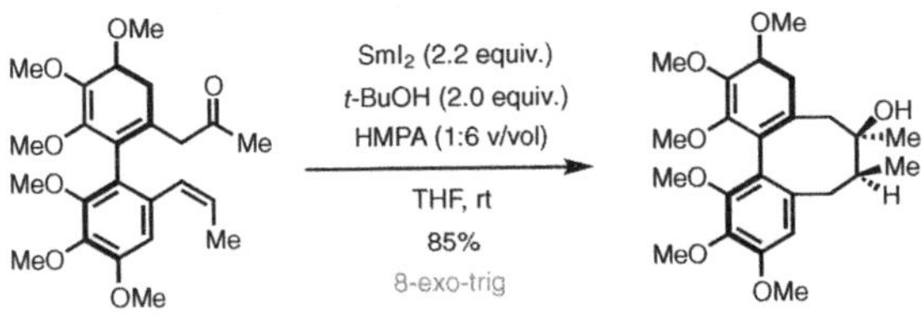

109

(+)-Isoschizandrin (110)
(dr >95:5)

Scheme 6.29 Synthesis of (+)-isoschizandrin using a 8-exo-trig cyclization by Molander and coworkers.

moiety of plumisclerin A (Scheme 6.28).[46] The reaction proved highly diastereoselective, affording the tricyclo[4,3,1,0^{1,5}]decane core of plumisclerin A at a 60% yield.

Finally, Molander and coworkers reported a clever SmI_2-promoted 8-*endo*-trig ketyl-alkene cyclization for the synthesis of (+)-isoschizandrin (Scheme 6.29).[47] The high diastereoselectivity observed was attributed to the concomitant presence of the (*Z*)-olefin, the pseudoequatorial orientation of the alkoxysamarium intermediate and the stereochemistry of the biaryl.

6.3 Conclusion

Since the pioneering work of Kagan, SmI_2 has emerged as a fantastic tool exhibiting high functional group tolerance and chemoselectivity as illustrated by the variety of coupling partners and the number of synthetic applications, in particular in the field of natural product synthesis. Nonetheless, many challenges still lie ahead, such as the development of efficient catalytic protocols and the development of asymmetric versions of the SmI_2-mediated cross-coupling reaction; however, considering the number of synthetic chemists around the world focused on these matters, these questions will undoubtedly be answered in the near future.

Acknowledgements

We gratefully thank the Agence Nationale de la Recherche for their financial support (NCiS project; ANR-2010-JCJC-715–1).

References

1 (a) Namy, J. L.; Girard, P.; Kagan, H. B. *New J. Chem.* **1977**, *1*, 5–7; (b) Girard, P.; Namy, J. L.; Kagan, H. B. *J. Am. Chem. Soc.* **1980**, *102*, 2693–2698.

2 For selected reviews of SmI_2-mediated reactions in organic synthesis, see: (a) Soderquist, J. A. *Aldrichimica Acta* **1991**, *24*, 15–23; (b) Molander, G. A. *Org. React.* **1994**, *46*, 211–367; (c) Imamoto, T. *Lanthanides in Organic Synthesis*, London: Academic Press, **1994**; (d) Molander, G. A.; Harris, C. R. *Chem. Rev.* **1996**, *96*, 307–338; (e) Skrydstrup, T. *Angew. Chem. Int. Ed.* **1997**, *36*, 345–347; (f) Molander, G. A.; Harris, C. R. *Tetrahedron* **1998**, *54*, 3321–3354; (g) Nomura, R.; Endo, T. *Chem. Eur. J.* **1998**, *4*, 1605–1610; (h) Krief, A.; Laval, A.-M. *Chem. Rev.* **1999**, *99*, 745–777; (i) Steel, P. G. *J. Chem. Soc. Perkin Trans.* **2001**, *1*, 2727–2751; (j) Agarwal, S.; Greiner, A. *J. Chem. Soc. Perkin Trans. 1* **2002**, 2033–2042; (k) Kagan, H. B. *Tetrahedron* **2003**, *59*, 10351–10372; (l) Berndt, M.; Gross, S.; Hölemann, A.; Reissig, H.-U. *Synlett* **2004**, 422–438; (m) Edmonds, D. J.; Johnston, D.; Procter, D. J. *Chem. Rev.* **2004**, *104*, 3371–3403; (n) Jung, D. Y.; Kim, Y. H. *Synlett* **2005**, 3019–3032; (o) Gopalaiah, K.; Kagan, H. B. *New J. Chem.* **2008**, *32*, 607–637; (p) Rudkin, I. M.; Miller, L. C.; Procter, D. J. *Organomet. Chem.* **2008**, *34*, 19–45; (q) Szostak, M.; Fazakerley, N. J.; Parmar, D.; Procter, D. J. *Chem. Rev.* **2014**, *114*, 5959 – 6039.

3 (a) Shabangi, M.; Flowers II, R. A. *Tetrahedron Lett.* **1997**, *38*, 1137–1140; (b) Flowers II, R. A. *Synlett* **2008**, 1427–1439.

4 Procter, D. J.; Flowers, R. A.; Skrydstrup, T. *Organic Synthesis using Samarium Diiodide: A Practical Guide*, London: Royal Society of Chemistry Publishing, **2010**.

5 For selected examples, see: (a) Inanaga, J.; Ishikawa, M.; Yamaguchi, M. *Chem. Lett.* **1987**, *16*, 1485–1486; (b) Cabrera, A.; Alper, H. *Tetrahedron Lett.* **1992**, *33*, 5007–5008; (c) McDonald, C. E.; Ramsey, J. D.; Sampsell, D. G.; Butler, J. A.; Cecchini, M. R. *Org. Lett.* **2010**, *12*, 5178–5181.

6 For a review, see: Molander, G. A. *Chem. Rev.* **1992**, *92*, 29–68.

7 For a comprehensive review on SmI_2 in total synthesis, see: Nicolaou, K. C.; Ellery, S. P.; Chen, J. S. *Angew. Chem. Int. Ed.* **2009**, *48*, 7140–7165.

8 For recent examples of the Reformatsky reaction, see: (a) Chattopadhyay, A.; Dubey, A. K. *J. Org. Chem.* **2007**, *72*, 9357–9359; (b) Parrish, J. D.; Shelton, D. R.; Little, R. D. *Org. Lett.* **2003**, *5*, 3615–3617; (c) Wolf, C.; Moskowitz, M. *J. Org. Chem.* **2011**, *76*, 6372–6376; (d) Moumne, R.; Lavielle, S.; Karoyan, P. *J. Org. Chem.* **2006**, *71*, 3332–3334.

9 For selected reviews of the Reformatsky reaction, see: (a) Fürstner, A. *Synthesis* **1989**, 571–590; (b) Ocampo, R.; Dolbier, W. R. *Tetrahedron* **2004**, *60*, 9325–9374.

10 Molander, G. A.; Hahn, G. *J. Org. Chem.* **1986**, *51*, 1135–1138.

11 (a) Molander, G. A.; Kenny, C. *Tetrahedron Lett.* **1987**, *28*, 4367–4370; (b) Molander, G. A.; Kenny, C. *J. Am. Chem. Soc.* **1989**, *111*, 8236–8246.

12 For selected examples, see: (a) Molander, G. A.; McKie, J. A. *J. Org. Chem.* **1995**, *60*, 872–822; (b) Curran, D. P.; Fevig, T. L.; Jaspersen, C. P.; Totleben, M. L. *Synlett* **1992**, 943–961; (c) Hutton, T. K.; Muir, K. W.; Procter, D. J. *Org. Lett.* **2003**, *5*, 4811–4814.

13 (a) Namy, J.-L.; Souppe, J.; Kagan, H. B. *Tetrahedron Lett.* **1983**, *24*, 765–766; (b) Souppe, J.; Danon, L.; Namy, J.-L.; Kagan, H. B. *J. Organomet. Chem.* **1983**, *250*, 227–236.

14 For selected reviews of the pinacol reaction, see: (a) Ladipo, F. T. *Curr. Org. Chem.* **2006**, *10*, 965–980; (b) Chatterjee, A.; Joshi, N. N. *Tetrahedron* **2006**, *62*, 12137–12158; (c) Hirao, T. *Top. Curr. Chem.* **2007**, *279*, 53–76.

15 Molander, G. A.; Kenny, C. *J. Org. Chem.* **1988**, *53*, 2132–2134.

16 For selected reviews of fragmentation reactions in organic chemistry, see: (a) Wakabayashi, T.; Watanabe, K. *J. Synth. Org. Chem. Jpn.* **1980**, *38*, 853–861; (b) Li, J. J. *Tetrahedron* **2001**, *57*, 1–24; (c) Bradley, D.; Williams G.; Blann, K.; Caddy, J. *Org. Prep. Proced. Int.* **2001**, *33*, 565–602; (d) Williams, D. B. G.; Caddy, J.; Blann, K. *Org. Prep. Proced. Int.* **2003**, *35*, 307–360.

17 (a) Moslin, R. M.; Jamison, T. F. *J. Am. Chem. Soc.* **2006**, *128*, 15106–15107; (b) Moslin, R. M.; Jamison, T. F. *J. Org. Chem.* **2007**, *72*, 9736 – 9745.

18 For linear approaches to the total synthesis of (+)-acutiphycin, see: (a) Smith III, A. B.; Chen, S. S.-Y.; Reichert, F. C.; Nelson, J. M.; Salvatore, B. A. *J. Am. Chem. Soc.* **1995**, *117*, 12013–12014; (b) Smith III, A. B.; Chen, S. S.-Y.; Nelson, F. C.; Reichert, J. M.; Salvatore, B. A. *J. Am. Chem. Soc.* **1997**, *119*, 10935–10946; (c) Hena, M. A.; Kim, C.-S.; Horiike, M.; Kiyooka, S.-I. *Tetrahedron Lett.* **1999**, *40*, 1161–1164; (d) Kiyooka, S.-I.; Hena, M. A. *J. Org. Chem.* **1999**, *64*, 5511–5523.

19 For selected examples, see: (a) Tabuchi, T.; Kawamura, K.; Inanaga, J. *Tetrahedron Lett.* **1986**, *27*, 3889–3890; (b) Inanaga, J.; Yokoyama, Y.; Handa, Y.; Yamaguchi, M. *Tetrahedron Lett.* **1991**, *32*, 6371–6372; (c) Corey, E. J.; Wu, Y. J. *J. Am. Chem. Soc.* **1993**, *115*, 8871–8872; (d) Inoue, M.; Sasaky, M.; Tachibana, K. *Angew. Chem. Int. Ed.* **1998**, *37*, 965–969; (e) Inoue, M.; Sasaky, M.; Tachibana, K. *J. Org. Chem.* **1999**, *64*, 9416–9429.

20 For different total synthesis of brevetoxin B, see: (a) Nicolaou, K. C.; Rutjes, F. P. J. T.; Theodorakis, E. A.; Tiebes, J.; Sato, M.; Untersteller, E. *J. Am. Chem. Soc.* **1995**, *117*, 1173–1174; (b) Nicolaou, K. C.; Theodorakis, E. A.; Rutjes, F. P. J. T.; Sato, M.; Tiebes, J.; Xiao, X.-Y. et al., *J. Am. Chem. Soc.* **1995**, *117*, 10239–10251.

21 For synthetic approaches toward brevetoxin B, see: (a) Hori, N.; Matsukura, H.; Matsuo, G.; Nakata, T. *Tetrahedron Lett.* **1999**, *40*, 2811–2814; (b) Matsuo,

G.; Hori, N.; Nakata, T. *Tetrahedron Lett.* **1999**, *40*, 8859–8862; (c) Matsuo, G.; Matsukura, H.; Hori, N.; Nakata, T. *Tetrahedron Lett.* **2000**, *41*, 7673–7676; (d) Matsuo, G.; Hori, N.; Matsukura, H.; Nakata, T. *Tetrahedron Lett.* **2000**, *41*, 7677–7680. (e) Matsuo, G.; Matsukura, H.; Hori, N.; Nakata, T. *Tetrahedron Lett.* **2000**, *41*, 7681–7684; (f) Matsukura, H.; Hori, N.; Matsuo, G.; Nakata, T. *Tetrahedron Lett.* **2002**, *58*, 1853–1864.

22 Matsuo, G.; Kawamura, K.; Hori, N.; Matsukura, H.; Nakata, T. *J. Am. Chem. Soc.* **2004**, *126*, 14374–14376.

23 Nicolaou, K. C.; Rutjes, F. P. J. T.; Theodorakis, E. A.; Tiebes, J.; Sato, M.; Untersteller, E. *J. Am. Chem. Soc.* **1995**, *117*, 10252–10253.

24 Su, J.-H.; Huang, H.-C.; Chao, C.-H.; Yan, L.-Y.; Wu, Y.-C.; Wu, C.-C. *et al.*, *Bull. Chem. Soc. Jpn.* **2005**, *78*, 877–879.

25 Clark, J. S.; Hayes, S. T.; Wilson, C.; Gobbi, L. *Angew. Chem. Int. Ed.* **2007**, *46*, 437–440.

26 (a) Nicolaou, K. C.; Huang, X.; Giuseppone, N.; Rao, P. B.; Bella, M.; Reddy, M. V.; *et al.*, *Angew. Chem. Int. Ed.* **2001**, *40*, 4705–4709; (b) Nicolaou, K. C.; Rao, P. B.; Hao, J.; Reddy, M. V.; Rassias, G.; Huang, X.; *et al.*, *Angew. Chem. Int. Ed.* **2003**, *42*, 1753–1758; (c) Nicolaou, K. C.; Snyder, S. A.; Giuseppone, N.; Huang, X.; Bella, M.; Reddy, M. V.; *et al.*, *J. Am. Chem. Soc.* **2004**, *126*, 10174–10182; (d) Nicolaou, K. C.; Hao, J.; Reddy, M. V.; Rao, P. B.; Rassias, G.; Snyder, S. A.; et al., *J. Am. Chem. Soc.* **2004**, *126*, 12897–12906.

27 For the first total synthesis of diazonamide A, see: (a) Nicolaou, K. C.; Bella, M.; Chen, D. Y.-K.; Huang, X.; Ling, T.; Snyder, S. A. *Angew. Chem. Int. Ed.* **2002**, *41*, 3495–3499; (b) Nicolaou, K. C.; Chen, D. Y.-K.; Huang, X.; Ling, T.; Bella, M.; Snyder, S. A. *J. Am. Chem. Soc.* **2004**, *126*, 12888–12896.

28 Lindquist, N.; Fenical, W.; Van Duyne, G. D.; Clardy, J. *J. Am. Chem. Soc.* **1991**, *113*, 2303–2304.

29 (a) Bode, J. W.; Carreira, E. M. *J. Am. Chem. Soc.* **2001**, *123*, 3611–3612; (b) Bode, J. W.; Carreira, E. M. *J. Org. Chem.* **2001**, *66*, 6410–6424.

30 (a) Höfle, G.; Bedorf, N.; Steinmetz, H.; Schomburg, D.; Gerth, K.; Reichenback, H. *Angew. Chem. Int. Ed.* **1996**, *35*, 1567–1569; (b) Gerth, K.; Bedorf, N.; Höfle, G.; Irschik, H.; Reichenback, H. J. *Antibiot.* **1996**, *49*, 560–563. For a review, see: Altmann, K.-H.; Pfeiffer, B.; Arseniyadis, S.; Pratt, B. A.; Nicolaou, K. C. *Chem. Med. Chem.* **2007**, 396–423.

31 Bode, J. W.; Carreira, E. M. *Org. Lett.* **2001**, *3*, 1587–1590.

32 (a) Pelletier, P. J.; Caventou, J. B. *Ann. Chim. Phys.* **1818**, *8*, 323; (b) Pelletier, P. J.; Caventou, J. B. *Ann. Chim. Phys.* **1819**, *10*,142.

33 Openshaw, H. T.; Robinson, R. *Nature* **1946**, *157*, 438.

34 Woodward, R. B.; Cava, M. P.; Ollis, W. D.; Hunger, A.; Daeniker, H. U.; Schenker, K. *J. Am. Chem. Soc.* **1954**, *76*, 4749–4751.

35 (a) Magnus, P.; Giles, M.; Bonnert, R.; Kim, C. S.; McQuire, L.; Merritt, A. *et al.*, *J. Am. Chem. Soc.* **1992**, *114*, 4403–4405; (b) Knight, S. D.; Overman, L. E.; Pairaudeau, G. *J. Am. Chem. Soc.* **1993**, *115*, 9293–9294; (c) Kuehne, M. E.; Xu, F.

J. Org. Chem. **1993**, *58*, 7490–7497; (d) Rawal, V. H.; Iwasa, S. *J. Org. Chem.* **1994**, *59*, 2685–2686; (e) Kuehne, M. E.; Xu, F. *J. Org. Chem.* **1998**, *63*, 9427–9433; (f) Solé, D.; Bonjoch, J.; García-Rubio, S.; Peidró, E.; Bosch, J. *Angew. Chem. Int. Ed.* **1999**, *38*, 395–397; (g) Ito, M.; Clark, C. W.; Mortimore, M.; Goh, J. B.; Martin, S. F. *J. Am. Chem. Soc.* **2001**, *123*, 8003–8010; (h) Eichberg, M. J.; Dorta, R.; Lamottke, L. K.; Vollhardt, K. P. C. *Org. Lett.* **2000**, *2*, 2479–2481; (i) Nakanishi, M.; Mori, M. *Angew. Chem. Int. Ed.* **2002**, *41*, 1934–1936; (j) Bodwell, G. J.; Li, J. *Angew. Chem. Int. Ed.* **2002**, *41*, 3261–3262; (k) Ohshima, T.; Xu, Y.; Takita, R.; Shimuzu, S.; Zhong, D.; Shibasaki, M. *J. Am. Chem. Soc.* **2002**, *124*, 14546–14547; (l) Kaburagi, Y.; Tokuyama, H.; Fukuyama, T. *J. Am. Chem. Soc.* **2004**, *126*, 10246–10247; (m) Zhang, H.; Boonsombat, J.; Padwa, A. *Org. Lett.* **2007**, *9*, 279–282; (n) Sirasani, G.; Paul, T.; Dougherty, Jr., W.; Kassel, S.; Andrade, R. B. *J. Org. Chem.* **2010**, *75*, 3529–3532; (o) Beemelmanns, C.; Reissig, H. U. *Angew. Chem. Int. Ed.* **2010**, *49*, 8021–8025; (p) Martin, D. B. C.; Vanderwal, C. D. *Chem. Sci.* **2011**, *2*, 649–651; (q) Jones, S. B.; Simmons, B.; Mastracchio, A.; MacMillan, D. W. C. *Nature* **2011**, *475*, 183–188; (r) Bonjoch, J.; Solé, D. *Chem. Rev.* **2000**, *100*, 3455–3482; (s) Mori, M. *Heterocycles* **2010**, *81*, 259–292; (t) Furst, L.; Stephenson C. R. *Nat. Chem. Biol.* **2011**, *7*, 582–583; (u) Beemelmanns, C.; Reissig, H. U. *Chem. Soc. Rev.* **2011**, *40*, 2199–2210; (v) Martin, D. B. C.; Nguyen, L. Q. C.; Vanderwal, D. *J. Org. Chem.* **2012**, *77*, 17–46.

36 Cannon, J. S.; Overman, L. E. *Angew. Chem. Int. Ed.* **2012**, *51*, 4288–4311.

37 (a) Beemelmanns, C.; Reissig, H.-U. *Angew. Chem. Int. Ed.* **2010**, *49*, 8021–8025; (b) Beemelmanns, C.; Reissig, H.-U. *Chem. Eur. J.* **2015**, *21*, 8416–8425.

38 Wani, M. C.; Taylor, H. L.; Wall, M. E.; Coggon, P.; McPhail, A. T. *J. Am. Chem. Soc.* **1971**, *93*, 2325–2327.

39 Schiff, P. B.; Fant, J. S.; Horwitz, B. *Nature* **1979**, *277*, 665–667.

40 (a) Holton, R. A.; Somoza, C.; Kim, H.-B.; Liang, F.; Biediger, R. J.; Boatman, P. D. et al., *J. Am. Chem. Soc.* **1994**, *116*, 1597 – 1598; (b) Nicolaou, K. C.; Yang, Z.; Liu, J. J.; Ueno, H.; Nantermet, P. G.; Guy, R. K. *et al.*, *Nature* **1994**, *367*, 630 – 634; (c) Masters, J. J.; Link, J. T.; Snyder, L. B.; Young, W. B.; Danishefsky, S. J. *Angew. Chem. Int. Ed.* **1995**, *34*, 1723 – 1726; (d) Wender, P. A.; Badham, N. F.; Conway, S. P.; Floreancig, P. E.; Glass, T. E.; Gränicher, C. *et al.*, *J. Am. Chem. Soc.* **1997**, *119*, 2755 – 2756; (e) Morihira, K.; Hara, R.; Kawahara, S.; Nishimori,T.; Nakamura, N.; Kusama, H. *et al.*, *J. Am. Chem. Soc.* **1998**, *120*, 12980 – 12981; (f) Mukaiyama,T.; Shiina, I.; Iwadare, H.; Saitoh, M.; Nishimura, T.; Ohkawa, N. *et al.*, *Chem. Eur. J.* **1999**, *5*, 121 – 161; (g) Doi, T.; Fuse, S.; Miyamoto, S.; Nakai, K.; Sasuga, D.; Takahashi, T. *Chem. Asian J.* **2006**, *1*, 370–383.

41 Fukaya, K.; Tanaka, Y.; Sato, A. C.; Kodama, K.; Yamazaki, H.; Ishimoto, T. et al., *Org. Lett.* **2015**, *17*, 2570–2573.

42 (a) Johnston, D.; McCusker, C. M.; Procter, D. J. *Tetrahedron Lett.* **1999**, *40*, 4913–4916; (b) McCusker, C. F.; Johnston, D.; Muir, K.; Procter, D. J. *J. Chem. Soc. Perkin Trans. 1* **2000**, 681–695; (c) Johnston, D.; Francon, N.; Edmonds,

D. J.; Procter, D. J. *Org. Lett.* **2001**, *3*, 2001–2004; (d) Johnston, D.; Couche, E.; Edmonds, D. J.; Muir, K. W.; Procter, D. J. *Org. Biomol. Chem.* **2003**, *1*, 328–337; (e) Baker, T. M.; Edmonds, D. J.; Hamilton, D.; O'Brien, C. J.; Procter, D. J. *Angew. Chem. Int. Ed.* **2008**, *47*, 5631–5633.

43 Xu, T.; Li, C. C.; Yang, Z. *Org. Lett.* **2011**, *13*, 2630–2633.

44 Izuchi, Y.; Kanomata, N.; Koshino, H.; Hongo, Y.; Nakata, T.; Takahashi, S. *Tetrahedron: Asymmetry* **2011**, *22*, 246–251.

45 Honda, T.; Matsukawa, T.; Takahashi, K. *Org. Biomol. Chem.* **2011**, *9*, 673–675.

46 Chen, J.-P., He, W., Yang, Z.-Y., Yao, Z.-J. *Org. Lett.* **2015**, *17*, 3379 – 3381.

47 Molander, G. A., George, K. M., Monovich, L. G. *J. Org. Chem.* **2003**, *68*, 9533–9540.

7

Asymmetric Organocatalysis in the Total Synthesis of Complex Natural Products

Gang Zhao[1], *Zheng Qing Ye*[1], *and Xiao Yu Wu*[2]

[1] *Shanghai Institute of Organic Chemistry, Chinese Academy of Sciences, Shanghai, China*
[2] *Department of Chemistry, Shanghai University, Shanghai, China*

7.1 Background

Natural products are the best sources of drugs and leading compounds in drug discovery, as it has been estimated that about 40% of medicines are derived directly or indirectly from natural products.[1] Total syntheses of natural products, especially those of biological activities, have long been an important part of organic chemistry. Almost all natural products possess some degree of complexity, and usually bear more than one chiral center. Therefore, how to introduce that chirality must be considered during the asymmetric total synthesis of these compounds. Over the years, strategies toward chiral target compounds have mostly relied on chiral pools, chiral resolution, and metal-based asymmetric catalysis.[2] At the turn of the new century, organocatalysis emerged as a new and efficient solution for the preparation of chiral targets (Scheme 7.1).[3] Organocatalysis, directly employing small chiral organic molecules as catalysts without involving metals as the central part, has been intensively investigated to date. And several different activation modes, including enamine catalysis, iminium catalysis, hydrogen bond catalysis (Brønsted acid and (thio)urea), Brønsted and Lewis base catalysis, phosphane catalysis, and carbenecatalysis have been developed to promote numerous types of asymmetric transformations.[4] Meanwhile, to demonstrate the practical utility of these organocatalytic reactions, more and more groups have adopted asymmetric organocatalysis as a key step in the total synthesis of natural products and biological active compounds.[5] In this chapter, we will describe some representative applications of asymmetric organocatalysis in total synthesis of relatively complex natural products in the past few years. Here, we do not

Efficiency in Natural Product Total Synthesis, First Edition. Edited by Pei-Qiang Huang, Zhu-Jun Yao, and Richard P. Hsung.

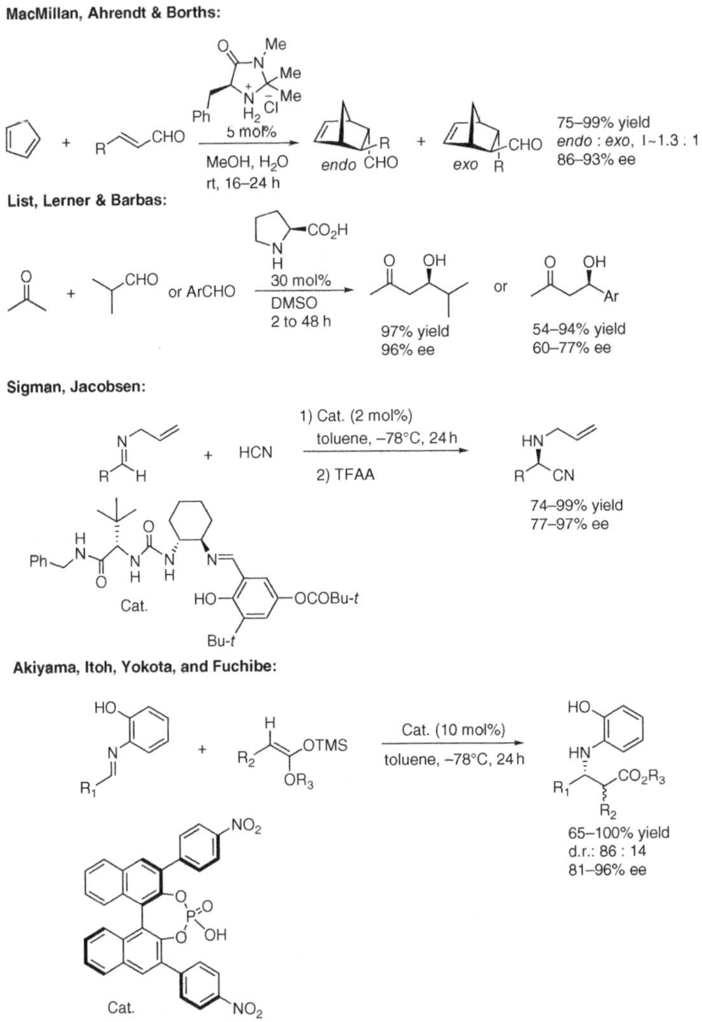

Scheme 7.1 Emerging of organocatalysis at the turn of the new century.

intend to cover all the examples in this field, as a more comprehensive review on this topic has been presented elsewhere.[5]

7.2 Total Synthesis of Alkaloids

7.2.1 Synthesis of (–)-Flustramine B

MacMillan's imidazolinone catalysts are probably one type of the most successful chiral secondary amine catalysts in asymmetric organocatalysis, and have been employed to catalyze various reactions by activation of carbonyl-containing substrates via the formation of either enamine or iminiumion.[4b,c] Furthermore, these catalysts are frequently recruited by MacMillan's group in their total synthesis of certain natural products.

In 2004, MacMillan *et al.* described a synthesis of alkaloid (−)-flustramine B, wherein a novel enantioselective organocatalytic cascade sequence between a tryptamine derivative and acrolein, leading to a chiral pyrroloindoline core, was the key step.[6] The cascade reaction commenced with the addition of tryptamine derivative **4** to iminium ion **3**, which was formed via condensation of catalyst **1** and α,β-unsaturated aldehyde **2**, leading to indoliumion **5** bearing a C(3)-quaternary chiral center (Figure 7.1). This catalytic cycle was driven by the intramolecular addition of an inbuilt nucleophile to the *in situ* generated iminium moiety. Thus, a 5-*exo*-heterocyclization of the pendant ethylamine nitrogen to iminium moiety produced the tricyclic system 7.

Following this strategy, the reaction between 6-bromo tryptamine derivative **8** and acrolein **9** proceeded smoothly under the promotion of imidazolidinone catalyst **1** using *p*-TSA as a cocatalyst, affording pyrroloindoline **10** at a yield of 78% with 90% *ee* (Scheme 7.2). Transformation of a 3-hydroxy propyl side chain into the corresponding prenyl and removal of the Boc protecting group, as well as reductive methylation, were achieved in five additional steps to deliver (−)-flustramine B.

7.2.2 Enantioselective Total Synthesis of (+)-Minfiensine

A strategy similar to the one in Section 7.2.1 was also used successfully to access a more complex molecule, (+)-minfiensine,[7] whose complex structure

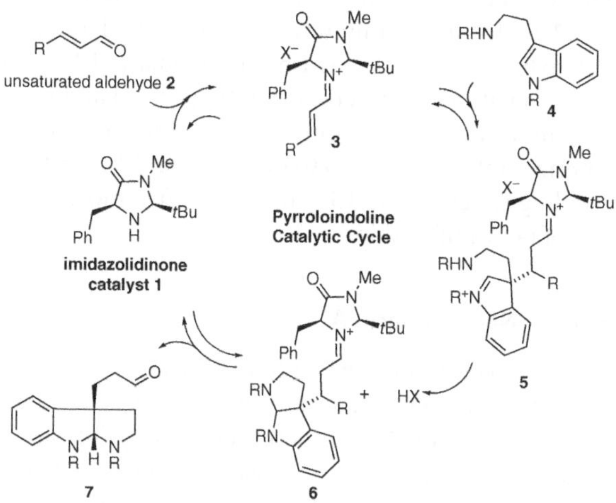

Figure 7.1 Catalytic cycle toward pyrroloindoline.

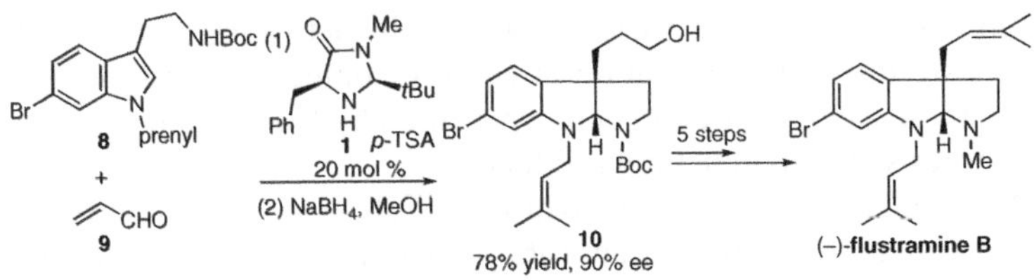

Scheme 7.2 Total synthesis of (–)-flustramine B.

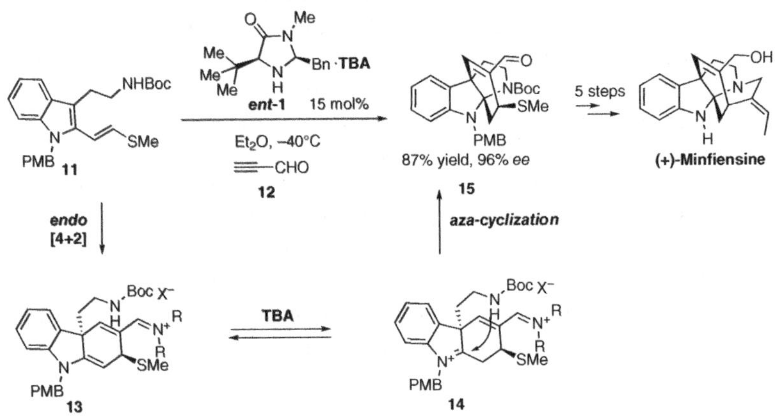

Scheme 7.3 Imidazolidinone catalyzed cascade as a key step in the total synthesis of (+)-minfiensine.

and potential biological activities made it an interesting and challenging target for total synthesis.[8] Toward this end, MacMillan's group developed a novel [4+2]-cyclization reaction cascade (Scheme 7.3). Firstly, catalyst *ent*-**1** condenses with propynal **12** to form an iminium ion, which undergoes a [4+2]-cyclization with 2-vinylindole **11** to furnish intermediate **13**. This cycloaddition reaction proceeds with high stereoselectivity (96% *ee*) and *endo* selectivity due to the rigid conformation of the iminium ion. Then, the counter-acid of the catalyst isomerizes intermediate **13** to iminium ion **14** to initiate a second *exo*-cyclization leading to tetracycle **15**, from which five more steps complete the total synthesis of (+)-minfiensine.

7.2.3 Concise Synthesis of (–)-Nakadomarin A

Hexacyclic alkaloid (–)-nakadomarin A contains an 8/5/5/5/15/6 ring system and four stereogenic centers, including one quaternary, which makes it a challenging and attractive target molecule for the synthetic community. Toward a short and highly stereoselective synthesis of (–)-nakadomarin A,[9] Dixon's group developed a diastereoselective Michael addition using a bifunctional

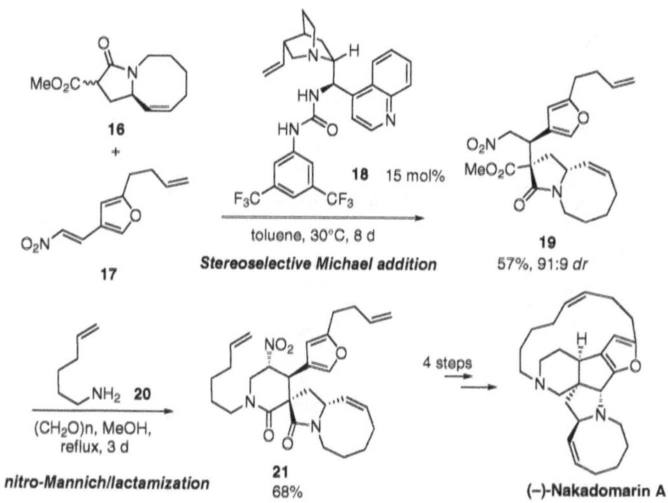

Scheme 7.4 Key steps in the total synthesis of (−)-nakadomarin A.

cinchona catalyst **18**[10] to deliver the desired compound **19** at a good yield as a 10:1 mixture of diastereomers.

Subjection of **19** to a three-component nitro-Mannich/lactamization cascade by following their previously reported protocol[10d,11] facilitated the construction of tetracycle **21** at a 68% yield (Scheme 7.4).

7.2.4 Collective Total Synthesis of Strychnine, Akuammicine, Aspidospermidine, Vincadifformine, Kopsinine, and Kopsanone

By means of organocatalytic cascades and collective natural product synthesis, the MacMillan group succeeded in the preparation of a range of structurally diverse alkaloids from a common scaffold **22**.[12] The alkaloids prepared from scaffold **22** included strychnine, aspidospermidine, vincadifformine, akuammicine, kopsanone, and kopsinine (Figure 7.2).

As shown in Figure 7.3, the key tetracyclic intermediate **22** could be accessed through a one-pot asymmetric Diels–Alder/elimination/conjugate addition organocascade sequence between a simple tryptamine-derived substrate **23** and propargyl aldehyde **12**. At first, exposure of the 2-(vinyl-1-selenomethyl) tryptamine system **23** to the iminium from condensation of propargyl aldehyde **12** with imidazolidinone catalyst **24** (20 mol%), an asymmetric Diels–Alder addition, occurred smoothly. In contrast to their previous studies with methylthio-substituted 2-vinyl indoles as the substrate in the synthesis of minfiensine, the incorporation of methylselanyl group on the diene in this study led to a new polycyclic system. Notably, the cycloadduct **25** could undergo

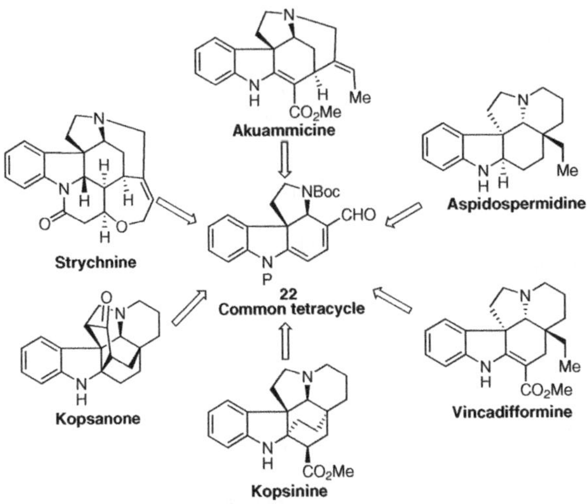

Figure 7.2 Diverse synthesis of alkaloids from a common intermediate tetracycle **22**.

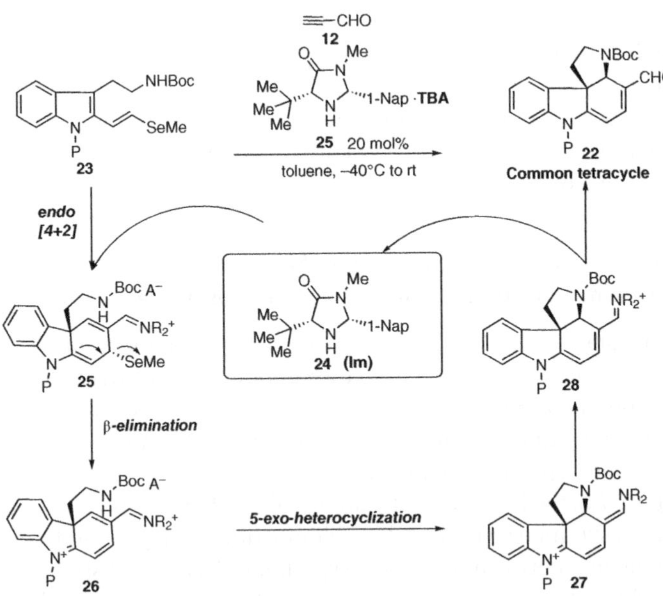

Figure 7.3 Proposed mechanism of an organocascade cycle toward key tetracyclic intermediate **22**.

facile *β*-elimination of methylselanyl group to furnish the unsaturated iminium ion **26**. Then conjugate addition of the pendant carbamate was expected to occur at the *δ*-position to the indolinium ion to deliver the enantio-enriched spiroindoline core **22**.

Key spiroindolines **29**, **32**, and ***ent*-32** were obtained with 1-naphthyl-substituted imidazolidinone catalyst **24** or ***ent*-24** in the presence of tribromoacetic acid at high yields and excellent enantioselectivities, providing a solid basis for the enantioselective synthesis of the six structurally diverse alkaloids mentioned previously (Scheme 7.5–7.7).

7.2.5 Asymmetric Synthesis of (−)-Lycoramine, (−)-Galanthamine, and (+)-Lunarine

The hydrodibenzofuran alkaloids contain a functionalized *cis*-hydrodibenzofuran nucleus with a crucial all-carbon quaternary stereogenic center, which is unique in their core molecular architecture. To meet the synthetic challenge posed by the hydrodibenzofuran alkaloids, Tu, Fan, and coworkers presented an elegant synthetic strategy for the catalytic enantioselective creation of the

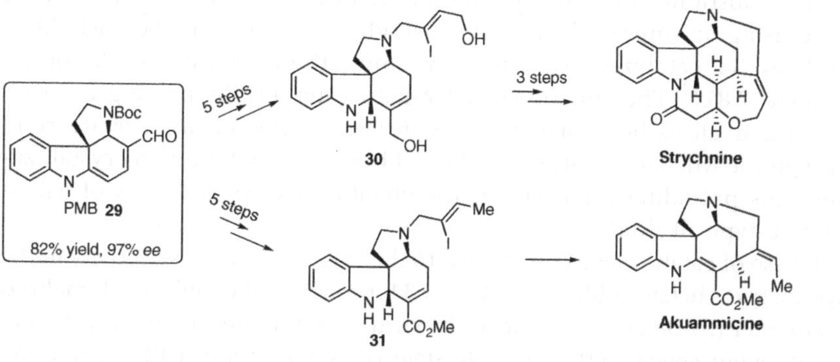

Scheme 7.5 Synthesis of strychnine and akuammicine from common key intermediate **29**.

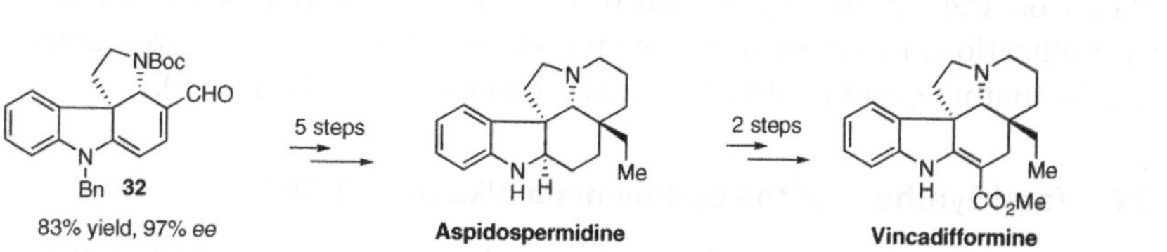

Scheme 7.6 Synthesis of aspidospermidine and vincadifformine from key intermediate **32**.

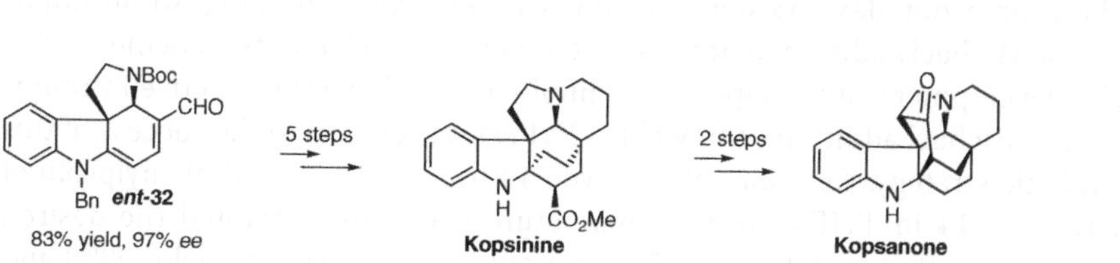

Scheme 7.7 Synthesis of kopsinine and kopsanone from key intermediate ***ent*-32**.

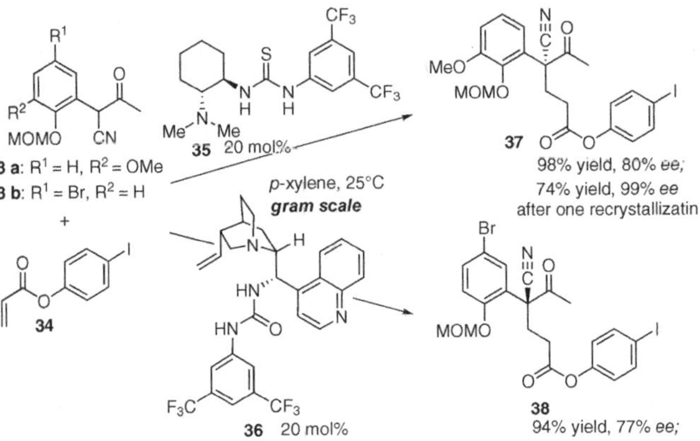

Scheme 7.8 Synthesis of key pseudoenantiomeric synthons **37** and **38**.

key aryl-substituted quaternary carbon center by a direct C–C bond forming tactic using an unprecedented Michael addition of α-cyanoketones **33** to acrylates **34** catalyzed by a tertiary amine-thiourea catalyst **35** or **36** (Scheme 7.8).[13] The functionalized δ-keto ester (*S*)-**38** bearing an aryl-substituted all-carbon quaternary center was obtained as a white solid through the Michael addition of **33a** to **34** in 99% *ee* after one recrystallization, thus providing the basis for the enantioselective synthesis of galanthamine-type alkaloids.

The use of catalyst **36** (20 mol%) led to the reversed enantioselectivity in the asymmetric Michael addition of **33b** to **34**, providing the multifunctionalized δ-keto ester (*R*)-**38** at 77% *ee* and a 94% yield. Notably, the enantiopurity could be further enriched to 99% *ee* by a late-stage recrystallization of **42** (Scheme 7.9) during the synthesis of (+)-lunarine.

Based on the careful exploration of this method as well as an additional recrystallization of compound **42**, enantioselective synthesis of (−)-lycoramine, (−)-galanthamine, and (+)-lunarine was accomplished (Scheme 7.9).

7.2.6 Total Synthesis of the Galbulimima Alkaloid (−)-GB17

The tetracyclic alkaloid GB17 contains a piperidine ring and a *trans*-decalincarbocylic core. Toward a stereoselective synthesis of GB17, the Thomson group developed an efficient synthetic route utilizing two intramolecular Michael additions to forge key stereochemical elements (Scheme 7.10).[14] The development of the diphenylprolinol-catalyzed enantioselective intramolecular Michael addition (99% yield, >20:1 *er*) was crucial to the success of this synthetic strategy. Exposure of aldehyde **43** to 5 mol% of the diphenylprolinol derivative **44** in THF at room temperature smoothly generated the desired *trans*-substituted cyclohexane **45** on a gram scale at excellent yield (99%) and

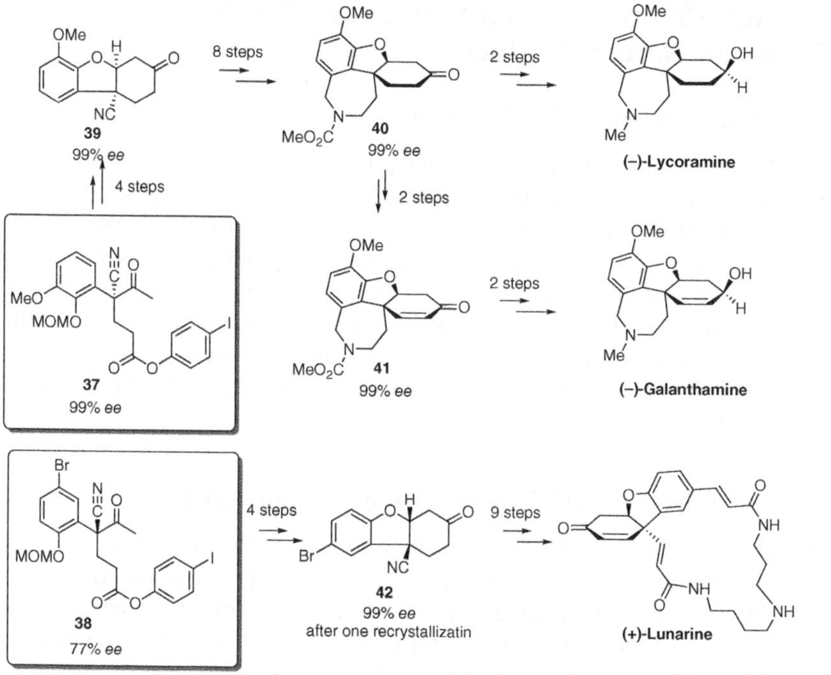

Scheme 7.9 Enantioselective synthesis of (–)-lycoramine, (–)-galanthamine and (+)-lunarine.

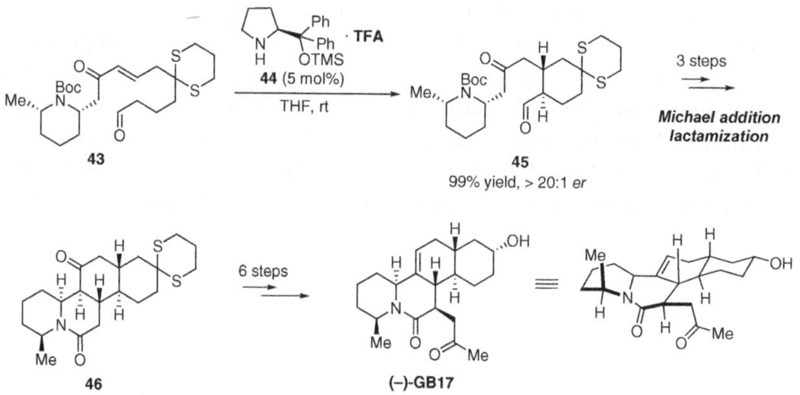

Scheme 7.10 Total synthesis of GB17.

high *er* value (>20:1). Interestingly, poor results were obtained when the intramolecular Michael reaction was conducted in the absence of the dithiane, which, as explained by the authors, might be ascribed to lack of the favorable Thorpe–Ingold effect.

7.3 Total Synthesis of Terpenoids and Related Multicyclic Natural Products

7.3.1 Total Synthesis of (+)-Hirsutene

With 1,4-cyclooctanediones as the substrates, List's group has developed an asymmetric, catalytic transannular aldol reaction that provides polycyclic products at good yields (53–84%) and high enantioselectivities (*er* 95:5–98:2). This methodology culminated in the shortest asymmetric total synthesis of (+)-hirsutene to date (Scheme 7.11).[15] Treatment of diketone **47** by *trans*-4-fluoro proline **48** (10 mol%) in DMSO at room temperature for 15 h, the transannular aldol reaction furnished *cis/anti/cisβ*-hydroxy ketone **49** at a 84% yield and with 98:2 *er*, which further afforded (+)-hirsutene after minor structural modification.

7.3.2 Total Synthesis of (–)-Brasoside and (–)-Littoralisone

Littoralisone features a stereochemically elaborate cyclobutane ring, an adjacent nine-membered lactone, and 14 stereocenters within a 24-carbon framework. In 2005, the MacMillan group accomplished the first total syntheses of littoralisone and brasoside in 13 steps (Scheme 7.12).[16] These syntheses demonstrate the capacity of proline catalysis to (1) overcome the inherent substrate-controlled stereo induction of enamine-Michael reactions to secure the formation of **53** with *cis*-selectivity, and (2) enable the two-step asymmetric construction of a polyol differentiated glucose coupling partner. Firstly, treatment of aldehyde **50** in the proline-catalyzed *R*-formyl oxidation protocol using nitrosobenzene produced the corresponding *R*-oxyamino aldehyde with a high *ee* value. Subsequent olefination of the resultant aldehyde and cleavage of the aminoxy bond provided *γ*-chiral *α,β*-unsaturated ester **51** in one pot. Secondly, exposure of **52** to *L*-proline in DMSO followed by *in situ* acetylation provided the **53** at a yield of 83% and with 10:1 *cis*-selectivity. Interestingly, this

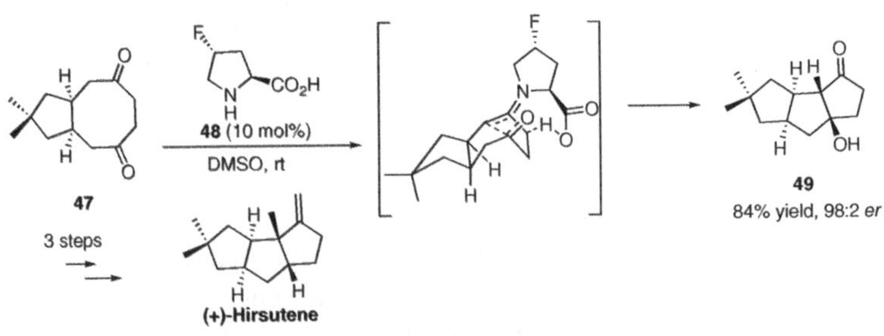

Scheme 7.11 Organocatalytic transannular aldol reaction in total synthesis of (+)-hirsutene.

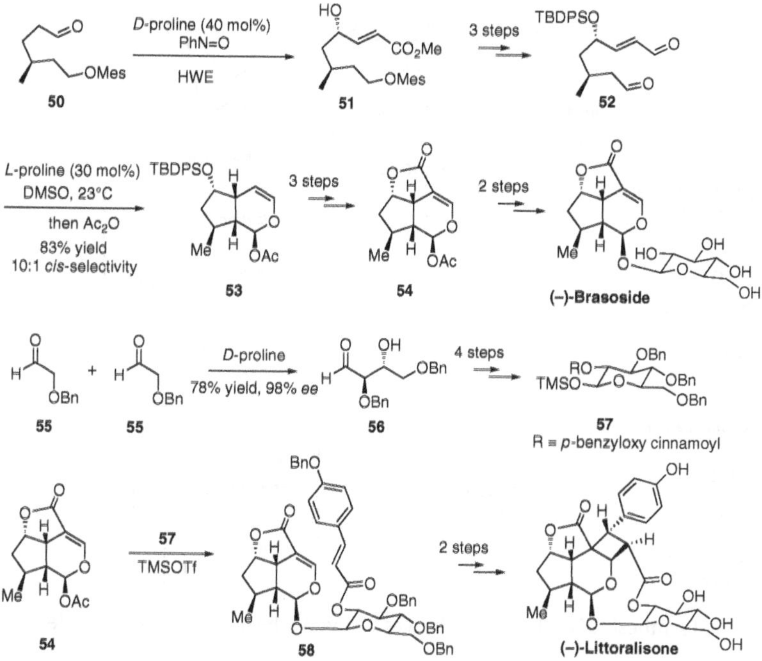

Scheme 7.12 Organocatalytic approach toward (−)-brasoside and (−)-littoralisone.

process can be accelerated either by heat or by the addition of water. Thirdly, preparation of the requisite glycosidic coupling partner was realized by a proline-catalyzed dimerization of benzyloxyacetaldehyde **55** to provide **56** with excellent enantioselectivity (98% *ee*). Furthermore, these syntheses confirm the relative and absolute stereochemistry of (−)-lrasoside and (−)-littoralisone and additionally provide a chemical support for a biosynthetic pathway between these two isolates.

7.3.3 Concise Synthesis of Ricciocarpin A

Ricciocarpin A, as a furanosesquiterpene lactone, exhibits high molluscicidal activity against the water snail *Biomphalariaglabtata*, a vector of schistosomiasis.[17] In 2009, the List group reported an efficient, highly enantioselective and diastereoselective total synthesis of ricciocarpin A.[18] The key feature of the synthesis is a one-pot, three-step, organocatalytic reductive Michael–Tishchenko cascade (Figure 7.4). After screening of reaction conditions, List's group found that a tandem conjugate reduction and Michael cyclization of substrate **60** gave the ketoaldehyde **61** using MacMillan catalyst ***ent*-1**. However, product **61** was obtained predominantly as the undesired *cis*-isomer

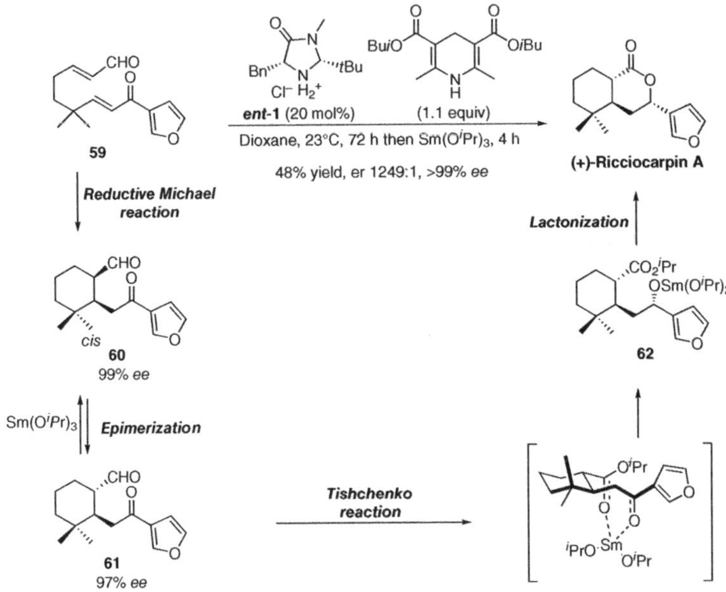

Figure 7.4 Total synthesis of ricciocarpin A by a reductive Michael–Tishchenko cascade.

(2:1, 79% yield) in excellent enantioselectivity (cis-*er* = 99, 99% *ee*, trans-*er* = 57, 97% *ee*). Interestingly, the *cis*-isomer slowly isomerized to give the thermodynamically more stable *trans*-isomer **62**, which underwent a rapid isomerization followed by a highly diastereoselective Tishchenko reaction promoted by $Sm(OiPr)_3$ to furnish the desired natural product (+)-ricciocarpin A as a single *trans*-diastereomer in excellent enantioselectivity (*er* = 249, 99% *ee*) at a 48% yield.

7.3.4 Total Synthesis and Absolute Stereochemistry of Seragakinone A

Seragakinone A features a densely oxygenated pentacyclic core containing an angular prenyl substituent. The Keisuke–Suzuki group reported the asymmetric total synthesis and confirmed absolute stereochemistry of Seragakinone A.[19] Noteworthy features of their synthesis include the benzoin-forming reactions, which worked well to effect two separate cyclizations, including one highly enantioselective reaction using the modified Rovis triazolium salt **64** (86% yield, 99% *ee*), and a Shi asymmetric epoxidation to afford the desired stereocenters (Scheme 7.13). First, when using the modified Rovis triazolium salt **64** (10 mol%) in the presence of triethylamine (10 mol%), the key benzoin

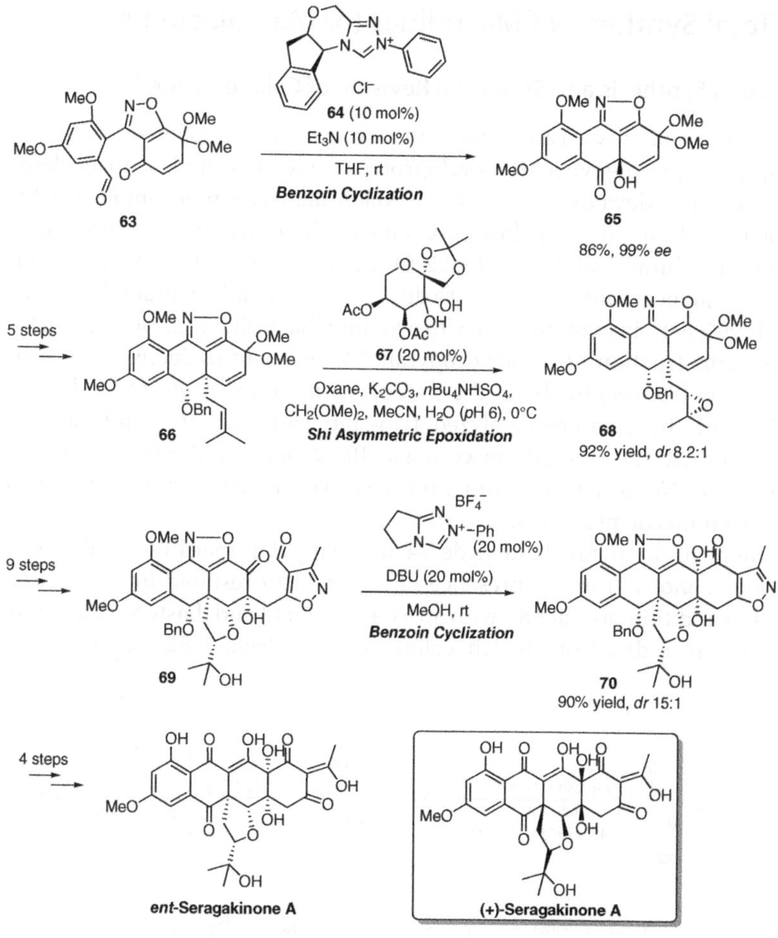

Scheme 7.13 Total synthesis and confirmation of absolute stereochemistry of seragakinone A.

cyclization of ketoaldehyde **63** proceeded with excellent enantioselectivity to give the cyclic ketol (*R*)-**65** at a high yield (86%, 99% *ee*). Second, Shi epoxidation catalyzed by ketonehydrate **67** (20 mol%) furnished the desired epoxide **68** in good diastereoselectivity (92% yield, d.r. 8.2:1). It is worth mentioning that benzoin cyclization of **69** proceeded smoothly using achiral triazoliumsalt (20 mol%) as a catalyst in the presence of DBU (20 mol%) in MeOH(rt, 30 min), to give ketol **70** at a 90% yield with excellent diastereoselectivity (d.r. 15:1), which indicates that the diastereoselectivity in this intramolecular reaction is entirely substrate controlled.

7.4 Total Synthesis of Macrolides (or Macrolactams)

7.4.1 Total Synthesis and Structural Revision of Callipeltoside C

Callipeltoside contains a 12-membered macrocycle comprising seven stereocenters, a unique dienynechlorocyclopropane sidechain and a novel deoxysugar 2-O-methylevalose. In 2008, the MacMillan group accomplished the first total synthesis of callipeltoside C and made a structural revision with respect to the chirality of the pendent 2-O-methylevalose carbohydrate.[20] This synthetic sequence features the use of a direct aldehyde-aldehyde aldol coupling (d.r. 12:1, 99% *ee*)[21] in combination with a Semmelhack alkoxycarbonylation for the rapid construction of tetrahydropyran **74**, an enantioselective formyl α-oxyamination (99% *ee*) to obtain the stereogenicity of the protected iodoalcohol 77[22] and the enamine-catalyzed two-step carbohydrate synthesis[23] to rapidly assemble the desired deoxy sugar **80** (99% *ee*) and ***ent*-80** (99% *ee*) (Scheme 7.14). Notably, proline functioned as an optimal organocatalyst for all of these asymmetric processes.

Treatment of the pyranyl aldehyde **74** with vinyl Grignard (derived *in situ* from iodoalcohol 77) in the presence of $MgBr_2{\cdot}Et_2O$ resulted in the desired *anti*-Felkin addition product **81** with an excellent yield and diastereoselectivity (98% yield, 16:1 dr), from which callipeltoside aglycone **82** was obtained

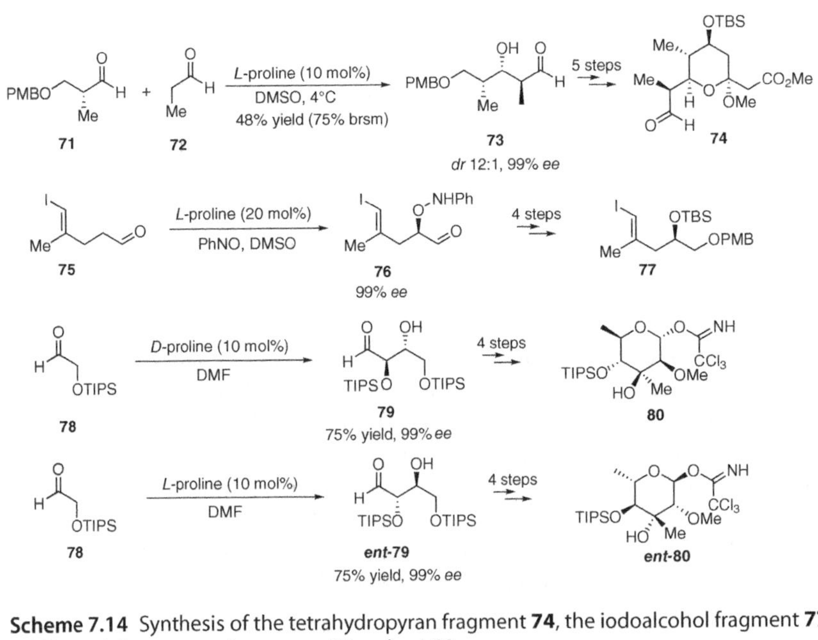

Scheme 7.14 Synthesis of the tetrahydropyran fragment **74**, the iodoalcohol fragment **77**, a differentially protected mannose **80** and *ent*-**80**.

successfully in nine additional steps. The coupling of the *D*-2-O-methylevalose hexose **80** and the aglycone **82** gave the reported structure of callipeltoside C; however, its spectra were inconsistent with those reported for the isolated natural callipeltoside C. Coupling the *L*-antipode of the callipeltose C sugar (***ent*-80**) with the macrocyclic core **82** produced the macrolide whose spectroscopic data were identical to those of the natural isolate (Scheme 7.15).

7.4.2 Total Synthesis of (+)-Cytotrienin A

Cytotrienin A comprises an *E,E,E*-triene motif within a 21-membered lactam, which also contains four chiral centers, including three contiguous chiral centers.

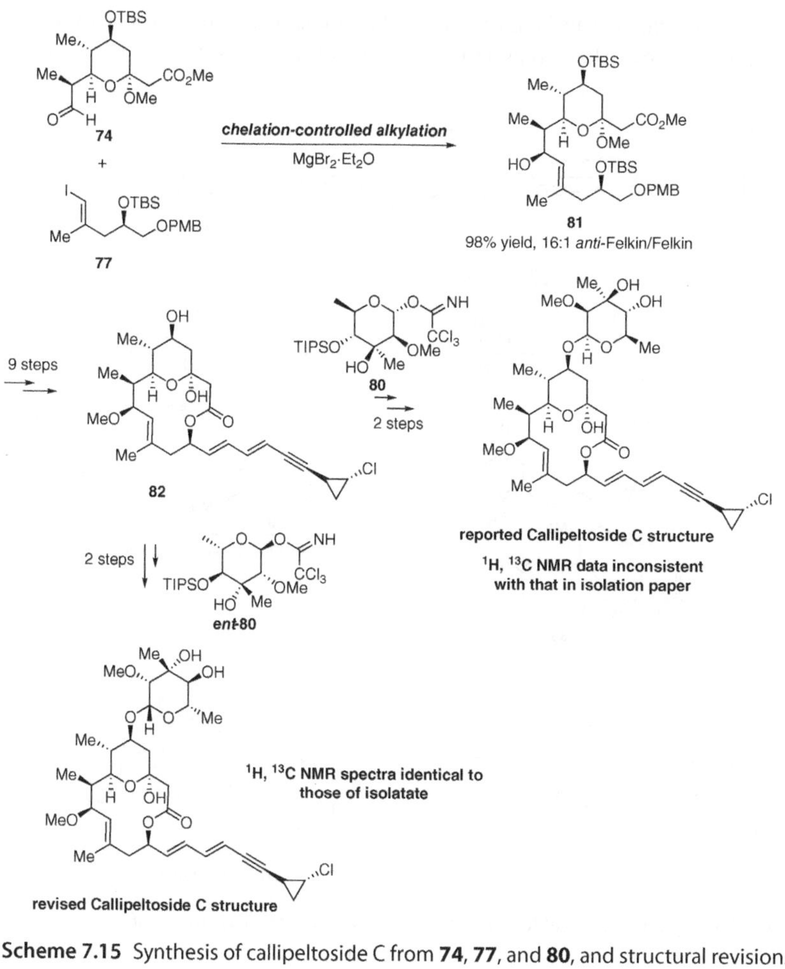

Scheme 7.15 Synthesis of callipeltoside C from **74**, **77**, and **80**, and structural revision.

To achieve the asymmetric total synthesis of (+)-cytotrienin A, the Hayashi group built up the stereocenters via an aldol reaction catalyzed by *L*-proline or **87** derived from 4-hydroxyproline (Scheme 7.16).[24] Firstly, proline-mediated α-aminoxylation of aldehyde **83** proceeded efficiently to provide **84**, which went all the way to afford carboxylic acid **85** in nine steps. More importantly, diol **88** could be obtained at a good yield, good d.r., and excellent *ee* (77% yield, *anti/syn* = 6.2:1, 96% *ee*) by using a surfactant-proline conjugated catalyst **87** without any solvent. Condensation between fragment **89** and carboxylicacid **85** in the presence of BOP-Cl afforded **91** at a yield of 79%. Finally, RCM forged the triene system and thus completed the total synthesis of (+)-cytotrienin A.

7.4.3 Total Synthesis of Diazonamide A

Through a seminal synthesis of the original proposed structure of diazonamide A, Harran and coworkers reassigned the structure in 2001.[25,26] The correct

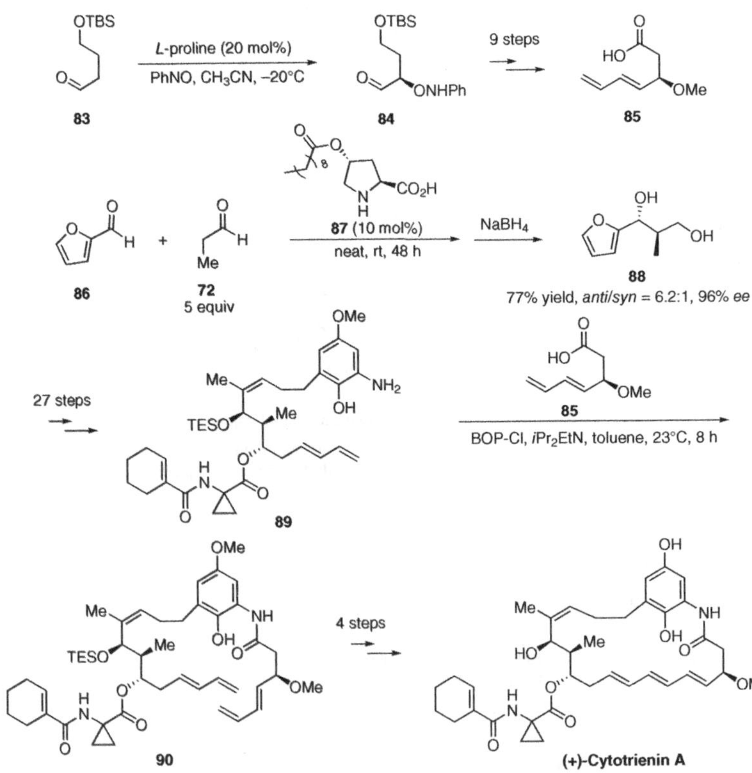

Scheme 7.16 Organocatalyticaldol reactions as key steps in total synthesis of (+)-cytotrienin A.

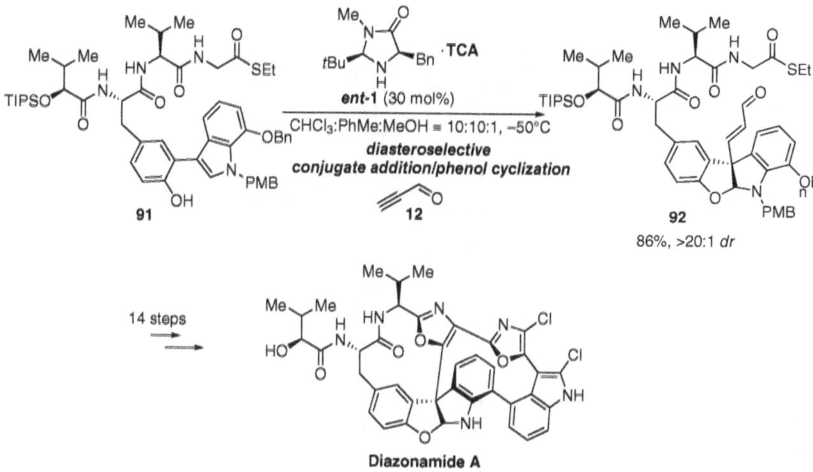

Scheme 7.17 Construction of the furanoindoline core for diazonamide A by organocatalysis.

structure of diazonamide A contains two 12-membered macrocycles that are connected by a fully functional quaternary carbon stereocenter embedded in an unusual furanoindoline core, which make it an exceptionally compact and rigid framework. During the enantioselective total synthesis of diazonamide A (Scheme 7.17), the MacMillan group developed a highly stereoselective organocatalytic conjugate addition/phenol cyclization cascade to construct the C(10) quaternary center and the central furanoindolinecore **92** (d.r. > 20:1).[27] Treatment of **91** with 30 mol% of imidazolidinone catalyst ***ent*-1·TCA** and excessive propynal **12** in the mixture of toluene, chloroform, and methanol delivered **92** at a 86% isolated yield and >20:1 d.r. after 24 h at −50 °C. Notably, use of racemic ***ent*-1·TCA** afforded a 1:1 mixture of diastereomers at C(10), indicating that the stereoselectivity is entirely catalyst controlled. Moreover, this example showed that complex substrate such as **91** could be employed as a reaction partner in organocatalysis for synthesis of challenging synthetic targets.

7.5 Total Synthesis of Peptide Natural Products

7.5.1 Total Synthesis of Chloptosin

Chloptosin consists of a biaryl motif connecting two identical cyclic hexapeptide subunits. Moreover, the two subunits contain exclusively nonproteinogenic amino acids in alternating enantiomeric forms, such as the unusual (*R*)- and (*S*)-piperazic acids, and a 6-chloropyrroloindole residue.[28,29] Toward the efficient preparation of the embedded piperazic acids, Ley's group has developed an organocatalytic, one-pot procedure for the enantioselective

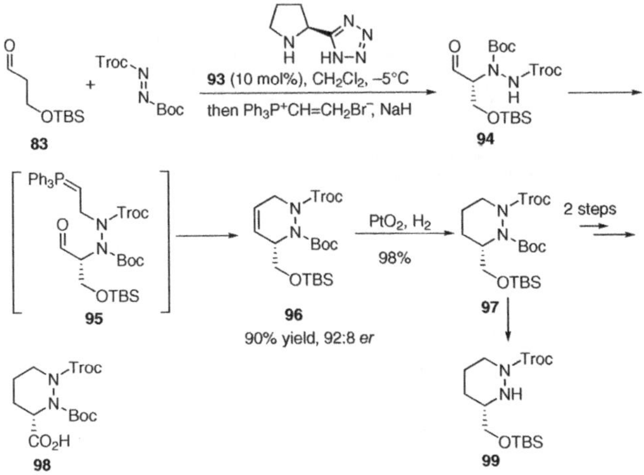

Scheme 7.18 Organocatalysis as key step in preparation of coupling partners **99** and **100**.

synthesis of 3,6-dihydropyridazines **96** from achiral aldehyde **83** using tetrazole catalyst **93**.[30,31] This process proceeds at a high yield (90%) and *er* value (92:8) through α-amination, a base-promoted conjugate addition to a vinylphosphonium salt, and a subsequent Wittig ring-closure to furnish desired 3,6-dihydropyridazine **97** as a sole product (Scheme 7.18). In this procedure, a differentially substituted (Boc and Troc) azodicarboxylate is essential to generate orthogonally protected piperazic acid building blocks. Additionally, ***ent*-98** and ***ent*-99** were obtained in similar way by using the corresponding tetrazole catalyst ***ent*-94**.

Coupling between reaction partners **99** and **100** performed well with no loss of enantiopurity. Thus, amidation of *ent*-**100** with piperazic acid **99** by employing Ghosez's reagent led to the desired dipeptide **100** at a 71% yield, which was transformed to tetrapeptide fragment **101** in seven steps, ready for the total synthesis of chloptosin (Scheme 7.19).

7.6 Summary of the Key Reactions and Tactics

Over the past decade, organocatalysis has witnessed a rapid growth and become increasingly important for the total synthesis of natural products. As we move into the future, organocatalysis will act as an efficient synthetic strategy providing access to myriad complex molecular architectures. In summary, there are several important synthetic applications in the field of organocatalytic synthesis of natural products. First, it provides chiral building blocks for natural product synthesis from simple starting materials via an aldol reaction, α-amination, α-oxyamination, Michael addition, and so on. Second,

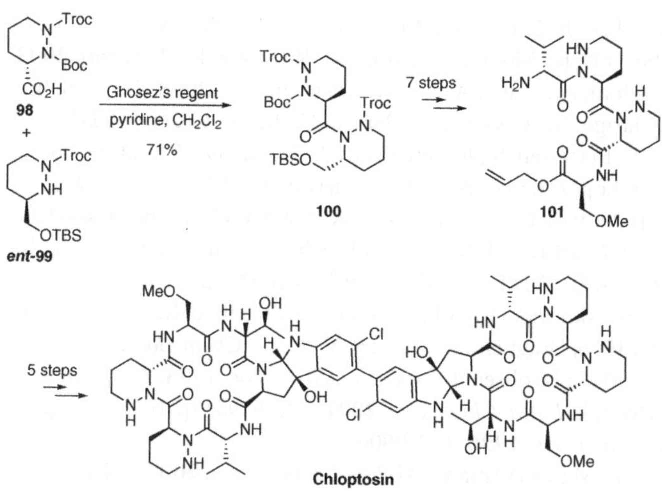

Scheme 7.19 Preparation of tetrapeptide fragment **101**.

organocatalytic reactions are employed to forge the complex frameworks for many challenging targets, toward which the traditional methods usually fail. Last, organocatalytic cascade reactions have been utilized in the total synthesis of natural products in an atom-, step-, and redox-economic fashion, which fits well to the end of "ideal synthesis." With the various recently developed modes of organocatalytic activation (iminium catalysis, enamine catalysis, hydrogen-bonding catalysis, SOMO catalysis, counterion catalysis) in the repertoire, alongside the concept of bifunctional organocatalysts and multicatalytic systems, numerous novel cascade sequences come into being and will find important applications in the total synthesis of natural products.

References

1 (a) Newman, D. J.; Cragg, G. M.; Snader, K. M. *J. Nat. Prod.* **2003**, *66*, 1022; (b) Harvey A. *Drug Discov. Today* **2000**, *5*, 294.

2 (a) Jacobsen, E. N.; Pfaltz, A.; Yamamoto, H. (Eds) *Comprehensive Asymmetric Catalysis*, 2nd edn, Berlin-Heidelberg: Springer-Verlag, **2004**; (b) Caprio, V.; Williams, J. M. J. (Eds) *Catalysis in Asymmetric Synthesis*, 2nd edn, Oxford: Wiley-Blackwell, **2009**; (c) Lin, G. Q.; Li, Y.-M; Chan, A. S. C. *Principles and Applications of Asymmetric Synthesis*, Chichester, UK: John Wiley & Sons, Ltd, **2001**.

3 (a) Ahrendt, A.; Borths, C. J.; MacMillan, D. W. C. *J. Am. Chem. Soc.* **2000**, *122*, 4243; (b) List, B.; Lerner, R. A.; Barbas III, C. F. *J. Am. Chem. Soc.* **2000**, *122*, 2395.

4 (a) Dalko, P. I.; Moisan, L. *Angew. Chem. Int. Ed.* **2004**, *43*, 5138; (b) Erkkilä, A.; Majander, I.; Pihko, P. M. *Chem. Rev.* **2007**, *107*, 5416; (c) Mukherjee, S.; Yang, J.

W.; Hoffmann, S.; List, B. *Chem. Rev.* **2007**, *107*, 5471; (d) MacMillan, D. W. C. *Nature* **2008**, *455*, 304; (e) Mohr, J. T.; Krout, M. R.; Stoltz, B. M. *Nature* **2008**, *455*, 323; (f) Dondoni, A.; Massi, A. *Angew. Chem. Int. Ed.* **2008**, *47*, 4638; (g) Melchiorre, P.; Marigo, M.; Carlone, A.; Bartoli, G. *Angew. Chem. Int. Ed.* **2008**, *47*, 6138; (h) Bertelsen, S.; Jørgensen, K. A. *Chem. Soc. Rev.* **2009**, *38*, 2178; (i) Marqués-López, E.; Herrera, R. P.; Christmann, M. *Nat. Prod. Rep.* **2010**, *27*, 1138; (j) Busacca, C. A.; Fandrick, D. R.; Song, J.-H. J.; Senanayake, C. H. *Adv. Synth. Catal.* **2011**, *353*, 1825; (k) Albrecht, Ł.; Jiang, H.; Jørgensen, K. A. *Angew. Chem. Int. Ed.* **2011**, *50*, 8492; (l) Pellissier, H. *Adv. Synth. Catal.* **2012**, *354*, 237; (m) Akiyama, T.; Itoh, J.; Yokota, K.; Fuchibe, K. *Angew. Chem. Int. Ed.* **2004**, *43*, 1566; Uraguchi, D.; Terada, M. *J. Am. Chem. Soc.* **2004**, *126*, 5356; (o) Sigman, M. S.; Jacobsen, E. N. *J. Am. Chem. Soc.* **1998**, *120*, 4901; (p) Tian, S. K.; Deng, L. *J. Am. Chem. Soc.* **2001**, *123*, 6195; (q) Tu, Y.; Wang, Z. X.; Shi, Y. *J. Am. Chem. Soc.* **1996**, *118*, 9806.

5 (a) de Figueiredo, R. M.; Christmann, M. *Eur. J. Org. Chem.* **2007**, 2575; (b) Marqués-López, E.; Herrera, R. P.; Christmann, M. *Nat. Prod. Rep.* **2010**, *27*, 1138.

6 Austin, J. F.; Kim, S.-G.; Sinz, C. F.; Xiao, W.-J.; MacMillan, D. W. C. *Proc. Natl Acad. Sci. USA* **2004**, *101*, 5483.

7 Jones, S. B.; Simmons, B.; MacMillan, D. W. C. *J. Am. Chem. Soc.* **2009**, *131*, 13606.

8 Massiot, G., Thepenier, P., Jacquier, M.-J., Le Men-Olivier, L.; Delaude, C. *Heterocycles* **1989**, *29*, 1435.

9 Jakubec, P.; Cockfield, D. M.; Dixon, D. J. *J. Am. Chem. Soc.* **2009**, *131*, 16632.

10 (a) Ye, J.; Dixon, D. J.; Hynes, P. S. *Chem. Commun.* **2005**, 4481; (b) McCooey, S. H.; Connon, S. J. *Angew. Chem. Int. Ed.* **2005**, *44*, 6367; (c) Okino, T.; Hoashi, Y.; Takemoto, Y. *J. Am. Chem. Soc.* **2003**, *125*, 12672; (d) Jakubec, P.; Halliwell, M.; Dixon, D. J. *Org. Lett.* **2008**, *10*, 4267.

11 Hynes, P.; Stupple, P. A.; Dixon, D. J. *Org. Lett.* **2008**, *10*, 1389.

12 Jones, S. B.; Simmons, B.; Mastracchio, A.; MacMillan, D. W. C. *Nature* **2011**, *475*, 183.

13 Chen, P.; Bao, X.; Zhang, L.-F.; Ding, M.; Han, X.-J.; Li, J.; *et al. Angew. Chem. Int. Ed.* **2011**, *50*, 8161.

14 Larson, R. T.; Clift, M. D.; Thomson, R. J. *Angew. Chem. Int. Ed.* **2012**, *51*, 2481.

15 Chandler, C. L.; List, B. *J. Am. Chem. Soc.* **2008**, *130*, 6737.

16 Mangion, I. K.; MacMillan, D. W. C. *J. Am. Chem. Soc.* **2005**, *127*, 3696.

17 (a) Wurzel, G.; Becker, H. *Phytochemistry* **1990**, *29*, 2565; (b) Wurzel, G.; Becker, H.; Eicher, T.; Tiefensee, K. *Planta Med.* **1990**, *56*, 444; (c) Zinsmeister, H. D.; Becker, H.; Eicher, T. *Angew. Chem. Int. Ed.* **1991**, *30*, 130.

18 Michrowska, A.; List, B. *Nature Chem.* **2009**, *1*, 225.

19 Takada, A.; Hashimoto, Y.; Takikawa, H.; Hikita, K.; Suzuki, K. *Angew. Chem. Int. Ed.* **2011**, *50*, 2297.

20 Carpenter, J.; Northrup, A. B.; Chung, de M.; Wiener, J. J. M.; Kim, S.-G.; MacMillan, D. W. C. *Angew. Chem. Int. Ed.* **2008**, *47*, 3568.

21 Northrup, A. B.; MacMillan, D. W. C. *J. Am. Chem. Soc.* **2002**, *124*, 6798.

22 (a) Brown, S. P.; Brochu, M. P.; Sinz, C. J.; MacMillan, D. W. C. *J. Am. Chem. Soc.* **2003**, *125*, 10808; (b) Zhong, G. *Angew. Chem. Int. Ed.* **2003**, *42*, 4247.

23 (a) Northrup, A. B.; Mangion, I. K.; Hettche, F.; MacMillan, D. W. C. *Angew. Chem. Int. Ed.* **2004**, *43*, 2152; (b) Northrup, A. B.; MacMillan, D. W. C. *Science* **2004**, *303*, 1752.

24 Hayashi, Y.; Shoji, M.; Ishikawa, H.; Yamaguchi, J.; Tamura, T.; Imai, H.; *et al. Angew. Chem. Int. Ed.* **2008**, *47*, 6657.

25 (a) Lindquist, N.; Fenical, W.; Van Duyne, G. D.; Clardy, J. *J. Am. Chem. Soc.* **1991**, *113*, 2303; (b) Fernandez, R.; Martin, M. J.; Rodriguez-Acebes, R.; Reyes, F.; Francesch, A.; Cuevas, C. *Tetrahedron Lett.* **2008**, *49*, 2282.

26 (a) Li, J.; Jeong, S.; Esser, L.; Harran, P. G. *Angew. Chem. Int. Ed.* **2001**, *40*, 4765; (b) Li, J.; Burgett, A. W. G.; Esser, L.; Amezcua, C.; Harran, P. G. *Angew. Chem. Int. Ed.* **2001**, *40*, 4770.

27 Knowles, R. R.; Carpenter, J.; Blakey, S. B.; Kayano, A.; Mangion, I. K.; Sinz, C. J.; *et al. Chem. Sci.* **2011**, *2*, 308.

28 Oelke, A. J.; France, D. J.; Hofmann, T.; Wuitschik, G.; Ley, S. V. *Angew. Chem. Int. Ed.* **2010**, *49*, 6139.

29 Oelke, A. J.; Antonietti, F.; Bertone, L.; Cranwell, P. B.; France, D. J.; Goss, R. J. M. *et al. Chem. Eur. J.* **2011**, *17*, 4183.

30 Oelke, A. J.; Kumarn, S.; Longbottom, D. A.; Ley, S. V. *Synlett* **2006**, *2548.*

31 Kumarn, S.; Oelke, A. J.; Shaw, D. M.; Longbottom, D. A.; Ley, S. V. *Org. Biomol. Chem.* **2007**, *5*, 2678.

8

Multicomponent Reactions in Natural Product Synthesis

Michael Smietana[1], Erica Benedetti[2], Cyril Bressy[3], and Stellios Arseniyadis[4]

[1] *Université de Montpellier, Montpellier, France*
[2] *Université Paris Descartes, Paris, France*
[3] *Aix Marseille Université, Marseille, France*
[4] *Queen Mary University of London, London, UK*

8.1 Background

Multicomponent reactions, usually referred to as MCRs, are particularly attractive as they allow easy and rapid access to large libraries of organic compounds in a one-pot fashion. Step-economical, this approach minimizes the number of synthetic operations and offers the possibility to build large libraries through high-throughput library-screening optimization. The concept itself in which three or more components are combined together in a single reaction vessel to produce multiple bond-forming events is truly bio-inspired as exemplified by the prebiotic synthesis of adenine from five molecules of HCN (Figure 8.1).

While some important MCRs were reported more than a century ago (see Strecker's synthesis of α-amino acids, 1850, Hantzsch's synthesis of dihydropyridines, 1882, Biginelli's synthesis of dihydropyrimidines, 1893), the general interest in MCRs has considerably increased in the past decade as witnessed by the exponential growth in the number of published papers.[1–9]

Obviously, the content of this chapter is not meant to cover all the literature in the field but to highlight important advances and strategies that were used in the total synthesis of various complex natural products.

Efficiency in Natural Product Total Synthesis, First Edition. Edited by Pei-Qiang Huang, Zhu-Jun Yao, and Richard P. Hsung.

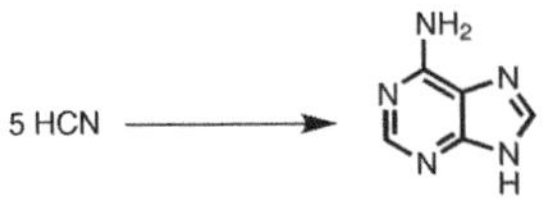

Figure 8.1 Prebiotic synthesis of adenine.

8.2 Multicomponent Reactions in Natural Product Synthesis

8.2.1 Synthesis of Martinelline by Powell and Batey

Martinelline **1** and martinellic acid **2** (Figure 8.2) bear a distinctive core constituted by a tetrahydroquinoline fused with a pyrrolidine ring and three pendant isoprenyl-derived guanidine moieties. Isolated in 1995 from the roots of the tropical plant *Martinella iquitosensis,* these two alkaloids exhibit potent bradykinin B antagonist activities.[10,11]

Attracted by its unique structures and activities, several research groups investigated a variety of synthetic methods for their preparation.[12–23] Among them, Batey and Powell devised an unprecedented three-component Povarov hetero Diels–Alder reaction to access the hexahydropyrrolo[3,2-c] quinoline core.

This Povarov reaction (Scheme 8.1) is an inverse electron-demand hetero-Diels–Alder between an *N*-arylimine and an electron-rich olefin.[24] The resulting [4+2] cycloaddition products generally obtained under acidic catalysis are tetrahydroquinolines, which can further be converted to substituted quinolines. Since the *N*-arylimine can be prepared *in situ* from an aniline and an aldehyde,

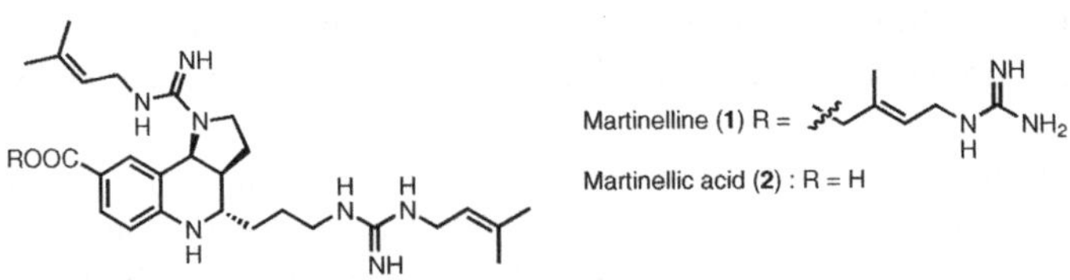

Figure 8.2 Structure of martinelline and martinellic acid.

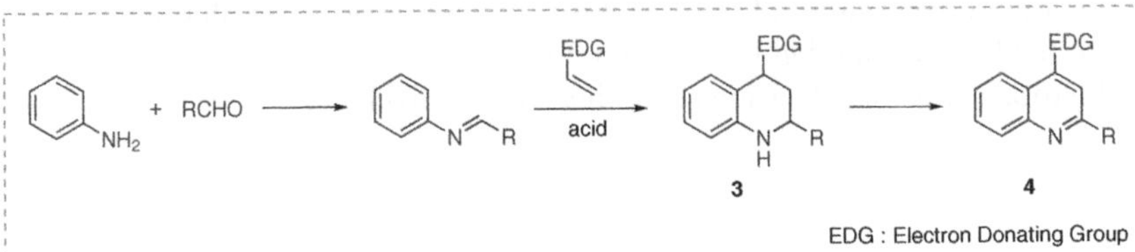

Scheme 8.1 General scheme for the Povarov reaction.

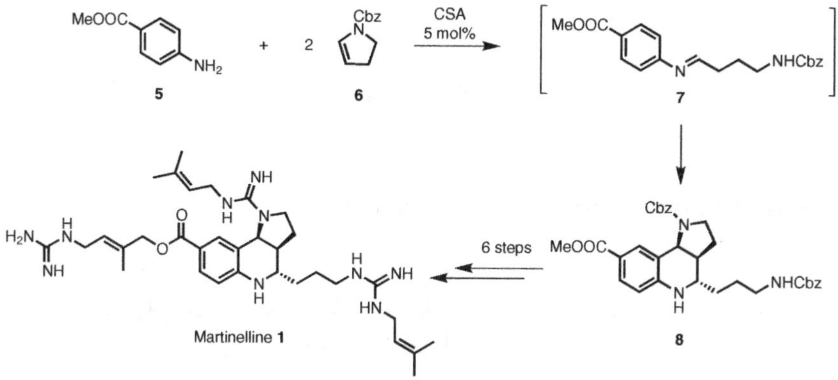

Scheme 8.2 Synthesis of martinelline by Powell and Batey.[28]

the Povarov reaction can potentially be performed in a one-pot fashion thus becoming an efficient three-component reaction (3-MCR).[25–27]

Batey and Powell used a modified Povarov reaction using two equivalents of enamine **6** serving here both as an aldehyde equivalent and as a dienophile to access the martinelline core structure.[28] The reaction was best performed in the presence of camphorsulfonic acid, leading to the aromatic imine intermediate **7**. A [4 + 2] cycloaddition followed by a rearomatization produced a 11:89 mixture of diastereomers in favor of the desired *exo*-cyclic adduct **8** (Scheme 8.2). After separation of the diastereomers by flash column chromatography, only six additional steps were required to complete the synthesis of martinelline **1**, which was achieved in nine steps and a 10% overall yield.

8.2.2 Synthesis of Eurystatin by Schmidt and Weinbrenner

Eurystatins A (**9**) and B (**10**) (Figure 8.3) are two prolyl endopeptidase inhibitors bearing a 13-membered macrocyclic core isolated from *Streptomyces eurythermus* R353–21.[29,30] The first total synthesis of this peptidomimetic compound constituted of leucine, ornithine, and α-ketoalaninamide subunits was reported as early as 1994 by Schmidt and Weinbrenner employing elegantly the Passerini reaction.[31]

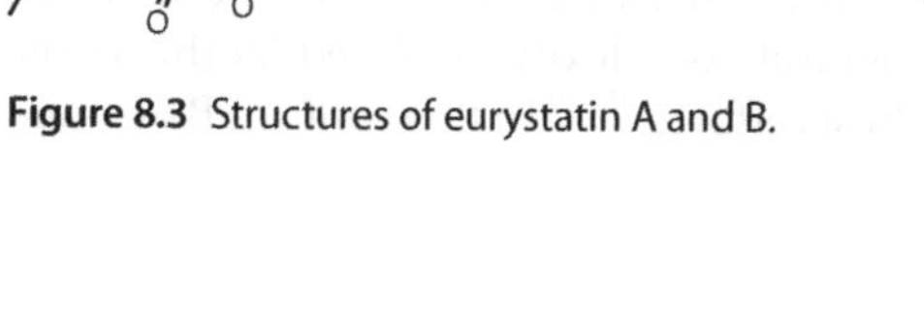

Figure 8.3 Structures of eurystatin A and B.

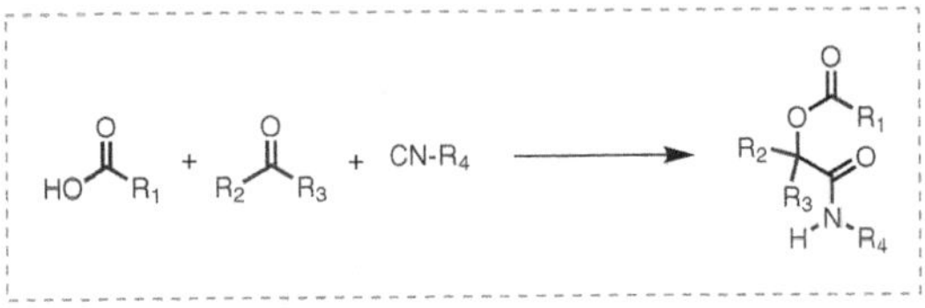

Scheme 8.3 General scheme for the Passerini reaction.

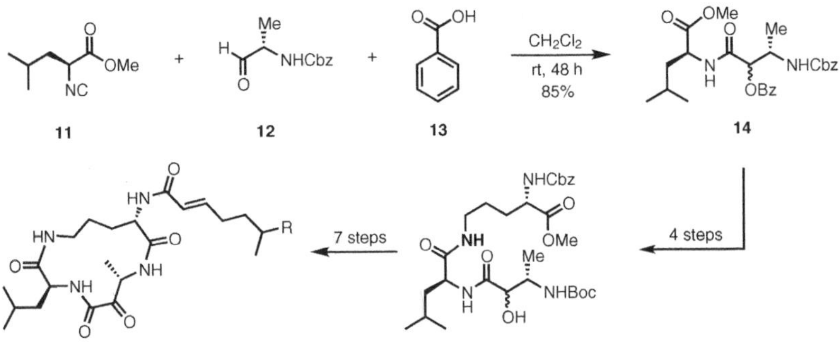

Scheme 8.4 Synthesis of eurystatin A by Schmidt and Weinbrenner.

The Passerini reaction is a three-component reaction combining a carboxylic acid, a carbonyl compound, and an isocyanide to access the corresponding α-acyloxycarboxamides in one step (Scheme 8.3).[32,33]

In their synthesis, Schmidt and Weinbrenner used methyl (*S*)-2-isocyano-4-methylpentanoate **11** with (*S*)-*Z*-alaninal **12** and benzoic acid **13** to obtain key fragment **14** as a diastereoisomeric mixture (Scheme 8.4). The total synthesis was eventually completed after a series of steps including the formation of a pentafluorophenyl ester intermediate for the peptide ring closure. Overall, the total synthesis of eurystatin A required 12 steps for a 12% overall yield.

Interestingly, in 2001, Semple *et al.* described a Passerini reaction/*N*-deprotection/acyl migration strategy to construct the acyclic skeleton of eurystatins, thus allowing to shorten its synthesis (Scheme 8.5).[34]

8.2.3 Synthesis of Motuporin by Bauer and Armstrong

Motuporin, a cyclic pentapeptide from the sponge *Theonella swinhoei* (Figure 8.4) was isolated in 1992 by Andersen and coworkers. It is one of the most potent inhibitors of phosphatase 1 known, inhibiting at subnanomolar concentrations and showing strong *in vitro* cytotoxicity against a number of cancer cell lines.[35] The biological activity and the structural complexity exhibited by this cyclic peptide stimulated many total synthesis efforts.[36–39] In this context, Bauer and

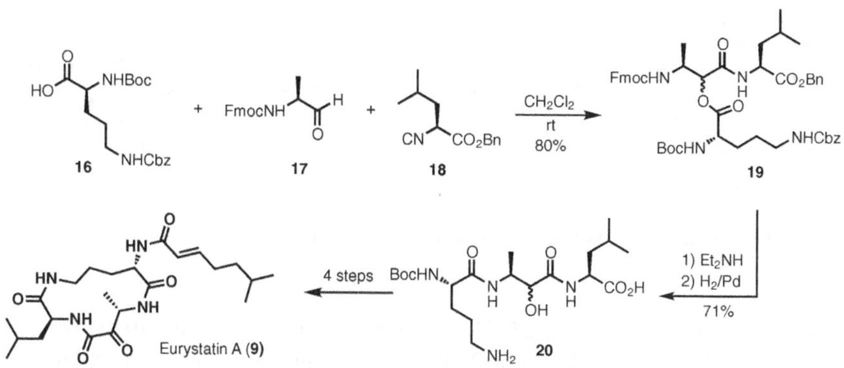

Scheme 8.5 Synthesis of eurystatin A by Semple, Owens, and coworkers.

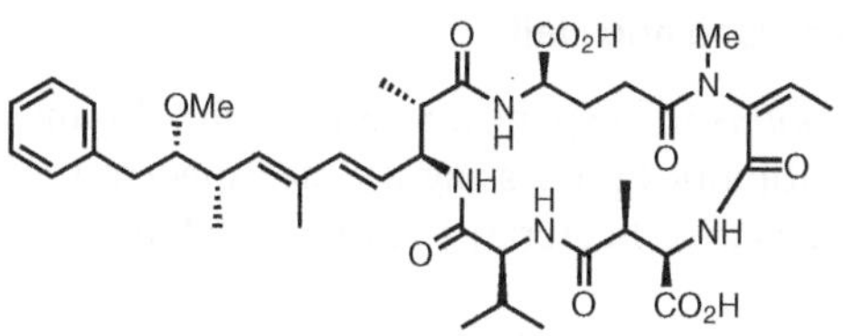

Motuporin (**21**)

Figure 8.4 Structures of motuporin.

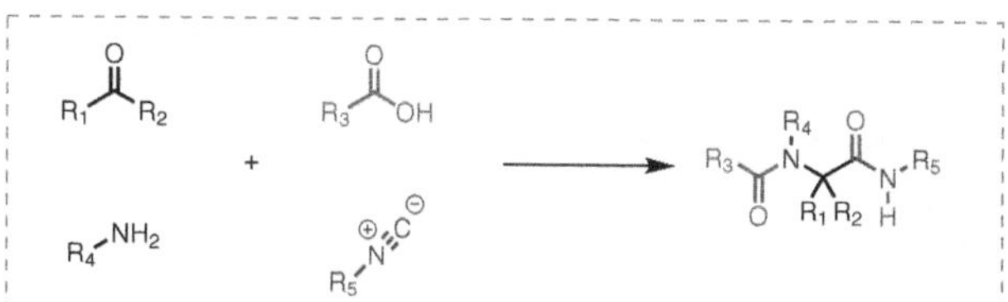

Scheme 8.6 General scheme of the Ugi reaction.

Armstrong synthesized the cyclic core of motuporin by means of a particularly elegant Ugi four-component reaction (U-4CR).

First published in 1959 by Ivar Karl Ugi, the Ugi reaction is a multicomponent reaction involving a carbonyl derivative, an amine, an isocyanide and a carboxylic acid to form a bis-amide (Scheme 8.6).[40,41] Principally exploited to develop chemical libraries, recent reports highlight its applications in total synthesis.[42–46]

Pioneering this tendency in their synthesis of motuporin, Armstrong and coworkers efficiently synthesized intermediate **26** via a four-component Ugi condensation using aldehyde **22**, protected glutamic acid **23**, methylamine **24**,

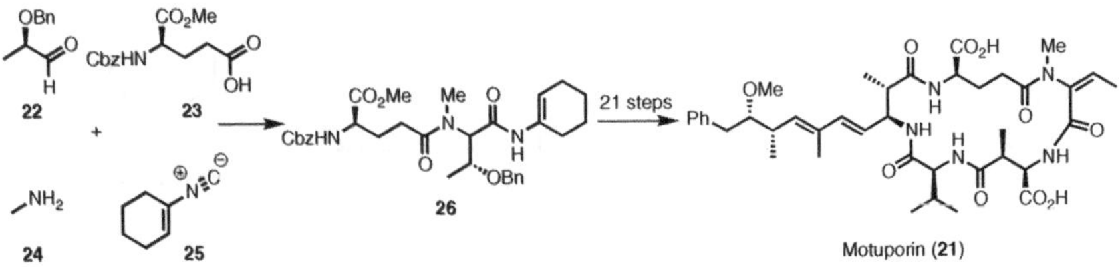

Scheme 8.7 Synthesis of motuporin by Bauer and Armstrong.

and cyclohexenyl isocyanide **25**. Hence, the 2-(*N*)-methylaminobutenyl residue was obtained at a 59% yield, while 21 more steps were necessary to complete the synthesis of this challenging macrocyclic natural product (Scheme 8.7).[47]

8.2.4 Synthesis of Thiomarinol H by Gao and Hall

Isolated from the bacterium *Alteromonas rava* sp. nov. SANK 73390, thiomarinol A **27**[48] and H **28** are potent antibiotics possessing a wider spectrum of activity (Gram-positive and Gram-negative bacteria) than the related pseudomonic acid A **29** (Figure 8.5).[49]

Based on an hetero-Diels–Alder variant of the Vaultier–Lallemand three-component reaction, Gao and Hall efficiently achieved the first total synthesis of thiomarinol A (**27**) [50].

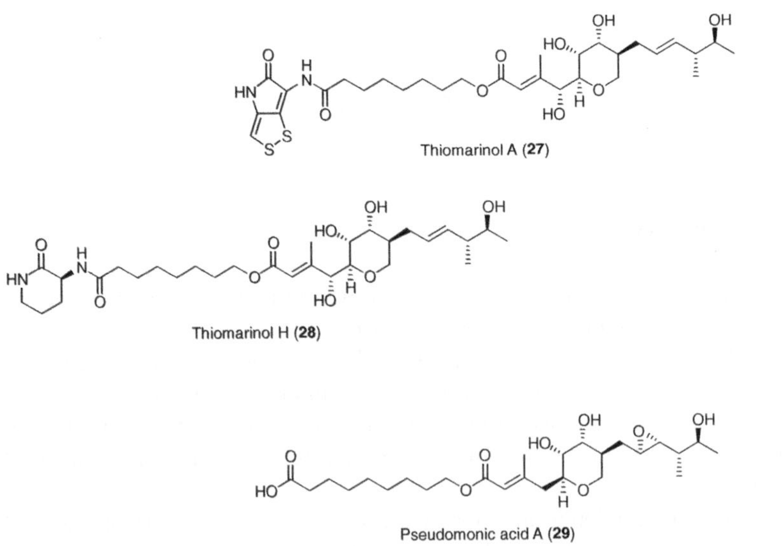

Figure 8.5 Structures of thiomarinol A (**27**) and H (**28**), and pseudomonic acid A (**29**).

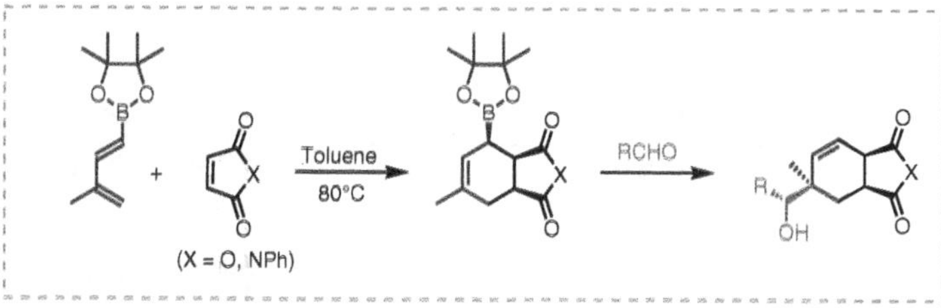

Scheme 8.8 Vaultier–Lallemand tandem cycloaddition/allylboration sequence.

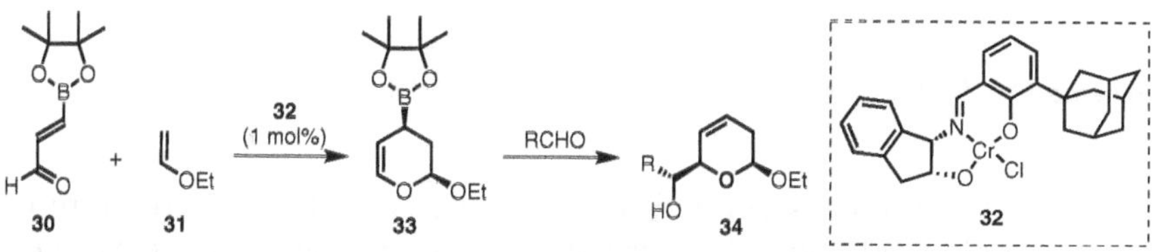

Scheme 8.9 Tandem hetero Diels–Alder cycloaddition/allylboration processes.

This Vaultier–Lallemand three-component reaction combines a Diels–Alder and an allylboration reaction. Hence, upon heating a 4-borono-1,3-butadiene with activated dienophiles such as maleic anhydride, maleimide, or methylacrylate, the resulting *cis*-cycloadducts bearing an allylboronate functionality then reacts with aldehydes to afford highly functionalized alcohols in a highly diastereoselective fashion (Scheme 8.8).[51–54]

Based on this elegant multicomponent reaction, Hall *et al.* and Carreaux *et al.* independently reported an enantioselective hetero Diels–Alder/allylboration reaction involving 3-boronoacrolein pinacolate **30**, ethyl vinyl ether **31**, and a variety of aldehydes catalyzed by Jacobsen's catalyst (**32**) to afford the corresponding α-hydroxyalkyl dihydropyrans **34** (Scheme 8.9).[55,56] The methodology was further applied by Carreaux *et al.* to the synthesis of various natural antitumour products belonging to the goniodiol family as well as to the enantioselective synthesis of (+)-iso-*exo*-brevicomin.[57–61]

This methodology was also applied by Gao and Hall to achieve the first total synthesis of thiomarinol H **28** (Scheme 8.10). The tandem hetero Diels–Alder cycloaddition/allylboration process was applied to 3-boronoacrolein pinacolate (**30**), a prenyl-substituted enol ether **35**, and aldehyde **37** as depicted in Scheme 8.10. Remarkably, while a 3:1 *Z*/*E* mixture of 2-substitued enol ether was used, the desired pyran **36** was obtained as a single diastereoisomer at a 76% yield and 95% *ee*. Pyran **36** was subsequently reacted in a one-pot fashion with aldehyde **37** to afford 2-hydroxyalkyldihydropyran **38**. It was assumed that Jacobsen's catalyst promoted the cycloaddition of the desired *Z* isomer faster than with the *E* isomer. The synthesis of thiomarinol H **29** was eventually completed after 14 additional steps, and the natural product was obtained at a 22% overall yield.

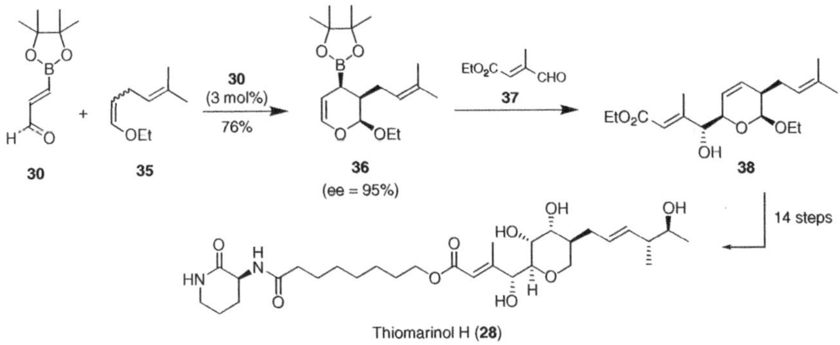

Scheme 8.10 Synthesis of thiomarinol H (**29**) by Gao and Hall.

Among other studies devoted to the synthesis of the core structure of pseudomonic acids and thiomarinols, the unified approach developed by Srihari *et al.*, which involves a Johnson-Claisen rearrangement, a Achmatowicz rearrangement and a Julia–Kocienski olefination as the key steps, is of particular interest.[62]

8.2.5 Synthesis of Minquartynoic Acid by Gung and Coworkers

Isolated from the stem bark of *Minquartia guianensis,* a traditional anthelmintic used by Ecuador's Amazonian, but also from south Asian *Ochanostachys amentacea* twigs, minquartynoic acid (**39**) along with some analogs **40** and **41** (Figure 8.6) displayed cytoxicity against various cell lines as well as anti-HIV activity.[63–65] Interested by these unusual natural products, Gung and Dickson devised a one-pot access to the tetrayne moiety through a three-component Cadiot–Chodkiewicz reaction.

The Cadiot–Chodkiewicz alkyne cross-coupling reaction allows the synthesis of bis-acetylene by coupling a terminal alkyne with a 1-bromoalkyne in the presence of a copper(I)-salt and an aliphatic amine.[66,67] Well-suited to the preparation of unsymmetrical diynes, the reaction mechanism involves the formation of a copper(I) acetylide followed by a cycle of oxidative addition and reductive elimination thus creating a new carbon-carbon bond while tolerating many oxygenated and nitrogenated functionalities (Scheme 8.11).[68]

Foreseeing the synthetic potential of this reaction, Gung and coworkers used a bidirectional synthon that could react with two different bromoalkynes successively.[69,70] Hence, under Cu(I)-catalysis, the desired cross-coupling product **46** was obtained at a 30% yield along with 29% of the unwanted symmetrical tetracetylenes **45** and **47**. After column chromatography separation, **46** was treated with HF•Pyr to afford minquartynoic acid, which was obtained in seven steps and a 17% overall yield (Scheme 8.12). While the yield

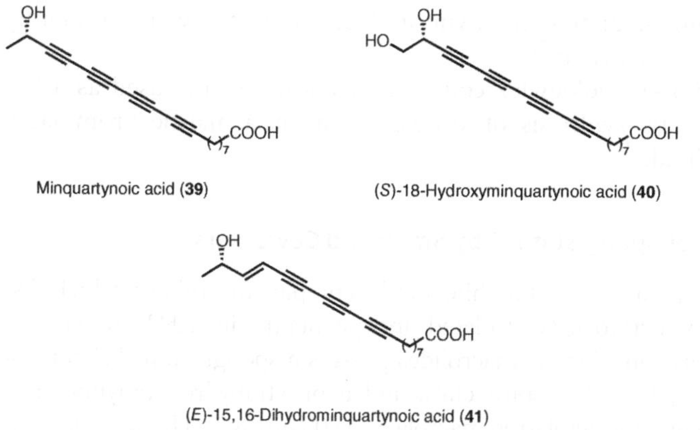

Figure 8.6 Structures of minquartynoic acid **39** and its analogs **40** and **41**.

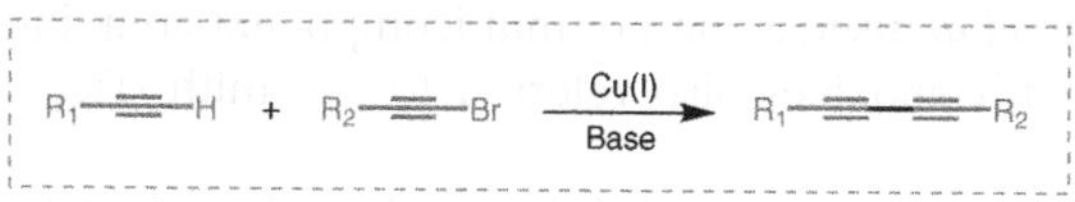

Scheme 8.11 Cadiot–Chodkiewicz heterocoupling.

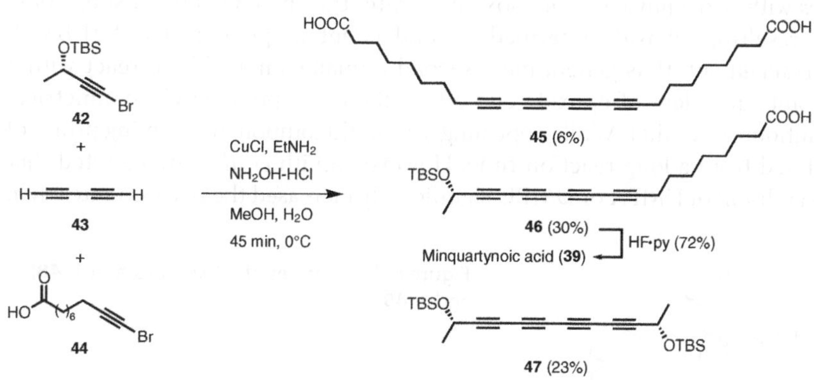

Scheme 8.12 3-MCR Cadiot–Chodkiewicz alkyne cross-coupling reaction.

of formation of **46** is modest, this MCR approach is by far the most convenient and efficient way to access these tetrayne units as all the attempts to assemble the acetylenic fonctions linearly have led to unstable intermediates.

It is worth pointing out that a more classical approach toward the synthesis of minquartynoic acid **39** and its analogs **40** and **41** using the Cadiot–Chodkiewicz reaction was reported in 2006 in which cross-coupling reactions of terminal diynes or enyne and bromodiynes were efficiently performed at a 70–78% yield.[71]

This same Cadiot–Chodkiewicz coupling was also recently used as a key step in the first total synthesis of xestospongenyne, a marine brominated polyunsaturated lipid.[72]

8.2.6 Synthesis of Spongistatin 2 by Smith and Coworkers

The spongistatins are a family of architecturally complex macrolides, which display extraordinary cytotoxicity. Isolated independently in 1993 by Pettit,[73] Kitagawa,[74] and Fusetani,[75] these macrolides possess a spongipyran skeleton, 24 stereogenic centers, two [6.6]-spiroketals, and a bistetrahydropyranylmethane moiety embedded in a 42-membered macrolactone (Figure 8.7). The spongistatins display remarkable cytotoxic activities[76–78] that triggered considerable interest in the synthetic community and led ultimately to gram-scale syntheses.[79–97]

In their second-generation synthesis of an advanced ABCD ring intermediate of the spongistatins, Smith *et al.* used a one-pot multicomponent coupling of 2-silylated-1,3-dithianes with epoxides also referred to as Smith–Tietze coupling.

This Smith–Tietze coupling uses 2-lithio-1,3-dithianes first described by Corey and Seebach[98] as a "carbonyl anion." In 1994, Tietze and coworkers described the symmetrical *bis*-alkylation of lithiated 2-trialkylsilyl-1,3-dithianes with two equivalents of epoxides.[99] After the opening of the first epoxide, the resulting alkoxide intermediate underwent a spontaneous [1,4]-Brook rearrangement, thus generating a second dithiane anion able to react with a second equivalent of epoxide and afford the corresponding C2-symmetrical enantiopure 1,5-diol. While appealing, this multicomponent coupling protocol suffered from a long reaction time. However, Smith *et al.* demonstrated that the addition of HMPA or DMPU significantly increased the reaction rate while

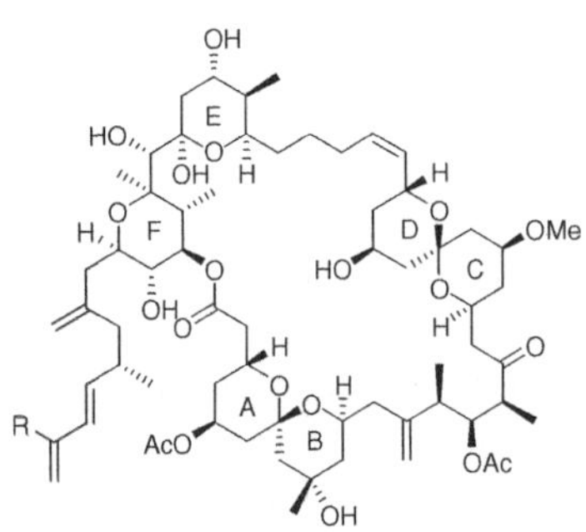

(+)-Spongistatin 1 (**48**, R = Cl)
(+)-Spongistatin 2 (**49**, R = H)

Figure 8.7 Structures of (+)-spongistatin 1 (**48**) and 2 (**49**).

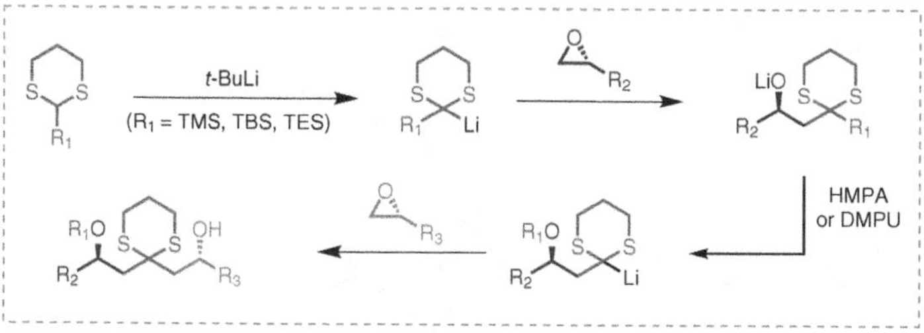

Scheme 8.13 Smith–Tietze multicomponent dithiane linchpin coupling.

allowing the coupling of two different epoxides to occur thanks to the additive-controlled Brook rearrangement (Scheme 8.13).[100] Moreover, the use of an enantiopure *bis*-epoxide allowed a one-pot *five-component linchpin coupling.*

The three-component dithiane linchpin coupling was the key bond-forming reaction in the second-generation synthesis of the advanced ABCD intermediate of the spongistatins by Smith *et al.*[87] as both the AB and the CD fragments were accessed by this multicomponent coupling. Interestingly, one of the epoxide components had to be added into the reaction mixture as its lithium alkoxide to avoid the formation of elimination products. Upon deprotection of the dithiane moiety, an *in situ* spiroketalization takes place.

In their approach toward the synthesis of the AB and CD bis-spiroketal units of the spongistatins, Smith *et al.* treated lithiated 2-triethylsilyl dithiane **50** with epoxide fragment **51**. The alkoxide intermediate was then submitted to HMPA, thus triggering the 1,4-Brook rearrangements to generate a second dithianyl anion, which was allowed to react with the second epoxide fragment **52** leading to the corresponding polyol derivative **53**. Deprotection of the dithiane moiety with Hg(II) promoted the *in situ* spiroketalization providing spiro derivative **54** as a single isomer, while eight additional steps were needed to access the elaborated AB fragment **55** (Scheme 8.14).

A similar strategy was used for the CD fragment with epoxides **56** and **57** (Scheme 8.15). Hence, hydrolysis of the dithiane moiety using CAN provided a mixture of spiroketals **59** and **60** in a 4:1 mixture in favor of the undesired spiroketal **59**. Dihydroxylation of the olefin moiety followed by perchlorate/perchloric acid equilibration and then 1,2-diol cleavage using $NaIO_4$ afforded a separable mixture of (+)-**61** and (−)-**62** (1:4), now favoring the desired spiroketal (−)-**62**. The two fragments **55** and **63** were then joined together through a stereoselective aldol reaction to afford the desired fragment bearing the ABCD rings. Overall, thanks to this elegant multicomponent reaction, (−)-**64** was obtained in 22 steps and a 4% overall yield, which is 15 steps less than their previous synthesis of the same fragment.[85] This multicomponent strategy, now referred to as Anion Relay Chemistry (ARC), has since been exploited in the synthesis of many complex natural and unnatural products.[101–106]

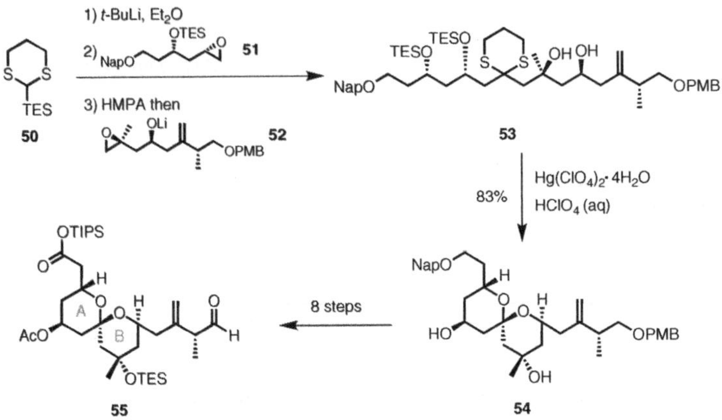

Scheme 8.14 Synthesis of the AB fragment **55**.

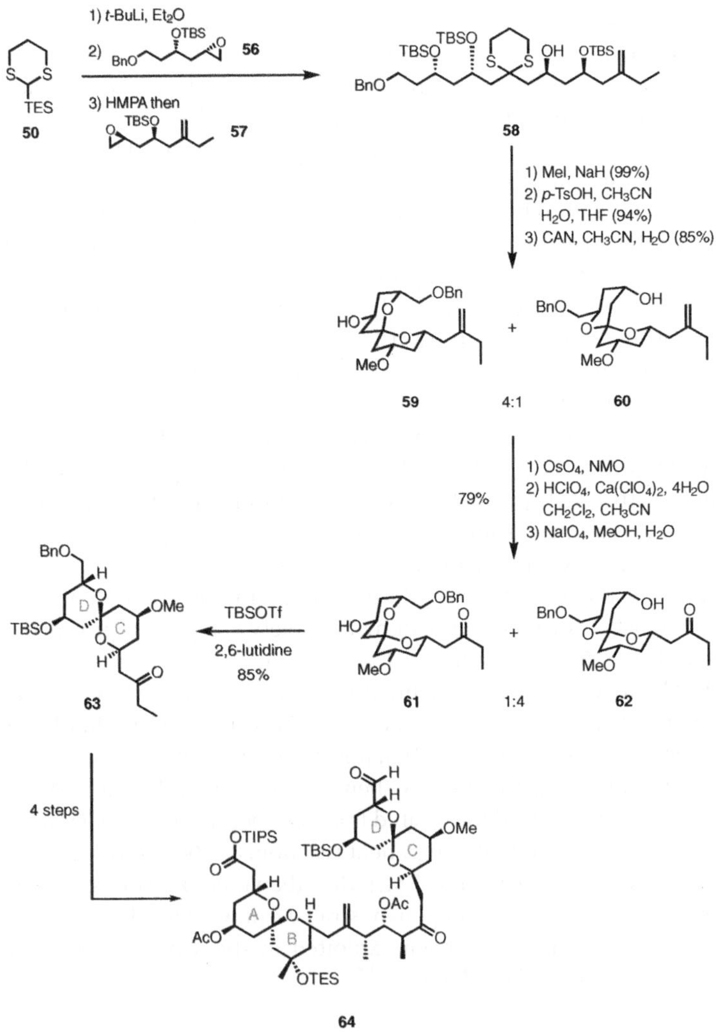

Scheme 8.15 Synthesis of the ABCD subunit of the spongistatins.

8.2.7 Synthesis of Vannusal A and B by Nicolaou and Coworkers

Vannusals A and B were isolated in 1999 from tropical strains of *Euplotes vannus*, an interstitial ciliate.[107] The two compounds were characterized as a unique C30 backbone arranged around six rings and 13 stereogenic centers (Figure 8.8). In fact, the total synthesis of these compounds has been a fantastic journey as the originally assigned structure of vannusal A (**65**) and B (**66**) was proven to be erroneous. The successful synthesis of the originally assigned structures by Nicolaou *et al.*, though disappointing, prompted an important synthetic effort that led to the synthesis of several alternative diastereomers (out of the 8192 possibilities), one of which matched the natural product.[108–111]

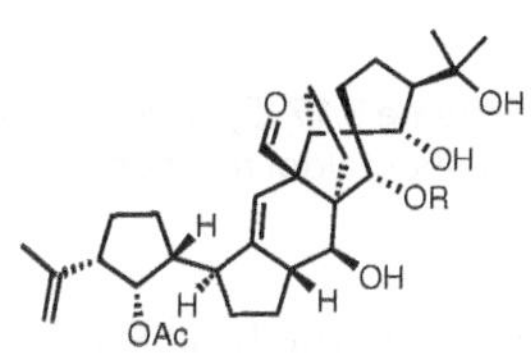

Vannusal A (R = Ac, **65**)
Vannusal B (R = H, **66**)

Figure 8.8 Structures of vannusal A and B.

The vannusal endeavor permitted the development of many remarkable methodologies including an elegant Noyori-type three-component reaction for the synthesis of the spirocyclic fragment. This original three-component reaction features a particularly appealing copper-catalyzed conjugate addition of a zinc species on a cyclic enone to afford a transient cyclic Zn-enolate that undergoes an aldol condensation in the presence of an aldehyde resulting in the corresponding aldol bearing three consecutive stereogenic centers (Scheme 8.16).[112]

This reaction was initially developed by Noyori to construct the prostanoic acid scaffold in a single step. Isolated from the prostate glands, prostaglandins are hormones derived from the C20 fatty acid arachidonic acid.[113] These compounds have in common a diversely substituted five-membered ring.[114–116]

This pioneering work used a tandem organo-copper conjugate addition/aldol reaction to introduce the ω side chain and an aldol condensations of the resulting enolate to attach the α side chain. The one-pot sequence allowed the construction of the functionalized prostaglandin skeleton at a 83% yield (Scheme 8.17).

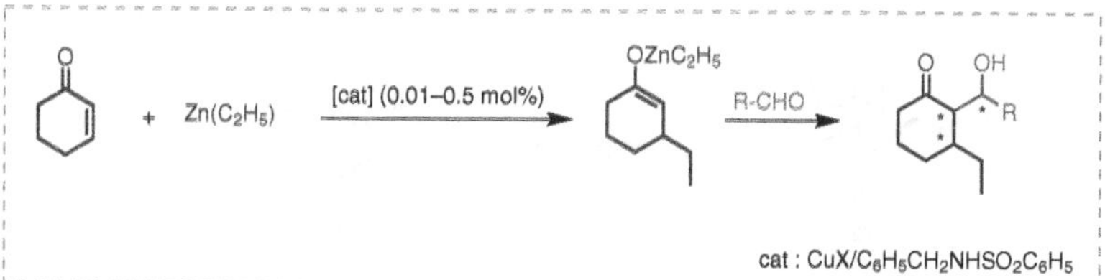

Scheme 8.16 General scheme for the Noyori three-component reaction.

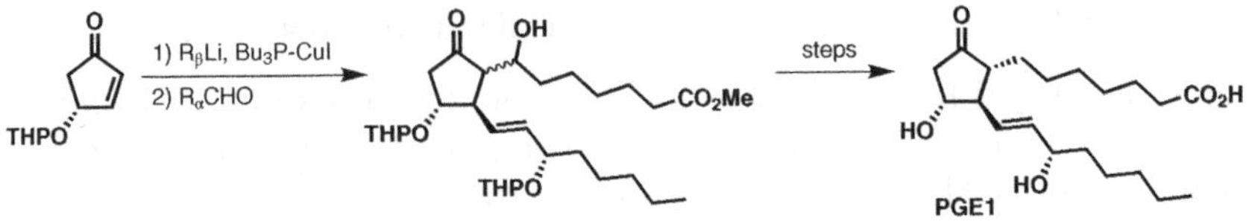

Scheme 8.17 Synthesis of prostaglandin PGE1 using a tandem organo-copper conjugate addition/aldol reaction.

Overall, the total synthesis of PGE1 was achieved in five steps and a 56% yield. Noyori and coworkers subsequently optimized the sequence using a more reactive zinc enolate intermediate. The tandem organo-copper conjugate addition/aldol reaction was subsequently applied to the synthesis of various prostaglandins, prostacyclin and related compounds.[117–120]

During their synthesis of the originally assigned structures of the vannusals, Nicolaou and coworkers reported a remarkable asymmetric Noyori-type three-component reaction in the presence of $[Rh(COD)Cl]_2$ using BINAP as ligand (Scheme 8.18). Based on the work of Oi and Inoue who reported in 2004 a rhodium-catalyzed addition of alkenyl zirconium species to enones,[121] the multi-component Noyori-type sequence was found to tolerate a wide range of cyclic enones and aldehydes leading to 2,3-disubstituted aldols with good to excellent enantioselectivities.[122]

Application of these conditions to cyclohexenone **67**, 4-siloxybutanal derivative **68** and the alkenyl zirconium derivative **69** provided aldol **70** at a 52% yield, which was converted into enone **71**. From this point, seven more steps were necessary to obtain the spirocyclic key intermediate **72**.[123]

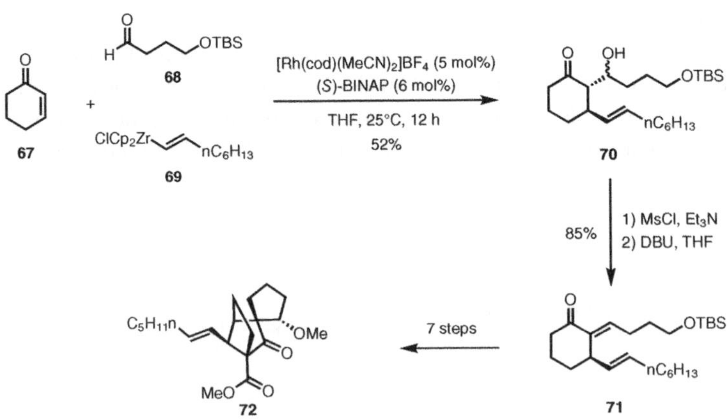

Scheme 8.18 Synthesis of key intermediate **72** in the total synthesis of vannusal A and B by Nicolaou *et al.*

8.2.8 Synthesis of Calystegine B-4 by Pyne and Coworkers

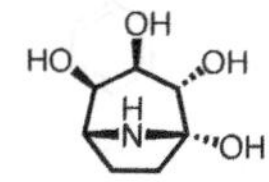

Calystegine B4 (**73**)

Figure 8.9 Structures of calystegine B4.

Calystegine B4 (**73**) belongs to the calystegines family of polyhydroxylated nortropane alkaloids that are subdivided into three groups based on the number of hydroxy groups present (Figure 8.9).[124–126] Isolated in 1996 from the root extracts of *Scopolia japonica,* calystegine B4 displays glycosidase inhibitory activities with an IC50 of 4.8 μM against pig kidney trehalase.[127]

Due to their promising biological activities, many synthetic routes toward the synthesis of the various calystegines have been reported over the years, most of them using a ring closing metathesis (RCM)/hydroboration/oxidation strategy that unfortunately led to mixtures of undesired regioisomers. To avoid this drawback, Pyne and coworkers disclosed a new strategy based on the Petasis–Borono–Mannich multicomponent reaction.

The Petasis–Borono–Mannich reaction, also referred to as the Petasis reaction was reported in 1993 by Petasis and coworkers.[128] Based on a modified Mannich reaction, vinylboronic acids act as the nucleophile in the presence of an amine and a carbonyl derivative to generate allylic amines, or benzylic amines when phenyl boronic acids are used (Scheme 8.19).[129] The process has since found many applications in natural product total synthesis[130–132] and for the preparation of various classes of compounds.[133]

The Petasis reaction was the key step in Pyne's synthesis of calystegine B4 (Scheme 8.20).[134] Hence, treatment of (–)-D-xylose (**74**) with benzylamine **75** and (*E*)-2-phenylvinyl boronic acid **76** generated the aminotetrol **77** at a 82% yield. A series of orthogonal protection and deprotection steps was necessary in order to obtain aldehyde **78**, which was treated with vinylmagnesium bromide to afford a 1:1 mixture of alcohols **79** and **80**. Both alcohols were separated and convergently transformed to calystegine B4 in three steps. **79** underwent a RCM/oxidation/reduction sequence whereas an oxidation/RCM/reduction was applied to **80**. Overall, the total synthesis of calystegine B4 was achieved in 10 steps and 4.7% (via **79**) and 3.4% (via **80**) overall yields.

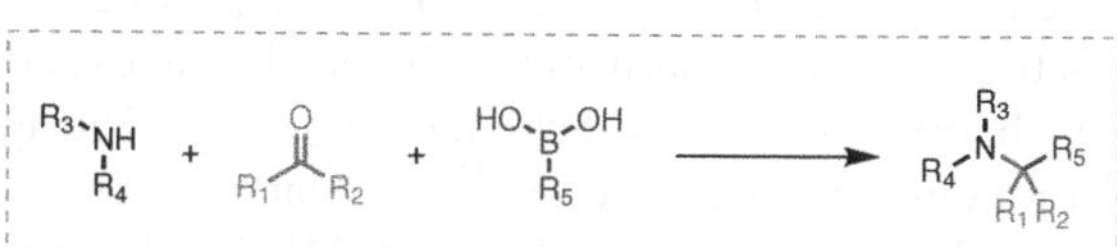

Scheme 8.19 General scheme for the Petasis–Borono–Mannich multicomponent reaction.

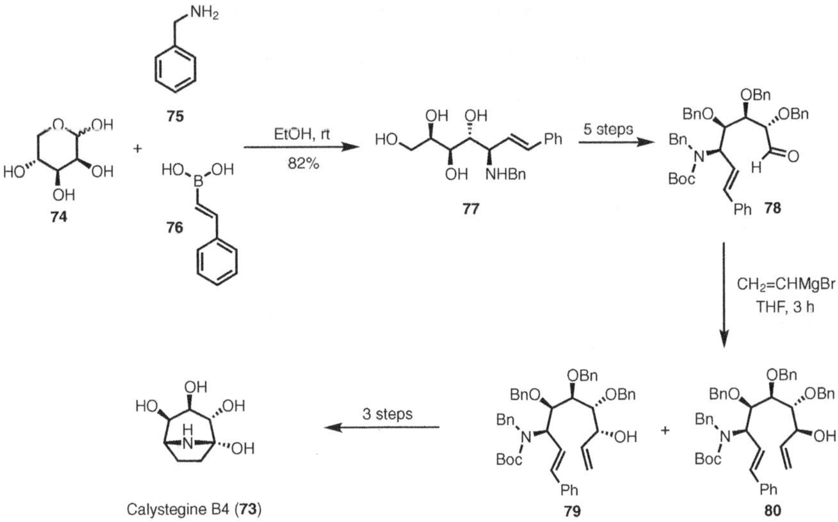

Scheme 8.20 Synthesis of calystegine B4 by Pyne and coworkers.[134]

8.2.9 Synthesis of Jerangolid D by Markó and Pospisil

The jerangolids are a family of antifungal compounds produced by the myxobacterium *Sorangium cellulosum*.[135] Isolated for the first time in 1987, the structures of these polyketides that display potent antifungal activities consist of a lactone and a dihydropyran moieties connected through a penta-1,4,dienyl linker (Figure 8.10). While the total synthesis of jerangolid D has been described in 2007 by Markó and Pospisil using an asymmetric three-component Hosomi–Sakurai reaction,[136] the synthesis of jerangolid D was achieved in 2010 by Hannessian *et al.* through a totally different approach.[137]

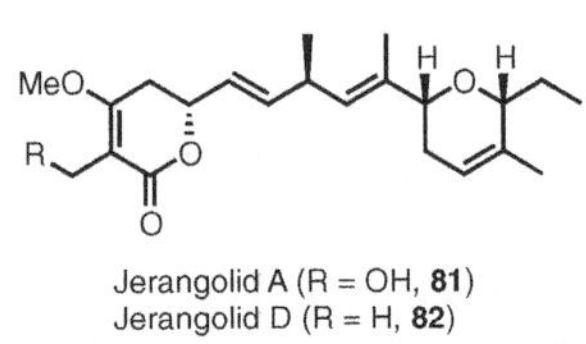

Figure 8.10 Structures of jerangolids A and D.

The Sakurai allylation reaction is a very useful reaction for carbon-carbon bond forming. Reported in 1976, this reaction allows the formation of homoallylic alcohols by treating allylsilanes with carbonyl derivatives in the presence of a $TiCl_4$.[138] Sakurai subsequently described a variant of this initial reaction reporting the Lewis acid catalyzed condensation of carbonyl derivatives with allylic silanes and alcohols for the formation of homoallylic ethers (Scheme 8.21).[139–141] Widely involved in the so-called silyl-modified Sakurai reaction,[142–152] Markó and cowworkers reported in 2006 the first chiral aldehyde based Sakurai MCR with complete diastereocontrol in the case of secondary and tertiary silyl ethers.[153]

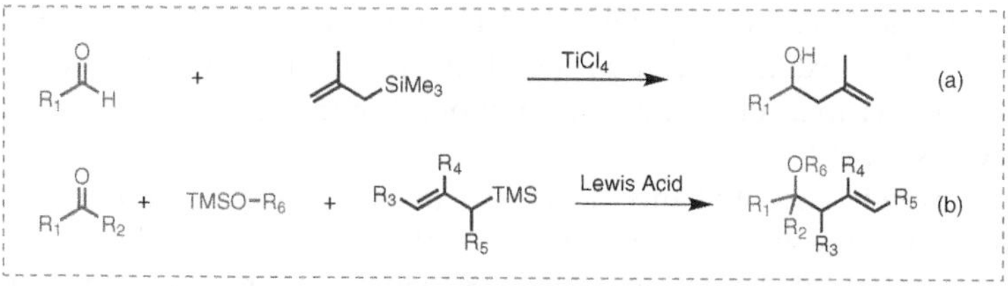

Scheme 8.21 The original Sakurai allylation (a) and the three-component Sakurai reaction (b).

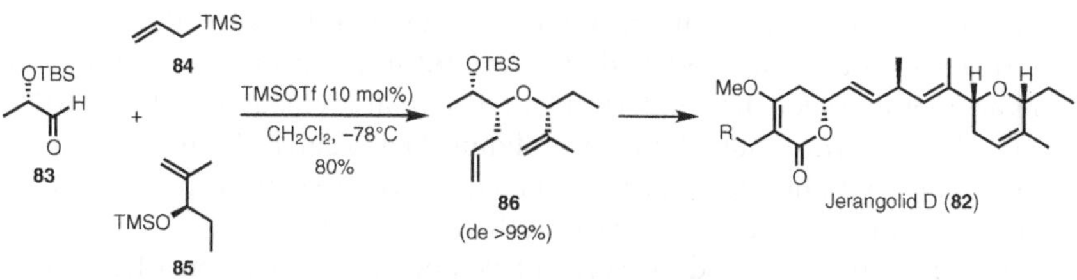

Scheme 8.22 Synthesis of jerangolid D by Pospisil and Markó.

Applying this methodology on chiral aldehyde **83**, allyl trimethyl silane **84** and chiral ether **85** in the presence of a catalytic amount of TMSOTf furnished the corresponding *syn,syn*-ether **86** as a single stereoisomer at an 80% yield (Scheme 8.22). After RCM, the dihydropyran moiety was obtained, derivatized, and finally coupled with the lactone moiety. Overall, the total synthesis of jerangolid D was completed in 22 steps and a 6.1% overall yield.[136]

8.2.10 Synthesis of (–)-Nakadomarin A by Young and Kerr

(–)-Nakadomarin A **87** is an alkaloid isolated for the first time in 1997 from the marine sponge *Amphimedon* sp. Its core structure features an astonishing hexacyclic 6/5/5/5/8/15 ring system one of them being a furan (Figure 8.11).[154,155] (–)-Nakadomarin A has been found to exhibit promising cytotoxic, antimicrobial, and antibacterial activities but its limited natural availability (6.0 mg, 0.0018% isolated yield) hampered its exploitation. As such, the total synthesis of nakadomarin A has been the subject of intense work for drug development.[156–170]

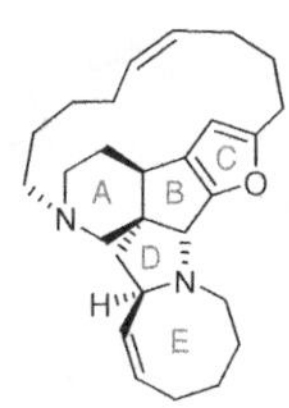

Nakadomarin A (**87**)

Figure 8.11 Structures of nakadomarin A.

Among the many strategies envisioned to access this remarkable core structure, a couple of approaches disclosed a multicomponent reaction involving a 1,3-dipolar cycloaddition. In these specific MCRs, the reactive 1,3-dipole is commonly generated *in situ* before reacting with the dipolarophile. Many examples can be found in the literature notably for the construction of pharmaceutically active compounds and the subject has been reviewed extensively.[171–174]

In 2004, Ahrendt and Williams devised a azomethine ylide 1,3-dipolar cycloaddition to access the ADE fragment of nakadomarin A.[175,176] The 1,3-dipolar cycloaddition of azomethine ylides **88** (which can be generated from aziridines, imines, iminiums, by *N*-metallation or by condensation of aldehydes to secondary amines) with alkenes **89** leads to the formation of the pyrrolidines **90** (Scheme 8.23).[177–181]

In their approach, Ahrendt and Williams generated the azomethine ylide by reacting secondary amine **91** and aldehyde **92** in the presence of enone **93**. This azomethine ylide then underwent a 1,3-dipolar cycloaddition-based MCR to afford the 2,5-*trans*-substituted pyrrolidine **94** as a single diastereoisomer at a 35% yield. Six additional steps were then necessary to convert pyrrolidine **95** to the ADE fragment of nakadomarin A (Scheme 8.24).

Another approach toward the core of nakadomarin A based on a 1,3-dipolar cycloaddition was also proposed by Young and Kerr. Their strategy relied on an elegant [3 + 2] cycloaddition between a *in situ* generated nitrone and a cyclopropane.[182–184] Indeed, while the 1,3-dipolar cycloaddition of nitrones **96** with alkenes **97** leading to the corresponding isoxazolidines **98** is well-known,[181] the ytterbium-catalyzed [3 + 2] cycloaddition between nitrones and strained

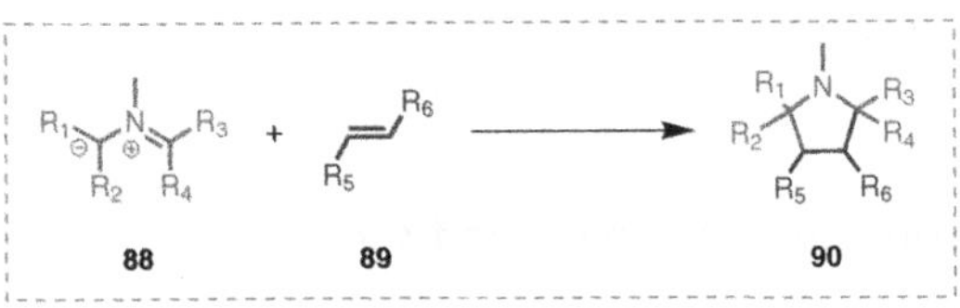

Scheme 8.23 1,3-Dipolar cycloaddition of azomethine ylides.

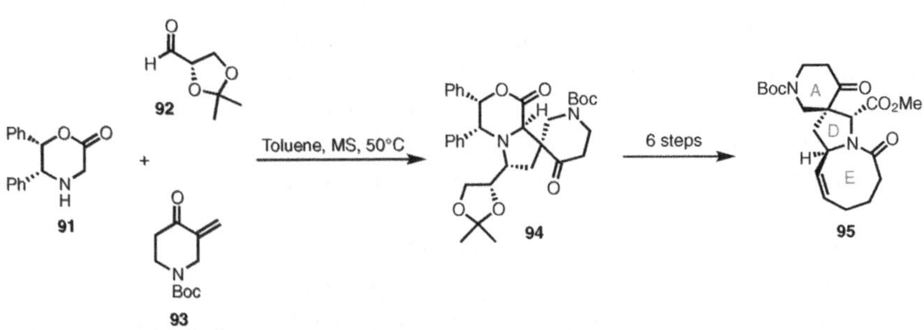

Scheme 8.24 Synthesis of the ADE fragment of nakadomarin A by Ahrendt and Williams.

cyclopropanes generating tetrahydro-1,2-oxazines **102** is more recent (Scheme 8.25).

Observing that these oxazines could be transformed into pyrrolidines after cleavage of the N-O bond, conversion of the resulting hydroxyl functionality to a leaving group and subsequent ring closure,[185] the authors devised a particularly straightforward strategy toward the tetracyclic core of nakadomarin A (Scheme 8.26).[186] Indeed, the 1,3-dipolar MCR between hydroxylamine **103**, furaldehyde **104**, and cyclopropane **105**, afforded the cyclic adduct **106** at an excellent yield and high regio- and stereocontrol. This oxazine was eventually converted to the tricyclic system **107** in seven steps, while 13 additional steps were necessary to complete the total synthesis of nakadomarin A, which was obtained in 23 steps and a 0.003% overall yield.

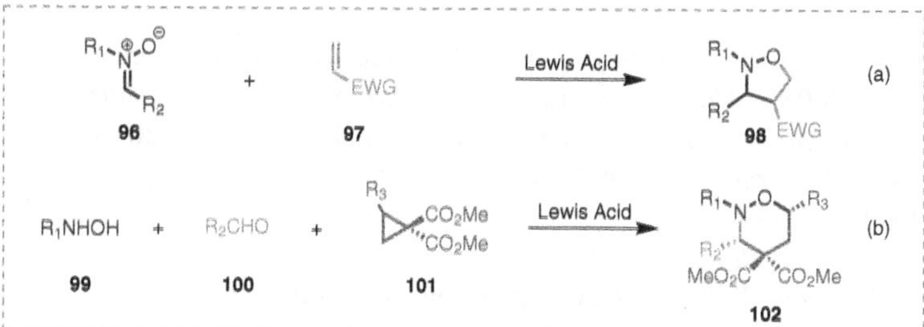

Scheme 8.25 (a) 1,3-Dipolar cycloaddition of nitrones with alkenes and (b) Young and Kerr's 3-MCR of nitrones with cyclopropanes.

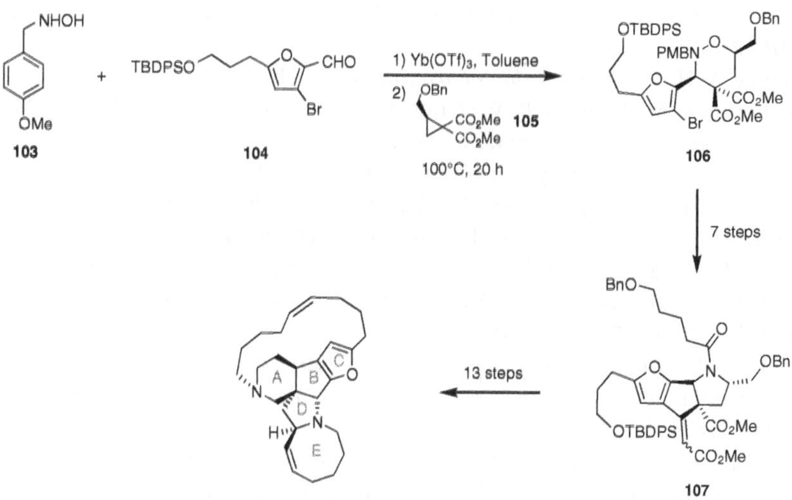

Scheme 8.26 Synthesis of nakadomarin A by Young and Kerr.

8.3 Conclusion

From the previous sections, it clearly appears that MCRs have become fantastic tools to access structural diversity in an effective and highly straightforward fashion. While the list of existing MCRs is in constant progress, it is evident that new synthetic applications will be obtained in the near future with implications in both industry and academia. As a matter of fact, the recent developments made in asymmetric multicomponent reactions as well as in transition metal-catalyzed MCRs are starting to play an important role in natural product synthesis and drug development.

References

1 (a) Zhu, J.; Bienaymé, H. *Multicomponent Reactions*, Weinheim, Germany: Wiley-VCH, **2005**. (b) Zhu, J.; Wang, Q.; Wang, M.-X. *Multicomponent Reactions in Organic Synthesis*. Wiley-VCH, Weinheim, Germany, **2015**.
2 (a) Pellissier, H. *Tetrahedron* **2007**, *63*, 3235. (b) Pellissier, H. *Adv. Synth. Catal.* **2011**, *353*, 189.
3 Pellissier, H. *Chem. Rev.* **2013**, *113*, 442.
4 Tietze, L. F.; Kinzel, T.; Brazel, C. C. *Acc. Chem. Res.* **2009**, *42*, 367.
5 Touré, B. B.; Hall, D. G. *Chem. Rev.* **2009**, *109*, 4439.
6 de Graaff, C.; Ruijter, E.; Orru, R. V. A. *Chem. Soc. Rev.* **2013**, *41*, 3969.
7 Ramón, D. J.; Yus, M. *Angew. Chem., Int. Ed.* **2005**, *44*, 1602.
8 Dömling, A.; Ugi, I. *Angew. Chem., Int. Ed.* **2000**, *39*, 3168.
9 Bonne, D.; Coquerel, Y.; Constantieux, T.; Rodriguez, J. *Tetrahedron: Asymmetry* **2010**, *21*, 1085.
10 Gentry, A. H.; Cook, K. *J. Ethnopharm.* **1984**, *11*, 337.
11 Witherup, K. M.; Ransom, R. W.; Graham, A. C.; Bernard, A. M.; Salvatore, M. J. *et al. J. Am. Chem. Soc.* **1995**, *117*, 6682.
12 Badarinarayana, V.; Lovely, C. J. *Tetrahedron Lett.* **2007**, *48*, 2607.
13 Ikeda, S.; Shibuya, M.; Iwabuchi, Y. *Chem. Commun*, **2007**, *504*.
14 He, Y.; Moningka, R.; Lovely, C. J. *Tetrahedron Lett.* **2005**, *46*, 1251.
15 Xia, C. F.; Heng, L. S.; Ma, D. *Tetrahedron Lett.* **2002**, *43*, 9405.
16 Miyata, O.; Shirai, A.; Yoshino, S.; Nakabayashi, T.; Takeda, Y.; Kiguchi, T. *et al. Tetrahedron* **2007**, *63*, 10092.
17 Snider, B. B.; Ahn, Y.; O'Hare, S. M. *Org. Lett.* **2001**, *3*, 4217.
18 He, Y.; Mahmud, H.; Moningka, R.; Lovely, C. J.; Dias, H. V. R. *Tetrahedron* **2006**, *62*, 8755.
19 Ma, D.; Xia, C.; Zhang, J.; Jiang, J.; Tang, W. *J. Org. Chem.* **2002**, *68*, 442.
20 Shirai, A.; Miyata, O.; Tohnai, N.; Miyata, M.; Procter, D. J.; Sucunza, D. *et al. J. Org. Chem.* **2008**, *73*, 4464.
21 Davies, S. G.; Fletcher, A. M.; Lee, J. A.; Lorkin, T. J. A.; Roberts, P. M.; Thomson, J. E. *Org. Lett.* **2013**, *15*, 2050.

22 Yoshitomi, Y.; Arai, H.; Makino, K.; Hamada, Y. *Tetrahedron* **2008**, *64*, 11568.
23 Ueda, M.; Kawai, S.; Hayashi, M.; Naito, T.; Miyata, O. *J. Org. Chem.* **2010**, *75*, 914.
24 Povarov, L. S. *Russ. Chem. Rev.* **1967**, *656*.
25 Dagousset, G.; Retailleau, P.; Masson, G.; Zhu, J. P. *Chem. Eur. J.* **2012**, *18*, 5869.
26 Vicente-Garcia, E.; Ramon, R.; Preciado, S.; Lavilla, R. *Beilstein J. Org. Chem.* **2011**, *7*, 980.
27 Vicente-Garcia, E.; Ramon, R.; Lavilla, R. *Synthesis* **2011**, *2237*.
28 Powell, D. A.; Batey, R. A. *Org. Lett.* **2002**, *4*, 2913.
29 Toda, S.; Numata, Y.; Obi, K.; Hamagishi, Y.; Tomita, K.; Komiyama, N. *et al.* *J. Antibiot.* **1992**, *45*, 1573.
30 Toda, S.; Kotake, C.; Tsuno, T.; Narita, Y.; Yamasaki, T.; Konishi, M. *J. Antibiot.* **1992**, *45*, 1580.
31 Schmidt, U.; Weinbrenner, S. *Chem. Commun* **1994**, *1003*.
32 Passerini, M. *Gazz. Chim. Ital.*, **1921**, *51*, 126.
33 Passerini, M. *Gazz. Chim. Ital.*, **1921**, *51*, 762.
34 Owens, T. D.; Araldi, G.-L.; Nutt, R. F.; Semple, J. E. *Tetrahedron Lett.* **2001**, *42*, 6271.
35 Desilva, E. D.; Williams, D. E.; Andersen, R. J.; Klix, H.; Holmes, C. F. B.; Allen, T. M. *Tetrahedron Lett.* **1992**, *33*, 1561.
36 Hu, T.; Panek, J. S. *J. Am. Chem. Soc.* **2002**, *124*, 11368.
37 Hu, T.; Panek, J. S. *J. Org. Chem.* **1999**, *64*, 3000.
38 Valentekovich, R. J.; Schreiber, S. L. *J. Am. Chem. Soc.* **1995**, *117*, 9069.
39 Samy, R.; Kim, H. Y.; Brady, M.; Toogood, P. L. *J. Org. Chem.* **1999**, *64*, 2711.
40 Ugi, I. M.; Fetzer, U.; Steinbrückner, C., *Angew. Chem.* **1959**, *71*, 373.
41 Ugi, I. *Angew. Chem., Int. Ed.* **1962**, *1*, 8.
42 Nakagawa, K.; Katayama, K.; Takeda, H.; Matsuda, A.; Ichikawa, S. *Org. Lett.* **2014**, *16*, 428.
43 Ieda, S.; Masuda, A.; Kariyama, M.; Fukuyama, T.; Wakimoto, T.; Asakawa, T. *et al.* *Heterocycles* **2012**, *86*, 1071.
44 Neves Filho, R. A. W.; Stark, S.; Westermann, B.; Wessjohann, L. A. *Beilstein J. Org. Chem.* **2012**, *8*, 2085.
45 Neves Filho, R. A. W.; Westermann, B.; Wessjohann, L. A. *Beilstein J. Org. Chem.* **2011**, *7*, 1504.
46 Breder, A.; Chinigo, G. M.; Waltman, A. W.; Carreira, E. M. *Chem. Eur. J.* **2011**, *17*, 12405.
47 Bauer, S. M.; Armstrong, R. W. *J. Am. Chem. Soc.* **1999**, *121*, 6355.
48 Shiozawa, H.; Kinoshita, T.; Kagasaki, T.; Domon, H.; Haruyama, H.; Utsui, Y. *et al.* *J. Antibiot.* **1993**, *46*, 1834.
49 Stierle, D. B.; Stierle, A. A. *Experientia* **1992**, *48*, 1165.
50 Gao, X.; Hall, D. G. *J. Am. Chem. Soc.* **2005**, *127*, 1628.
51 Vaultier, M.; Truchet, F.; Carboni, B.; Hoffmann, R. W.; Denne, I. *Tetrahedron Lett.* **1987**, *28*, 4169.
52 Lallemand, J. Y.; Six, Y.; Ricard, L. *Eur. J. Org. Chem.* **2002**, *503*.
53 Six, Y.; Lallemand, J. Y. *Tetrahedron Lett.* **1999**, *40*, 1295.

54 Renard, P. Y.; Six, Y.; Lallemand, J. Y. *Tetrahedron Lett.* **1997**, *38*, 6589.
55 Gao, X.; Hall, D. G. *J. Am. Chem. Soc.* **2003**, *125*, 9308.
56 Deligny, M.; Carreaux, F.; Toupet, L.; Carboni, B. *Adv. Synth. Catal.* **2003**, *345*, 1215.
57 Favre, A.; Carreaux, F.; Deligny, M.; Carboni, B. *Eur. J. Org. Chem.* **2008**, *4900*.
58 Deligny, M.; Carreaux, F.; Carboni, B. *Synlett* **2005**, *1462*.
59 Gao, X. R.; Hall, D. G.; Deligny, M.; Favre, A.; Carreaux, F.; Carboni, B. *Chem. Eur. J.* **2006**, *12*, 3132.
60 Eberlin, L.; Tripoteau, F.; Carreaux, F.; Whiting, A.; Carboni, B. *Beilstein J. Org. Chem.* **2014**, *10*, 237.
61 Bouziane, A.; Régnier, T.; Carreaux, F.; Carboni, B.; Bruneau, C.; Renaud, J.-L. *Synlett* **2010**, *207*.
62 Sridhar, Y.; Srihari, P. *Org. Biomol. Chem.* **2014**, *12*, 2950.
63 Marles, R. J.; Farnsworth, N. R.; Neill, D. A. *J. Nat. Prod.* **1989**, *52*, 261.
64 Rashid, M. A.; Gustafson, K. R.; Cardellina, J. H.; Boyd, M. R. *Nat. Prod. Lett.* **2001**, *15*, 21.
65 Ito, A.; Cui, B. L.; Chavez, D.; Chai, H. B.; Shin, Y. G.; Kawanishi, K. *et al.* *J. Nat. Prod.* **2001**, *64*, 246.
66 Chodkiewicz, W. *Ann. Chim.* **1957**, *2*, 819.
67 Chodkiewicz, W.; Cadiot, P. *Hebd. Seances Acad. Sci.* **1955**, *241*, 1055.
68 Siemsen, P.; Livingston, R. C.; Diederich, F. *Angew. Chem., Int. Ed.* **2000**, *39*, 2633.
69 Gung, B. W. *C. R. Chimie* **2009**, *12*, 489.
70 Gung, B. W.; Dickson, H. *Org. Lett.* **2002**, *4*, 2517.
71 Sabitha, G.; Reddy, C. S.; Yadav, J. S. *Tetrahedron Lett.* **2006**, *47*, 4513.
72 Gong, J.-X.; Wang, H.-Y.; Yao, L.-G.; Li, X.-W.; Guo Y.-W. *Synlett* **2016**, *391*.
73 Pettit, G. R.; Chicacz, Z. A.; Gao, F.; Herald, C. L.; Boyd, M. R.; Schmidt, J. M. *et al.* *J. Org. Chem.* **1993**, *58*, 1302.
74 Kobayashi, M.; Aoki, S.; Sakai, H.; Kawazoe, K.; Kihara, N.; Sasaki, T. *et al.* *Tetrahedron Lett.* **1993**, *34*, 2795.
75 Fusetani, N.; Shinoda, K.; Matsunaga, S. *J. Am. Chem. Soc.* **1993**, *115*, 3977.
76 Pettit, G. R.; Cichacz, Z. A.; Gao, F.; Herald, C. L.; Boyd, M. R. *Chem. Commun* **1993**, *1166*.
77 Pettit, G. R.; Cichacz, Z. A.; Herald, C. L.; Gao, F.; Boyd, M. R.; Schmidt, J. M. *et al.* *Chem. Commun* **1994**, *1605*.
78 Bai, R.; Taylor, G. F.; Cichacz, Z. A.; Herald, C. L.; Kepler, J. A.; Pettit, G. R. *et al.* *Biochemistry* **1995**, *34*, 9714.
79 Evans, D. A.; Coleman, P. J.; Dias, L. C. *Angew. Chem., Int. Ed.* **1997**, *36*, 2738.
80 Evans, D. A.; Trotter, B. W.; Côté, B.; Coleman, P. J. *Angew. Chem., Int. Ed.* **1997**, *36*, 2741.
81 Evans, D. A.; Trotter, B. W.; Côté, B.; Coleman, P. J.; Dias, L. C.; Tyler, A. N. *Angew. Chem., Int. Ed.* **1997**, *36*, 2744.
82 Evans, D. A.; Trotter, B. W.; Coleman, P. J.; Côté, B.; Dias, L. C.; Rajapakse, H. A. *et al.* *Tetrahedron* **1999**, *55*, 8671.

83 Guo, J.; Duffy, K. J.; Stevens, K. L.; Dalko, P. I.; Roth, R. M.; Hayward, M. M. *et al. Angew. Chem., Int. Ed.* **1998**, *37*, 187.

84 Hayward, M. M.; Roth, R. M.; Duffy, K. J.; Dalko, P. I.; Stevens, K. L.; Guo, J.; *et al. Angew. Chem., Int. Ed.* **1998**, *37*, 190.

85 Smith III, A. B.; Doughty, V. A.; Lin, Q.; Zhuang, L.; McBriar, M. D.; Boldi, A. M. *et al. Angew. Chem., Int. Ed.* **2001**, *40*, 191.

86 Smith III, A. B.; Lin, Q.; Doughty, V. A.; Zhuang, L.; McBriar, M. D.; J. K. Kerns, *et al. Angew. Chem., Int. Ed.* **2001**, *40*, 196.

87 Smith III, A. B.; Doughty, V. A.; Sfouggatakis, C.; Bennett, C. S.; Koyanagi, J.; Takeuchi, M. *Org. Lett.* **2002**, *4*, 783.

88 Smith III, A. B.; Zhu, W.; Shirakami, S.; Sfouggatakis, C.; Doughty, V. A.; Bennett, C. S. *et al. Org. Lett.* **2003**, *5*, 761.

89 Paterson, I.; Chen, D. Y. K.; Coster, M. J.; Aceña, J. L.; Bach, J.; Gibson, K. R. *et al. Angew. Chem., Int. Ed.* **2001**, *40*, 4055.

90 Crimmins, M. T.; Katz, J. D.; Washburn, D. G.; Allwein, S. P.; McAtee, L. F. *J. Am. Chem. Soc.* **2002**, *124*, 5661.

91 Hubbs, J. L.; Heathcock, C. H. *J. Am. Chem. Soc.* **2003**, *125*, 12836.

92 Heathcock, C. H.; McLaughlin, M.; Medina, J.; Hubbs, J. L.; Wallace, G. A.; Scott, R. *et al. J. Am. Chem. Soc.* **2003**, *125*, 12844.

93 Ball, M.; Gaunt, M. J.; Hook, D. F.; Jessiman, A. S.; Kawahara, S.; Orsini, P. *et al. Angew. Chem., Int. Ed.* **2005**, *44*, 5433.

94 Kraus, H.; Francais, A.; O'Brien, M.; Frost, J.; Polara, A.; Dieguez-Vazquez, A. *et al. Chem. Sci.* **2013**, *4*, 1989.

95 Reznik, S. K.; Leighton, J. L. *Chem. Sci.* **2013**, *4*, 1497.

96 Smith III, A. B.; Sfouggatakis, C.; Risatti, C. A.; Sperry, J. B.; Zhu, W.; Doughty, V. A. *et al. Tetrahedron* **2009**, *65*, 6489.

97 Smith III, A. B.; Lin, Q.; Doughty, V. A.; Zhuang, L.; McBriar, M. D.; Kerns, J. K. *et al. Tetrahedron* **2009**, *65*, 6470.

98 Corey, E. J.; Seebach, D. *Angew. Chem., Int. Ed.* **1965**, *4*, 1075.

99 Tietze, L. F.; Geissler, H.; Gewert, J. A.; Jakobi, U. *Synlett* **1994**, *511*.

100 Smith III, A. B.; Boldi, A. M. *J. Am. Chem. Soc.* **1997**, *119*, 6925.

101 Smith III, A. B.; Adams, C. M. *Acc. Chem. Res.* **2004**, *37*, 365.

102 Smith III, A. B.; Wuest, W. M. *Chem. Commun* **2008**, *5883*.

103 Smith III, A. B.; Foley, M. A.; Dong, S.; Orbin, A. *J. Org. Chem.* **2009**, *74*, 5987.

104 Chen, J. L.-Y.; Brimble, M. A. *Chem. Commun* **2010**, *46*, 3967.

105 Chen, J. L.-Y.; Brimble, M. A. *J. Org. Chem.* **2011**, *76*, 9417.

106 Spangenberg, T.; Aubry, S.; Kishi, Y. *Tetrahedron Lett.* **2010**, *51*, 1782.

107 Guella, G.; Pietra, F.; Dini, F. *Angew. Chem., Int. Ed.* **1999**, *38*, 1134.

108 Nicolaou, K. C.; Ortiz, A.; Zhang, H. *Angew. Chem., Int. Ed.* **2009**, *48*, 5648.

109 Nicolaou, K. C.; Zhang, H.; Ortiz, A. *Angew. Chem., Int. Ed.* **2009**, *48*, 5642.

110 Nicolaou, K. C.; Ortiz, A.; Zhang, H.; Dagneau, P.; Lanver, A.; Jennings, M. P. *et al. J. Am. Chem. Soc.* **2010**, *132*, 7138.

111 Nicolaou, K. C.; Ortiz, A.; Zhang, H.; Guella, G. *J. Am. Chem. Soc.* **2010,** *132,* 7153.

112 Kitamura, M.; Miki, T.; Nakano, K.; Noyori, R. *Tetrahedron Lett.* **1996,** *37,* 5141.

113 Bergström, S. Samuelsson, B. *J. Biol. Chem.* **1962,** *237,* 3005.

114 Suzuki, M.; Suzuki, T.; Kawagishi, T.; Noyori, R. *Tetrahedron Lett.* **1982,** *23,* 4057.

115 Noyori, R.; Suzuki, M. *Angew. Chem., Int. Ed. Engl.* **1984,** *23,* 847.

116 Suzuki, M.; Yanagisawa, A.; Noyori, R. *J. Am. Chem. Soc.* **1988,** *110,* 4718.

117 Furstner, A.; Mathes, C.; Lehmann, C. W. *J. Am. Chem. Soc.* **1999,** *121,* 9453.

118 Lipshutz, B. H.; Wood, M. R. *J. Am. Chem. Soc.* **1994,** *116,* 11689.

119 Wrobleski, A.; Sahasrabudhe, K.; Aubé, J. *J. Am. Chem. Soc.* **2002,** *124,* 9974.

120 Kuramochi, A.; Usuda, H.; Yamatsugu, K.; Kanai, M.; Shibasaki, M. *J. Am. Chem. Soc.* **2005,** *127,* 14200.

121 Oi, S.; Sato, T.; Inoue, Y. *Tetrahedron Lett.* **2004,** *45,* 5051.

122 Nicolaou, K. C.; Tang, W. J.; Dagneau, P.; Faraoni, R. *Angew. Chem., Int. Ed.* **2005,** *44,* 3874.

123 Nicolaou, K. C.; Jennings, M. P.; Dagneau, P. *Chem. Commun.* **2002,** *2480.*

124 Ducrot, P. H.; Lallemand, J. Y. *Tetrahedron Lett.* **1990,** *31,* 3879.

125 Asano, N.; Kato, A.; Yokoyama, Y.; Yamamoto, M.; Miyauchi, M.; Kizu, H. *et al. Carbohydr. Res.* **1996,** *284,* 169.

126 Drager, B. *Nat. Prod. Rep.* **2004,** *21,* 211.

127 Asano, N.; Kato, A.; Kizu, H.; Matsui, K.; Watson, A. A.; Nash, R. J. *Carbohydr. Res.* **1996,** *293,* 195.

128 Petasis, N. A.; Akritopoulou, I. *Tetrahedron Lett.* **1993,** *34,* 583.

129 Petasis, N. A.; Zavialov, I. A. *J. Am. Chem. Soc.* **1998,** *120,* 11798.

130 Arai, S.; Sugiyama, S.; Ishii, K. *Tetrahedron: Asymm.* **2004,** *15,* 3149.

131 Batey, R. A.; MacKay, D. B. *Tetrahedron Lett.* **2000,** *41,* 9935.

132 Hong, Z. Y.; Liu, L.; Hsu, C. C.; Wong, C. H. *Angew. Chem., Int. Ed.* **2006,** *45,* 7417.

133 Candeias, N. R.; Montalbano, F.; Cal, P. M. S. D.; Gois, P. M. P. *Chem. Rev.* **2010,** *110,* 6169.

134 Moosophon, P.; Baird, M. C.; Kanokmedhakul, S.; Pyne, S. G. *Eur. J. Org. Chem.* **2010,** *3337.*

135 Gerth, K.; Washausen, P.; Hofle, G.; Reichenbach, H.; Irschik, H. *J. Antibiotics* **1996,** *49,* 71.

136 Pospisil, J.; Markó, I. E. *J. Am. Chem. Soc.* **2007,** *129,* 3516.

137 Hanessian, S.; Focken, T.; Oza, R. *Org. Lett.* **2010,** *12,* 3172.

138 Hosomi, A.; Sakurai, H. *Tetrahedron Lett.* **1976,** *1295.*

139 Sakurai, H. *Pure Appl. Chem.* **1982,** *54,* 1.

140 Sakurai, H.; Sasaki, K.; Hayashi, J.; Hosomi, A. *J. Org. Chem.* **1984,** *49,* 2808.

141 Tietze, L. F.; Kinzel, T.; Brazel, C. C. *Acc. Chem. Res.* **2009,** *42,* 367.

142 Leroy, B.; Markó, I. E. *J. Org. Chem.* **2002,** *67,* 8744.

143 Markó, I. E.; Dumeunier, R.; Leclercq, C.; Leroy, B.; Plancher, J. M.; Mekhalfia, A. *et al. Synthesis* **2002**, *958*.
144 Markó, I. E.; Bayston, D. J. *Synthesis* **1996**, *297*.
145 Markó, I. E.; Bailey, M.; Murphy, F.; Declercq, J. P.; Tinant, B.; Feneaudupont, J. *et al. Synlett* **1995**, *123*.
146 Markó, I. E.; Bayston, D. J. *Tetrahedron* **1994**, *50*, 7141.
147 Markó, I. E.; Bayston, D. J. *Tetrahedron Lett.* **1993**, *34*, 6595.
148 Markó, I. E.; Bayston, D. J.; Mekhalfia, A.; Adams H. *Bull. Soc. Chim. Belges* **1993**, *102*, 655.
149 Markó, I. E.; Mekhalfia, A.; Bayston, D. J.; Adams H. *J. Org. Chem.* **1992**, *57*, 2211.
150 Markó, I. E.; Mekhalfia, A. *Tetrahedron Lett.* **1992**, *33*, 1799.
151 Mekhalfia, A.; Markó, I. E. *Tetrahedron Lett.* **1991**, *32*, 4779.
152 Mekhalfia, A.; Markó, I. E.; Adams H. *Tetrahedron Lett.* **1991**, *32*, 4783.
153 Pospisil, J.; Kumamoto, T.; Markó, I. E. *Angew. Chem., Int. Ed.* **2006**, *45*, 3357.
154 Kobayashi, J.; Watanabe, D.; Kawasaki, N.; Tsuda, M. *J. Org. Chem.* **1997**, *62*, 9236.
155 Kobayashi, J.; Tsuda, M.; Ishibashi, M. *Pure App. Chem.* **1999**, *71*, 1123.
156 Bonazzi, S.; Cheng, B.; Wzorek, J. S.; Evans, D. A. *J. Am. Chem. Soc.* **2013**, *135*, 9338.
157 Jakubec, P.; Kyle, A. F.; Calleja, J.; Dixon, D. J. *Tetrahedron Lett.* **2011**, *52*, 6094.
158 Cheng, B.; Wu, F.; Yang, X.; Zhou, Y.; Wan, X.; Zhai, H. *Chem. Eur. J.* **2011**, *17*, 12569.
159 Kyle, A. F.; Jakubec, P.; Cockfield, D. M.; Cleator, E.; Skidmore, J.; Dixon, D. J. *Chem. Commun.* **2011**, *47*, 10037.
160 Nilson, M. G.; Funk, R. L. *Org. Lett.* **2010**, *12*, 4912.
161 Inagaki, F.; Kinebuchi, M.; Miyakoshi, N.; Mukai, C. *Org. Lett.* **2010**, *12*, 1800.
162 Stockman, R. A.; McDermott, P. J.; Newton, A. F.; Magnus, P. *Synlett* **2010**, *559*.
163 Martin, D. B. C.; Vanderwal, C. D. *Angew. Chem., Int. Ed.* **2010**, *49*, 2830.
164 Jakubec, P.; Cockfield, D. M.; Dixon, D. J. *J. Am. Chem. Soc.* **2009**, *131*, 16632.
165 Young, I. S.; Kerr, M. A. *J. Am. Chem. Soc.* **2007**, *129*, 1465.
166 Nakagawa, M.; Nagata, T.; Ono, K.; Nishida, A. *J. Synth. Org. Chem. Jpn.* **2005**, *63*, 200.
167 Ono, K.; Nakagawa, M.; Nishida, A. *Angew. Chem., Int. Ed.* **2004**, *43*, 2020.
168 Nagata, T.; Nakagawa, M.; Nishida, A. *J. Am. Chem. Soc.* **2003**, *125*, 13618.
169 Nagata, T.; Nakagawa, M.; Nishida, A. *J. Am. Chem. Soc.* **2003**, *125*, 7484.
170 Yu, M.; Wang, C.; Kyle, A. F. P.; Jakubec, P.; Dixon, D. J.; Schrock, R. R. *et al. Nature* **2011**, *479*, 88.
171 de Graaff, C.; Ruijter, E.; Orru, R. V. A. *Chem. Soc. Rev.* **2013**, *41*, 3969.
172 Pellissier, H. *Chem. Rev.* **2013**, *113*, 442.
173 Pellissier, H. *Adv. Synth. Catal.* **2011**, *353*, 189.

174 Pellissier, H. *Tetrahedron* **2007**, *63*, 3235.
175 Ahrendt, K. A.; Williams, R. M. *Org. Lett.* **2004**, *6*, 4539.
176 Ahrendt, K. A.; Williams, R. M. *Org. Lett.* **2005**, *7*, 957.
177 Pinho e Melo, T. M. V. D. *Eur. J. Org. Chem.* **2006**, *2873*.
178 Najera, C.; Sansano, J. M. *Curr. Org. Chem.* **2003**, *7*, 1105.
179 Padwa, A.; Dean, D. C.; Osterhout, M. H.; Precedo, L.; Semones, M. A. *J. Org. Chem.* **1994**, *59*, 5347.
180 Dauban, P.; Malik, G. *Angew. Chem., Int. Ed.* **2009**, *48*, 9026.
181 Gothelf, K. V.; Jorgensen, K. A. *Chem. Rev.* **1998**, *98*, 863.
182 Young, I. S.; Kerr, M. A. *Angew. Chem., Int. Ed.* **2003**, *42*, 3023.
183 Carson,C. A.; Young, I. S.; Kerr, M. A. *Synthesis* **2008**, *485*.
184 Young, I. S.; Kerr, M. A. *Org. Lett.* **2004**, 6, 139.
185 Pulz, R.; Al-Harrasi, A.; Reissig, H.-U. *Org. Lett.* **2002**, *4*, 2353.
186 Young, I. S.; Williams, J. L.; Kerr, M. A. *Org. Lett.* **2005**, *7*, 953.

9

Renewable Resource-Based Building Blocks/Chirons for the Total Synthesis of Natural Products

Wai-Lung Ng[1,2,3], *Anthony W. H. Wong*[3], *and Tony K. M. Shing*[3,4]

[1] *Department of Medicine, Harvard Medical School, Boston, MA, USA*
[2] *Department of Cancer Biology, Dana-Farber Cancer Institute, Boston, MA, USA*
[3] *Department of Chemistry, The Chinese University of Hong Kong, Shatin, New Territories, Hong Kong SAR, China*
[4] *Department of Chemistry, Faculty of Science and Technology, Keio University, Yokohama, Kanagawa, Japan*

9.1 Introduction

9.1.1 The Chiron Approach Toward the Total Synthesis of Natural Products

The use of enantiopure small molecules as chiral building blocks in total synthesis was coined the **chiron approach** by Hanessian.[1] The term "chiron" is a chiral synthon,[1] and in this chapter, we will focus our discussion on using renewable resource-based chirons in natural product total synthesis. For further reading, readers may refer to the comprehensive text on both natural and non-natural chiron-based syntheses written by Hanessian *et al.*[1d]

The two major strategies toward the total synthesis of optically active natural products are the chiron approach and the **asymmetric synthesis approach**.[2] The asymmetric synthesis approach offers the advantage that the stereochemical outcome of reactions can be controlled through judicial choice of a chiral ligand/catalyst. On the other hand, the chiron approach offers facile establishment of desired stereochemistry and functionalities in addition to the carbon skeleton of the target molecules. This chapter aims to introduce the latter approach with the aid of some selected literature examples.

9.1.2 General Survey of Natural Chirons

Mother Nature provides synthetic organic chemists with an abundant source of chiral building blocks. This chiral pool mainly consists of five types of small

Efficiency in Natural Product Total Synthesis, First Edition. Edited by Pei-Qiang Huang, Zhu-Jun Yao, and Richard P. Hsung.

and optically active molecules, namely carbohydrates, α-amino acids, terpenes, α-hydroxy acids, and cyclitols (Figure 9.1).

Carbohydrates have frequently been employed in the total synthesis of complex natural products. They are commercially available in large quantities and are inexpensive. The availability of D- and L-sugars with different chain lengths and stereochemistry makes them excellent candidates for the construction of tetrahydrofuran-type moieties with densely functionalized stereogenic centers. However, significant modifications of the carbon skeleton and extensive protection/deprotection steps should be minimized in order to achieve an efficient and economical synthetic route.

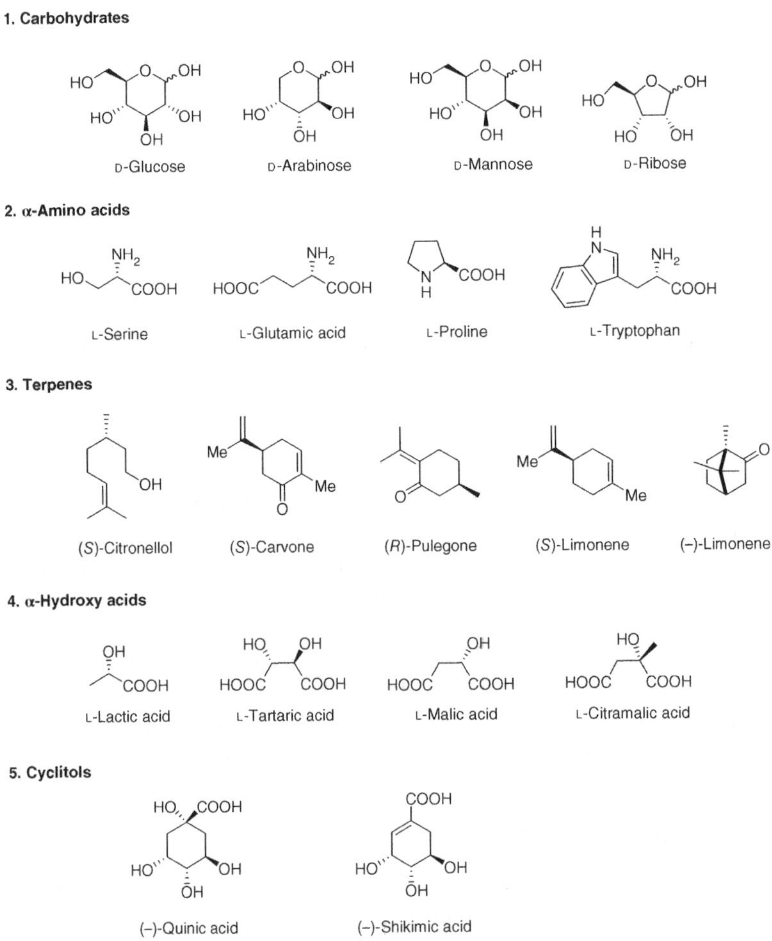

Figure 9.1 Some representative and easily available natural chirons.

α-Amino acids constitute another important class of natural chirons. The 21 natural L-α-amino acids consist of side-chains with diverse functionality and chain lengths, making them ideal building blocks in natural product total synthesis. In particular, amino acids like L-serine and L-glutamic acid have been popular starting materials in the total synthesis of alkaloids. Terpenes such as citronellol, carvone, and camphor are naturally occurring hydrocarbons that possess synthetic handles like carbonyl, hydroxyl, and alkene moieties. They are frequently used in the total synthesis of terpenoid natural products. With careful choice of acyclic, cyclic, or bicyclic terpenes, synthetic chemists can rapidly construct the carbon skeleton of the target molecules with stereocenters pre-installed. α-Hydroxy acids are also nature's gifts that enabled synthetic chemists to complete many total syntheses. Within this class of molecules, D- and L-tartaric acids possess a C_2 symmetry that makes them valuable building blocks.[3]

Cyclitols are a class of primary metabolites that contain multiple hydroxyl groups and a six-membered carbocycle. Examples of cyclitols, include (−)-quinic acid and (−)-shikimic acid, are biosynthetically derived from D-glucose. The existing polyhydroxylated cyclohexane ring in cyclitols can provide a facile entry to the highly functionalized carbon framework in complex natural products.

In this chapter, we discuss recent literature examples that have employed the chiron approach in the total syntheses of alkaloids, terpenoids, and other natural products like polyketides and peptide-based small molecules. The key reactions involved in each synthesis will be highlighted.

9.2 Total Synthesis of Alkaloids

9.2.1 Amino Acids as Starting Chirons

Hanessian *et al.* employed D-serine as the starting chiron to synthesize the enantiomer of cyclizidine (**1**).[4] The key reactions involved the stereocontrolled construction of the functionalized indolizidine ring and the installation of the cyclopropyl dienyl side chain. First, mesylate **3** was constructed from the readily available *N*-Boc-D-serine (**2**) in 21 steps. After that, tandem removal of fluorenylmethoxycarbonyl (Fmoc) group and cyclization gave bicyclic compound **4** at an excellent yield. The cyclopropyl diene unit was constructed in four steps to obtain the (+)-cyclizidine (**1**) (Scheme 9.1).

N-Boc-D-Serine (**7**) was also employed by Knight *et al.* in the synthesis of D-mannolactam (**5**) and deoxymannojirimycin (**6**).[5] The key step was a palladium-catalyzed decarboxylative carbonylation of a serine-derived 5-vinyloxazolidin-2-one (**8**) to give 6-(tertbutyldimethylsilyloxymethyl)-3,6-dihydro-1H-pyridin-2-one (**9**), which was ultimately converted into the target compounds (Scheme 9.2).

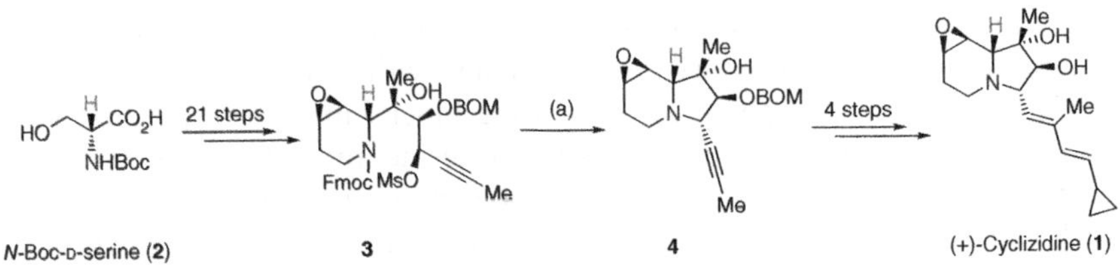

Scheme 9.1 Reagents and conditions: (a) Piperidine, CH_3CN, rt, 100%.

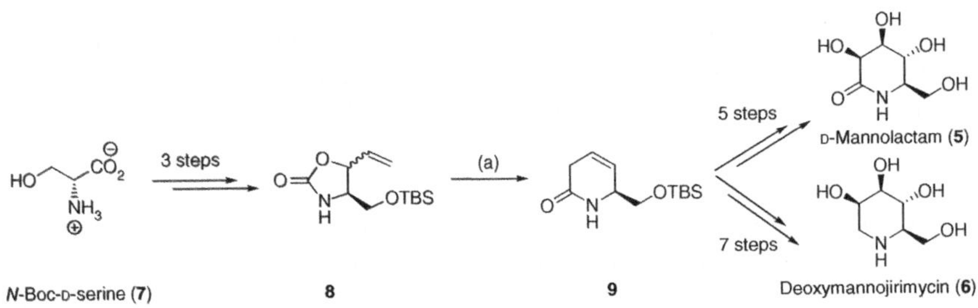

Scheme 9.2 Reagents and conditions: (a) $PdCl_2(PPh_3)_2$ (10 mol%), CO (65 atm), EtOH, 60°C, 81%.

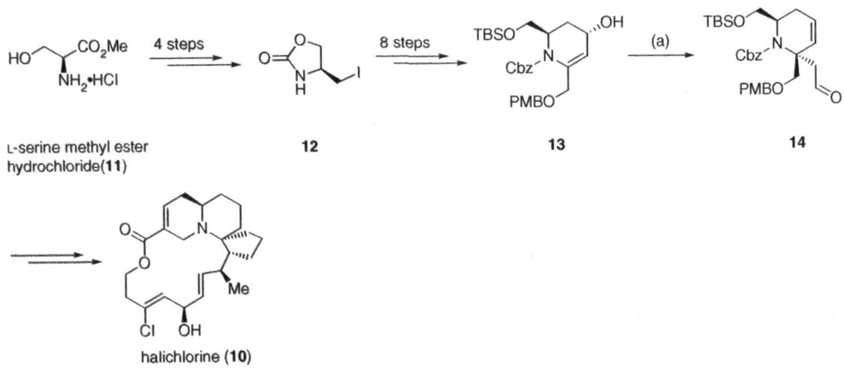

Scheme 9.3 Reagents and conditions: (a) Butyl vinyl ether, $Hg(OAc)_2$, Et_3N, 110°C, 36 h, 79%.

In the total synthesis of the marine alkaloid halichlorine (**10**),[6] L-serine methyl ester hydrochloride (**11**) was converted into iodide **12** in four steps,[7] which was subsequently converted into alcohol **13** (Scheme 9.3). Alcohol **13** was subjected to the key Claisen rearrangement by heating in butyl vinyl ether in the presence of $Hg(OAc)_2$ and Et_3N, furnishing an unstable aldehyde **14**. Further transformations yielded (+)-halichlorine (**10**). The highlight of this Claisen rearrangement is that the required conditions were mild compared with the usual conditions,[8] possibly due to the involvement of mercuric ion catalysis.[9] No previous examples of Claisen rearrangement in which the distal terminus of the allylic double bond system carries a nitrogen atom were reported.

In the study of the total synthesis of (+)-chaetocins A (**15**) (Scheme 9.4) and C (**16**) and (+)-12,12′-dideoxychetracin A (**17**) (Scheme 9.5), Kim *et al.* used commercially available *N*-Boc-L-tryptophan (**18**) and L-serine methyl ester hydrochloride (**19**) to synthesize the key intermediate diketopiperazine (−)-**19** followed by bromocyclization to afford *endo*-tetracyclic bromide (+)-**20** in 59% with high diastereoselectivity (Scheme 9.4).[10] The key intermediate (+)-**22** was obtained by $CoCl(PPh_3)_3$-mediated reductive radical dimerization of (+)-**21** in 49% yield.[11] The innate reactivity differences between the C-11 and C-15 hemiaminals permitted a regio- and stereoselective introduction of the critical sulfur moiety of *epi*di-, *epi*tri-, and *epi*tetra-thiodiketopiperazines. Reacting tetraol (−)-**23** with trifluoroacetic acid in hydrogen sulfide-saturated nitromethane gave the sensitive bisthiohemiaminal **24** in a highly diastereoselective fashion. Dithioisobutyrate (+)-**25** was afforded at a 53% yield in two steps by concentrating the reaction mixture followed by the addition of isobutyryl chloride. The isobutyrates served the dual role of preventing detrimental hemiaminal opening under polar protic conditions and activating the C-15 hemiaminal function in an anticipated mild ionization. The unique challenge of C-11 ionization was solved by irradiating a solution of (+)-**25** with a black-light phosphor-coated lamp in the presence of the photosensitizer 1,4-dimethoxynaphthalene. The terminal reductant L-ascorbic acid cleanly provided the desired diaminodithioisobutyrate (+)-**26** at a 51% yield.[12] Subsequent transformations delivered the iminium ion **29**. Efficient cyclization with concomitant loss of a triphenylmethyl cation was accomplished (82% yield) to give (+)-chaetocin A diacetate (**30**). Deacetylation gave the natural (+)-chaetocin A (**15**). Syntheses of (+)-chaetocin C (**16**) and (+)-12,12′-dideoxychetracin A (**17**) are outlined in Scheme 9.5, which was accomplished by using the same strategy as shown in Scheme 9.4.

L-Serine methyl ester (**11**) was also employed in the total synthesis of (+)-malbrancheamide B (**35**) and (+)-premalbrancheamide (**36**) (Scheme 9.6) by Laws *et al.*[13] The key of the synthesis was the domino reaction sequence including aldol condensation, alkene isomerization, and an intramolecular Diels–Alder (IMDA) reaction. The domino reaction of compound **37**, which was obtained in four steps from L-serine methyl ester (**11**)[14], and indole **38**, which was derived from 6-chloroindole in four steps,[15] gave diastereomeric cycloadducts **39** and **40** in a 1:2 ratio and an 85% yield (Scheme 9.7). Further transformation on cycloadduct **40** afforded (+)-malbrancheamide B (**43**).

The racemic (±)-malbrancheamide B **35** and **41** were obtained by a domino reaction of indole **43** with compound **44**, which was derived from L-proline methyl ester (**42**) in three steps (Scheme 9.7). This reaction sequence can be rationalized by initial enolization of compound **44** followed by addition to the aldehyde in indole **43** to give the intermediate, the aldol condensation product **45**. Isomerization of the exocyclic alkene in **45** to the reactive endocyclic azadiene **46** was effected under basic reaction conditions. IMDA cycloaddition of the azadiene terminated the reaction sequence, yielding a mixture of **47**–**49**.

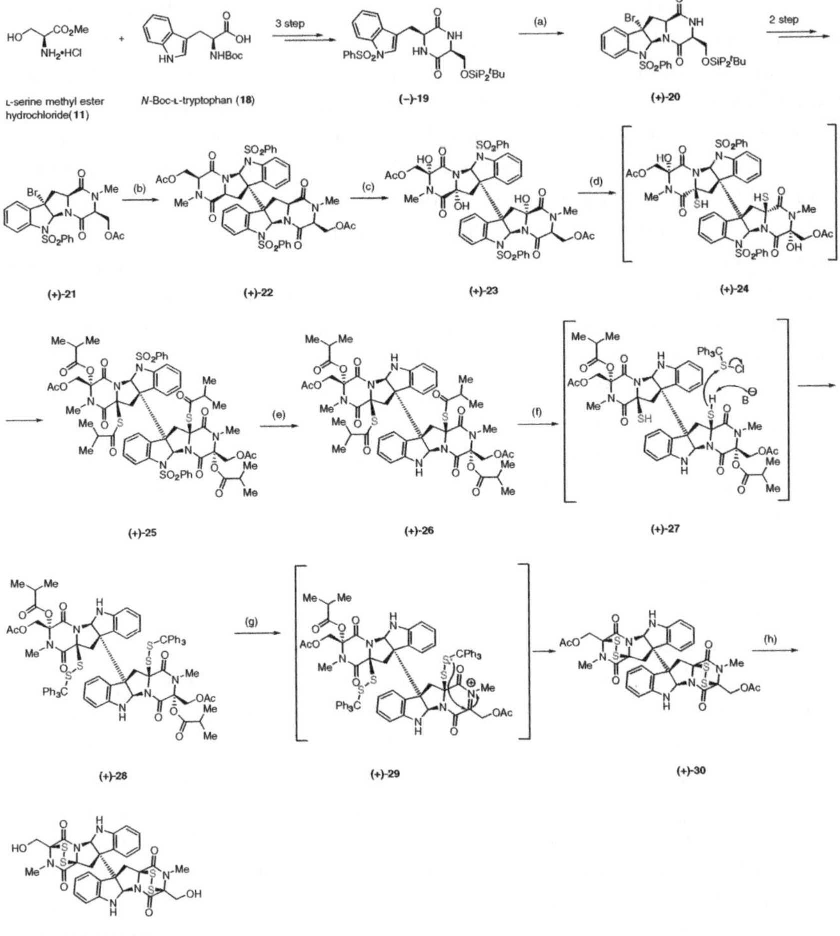

Scheme 9.4 Reagents and conditions: (a) Br_2, benzene, 59%; (b) $CoCl(PPh_3)_3$, CH_2Cl_2, 49%; (c) Py_2AgMnO_4, CH_2Cl_2, 55%; (d) H_2S, TFA, $MeNO_2$; ^iPrCOCl, CH_2Cl_2, 53% (two steps); (e) *hυ* (350 nm), L-ascorbic acid, 1,4-dimethoxynaphthalene, H_2O, CH_3CN, 51%; (f) N_2H_4, THF, 0°C; NaH, Ph_3CSCl, 90%; (g) BF_3•OEt_2, 2,6-di-^{t}Bu-4-Me-pyridine, Et_3SiH, CH_2Cl_2, 82%; (h) Otera's cat., MeOH, toluene, 85°C, 92%.

Diketopiperazine product **49** arose from lactam hydrolysis of cycloadduct **48** during the course of the reaction. Compound **48** and diketopiperazine product **49** were converted into (±)-malbrancheamide B (**35**) and (**41**).

In an improved total synthesis of (+)-macroline (**50**) and (−)-alstonerine (**51**) reported by Liao *et al.*, D-tryptophan methyl ester (**52**) was converted into azabicyclononone **53** in greater than 98% *ee* in a three-pot process.[16,17] Azabicyclononone **53** was then converted into the coupled

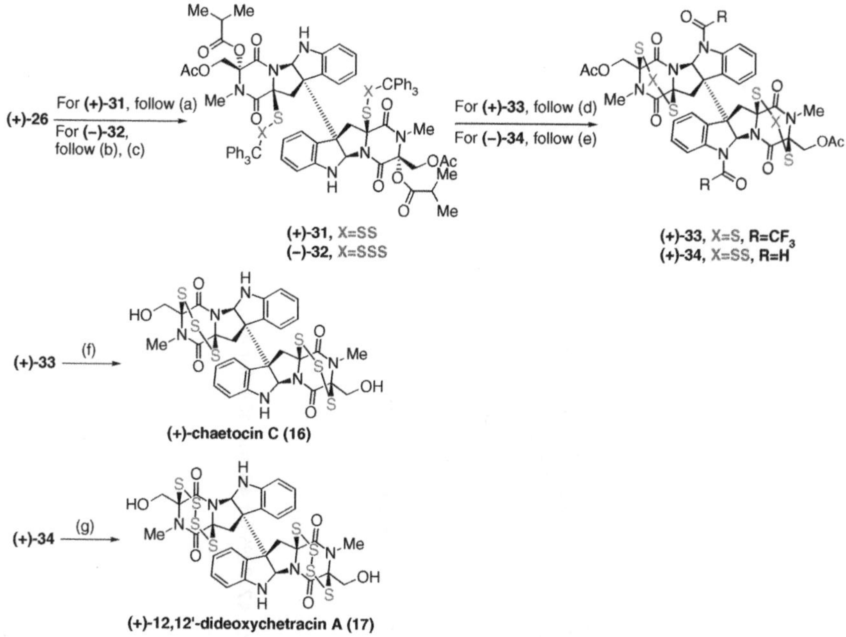

Scheme 9.5 Reagents and conditions: (a) N_2H_4, THF, 0°C; TrSSCl, Et_3N, 86%; (b) N_2H_4, THF, 0°C, 93%; (c) TrSSSCl, Et_3N, 80%; (d) TFAA, 2,6-di-^{t}Bu-4-Me-pyridine, CH_3CN; BF_3•OEt_2, 91%. (e) HCO_2Ac; CH_3CN, BF_3•OEt_2, 60%. (f) Otera's cat., MeOH, toluene, 90°C; N_2H_4, 95%. (g) HCl, MeOH, 52%.

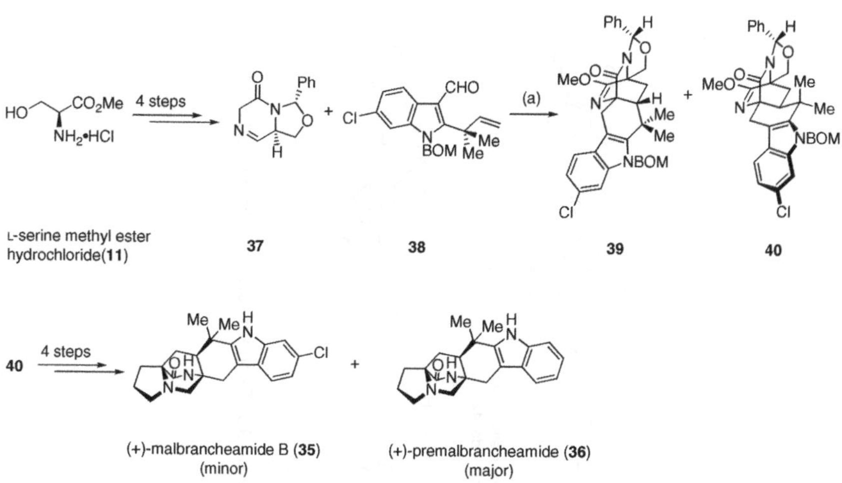

Scheme 9.6 Reagents and conditions: (a) NaOMe, MeOH, 65°C, 85%.

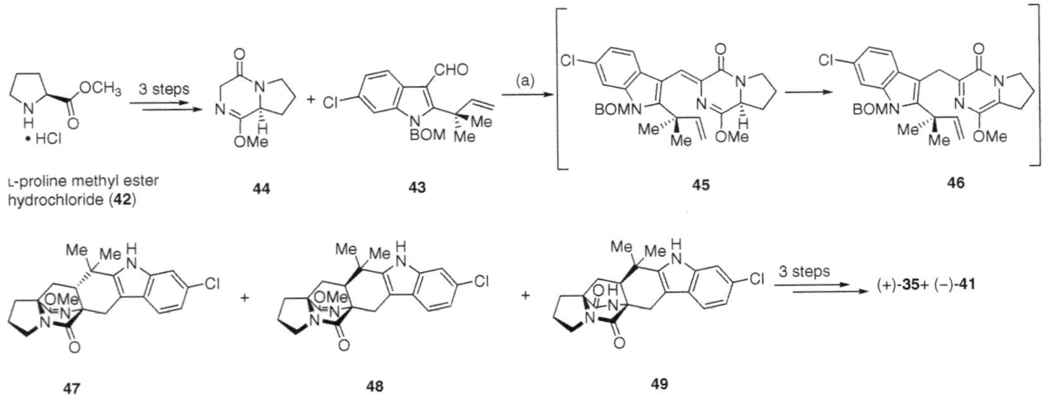

Scheme 9.7 Reagents and conditions: (a) NaOMe, MeOH, 65°C, 99% (29% for **47**, 22% for **48** and 48% for **49**).

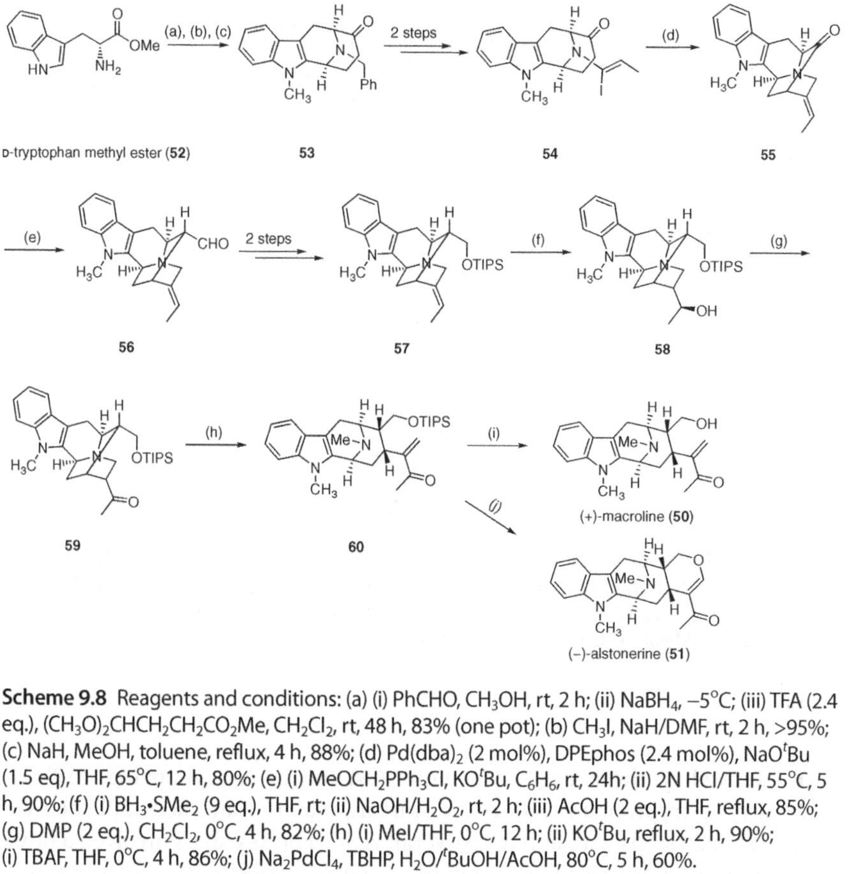

Scheme 9.8 Reagents and conditions: (a) (i) PhCHO, CH_3OH, rt, 2 h; (ii) $NaBH_4$, −5°C; (iii) TFA (2.4 eq.), $(CH_3O)_2CHCH_2CH_2CO_2Me$, CH_2Cl_2, rt, 48 h, 83% (one pot); (b) CH_3I, NaH/DMF, rt, 2 h, >95%; (c) NaH, MeOH, toluene, reflux, 4 h, 88%; (d) $Pd(dba)_2$ (2 mol%), DPEphos (2.4 mol%), NaO^tBu (1.5 eq), THF, 65°C, 12 h, 80%; (e) (i) $MeOCH_2PPh_3Cl$, KO^tBu, C_6H_6, rt, 24h; (ii) 2N HCl/THF, 55°C, 5 h, 90%; (f) (i) $BH_3{\bullet}SMe_2$ (9 eq.), THF, rt; (ii) $NaOH/H_2O_2$, rt, 2 h; (iii) AcOH (2 eq.), THF, reflux, 85%; (g) DMP (2 eq.), CH_2Cl_2, 0°C, 4 h, 82%; (h) (i) MeI/THF, 0°C, 12 h; (ii) KO^tBu, reflux, 2 h, 90%; (i) TBAF, THF, 0°C, 4 h, 86%; (j) Na_2PdCl_4, TBHP, $H_2O/^tBuOH/AcOH$, 80°C, 5 h, 60%.

product **54** using the newly discovered Pd(0)-catalyzed α-vinylation reaction, followed by Wittig reaction to obtain the key intermediate N_a-methylvellosimine **55** (Scheme 9.8).[18,19] (+)-Macroline (**50**) was then obtained via a selective hydroboration-oxidation and retro-Michael ring-opening sequence. In addition, the intramolecular Tsuji–Wacker oxidation[20] was employed to convert the protected (+)-macroline **60** into (−)-alstonerine **51** (Scheme 9.8), which was further elaborated into (−)-talcarpine[16] and (−)-anhydromacrolsalhinemethine.[21]

(+)-Vellosimine **64**, which was derived from D-tryptophan methyl ester (**52**),[16,22] was also employed for the synthesis of (+)-polyneuridine (**61**), (+)-polyneuridine aldehyde (**62**), and macusine A (**63**). The key oxidation was accomplished with the Corey–Kim reagent to provide the important biosynthetic intermediate, polyneuridine aldehyde (**62**) (Scheme 9.9).[22]

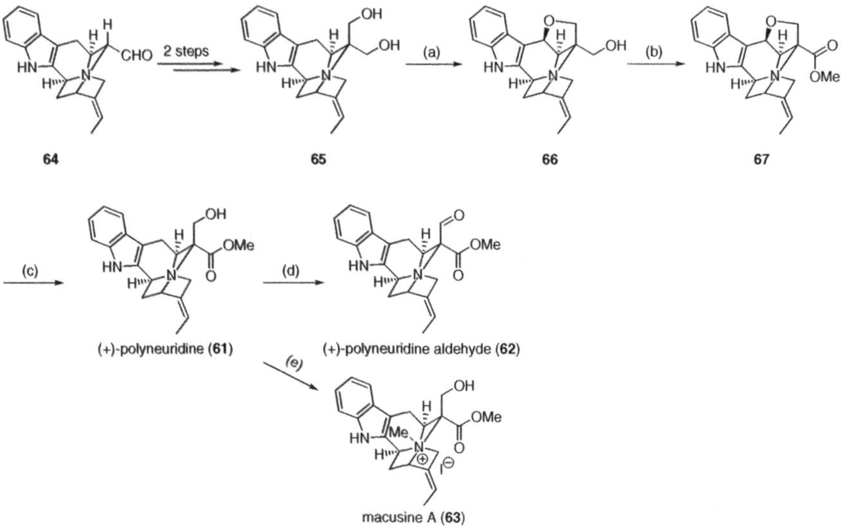

Scheme 9.9 Reagents and conditions: (a) DDQ, THF, −78°C, 30 mins, 90%; (b) (i) NCS, SMe_2, CH_2Cl_2, −78°C, 2 h then Et_3N, −78°C, 1 h; (ii) 85% KOH, I_2, MeOH, 0°C, 6 h, 90% from **71**; (c) Et_3SiH, TFA, CH_2Cl_2, rt, 12 h, 87%; (d) (i) NCS, SMe_2, CH_2Cl_2, −78°C, 2 h then Et_3N, −78°C, 1 h; (ii) 85% KOH, I_2, MeOH, 0°C, 6 h, 75%; (e) MeI, THF, 0°C, 8 h, 81%.

In the total synthesis of (−)-ardeemin (**68**) and its *N*-acyl analogs (**69**), He *et al.* used L-tryptophan (**70**) as the starting material and accomplished the synthesis in 20 steps with about 2% overall yield. The key steps involved a novel three-step one-pot cascade reaction of **71** with diazoester **72** via an intermolecular cyclopropanation, ring opening, and a ring closure to assemble the chiral 3-substituted hexahydropyrrolo[2,3-*b*]indole **76** (Scheme 9.10).[23]

L-Proline is another widely used chiron for syntheses of alkaloids. In 2010, Niphakis *et al.* used Boc-L-proline (**79**) to synthesize (+)-ipalbidine (**77**) and (+)-antofine (**78**) (Scheme 9.11).[24] The key features of the synthesis were the 6-*endo-trig* cyclization to obtain enone **82** and the direct C-H arylation reaction to afford the arylindolizidine **83**. Negishi cross-coupling reaction[25] was used to install the 3,4-dimethoxyphenyl group in the synthesis of (+)-antofine (**78**).

Derivatives of L-proline are also popular chirons for alkaloid syntheses. In 2007, Sánchez-Izquierdo *et al.* reported the total synthesis of the putative structure of (−)-stemonidine (**85**) which converted commercially available L-prolinol (**86**) into nitrone **89**, followed by an 1,3-dipolar cycloaddition with olefin **90**.[26] Another key step for this synthesis was the spirolactonization of ketone **92** and aldehyde **93** (Scheme 9.12).[26] The original configuration of the

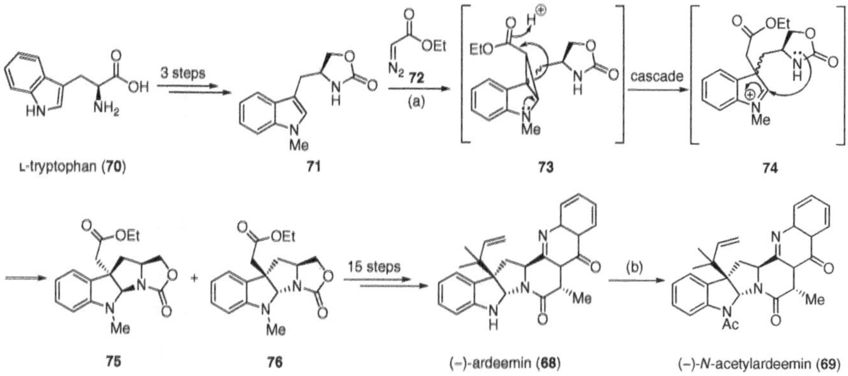

Scheme 9.10 Reagents and conditions: (a) Cu(OTf)-toluene, toluene, 25°C, 45% yield of **75** and 28% yield of **76**; (b) Ac_2O, DIPEA, 60°C, C_6H_6, 36 h, 72%.

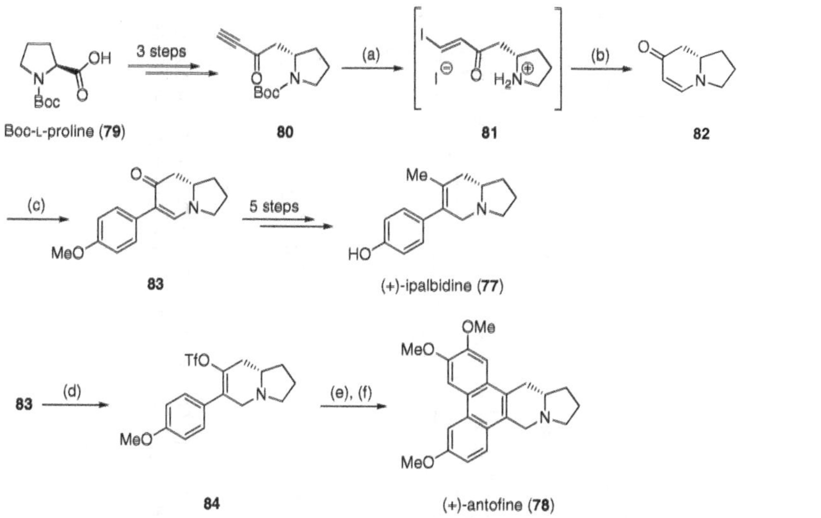

Scheme 9.11 Reagents and conditions: (a) NaI, HCO_2H, rt; (b) K_2CO_3, MeOH, 1 h, 96% from **80**; (c) $Pd(OAc)_2$ (30 mol%), $Cu(OAc)_2$ (3 eq.), PMP-BF_3K, K_2CO_3, ^{t}BuOH/AcOH/DMSO (20:5:1), 60°C, 70%; (d) L-selectride, THF, −78 to 0°C; then Comins' reagent, −78 to 0°C, 76%; (e) $Pd(PPh_3)_4$ (10 mol%), 3,4-dimethoxyphenylzinc bromide, THF, 60°C, 96%; (f) $PhI(O_2CCF_3)_2$, BF_3•Et_2O, CH_2Cl_2, −78°C, 70%.

natural (−)-stemonidine (**85**) was incorrectly assigned by Xu and coworkers[27,28] and the new assignment should be (−)-*epi*-stemonidine (**98**), which was also named as stemospironine.

In the total synthesis of (+)-dibromophakellin (**99**) (Scheme 9.13) and (+)-dibromophakellstation (**100**) (Scheme 9.14) reported by Imaoka and

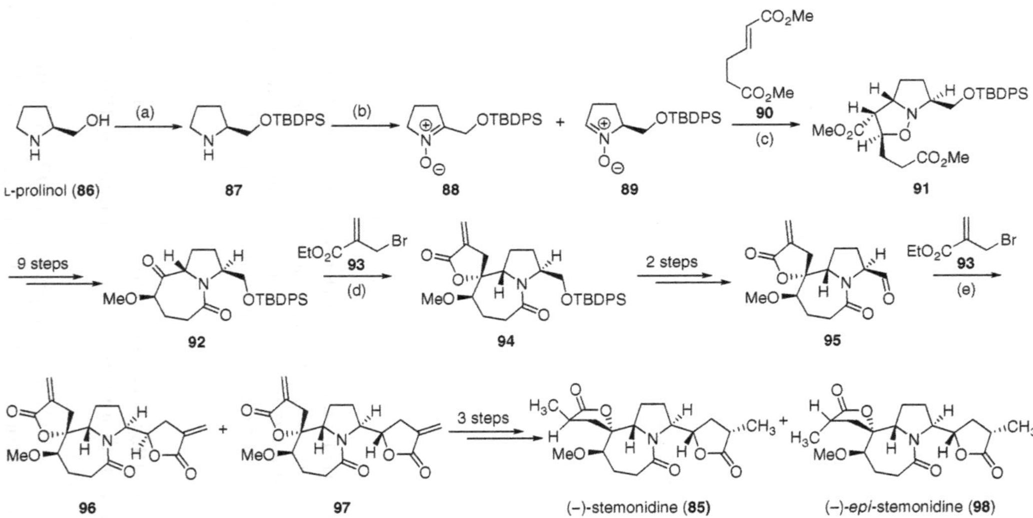

Scheme 9.12 Reagents and conditions: (a) TBDPSCl, Im, THF, 82%; (b) Oxone®, EDTA, $NaHCO_3$, THF/CH_3CN, 0°C, 90% (**91**:**92** = 1:1.3); (c) **90**, toluene, reflux, 78%; (d) **93**, Zn, THF, reflux, 86%; (e) **93**, Zn, THF, reflux, 38% of **96**, 8% of mixture of **96** and **97** and 27% of **97**.

coworkers, commercial available *trans*-4-hydroxyl-L-proline (**101**) was selected as the starting material. The key step involved a new enamide-type Overman rearrangement, yielding the quaternary chiral aminals (Scheme 9.13).[29] Surprisingly, only the β-isomer of **102** was converted into the target aminal. It was proposed that the acetate group in the α-isomer was sterically repulsive with the imidate group, resulting in an elimination rather than a rearrangement reaction.

In recent years, the readily available L-pyroglutamic acid (**110**) has been another commonly used chiron for alkaloids synthesis. In 2009, Lesma and coworkers developed a versatile route to diversely functionalized indolizidine and pyrrolizidine natural and non-natural alkaloids using cross-metathesis and

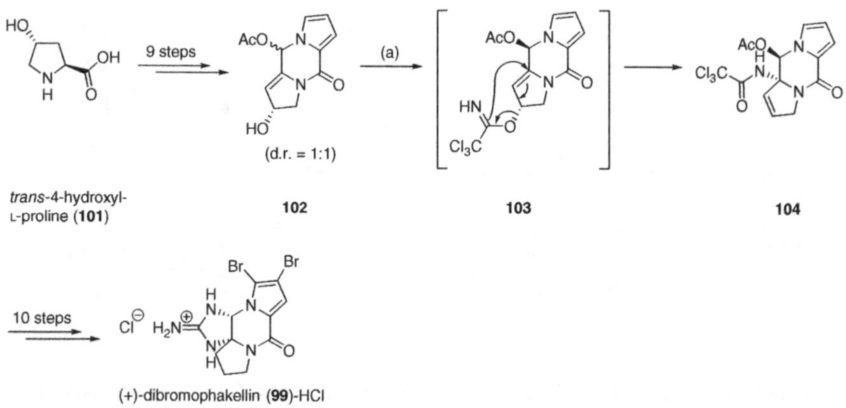

Scheme 9.13 Reagents and conditions: (a) Cl_3CCN, DBU, CH_2Cl_2, 0–25°C, 70%.

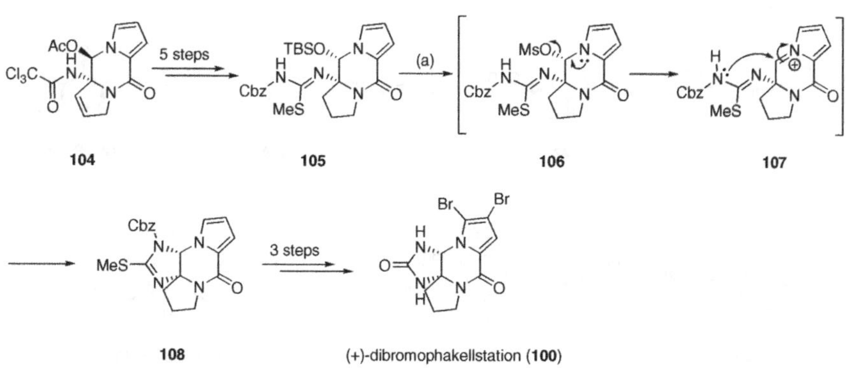

Scheme 9.14 Reagents and conditions: (a) (i) TBAF, 4 Å MS, THF/CH_2Cl_2, −40°C; (ii) MsCl, −40°C, 57%.

ring-closing metathesis as the key steps. It was further utilized on the total synthesis of (+)-monomorine (**109**) (Scheme 9.15).[30]

A new synthetic approach to (+)-batzellaside B (**121**) from naturally abundant L-pyroglutamic acid (**110**) was reported by Wierzejska *et al.* in 2012 (Scheme 9.16).[31] The key synthetic step involved Sharpless asymmetric dihydroxylation of allylic acetate **123** to introduce two chiral centers diastereoselectively. Diol **125** was then further converted into heterocyclic hemiaminal **126** in nine steps. Formal synthesis of the target natural product was accomplished in eight steps.[33]

In the total synthesis of (+)-septicine (**127**) reported by Reddy and coworkers, L-glutamic acid (**129**) was converted into the key intermediate diketone **131**, which underwent a intramolecular McMurry coupling to give compound **132**. Reduction of amide **132** with $LiAlH_4$ furnished (+)-septicine (**127**)

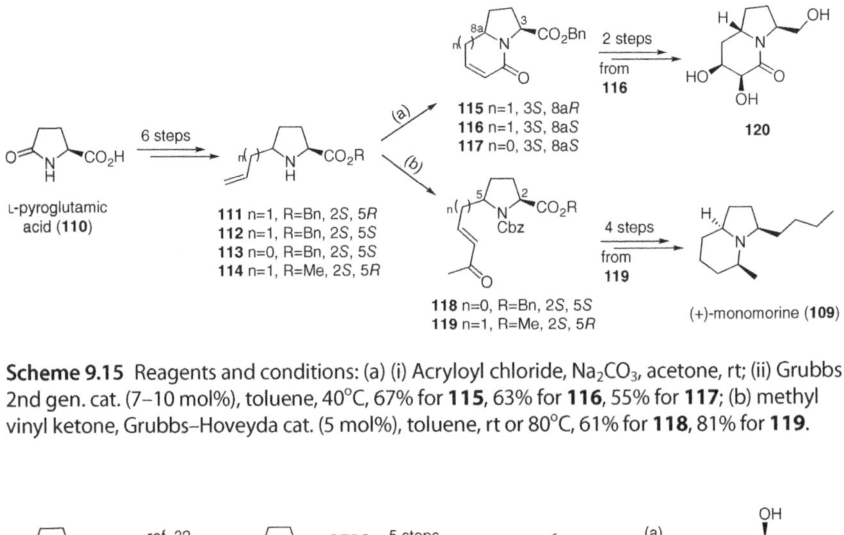

Scheme 9.15 Reagents and conditions: (a) (i) Acryloyl chloride, Na_2CO_3, acetone, rt; (ii) Grubbs 2nd gen. cat. (7–10 mol%), toluene, 40°C, 67% for **115**, 63% for **116**, 55% for **117**; (b) methyl vinyl ketone, Grubbs–Hoveyda cat. (5 mol%), toluene, rt or 80°C, 61% for **118**, 81% for **119**.

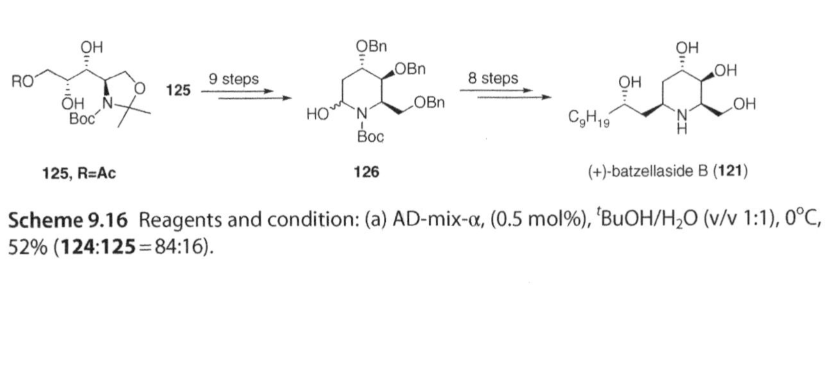

Scheme 9.16 Reagents and condition: (a) AD-mix-α, (0.5 mol%), $^tBuOH/H_2O$ (v/v 1:1), 0°C, 52% (**124**:**125** = 84:16).

(Scheme 9.17).[34] Lastly, (+)-septicine (**127**) was readily converted into (+)-tylophorine (**128**) using the reported protocol.[35]

The starting chiron **138**, which was derived from D-glutamic acid (**137**) in four steps,[37] was used by the same research group to synthesize 9-*epi*-sessilifoliamide J (**133**) (Scheme 9.18),[38] the potent immunosuppressant FR901483 (**135**) and its 8-epimer (**136**) (Scheme 9.19).[39] The key steps in Scheme 9.18 involved a *threo*-selective vinylogous Mannich reaction on bicyclic *N,O*-acetal **139** and a Ley oxidation-SmI_2-mediated coupling lactonization on alcohol **143**. 9-*epi*-sessilifoliamide J (**133**) was obtained as the major product while the natural product (**134**) was the minor component.

Meanwhile, in Scheme 9.19, the key feature was the one-pot amide reductive *bis*-alkylation method to construct the chiral aza-quaternary center ($dr = 9{:}1$) in amine **148**. The stereochemistry of the major diastereomer was deduced to be *trans* based on mechanistic consideration that the vinyl group would approach the *N*-acyliminium ion intermediate from the α-side opposing the benzyloxy group. Further transformations of **148** furnished ketone **149**, which underwent a regio- and diastereoselective intramolecular aldol reaction to build the bridged ring in **150**. The advanced diene **151** underwent ring-closing metathesis (RCM) to form the 3-pyrrolin-2-one ring, and the cyclized product **152** was harvested at a good yield. The common intermediate **153** of both target molecules were synthesized in eight more steps. Finally, FR901483 (**135**) or its 8-*epi*-FR901483 (**136**) were successfully obtained in three more steps.

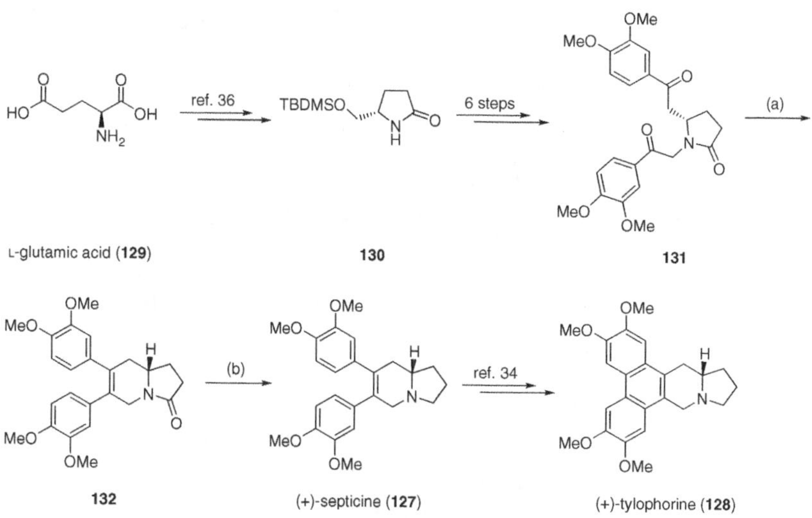

Scheme 9.17 Reagents and conditions: (a) anhyd., $TiCl_4$, Zn, THF, reflux, 24 h, 81%; (b) $LiAlH_4$, THF, 0°C to rt, 1 h, 91%.

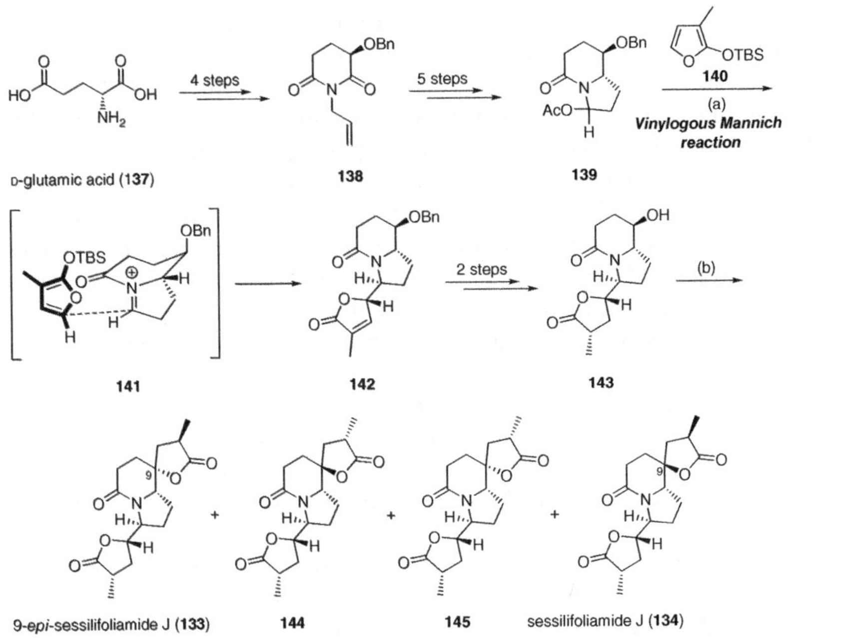

Scheme 9.18 Reagents and conditions: (a) TMSOTf, CH_2Cl_2, −78°C to rt, 61%; (b) (i) TPAP, 4Å MS, NMO, CH_2Cl_2, 0°C to rt; (ii) methyl methacrylate, ^{t}BuOH, SmI_2, THF, 0°C, 70% (**133**:**144**:**145**:**134** = 53:21:20:6).

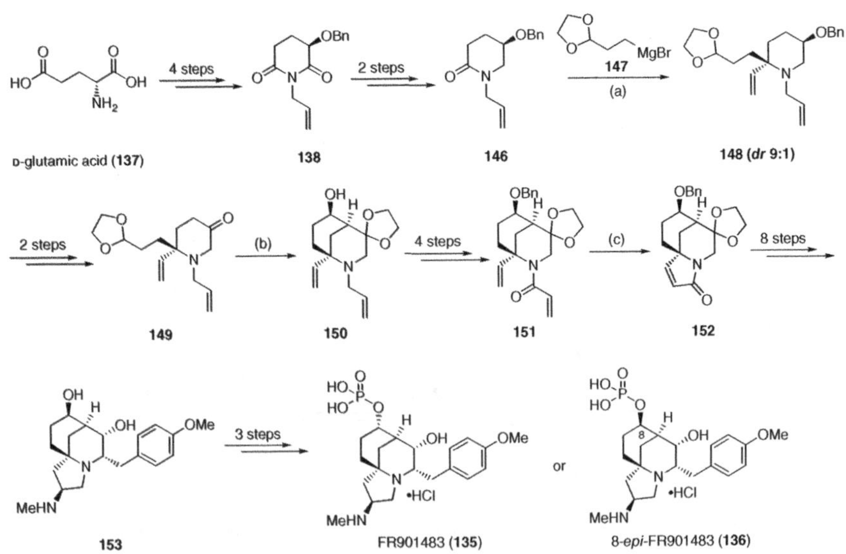

Scheme 9.19 Reagents and conditions: (a) Tf_2O, DTBMP, **147**, vinylmagnesium bromide, CH_2Cl_2, −78°C, 75% (one-pot); (b) 4 N HCl, THF, 50°C, ethylene glycol, CSA, toluene, 90°C, 53% (one-pot); (c) $Ti(O^iPr)_4$ (30 mol%), Grubbs II (25 mol%), toluene, 85°C, 92%.

There are four chiral centers in (−)-renieramycin G (**154**). Two of them (C-3 and C-13) came from the chiron L-tyrosine methyl ester (**155**) while the other two (C-1 and C-11) were introduced through an intermolecular and intramolecular Pictet–Spengler reaction, respectively (Scheme 9.20).[40]

9.2.2 Carbohydrates as Starting Chirons

Apart from amino acids, carbohydrates are another class of chirons to synthesize alkaloids. In 2011, Shing *et al.* reported the use of D-(−)-ribose (**164**) to synthesize natural (−)-cocaine (**163**) and its analogs (**173, 174, 177** and **178**), using an intramolecular nitrone-alkene cycloaddiontion (INAC) as the key step (Scheme 9.21).[41] D-(−)-Ribose was converted into aldehyde **165** in six steps, which was reacted with *N*-methylhydroxylamine to generate nitrone **166** *in situ* and cyclized to give *endo*-cycloadduct **167** and six-membered *exo*-cycloadduct **168** in a ratio of 20:1. The *endo*-cycloadduct **167** was further transformed into **168,** and Raney-Nickel mediated hydrogenolysis yielded azetidine-diol **169**. The N-O bond of **168** was first cleaved to give the corresponding amine, which opened the epoxy ring at C-6 to afford bicyclo[4.1.1] diol **170**. The formation of a bicyclo[4.1.1] skeleton instead of the tropane structure (by attacking C-5) might be rationalized by Baldwin's rule,[42] in which the 4-*exo-tert* cyclization (leading to a bicyclo[4.1.1] skeleton) is more favored than the 5-*endo-tert* cyclization (leading to a tropane skeleton). This azetidine-diol **169** was then transformed into (6*S*,7*R*)-6-chloro-7-benzyloxy cocaine (**173**) by a benzoylation-mesylation sequence at an excellent overall yield. The C-3 hydroxyl group of **169** was first benzoylated, and the free C-5 hydroxyl group in **170** was then mesylated. As the aza-bridge is *anti* to the C-5 –OMs group, neighboring-group participation would assist the mesylate ion in **171** to dissociate easily, giving ammoniumion **172.**[43] Nucleophilic attack of chloride ion at C-6α position of **172** furnished the novel (6*S*)-chloride cocaine analog **173**. The release of ring strain from the bicyclo-[4.1.1] skeleton to the tropane ring was probably the driving force for the attack at C-6 instead of C-5. Hydrogenolysis of **173** in the presence of Raney nickel provided the first synthesis of another cocaine analog, (7*S*)-hydroxy cocaine **174**, which was the 7-epimer of the known (7*R*)-hydroxy cocaine.[44] The third and fourth cocaine analogs, (7*S*)-chloride cocaine **177** and (7*S*)-iodide cocaine **178**, were synthesized from **174** in good yields. Finally, hydrogenolysis of **177** afforded natural (−)-cocaine (**163**).

In the formal total synthesis of the stemofoline alkaloids, Shanahan and coworkers employed 2-deoxy-ribose as starting material (Scheme 9.22).[45] The key step in the synthesis was a novel dipolar cycloaddition cascade that was initiated by cyclization of a rhodium-derived carbene **186**. The resulting azomethine ylide then underwent spontaneous cyclization via a dipolar cycloaddition. The synthesis also featured several other interesting reactions, including a

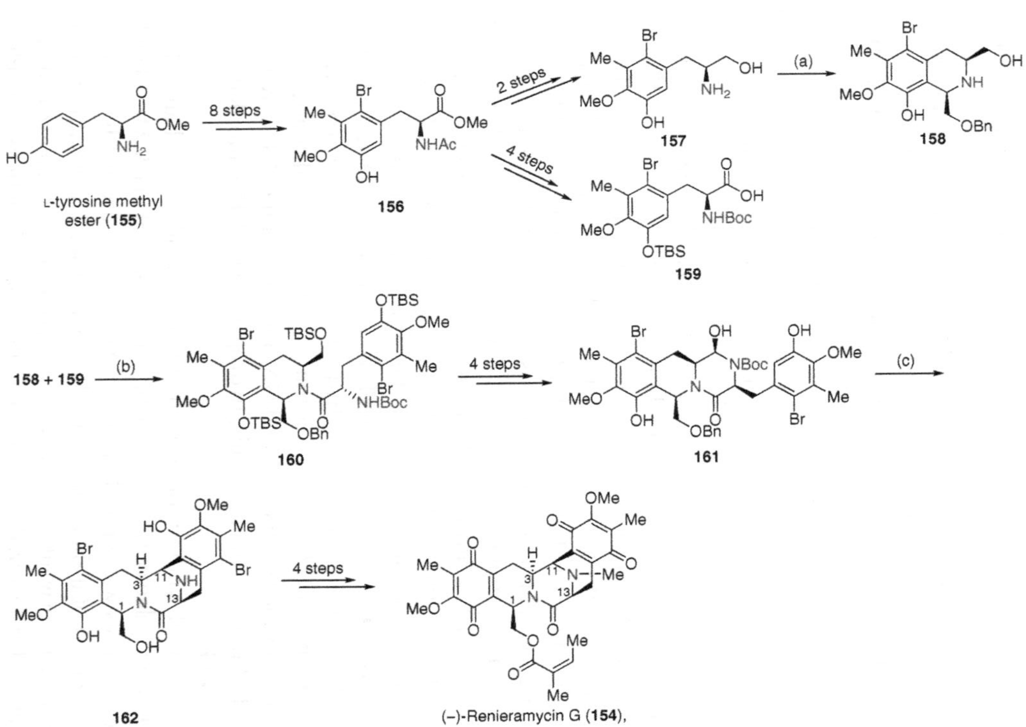

Scheme 9.20 Reagents and conditions: (a) $BnOCH_2CHO$, AcOH, 4Å MS, CH_2Cl_2/CF_3CH_2OH, 0°C, 87%; (b) BOPCl, Et_3N, CH_2Cl_2, rt, 88%; (c) CF_3SO_3H, rt, 82%.

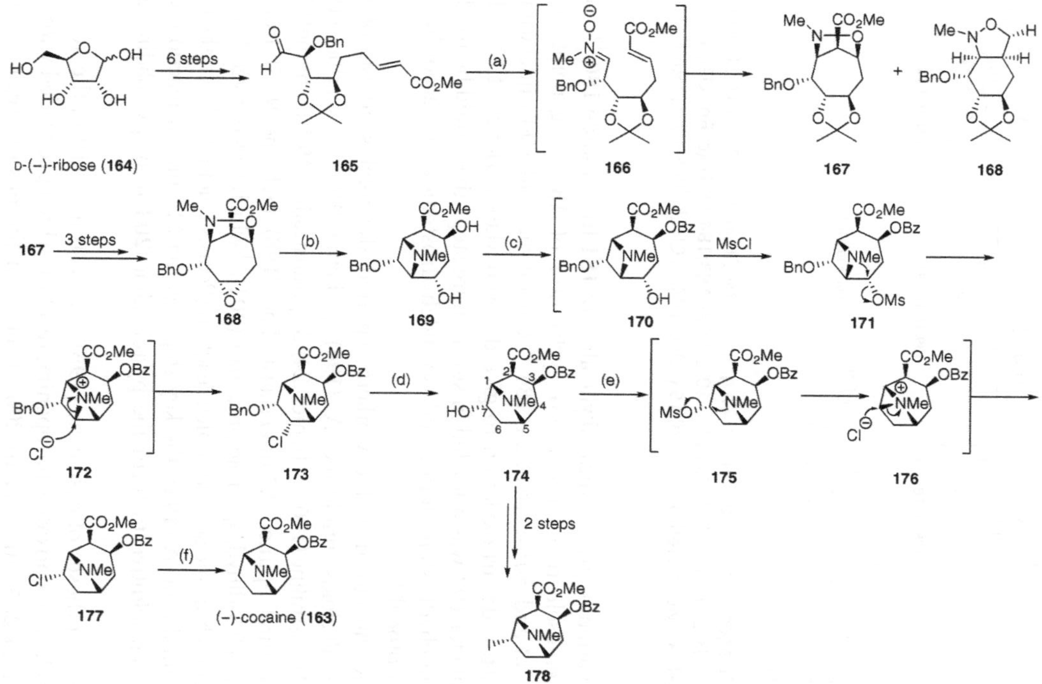

Scheme 9.21 Reagents and conditions: (a) MeNHOH, toluene, reflux, 75% (**167**:**168** = 20:1); (b) H_2, Raney® Ni, AcOH, MeOH, rt to 60°C, 84%; (c) (i) BzCl, Et_3N, DMAP, CH_2Cl_2, 0°C; (ii) MsCl, rt, 89%; (d) H_2, Raney® Ni, MeOH, 50°C, 74%; (e) MsCl, Et_3N, DMAP, seal tube, CH_2Cl_2, rt then 70°C, 91%; (f) H_2, Raney® Ni, MeOH, rt, 76%.

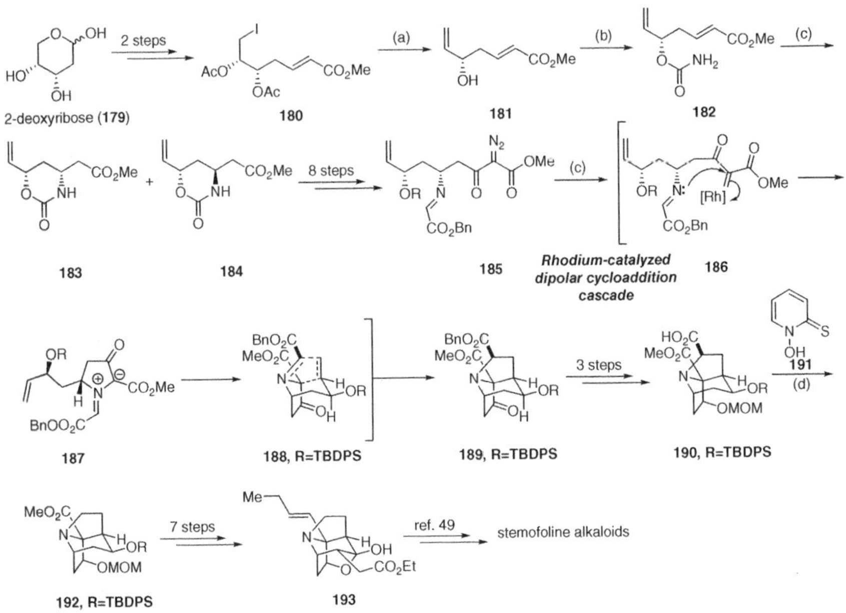

Scheme 9.22 Reagents and conditions: (a) Zn granules, MeOH, reflux, 62%; (b) $ClSO_2NCO$, CH_2Cl_2, 0°C, then H_2O, 99%; (c) NaH, CH_2Cl_2, −10°C, 80% (**183**:**184** = 8:1); (c) $Rh_2(OAc)_4$ (3 mol%), NEt_3•TFA, xylenes, reflux, 75%; (d) **191**, DCC, DMAP, $CHCl_3$, ^{t}BuSH, *hυ* (250W), 71%.

Boord elimination to give a chiral allylic alcohol **181** from iodide **180**,[46] and a highly diastereoselective Hirama–Itô cyclization of **182** to generate *syn*-carbamate **183** and *anti*-carbamate **184**.[47] Carboxylic acid **190** underwent a modified Barton decarboxylation to give decarboxylated tricycle **192**.[48] The known hemiketal **193** was afforded in seven more steps that could be used to synthesize didehydrostemofoline and iso-didehydrostemofoline according to Overman's protocol.[49]

In 2010, Ritthiwigrom and coworkers reported the total synthesis of the pyrrolizidine alkaloids uniflorine A (**194**) (Scheme 9.23), casuarine (**195**) (Scheme 9.24), australine (**196**) (Scheme 9.24), 3-*epi*-australine (**197**) (Scheme 9.25), and the unnatural epimer 3,7-di-*epi*-australine (**198**) (Scheme 9.25) from a common chiral 2,5-dihydropyrrole precursor (**201**),[50] which was derived from L-xylose (**199**) by a boronic acid-Mannich reaction (Petasis reaction).[51,52] In the total synthesis of uniflorine A (**194**), the key step was the Mitsunobu reaction with Et_3N•HCl as an additive to form the pyrrolizidine **203** from pyrrolidine **202** (Scheme 9.23).[52]

In the total syntheses of casuarine (**195**) and australine (**196**), the key step was a regioselective epoxide ring-opening reaction which proceeded with a *dr* of 92:8 (**204** to **205**, and **204** to **206**) in favor of the desired regioisomeric

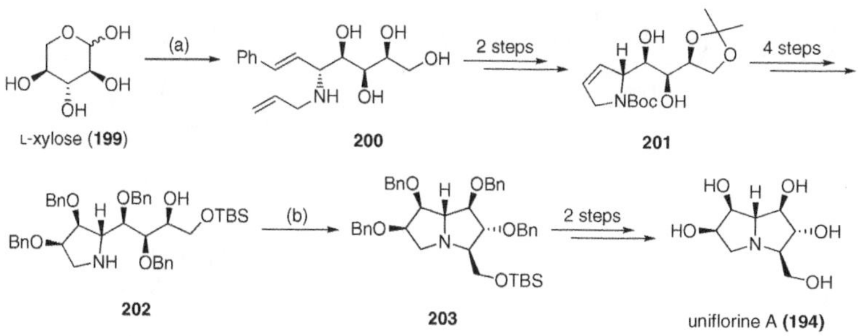

Scheme 9.23 Reagents and conditions: (a) $AllNH_2$, (*E*)-styrene boronic acid, EtOH, rt, ion-exchange, 92%; (b) PPh_3, DIAD, NEt_3•HCl, pyridine, rt, 92%.

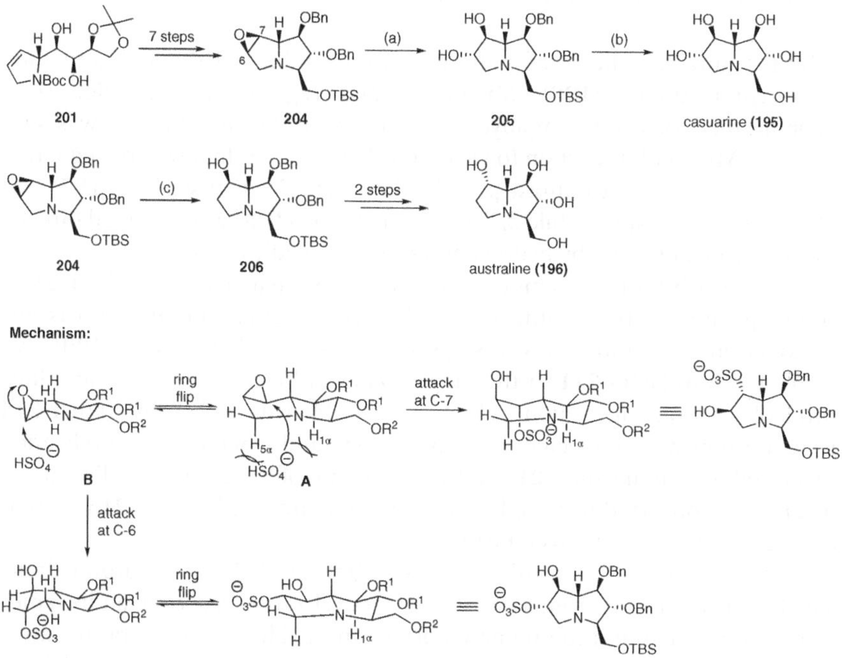

Scheme 9.24 Reagents and conditions: (a) $NaHSO_4$, CH_2Cl_2, reflux, 2 d then H_2O, rt, 47%; (b) $PdCl_2$, H_2, MeOH, rt, ion-exchange, 93%; (c) $LiAlH_4$, THF, 0°C, 59%.

products (Scheme 9.25). For *trans*-1,2-diaxial ring opening of epoxide **204** by HSO_4^-, the two reactive conformations **A** and **B** are possible. Attack on conformation **A** at C-7 was inhibited by 1,3-diaxial interactions between the nucleophile (HSO_4^-) and the pseudoaxial protons H-1α and H-5α. Thus, nucleophilic attack on conformation B at C-6 predominated and resulted in the formation of **205** as the major regioisomeric product.

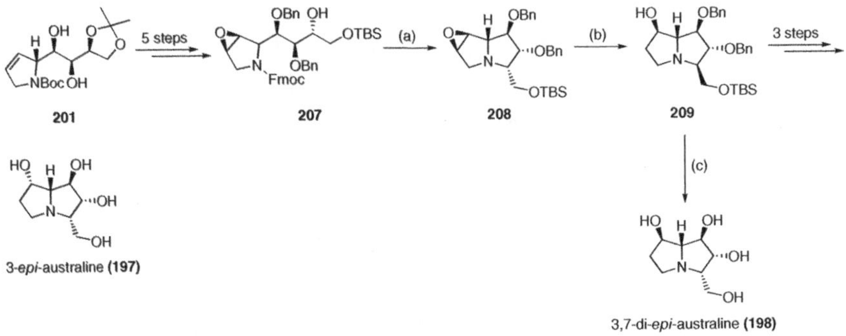

Scheme 9.25 Reagents and conditions: (a) PPh_3, DIAD, toluene, 80°C, 70%; (b) $LiAlH_4$, THF, 0°C, 41%; (c) $PdCl_2$, H_2, MeOH, rt, ion-exchange, 88%.

In the total synthesis of natural 3-*epi*-australine (**197**) and unnatural 3,7-di-*epi*-australine (**198**), Mitsunobu and regioselective epoxide ring-opening reactions were employed (Scheme 9.26). Pyrrolidine **207** was subjected to Mitsunobu reaction to yield pyrrolizidine **208**, followed by reduction of the epoxide ring to afford pyrrolizidine **209**. 3-*epi*-australine (**197**) was obtained in three steps while 3,7-di-*epi*-australine (**198**) was obtained directly from pyrrolizidine **209** by hydrogenolysis under acidic conditions.

L-Xylose (**199**) was also employed by Chandrasekhar *et al.* to synthesize (−)-codonopsinine (**210**), featuring a Julia *trans*-olefination and a cascade epoxidation-cyclization as key steps (Scheme 9.26).[53] L-Xylose (**199**) was converted into lactol **211** in five steps, followed by glycol cleavage and Julia *trans*-olefination with sulfone **212**[54] to yield aldehyde **213**. Allylic ether **214** was subsequently subjected to epoxidation, and concomitant cyclization furnished pyrrolidine diols **215** and **216** in a ratio of 9:1. Finally, the Boc group in **215** was converted to a methyl group by refluxing Red-Al and **215** in toluene[55] to yield (−)-codonopsinine (**210**).

In 2004, Bohno *et al.* synthesized (+)-vittatine (**217**) from carbohydrate. The key reactions involved a Ferrier's carbocyclization reaction, an intramolecular aminomercuration-demercuration, and a Chugaev reaction to introduce the hexahydroindole skeleton.[56] The chiron (**219**) was derived from commercially available methyl 4,6-*O*-benzylidene-α-glucopyranoside (**218**) in two steps (75% overall yield), which was followed by a catalytic Ferrier's carbocyclization reaction[57] to afford chiral cyclohexenone **220**. Allylic alcohol **221** was subjected to Clasien rearrangement with triethyl orthoacetate to construct the quaternary carbon, giving the rearranged product **222** at a 60% yield.[58] The perhydroindole skeleton in amine **224** was successfully introduced by intramolecular aminomercuration-demercuration of **223**.[58e,f,59] (+)-Vittatine (**217**) was then successfully obtained from **224** in three steps (Scheme 9.27).

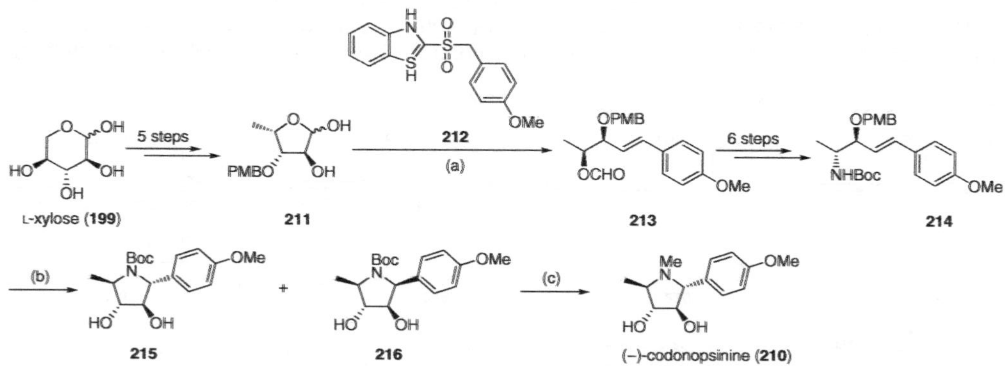

Scheme 9.26 Reagents and conditions: (a) (i) $NaIO_4$, MeOH/H_2O; (ii) **212**, NaHMDS, THF, −78°C, 72% for two steps; (b) *m*CPBA, CH_2Cl_2, 0°C, 89%; (c) Red-Al, toluene, reflux, 83%.

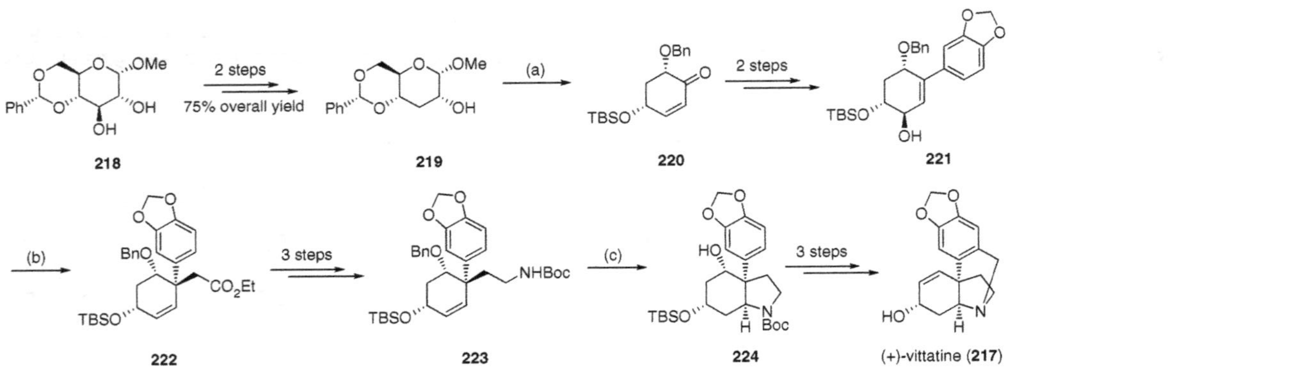

Scheme 9.27 Reagent and conditions: (a) (i) DIBAL-H, toluene, rt; (ii) I_2, Ph_3P, Im, toluene; (iii) TBSCl, Im, DMF; (iv) ^{t}BuOK, THF; (v) $Hg(OCOCF_3)_2$ (30 mol%), acetone/acetate buffer (1:1), rt, then MsCl, Et_3N, CH_2Cl_2, 50% for five steps; (b) $CH_3C(OEt)_3$, cat. EtCOOH, 4 Å MS, 130°C, seal tube, 60%; (c) $Hg(OCOCF_3)_2$, THF, rt, then $NaBH_4$, 0.5 M aq. NaOH/MeOH, rt, 75%.

(+)-Castanospermine (**225**) was first isolated in 1981, and was proved to be a powerful inhibitor of mammalian and plant α- and β-glucosidases.[60] The alkaloid is also a potent antiviral agent[61] and once entered the phase II clinical trial. Many research groups devoted to study its synthesis using sugar as the starting chirons.[62] The most efficient route reported to date is the work of Jensen *et al.* who successfully synthesized the natural product in nine steps with a 22% overall yield, using commercial readily available methyl α-D-glucopyranoside (**226**) as the starting material.[63] The key features of the synthesis involved a zinc-mediated fragmentation of the benzyl-protected 6-iodoglucopyranoside, RCM, and strain-release transannular cyclization to afford the indolizidine skeleton (Scheme 9.28). Methyl α-D-glucopyranoside (**226**) was converted into 6-iodoglucopyranoside **227** in two steps, followed by zinc-mediated fragmentation to furnish aldehyde **228**. Aldehyde **228** was converted into diene **230** in two steps, which was the precursor for RCM. It was found that Grubbs catalyst **231** was the best catalyst for the RCM of diene **230**, affording the nine-membered heterocycle **232** at a good yield. Epoxidation and transannular cyclization of **232** gave the desired benzyl-protected castanospermine **233**, which was deprotected to give the natural product.

The analog of natural castanospermine, 6-*O*-butanoyl castanospermine, is currently undergoing phase II clinical trial for treating chronic HCV. It could be obtained from natural castanospermine by a one-pot procedure reported by Furneaux *et al.*[64]

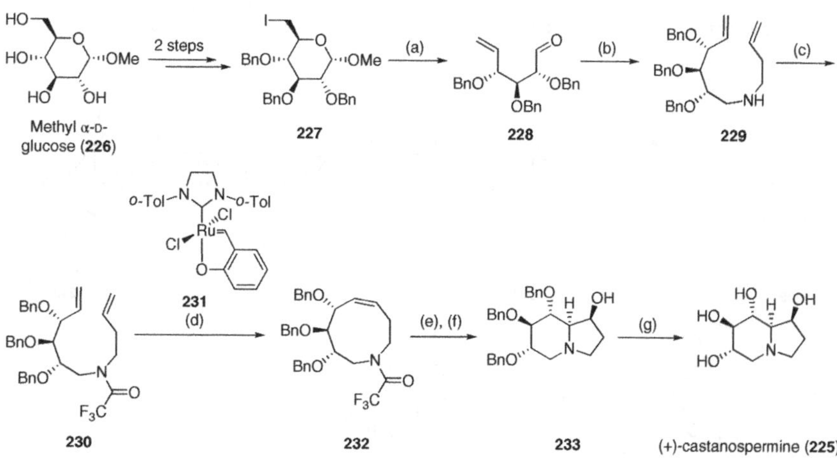

Scheme 9.28 Reagents and conditions: (a) Zn, THF/H_2O, ultrasound, 40°C, 99%; (b) homoallylamine, 4 Å MS, AcOH, $NaCNBH_3$, THF, 0–22°C, 89%; (c) $(CF_3CO)_2O$, Et_3N, CH_2Cl_2, 0°C, 93%; (d) Catalyst **231**, benzene, 80°C, 78%; (e) CF_3COCH_3, oxone, $NaHCO_3$, Na_2EDTA, CH_3CN/H_2O, −10–0°C; (f) ^{t}BuOK, H_2O, Et_2O, 0–22°C, 44% for two steps; (g) H_2, Pd/C, HCl, MeOH, 22°C, 94%.

9.2.3 Terpene and α-Hydroxyl Acid as Starting Chirons

Beside amino acids and carbohydrates, α-hydroxyl acid and terpenes were also used for the syntheses of alkaloids. In 2006, Garg *et al.* reported the first total synthesis of (−)-sarain A (**234**) from D-diethyl tartrate (**235**) (Scheme 9.29).[65] The synthesis featured the use of an intramolecular Stille reaction to fashion the 14-membered ring, yielding the skipped triene functionality. D-Diethyl tartrate (**235**) was first converted into oxazoline **236** in five straightforward steps, followed by Michael addition with α,β-unsaturated ester **237** to give adduct **238**. The key intermediate **239** was subjected to the critical transformation to install the quaternary carbon stereocenter and construct the core of the target molecule. Exposure of **239** to BCl_3 facilitated the intramolecular enoxysilane/*N*-sulfonylimmnium ion cyclization, affording the tetracyclic aldehyde **241**. Subsequent transformations on **241** furnished diene **242**, which underwent RCM to yield the macrocycle in **243**. In the presence of catalytic amount of $[Pd(PPh_3)_4]$ and excess LiCl, stannane iodide **243** cyclized at ambient temperature to assemble the 14-membered triene ring. After reduction of the *N,O*-acetal functionality of the hexacyclic product with *i*-Bu_2AlH, pentacyclic alcohol **244** was obtained. Finally, oxidation and deprotection yielded (−)-sarain A (**234**).

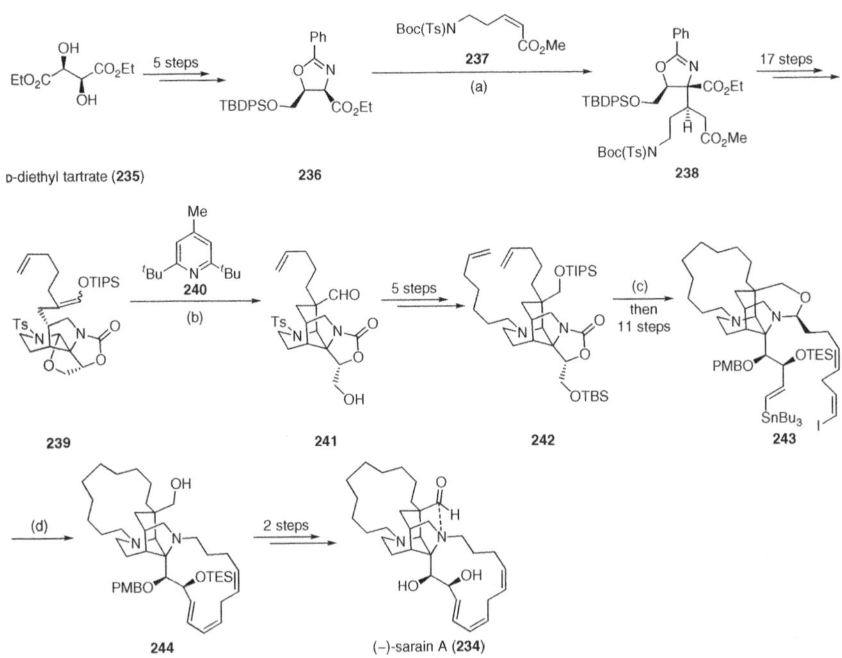

Scheme 9.29 Reagents and conditions: (a) **237**, LHMDS, DME, −78°C, 71%; (b) **240**, BCl_3, CH_2Cl_2, 0–23°C, 92% (*dr* > 20:1); (c) Grubbs II, CH_2Cl_2, reflux then H_2, Pd/C; (d) (i) $Pd(PPh_3)_4$, LiCl, THF, 23°C; (ii) DIBAL-H, toluene, −78–0°C, 62% for two steps.

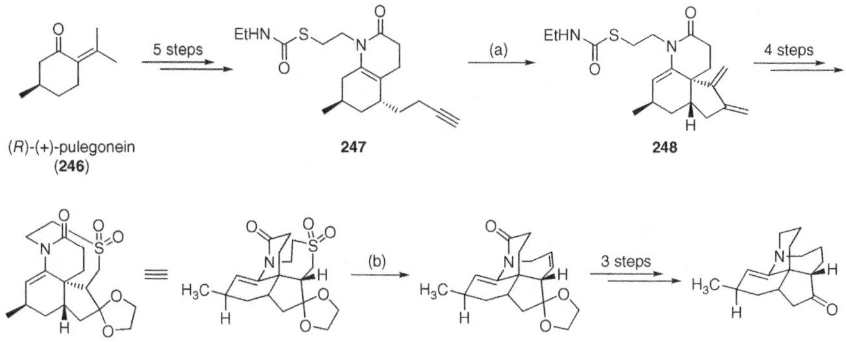

Scheme 9.30 Reagents and conditions: (a) $PtCl_2$ (10 mol%), toluene, 90°C, 87%; (b) CBr_2F_2, KOH•alumina, ^{t}BuOH, 46%.

Kozak *et al.* reported the total synthesis of (+)-fawcettidine (**245**) from (*R*)-(+)-pulegonein (**246**) in 16 steps. The key features were platinum(II)-catalyzed annulation reaction of a highly functionalized enamide, and a one-pot Ramberg–Bäcklund reaction to form a seven-membered ring (Scheme 9.30).[66] (*R*)-(+)-pulegonein (**237**) was firstly converted into enamide **247** in seven steps. Enamide **247** was considered as a potentially disastrous substrate for a reaction involving platinum catalysis, due to the affinity of sulfur atoms toward platinum. Fortunately, in this case, the reaction proceeded smoothly to afford tricycle **248** at a good yield, even on a gram scale. Further functionalization on tricycle **248** furnished sulfone **249**, which was subsequently subjected to Ramberg–Bäcklund reaction to yield alkene **250**. Interestingly, the two-step procedure failed but the one-pot procedure disclosed by Chan *et al.* turned out to be successful.[67] Finally, the hydrogenation of alkene **250** over Pd/C, reduction of the enamide unit with $LiAlH_4$, and the deprotection of ketone successfully afforded the target molecule.

9.3 Total Synthesis of Terpenoids

9.3.1 Terpene as a Starting Chiron

In 2013, Baran and coworkers[68] reported an elegant total synthesis of (+)-ingenol (**251**) from (+)-3-carene (**252**) in just 14 steps, representing the shortest total synthesis to date (Scheme 9.31).[69] (+)-Ingenol is a biologically important plant-derived diterpenoid possessing a unique *in,out*-[4.4.1]bicycloundecane motif.[68,70] Inspired by the biosynthetic pathway of ingenane terpenoids, Baran *et al.* designed a two-phase synthesis consisting of a cyclase phase and an oxidase phase toward the total synthesis of (+)-ingenol.

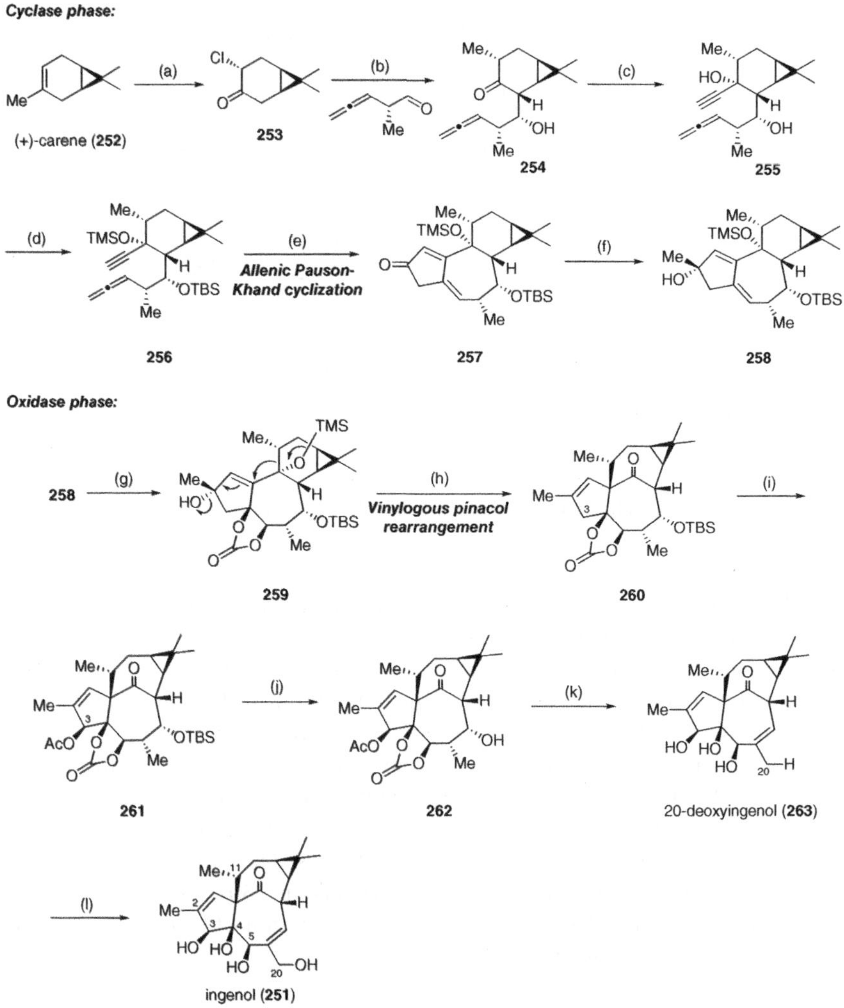

Scheme 9.31 Reagents and conditions: (a) (i) NCS, DMAP (10 mol%), rt; (ii) O_3, thiourea, CH_2Cl_2/MeOH, 48% over two steps; (b) LiNap, HMPA, MeI, THF, −78°C to rt, then LiHMDS, **N**, −78°C, 44% (one-pot); (c) ethynylmagnesium bromide, THF, −78– −15 °C, 81%, *dr* = 10:1; (d) TBSOTf, Et_3N, CH_2Cl_2, 0°C, then TMSOTf, Et_3N, CH_2Cl_2, 71%; (e) $[RhCl(CO)_2]_2$ (10 mol%), CO, *p*-xylene, 140°C, 72%; (f) MeMgBr, THF, −78–0 °C, 80%; (g) (i) OsO_4, pyridine, aq. Na_2SO_3/THF, rt; (ii) CDI, DMAP (10 mol%), CH_2Cl_2, rt, 64% over two steps; (h) $BF_3{\cdot}Et_2O$, CH_2Cl_2, −78– −40°C, then Et_3N/MeOH, 80%; (i) SeO_2, dioxane, 80°C, then Ac_2O, DMAP (10 mol%), pyridine, rt, 59%; (j) HF, CH_3CN, 50°C, 90%; (k) Martin's sulfurane, $CHCl_3$, 80°C, then NaOH, THF, 81%; (l) SeO_2, dioxane/HCO_2H = 1:1, 80°C, 76%.

The cyclase phase, which consisted of seven steps and established seven C–C bonds, helped construct the bicyclo[4.4.1]undecane core. It began with chlorination of (+)-3-carene (**252**) by using *N*-chlorosuccinimide (NCS). Subsequent ozonolysis afforded chloroketone **253** in 48% yield (over two steps). With **253** in hand, a one-pot reductive methylation-aldol reaction with allenic aldehyde was carried out to give aldol adduct **254** diastereoselectively. Grignard addition of ethynyl magnesium bromide to ketone **254** delivered diol **255** with a *dr* of 10:1. Diol **255**, which was synthesized in gram scale, was the key intermediate for the subsequent Pauson–Khand cyclization. However, it was found that the hydroxyl groups of diol **255** led to the failure of the cyclization. Thus, the hydroxyl functionalities were protected as silyl ethers. Allenic Pauson–Khand cyclization of **256** was catalyzed by $[RhCl(CO)_2]_2$ under a CO atmosphere to yield dienone **257**. The cyclase phase was completed by the Grignard addition of methylmagnesium bromide to dienone **257**.

The oxidase phase, which consisted of seven steps and established four C–O bonds, secured the *in, out*-stereochemistry. It began with a stoichiometric OsO_4 dihydroxylation to afford a diol intermediate, which was then protected as carbonate **259**. The key vinylogous pinacol rearrangement, which expanded the cyclohexane ring and established the *in/out* topology, was achieved by treatment of **259** with $BF_3{\bullet}Et_2O$ at −78°C. Quenching the reaction with a 1:1 mixture of $MeOH/Et_3N$ at −40°C was critical to obtain a high yield of the desired rearrangement product **260**.[70] The C-3 hydroxyl group was installed by SeO_2-mediated allylic oxidation, followed by *in situ* protection as acetate **261**. Deprotection of the TBS ether in **261** delivered alcohol **262**, which was treated with Martin's sulfurane and then with NaOH to give the highly crystalline 20-deoxyingenol (**263**). The oxidase phase, and thus the total synthesis, was completed by Shibuya allylic oxidation[71] to secure the oxidation state at C-20.

In summary, the naturally occurring terpene (+)-3-carene (**252**) served as chiron in the total synthesis of (+)-ingenol (**251**), providing a short, efficient and scalable synthetic route toward this biologically active diterpenoid. It is noteworthy that dimethyl cyclopropane moiety of (+)-3-carene effectively governed the stereochemical course so no chiral auxiliaries or chiral catalysts were employed in this synthesis.

Shing and Yeung completed the first total synthesis of (−)-samaderine Y (**264**), a pentacyclic quassinoid, in 2005 by utilizing (+)-carvone (**265**) as the chiron (Scheme 9.32).[72] The synthesis started with the formation of enone **266** in two steps according to their previous endeavor.[73] A regioselective allylic oxidation at C-11 yielded ene-dione **267**, which was reduced regio- and stereoselectively by Luche reduction to give an intermediate allylic alcohol. The alcohol was protected as TBS ether **268**, and subsequent epoxdiation and chelation-controlled hydride reduction afforded alcohol **269**. The charactistic ether bridge was formed by a one-pot acetonide migration-intramolecular epoxide ring opening. The C-12 hydroxyl group in the thermodynamically

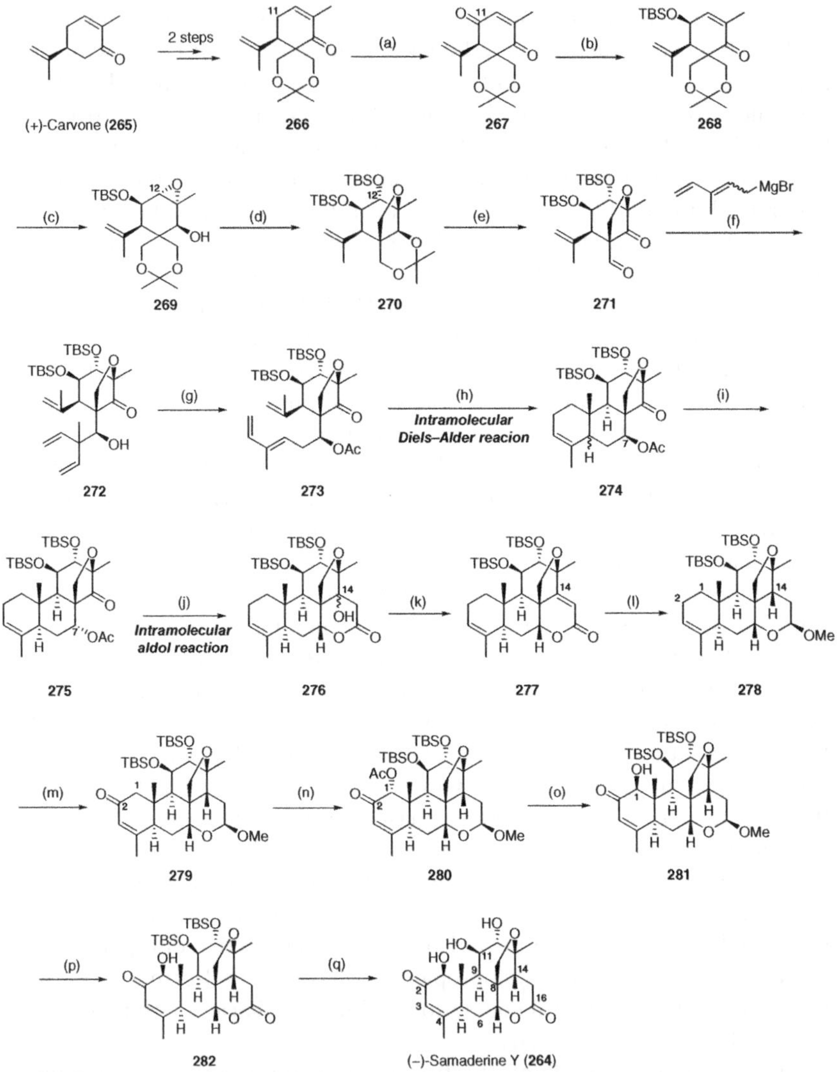

Scheme 9.32 Reagents and conditions: (a) CrO_3, 3,5-dimethylpyrazole, CH_2Cl_2, reflux, 70%; (b) (i) $NaBH_4$, $CeCl_3$•$7H_2O$, MeOH, 0°C; (ii) TBSOTf, Et_3N, CH_2Cl_2, rt, 87% (2 steps); (c) (i) TBHP, NaOH, MeOH, 40°C; (ii) $NaBH_4$, $CeCl_3$•$7H_2O$, MeOH, 0°C; (d) (i) TFA, CH_2Cl_2, rt; (ii) 2,2-DMP, *p*-TsOH, rt; (iii) TBSOTf, Et_3N, CH_2Cl_2, 73% (3 steps); (e) (i) TFA, H_2O, CH_2Cl_2, rt, 92%; (ii) NMO, TPAP, 3Å MS, CH_2Cl_2, rt, 85%; (f) Et_2O, 0°C, 78%; (g) (i) NaH, 4-methylbenzo[15]crown-5, THF, rt; (ii) Ac_2O, Et_3N, DMAP, CH_2Cl_2, rt, 83% (2 steps); (h) methylene blue, toluene, 180°C, 92% (*trans:cis*=2:1); (i) (1) NaOH, MeOH, rt; (2) Tf_2O, pyridine, DMAP, CH_2Cl_2, rt; (3) $^nBu_4NOAc$, THF, rt, 65%, 3 steps; (j) LDA, THF, –78°C, 88%; (k) $SOCl_2$, pyridine, CH_2Cl_2, 45°C, 94%; (l) (i) $NaBH_4$, $NiCl_2$•$6H_2O$, MeOH, rt; (ii) conc. HCl, MeOH, rt, 78% (2 steps); (m) $Mn(OAc)_3$•$2H_2O$ (10 mol%), TBHP, 3Å MS, EtOAc, rt, 72%; (n) $Mn(OAc)_3$•$2H_2O$, benzene, reflux, 68%; (o) (i) K_2CO_3, MeOH, rt; (ii) Dess-Martin periodinane, CH_2Cl_2, rt; (iii) $NaBH_4$, THF, MeOH, 0°C, 72% (3 steps); (p) (i) conc. HCl, H_2O, THF, 45°C; (ii) $AgCO_3$/celite, benzene, reflux, 68% (2 steps); (q) conc. HCl, TFA, rt, 61%.

stable acetonide was protected, yielding disilylated ether **270**. Acid hydrolysis of acetonide **270** followed by Ley Oxidation[74] delivered dialdehyde **271**, which was then subjected to the nucleophilic addition of an allylic Grignard reagent to give the rearranged diene **272** diastereoselectively. Although the undesired 1,4-diene **272** was obtained, Shing and Yeung realized that a cationic [1,3]-sigmatropic rearrangement[75] could give the desired 1,3-diene. This rearrangement was achieved in the presence of NaH and a crown ether, and the resulted homoallylic alcohol was protected as acetate **273**.

The key step that constructed the AB ring was the intramolecular Diels–Alder reaction of **273**, affording the tetracyclic intermediate **274** with a *dr* of 2:1 in favor of the *trans*-fused cycloadduct. To invert the chiral center at C-7, three sequential transformations were carried out which included: (1) base-promoted hydrolysis of the acetate group at C-7; (2) formation of a triflate intermediate; and (3) S_N2 reaction with $^{n}Bu_4NOAc$. The intramolecular aldol reaction of acetate **275** effectively constructed the D ring, leading to the formation of lactone **276**. Removal of the unnecessary hydroxyl group at C-14 furnished the unsaturated lactone **277**, which was then subjected to conjugated reduction and glycosidation to give acetal **278**. The oxidation state at C-2 was set by a newly discovered Mn-catalyzed allylic oxidation[76] while that at C-1 was set by a Mn-promoted α-keto acetoxylation.[77] The stereocenter at C-1 was inverted by a three-step sequence that established the C-1 β-hydroxyl group as in the target molecule. With methyl acetal **281** in hand, acid hydrolysis and the subsequent Ag-mediated oxidation were performed to give lactone **282**. The synthesis (−)-samaderine Y (**264**) was completed by silyl ether deprotection of **282** under acidic conditions.

In summary, (+)-carvone (**265**) served not only as the precursor of the C ring, it also served as a stereocontrolling element, which allowed a multitude of stereoselective transformations. Thus, this total synthesis of (−)-samaderine Y (**264**) again demonstrated the advantages of the chiron approach toward the syntheses of complex terpenoids.

Ghosh and Xi reported the synthesis of the oxatetracyclic core of (−)-platensimycin (**283**) from the chiron (+)-carvone (**265**) (Scheme 9.33).[78] First, lactone **284** was obtained in four steps from (+)-carvone according to modified literature procedures.[79] Baeyer–Villiger oxidation of the ketone moiety of **285** led to the formation of an ester intermediate which was subjected to base-promoted hydrolysis to yield alcohol **285**. After protecting the hydroxyl group as TBS ether, Petasis olefination[80] was carried out to give an enol ether, which was then converted into primary alcohol **286** by hydroboration. Its primary hydroxyl group was protected as TBDPS ether, while the secondary TBS ether was cleaved by DDQ. Subsequent Swern oxidation delivered ketone **287**. An *E*-selective Horner–Emmons olefination was achieved using the chiral phosphonate **288**, which was reduced immediately to yield allylic alcohol **288**. Simple protecting group manipulations afforded primary alcohol **289**, which was oxidized and elongated by a second Horner–Emmons olefination.

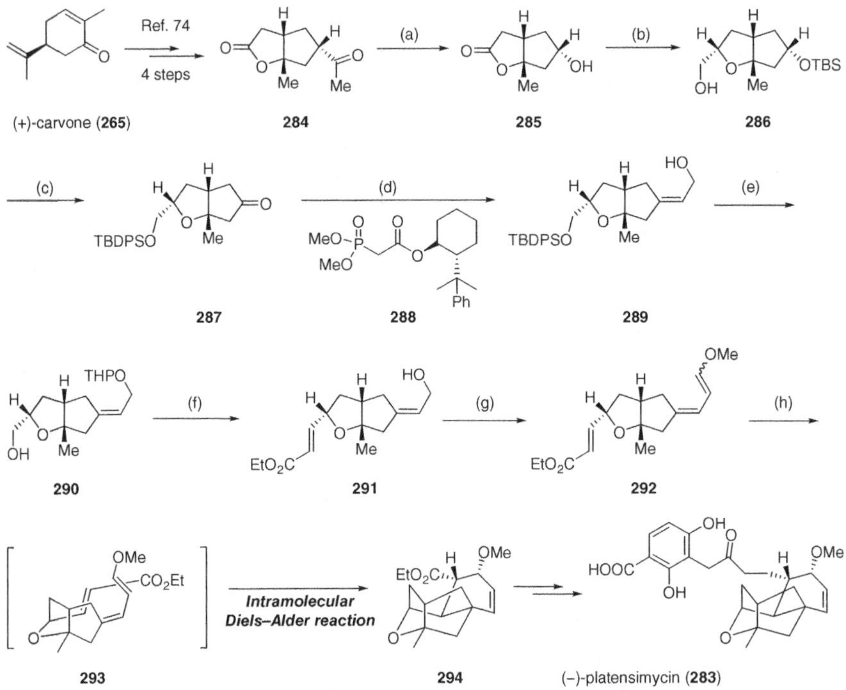

Scheme 9.33 Reagents and conditions: (a) (i) CF_3CO_3H, CH_2Cl_2, 0°C; (ii) K_2CO_3, MeOH, 80%; (b) (i) TBSCl, imidazole; (ii) Cp_2TiMe_2, toluene, 90°C; (iii) 9-BBN, THF, 81% (3 steps), *dr*=2:1; (c) (i) TBDPSCl, imidazole; (ii) DDQ, THF; (iii) Swern Ox., 87% (3 steps); (d) (i) **288**, ^{n}BuLi, −78–−20°C; (ii) DIBAL-H, −78°C, CH_2Cl_2, 61% (2 steps); (e) (i) DHP, PPTS; (ii) TBAF, THF, 99% (2 steps); (f) (i) Swern Ox.; (ii) LHMDS, $(EtO)_2P(O)CH_2CO_2Et$; (iii) CSA, EtOH, 65% (3 steps); (g) (i) Dess–Martin; (ii) Ph_3P=CHOMe, THF, −78°C, (*E*):(*Z*)=1:1, 77% (2 steps); (h) PhCl, BHT, 200°C, 39%.

Acid hydrolysis of the THP group then secured primary allylic alcohol **291**. Dess–Martin oxidation followed by Wittig olefination gave the key intermediate triene **292** as a 1:1 mixture of *E/Z* enol ether. The intramolecular Diels–Alder reaction of the *E*-isomer was achieved under high-dilution conditions, yielding the cycloadduct **294**, which served as the oxatetracyclic core of (−)-platensimycin.

Thus, (+)-carvone (**265**) again provided a facile entry to this kind of natural product. The key intramolecular Diels–Alder reaction also effectively established the molecular complexity.

In 2005, MacMillan and Mangion showed that (−)-citronellol (**297**) was a suitable chiron for the total syntheses of (−)-brasoside (**295**) and (−)-littoralisone (**296**) (Schemes 9.34 and 9.35).[81] The syntheses featured an organocatalytic intramolecular Michael addition that constructed the *cis*-fused bicyclic core, and a photolytic [2+2] cycloaddition. With the bicyclic core established, glycosidation of **302** followed by a global deprotection gave (−)-brasoside (**295**) in just 13 steps. On the other hand, glycosidation of **302** with the differentially protected glucose **307** yielded glycoside **304**, which was the key intermediate of the subsequent

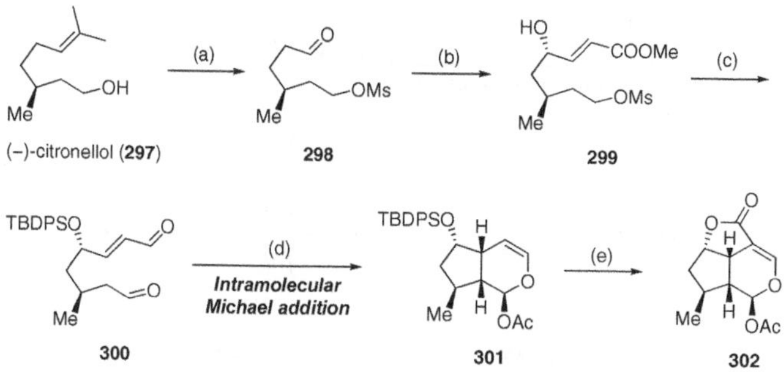

Scheme 9.34 Reagents and conditions: (a) (i) MsCl, DMAP, pyridine, CH_2Cl_2; (ii) O_3, MeOH, CH_2Cl_2, −78°C, 95% (2 steps); (b) (i) PhNO, D-proline (40 mol%), DMSO; then $(EtO)_2P(O)CH_2CO_2Me$, LiCl, DBU; (ii) NH_4Cl, MeOH, 56%; (c) (i) TBDPSCl, imidazole, DMF; (ii) DIBAL-H, Et_2O, −78°C; (iii) DMP, CH_2Cl_2, 92% (3 steps); (d) L-proline, DMSO, 40°C; Ac_2O, DMAP, pyridine, 0°C, 83%; (e) (i) $POCl_3$, DMF, 40°C; (ii) $NaClO_2$, NaH_2PO_4, *t*-BuOH; (iii) HF/pyridine, THF; (iv) DCC, CH_2Cl_2, 56% (4 steps).

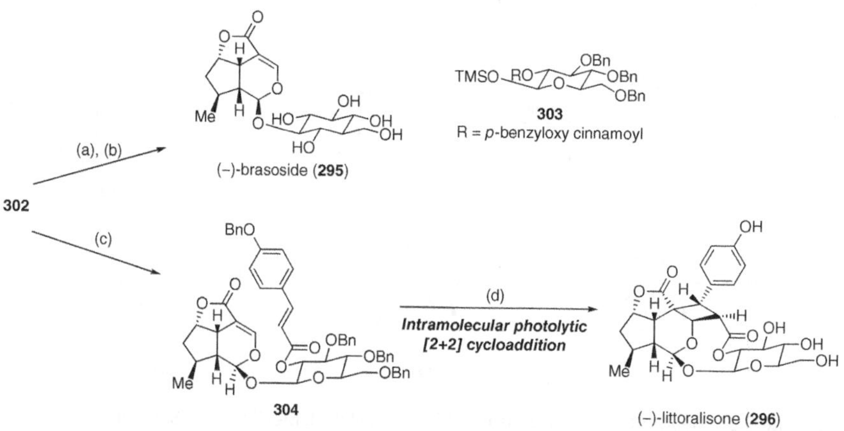

Scheme 9.35 Reagents and conditions: (a) 1-*O*-TMS-*β*-D-glucose tetraacetate, TMSOTf (0.4 eq.), CH_3CN, −30°C. (b) Et_3N, H_2O, MeOH, CH_2Cl_2, −15°C, 82% (2 steps); (c) TMSOTf, **303**, 74%; (d) (ii) *hυ* 350 nm; (ii) H_2, Pd/C, 84% (2 steps).

intramolecular [2+2] cycloaddition. Global deprotection of cycloadduct **304** then furnished (−)-littoralisone (**296**), completing this 13-step-total synthesis.

Ley and coworkers once again employed the chiron (+)-carvone (**265**) in their total syntheses of the sesquiterpene lactones trilobolide (**305**), nortrilobolide (**306**), and thapsivillosin F (**307**) (Scheme 9.36).[82] Their synthetic work featured (1) a Favorskii rearrangement from ketone **308** to ester **310**; (2) RCM of diene **315** for the construction of cycloheptene **316**; (3) Sharpless dihydroxylation of **316** and **320**; and (4) Se-mediated [2,3]-sigmatropic rearrangement of **327** and the subsequent elimination.[83] Esterification of the key intermediate **330** with different acid anhydrides delivered the three natural products.

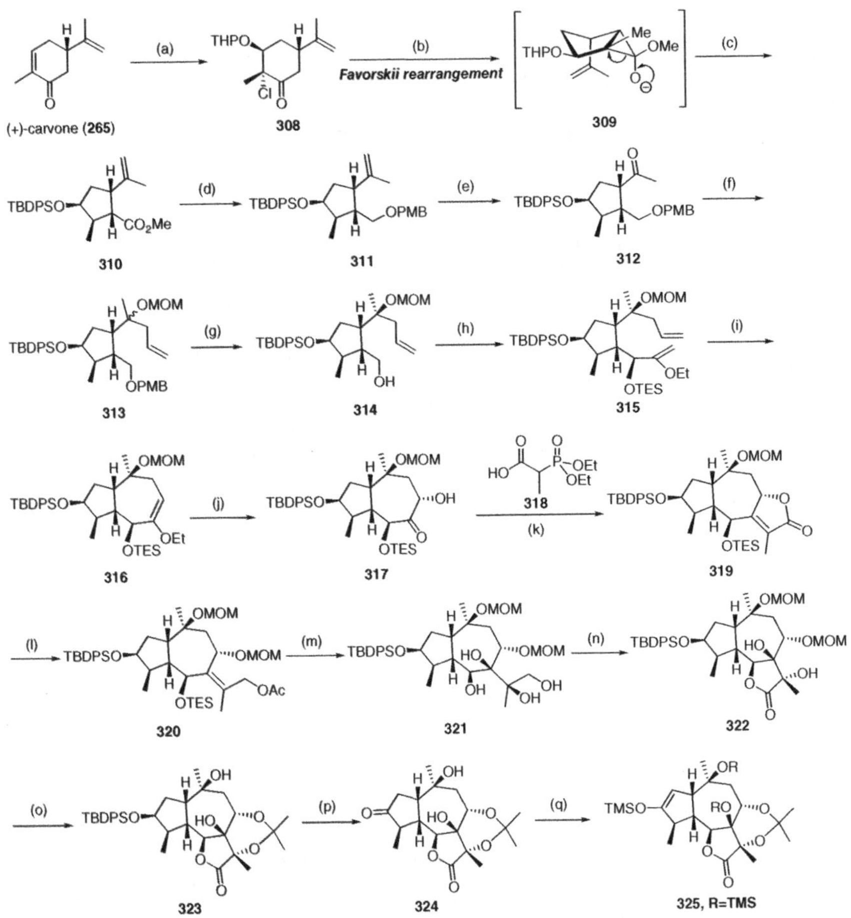

Scheme 9.36 Reagents and conditions: (a) (i) H_2O_2, NaOH, MeOH, 10°C, 88%; (ii) LiCl, TFA, 95%; (iii) DHP, PPTS (cat.), CH_2Cl_2, rt, 87%; (b) NaOMe, MeOH, 0°C, 95%, *dr* > 95:5; (c) (i) PPTS (cat.), MeOH, 40°C, 84%; (ii) TBDPSCl, imidazole, DMF, rt, 98%; (d) (i) LAH, THF, 0°C; (ii) NaH, PMBCl, DMF, rt; (e) (i) OsO_4 (cat.), NMO, acetone/H_2O, rt; (ii) $NaIO_4$, rt, 74% over four steps; (f) (i) AllylMgBr, THF, −78°C, 99%, *dr* = 3.5:1; (ii) MOMCl, DIPEA, DMAP (cat.), CH_2Cl_2, rt, 88%; (g) DDQ, pH 7 phosphate buffer, CH_2Cl_2, rt, 72%; (h) (i) TPAP (cat.), NMO, 4Å MS, CH_2Cl_2, rt, 90%; (ii) CH_2=CHOEt, ^{t}BuLi, THF, −78°C, 96%, *dr* > 95:5; (iii) TESCl, imidazole, DMF, rt, 95%; (i) Grubbs-2nd catalyst (2.5 mol%), CH_2Cl_2, reflux, 92%; (j) $K_2OsO_2(OH)_4$, $K_3Fe(CN)_6$, $NaHCO_3$, $MeSO_2NH_2$, K_2CO_3, ^{t}BuOH, H_2O, rt, 90%, *dr* = 16:1; (k) (i) **318**, EDCI, CH_2Cl_2, rt; (ii) NaH, THF, reflux, 79% over two steps; (l) (i) $LiBH_4$, THF, reflux, 92%; (ii) Ac_2O, DMAP, 2,6-lutidine, CH_2Cl_2, rt, 95%; (iii) MOMCl, DIPEA, DMAP, CH_2Cl_2, rt, 92%; (m) (i) $K_2OsO_2(OH)_4$ (20 mol%), quinuclidine, K_2CO_3, $MeSO_2NH_2$, $K_3Fe(CN)_6$, ^{t}BuOH, H_2O, rt, *dr* > 95:5; (ii) K_2CO_3, MeOH, rt, 85% over two steps; (n) TPAP (10 mol%), NMO, 4Å MS, CH_3CN, rt, 73%; (o) Amberlyst-15, acetone, rt, 85%; (p) (i) TBAF, THF, rt, 98%; (ii) TPAP (cat.), NMO, 4Å MS, CH_2Cl_2, rt, 93%; (q) TMSCl, Et_3N, DMF, 120°C, 88%; (r) dimethyldioxirane, acetone, CH_2Cl_2, 99%, 0°C, *dr* > 95:5; (s) TMSCl, Et_3N, DMF, 150°C, 90%; (t) PhSeBr (10 mol%), CH_2Cl_2, 0°C to rt, 94%; (u) (i) $NaBH_4$, MeOH, 0°C, 86%, *dr* = 4:1; (ii) angelic acid, Et_3N; then 2,4,6-trichlorobenzoyl chloride, toluene, 80°C, 76%; (v) (i) isopropenyl acetate, PS-TsOH, CH_2Cl_2, rt, 68%; (ii) $HCl_{(aq)}$, MeOH, 40°C; (w) butyric anhydride, DMAP, CH_2Cl_2, rt, 72% over two steps; (x) (*S*)-2-methylbutyric anhydride, DMAP, CH_2Cl_2, rt, 78% over two steps; (y) senecioic anhydride, DMAP, CH_2Cl_2, rt, 73% over two steps.

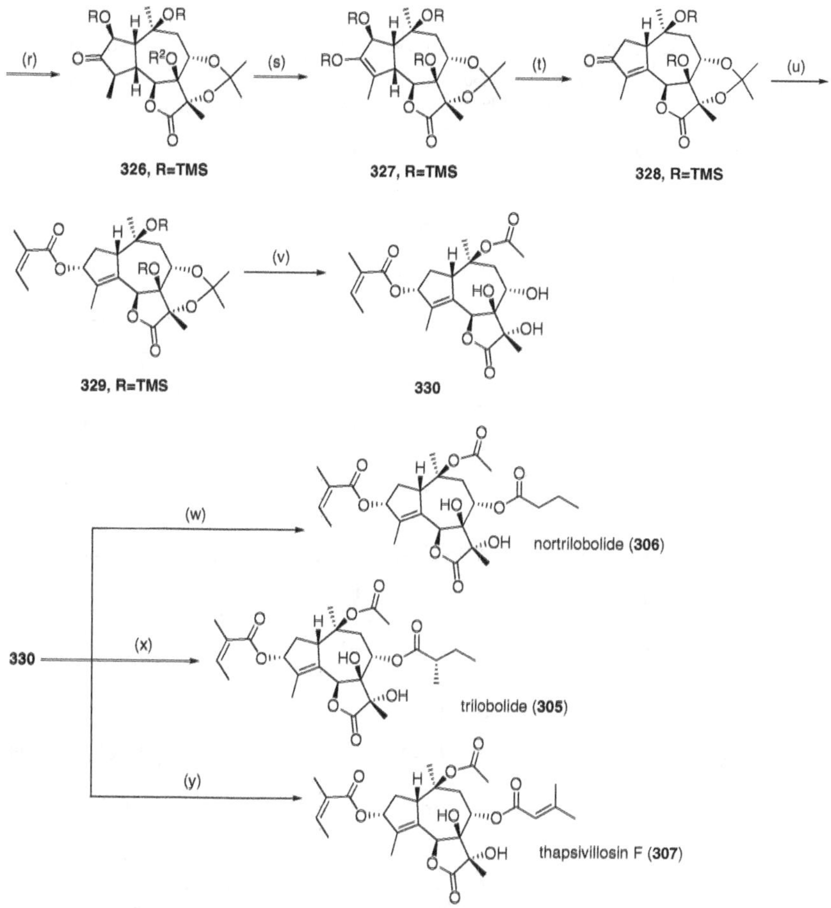

Scheme 9.36 (Continued)

In 2007, Danishefsky and coworkers reported the total synthesis of the sesquiterpenoid peribysin E **331** from the chiron (+)-carvone (**265**) (Scheme 9.37).[84] The key steps in this total synthesis were (1) a Lewis acid-catalyzed Diels–Alder reaction between (+)-carvone and diene **332**; (2) Wittig–Levine methoxymethenylation[85] of **335**, and the subsequent hydrolysis to yield **337**; (3) a vinyl-vinyl Suzuki-coupling reaction between **342** and **343**; and (4) a late-stage Ti-mediated semi-pinacol rearrangement of **346** to **347**.

Paquette and coworkers developed an asymmetric synthesis of the sesquiterpene (+)-fomannosin (**348**),[86] starting from the inexpensive and readily available chiron D-glucose (**349**). The glucose-derived acetal **350** was subjected to a zirconocene-promoted ring contraction reaction to form the cyclobutane core (Scheme 9.38). The resulted alcohol was protected as TBS ether **351**, in which the vinyl alkene moiety was reductively cleaved to give aldehyde **352**. Nucleophilic addition of an organolithium reagent delivered

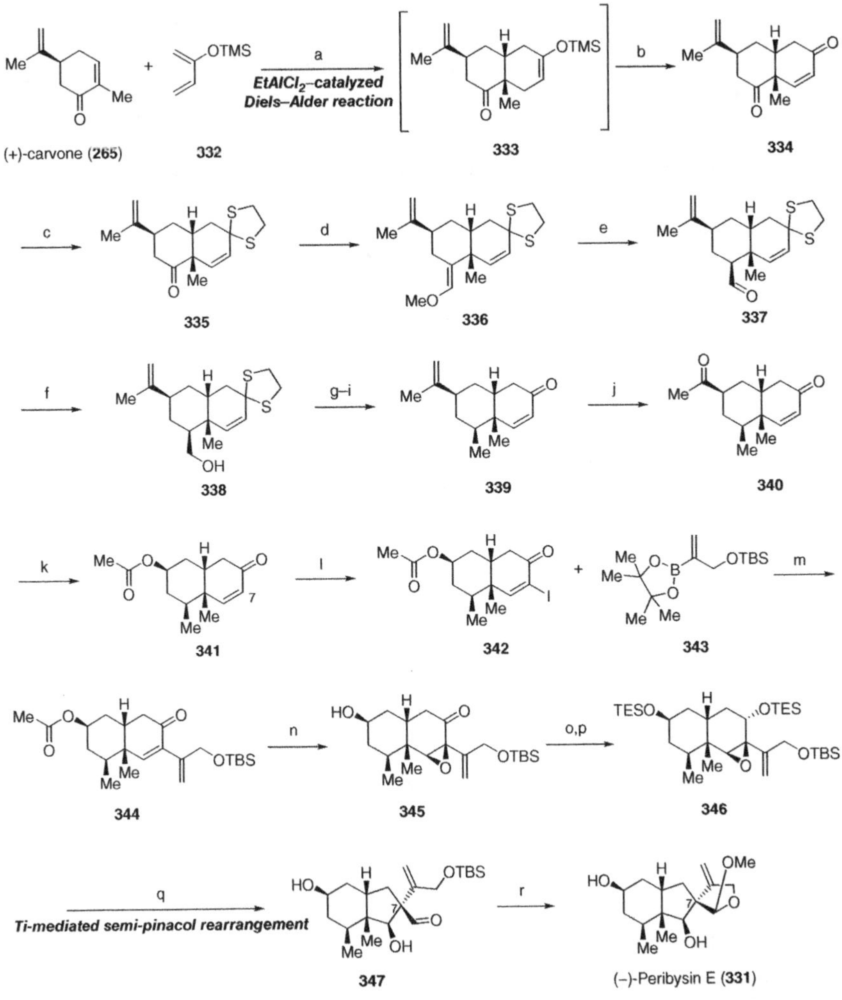

Scheme 9.37 Reagents and conditions: (a) $EtAlCl_2$, toluene, 0°C to rt; (b) $Pd(OAc)_2$, CH_3CN, rt, 63% over two steps, *cis/trans* = 19:1; (c) 1,2-ethanedithiol, MeOH, $BF_3{\cdot}OEt_2$, 0°C, 83%; (d) $Ph_3PCH_2OCH_3Cl$, $KN(SiMe_3)_2$, THF, −30–0°C to rt; (e) 4 N HCl, MeOH, THF, 0°C, 89% over two steps, $\beta/\alpha = 13:1$; (f) $NaBH_4$, MeOH, THF, 0°C to rt, 90%; (g) MsCl, Et_3N, CH_2Cl_2, 0°C to rt; (h) $LiBHEt_3$, THF, 0°C to rt, 71% over two steps; (i) $(CF_3CO_2)_2IPh$, MeOH, H_2O, CH_2Cl_2, rt, 87%; (j) OsO_4, H_2O, $NaIO_4$, 2,6-lutidine, dioxane, rt, 85%; (k) *m*CPBA, CH_2Cl_2, 0°C to rt, 45% (80% based on recovered starting material); (l) $TMSN_3$, I_2, pyridine, CH_2Cl_2, 71% (100% based on recovered starting material); (m) $[Pd(PhCN)_2Cl_2]$, THF/H_2O, Ag_2O, Ph_3As, rt, 89%; (n) H_2O_2, NaOH, MeOH, 0°C, 85%; (o) $NaBH_4$, MeOH, THF, 0°C, 91%, $\alpha/\beta = 7:1$; (p) TESCl, imidazole, DMF, rt, 93%; (q) $TiCl_4$, CH_2Cl_2, −78°C, 50%; (r) HCl, MeOH, 0°C, 80%.

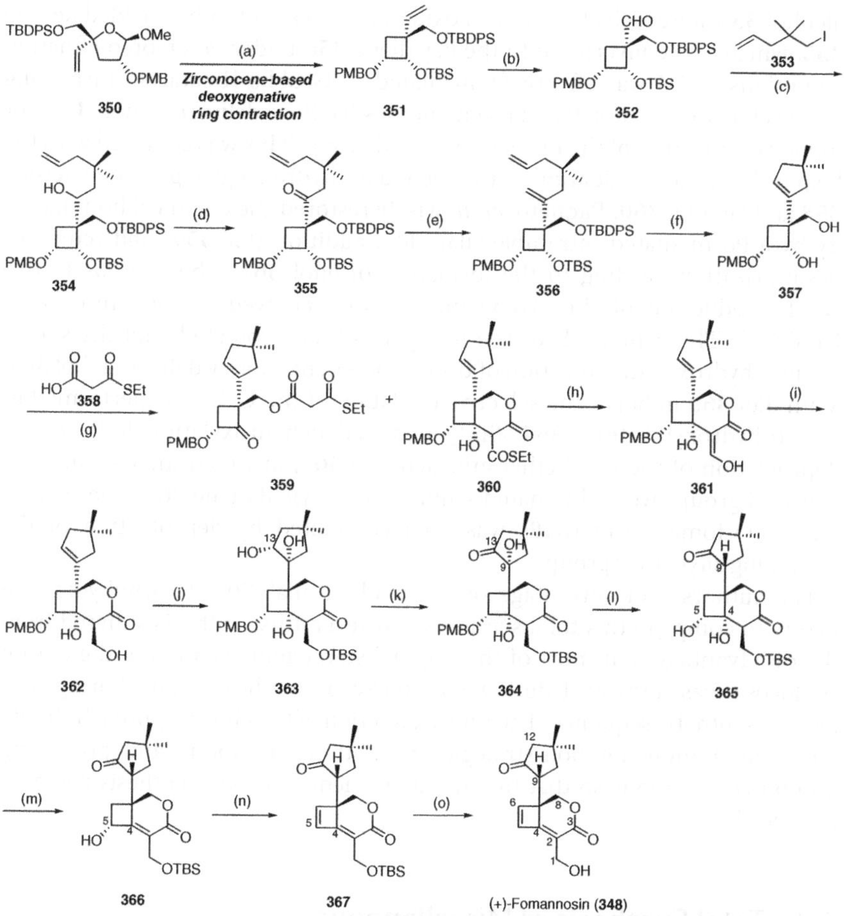

Scheme 9.38 Reagents and conditions: (a) (i) $[Cp_2ZrCl_2]$, ^nBuLi, THF, –78°C to rt, 60%; (ii) TBSCl, imidazole, CH_2Cl_2, rt, 90%; (b) O_3, Sudan III, CH_2Cl_2, –78°C, then PPh_3, 91%; (c) **353**, ^tBuLi, THF, –78°C, 98%; (d) PDC, 4Å MS, CH_2Cl_2, rt, 24 h, 83%; (e) (i) $(CH_3)_3SiCH_2Li$, pentane/toluene (1:1), –78 °C; (ii) PTSA, benzene, rt, 83% (2 steps); (f) (i) Grubbs-II, benzene, reflux, 91%; (ii) TBAF, THF, rt, 76%; (g) (i) **358**, EDCI, CH_2Cl_2, –40°C to rt; (ii) IBX, DMSO, rt, 44% (2 steps); (h) Pd/C, Et_3SiH, CH_2Cl_2; then silica gel, 74%; (i) $NaBH_4$, KH_2PO_4, AcOH, MeOH, 0°C, 43%, *dr* = 2:1; (j) (i) TBSOTf, 2,6-lutidine, CH_2Cl_2, –78°C, 89%; (ii) OsO_4, THF/pyridine = 4:1, 0°C; then H_2S, 76%; (k) ClCOCOCl, DMSO, Et_3N, CH_2Cl_2, –78°C to rt, 78%; (l) (i) SmI_2, ^tBuOH/THF = 1:4, 64%; (ii) TFA, CH_2Cl_2, rt, 50% (2 steps); (m) (i) $SOCl_2$, Et_3N, CH_2Cl_2, rt; (ii) DBU, CH_2Cl_2, 0°C, 31% (2 steps); (n) (i) Tf_2O, CH_2Cl_2, 0°C; (ii) DBU, benzene, rt, 35% (over two steps); (o) TBAF, THF, 0°C, 86%.

alcohol **354** stereoselectively. PDC oxidation of **354** smoothly yielded ketone **355**, which was converted into the key diene **356** under Peterson olefination conditions. RCM was effectively mediated by Grubbs-II catalyst, furnishing the cyclopentene moiety. Subsequent desilylation delivered diol **357** for monoesterification of the primary hydroxyl group. This was achieved with the help of EDCI, and oxidation of the secondary hydroxyl group yielded ketone **358** and lactone **360**. Paquette *et al.* wisely restored the chemical homogeneity by a Pd-mediated intramolecular aldol addition (for **359**) and reductive desulfurization, leading to the formation of enol **361**. The chemoselective $NaBH_4$ reduction of the alkene moiety was achieved in the presence of KH_2PO_4.[87] The primary hydroxyl group was then protected, and the subsequent dihydroxylation reaction of the cyclopentene moiety delivered diol **363**. With diol **363** in hand, the selective oxidation of the C-13 was performed to give α-hydroxyl ketone **364**. SmI_2-mediated dehydroxylation followed by deprotection of the PMB ether afforded diol **365**, in which the C-4 and C-5 hydroxyl groups were eliminated sequentially to yield diene **367**. The synthesis of (+)-fomannosin (**348**) was then completed by deprotection of the remaining silyl ether group.

The authors took advantage of the ready availability of carbohydrate in nature to complete this total synthesis. However, this synthesis suffered from the disadvantage that most of the original stereogenic centers of the chiron D-glucose was destroyed during the course of synthesis, contributing to a lengthy synthetic sequence. The future direction of the chiron approach should focus on minimizing non-strategic redox transformations and protecting group manipulations so that the overall efficiency of total synthesis could be enhanced.

9.4 Total Synthesis of Miscellaneous Natural Products

9.4.1 Amino Acids as Starting Chirons

In the total synthesis of pactamycin (**368**) and pactamycate (**369**), Hanessian and coworkers used L-theronine (**370**) as the starting chiron with was subsequently converted into methyl ketone **371** in 8 steps (Scheme 9.39).[88] Ozonolytic cleavage of the exocyclic methylene group, followed by a highly stereoselective Mukaiyama-type intramolecular aldol condensation afforded cyclopentanone **372**. Cyclopentanone **372** was subsequently converted into α-epoxide **373** in six steps, followed by the key "epoxide-inversion" reaction with $Zn(OTf)_2$ in AcOH to give triol **375** with C-4 inversion. Presumably, this arose from the spiroepoxide **374** that underwent solvolysis to afford the primary acetate as in **375**. A two-step sequence restored the robust TBDPS ether group,

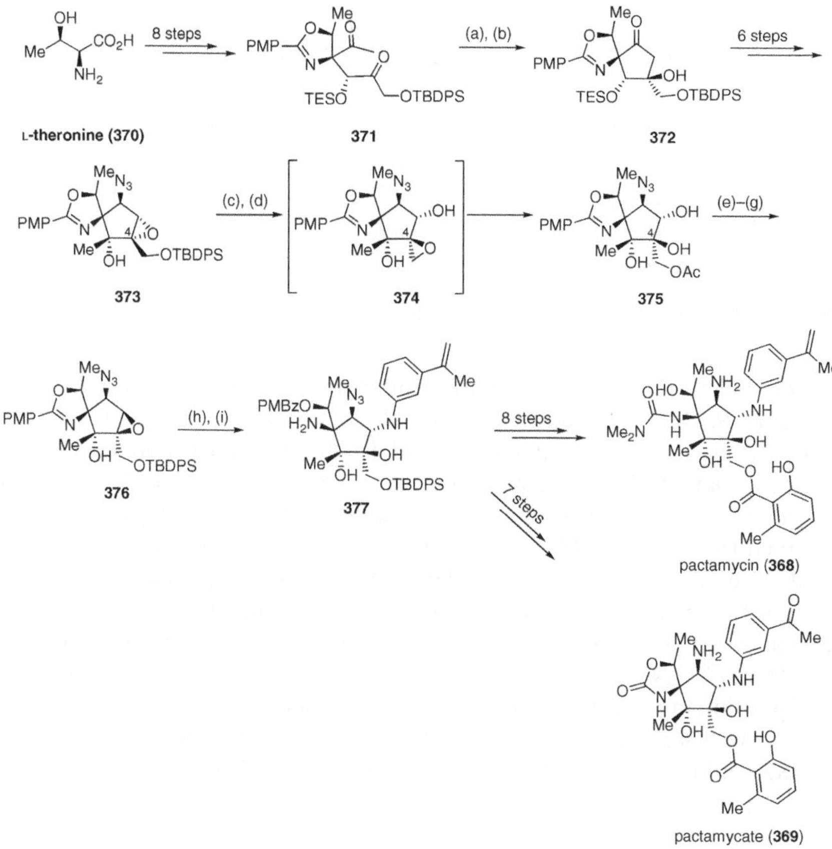

Scheme 9.39 Reagents and conditions: (a) O_3, CH_2Cl_2, −78°C, then DMS, 84%; (b) $TiCl_4$, CH_2Cl_2, DIPEA, TMSCl, 0°C, 85%; (c) TBAF, THF, 0°C, 86%; (d) $Zn(OTf)_2$, AcOH, 80°C; (e) K_2CO_3, MeOH, rt; (f) TBDPSCl, TEA, DMAP, rt, 85% (over 3 steps); (g) Tf_2O, pyridine, CH_2Cl_2, −78–0°C, 96%; (h) 3-(prop-1-en-2-yl)aniline, $Yb(OTf)_3$, toluene, 80°C, 81%, (91% brsm); (i) 2 N HCl, THF, rt, 63%, (83% after two cycles).

and the resulting triol was converted *in situ* into the epoxide **376** through the secondary triflate. Highly stereoselective epoxide opening at C3 with the 3-(prop-1-en-2-yl)aniline in the presence of $Yb(OTf)_3$,[89] followed by cleavage of the oxazoline moiety afforded the common intermediate **377**. Pactamycin (**368**) and pactamycate (**369**) were then obtained from **377** in eight and seven steps, respectively.

In the revision of the structure of aeruginosins 205A and 205B, Hanessian and coworkers synthesized the enantiopure aeruginosin 205B (**378**) from dimethyl *N*-Cbz-L-glutamate (**379**) and D-xylose.[90] The 2-carboxy-5-hydroxyoctahydroindole core **374** (Choi subunit) was obtained from *N*-Cbz-L-glutamate

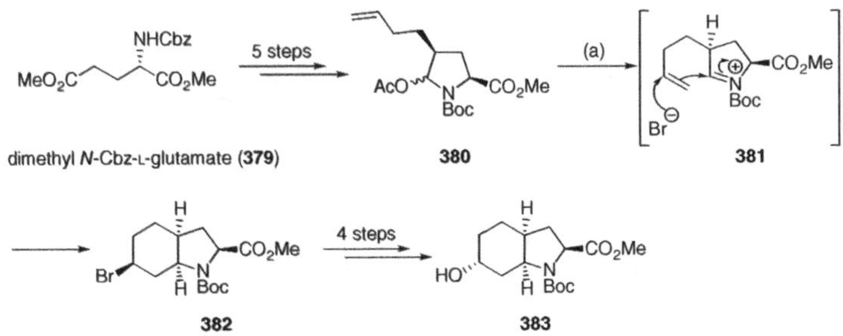

Scheme 9.40 Reagents and condition: (a) $SnBr_4$, CH_2Cl_2, $-78°C$, 90%.

(379), which was converted into hemiaminal acetate **380** in five steps (Scheme 9.40).[91] Subsequent key aza-Prins cyclization afforded the key aza[4.3.0]-bicyclic motif **382** in good yield. Further transformations yielded the *N*-Boc Choi unit **383**. The glycosyl unit in aeruginosin 205B was derived from D-xylose (**384**), which was converted to compound **385** in five steps. Treatment of a mixture of **383** and **385** in ether-DCM with AgOTf in the presence of tetramethylurea as an acid scavenger[91] led to the desired *R*-anomer **386** (43%), which was easily separable from the *S*-anomer (53%) (Scheme 9.41). The presumed aeruginosin 205B (**378**) was obtained in six steps from **386**. After comparing the ^1H and ^{13}C NMR data, it was discovered that the sulfate unit should be at C-4' position instead of C-3' position of the glycosol unit.

In 2006, Nakamura *et al.* reported the first total synthesis of polygalolides A (**387**) and B (**388**) from D-arabinose (**389**). The key feature of this synthesis

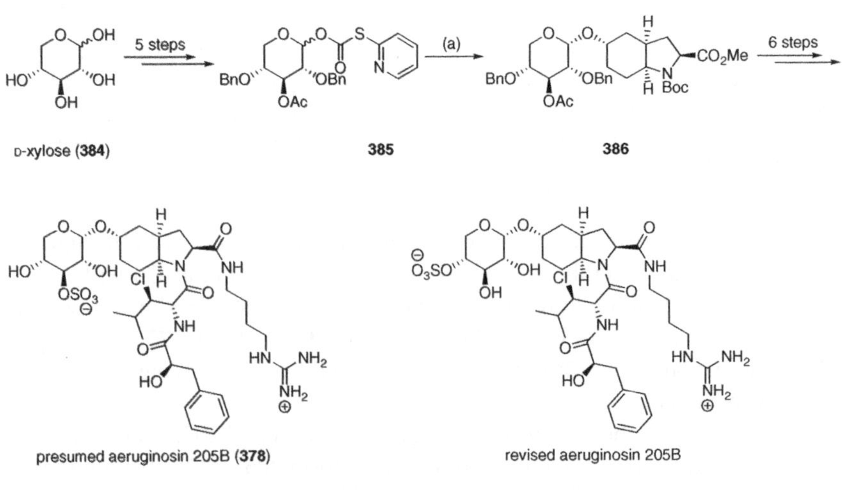

Scheme 9.41 Reagents and conditions: (a) (i) **383**, tetramethylurea, MS, rt; (ii) AgOTf, rt, dark, Et_2O/CH_2Cl_2 (v/v 10:1), 96% (α:β = 1:1.2).

was the tandem carbonyl ylide formation/1,3-dipolar cycloaddition to construct the unprecedented 5,10-dioxatricyclo[6.2.1.0]undecane core structure in a single step (Scheme 9.42).[92] D-arabinose was firstly converted into alcohol **390** according to Wong's protocol,[93] followed by subsequent transformation to alcohol **391**. The alkene moiety was installed by a three-step sequence to afford alkene **392**. α-Diazoketone **393** was obtained from alkene **392** in five steps, followed by the critical carbonyl ylide formation/intramolecular 1,3-dipolar cycloaddition reaction sequence with Rh catalyst. After extensive experimentation, it was found that $Rh_2(OAc)_4$ in benzotrifluoride[94] was the optimal system to afford cycloadduct **395**. The key intermediate **396** was obtained from cycloadduct **395** in five steps. Polygalolides A (**387**) and B (**388**) were then furnished in three and four more steps, respectively.

In 2010, an enantioselective synthesis of the antifungal natural product (+)-ambruticin S (**397**) was accomplished by Hanessian *et al.* The synthesis started with the readily available methyl α-D-glucopyranoside (**226**), (*R*)-Roche ester (**398**), and (*S*)-glycidol (**399**) as chirons, which encompassed seven of the 10 stereogenic centers of the target molecule. The remaining three centers were set by a highly diastereoselective cyclopropanation

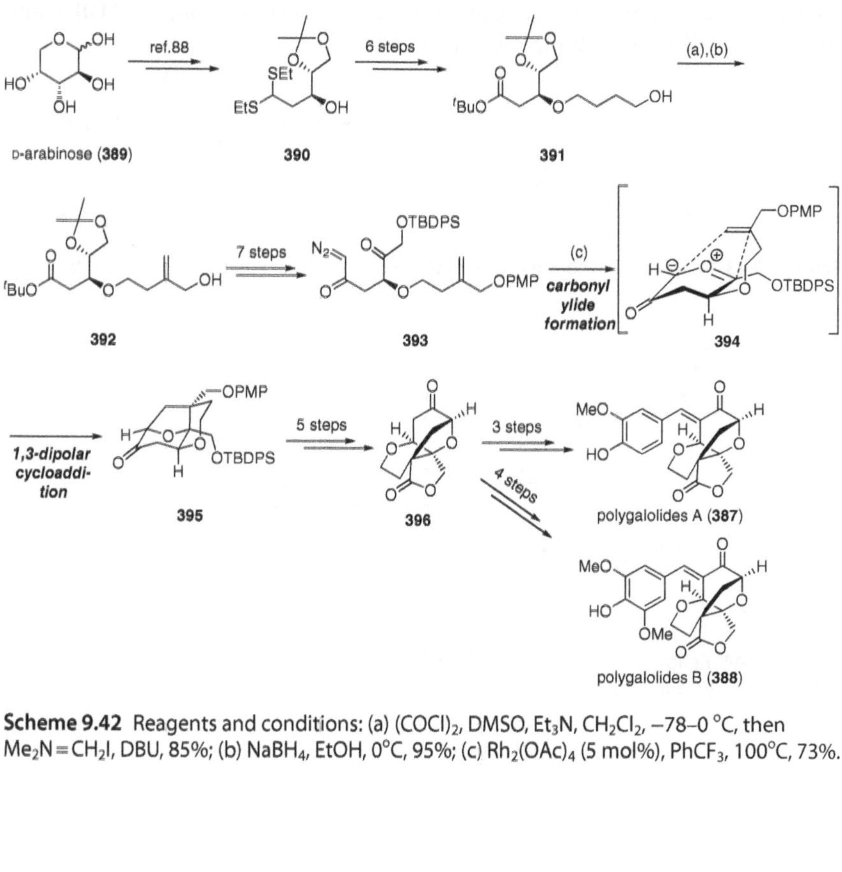

Scheme 9.42 Reagents and conditions: (a) $(COCl)_2$, DMSO, Et_3N, CH_2Cl_2, −78–0 °C, then $Me_2N{=}CH_2I$, DBU, 85%; (b) $NaBH_4$, EtOH, 0°C, 95%; (c) $Rh_2(OAc)_4$ (5 mol%), $PhCF_3$, 100°C, 73%.

employing a chiral phosphonamide reagent. The strategy for the construction of the dihydropyran subunit involved a highly *syn*-selective Lewis acid-catalyzed 6-*endo-trig* cyclization. Other key steps in the synthesis featured an epoxide opening with a dithiane anion, two efficient phosphonamide-anion based olefinations, and a late-stage *C*-glycosylation.[95] The synthesis started from ring A in (+)-ambruticin S in which commercial available methyl α-D-glucopyranoside (**226**) was converted to chloride **400** in four steps. Selective substitution of the C-6 chloride was discovered and treatment of chloride **400** with allyltriphenylstannane in the presence of ACCN afforded alkene **401**. Alkene **401** was then converted to lactone **402** in six steps (Scheme 9.43).

On the other hand, ring C was constructed from (*S*)-glycidol (**399**), which was converted to its benzyl ether derivative **403** and underwent the key epoxide opening with the dithiane anion **404** (9:1 ratio of *E*/*Z* isomers).[96] Interestingly, only the *E*-isomer reacted with the epoxide to give adduct **405**. Further transformations gave the key intermediate alkene **406**, which underwent a *syn*-selective Lewis acid-catalyzed 6-*endo-trig* cyclization to yield dihydropyrane **407**. The desired ketone **408** was then obtained in four more steps (Scheme 9.44).

The third task was to construct ring B, which was accomplished through a highly stereoselective cyclopropanation of *tert*-butyl crotonate (**410**) using *trans*-chloroallyl phosphonamide **409** as chiral auxiliary to give the ideal

methyl α-D-arabinose (226) 400 401 402

Scheme 9.43 Reagents and condition: (a) Allyl-SnPh$_3$, ACCN, toluene, 150°C, 52% (91% brsm).

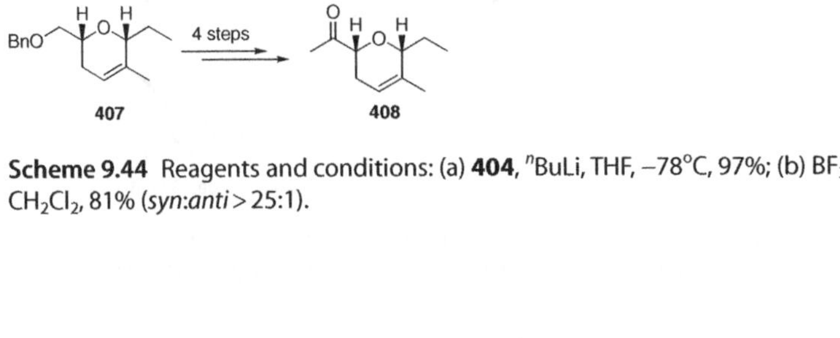

Scheme 9.44 Reagents and conditions: (a) **404**, ^{n}BuLi, THF, −78°C, 97%; (b) BF$_3$•OEt$_2$ (cat.), CH$_2$Cl$_2$, 81% (*syn*:*anti* > 25:1).

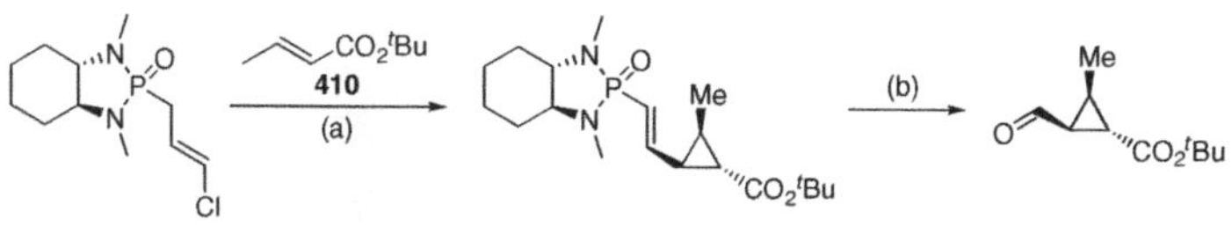
409 411 412

Scheme 9.45 Reagents and conditions: (a) **410**, ^{n}BuLi, THF, −78°C, 89% (*dr* = 99:1); (b) O_3, CH_2Cl_2, −78°C then DMS, 70%.

cyclopropane adduct **411** as a single diastereomer. Removal of the chiral auxiliary by ozonolysis yielded aldehyde **412** (Scheme 9.45).

With all the rings in hand, the synthesis of the middle segment started with (*R*)-Roche ester (**398**), which was converted to iodide **413** using Marshall's protocol.[97] Reacting **413** with phosphonamide **414**, which was derived from alkylation of 1,3-dimethyl-2-oxo-1,3,2-diazaphospholidine,[98,99] afforded phosphonamide **415**. Cyclopropane **416** was obtained via they key phosphonamide-anion based olefinations with aldehyde **412** (Scheme 9.46). Cyclopropane **416** was converted into phosphonamide **417** in seven steps, which underwent another phosphonamide-anion based olefinations with ketone **408** to give triene **418** as a separable 6:1 mixture of *E*/*Z* ratio isomer. Deprotection of the acetylene gave alkyne **419** that underwent the key *C*-glycosylation with lactone **402** to afford glycoside **420**. (+)-Ambruticin S (**397**) was afforded in four more steps from glycoside **420**.

9.5 Conclusions and Perspectives

In summary, renewable resource-based chirons have provided a facile entry to the total synthesis of many complex natural products including alkaloids and terpenoids. Carbohydrates, α-amino acids, terpenes, α-hydroxy acids, and cyclitols are the key building blocks utilized by synthetic organic chemists. This chapter only highlighted the most recent literature examples in the past decade. To gain a more detailed understanding on this subject, readers are highly recommended to pursue the recent text written by Hanessian *et al.*[1d]

We envision that these chirons will continue to play important roles in organic synthesis – not only as building blocks but also as chiral auxiliaries/catalysts. Combining with redox economic synthetic manipulations,[100] the chiron approach can serve as a powerful tool in the new era of total synthesis. The recent concept of chemomics also relies heavily on the use of chiral building blocks that can accelerate drug discovery.[101]

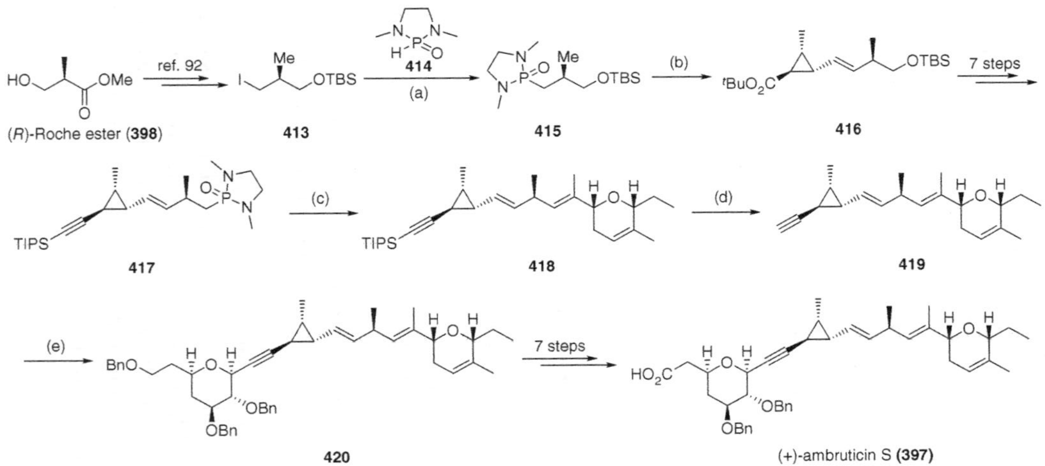

Scheme 9.46 Reagents and conditions: (a) **414**, NaH, DMF, THF, 0°C, 95%; (b) **412**, ^{n}BuLi, THF, −78°C, 72% (E/Z > 25:1); (c) **408**, ^{n}BuLi, THF, −78°C, 92% brsm (E/Z > 6:1); (d) TBAF, THF, 88%; (e) (i) **402**, ^{n}BuLi, THF, −78°C; (ii) BF_3•OEt_2, Et_3SiH, CH_3CN, CH_2Cl_2, −40°C, 66% (98% brsm).

References

1 (a) Hanessian, S. *Aldrichimica Acta* **1989**, *22*, 3–16; (b) Hanessian, S. In *Total Synthesis of Natural Products: The Chiron Approach*; Baldwin, J. E. (ed.); Oxford: Pergamon, **1983**; (c) Hanessian, S.; Franco, J.; Larouche, B. *Pure. Appl. Chem.* **1990**, *62*, 1887–1910; (d) Hanessian, S.; Giroux, S.; Mernet, B. L. In *Design and Strategy in Organic Synthesis: The Chiron Approach to Catalysis*; Hanessian, S. (ed.), Weinheim: Wiley-VCH, **2013**.

2 Walsh, P. J.; Kozlowski, M. C. In *Fundamentals of Asymmetric Catalysis*; Murdzek, J. (ed.) Sausalito: University Science Books, **2009**.

3 Coppola, G. M.; Schuster, H. F. In *α-Hydroxy Acids in Enantioselective Syntheses*; Walter, G. (ed.), Weinheim: Wiley-VCH, **1997**.

4 Hanessian, S.; Soma, U.; Dorich, S.; Deschênes-Simard, B. *Org. Lett.* **2011**, *13*, 1048–1051.

5 Knight, J. G.; Tchabanenko, K. *Tetrahedron* **2003**, *59*, 281–286.

6 Liu, D.; Acharya, H. P.; Yu, M.; Wang, J.; Yeh, V. S. C.; Kang, S.; *et al. J. Org. Chem.* **2009**, *74*, 7417–7428.

7 Sibi, M. P.; Rutherford, D.; Renhowe, P. A.; Li, B. J. *Am. Chem. Soc.* **1999**, *121*, 7509–7516.

8 Rhoads, S. J.; Raulins, N. R. *Org. React.* **1975**, *22*, 1–252.

9 (a) Overman, L. E. *Angew. Chem., Int. Ed. Engl.* **1984**, *23*, 579–586; (b) Lutz, R. P. *Chem. Rev.* **1984**, *84*, 205–247.

10 Kim, J.; Movassaghi, M. *J. Am. Chem. Soc.* **2010**, *132*, 14376–14378.

11 (a) Movassaghi, M.; Schmidt, M. A.; Ashenhurst, J. A. *Angew. Chem., Int. Ed.* **2008**, *47*, 1485–1487. (b) Movassaghi, M.; Schmidt, M. A. *Angew. Chem., Int. Ed.* **2007**, *46*, 3725–3728.

12 Hamada, T.; Nishida, A.; Yonemitsu, O. *J. Am. Chem. Soc.* **1986**, *108*, 140–145.

13 Laws, S. W.; Scheerer, J. R. *J. Org. Chem.* **2013**, *78*, 2422–2429.

14 Margrey, K. A.; Chinn, A. J.; Laws, S. W.; Pike, R. D.; Scheerer, J. R. *Org. Lett.* **2012**, *43*, 2458 - 2461.

15 (a) Miller, K. A.; Welch, T. R.; Greshock, T. J.; Ding, Y. S.; Sherman, D. H.; Williams, R. M. *J. Org. Chem.* **2008**, *73*, 3116 - 3119. (b) Valente, M. W. N.; Williams, R. M. *Heterocycles* **2006**, *70*, 249 - 259.

16 Yu, P.; Wang, T.; Li, J.; Cook, J. M. *J. Org. Chem.* **2000**, *65*, 3173–3191.

17 Li, J.; Cook, J. M. *J. Org. Chem.* **1998**, *63*, 4166–4167.

18 Liao, X.; Zhou, H.; Yu, J.; Cook, J. M. *J. Org. Chem.* **2006**, *71*, 8884–8890.

19 Yu, J.; Wang, T.; Liu, X.; Deschamps, J.; Flippen-Anderson, J.; Liao, X.; Cook, J. M. *J. Org. Chem.* **2003**, *68*, 7565–7581.

20 Liao, X.; Zhou, H.; Wearing, X.; Ma, J.; Cook, J. M. *Org. Lett.* **2005**, *7*, 3501–3504.

21 Gan, T.; Cook, J. M. *J. Org. Chem.* **1998**, *63*, 1478–1483.

22 (a) Yin, W.; Ma, J.; Rivas, F. M.; Cook, J. M. *Org. Lett.* **2007**, *9*, 295–298; (b) Yin, W.; Kabir, S.; Wang, Z.; Rallapalli, S. K.; Ma, J.; Cook, J. M. *J. Org. Chem.* **2010**, *75*, 3339–3349.

23 He, B.; Song, H.; Du, Y.; Qin, Y. *J. Org. Chem.* **2009**, *74*, 298–304.
24 Niphakis, M. J.; Georg, G. I. *J. Org. Chem.* **2010**, *75*, 6019–6022.
25 Comins, D. L.; Chen, X.; Morgan, L. A. *J. Org. Chem.* **1997**, *62*, 7435–7438.
26 Sánchez-Izquierdo, F.; Blanco, P.; Busqué, F.; Alibés, R.; De March, P.; Figueredo, M.; *et al. Org. Lett.* **2007**, *9*, 1769–1772.
27 Xu, R.; Lu, Y.; Chu, J.; Iwashita, T.; Naoki, H.; Naya, Y.; *et al. Tetrahedron* **1982**, *38*, 2667–2670.
28 He, X.; Lin, W.; Xu, R. *Huaxue Xuebao* **1990**, *48*, 694–699.
29 Imaoka, T.; Akimoto, T.; Iwamoto, O.; Nagasawa, K. *Chem. Asian. J.* **2010**, *5*, 1810–1816.
30 Lesma, G.; Colombo, A.; Sacchetti, A.; Silvani, A. *J. Org. Chem.* **2009**, *74*, 590–596.
31 Wierzejska, J.; Motogoe, S.; Makino, Y.; Sengoku, T.; Takahashi, M.; Yoda, H. *Beilstein J. Org. Chem.* **2012**, *8*, 1831–1838.
32 Sengoku, T.; Satoh, Y.; Oshima, M.; Takahashi, M.; Yoda, H. *Tetrahedron* **2008**, *64*, 8052–8058.
33 Wierzejska, J.; Ohshima, M.; Inuzuka, T.; Sengoku, T.; Takahashi, M.; Yoda, H. *Tetrahedron Lett.* **2011**, *52*, 1173–1175.
34 Reddy, K. U.; Reddy, A. P. *Synth. Commun.* **2013**, *43*, 2678–2684.
35 Comins, D. L.; Chen, X.; Morgan, L. A. *J. Org. Chem.* **1997**, *62*, 7435–7438.
36 (a) Ackermann, J.; Matthes, M.; Tamm, C. *Helv. Chim. Acta* **1990**, *73*, 122–132; (b) Oliveria, D. J.; Coelho, F. *Synth. Commun.* **2000**, *30*, 2143–2159.
37 Fu, R.; Du, Y.; Li, Z.-Y.; Xu, W.-X.; Huang, P.-Q. *Tetrahedron* **2009**, *65*, 9765–9771.
38 Tuo, S.-C.; Ye, J.-L.; Wang, A.-E; Huang, S.-Y.; Huang, P.-Q. *Org. Lett.* **2011**, *13*, 5270–5273.
39 Huo, H.-H.; Xia, X.-E.; Zhang, H.-K.; Huang, P.-Q. *J. Org. Chem.* **2013**, *78*, 455–465.
40 Liao, X. W.; Liu, W.; Dong, W. F.; Guan, B. H.; Chen, S. Z.; Liu, Z. Z. *Tetrahedron* **2009**, *65*, 5709–5715.
41 Shing, T. K. M.; So, K. H. *Org. Lett.* **2011**, *13*, 2916–2919.
42 Baldwin, J. E. *J. Chem. Soc., Chem. Commun.* **1976**, 734–736.
43 Archer, S.; Lewis, T. R.; Bell, M. R.; Schulenberg, J. W. *J. Am. Chem. Soc.* **1961**, *83*, 2386–2387.
44 (a) Kozikowski, A. P.; Simoni, D.; Manfredini, S.; Roberti, M.; Stoelwinder, J. *Tetrahedron Lett.* **1996**, *37*, 5333–5336; (b) Dickerson, T. J.; Janda, K. D. *Bioorg. Med. Chem. Lett.* **2007**, *17*, 4280–4283.
45 Shanahan, C. S.; Fang, C.; Paull, D. H.; Martin, S. F. *Tetrahedron* **2013**, *69*, 7592–7607.
46 Swallen, L. C.; Boord, C. E. *J. Am. Chem. Soc.* **1930**, *52*, 651–660.
47 (a) Hirama, M.; Shigemoto, T.; Itô, S. *Tetrahedron Lett.* **1985**, *26*, 4137–4140; (b) Hirama, M.; Shigemoto, T.; Yamazaki, Y.; Itô, S. *Tetrahedron Lett.* **1985**, *26*, 4133–4136; (c) Hirama, M.; Shigemoto, T.; Yamazaki, Y.; Itô, S. *J. Am. Chem. Soc.* **1985**, *107*, 1797–1798; (d) Hirama, M.;

Nishizaki, I.; Shigemoto, T.; Itô, S. *J. Chem. Soc., Chem. Commun.* **1986**, 393–394; (e) Hirama, M.; Shigemoto, T.; Itô, S. *J. Org. Chem.* **1987**, *52*, 3342–3346.

48 Ko, E. J.; Savage, G. P.; Williams, C. M.; Tsanaktsidis, J. *Org. Lett.* **2011**, *13*, 1944–1947

49 Brüggemann, M.; McDonald, A. I.; Overman, L. E.; Rosen, M. D.; Schwink, L.; Scott, J. P. *J. Am. Chem. Soc.* **2003**, *125*, 15284–15285.

50 Ritthiwigrom, T.; Willis, A. C.; Pyne, S. G. *J. Org. Chem.* **2010**, *75*, 815–824.

51 Davis, A. S.; Pyne, S. G.; Skelton, B. W.; White, A. H. *J. Org. Chem.* **2004**, *69*, 3139–3143.

52 Petasis, N. A.; Zavialov, I. A. *J. Am. Chem. Soc.* **1998**, *120*, 11798–11799.

53 Chandrasekhar, S.; Saritha, B.; Jagadeshwar, V.; Prakash, S. J. *Tetrahedron: Asymmetry* **2006**, *17*, 1380–1386.

54 Blakemore, P. R.; Cole, W. J.; Kocienski, P. J.; Morley, A. *Synlett* **1998**, 26–28.

55 Tsuzuki, Y.; Tomita, K.; Sato, Y.; Kashimoto, S.; Chiba, K. *J. Med. Chem.* **2004**, *47*, 2097–2109.

56 Bohno, M.; Imase, H.; Chida, N. *Chem. Commun.* **2004**, 1086–1087.

57 Chida, N.; Ohtsuka, M.; Ogura, K.; Ogawa, S. *Bull. Chem. Soc. Jpn.* **1991**, *64*, 2118–2121.

58 (a) Kano, S.; Yokoyama, T.; Yuasa, Y.; Shibuya, S. *Chem. Lett.* **1982**, *11*, 1915–1918; Strauss, H. F.; (b) Wiechers, A. *Tetrahedron Lett.*, **1979**, *20*, 4495–4496; (c) Fujioka, H.; Annoura, H.; Murano, K.; Kita Y.; Tamura, Y. *Chem. Pharm. Bull.*, **1989**, *37*, 2047–2052; (d) Bauermeister S.; Gouws, I. D.; Strauss H. F.; Venter, E. M. M. *J. Chem. Soc., Perkin Trans. 1*, **1991**, 561–565; (e) Chida, N.; Sugihara K.; Ogawa, S. *J. Chem. Soc., Chem. Commun.*, **1994**, 901–902; (f) Chida, N.; Sugihara, K.; Amano, S.; Ogawa, S. *J. Chem. Soc., Perkin Trans. 1* **1997**, 275–280.

59 Saitoh, Y.; Moriyama, Y.; Hirota, H.; Takahashi T.; Khuong-Huu, Q. *Bull. Chem. Soc. Jpn.* **1981**, *54*, 488–492.

60 de Melo, E. B.; da Silveira Gomes, A.; Carvalho, I. *Tetrahedron* **2006**, *62*, 10277–10302.

61 Whitby, K.; Pierson, T. C.; Geiss, B.; Lane, K.; Engle, M.; Zhou, Y.; Doms, R. W.; Diamond, M. S. *J. Virol.* **2005**, *79*, 8698–8706.

62 (a) Machan, T.; Davis, A. S.; Liawruangrath, B.; Pyne, S. G. *Tetrahedron* **2008**, *64*, 2725–2732. (b) Kang, S. H.; Kim, J. S. *Chem. Commun.* **1998**, 1353–1354. (c) Furneaux, R. H.; Gainsford, G. J.; Mason, J. M.; Tyler, P. C. *Carbohydr. Res.* **2004**, *339*, 1747–1751.

63 Jensen, T.; Mikkelsen, M.; Lauritsen, A.; Andresen, T. L.; Gotfredsen, C. H.; Madsen, R. *J. Org. Chem.* **2009**, *74*, 8886–8889.

64 Furneaux, R. H.; Gainsford, G. J.; Manson, J. M.; Tyler, P. C. *Tetrahedron* **1994**, *50*, 2131–2160.

65 Garg, N. K.; Hiebert, S.; Overman, L. E. *Angew. Chem. Int. Ed.* **2006**, *45*, 2912–2915.

66 Kozak, J. A.; Dake, G. R. *Angew. Chem. Int. Ed.* **2008**, *47*, 4221–4223.

67 Chan, T.-L.; Fong, S.; Li, Y.; Man, T.-O.; Poon, C.-D. *J. Chem. Soc. Chem. Commun.* **1994**, 1771–1772.

68 Jørgensen, L.; McKerrall, S. J.; Kuttruff, C. A.; Ungeheuer, F.; Felding, J.; Baran, P. S. *Science* **2013**, *341*, 878–882.

69 aWinkler, J. D.; Rouse, M. B.; Greaney, M. F.; Harrison, S. J.; Jeon, Y. T. *J. Am. Chem. Soc.* **2002**, *124*, 9726 - 9728. (b) Tanino, K.; Onuki, K.; Asano, K.; Miyashita, M.; Nakamura, T.; Takahashi, Y.; *et al. J. Am. Chem. Soc.* **2003**, *125*, 1498 - 1500. (c) Nickel, A.; Maruyama, T.; Tang, H.; Murphy, P. D.; Greene, B.; Yusuff, N.; *et al. J. Am. Chem. Soc.* **2004**, *126*, 16300 - 16301.

70 McKerrall, S. J.; Jørgensen, L.; Kuttruff, C. A.; Ungeheuer, F.; Baran, P. S. *J. Am. Chem. Soc.* **2014**, *136*, 5799 - 5810.

71 Shibuya, K. *Synth. Commun.* **1994**, *24*, 2923–2941.

72 Shing, T. K. M.; Yeung, Y.-Y. *Angew. Chem. Int. Ed.* **2005**, *44*, 7981–7984.

73 (a) Shing, T. K. M.; Zhu, X. Y.; Yeung, Y. Y. *Chem. Eur. J.* **2003**, *9*, 5489–5500; (b) Shing, T. K. M.; Jiang, Q. *Tetrahedron Lett.* **2001**, *42*, 5271–5273.

74 Ley, S. V.; Norman, J.; Griffith, W. P.; Marsden, S. P. *Synthesis* **1994**, *7*, 639–666.

75 Wilson, S. R.; Mao, D. T.; Jernberg, K. M.; Ezmirly, S. T. *Tetrahedron Lett.* **1977**, *18*, 2559–2562.

76 Shing, T. K. M.; Yeung, Y.-Y.; So, P. L. *Org. Lett.* **2006**, *8*, 3149– 3151.

77 Cambie, R. C.; Hay, M. P.; Larsen, L.; Rickard, C. E. F.; Rutledge, P. S.; Woodgate, P. D. *Aust. J. Chem.* **1991**, *44*, 821–842.

78 Ghosh, A. K.; Xi, K. *Org. Lett.* **2007**, *9*, 4013–4016.

79 (a) Srikrishna, A.; Hemamalini, P. *J. Org. Chem.* **1990**, *55*, 4883–4887. (b) Weinges, K.; Reichert, H. *Synlett* **1991**, 785–786. (c) Weinges, K.; Reichert, H.; Huber-Patz, U.; Irngartinger, H. *Liebigs Ann.Chem.* **1993**, 403–411. (d) Weinges, K.; Reichert, H.; Braun, R. *Chem. Ber.* **1994**, *127*, 549–550.

80 (a) Petasis, N. A.; Bzowej, E. I. *J. Am. Chem. Soc.* **1990**, *112*, 6392–6394. (b) Dollinger, L. M.; Ndakala, A. J.; Hasshemzadeh, M.; Wang, G.; Wang, Y.; Martinez, I.; Arcari, J. T.; Galluzzo, D. J.; Howell, A. R. J. *Org. Chem.* **1999**, *64*, 7074–7080.

81 Mangion, I. K.; MacMillan, D. W. C. *J. Am. Chem. Soc.* **2005**, *127*, 3696–3697.

82 Oliver, S. F.; Hogennauer, K.; Simic, O.; Antonello, A.; Smith, M. D.; Ley, S. V. *Angew. Chem. Int. Ed.* **2003**, *48*, 5996–6000.

83 Ley, S. V.; Antonello, A.; Balskus, E. P.; Booth, D. T.; Christensen, S. B.; Cleator, E.; *et al. Proc. Nat. Acad. Sci. U.S.A.* **2004**, *101*, 12073–12078.

84 Angeles, A. R.; Dorn, D. C.; Kou, C. A.; Moore, M. A. S.; Danishefsky, S. J. *Angew. Chem., Int. Ed.* **2007**, *46*, 1451–1454.

85 Levine, S. G. J. *Am. Chem. Soc.* **1958**, *80*, 6150–6151.

86 Paquette, L. A.; Peng, X.; Yang, J. *Angew. Chem. Int. Ed.* **2007**, *46*, 7817–7819.

87 (a) Fujii, H.; Hirano, N.; Uchiro, H.; Kawamura, K.; Nagase, H. *Chem. Pharm. Bull.* **2004**, *52*, 747–750. (b) Rabiczko, J.; Urbanczyk-Lipkowska, Z.; Chemielewski, M. *Tetrahedron* **2002**, *58*, 1433–1441.

88 (a) Hanessian, S.; Vakiti, R. R.; Dorich, S.; Banerjee, S.; Lecomte, F.; Del Valle, J. R.; *et al. Angew. Chem. Int. Ed.* **2011**, *50*, 3497–3500; (b) Hanessian, S.; Vakiti, R. R.; Dorich, S.; Banerjee, S.; Deschênes-Simard, B. *J. Org. Chem.* **2012**, *77*, 9458–9472.

89 Hanessian, S.; Wang, X.; Ersmark, K.; Del Valle, J. R.; Klegraf, E. *Org. Lett.* **2009**, *11*, 4232–4235.

90 Hanessian, S.; Tremblay, M.; Petersen, J. F. W. *J. Am. Chem. Soc.* **2004**, *126*, 6064–6071.

91 (a) Lou, B.; Huynh, H. K.; Hanessian, S. In *Preparative Carbohydrate Chemistry*; Hanessian, S. (ed.), New York: Marcel Dekker, **1997**; Ch.19, p. 431. (b) Hanessian, S.; Mascitti, V.; Rogel, O. *J. Org. Chem.* **2002**, *67*, 3346–3354.

92 Nakamura, S.; Sugani, Y.; Kikuchi, F.; Hashimoto, S. *Angew. Chem. Int. Ed.* **2006**, *45*, 6532–6535.

93 Wong, M. Y. H.; Gray, G. R. *J. Am. Chem. Soc.* **1978**, *100*, 3548–3553.

94 Kitagaki, S.; Anada, M.; Kataoka, O.; Matsuno, K.; Umeda, C.; Watanabe, N.; *et al. J. Am. Chem. Soc.* **1999**, *121*, 1417–1418.

95 Hanessian, S.; Focken, T.; Mi, X.; Oza, R.; Chen, B.; Ritson, D.; *et al. J. Org. Chem.* **2010**, *75*, 5601–5618.

96 Ziegler, F. E.; Fang, J. M.; Tam, C. C. *J. Am. Chem. Soc.* **1982**, *104*, 7174–7181.

97 Marshall, J. A.; Grote, J.; Audia, J. E. *J. Am. Chem. Soc.* **1987**, *109*, 1186–1194.

98 Pudovik, M. A.; Pudovik, A. N. *Zh. Obshch. Khim.* **1973**, *43*, 2147–2149.

99 Koeller, K. J.; Spilling, C. D. *Tetrahedron Lett.* **1991**, *32*, 6297–6300.

100 Burns, N. Z.; Baran, P. S.; Hoffmann, R. W. *Angew. Chem. Int. Ed.* **2009**, *48*, 2854–2867.

101 Xu, J.; Gu, Q.; Liu, H. B.; Zhou, J. J.; Bu, X. Z.; Huang, Z. S.; *et al. Sci. China: Chem.* **2013**, *56*, 71 – 85.

10

Natural Product Synthesis for Drug Discovery and Chemical Biology

Zhu-Jun Yao[1] *and Wan-Guo Wei*[2]

[1] *School of Chemistry and Chemical Engineering, Nanjing University, Nanjing, China*
[2] *Shanghai Advanced Research Institute, Chinese Academy of Sciences, Shanghai, China*

10.1 The Importance of Bioactive Natural Products in Biological Investigation

For over three billion years, microorganisms have evolved to produce nucleic acids, proteins, and other biological macromolecules alongside smaller metabolites.[1] It is thus hardly surprising that there exists a small metabolite, or natural-product, ligand for many of the distinct macromolecular targets that exist within today's cells. Natural products are biosynthesized by protein catalysts, and as such natural-product and protein structures must necessarily have co-evolved to bind to one another selectively.[2] Indeed, it is likely that during evolution, there has existed a natural-product ligand that could bind to all possible protein folds. In addition to this, many metabolites are known to regulate their own biosynthesis or degradation through interaction with the genes, particularly mRNAs, which encode the enzymes responsible for their metabolism.[3,4] Thus, while primary metabolites serve defined roles in the normal growth and development of living organisms, it seems likely that the secondary metabolites have evolved specifically to interact with and modulate the function of biological macromolecules.

Most secondary metabolites are produced by bacteria, fungi, plants, and marine organisms. Some of these molecules are known to provide a selective advantage against microbial attack or defense against infestation and disease,[5] while others function as signaling molecules in quorum sensing,[6] as pheromones or pigments facilitating reproduction. However, the role of many of these molecules and their specific molecular targets within the producing cell or local environment remains unclear. In fact, it has been suggested that the

Efficiency in Natural Product Total Synthesis, First Edition. Edited by Pei-Qiang Huang, Zhu-Jun Yao, and Richard P. Hsung.

majority of natural products do not have any significant biological activity or functional role.[7] However, evolution could still favor their retention, as part of the wider chemical diversity. Through the presence of promiscuous enzymes and mutations within biosynthetic pathway, this chemical diversity can evolve, until a natural product with potent biological activity is selected, which would increase the fitness of the population. Despite this, specific targets have been discovered for many secondary metabolites within human cells. Indeed, a significant portion of the new pharmaceuticals released for clinical use are derived from or inspired by natural products, and this has been one of the major cornerstones of pharmaceutical drug development.[8] The fact that many natural products can selectively bind to and modulate the function of many targets within the human cell is not surprising. Most macromolecular targets in humans have functional counterparts, possessing sequence similarity, in lower eukaryotes and prokaryotes. As a result, the biological activity of natural products in human cells is a consequence of their co-evolution alongside similar targets within the producing or neighboring organism in the environment.

In addition to pharmaceutical drug development, the enormous potential of natural products as modulators of biomolecular function has become increasingly evident with the advent of chemical genomics and increasingly powerful high-throughput screening methodologies. Unlike pharmaceutical research, which focuses on only those cellular targets that are deemed therapeutically relevant, chemical genomics aims to uncover small molecules that can specifically modulate every function of all gene products in a given cell.[9] In this respect, natural products have proved invaluable constituents of compound libraries in chemical genomics. Moreover, this research has uncovered a plethora of hitherto unidentified targets to which natural products interact with exquisite selectivity. As a result of this and the continuing need for improved pharmaceuticals, there has been an increased drive to expand the number and structural diversity of available natural-product compound libraries.

Access to structurally complex and diverse small molecules through synthesis is driving recent efforts to dissect biological pathways in ways analogous to those of genetics, where random mutations are first generated and then screened in search of a specific cellular or organism phenotype. Finding small molecules or mutations that affect a specific pathway and identifying the cellular target of the small molecule or the molecular sequence of the mutant gene can shed light on the pathway. From the perspective of drug discovery, the small-molecule approach offers the means for the simultaneous identification of proteins that may serve as targets for therapeutic intervention and small molecules that can modulate the functions of these therapeutic targets. The structures of the small-molecule modulators provide leads for the drug discovery process. The overall process differs from the current primary means of drug discovery where biological methods are first used to select protein targets for therapeutic intervention, followed by chemical efforts to determine whether the protein target can be modulated by small molecules. The latter process involves screening small molecules for their ability to bind the preselected protein target. Biologically active natural products can be considered "privileged" scaffolds that have been evolutionarily selected for binding to particular domains of biological macromolecules. They could potentially address poorly populated, underexplored chemical space. Natural products are central to biology and medicine, serving as pharmaceutical leads, drugs, and powerful reagents for studying cell biology.

10.2 Bioactive Natural-Product-Inspired Chemical Biology

Chemical biology uses small molecules to study and manipulate biological systems. By altering an organisms' normal state, and thereby affecting growth or development, we can learn about the contributions of the perturbed processes to the organisms' fitness, physiology, and homeostasis. The approach is analogous to genetic manipulation to produce an observable phenotype. Small molecules, in addition to complementing genetic perturbations, have several advantages: They can be applied at varied concentrations, during different stages of development, and on specific tissues to different organisms, and their rapid reversibility can be used to modulate dynamic processes. Natural products embody inherent structural complexity and biological activity that often leads to new targets, pathways, or modes of action. As it turns out, the underlying biological effects and mechanisms of natural products are fundamental, and often go far beyond what scientists can design alone. Therefore, in addition to its crucial role in drug discovery, the natural product has played a special role in chemical biology. Owing to their diversity, target affinity, and specificity, natural products have demonstrated enormous potential as modulators of biomolecular function.[10] Numerous natural products, including erythromycin, cyclosporine A, rapamycin, geldanamycin, trapoxin A, FTY720, and diazonamide A have been used for the study of important signaling pathways (Figure 10.1). The following examples demonstrate the essential role of natural-product synthesis in drug discovery and probe development for chemical biology.

The mechanistic studies of diazonamide A reported by Harran and coworkers is one of the classic examples highlighting the importance of natural-product synthesis in understanding biological processes. Diazonamide A was isolated from the colonial marine ascidian *Diazona angulata*,[11] and attracted a considerable amount of attention from the organic chemistry community due to its potent cytotoxicity against various types of human cancer cell lines. The initial NCI's COMPARE screen suggested that the antitumor activity of diazonamide A comes from its microtubule binding activity. However, detailed

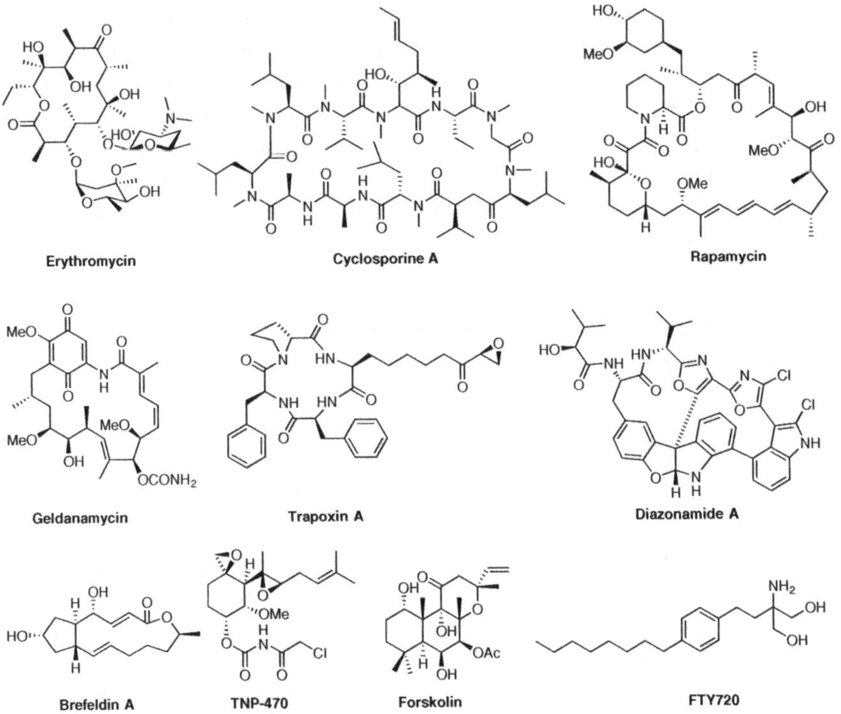

Figure 10.1 Examples of natural products inspired chemical biology.

mechanistic studies showed that diazonamide A does not compete with other tubulin-binding agents, such as colchicine or vinblastine. Based on the intriguing bioactivity and remarkable molecular structure of diazonamide A, Harran and coworkers completed the total synthesis of diazonamide A, paving the way for structural modification of mode of action studies.[12] Remarkably, these studies also served to correct the X-ray misassigned structure of diazonamide A and led to Harran's discovery of the misassignment, its origin, and his proposed and synthetically confirmed key structural reassignment.

Using a biotinylated derivative of diazonamide diazonamide A, Harran, Wang, and coworkers identified ornithine δ-amino transferase (OAT), a mitochondrial matrix enzyme, as the molecular target of the natural product (Figure 10.2).[13] To isolate diazonamide-binding protein(s), HeLa extracts were fractionated by Hitrap Q anionexchange chromatography. Each eluted column fraction was then probed with affinity matrices prepared from both compounds **3** (active) and **4** (inactivated control). Affinity-purified proteins were visualized by silver-stained SDS/PAGE wherein two bands ~50 kDa in size were recovered specifically by the active compound matrix. Mass spectrometry identified both as OAT. These mechanistic studies suggested a unique mode of

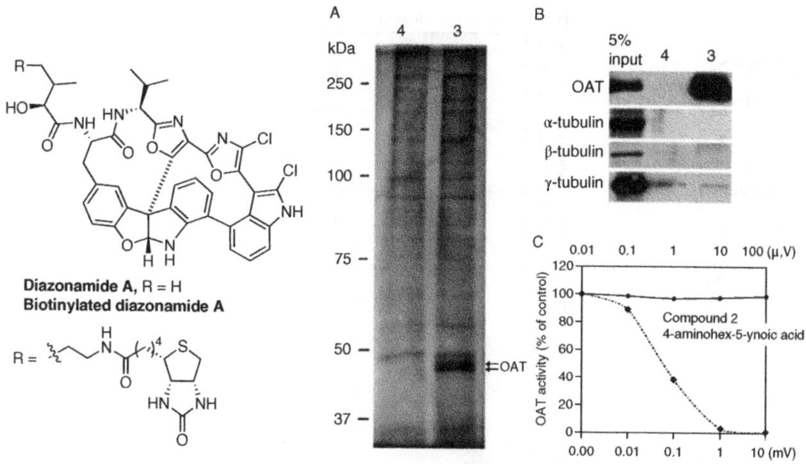

Figure 10.2 Identification of ornithine δ-amino transferase (OAT) as a diazonamide A binding protein.

action involving OAT and identified the protein as a target for chemotherapeutic drug development. Here, the total synthesis of diazonamide A proved to be the key milestone that led to the discovery of an unanticipated, paradoxical function of OAT in mitotic cell division.

Diterpenoids form a large and structurally diverse class of secondary metabolites that are isolated from plants and have a wide spectrum of important biological activities.[14,15] The identification of the cellular activities and target molecules of diterpenoids have attracted considerable interest. In 2012, Chen and coworkers reported that adenanthin, a diterpenoid isolated from the leaves of *Rabdosia adenantha*, induces differentiation of acute promyelocytic leukemia (APL) cells (Figure 10.3).[16] Detailed mechanism studies show that adenanthin directly targets the conserved resolving cysteines of Prx I and Prx II and inhibits their peroxidase activities. By installation of a biotin tag to the C3 hydroxyl group to yield a biotin-adenanthin derivative that retained the ability to induce NB4 cell differentiation, they were able to conduct the pull-down experiment. Then, NB4 cell lysates were incubated with biotin-adenanthin or free biotin, and the mixtures were precipitated with streptavidin-coated agarose beads, followed by gel electrophoresis and silver staining. Only one band, with a molecular mass of ~23 kDa, was clearly precipitated by biotin-adenanthin but not by free biotin. The band was competed away by high concentrations of unlabeled adenanthin (Figure 10.3c), indicating that the protein that bound biotin-adenanthin also bound adenanthin. Mass spectrometry revealed that the adenanthin-bound protein is Prx I and Prx II. Together with other biomedical proofs, these mechanism studies revealed that Prx I and Prx II are direct targets of adenanthin's induction of APL cell differentiation.

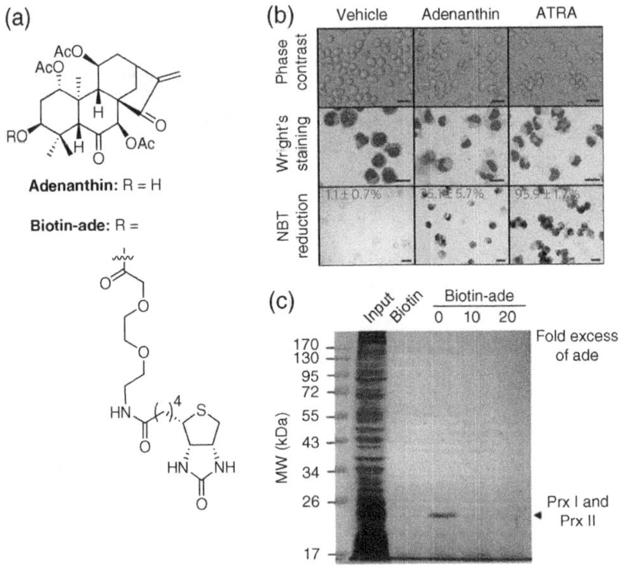

Figure 10.3 Differentiation induction of ATRA-sensitive NB4 cells by adenanthin.

The genus *Ainsliaea* is an important herbal resource in traditional Chinese medicine for treating various diseases, including rheumatism, traumatic injuries, edema, stomach ache, and anorexia.[17] Phytochemistry studies have shown that the major secondary metabolites isolated from these plant species are monomeric, dimeric, and trimeric guaianolides.[18] (−)-Ainsliadimer A was an architecturally complex and unique trimeric sesquiterpenoid, which not only displayed potent cytotoxicity against several cancer cell lines, but also demonstrated that the activity resulted from the induction of apoptosis. However, the exact mode of anticancer action remained elusive. Most recently, Lei's group[19] reported a combined strategy of diverted total synthesis (DTS) and bioorthogonal ligation (TQ ligation) to facilitate identification of the target of this complex natural-product (Figure 10.4). The authors developed a novel bioorthogonal ligation strategy to obtain a chemical probe named TV-modified ainsliatrimer A. Gratifying, this probe possesses equipotent activity at approximately double the concentration of ainsliatrimer A, and retained its apoptosis-inducing ability, which makes it a suitable probe for bioimaging studies in living cells. Pretarget imaging experiments indicated that the potential target(s) are localized in the nucleus. An affinity pull-down assay was also performed with two finely designed chemical probes, and finally led to the identification of PPARg as a functional target of (−)-ainsliadimer A.

Bioactive natural products offer a "privileged" starting point in the search for highly specific and potent modulators of biomolecular function. Natural products

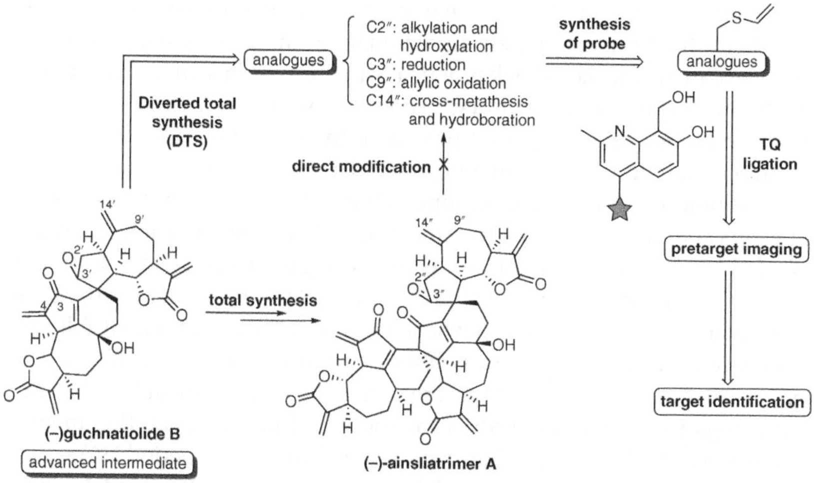

Figure 10.4 Target identification of (−)-ainsliatrimer A enabled by DTS and bioorthogonal ligation (TQ ligation).

have been the richest sources for novel modulators of biomolecular function. Cell-based phenotypic screenings of natural products are followed by target identification, which is often among the most difficult process in chemical biology research. However, target identification of biologically active natural products can be extremely fruitful when their target molecules are drugable. Therefore, by choosing the right target and using both efficient and innovative synthetic technology, there is no doubt that natural-product synthesis will remain not only relevant, but also essential to progress in drug discovery and chemical biology.

10.3 Natural Products in Drug Discovery

Natural products, obtained from plants, animals and marine organisms, or microorganisms are shaped by evolution. Through evolutionary time, the biosynthetic "engine" of nature has produced a myriad of natural products with huge chemical diversity and often distinct biological properties. Such natural products are often stereochemically complex molecules annotated with diverse functional groups, interacting with high specificity with biological targets. Natural products have played a vital role in the treatment of human ailments for thousands of years[20,21] and continue to play a big role in the modern discovery of new agents for the treatment of diseases today. In certain therapeutic areas, natural products account for almost all key modern medicine used today. Many drugs are formulated and used directly as they are found in nature, some are derived directly from natural products by semisynthesis, and others are modeled after natural products. The importance of natural products in modern medicine has been extensively discussed in recent reviews and reports.[22–27]

History of medicine dates back practically to the existence of human civilization. Before the twentieth century, crude and semipure extracts of plants, animals, microbes, and minerals represented the only medication available to treat human and domestic animal illnesses. The isolation of strychnine, morphine, atropine, colchicine, and quinine in the early 1800s from the commonly used plants and their use for the treatment of certain ailments might constitute the early idea of "pure" compounds as drugs (Figure 10.5). E. Merck isolated and commercialized morphine as the first pure natural-product for the treatment of pain.[28–30] Isolation of salicylic acid as the active component followed by acetylation produced the semisynthetic product called "Aspirin" that was commercialized by Bayer in 1899 as a remedy for rheumatism and back pain.[31] The investigation of natural products as sources of novel human therapeutics reached its peak in the western pharmaceutical industry in the period 1970–1980, which resulted in a pharmaceutical landscape heavily influenced by non-synthetic molecules. By 1990, about 80% of drugs were either natural products or analogs inspired by them. Antibiotics (e.g., penicillin, tetracycline, erythromycin), antiparasitics (e.g., avermectin), antimalarials (e.g., quinine, artemisinin), lipid control agents (e.g., lovastatin and analogs), immunosuppressants for organ transplants (e.g., cyclosporine, rapamycins), and anticancer drugs (e.g., taxol, doxorubicin) revolutionized medicine.[32]

The twentieth century revolutionized the thinking in the use of drugs, as the receptor theory of drug action. The idea that effects of drugs in human body are mediated by specific interactions of the drug molecule with biological macromolecules (proteins or nucleic acids in most cases) led scientists to the conclusion that individual chemical compounds in extracts, rather than some mystical "power of life," are the factors required for the biological activity of the drug. This led to the beginning of a totally new era in pharmacology, as pure,

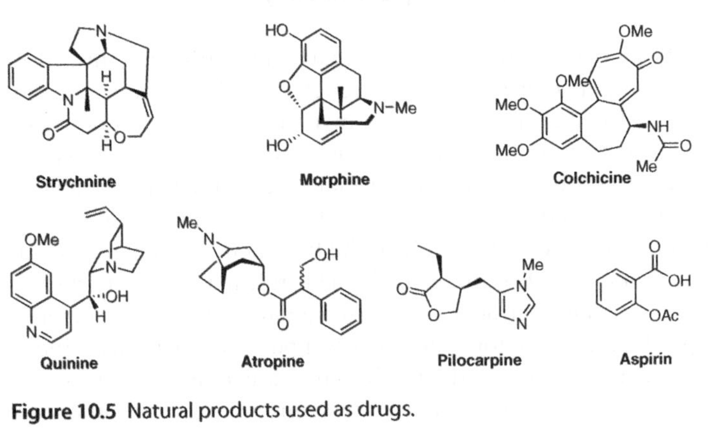

Figure 10.5 Natural products used as drugs.

isolated chemicals instead of extracts became the standard treatments for diseases. There is no doubt that plants are among the most perfect "natural laboratories" for the synthesis of various molecules ranging from a simple skeleton to highly complex chemical structures. If secondary metabolites are compared with randomly synthesized compounds, natural metabolites are superior in terms of biological and pharmacological activities. Present drug discovery from medicinal plants has mainly relied on biological activity guided isolation methods, which, for example, have led to the isolation, identification, and the discovery of important drugs.

The development of high-throughput screens based on molecular targets had led to demand for the generation of large libraries of compounds to satisfy the enormous capacities of these screens. One of the next breakthroughs was the use of a mechanism-based screening for bioassay-guided fractionation. Through continual improvement of screening formats, reagent production, robotics, and data management, mechanism-based screening has since become the mainstay of high-throughput screening (HTS). Some of the first compounds identified in the early 1970s using mechanism-based screening methods included the β-lactamase inhibitor clavulanic acid from *Streptomyces clavuligerus*[33] and the HMG-CoA reductase inhibitor mevastatin from *Penicillium citrinum*.[34] Mevastatin was also reported as an antifungal agent from *P. brevicompactum*.[35] A mixture of clavulanic acid and amoxicillin is still being used today as a front-line antibiotic, while mevastatin and lovastatin were the lead compounds for a series of antilipidemic drugs collectively known as the "statins" (Figure 10.6).[36,37]

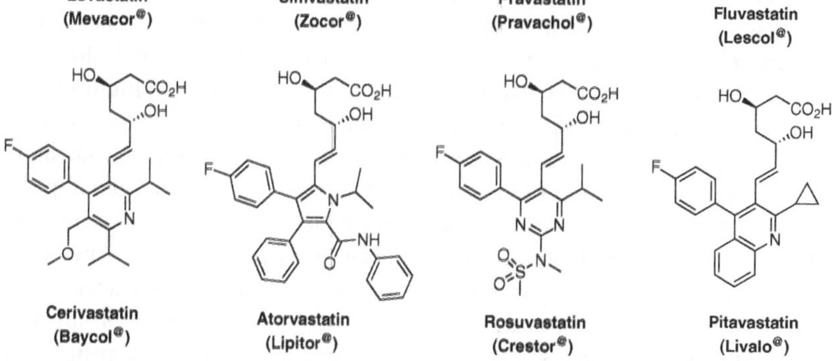

Figure 10.6 Statin class of antilipidemic drugs derived from lovastatin.

Statins are some of the most widely prescribed drugs in the world used to lower cholesterol levels by inhibiting the enzyme HMG-CoA reductase, which plays a central role in the production of cholesterol in the liver. HMG-CoA reductase (HMGCR) catalyzes the conversion of HMG-CoA to mevalonate, the rate-limiting step in cholesterol synthesis. The statin pharmacophore is a modified 3,5-dihydroxyglutaric acid (DHGA) moiety, which is structurally similar to the endogenous substrate HMG-CoA and the mevaldyl CoA transition-state intermediate. The statin pharmacophore binds to the same active site as the substrate HMG-CoA and inhibits the HMGCR enzyme. So, all statins must have the required *3R,5R* stereochemistry to be effective inhibitors. Traditionally, structure-activity relationships in bioactive natural products were examined by simple chemical transformations. For example, methylation of the side chain of the cholesterol-lowering agent lovastatin produces simvastatin (Zocor), an improved drug that had sales in excess of $4.3 billion in 2006 (before the loss of patent protection). The developing understanding of secondary metabolite biogenesis also allows use of biosynthetic enzymes or genetically altered organisms to generate derivatives of drugs. Tang and coworkers used an esterase (LovD), which normally attaches the side chain of lovastatin, to make a series of analogs, including simvastatin.

Recent advances in genomics and structural biology during the past 5–10 years are painting a clearer picture of the diversity of proteins targeted by natural-product molecules. The sequencing of the human genome has promised a revolution in medicine. The genome encodes 20,000–25,000 human genes and thousands more proteins as a result of alternative gene splicing. Many of these hold the keys to treating diseases, especially numerous enzymes of thus far undefined physiologic functions.[38] Out of 1200 registered drugs, over 300 act as enzyme inhibitors. Most of them are simple analogs of substrates of certain enzymatic reactions. In 1995, there were transition-state analogs for at least 130 enzymes.[39] The most successful structure-based anti-influenza drug Oseltamivir (Tamifliu)[40] and Zanamivir (Renenza)[41] are sialic acid–mimicking inhibitors of neuraminidase (NA) targeting the sialidase function.

Sialic acids play an important role in a variety of biological processes. They are usually attached to the terminal positions of glycoproteins, glycolipids, and oligosaccharides. From more than 100 different sialic acids, *N*-acetylneuraminic acid (NeuAc) is the most abundant one. Sialidases or neuraminidases are a family of exo-glycoside hydrolases that catalyze the cleavage of terminal sialic acid residues from sialylated oligosaccharides, glycoproteins, and glycolipids. The sialic acid binding pocket in neuraminidase (NA) was described when the initial crystal structure of neuraminidase (NA) was reported.[42,43] The non-specific transition-state analog of sialidase, 2-deoxy-2,3-dehydro-*N*-acetylneuraminic acid (DANA) is a good starting point for the synthesis of specific inhibitors of human enzymes. During recent decades, chemical synthesis of sialic acids,[44] as well as structural modification of a sialic acid backbone[45] based on structural information has been made at every position, mostly targeting Neu5Ac. Three inhibitors of neuraminidase have been successfully introduced as anti-influenza drugs, all of them being transition-state analog inhibitors (Figure 10.7). They were designed by systematic reduction of DANA structure using crystallographic data and computer-aided methods.[46] Relenza (zanamivir) was the first inhibitor to be synthesized, which specifically inhibited neuraminidases of both Type A and Type B influenza viruses and is effective in controlling influenza infections. Tamiflu (oseltavimir) was developed in 1997 as a neuraminidase inhibitor, which was orally bioavailable. The third drug is permavir, which has been authorized for the emergency use of treatment of certain hospitalized patients with known or suspected 2009 H1N1 influenza. All three drugs soon became lead structures for the design and preparation of new, presumably more effective neuraminidase drugs.

The great success of natural products in drug discovery comes from their structural diversity. General differences among biologically active natural products, combinatorial chemistry libraries, and existing synthetic drugs on the basis of structural and physicochemical properties have been extensively reviewed.[47] In particular, drugs that originated from natural products have been highly successful for protein–protein interactions, nucleic acid complexes, and antibacterial targets.[48–55] They are classically challenging targets for standard small-molecule drug discovery due to the involvement of large, flat interfaces, and the macromolecular nature of their interactions. For many years, nature has evolved to produce small ligands (or natural products) for macromolecular targets within living organisms that contain structural domains similar to many human proteins.[56] As a result of the natural selection process, natural products possess a unique and vast chemical diversity with optimal interactions with biological macromolecules. Due to this diversity and

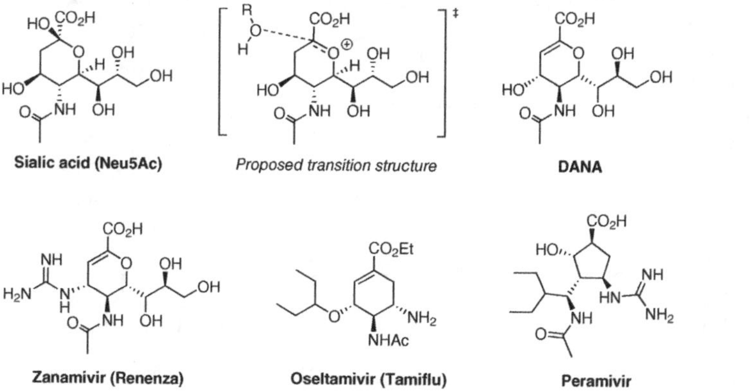

Figure 10.7 Selected influenza neuraminidase inhibitors inspired from sialic acid.

specificity, natural products have proven to be by far the richest sources of new drug development. Of the 1355 New Chemical Entities (NCEs) reported in 1981–2010 that 540 (40%) NCEs were either natural products or natural-product derived. In particular, 63 of the 99 (64%) small-molecule anticancer drugs and 78 of the 104 (75%) antibiotics developed from 1981 to 2010 have originated from natural products.[57] Representative examples include penicillin V (antibiotic), erythromycin (antibiotic), paclitaxel (anticancer), artemether (antimalaria), and galantamine (treatment of Alzheimer's disease; Figure 10.8).

The success of natural products is related to the forces of natural product chemistry, molecular, and cellular biology, synthetic and analytical chemistry, biochemistry, and pharmacology to exploit the vast diversity of chemical structures and biological activities of these products. Natural products are important sources for new drugs and are also good lead compounds suitable for further modification during drug development. The large proportion of natural products in drug discovery has stemmed from the diverse structures and the intricate carbon skeletons of natural products. Since secondary metabolites from natural sources have been elaborated within living systems, they are often perceived as showing more "drug-likeness and biological

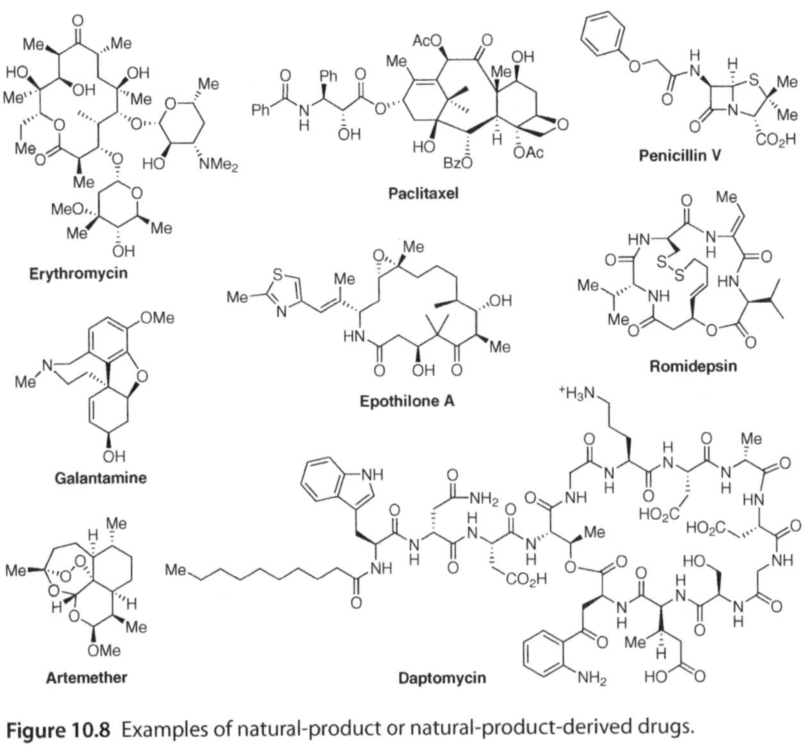

Figure 10.8 Examples of natural-product or natural-product-derived drugs.

friendliness than totally synthetic molecules," making them good candidates for further drug development.

However, not all natural products can be fully synthesized and many natural products have very complex structures that are too difficult and expensive to synthesize on an industrial scale. These include drugs such as penicillin, morphine, and formerly paclitaxel. Such compounds can only be harvested from their natural source; a process that can be tedious, time consuming, and expensive, as well as being potentially unsustainable for the resource. Furthermore, the number of structural analogs that can be obtained from harvesting is severely limited. In spite of the success of natural products as a chemical resource for drug leads, during the last two decades the drug discovery industry has de-emphasized or completely abandoned natural product-based drug discovery in favor of high-throughput synthetic compound library screening and computational design of drug leads. However, the high expectations of a rich drug pipeline from synthetic compound library screening have not materialized. Many experts believe that the waning numbers of NCEs during the last decade may be a direct result of the emphasis on natural products as leads for drug discovery. In the meantime, major technological breakthroughs; including microscale analytical techniques for structure determination, advanced strategies and methods for total synthesis, structure-activity relationship studies of even highly complex natural products, advances in microbial genomics, accessing natural products from unculturable microorganisms via mega-genomic approaches and improved culturing strategies, engineering unnatural natural products via biocombinatorial chemistry, and novel approaches for natural-product cellular target identification including genome-wide target-based screening platforms leading to novel disease targets are all rapidly changing how we approach natural product-based drug discovery.

10.3.1 Natural Products as Antibody-Drug Conjugate (ADC) Payloads

Natural products have had a profound impact on the treatment of cancer. Their ability to eradicate cancer cells through a variety of mechanisms has resulted in many successful drugs.[58,59] Yet the capacity to select for cancer cells in the presence of normal cells has been and will continue to be a significant challenge. This limited selectivity is the cause of many side effects associated with cancer chemotherapy as well as the inability to eliminate the tumor at tolerated doses in many cases. Conversely, monoclonal antibodies are able to exquisitely differentiate between cancer cells and normal cells by binding to tumor-associated antigens that are nonexistent or present only at low levels on normal cells. While several monoclonal antibodies have been approved for the treatment of cancer,[60] many others lack therapeutic activity. Conjugation of toxic natural molecules to monoclonal antibodies or

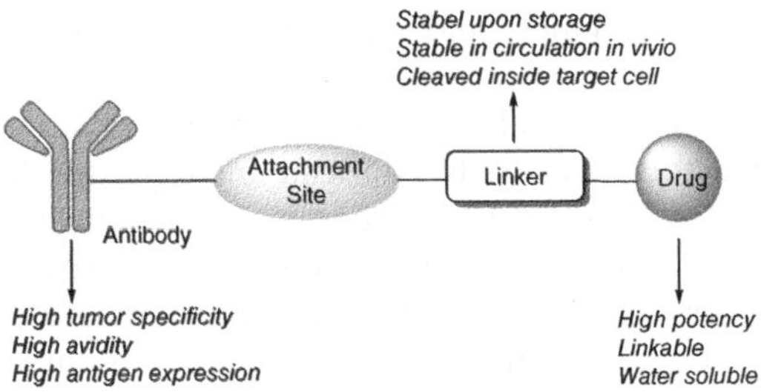

Figure 10.9 Antibody-drug conjugate (ADC) and its key characteristics.

polymeric carriers specifically targeted to epitopes on tumors of interest can also lead to the development of efficacious targeted therapies.

The ADC is a therapeutic modality consisting of a monoclonal antibody attached to a cytotoxic, small-molecule payload through a linker (Figure 10.9). Typically, there are three key components of an ADC: The monoclonal antibody, the cytotoxic drug, and the linker connecting the drug to the antibody. Ideally, the mAb will specifically bind to an antigen with substantial expression on tumor cells but limited expression on normal tissues. Specificity allows the utilization of drugs that otherwise would be too toxic for clinical application. Thus, most of the recent work in this field has centered on the use of highly potent cytotoxic agents. This requires the development of linker technologies that provide conditional stability, so that drug release occurs after tumor binding, rather than in circulation. Finally, the method of ADC conjugation, which determines the drug loading stoichiometry and homogeneity, has been shown to play roles not only in pharmacokinetics, but also in activity, potency, and tolerability. Optimization of the parameters influencing ADC activity has led to the development of new agents with promising activities. Recent clinical results have sparked significant additional interest and the number of ADC candidates in clinical development is growing rapidly. Use of highly potent drugs, stable linkers, optimized drug stoichiometry, and appropriate antigen target selection, have been crucial in advancing the technology to where it is today.

The explosive increase in ADC research and development likely presages the entry of a second wave of natural products into the oncology clinics – as ADC payloads, rather than as free drugs. ADCs are one of the hottest areas of cancer research and there are currently over 30 ADCs in clinical development. Natural products account for all ADC payloads currently approved for clinical use or in advanced clinical development. Hallmark features of natural-product cytotoxins include their unique mechanisms of action (MOAs), exquisite potencies, and structural diversity, even within the same target class.[61,62]

The most commonly used toxins for ADCs in clinical development are maytansine,[63–65] calicheamicin,[66,67] and auristatin.[68,69] Auristatins and

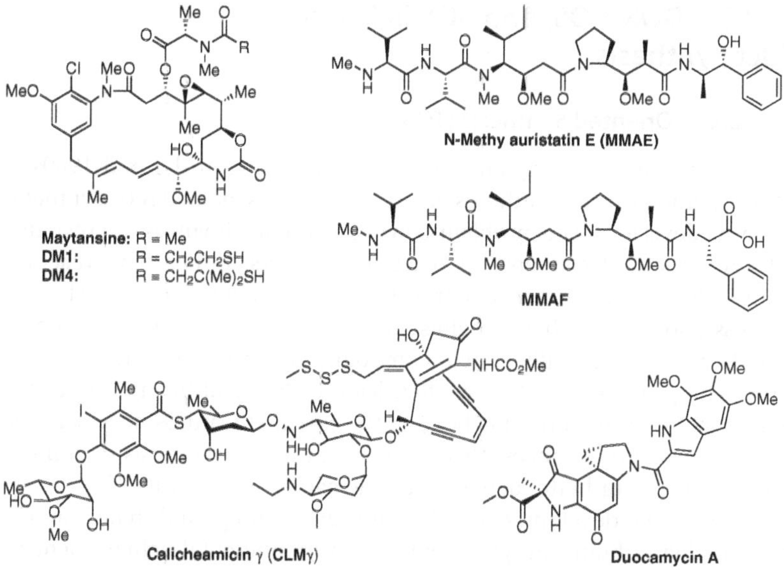

Figure 10.10 Representative natural products as ADC payloads.

maytasines both exert their cytotoxic effects by binding to tubulin, causing G2/M cell cycle arrest, and subsequently leading to apoptosis. The other natural-product payloads under clinical investigation are derived from the DNA alkylating agent duocarmycin/CC-106533,[70] the ansamycin intercalator and/topoisomerase II inhibitor doxorubicin,[71] and the topoisomerase I inhibitor SN-38 (derived from the plant metabolite camptothecin) (Figure 10.10).[72,73]

Currently, approximately 80 ADCs are being developed in oncology and about 30 are undergoing clinical testing. Natural-product payloads, including tubulin inhibitors (auristatins and maytansins) and DNA-damaging agents (calicheamicin and duocarmycin) are most of the ADCs now in clinical development. Recent clinical pronouncements, most notably the approval of T-DM1 for the treatment of breast cancers that are refractory to trastuzumab plus paclitaxel, have provoked an explosion of interest in the development of newer-generation ADCs for the treatment of both hematologic and solid malignancies. Although conventional cytotoxics, many of which are natural products, have in recent years retreated into the background in terms of cancer drug development, the emergence of ADCs as anticancer agents portends a new era of cytotoxic drug development with a consequent reinvigoration of interest in natural-product drugs. The next several years will witness an explosion of interest in ADCs, and natural products will play major roles in this revolutionary shift in cancer therapy.

10.4 TOS, DOS, FOS, and BOS in Natural Product Synthesis

10.4.1 Target-Oriented Synthesis (TOS)

TOS has a long history in organic chemistry. Beginning in the mid-1960s, a systematic method to plan syntheses of target molecules, named retrosynthetic analysis, was devised.[74] This problem-solving technique involves the recognition of key structural elements in reaction products, rather than reaction substrates, which code for synthetic transformations. Repetitive application of this process allows a synthetic chemist to start with a structurally complex target and find a structurally simple compound that can be used to start a synthesis. The process of TOS can often lead to the identification of a substructure portion of the molecule bearing the essential features necessary for activity and, in some instances, this has resulted in the synthesis of simpler analogs with similar or better activity than the natural-product itself.

The discovery and development of then antitumor compounds halichondrin B and its analog eribulin mesylate represent a monumental achievement in both synthetic organic chemistry and drug discovery (Figure 10.11). Since the discovery and structure elucidation of halichondrin B in 1986 by Uemura and coworkers, its complex structure and biological activity have been sources of interest for both chemists and biologists. Halichondrin B was isolated from the marine sponge *Halichondria okadai* and identified as the most potent congener of a family of compounds active on B-16 melanoma cell lines ($IC_{50} = 0.09$ nM).

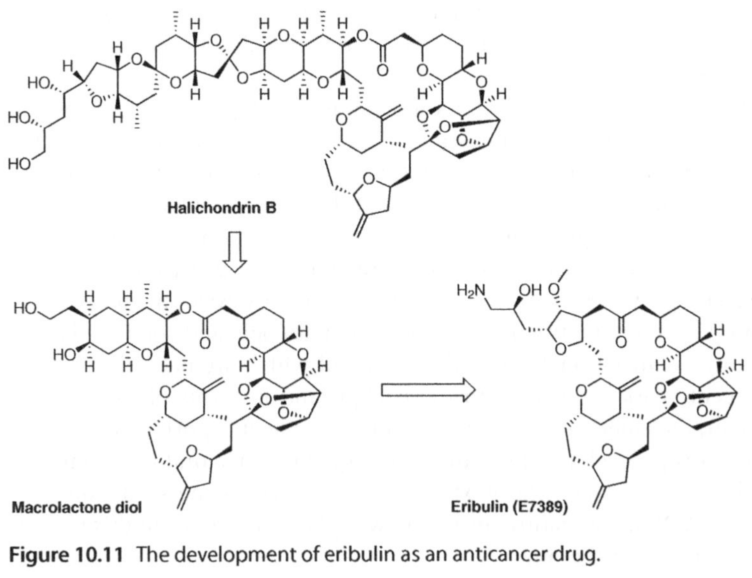

Figure 10.11 The development of eribulin as an anticancer drug.

The exceptional potency of halichondrin B led to additional studies that showed that the compound exerts its biological activity by the destabilization of tubulin in a mechanism different from those of other antimitotic agents. The antimitotic activity of halichondrin B translated to *in vivo* efficacy on chemoresistant solid tumors in xenograft models and lead to the preclinical evaluation of halichondrin B.

Preclinical development as a new anticancer chemotherapeutic, however, was hampered by extremely limited supply of this agent isolated from natural sources as the total synthesis of the molecule had not yet been achieved. The first total synthesis of halichondrin B was reported by Kishi and coworkers in 1992.[75] The breakthrough occurred during the total synthesis when samples of intermediates were also being submitted for biological testing. The active intermediate identified was a macrolactone diol and was nearly equipotent to halichondrin B on human colon cancer cell lines ($IC_{50} = 4.6$ nM). Subsequent studies on this intermediate found that the western fused pyran moiety could be replaced with simpler pyran or furan fragments. The discovery that macrolactone right-half synthetic intermediates also exhibited potent cell growth inhibitory activity *in vitro* provided an exciting and compelling starting point for a drug discovery effort around this important class of antitumor agents. Several synthetic routes for the preparation of eribulin have been disclosed since then,[76–79] each of which utilizes the same strategy described by Kishi and coworkers for the total synthesis of halichondrin B.[80]

The synthesis of eribulin represents a significant accomplishment in the field of total synthesis and brings a novel chemotherapeutic option to cancer patients. As depicted in Scheme 10.1, the strategy to prepare eribulin mesylate employs a convergent synthesis featuring the following: The late-stage coupling of sulfone **5** and aldehyde **6** followed by macrocyclization under Nozaki–Hiyami–Kishi coupling conditions, formation of a challenging cyclic ketal, and installation of the primary amine. Eribulin mesylate as approved by the U.S. Food and Drug Administration in 2010 to treat patients with metastatic breast cancer who have received at least two prior chemotherapy regimens for late-stage disease, including both anthracycline- and taxane-based chemotherapies.

10.4.2 Diversity-Oriented Synthesis (DOS)

To continue to be competitive with other drug discovery methods, natural-product research needs to continually improve the speed of the screening, isolation, and structure elucidation processes, as well addressing the suitability of screens for natural-product extracts and dealing with issues involved with large-scale compound supply. An integrative approach by combining the various discovery tools and the new discipline of integrative biology will provide the key for success in natural-product drug discovery and development.

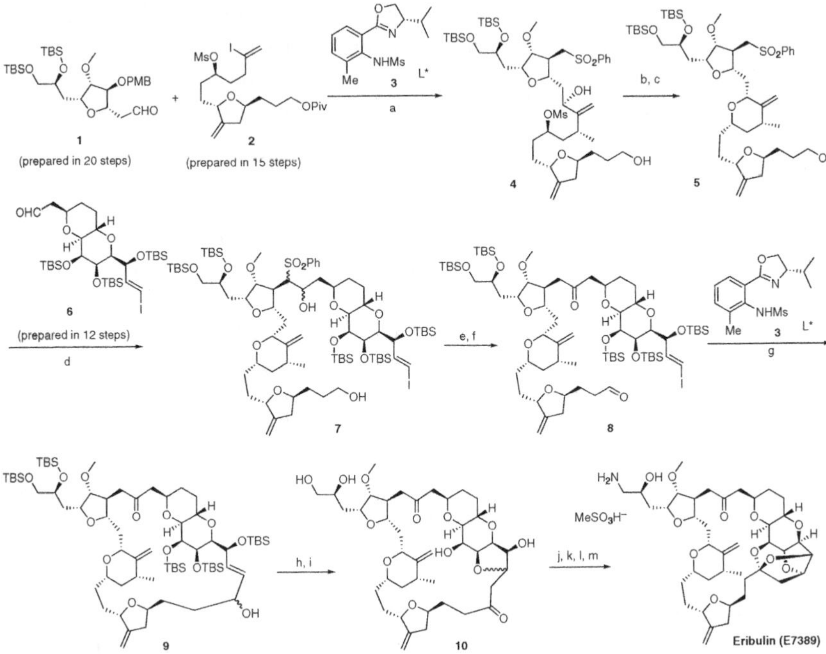

Scheme 10.1 The commercial synthesis of eribulin mesylate. Reagents and conditions: (a) $CrCl_2$, $NiCl_2$, **3**, Et_3N, THF; (b) KHMDS, THF, −14°C; (c) DIBAL-H, CH_2Cl_2, −78°C, 92%; (d) n-BuLi, THF, −75°C, 84%; (e) DMP, CH_2Cl_2, rt, 90%; (f) SmI_2, MeOH, THF, −78°C, 85%; (g) NiCl, $CrCl_2$, **3**, Et_3N, CH_3CN, THF, 70%; (h) TCAA, DMSO, Et_3N, $PhCH_3$, −15–0°C, 91%; (i) TBAF, imidazole-HCl, THF; (j) PPTS, CH_2Cl_2; (k) Ts_2O, collodine, Py (cat.), CH_2Cl_2, −20°C; (l) NH_4OH, *i*PrOH, rt; (m) $MeSO_3H$, aq. NH_4OH, 15°C, 84% over three steps.

Although various chemical libraries are now available commercially, these remain focused primary on so-called "drug-like" compounds. Because these libraries are concentrated in a relatively narrow region of chemical structure space, it seems unlikely that they will provide useful probes for all biological targets of interest. To address this important need, DOS has emerged as a valuable approach to generating libraries that explore untapped or under-represented regions of chemical structure space.[81–83] In contrast to target-oriented syntheses, diversity-oriented syntheses are not aimed at one particular target, and they are instead aimed at a collection of many compounds having structural complexity and diversity. Complexity is important because many biological processes are critically dependent on protein–protein interactions, and many of the small molecules known to disrupt these interactions are structurally complex natural products. Increasing the size and number of rigidifying and protein-binding elements in small molecules is generally viewed as essential in order for these compounds to bind tightly to sites of protein–protein interactions, which tend to be relatively flat in comparison with the concave topography characteristic of enzyme active sites.

The pioneering work of Schreiber, published in 1999, still remains one of the most important examples of a striking DOS synthesis for chemical genetics and cell-based assays.[84,85] Starting from the simple shikimic acid **11**, Schreiber and coworkers were able to generate the tetracyclic natural-product-like scaffold **13**, using a tandem acylation/1,3-dipolar cycloaddition with an array of nitrone-carboxylic acids (Figure 10.12).[86]

As shown in Figure 10.13, the tetracyclic template **13** is a rigid and densely functionalized compound, which can be decorated with a variety of appendages. The introduction ofbenzyl iodide substituents on the isoxazoline nitrogen allows for the application of a wide number of different palladium-catalyzed reactions (e.g., cross-coupling, amination, etherification, carbonylation), even though the authors demonstrated that only Suzuki,[87] Stille,[88] and Sonogashira/Castro–Stephens[89,90] reactions were successfully performed on template **13** (Figure 10.13). Additionally, the electrophilic lactone and epoxide moieties can react with nucleophiles like amines and nitriles, unmasking alcohol functionalities suitable for further reactions with acid chlorides, anhydrides, sulfonyl chlorides, chloroformates, carbamoyl chlorides, and isocyanates.

Starting from these possibilities, the authors chose the most efficient reactions compatible with solid-phase split-and-pool technique as a synthetic strategy of choice to provide the greatest number of small molecules as possible.[91,92] By utilizing a solid-phase split-and-pool technique, Schreiber and coworkers were able to construct a DOS library of more than 2 million densely functionalized compounds starting form shikimic acid-derived epoxycyclohexenol (Figure 10.14). A convergent tandem acylation/1,3-dipolar cycloaddition of both enantiomers of epoxycyclohexenol with substituted nitro carboxylic acids provided products with stereochemical diversity and handles for appendage diversity. These rigid polycyclic scaffolds then underwent a variety of appendage and skeletal diversity generating reactions to generate bicyclic and tricyclic cores. From this library, several compounds were discovered that activate the TGF-β-responsive reporter gene in mammalian cells in a chemical genetics assay with the EC_{50} of the most active being 50 μM.

Natural products exhibit tremendous structural diversity and often have increased size and structural complexity compared to synthetic drugs. They

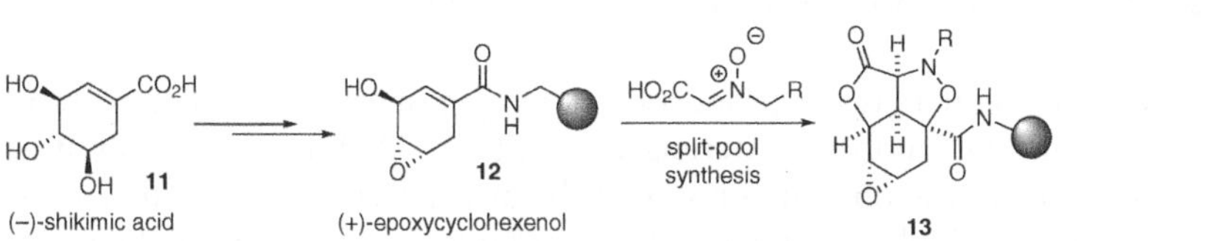

Figure 10.12 synthesis of tetracyclic **13** with tandem acylation/1,3-dipolar cycloaddition.

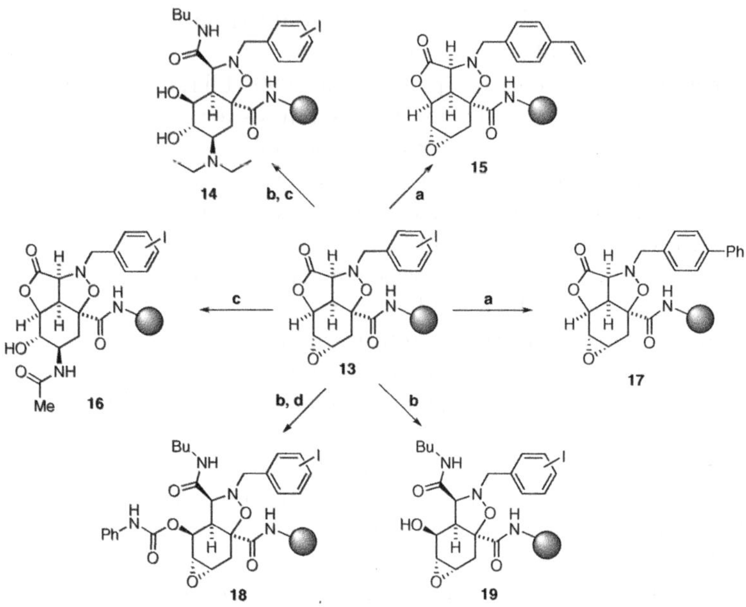

Figure 10.13 Possible diversification points and examples of derivatives of scaffold **13** by (a) cross-coupling reactions, (b) aminolysis of lactone moiety, (c) epoxide ring opening, and (d) alcohol esterification.

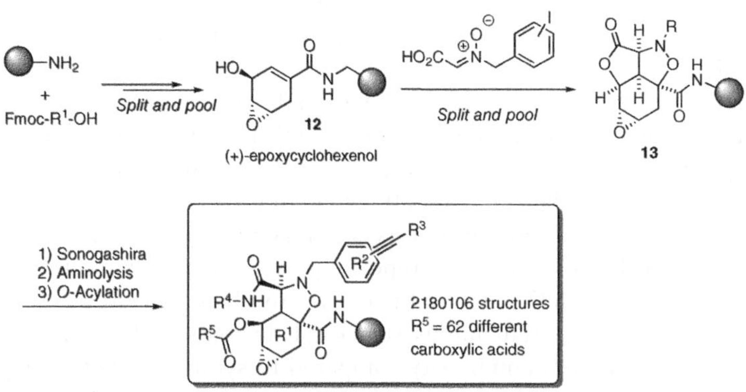

Figure 10.14 Split-and-pool DOS library based on scaffold **13**.

frequently contain a greater proportion of oxygen than nitrogen heteroatoms and a significant number of stereogenic centers.[93] Although clinically used natural products are sometimes not orally bioavailable, they are able to address a wider range of biological targets than synthetic drugs. For example, rather than acting as ligands that bind in a protein pocket, glycopeptide antibiotics such as vancomycin act as receptors for the C-terminal D-ala-D-ala of bacterial peptidoglycan precursors. Moreover, protein–protein interactions, which have historically been very difficult targets for synthetic drugs,[94,95] can often be modulated with natural products. The natural-product anticancer drugs paclitaxel (Taxol) and vincristine are examples that modulate tubulin protein–protein interactions.

DOS of "natural-product-like" libraries has recently emerged as a major area of current interest for use in biological screens, which aid both biological and drug discoveries.[96,97] Library design strategies have been divided into three general categories, according to the degree of structural similarity to natural products[98,99]: (1) Libraries based on the core scaffold of an individual natural-product, (2) libraries based on specific structural motifs that are found across a class of natural products, and (3) libraries that emulate the structural characteristics of natural products in a more general sense. Each strategy balances the degree of connection to natural-product structure space against the accessibility of structural diversity that is likely required to address multiple different biological targets. Interestingly, some structural motifs originally found in natural products have subsequently been identified as privileged structures and have been used in synthetic drugs.

Shair and coworkers reported an effective short method for the solid-phase library generation of complex polycyles involving Ugi four-component coupling, intramolecular Diels–Alder, and ring-opening-closing olefin metathesis as the key steps on alkylsily macrobeads to generate a 10,000-membered DOS library by using the natural-product carpanone as a starting point (Figure 10.15).[100] Carpanone contains a rigid polycyclic core with five contiguous stereocenters

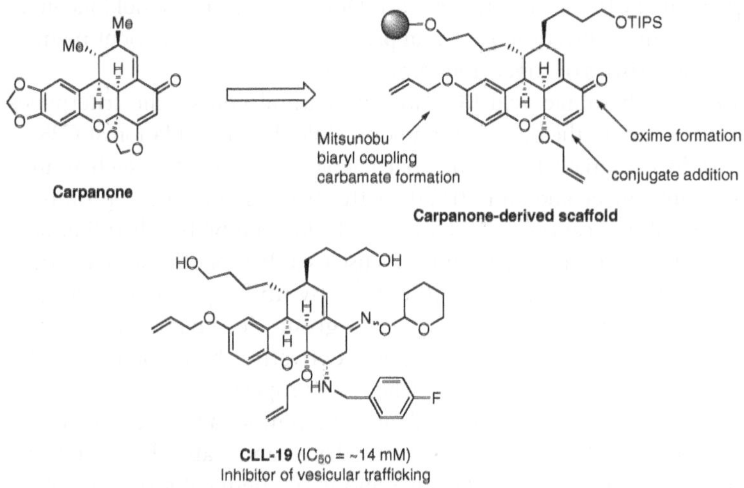

Figure 10.15 A DOS library based on carpanone's core structure.

and enone and phenol functionalities for the introduction of appendage diversity. In a convergent manner, a dimerization strategy involving a [4+2] cycloaddition was used to assemble the core. Screening of this library in a phenotypic assay resulted in the identification of several hits that perturb the secretory pathway by inhibiting the exit of proteins from the Golgi apparatus. A powerful aspect of this approach was that structure-activity relationships (SAR) could be determined directly from the screening results. The most active compounds, CLL-19 blocked protein trafficking at ~14 μM, while carpanone itself had no effect at 300 μM.

DOS has proven to be a successful strategy for the rapid access to chemical libraries of natural-product-like three-dimensional molecular architectures from simple starting materials.[101–106] However, it was rarely applied to natural-product total synthesis in parallel with skeletal diversification. As shown in Figure 10.16, Lei's group recently reported a typical build/couple/pair strategy to obtain skeletally diverse compounds with complexities comparable to natural products.[107] Inspired by the functional group pairing patterns hidden in lycopodium alkaloids, they efficiently and in parallel construct four natural products, (+)-serratezomine A, (−)-serratinine, (+)-*8a*-hydroxyfawcettimine and (−)-lycoposerramine-U, as well as six different unnatural scaffolds, following the advanced build/couple/pair (B/C/P) algorithm.[108] In this strategy, complex molecules are synthesized by the following steps that take advantage of the B/C/P algorithm: (1) identifying the hidden functional group pairing patterns (FGPPs) in natural products and then "depairing" them to afford a densely functionalized precursor; (2) performing reactions to provide this precursor through B/C phases; and (3) exploring a variety of functional group pairing reactions sequentially to afford the natural products and distinct natural-product-like scaffolds in parallel. This approach is expected to be complementary to the current standard DOS strategy and should facilitate the synthesis and diversification of natural products in a highly efficient manner due to the characteristics of concurrent branch points.

Betulin (**32**) and betulinic acid (**33**) are two natural pentacyclic triterpenes isolated from the bark of the birch tree and only differ by the oxidation at C28.[109] Betulinic acid displays many intriguing pharmacological properties such as anti-inflammatory, anticancer, and anti-HIV, the latter two being the most promising for pharmaceutical applications.[110,111] However, the low solubility of betulinic acid in H_2O constitutes a serious limitation for its use as a therapeutic, since it would make formulation for oral delivery difficult.[112] By utilizing a dual chemical/enzymatic C–H oxidation strategies, Baran's group developed the diverse-oriented synthesis of betulin and betulinic acid derivatives based on the betulin skeleton in order to improve their physicochemical properties (Figure 10.17).

DOS is a powerful new approach to identifying new small-molecule probes to dissect complex biological systems. Both drug- and natural-product-like libraries that target biologically relevant regions of chemical structure space have proven useful for discovering such probes. New synthetic planning

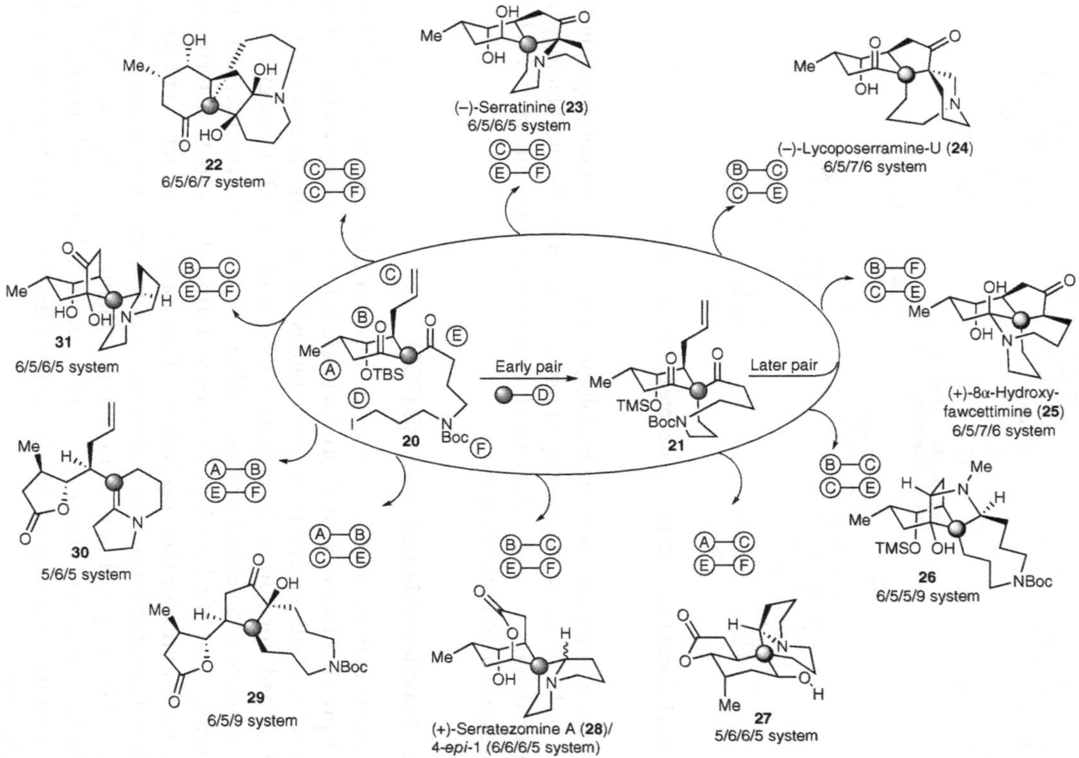

Figure 10.16 DOS of lycopodium alkaloids.

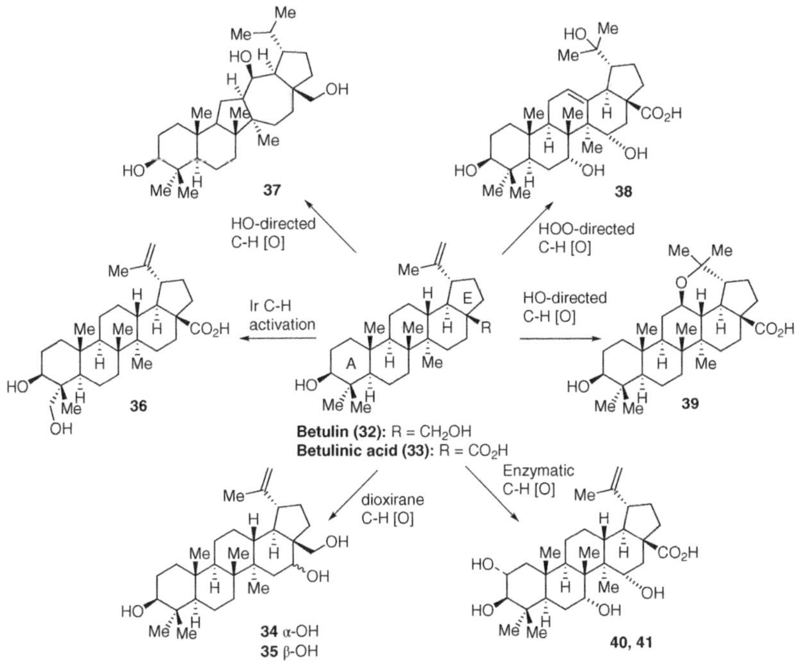

Figure 10.17 Diversification of the lupane core by C-H oxidation.

strategies and new chemical methodologies have also been developed in the context of DOS. Thus, the exciting potential of DOS in chemical biology has now been demonstrated clearly. Further evolution and refinement of this field can be expected in the coming years.

10.4.3 Function-Oriented Synthesis (FOS)

The aim of FOS is to address several common problems associated with natural-product leads: Natural products are often too complex to be made in a practical fashion, and many carry undesired side effects given they are not "designed" for human therapeutic use. Use of the FOS strategy focuses on the function of a natural-product, rather than its total structure, to design structurally simplified analogs by incorporating the activity-determining structural features (or their equivalent) of the natural product. Employment of FOS yields natural-product analogs with comparable or superior function that could be prepared in a step-economic manner. Also, by focusing on a specific function, the FOS can be used to enhance beneficial activities and also to minimize off-target activities better suited for biomedical research or medicinal use.

The central principle of FOS is that the function of a biologically active lead structure can be emulated, tuned, or even improved by replacement with simpler scaffolds designed to incorporate the activity-determining structural features (or their equivalent) of the lead compound. This approach allows for the step-economical synthesis of novel structures with improved or new activities. FOS addresses several concerns related to drug discovery. First natural products are not "designed" for human therapeutic use and as a consequence often have undesired side effects. By focusing on target-specific function, FOS can be used to minimize off-target activities and to enhance beneficial activities. Second, FOS can be used to optimize formulation, ADME, and pharmacokinetic performance, thereby avoiding problems exhibited by many natural products (e.g., the formulation of taxol). Third, because the focus of FOS is function rather than structure, it can address the concern that some natural products are too complex to make in a practical fashion.

The bryostatins are significantly complex molecules that have attracted attention as anticancer agents due to a unique range of activities, including their ability to induce apoptosis,[113] reverse multidrug resistance,[114] and modulate the immune system.[115] Bryostatin also synergizes with other antineoplastic agents and is currently in clinical trials for the treatment of cancer.[116] Exciting recent work has also shown that bryostatin can enhance memory and learning in animals, with potential implications for the treatment of Alzheimer's disease.[117] Unfortunately, access to bryostatin is extremely limited, as it is obtained in a low yield of only 0.00014% from marine bryozoa.[118] Before Wender applied FOS to the synthesis of bryostatin analogs, total syntheses of bryostatins had required more than 70 steps.[119] Recently, Trost's group completed a similar synthesis in 39 steps. The problems and opportunities presented by bryostatin are ideally suited for FOS because its therapeutic activity is connected to only a subset of its structure and that pharmacophore could be designed into a simplified target more readily accessed through synthesis (Figure 10.18).

Bryostatin potently binds to the C1 domain of protein kinase C (PKC, $Ki = 1.4$ nM). The C1 carbonyl and C19 and C26 alcohols as well as a corresponding lipophilic region of bryostatin were hypothesized as key binding elements.[120] These have met the requirements for the development of a convergent synthesis. Retrosynthetic analysis of **43** led to the C1–C13 spacer domain **44** and C15–C26 recognition domain **45**, which featured Yamaguchi esterification and macrotransacetalization. The FOS approach led to the realization of bryostatin analogs that bind to protein kinase C (PKC) with affinities comparable, and in some cases superior, to bryostatins (*Ki* value of 0.3 versus 1.4 nM). Moreover, these analogs can be made in 29 steps, significantly shorter than the previous pathway for the total synthesis of bryostatins. Further modification of bryostatins on the fragments of the A-, B-, and C-rings led to the efficient synthesis of a library of 31 analogs. Importantly, the analogs show greater potency than bryostatin when tested for growth inhibitory activity against the US National Cancer Institute panel of human cancer cell lines.

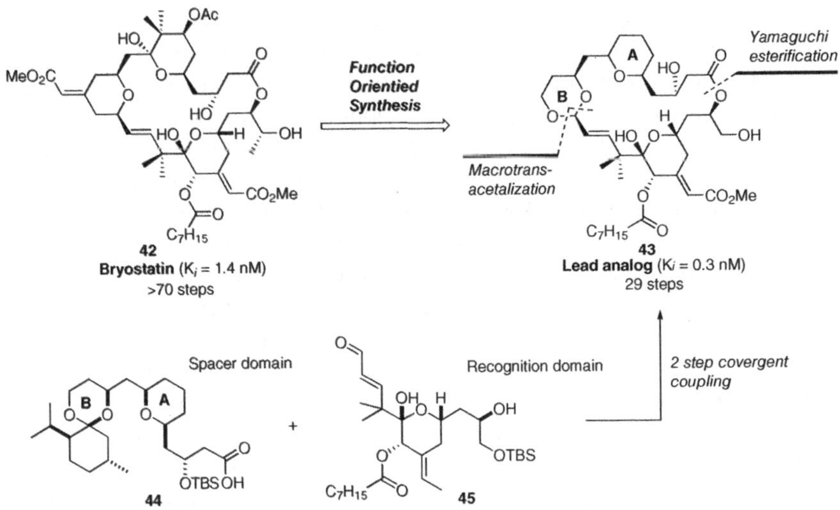

Figure 10.18 FOS of bryostatin analogs.

In summary, FOS led to more potent analogs that can be supplied in quantity and tuned for performance and at the same time allowed for the development of effective methods for convergent macrolide formation. An advantage of FOS was demonstrated in this study; namely that superior function can be achieved in fewer steps with simplified structures while enabling synthetic innovation.

10.4.4 Biology-Oriented Synthesis (BIOS)

Waldmann has developed a new concept for the design of combinatorial libraries based on natural products that he calls BIOS.[121–123] This concept is based on the recognition of fundamental and complementary properties of natural products and their protein targets. In order to guide the scaffold selection of library scaffold, the notion of taking advantage of privileged structures has inspired attempts to provide a structural classification of natural products (SCONP). In BIOS, the analysis of evolutionarily selected natural-product structures is employed to define prevalidated, educated-guess starting points for the synthesis and further development of biologically relevant small molecules for chemical biology and medicinal chemistry research.

BIOS is based upon two concepts developed by the Waldmann group. The scaffolds of natural products can be mapped in a hierarchical manner to create a scaffold tree, a "SCONP"[124,125] in a tree-like arrangement. This allows for logical pathways for the structural simplification of scaffolds. In the second concept, "protein structure similarity clustering" (PSSC), proteins are clustered by three-dimensional shape around the ligand-binding sites, regardless

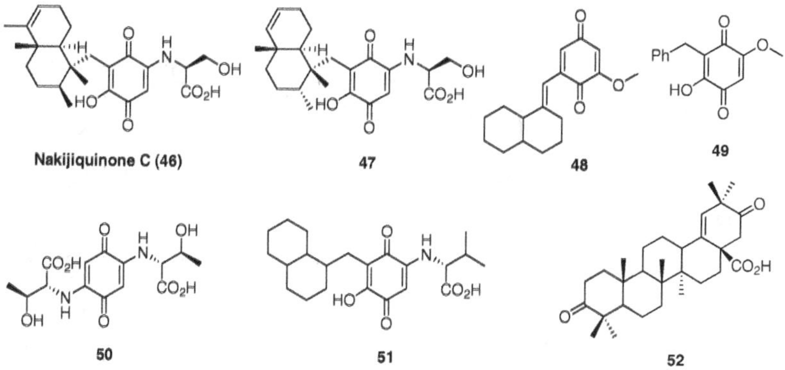

Figure 10.19 Protein kinase inhibitors discovered by BIOS.

of sequence similarity.[126–128] Merging these two concepts led to the BIOS approach. The ligand of any member of a PSSC could be expected to exhibit some degree of complementarity toward other members of the PSSC and, thus, serve as a starting point for the development of modulators of the other members.

The Waldman group successfully utilized BIOS to search for kinase inhibitors. Nakijiquinone C (**46**; Figure 10.19), isolated from a marine sponge and first reported by Kobayashi *et al.* in 1995,[129] was shown to be an inhibitor of epidermal growth factor receptor (EGFR), c-ErbB2, and protein kinase C (PKC), in addition to having cytotoxic activity against L1210 and KB cell lines. Testing of a library of 74 compounds, built around the basic nakijiquinone C structure, against a battery of kinases with similar protein domain folds, yielded seven new inhibitors with low micromolar activity *in vitro*, including one VEGFR-2 inhibitor **47** and four inhibitors of Tie-2 kinase **48–51**, a protein intimately involved in angiogenesis and for which, at the beginning of the study, no inhibitors were known.[130] During the study, the first natural-product inhibitor **52** of Tie-2 kinase was reported from the plant *Acacia aulacocarpa*,[131] and a set of four papers from another research group demonstrated the activity of synthetic pyrrolo[2,3-*d*]pyrimidines as inhibitors of the same class of kinases.[132–135]

A natural-product inspired synthesis of dysidiolide-like combinatorial library was developed to identify biologically active analogs of the natural phosphatase inhibitor dysidiolide (**58**; Figure 10.20).[136] To this end, chiral dienophile **54** was employed to enhance the stereodirecting influence of the resin-bound chiral diene **53**. The bicyclic scaffold **55** was built up by Diels–Alder reaction of diene **53** with acetal **54** derived from tiglic aldehyde and displayed an endo/exo ratio of 91:1 and a selectivity of 95:5 in favor of the desired endo isomer. The cycloadduct **55** was released by a ring-closing

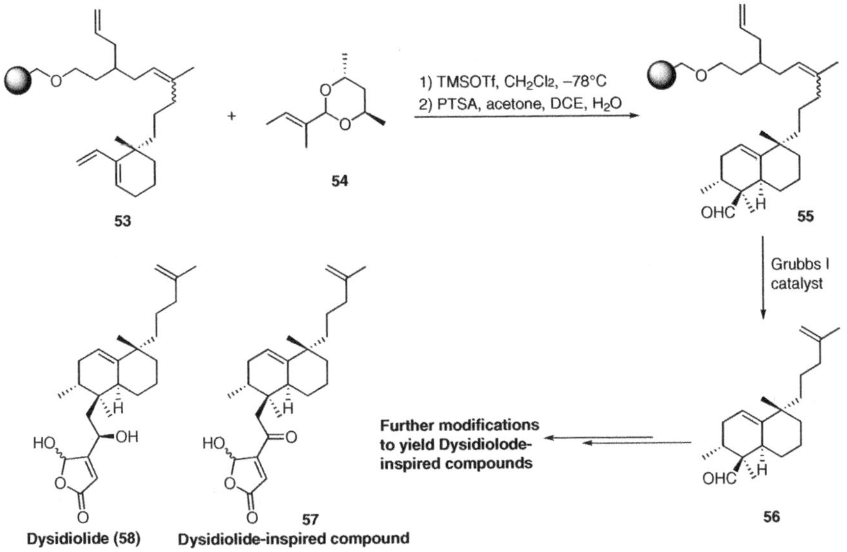

Figure 10.20 Solid-phase synthesis of dysidiolide-inspired compounds.

metathesis (RCM) reaction. Further modifications of the cycloadduct **56** provided analogs of dysidiolide. Biological evaluation of this focused small library revealed inhibitors of phosphatases and cytotoxic activity against different cancer cell lines, with dysidiolide-like molecule **57** being the most potent inhibitor of the phosphatase Cdc25C with an IC_{50} value of 0.8 μm. In addition, other members of the library were identified with low micromolar activities against the enzymes acetylcholinesterase and 11*β*-hydroxysteroid dehydrogenase type 1, which fall within the same PSSC as Cdc25A.[137]

BIOS represents a refinement of combinatorial libraries based on natural-product scaffolds by focusing on the most biologically relevant chemical space for the target. Furthermore, it allows the transfer of knowledge about the modulation of a target by a natural product to a whole cluster of structurally related proteins, even when those proteins catalyze mechanistically different reactions. These examples illustrate how BIOS may allow the identification of structurally simpler starting points for library design while providing new classes of inhibitors. The successful development of the BIOS approach paves the way to employ the biological prevalidation of natural product structure by evolution in chemical biology and medicinal chemistry research, thereby overcoming limitations of synthetic tractability or accessibility of natural products and suggests that natural products, and compound collections inspired by them should be reconsidered in future drug discovery efforts.

10.5 Semisynthesis

Semisynthesis or partial chemical synthesis is a type of chemical synthesis that uses compounds isolated from natural sources (e.g., plant material or bacterial or cell cultures) as starting materials. These natural biomolecules are usually large and complex molecules. Many natural products have been shown to occupy different and sometimes difficult to access chemical space compared to synthetic compounds. The uniqueness of many natural-product core structures makes these compounds of interest for use as starting points for semisynthesis and total synthesis.

Although the natural product of interest may not be readily available from biomass, sometimes another natural product that can serve as a starting material for the semisynthesis of the target is readily available. Large numbers of analogs can be rapidly generated by such semisynthetic approaches; however, many desired transformations cannot be accomplished due to incompatibilities with pre-existing functional groups or the lack of a feasible reaction. Thus, the structural diversity of the analogs accessible by derivatization is limited. There are numerous examples of this approach including taxanes,[138,139] camptothecins,[140] and combretastatins.[141]

Alkaloid ecteinascidin 743 (Et-743, Yondelis), discovered from the colonial tunicate *Ecteinascidia turbinate.*[142,143] Ecteinascidin 743 was developed as is an antitumor drug, sold by Zeltia and Johnson and Johnson under the brand name Yondelis. It is approved for use in Europe, Russia, and South Korea for the treatment of advanced soft tissue sarcoma. It is also undergoing clinical trials for the treatment of breast, prostate, and pediatric sarcomas. The issue of compound supply for advanced studies was solved by the development of a semisynthetic route from the microbial product cyanosafracin B.[144] Safracin B is an antibiotic of bacterial origin, available through fermentation of the bacteria *Pseudomonas fluorescens.*[145] Optimization of the fermentation process has allowed for the synthesis of cyano derivative cyanosafracin B on a kilogram scale providing a robust, sophisticated, and cheap starting material for the synthesis of ecteinascidin compounds.

The synthesis of ecteinascidin 743 (Et-743) from cyanosafracin B is outlined in Scheme 10.2. Starting from cyanosafracin B, its amino and phenol groups were protected as the BOC and MOM derivatives, respectively. Hydroxylation of the methoxy-*p*-quinone with NaOH afforded **58**. Hydrogenation of the quinine to give an unstable hydroquinone, which was treated with bromochloromethane and Cs_2CO_3 to yield **59**. Alkylation of the phenol, deprotection of MOM and BOC groups, cleavage of the amide and treatment with HCl/dioxane gave **60**. Temporary protection of the primary amine as the TROC carbamate, followed by reaction with MOMBr and $i Pr_2NEt$, removal of the TROC with Zn/HOAC, and treatment $NaNO_2$/HOAc transformed primary amine to alcohol **61**, a key intermediate in the total synthesis of ecteinascidin by Corey,[146] thus conducting the formal synthesis of the natural-product.

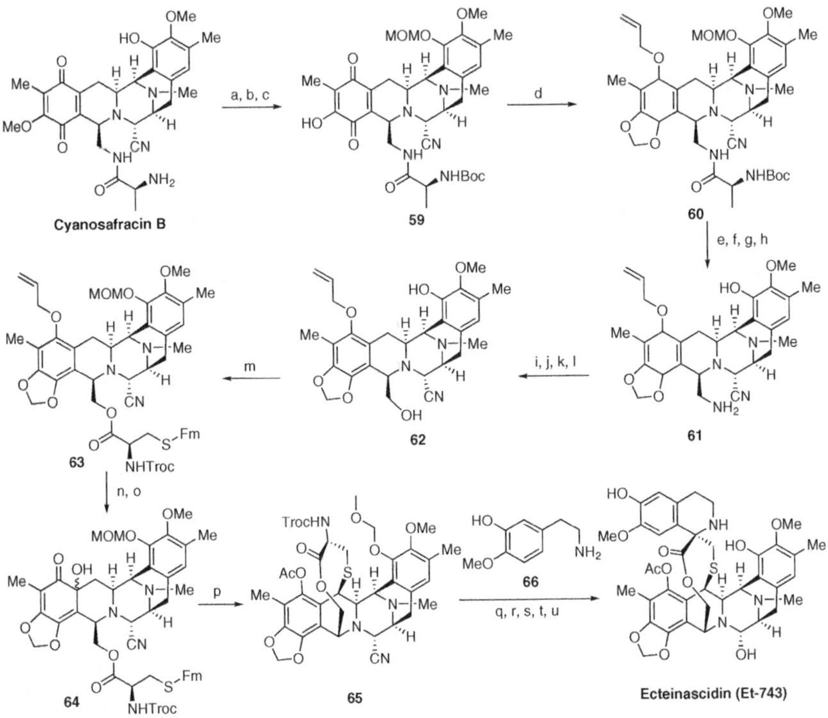

Scheme 10.2 Semisynthesis of ecteinascidin 743 (Et-743). Reagents and conditions: (a) Boc_2O, EtOH, 23°C, 81%; (b) MOMBr, *i*-Pr_2NEt, DMAP, CH_3CN, 40°C, 83%; (c) 1 M NaOH, MeOH, 20°C, 68%; (d) H_2, 10% Pd/C, 23°C; $ClBrCH_2$, Cs_2CO_3, DMF, 110°C; (e) allyl bromide, Cs_2CO_3, DMF, 23°C, 56% for two steps; (f) TFA, CH_2Cl_2, 23°C, 95%; (g) phenyl isothiocyanate, CH_2Cl_2, 23°C, 87%; (h) HCl/dioxane 4.3 M, 23°C, 82%; (i) TrocCl, pyridine, CH_2Cl_2, 0°C, 1 h, 98%; (j) MOMBr, *i*-Pr_2NEt, DMAP, CH_3CN, 40°C, 88%; (k) Zn, AcOH aq., 23°C, 83%; (l) $NaNO_2$, AcOH, THF, H_2O, 0°C, 50%; (m) (*S*)-*N*-[(trichloroethoxy)carbonyl]-*S*-(9-fluorenylmethyl) cysteine, EDC-HCl, DMAP, CH_2Cl_2, 23°C, 95%; (n) Bu_3SnH, $(PPh_3)_2PdCl_2$, AcOH, CH_2Cl_2, 23°C, 90%; (o) $(PhSeO)_2O$, CH_2Cl_2, −10°C, 91%; (p) DMSO, Tf_2O, −40°C; *i*-Pr_2NEt, 0°C; *t*-BuOH, 0°C; $(CH_3N)_2CdN$-*t*-Bu, 23°C, n; Ac_2O, 23°C, 58%; (q) TMSCl, NaI, CH_2Cl_2, CH_3CN, 23°C; (r) Zn, AcOH aq., 70°C, 77% for two steps; (s) [*N*-methylpyridinium-4-carboxaldehyde]$^+I^-$, DBU, $(CO_2H)_2$, 23°C, 57%; (t) **66**, silica gel, EtOH, 23°C, 90%; (u) $AgNO_3$, CH_3CN, H_2O, 23°C, 90%.

The cortistatins belong to a family of marine 9-(10,19)-*abeo*-androstane steroids that were isolated from the sponge corticium simplex in 2006 by Kobayashi.[147] They immediately attracted wide attention due to their potent antiangiogenic properties, rendering them promising therapeutics to combat tumor progression. (+)-Cortistatin A, the most highly functionalized and potent member of the family, which showed antiproliferative activity against human umbilical vein endothelial cells (HUVECs) at a low concentration with an $IC_{50} = 1.8$ nM, while it demonstrated a selectivity index of more than 3000-fold against HUVECs in comparison with NHDF (normal human dermal

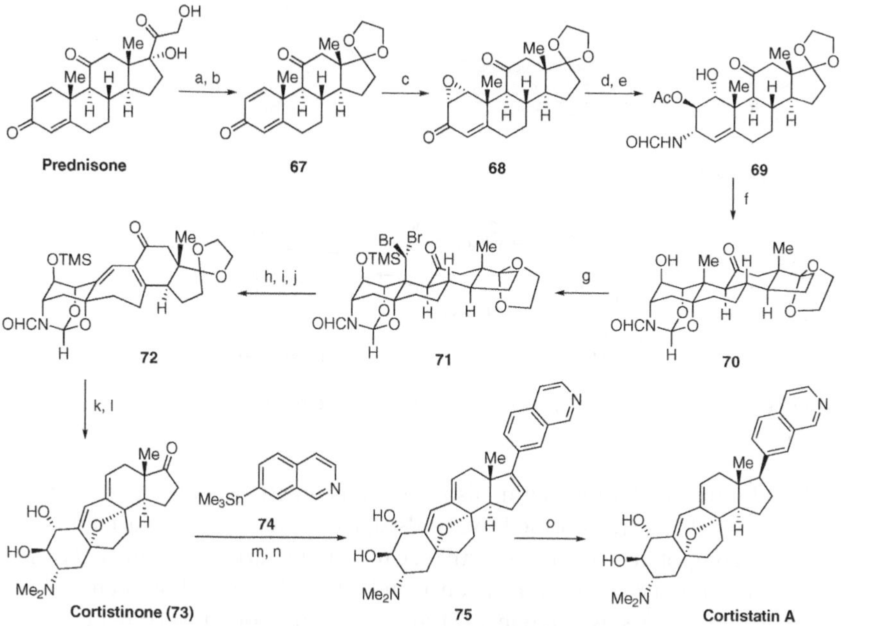

Scheme 10.3 Synthesis of cortistatin A from prednisone. Reagents and conditions: (a) BH_3-THF, $NaIO_4$, THF; (b) *p*-TsOH, ethylene glycol, toluene, reflux, 92% over two steps; (c) *t*-BuOOH, DBU, THF, 82%; (d) $Ti(O^iPr)_4$, NH_3, then $NaBH_4$; HCO_2H, CDMT, NMM, DMAP, 95%; (e) Et_3N, HOAc, 84%; (f) $Co(acac)_2$, $PhSiH_3$, O_2; $HC(OEt)_3$, TsOH, MeOH, K_2CO_3, 65%; (g) $PhI(OAc)_2$, Br_2, *hv*, TMSCl, Imidazole, 57%; (h) DBU, LiCl, 85%; (i) SmI_2, TBCHD; (j) LiBr, Li_2CO_3, 65% over two steps; (k) AlH_3; (l) $BiCl_3$, 62% over two steps; (m) N_2H_4, I_2, Et_3N; (n) **74**, $Pd(PPh_3)_4$, 55% over two steps; (o) Raney Ni, 50%.

fibroblast), KB3–1 (KB epidermoid carcinoma cells), K562 (human chronic myelogenous leukemia cells), and Neuro2A (murine neuroblastoma cells), indicating that cortistatin A selectively inhibits angiogenesis but has no apparent cytotoxicity toward cancer cells.

In 2011, Baran's group reported a semisynthesis from readily available prednisone (Scheme 10.3).[148] In the forward direction, prednisone was transformed into **67** in six steps. In one of the key steps of the sequence, the treatment of **70** with $PhI(OAc)_2$ and bromine under photochemical irradiation effected the alcohol-directed, geminal dihalogenation of an unactivated methyl group. Compound **71** was then elaborated in five steps and at a gram scale to cortistatinone, the key intermediate and branching point of the synthesis. Cortistatinone was finally converted into (+)-cortistatin A and a variety of analogs that were subsequently tested for bioactivity. The ready availability of synthetic analogs enabled by this synthesis has led to a number of collaborations with both pharmaceutical companies and academic groups.

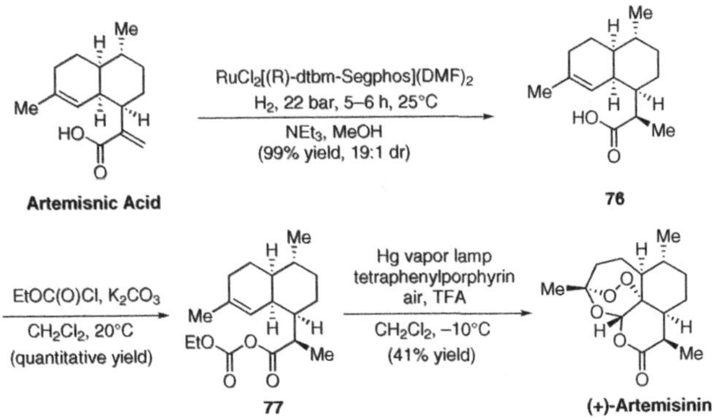

Scheme 10.4 Improved semisynthetic process of (+)-artemisinin by Sanofi.

The discovery of artemisinin, also known as qinghaosu, and its derivatives met this need and transformed the landscape of modern antimalarial therapy, as these compounds remain effective against multi-drug-resistant *P. falciparum*, with minimal report of resistance and efforts to control resistance.[149–151] Artemisinin was first isolated in 1971 in China as part of a government-backed initiative to discover antimalarial drugs for North Vietnamese troops during the Vietnam War.[152,153] Artemisinin is a structurally fascinating natural product containing a complex tetracyclic core. Embedded within this core is a unique endoperoxide bridge. Furthermore, artemisinin contains seven stereocenters, six of which are contiguous, and an acetal linkage between a seven-membered and six-membered ring. Production of this compound has been a topic of interest for several years and, fortunately, there have been many efforts from different areas of science to solve the supply problem.[154–156]

In 2013, Sanofi announced the large-scale production of artemisinin from microbially derived artemisinic acid at its Garessio site in Italy using the technology discovered by Amyris.[157] The semisynthetic transformations of artemisinic acid to artemisinin are related to those first described by Acton and Roth over 25 years ago and mimic what is assumed to occur in *A. annua*. In 2014, Sanofi described the technical details of their highly optimized semisynthesis (Scheme 10.4).[158] Related to the approach described by Amyris,[159] the first step consisted of transition metal-catalyzed diastereoselective hydrogenation of artemisinic acid to dihydroartemisinic acid. By using a chiral ruthenium catalyst, $RuCl_2[(R)$-dtbm-Segphos]$(DMF)_2$, the diastereoselectivity of the reaction was improved to 19:1 favoring **76**, reduction of the temperature in the reaction from 80 to 25°C, reduction in the time from 19 h to 5–6 h, and the use of methanol as a green solvent alternative to toluene, all while drastically improving the s/c ratio from 100/1 to over 4000/1. Subsequently, dihydroartemisinic acid was converted to mixed anhydride 77 at a quantitative yield, which was served as the precursor for the oxygenation/acid-mediated ring closure sequence. Despite the tremendous challenges associated with performing photochemical processes on a large scale, Sanofi employed a singlet oxygen ene reaction involving light, air, and tetraphenylporphyrin (TPP). This impressive semisynthesis proceeds in an overall 55% yield of crystalline artemisinin and is one of few large-scale photochemical reactions practiced industrially.

10.6 Representative Natural-Product Drugs and Their Synthesis

The total synthesis of complex natural products has long posed challenges to the top synthetic chemistry groups worldwide and has led to dramatic advances in the field of organic chemistry. Significant strides have been made in the synthesis and structural modification of drugs that are difficult to isolate in sufficient quantities for development. Adequate supply can be a serious limiting factor in the preclinical and clinical development of some naturally derived drugs, and the focus of many top synthetic groups on devising economically feasible synthetic strategies is a very welcome development for both clinicians conducting clinical trials and patient populations. In this section, we have chosen to highlight a few natural-product drugs and synthetic strategies.

10.6.1 Nicolaou and Yang's Synthesis of Taxol

Taxol (Paclitaxel), one of the most celebrated natural products, was isolated from the Pacific yew tree and its structure was reported in 1971.[160] Its arduous journey to the clinic took more than 20 years, being approved by the Food and Drug Administration (FDA) in 1992 for the treatment of refractory ovarian cancer,[161] and subsequently being approved for the treatment of metastatic breast and lung cancers, and Kaposi's sarcoma. The unique action of taxol also spurred the development of a second-generation semisynthetic taxane, docetaxel (Taxotere), approved in 1996 for anthracycline-refractory advanced breast cancer and now also used in lung cancer regimens. The events surrounding the discovery and development of these drugs provide numerous examples of the power of natural products to uncover new therapeutic agents and define novel drug targets. However, the scarcity of taxol and the ecological impact of harvesting it have prompted extensive searches for alternative sources including semisynthesis, cellular culture production, and chemical synthesis. In 1994, two essentially simultaneous reports[162,163] described two distinctly different total syntheses of taxol. These first two syntheses, by Nicolaou's group[162] and that of Holton,[163] were followed by those of Danishefsky,[164] Wender,[165] Mukaiyama,[166] and Kuwajima.[167] All these

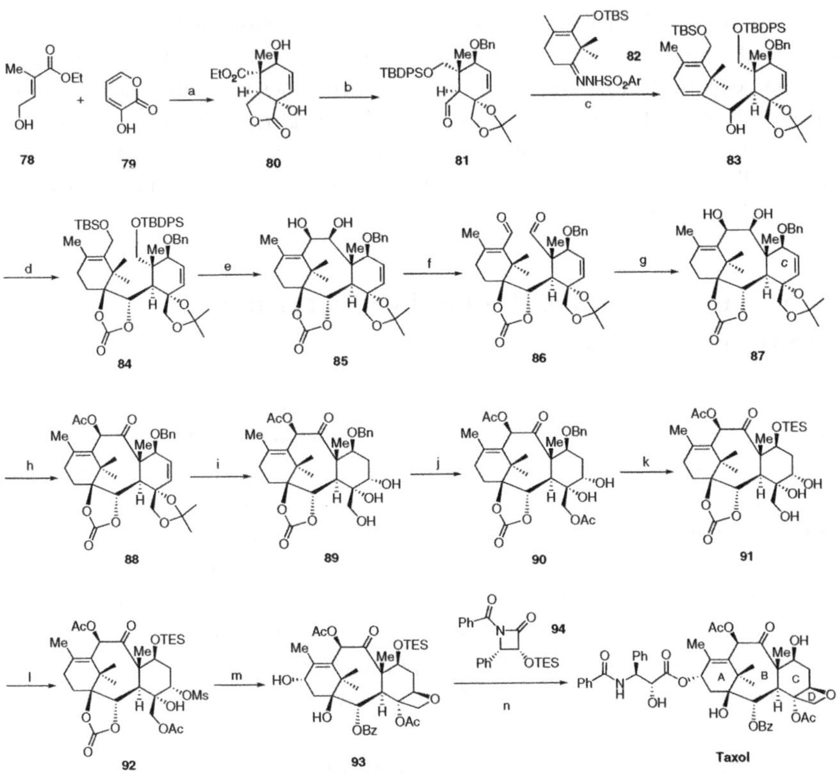

Scheme 10.5 *Reagents and Conditions*: (a) $PhB(OH)_2$, PhMe, reflux; then $Me_2C(CH_2OH)_2$; (b) 1. TBSOTf, 2,6-lutidine, DMAP, CH_2Cl_2, 0°C, 95%; 2. $LiAlH_4$, Et_2O, 0°C, 94%; 3. CSA, MeOH, CH_2Cl_2; then t-$BuPh_2SiCl$, imidazole, DMF, 92%; 4. KH, Et_2O, n-Bu_4NI, BnBr, rt, 87%; 5. $LiAlH_4$, Et_2O, 80%; 6. DMP, CSA, CH_2Cl_2, 82%; 7. TPAP, NMO, CH_3CN, 95%; (c) **82,** n-BuLi, THF, −78°C to rt, 82%; (d) $VO(acac)_2$, t-BuOOH, molecular sieves (cat.), benzene, 87%; (e) $LiAlH_4$, Et_2O, 76%; (f) 1. KH, HMPA/Et_2O, $COCl_2$, rt, 48%; 2. TBAF, THF, 82%; 3. TPAP (cat.), NMO, CH_3CN/CH_2Cl_2, 82%; (g) $(TiCl_3)_2$-$(DME)_3$, Zn-Cu, DME, 23%; (h) 1. Ac_2O, DMAP, CH_2Cl_2, 95%; 2. TPAP, NMO, CH_3CN, 93%; (i) 1. $BH_3{\cdot}THF$, THF, then H_2O_2, $NaHCO_3$, 55%; 2. *conc.* HCl, MeOH, 80%; (j) Ac_2O, DMAP, CH_2Cl_2; (k) 1. H_2, 10% $Pd(OH)_2/C$, EtOAc, 97%; 2. TMSCl, Py, 85%; 3. K_2CO_3, MeOH; (l) 1. TMSCl, Py, CH_2Cl_2, 70%; 2. Tf_2O, iPr_2NEt, CH_2Cl_2; 3. CSA, MeOH, rt, then silica gel, CH_2Cl_2, 60%; (m) Ac_2O, DMAP, CH_2Cl_2, 94%; (n) 1. PhLi, THF, −78°C, 80%; 2. PCC, NaOAc, celite, benzene, 75%; 3. $NaBH_4$, MeOH, 83%; 4. LiHMDS, **94,** THF, 0°C, 87%; 5. HF-Py, THF, rt, 80%.

synthesis, which are characterized by novel strategies and brave tactics, contributed enormously to the advancement of total synthesis and enabled investigations in biology and medicine. Among the most notable features of Nicolaou's total synthesis of taxol (Scheme 10.5) are the boron-mediated Diels–Alder reaction to construct the highly functionalized C ring, the application of the Shapiro and McMurry coupling reactions, and the selective manner in which the oxygen functionalities were installed onto the eight-membered ring of the molecule. Nicolaou's synthesis started with two key intermediates **81**[168] and **82**[169,170] and synthesized taxol by a convergent strategy, which opens a chemical pathway for the production of both the natural product itself and a variety of designed taxoids.

10.6.2 Danishefsky's Synthesis of Epothilone A

Epothilones A and B were isolated from the myxobacteria of the genus *Sorangium cellulosum*, and have been shown to promote microtubule assembly but without the risk of resistance from cellular export by the MDR efflux pump. The full structures of these compounds were determined from an X-ray crystallographic analysis by Hofle.[171] During the mid-1990s, epothilones A and B stimulated intense research activities in several laboratories.[172] The impetus for their total synthesis came not so much from their modestly complex macrolide structures but more so from their potent tubulin-binding properties and their potential to overshadow taxol as superior anticancer agents. The principal mechanism of the epothilone class is inhibition of microtubule function. Due to the high potency and clinical need for cancer treatments, epothilones have been the target of many total synthesis. The first total synthesis of epothilone A came from the Danishefsky laboratories in 1996[173] and was followed shortly thereafter by syntheses from Nicolaou's lab[174] and from those by Schinzer.[175] Danishefsky's first total synthesis of epothilone A (Scheme 10.6) featured a Suzuki coupling reaction to form a crucial C–C bond and an intramolecular ester enolate-aldehyde condensation to form the 16-membered macrocyclic lactone. This method as well as others allowed the Danishefsky group to synthesize several additional natural and designed members of the epothilone family, including epothilone B,[176] for extensive biological investigations.

10.6.3 Smith's Synthesis of Kendomycin

Kendomycin is an antitumor antibiotic first isolated in 1996 from *Streptomyces violaceoruber*, possesses potent activity as both an endothelin receptor antagonist[177] and an antiosteoporotic agent.[178] It also bears significant activity as an antibacterial, against multi-resistant strains of *Staphylococcus aureus*, as well as remarkable cytotoxicity, having potency similar to that of the clinically important drugs doxorubicin and cisplatin, against a series of human tumor cell lines (HMO2, HEP G2, MCF7, $GI_{50} < 0.1$ μM).[179] The structure of kendomycin comprises a unique quinone-methide-lactol chromophore attached to a densely substituted tetrahydropyran ring in conjunction with an aliphatic *ansa* ring. The first total synthesis of kendomycin was reported by Lee's group[180] and considerable synthetic effort has been made since then.[181] In 2005, Smith's group reported an elegant synthesis of kendomycin exploiting a Petasis–Ferrier rearrangement/ring-closing olefin metathesis strategy (Scheme 10.7).[182]

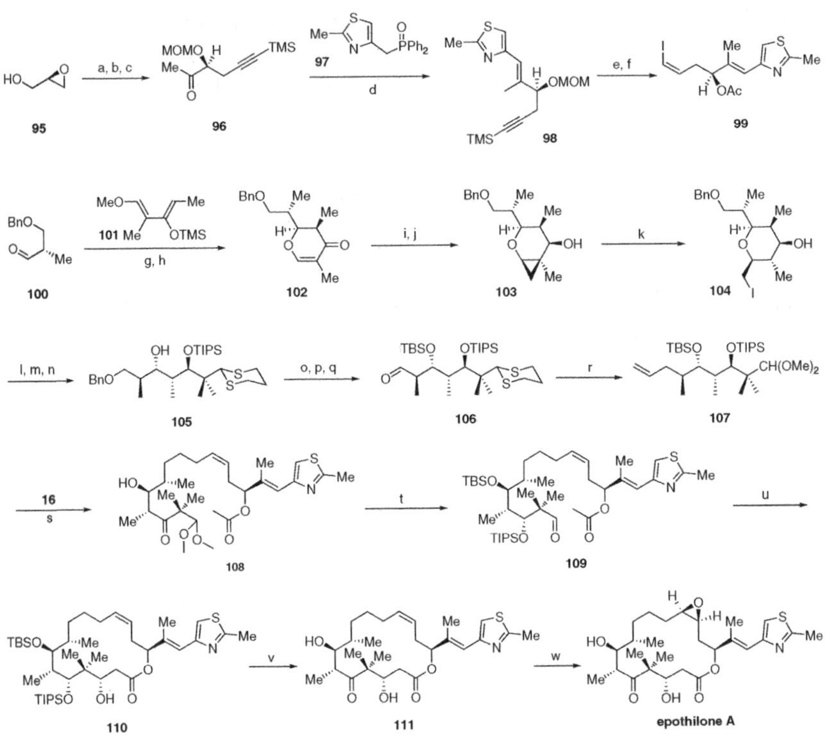

Scheme 10.6 *Reagents and conditions:* (a) DHP, PPTS, CH_2Cl_2, rt; (b) 1. TMSCCLi, $BF_3{\cdot}OEt_2$, THF, −78°C; 2. MOMCl, *i*Pr_2NEt, DCE, 55°C; 3. PPTS, MeOH, rt; (c) 1. $(COCl)_2$, DMSO, CH_2Cl_2, −78°C then Et_3N, −78°C to rt; 2. MeMgBr, Et_2O, 0°C to rt; 3. TPAP, NMO, 4Å mol. Sieves, CH_2Cl_2, 0°C to rt; (d) **97**, *n*-BuLi, THF, −78°C, then **96**, THF, −78°C to rt; (e) 1. NIS, $AgNO_3$, $(CH_3)_2CO$; 2, Cy_2BH, Et_2O, AcOH; (f) PhSH, $BF_3{\cdot}OEt_2$, CH_2Cl_2, rt; 2. Ac_2O, Py, DMAP, CH_2Cl_2, rt; (g) $TiCl_4$, CH_2Cl_2, −78°C; (h) TFA, CH_2Cl_2, rt; (i) $LiAlH_4$, Et_2O, −78°C; (j) Et_2Zn, CH_2I_2, Et_2O, rt; (k) NIS, MeOH, rt; (l) *n*-Bu_3SnH, AIBN, PhH, reflux; (m) Ph_3SiCl, imid. DMF, rt; (n) 1,3-propanedithiol, $TiCl_4$, CH_2Cl_2, −78°C to −40°C; (o) TBSOTf, 2,6-lutidine, CH_2Cl_2; (p) DDQ, CH_2Cl_2/H_2O; (q) $(COCl)_2$, DMSO, CH_2Cl_2, then Et_3N, −78°C to rt; (r) 1. PTSA, dioxane/H_2O; 2. CH_3PPh_3Br, NaHMDS, $PhCH_3$, 0°C to rt; 3. $PhI(OCOCF_3)_2$, MeOH/THF, rt; (s) **107**, 9-BBN, THF, rt, then $PdCl_2(dppf)_2$, Cs_2CO_3, Ph_3As, H_2O, DMF, **99**, rt, 71%; (t) *p*-TsOH, dioxane/H_2O, 50°C; (u) KHMDS, THF, −78°C, 51%; (v) 1. HF·Py, Py, THF, rt, 97%; 2. TBSOTf, 2,6-lutidine, CH_2Cl_2, −25°C, 93%; 3. DMP, CH_2Cl_2, 87%; 4. HF·Py, THF, rt, 99%; (w) dimethyl dioxirane, CH_2Cl_2, −50°C, 45% (≧20:1).

10.6.4 Yao's Synthesis of Camptothecin

The pentacyclic natural alkaloid camptothecin (CPT) was first isolated by Wall *et al.* from the Chinese medicinal plant *Camptotheca acumunata* in 1966 and is used as traditional Chinese medicine.[183] Although CPT shows potent antitumor activities by interfering with the interaction between DNA and topoisomerase I (Top I) to initiate a cytotoxic cascade to kill cancer cells, it could not be used in clinical treatments due to its poor aqueous solubility and high toxicity.[184–186]

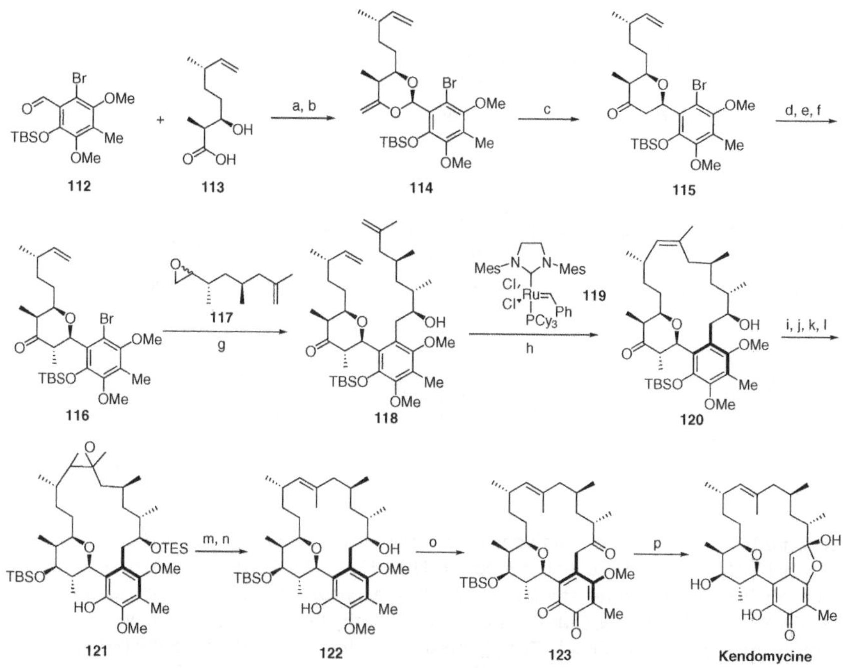

Scheme 10.7 *Reagents and conditions:* (a) *i*PrOTMS, TMSOTf, CH_2Cl_2, –78°C, 77%; (b) Cp_2TiMe_2, THF, 60–63°C, 85%; (c) Me_2AlCl, CH_2Cl_2, –78°C, 85%; (d) LiHMDS, MeI, THF, –78°C, 70%; (e) $NaBH_4$, EtOH, –78°C, 97%, 5:1 dr; (f) TBSOTf, 2,6-lutidine, CH_2Cl_2, –10°C, 95%; (g) *t*-BuLi, THF, –78°C, **116**, BF_3·Et_2O, THF, 60%, 2:1 dr; (h) Grubbs II (**119**), CH_2Cl_2, reflux, 57%; (i) TESOTf, DMAP, 2,6-lutidine, Py, 0°C, 89%; (j) OsO_4, Py, THF, 0°C, 95%; (k) MsCl, Py, CH_2Cl_2, 0°C, 95%; (l) $BnNMe_3OH$, MeOH/THF, 0°C, 84%; (m) WCl_6, BuLi, THF, 0°C to rt, 71%; (n) PPTS, MeOH, 0°C, 95%; (o) Dess–Martin Peridinane, Py, CH_2Cl_2, 0°C to rt, 69%; (p) aq. HF, CH_3CN, rt, 40%.

Despite being dropped from clinical trials by NCI in the 1970s due to severe bladder toxicity. During the past several decades, continuous structural modification of camptothecin-family alkaloids has led to quite a number of pharmaceutically useful camptothecin derivatives, such as topotecan (Hycamtin) and irinotecan (CPT-11; Camptosar), which have been approved and are used in cancer chemotherapy today.[187] Despite the rapid growth of the drug market, industrial sources of CPT and 10-OH CPT are still highly dependent on extraction from the medical plants *C. acuminata* and *Nothapodytes fetida*. Newer analogs with improved and more predictable bioavailability relative to these two existing agents are certain to expand the use of this natural-product drug class. Moreover, recent advances in the basic and clinical pharmacology of the camptothecins have renewed enthusiasm on a number of fronts. Many impressive total syntheses of CPT and its analogs have emerged from numerous research groups[188,189] since the first total synthesis of *rac*-camptothecin by Stock and Schultz in 1971.[190] In 2008, Yao and

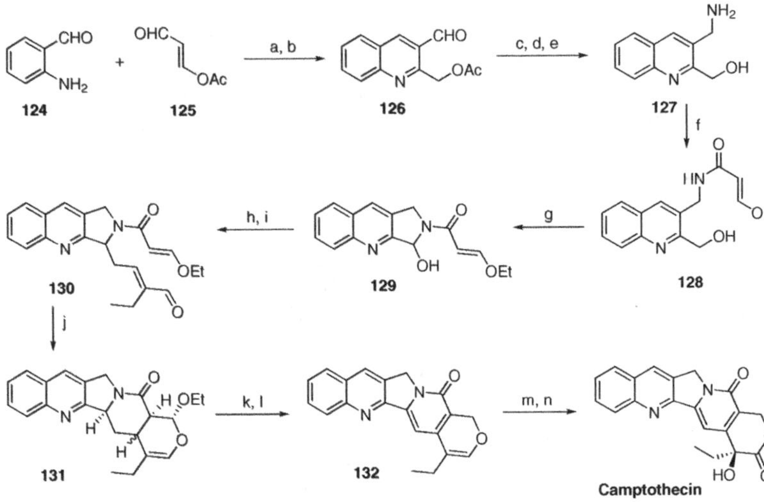

Scheme 10.8 *Reagents and conditions:* (a) benzolic acid, Pyrrolidine (cat.), CH_2Cl_2, then silica gel; (b) MnO_2, CH_2Cl_2, 75% over two steps; (c) NH_2OH, HCl, EtOH; (d) K_2CO_3, MeOH; (e) 10% Pd-C, MeOH, 79% over three steps; (f) (*E*)-3-ethoxyacryloyl chloride, DMF, 85%; (g) MnO_2, CH_2Cl_2, 84%; (h) Ac_2O, Et_3N, DMAP, CH_2Cl_2, (i) (*Z*)-((2-ethylbuta-1,3-dien-1-yl)oxy) trimethylsilane, $BF_3{\cdot}OEt_2$, CH_2Cl_2, 72%; (j) mesitylene, 160°C, sealed tube, 83%; (k) DDQ, cat. HOAc, dioxane; (l) Et_3SiH, $BF_3{\cdot}Et_2O$, CH_2Cl_2, 76% over two steps; (m) $(DHQD)_2$-PYR (cat.), $K_2OsO_2(OH)_4$ (cat.), $K_3Fe(CN)_6$, K_2CO_3, $MeSO_2NH_2$, t-BuOH/H_2O (1:1), 0°C; (n) I_2, $CaCO_3$, CH_3OH/H_2O (2:1), 83%, 95% *ee*.

coworkers reported a short and efficient route to the synthesis of camptothecin.[191] They developed a cascade reaction consisting of a pyrrolidine-catalyzed Michael addition, an intramolecular aldol condensation, and an oxidative aromatization, the intramoculecular oxa Diels–Alder cycloaddition, and the Sharpless asymmetric dihydroxylation making camptothecin synthesis more efficient and straightforward. The achieved syntheses and newly developed methodologies will be valuable in future endeavors for an economic industrial process of CPT alkaloids and their derivatives (Scheme 10.8).[192]

10.6.5 Nicolaou and Li's Synthesis of Platensimycin

Platensimycin, a metabolite of *Streptomyces platensis* (MA 7339) from a soil sample collected in South Africa,[193] which is an excellent example of a unique structural class of natural antibiotics, has been demonstrated to be a breakthrough in recent antibiotic research due to its unique functional pattern and significant antibacterial activity.[194] Platensimycin is a selective FabF inhibitor and exhibits *in vitro* activity against several drug-resistant bacteria. The novel and intricate chemical structure and vitally important biological activity of platensimycin made it an attractive target for total synthesis, and a number of groups have reported their synthetic studies toward this end.[195] Nicolaou and his team were the first to describe the total synthesis of platensimycin.[196] Their strategy involved the synthesis of the aromatic portion of platensimycin (3-amino-2,4-dihydroxybenzoic acid) in a protected form (**149**) starting from 2-nitroresorcinol in five steps. One of the asymmetric approaches to another key intermediate **143** was devised and featured by a well-known asymmetric alkylation reaction and a novel de-aromatizing cyclization reaction of a phenolic allyl silane. The reaction was developed as a viable and general method for the oxidative cyclization of phenolic allyl silanes to spiro hemiquinones (Scheme 10.9). Efficient coupling of each half of the molecule was readily achieved and subsequent deprotection furnished platensimycin.

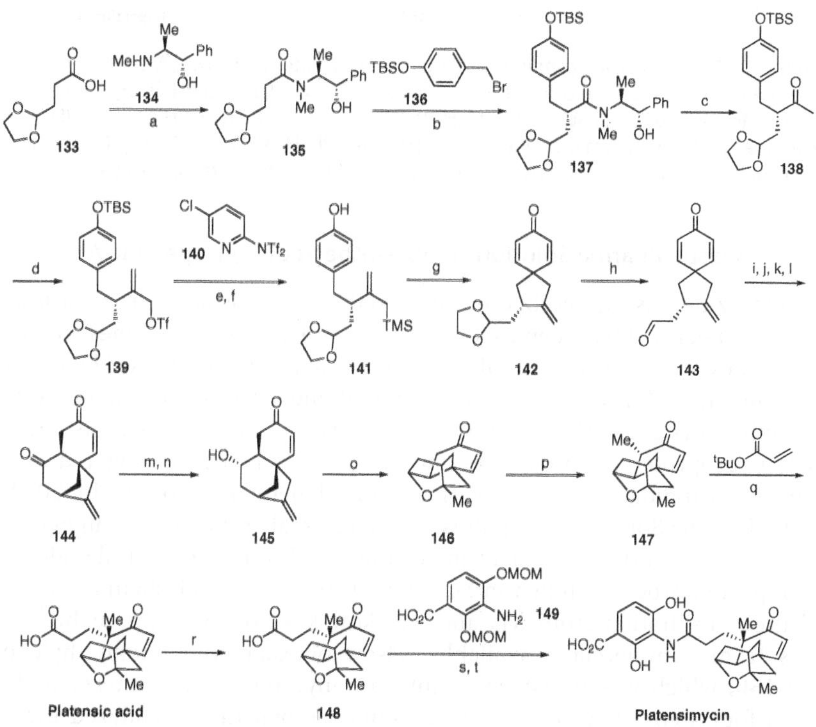

Scheme 10.9 *Reagents and conditions:* (a) PivCl, Et_3N, CH_3CN, 0°C, then **134**, Et_3N, THF; (b) LDA, LiCl, **136**, −78°C to 0°C, 87%; (c) MeLi, THF, −78°C to −22°C, 91%; (d) KHMDS, **140**, THF, −78°C to 0°C, 92%; (e) $TMSCH_2MgCl$, LiCl, $Pd(PPh_3)_4$, THF, 97%; (f) NaOH, MeOH, 100%; (g) $PhI(OAc)_2$, TFE, −1°C; (h) 1N aq. HCl/THF, 90%; (i) TMSCN, Et_3N, CH_2Cl_2, then aq. HCl; (j) vinyl ethyl ether/CH_2Cl_2, PPTS, 0°C, 99%; (k) KHMDS, THF, 0°C, 86%; (l) 1N aq. HCl/THF, 0°C, then aq. NaOH, 0°C, 89%; (m) DIBAL-H; (n) MnO_2, EtOAc, 82%; (o) TFA, CH_2Cl_2, 0°C, 87%; (p) KHMDS, MeI, THF/HMPA, −78°C to −10°C, 95%; (q) *t*-BuOK, *tert*-butyl acrylate, THF, −10°C, 84%; (r) TFA, CH_2Cl_2, 0°C, 100%; (s) **149**, HATU, Et_3N, 86%; (t) aq. LiOH, then aq. HCl, 86% two steps.

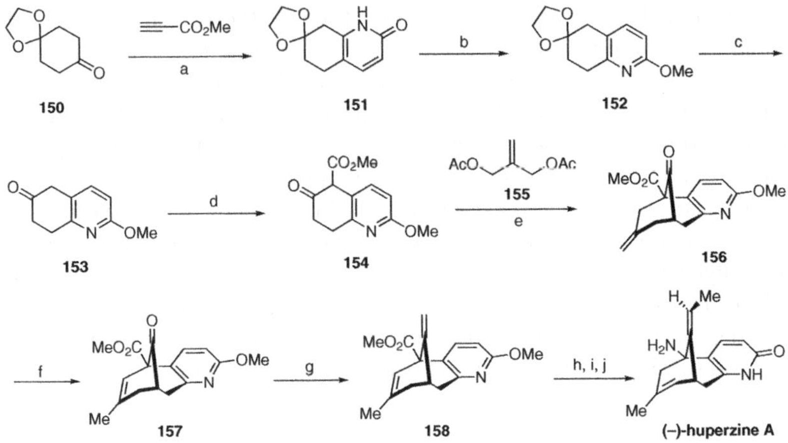

Scheme 10.10 *Reagents and conditions:* (a) methyl propiolate, *i*PrOH, NH_3, MeOH, 43%; (b) benzyltriethylammonium chloride, NaH, then Ag_2CO_3, MeI, 92%; (c) H_3PO_4, H_2O, 98%; (d) NaH, $(MeO)_2CO$, 65%; (e) SL-T002–1, $[C_3H_5PdCl]_2$, TMG, 82%; (f) TfOH, CH_2Cl_2, 80%; (f) ethyltriphenylphosphonium bromide, n-BuLi, THF, then PhSH, AIBN, Zn, 67%; (g) NaOH, THF, MeOH; (h) diphenylphosphoryl azide; (i) MeOH; (j) TMSCl, NaI, 40% over four steps.

10.6.6 Shasun Pharma Solutions Ltd's Synthesis of (−)-Huperzine A

(−)-Huperzine A is a quinolizidine alkaloid that can be isolated from various Huperzia species. It has been used for centuries in traditional Chinese medicine, mainly for the treatment of schizophrenia, fever, blood disorders, and loss of memory.[197] As one of the most potent and selective acetylcholinesterase inhibitors and NMDA receptor antagonists, (−)-huperzine A also possesses neuroprotective properties and is currently being investigated as a promising candidate for the treatment of Alzheimer's disease. In 2012, Shasun Pharma Solutions Ltd published a large-scale route to (−)-huperzine A that could be performed safely and economically on an industrial scale.[198] The reported process, shown in Scheme 10.10, required 11 chemical steps and drew inspiration from Xia and Kozikowski's previous approach.[199] It starts from commercially available 1,4-cyclohexanedione monoethylene ketal **150**, which was transformed into tetrahydroquinalone intermediate **154** in four steps. The subsequent asymmetric palladium-catalyzed bicycloannulation of β-ketoester **154** to **156** was carried out under optimized conditions. The conversion of compound **154** into (−)-huperzine A required six more steps. Using this route, 275 g of (−)-huperzine A were synthesized under cGMP conditions. Notably, all the reactions were carried out on a kilo- or decagram-scale in a chromatography-free process that is capable of providing large quantities of (−)-huperzine A for the pharmaceutical or nutraceutical industries.

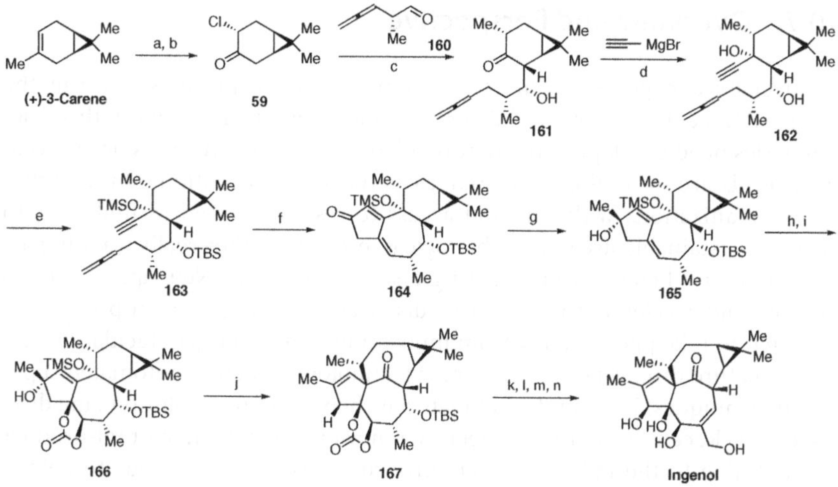

Scheme 10.11 *Reagents and conditions:* (a) NCS, DMAP, CH_2Cl_2, rt; (b) O_3, CH_2Cl_2, MeOH, −78°C, then thiourea, rt, 48% over two steps; (c) LiNap, MeI, HMPA, THF, −78°C, then LiHMDS, **160**, −78°C, 44%; (d) HC≡CMgBr, THF, −78–0°C, 81%, 10:1 *d.r.*; (e) TBSOTf, Et_3N, CH_2Cl_2, 0°C, then TMOTf, Et_3N, 71%; (f) $[RhCl(CO)_2]_2$, CO, xylenes, 140°C, 72%; (g) MeMgBr, THF, −78 to 0°C, 80%; (h) OsO_4 (1.0 equiv.), pyridine, rt, then saturated aq. Na_2SO_3, THF; (i) CDI, DMAP, CH_2Cl_2, rt; (j) $BF_3 \cdot Et_2O$, CH_2Cl_2, −78°C to −40°C, then Et_3N, MeOH, 80%; (k) SeO_2, dioxane, 80°C, then Ac_2O, DMAP, rt, 59%; (l) HF, CH_3CN, 50°C, 90%; (m) Martin sulfurane, $CHCl_3$, 60°C, then 10% aq. NaOH, THF, rt; (n) SeO_2, HCO_2H, dioxane, 80°C, 76%.

10.6.7 Baran's Synthesis of Ingenol

Ingenol is a diterpenoid first isolated in 1966 by Hecker and coworkers from *Euphorbia ingens*[200] and its structure was determined in 1967.[201] Ingenol esters display remarkable biological properties such as anticancer and anti-HIV activities, derived from their interaction with protein kinase C.[202] In particular, ingenol mebutate (commercially available as Picato, developed by LEO Pharma) was recently approved by the FDA as a first-in-class treatment for actinic keratosis, a pre-cancerous skin condition,[203] and has completed Phase II clinical trials for the topical treatment of basal cell carcinomas.[204] In 2013,[205,206] Baran's group reported their synthesis of ingenol started from inexpensive 3-carene, which was elaborated in five steps into **163** (Scheme 10.11). The key allenic Pauson–Khand reaction delivered dienone **164**, which was further transformed into carbinol **165**. Intermediate **165** represents the "cyclase phase" endpoint of the synthesis and at the same time a branching point for the synthesis of a variety of novel analogs of Picato®. Most notably, all the steps of the cyclase phase were carried out on a gram scale, and the route is currently being scaled up, in conjunction with LEO Pharma, to provide larger quantities of the Pauson–Khand product **164**. In a sequence of four more steps, the target ingenol was completed, resulting in the shortest, most scalable, and diversifiable synthesis of ingenol thus far.

10.7 Overview and Perspective

Biodiversity and genomics are two key words governing bioscience in the twenty-first century. Biological diversity is increasingly important in this field. Industries need to adapt to the new regulations and have clear visions of what their needs are for biological resources as tools for innovation. On the other hand, organizations with title and access to resources need to find ways to facilitate sustainable use of their biological resources if they really want to participate in and benefit from the drug discovery process. New approaches to improve and accelerate the joint drug discovery and development process are expected to take place mainly from innovation in drug target elucidation and lead structure discovery. Powerful new technologies such as automated separation techniques, HTS, and combinatorial chemistry are revolutionizing drug discovery. Recently, there has been a renewed interest in natural-product research due to the failure of alternative drug discovery methods to deliver many lead compounds in key therapeutic areas such as immune-suppression, anti-infectives, and metabolic diseases. Natural products extracted from tissues of terrestrial plants, marine organisms, or microorganism fermentation broths are the evolutionary shaped molecules with a profound impact on human health. Moreover, natural products have been, and will be, important sources of new pharmaceutical compounds. It is clear that demand for drugs, disposable consumer products, biological agents, and insecticides will continue to increase for the foreseeable future.

Acknowledgements

The authors thank National Natural Science Foundation (No. 21532002, 21778031) and Double First-class University Construction Plan of Nanjing University for the financial support.

References

1 Knoll, A. H. *Life on a Young Planet, The First Three Billion Years of Evolution on Earth*, Princeton, NJ: Princeton University Press, **2003**.
2 Breinbauer, R.; Vetter, I. R.; Waldmann, H. *Angew. Chem. Int. Ed.* **2002**, *41*, 2878–2890.
3 Mandal, M.; Breaker, R. R. *Nat. Rev. Mol. Cell Biol.* **2004**, *5*, 451–463.
4 Vitreschak, A. G.; Rodionov, D. A.; Mironov, A. A.; Gelfand, M. S. *Trends Genet.* **2004**, *20*, 44–50.
5 Dixon, R. A. *Nature* **2001**, *411*, 843–847.
6 Camilli, A.; Bassler, B. L. *Science* **2006**, *311*, 1113–1116.

7 Firn, R. D.; Jones, C. G. *Nat. Prod. Rep.* **2003**, *20*, 382–391.
8 Newman, D. J.; Cragg, G. M.; Snader, K. M. *J. Nat. Prod.* **2003**, *66*, 1002–1037.
9 Spring, D. R. *Chem. Soc. Rev.* **2005**, *34*, 472–482.
10 Dixon, N.; Wong, L.-S.; Geerlings, T.-H.; Micklefield, J. *Nat. Prod. Rep.* **2007**, *24*, 288–1310.
11 Lindquist, N.; Fenical, W.; Vanduyne, G. D.; Clardy, J. *J. Am. Chem. Soc.* **1991**, *113*, 2303 –2304.
12 Burgett, A. W.; Li, Q.; Wei, Q.; Harran, P. G. *Angew. Chem. Int. Ed.* **2003**, *42*, 4961–4966.
13 Wang, G.; Shang, L.; Burgett, A. W.; Harran, P. G.; Wang, X. *Proc. Natl. Acad. Sci. U.S.A.* **2007**, *104*, 2068 –2073.
14 Hanson, J. R. *Nat. Prod. Rep.* **2009**, *26*, 1156–1171.
15 Roberts, S. C. *Nat. Chem. Biol.* **2007**, *3*, 387–395.
16 Liu, C.-X.; Yin Q.-Q.; Zhou H.-C.; Wu Y.-L.; Pu J.-X.; Xia, L.; *et al. Nat. Chem. Biol.* **2012**, *8*, 486–493.
17 Luo, Y.-P.; Zhao, X.-T.; Yang, X.-D.; Zhao, J.-F.; Li, L. *Asia-Pacific Trad. Med.* **2009**, *5*, 158–61.
18 (a) Zhan, Z.-J.; Ying, Y.-M.; Ma, L.-F.; Shan, W.-G. *Nat. Prod. Rep.* **2011**, *28*, 594–629; (b) Fraga, B. M. *Nat. Prod. Rep.* **2010**, *27*, 1681–1708; (c) Fraga, B. M. *Nat. Prod. Rep.* **2008**, *25*, 1180–1209.
19 Li, C.; Dong, T.; Li, Q.; Lei, X.-G. *Angew. Chem. Int. Ed.* **2014**, *53*, 12111–12115.
20 Lahlou, M. *Exp. Opin. Drug Disc.* **2007**, *2*, 697–705.
21 Patwardhan, B.; Vaidya, A. D. B.; Chorghade, M. *Curr. Sci.* **2004**, *86*, 789–799.
22 Newman, D. J.; Cragg, G. M.; Snader, K. M. *Nat. Prod. Rep.* **2000**, *17*, 215–234.
23 Newman, D. J.; Cragg, G. M.; Snader, K. M. *J. Nat. Prod.* **2003**, *66*, 1022–1037.
24 Koehn, F. E.; Carter, G. T. *Nat Rev Drug Discov.* **2005**, *4*, 206–220.
25 Paterson, I., Anderson, E. A. *Science* **2005**, *310*, 451–455.
26 Balunas, M. J.; Kinghorn, A. D. *Life Sci.* **2005**, *78*, 431–441.
27 Jones, W.-P.; Chin, Y.-W.; Kinghorn, A. D. *Curr. Drug Target* **2006**, *7*, 247–264.
28 Butler, M. *J. Nat. Prod.* **2004**, *67*, 2141–2153.
29 Clardy, J.; Walsh, C. T. *Nature* **2004**, *432*, 829–836.
30 Newman, D. J.; Cragg, G. M.; Snader, K. M. *J. Nat. Prod.* **2003**, *66*, 1022–1037.
31 Newman, D. J.; Cragg, G. M.; Snader, K. M. *Nat. Prod. Rep.* **2000**, *17*, 215–234.
32 Li, W. H. L.; Vederas, J. C. *Science* **2009**, *325*, 161–165.
33 (a) Brown, A. G.; Butterworth, D.; Cole, M.; Hanscomb, G.; Hood, J. D.; Reading, C.; *et al. J. Antibiot.* **1976**, *29*, 668–669. (b) Howarth, T. T.; Brown, A. G.; King, T. J. *J. Chem. Soc. Chem. Commun.* **1976**, 266–267. (c) Brown, A. G. *Drug Des. Delivery* **1986**, *1*, 1–21.
34 (a) Endo, A.; Kuroda, M.; Tsujita, Y. *J. Antibiot.* **1976**, *29*, 1346–1348. (b) Endo, A. *J. Lipid Res.* **1992**, *33*, 1569–1582.
35 Brown, A. G.; Smale, T. C.; King, T. J.; Hasenkamp, R.; Thompson, J. *J. Chem. Soc., Perkin Trans. 1* **1976**, 1165–1170.

36 Tobert, J. A. *Nature Rev. Drug Discovery* **2003**, *2*, 517–526.

37 Gaw, A., Packard, C. J., Shepherd J. (eds). *Statins: The HMG CoA Reductase Inhibitors in Perspective*, 2nd Edn, Martin Dunitz: London, **2004**.

38 Gonzaga-Jagureui, C.; Lupski, J. R.; Gibbs, R. A. *Annu. Rev. Med.* **2012**, *63*, 35–61.

39 Radzicka, A.; Wolfenden, R. *Methods Enzymol.* **1995**, *249*, 284–312

40 Kim, C. U. *et al. J. Am. Chem. Soc.* **1997**, *119*, 681–690.

41 Von Itzstein, M. *Nature* **1993**, *363*, 418–423.

42 Varghese, J. N., Laver, W. G., Colman, P. M. *Nature* **1983**, *303*, 35–40.

43 Colman, P. M., Varghese, J. N., Laver, W. G. *Nature* **1983**, *303*, 41–44.

44 Liu, K.-G.; Yan, S.; Wu, Y.-L.; Yao, Z.-J., *J. Org. Chem.* **2002**, *67*, 6758–6763.

45 (a) Gao, Z.-X.; Wang, M.; Wang, S.-Z.; Yao, Z.-J. *Org. Lett.* **2009**, *11*, 3678–3681; (b) Cong, X.; Yao, Z.-J. *J. Org. Chem.* **2006**, *71*, 5365–5368; (c) Liu, K.-G.; Zhou, H.-B.; Wu, Y.-L.; Yao, Z.-J. *J. Org. Chem.* **2003**, *68*, 9528–9531.

46 Wei, D.-Q.; Du, Q.-S.; Sun, H.; Chou, K. H. *Biochem. Biophys. Res. Commun.* **2006**, *344*, 1048–1055;

47 Feher, M.; Schmidt, J. M. *J. Chem. Inf. Comput. Sci.* **2003**, *43*, 218–227.

48 Bauer, R. A.; Wurst, J. M.; Tan, D. S. *Curr. Opin. Chem. Biol.* **2010**, *14*, 308–314.

49 Zhang, M.Q.; Wilkinson, B. *Curr. Opin. Biotechnol.* **2007**, *18*, 478–488.

50 Reayi, A.; Arya, P. *Curr. Opin. Chem. Biol.* **2005**, *9*, 240–247.

51 Wells, J. A.; McClendon, C. L. *Nature* **2007**, *450*, 1001–1009.

52 Whitty, A.; Kumaravel, G. *Nat. Chem. Biol.* **2006**, *2*, 112–118.

53 Berg, T. *Curr. Opin. Chem. Biol.* **2008**, *12*, 464–471.

54 Arndt, H. D. *Angew. Chem. Int. Ed.* **2006**, *45*, 4552–4560.

55 Payne, D. J.; Gwynn, M. N.; Holmes, D. J.; Pompilano, D. L. *Nat. Rev. Drug Discov.* **2007**, *6*, 29–40.

56 Driggers, E. M.; Hale, S. P.; Lee, J.; Terrett, N. K. *Nat. Rev. Drug Discov.* **2008**, *7*, 608–624.

57 Schreiber, S. L.; Crabtree, G. R. *Immunol. Today* **1992**, *13*, 136–142.

58 Mann, J. *Nat. Rev. Cancer* **2002**, *2*, 143 – 148.

59 Cragg, G. M.; Grothaus, P. G.; Newman, D. J. *Chem. Rev.* **2009**, *109*, 3012 – 3043.

60 Scott, A. M.; Wolchok, J. D.; Old, L. *Nat. Rev. Cancer* **2012**, *12*, 278 – 287.

61 Dancik, V.; Seiler, K. P.; Young, D. W.; Schreiber S. L.; Clemons, P. A. *J. Am. Chem. Soc.* **2010**, *132*, 9259–9261.

62 Mayer, A. M. S.; Gustafson, K. R. *Eur. J. Cancer.* **2008**, *44*, 2357–2387.

63 Lewis Phillips, G. D.; Li, G.; Dugger, D. L., *et al. Cancer Res.* **2008**, *68*, 9280–9290.

64 Tolcher, A. W., Ochoa, L.; Hammond, L. A. *et al. J. Clin. Oncol.* **2003**, *21*, 211–222.

65 Goff, L. W.; Papadopoulos, K.; Posey, J. A. *et al. J. Clin. Oncol.* **2009**, *27*, 15625.

66 Hamann, P. R.; Hinman, L. M.; Hollander, I. *et al. Bioconjug. Chem.* **2002**, *13*, 47–58.

67 DiJoseph, J. F.; Armellino, D. C.; Boghaert, E. R. *et al. Blood* **2004**, *103*, 1807–1814.

68 Doronina, S. O.; Mendelsohn, B. A.; Bovee, T. D. *et al. Bioconjug. Chem.* **2006**, *17*, 114–124.

69 Tse, K. F.; Jeffers, M., Pollack, V. A. *et al. Clin. Cancer Res.* **2006**, *12*, 1373–1382.

70 Tietze, L. F.; Krewer, B.; von Hof, J. M.; Frauendorf, H.; Schuberth, I. *Toxins* **2009**, *1*, 134–150.

71 Stein, R.; Smith, M. R.; Chen, S.; Zalath, M.; Goldenberg, D. M., *Clin. Cancer Res.* **2009**, *15*, 2808–2817.

72 Cardillo, T. M.; Govindan, S. V.; Sharkey, R. M.; Trisal, P.; Goldenberg, D. M. *Clin. Cancer Res.* **2011**, *17*, 3157–3169.

73 Tsang, R. Y.; Finn, R. S. *Br. J. Cancer* **2012**, *106*, 6–13.

74 Corey, E. J.; Cheng, X.-M. *The Logic of Chemical Synthesis*, New York: John Wiley & Sons, Inc., **1989**.

75 Aicher, T. D.; Buszek, K. R.; Fang, F. G.; Forsyth, C. J.; Jung, S. H.; Kishi, Y.; *et al. J. Am. Chem. Soc.* **1992**, *114*, 3162.

76 Chiba, H.; Tagami, K. *J. Synth. Org. Chem. Jpn.* **2011**, *69*, 600.

77 Choi, H.; Demeke, D.; Kang, F.-A.; Kishi, Y.; Nakajima, K.; Nowak, P.; *et al. Pure Appl. Chem.* **2003**, *75*, 1.

78 Kishi, Y.; Fang, F.; Forsyth, C. J.; Scola, P. M.; Yoon, S. K. WO 9317690 A1, **1993**.

79 Littlefield, B. A.; Palme, M.; Seletsky, B. M.; Towle, M. J.; Yu, M. J.; Zheng, W. WO 9965894 A1, **1999**.

80 Jackson, K. L.; Henderson, J. A.; Motoyoshi, H.; Phillips, A. J. *Angew. Chem. Int. Ed.* **2009**, *48*, 2346–2350.

81 Spring, D. R. *Org. Biomol. Chem.* **2003**, *1*, 3867.

82 Spandl, R. J.; Spring, D. R.; Bender, A. *Org. Biomol. Chem.* **2008**, *6*, 1149.

83 Burke, M. D.; Schreiber, S. L. *Angew. Chem. Int. Ed.* **2004**, *43*, 46.

84 Tan, D. S.; Foley, M. A.; Shair, M. D.; Schreiber, S. L. *J. Am. Chem. Soc.* **1998**, *120*, 8565–8566.

85 Tan, D. S.; Foley, M. A.; Stockwell, B. R.; Shair, M. D.; Schreiber, S. L. *J. Am. Chem. Soc.* **1999**, *121*, 9073–9087.

86 Tamura, O.; Okabe, T.; Yamaguchi, T.; Gotanda, K.; Noe, K.; Sakamoto, M. *Tetrahedron* **1995**, *51*, 107–118.

87 Miyaura, N.; Suzuki, A. *Chem. Rev.* **1995**, *95*, 2457–2483.

88 Stille, J. K. *Angew. Chem. Int. Ed. Engl.* **1986**, *25*, 508–524.

89 Sonogashira, K.; Tohda, Y.; Hagihara, N. *Tetrahedron Lett.* **1975**, *50*, 4467–4470.

90 Stephens, R. D.; Castro, C. E. *J. Org. Chem.* **1963**, *28*, 3313–3315.

91 Lam, K. S.; Salmon, S. E.; Hersh, E. M.; Hruby, V. J.; Kazmierski, W. M.; Knapp, R. J. *Nature* **1991**, *354*, 82–84.

92 Furka, A.; Sebestyen, F.; Asgedom, M.; Dibo, G. *Int. J. Pept. Protein Res.* **1991**, *37*, 487–493.

93 Feher, M.; Schmidt, J. M. *J. Chem. Inf. Comput. Sci.* **2003**, *43*, 218–227.
94 Arkin, M. R.; Wells, J. A. *Nat. Rev. Drug Discov.* **2004**, *3*, 301–317.
95 Pagliaro, L.; Felding, J.; Audouze, K.; Nielsen, S. J.; Terry, R. B.; Krog-Jensen, C., *et al. Cur. Opin. Chem. Biol.* **2004**, *8*, 442–449.
96 Tan, D. S. *Nat. Chem. Rev.* **2005**, *1*, 74–84.
97 Arya, P.; Joseph, R.; Gan, Z.; Rakic, B. *Chem. Biol.* **2005**, *12*, 163–180.
98 Shang, S.; Tan, D. S. *Curr. Opin. Chem. Biol.* **2005**, *9*, 248–258.
99 Tan, D. S. *Comb. Chem. High Throughput Screen* **2004**, *7*, 631–643.
100 Goess, B. C.; Hannoush, R. N.; Chan, L. K.; Kirchhausen, T.; Shair, M. D. *J. Am. Chem. Soc.* **2010**, *128*, 5391–5403.
101 Morton, D.; Leach, S.; Cordier, C.; Warriner, S.; Nelson, A. *Angew. Chem. Int. Ed.* **2009**, *48*, 104–109.
102 Robbins, D. *et al. Chem. Sci.* **2011**, *2*, 2232–2235.
103 Ibbeson, B. M. *et al. Nat. Commun.* **2014**, *5*, 3155.
104 Oguri, H. *et al. J. Am. Chem. Soc.* **2011**, *133*, 7096–7105.
105 Beckmann, H. S. G. *et al. Nat. Chem.* **2013**, *5*, 861–867.
106 McLeod, M. C. *et al. Nat. Chem.* **2014**, *6*, 133–140.
107 Zhang, J.; Wu, J.-B.; Hong, B.-K.; Ai, W.-Y.; Wang, X.-M.; Li, H.-H.; *et al. Nat. Commun.* **2014**, *5*, 4614.
108 Nielsen, T. E.; Schreiber, S. L. *Angew. Chem. Int. Ed.* **2008**, *47*, 48–56.
109 (a) Lowitz, J. T. *Crell's Chem. Ann.* **1788**, *1*, 312; (b) Bruckner, V.; Kovacs, J.; Koczka, I. *J. Chem. Soc. Abstr.* **1948**, *948*; (c) Hayek, E. W. H.; Jordis, U.; Moche, W.; Sauter, F. *Phytochemistry* **1989**, *28*, 2229.
110 (a) Cichewicz, R. H.; Kouzi, S. A. *Med. Res. Rev.* **2003**, *24*, 90; (b) Alakurtti, S.; Makela, T.; Koskimies, S.; Yli-Kauhaluoma, J. *Eur. J. Pharm. Sci.* **2006**, *29*, 1.
111 (a) Ruzicka, L.; Lamberton A. H.; Christie, E. W. *Helv. Chim. Acta* **1938**, *21*, 1706; (b) Robertson, A.; Soliman, G.; Owen, E. C. *J. Chem. Soc.* **1939**, *1267*; (c) Kim, D. S. H. L.; Chen, Z.; Nguyen, V. T.; Pezzuto, J. M.; Qiu, S.; Lu, Z. *Synth. Commun.* **1997**, *27*, 1607.
112 (a) Pisha, E.; Chai, H.; Lee, I. S.; Chagwedera, T. E.; Farnsworth, N. R.; Cordell, G. A.; *et al. Nat. Med.* **1995**, *1*, 1046; (b) De Clercq, E. *Med. Res. Rev.* **2000**, *20*, 323.
113 Wall, N. R.; Mohammad, R. M.; Al-Katib, A. M. *Leuk. Res.* **1999**, *23*, 881–888.
114 Elgie, A. W.; Sargent, J. M.; Alton, P.; Peters, G. J.; Noordhuis, P.; Williamson, C. J.; *et al. Leuk. Res.* **1998**, *22*, 373–378.
115 Oz, H. S.; Hughes, W. T.; Rehg, J. E.; Thomas, E. K. *Microb. Pathog.* **2000**, *29*, 187–190.
116 Ajani, J. A.; Jiang, Y. X.; Faust, J.; Chang, B. C. B.; Ho, L.; Yao, J. C.; et al. *Invest. New Drugs* **2006**, *24*, 353–357.
117 Alkon, D. L.; Epstein, H.; Kuzirian, A.; Bennett, M. C.; Nelson, T. J. *Proc. Natl. Acad. Sci. U.S.A.* **2005**, *102*, 16432–16437.

118 Schaufelberger, D. E.; Koleck, M. P.; Beutler, J. A.; Vatakis, A. M.; Alvarado, A. B.; Andrews, P.; *et al. J. Nat. Prod.* **1991**, *54*, 1265–1270.

119 (a) Ohmori, K.; Ogawa, Y.; Obitsu, T.; Ishikawa, Y.; Nishiyama, S.; Yamamura, S. *Angew. Chem. Int. Ed.* **2000**, *39*, 2290–2294. (b) Evans, D. A.; Carter, P. H.; Carreira, E. M.; Charette, A. B.; Prunet, J. A.; Lautens, M. *J. Am. Chem. Soc.* **1999**, *121*, 7540–7552. (c) Kageyama, M.; Tamura, T.; Nantz, M. H.; Roberts, J. C.; Somfai, P.; Whritenour, D. C. *et al. J. Am. Chem. Soc.* **1990**, *112*, 7407–7408. (d) Keck, G. E.; Welch, D. S.; Poudel, Y. B. *Tetrahedron Lett.* **2006**, *47*, 8267–8270.

120 Wender, P. A.; DeBrabander, J.; Harran, P. G.; Jimenez, J. M.; Koehler, M. F. T.; Lippa, B. *et al. Proc. Natl. Acad. Sci. U.S.A.* **1998**, *95*, 6624–6629.

121 Breinbauer, R.; Manger, M.; Scheck, M.; Waldmann, H. *Curr. Med. Chem.* **2002**, *9*, 2129.

122 Breinbauer, R.; Vetter, I. R.; Waldmann, H. *Angew. Chem. Int. Ed.* **2002**, *41*, 2879.

123 Kaiser, M.; Wetzel, S.; Kumar, K.; Waldmann, H. *Cell. Mol. Life Sci.* **2008**, *65*, 1186.

124 Koch, M. A.; Schuffenhauer, A.; Scheck, M.; Wetzel, S.; Casaulta, M.; Odermatt, A. *et al. Proc. Natl. Acad. Sci. U.S.A.* **2005**, *102*, 17272.

125 Schuffenhauer, A.; Ertl, P.; Roggo, S.; Wetzel, S.; Koch, M. A.; Waldmann, H. *J. Chem. Inf. Model* **2007**, *47*, 47.

126 Dekker, F. J.; Koch, M. A.; Waldmann, H. *Curr. Opin. Chem. Biol.* **2005**, *9*, 232.

127 Koch, M. A.; Breinbauer, R.; Waldmann, H. *Biol. Chem.* **2003**, *384*, 1265.

128 Koch, M. A.; Waldmann, H. *Drug Disc. Today* **2005**, *10*, 471.

129 Kobayashi, J.; Madono, T.; Shigemori, H. *Tetrahedron* **1995**, *51*, 10867.

130 Stahl, P.; Kissau, L.; Mazitschek, R.; Giannis, A.; Waldmann, H. *Angew. Chem. Int. Ed.* **2002**, *41*, 1174.

131 Zhou, B. N.; Johnson, R. K.; Mattern, M. R.; Fisher, P. W.; Kingston, D. G. I. *Org. Lett.* **2001**, *3*, 4047.

132 Arnold, L. D.; Calderwood, D. J.; Dixon, R. W.; Johnston, D. N.; Kamens, J. S.; Munschauer, R.; *et al. Bioorg. Med. Chem. Lett.* **2000**, *10*, 2167.

133 Burchat, A. F.; Calderwood, D. J.; Hirst, G. C.; Holman, N. J.; Johnston, D. N.; Munschauer, R.; *et al. Bioorg. Med. Chem. Lett.* **2000**, *10*, 2171.

134 Burchat, A. F.; Calderwood, D. J.; Friedman, M. M.; Hirst, G. C.; Li, B.; Rafferty, P.; *et al. Bioorg. Med. Chem. Lett.* **2002**, *12*, 1687.

135 Calderwood, D. J.; Johnston, D. N.; Munschauer, R.; Rafferty, P. *Bioorg. Med. Chem. Lett.* **2002**, *12*, 1683.

136 Brohm, D.; Metzger, S.; Bhargava, A.; Muller, O.; Lieb, F.; Waldmann, H. *Angew. Chem. Int. Ed.* **2002**, *41*, 307.

137 Koch, M. A.; Wittenberg, L. O.; Basu, S.; Jeyaraj, D. A.; Gourzoulidou, E.; Reinecke, K. *et al. Proc. Natl. Acad. Sci. U.S.A.* **2004**, *101*, 16721.

138 Kingston, D. G. In *Anticancer Agents from Natural Products*. Cragg, G. M.; Kingston, D. G. I.; Newman, D. J. (eds); Boca Raton, FL: CRC Press LLC, **2005**, p. 89.

139 Kingston, D. G. I. *J. Org. Chem.* **2008**, *73*, 3975.
140 Rahier, N. J.; Thomas, C. J.; Hecht, S. M. In *Anticancer Agents from Natural Products*; Cragg, G. M.; Kingston, D. G. I.; Newman, D. J. (eds); Boca Raton, FL: CRC Press LLC, **2005**, p. 5.
141 Pinney, K. G.; Jelinek, C.; Edvardsen, K.; Chaplin, D. J.; Pettit, G. R. In *Anticancer Agents from Natural Products*; Cragg, G. M.; Kingston, D. G. I.; Newman, D. J. (eds); Boca Raton, FL: CRC Press LLC, **2005**, p. 23.
142 Rinehart, K. L.; Holt, T. G.; Fregeau, N. L.; Stroh, J. G.; Keifer, P. A.; Sun, F.; *et al. J. Org. Chem.* **1990**, *55*, 4512.
143 Wright, A. E.; Forleo, D. A.; Gunawardana, G. P.; Gunasekera, S. P.; Koehn, F. E.; McConnell, O. J. *J. Org. Chem.* **1990**, *55*, 4508.
144 Cuevas, C.; Perez, M.; Martın, M. J.; Chicharro, J. L.; Fernandez-Rivas, C.; Flores, M.; *et al. Org. Lett.* **2000**, *2*, 2545.
145 Ikeda, Y.; Idemoto, H.; Hirayama, F.; Yamamoto, K.; Iwao, K.; Asao, T.; *et al. J. Antibiot.* **1983**, *36*, 1279.
146 Corey, E. J.; Gin, D. Y.; Kania, R. S. *J. Am. Chem. Soc.* **1996**, *118*, 9202–9203.
147 Aoki, S.; Watanabe, Y.; Sanagawa, M.; Setiawan, A.; Kotoku, N.; Kobayashi, M. *J. Am. Chem. Soc.* **2006**, *128*, 3148–3149.
148 (a) Shenvi, R. A.; Guerrero, C. A.; Shi, J.; Li, C.-C.; Baran, P. S. *J. Am. Chem. Soc.* **2008**, *130*, 7241–7243; (b) Shi, J.; Shigehisa, H.; Guerrero, C. A.; Shenvi, R. A.; Li, C.-C.; Baran, P. S. *Angew. Chem. Int. Ed.* **2009**, *48*, 4328–4331; (c) Shi, J.; Manolikakes, G.; Yeh, C.-H.; Guerrero, C. A.; Shenvi, R. A.; Shigehisa, H.; *et al. J. Am. Chem. Soc.* **2011**, *133*, 8014–8027.
149 *Guidelines for the Treatment of Malaria*, World Health Organization, Geneva, 2nd Edn, **2010**.
150 Dondorp, A. M.; Fairhurst, R. M.; Slutsker, L.; MacArthur, J. R.; Breman, J. G.; Guerin, P. J. *et al. New Engl. J. Med.* **2011**, *365*, 1073–1075.
151 Ariey, F.; Witkowski, B.; Amaratunga, C.; Beghain, J.; Langlois, A.-C.; Khim, N. *et al. Nature* **2014**, *505*, 50–55.
152 White, N. J. *Science* **2008**, *320*, 330–334.
153 Tu, Y. *Nat. Med.* **2011**, *17*, 1217–1220.
154 Abdin, M. Z.; Israr, M.; Rehman, R. U.; Jain, S. K. *Planta Med.* **2003**, *69*, 289–299.
155 Covello, P. S. *Phytochemistry* **2008**, *69*, 2881–2885.
156 Farhi, M.; Marhevka, E.; Ben-Ari, J.; Algamas-Dimantov, A.; Liang, Z.; Zeevi, V. *et al. Nat. Biotechnol.* **2011**, *29*, 1072–1074.
157 Sanofi. Cashin-Garbutt, A.; Online: www.news-medical.net/news/20150628/Artemisinin-and-the-fight-against-malaria-an-interview-with-Dr-Robert-Sebbag-Sanofi.aspx (accessed March 2018).
158 Turconi, J.; Griolet, F.; Guevel, R.; Oddon, G.; Villa, R.; Geatti, A. *et al. Org. Process Res. Dev.* **2014**, *18*, 417–422.
159 Paddon, C. J.; Westfall, P. J.; Pitera, D. J.; Benjamin, K.; Fisher, K.; McPhee, D. *et al. Nature* **2013**, *496*, 528–532.

160 Wani, M. C.; Taylor, H. L.; Wall, M. E.; Coggon, P.; McPhail, A. T. *J. Am. Chem. Soc.* **1971**, *93*, 2325–2327.

161 Nicolaou, K. C.; Dai, W.-M.; Guy, R. K. *Angew. Chem. Int. Ed. Engl.* **1994**, *33*, 15–44;

162 (a) Nicolaou, K. C.; Yang, Z.; Liu, J.-J.; Ueno, H.; Nantermet, P. G.; Guy, R. K. *et al. Nature* **1994**, *367*, 630–634; (b) Nicolaou, K. C.; Nantermet, P. G.; Ueno, H.; Guy, R. K.; Couladouros, E. A.; Sorensen, E. J. *J. Am. Chem. Soc.* **1995**, *117*, 624–633; (c) Nicolaou, K. C.; Liu, J.-J.; Yang, Z.; Ueno, H.; Sorensen, E. J.; Claiborne, C. F. *et al. J. Am. Chem. Soc.* **1995**, *117*, 634–644; (d) Nicolaou, K. C.; Yang, Z.; Liu, J.-J.; Nantermet, P. G.; Claiborne, C. F.; Renaud, J. *et al. J. Am. Chem. Soc.* **1995**, *117*, 645–652; (e) Nicolaou, K. C.; Ueno, H.; Liu, J.-J.; Nantermet, P. G.; Yang, Z.; Renaud, J. *et al. J. Am. Chem. Soc.* **1995**, *117*, 653–659; (f) Nicolaou, K. C.; Guy, R. K. *Angew. Chem. Int. Ed. Engl.* **1995**, *34*, 2079–2090.

163 (a) Holton, R. A.; Somoza, C.; Kim, K. B.; Liang, F.; Biediger, R. J.; Boatman, P. D. *et al. J. Am. Chem. Soc.* **1994**, *116*, 1597–1598; (b) Holton, R. A.; Kim, K. B.; Somoza, C.; Liang, F.; Biediger, R. J.; Boatman, P. D. *et al. J. Am. Chem. Soc.* **1994**, *116*, 1599–1600.

164 Masters, J. J.; Link, J. T.; Snyder, L. B.; Young, W. B.; Danishefsky, S. J. *Angew. Chem. Int. Ed. Engl.* **1995**, *34*, 1723–1726.

165 (a) Wender, P. A.; Badham,N. F.; Conway, S. P.; Floreancig, P. E.; Glass, T. E.; Houze, J. B. *et al. J. Am. Chem. Soc.* **1997**, *119*, 2757–2758; (b) Wender, P. A.; Badham, N. F.; Conway, S. P.; Floreancig, P. E.; Glass, T. E.; Graenicher, C. *et al. J. Am. Chem. Soc.* **1997**, *119*, 2755–2756. (c) Wender, P. A.; Marquess, D. G.; McGrane, L. P.; Taylor, R. E. *Chemtracts-Org. Chem.* **1994**, *7*, 160–171.

166 (a) Shiina, I.; Saitoh, K.; Frechard-Ortuno, I.; Mukaiyama, T. *Chem. Lett.* **1998**, 3–4; (b) Mukaiyama, T.; Shiina, I.; Iwadare, H.; Saitoh, M.; Nishimura, T. Ohkawa, N. *et al. Chem. Eur. J.* **1999**, *5*, 121–161.

167 Morihira, K.; Hara, R.; Kawahara, S.; Nishimori, T.; Nakamura, N.; Kusama, H. *et al. J. Am. Chem. Soc.* **1998**, *120*, 12980–12981.

168 Nicolaou, K. C., Liu, J. J.; Hwang, C.-K.; Dai, W.-M.; Guy, R. K., *J. Chem. Soc. Chem. Commun.* **1992**, 1118–1120.

169 Nicolaou, K. C., Hwang, C.-K., Sorensen, E. J.; Claiborne, C. F., *J. Chem. Soc. Chem. Commun.* **1992**, 1117–1118.

170 Nicolaou, K. C., Yang, Z., Sorensen, E. J., Nakada, M. *J. Chem. Soc. Chem. Commun.* **1993**, 1024–1026.

171 Hdfle, G.; Bedorf, N.; Steinmetz, H.; Schomberg, D.; Gertb, K.; Reichenbdch, H. *Angew. Chem. Int. Ed. Engl.* **1996**, *35*. 1567.

172 Nicolaou, K. C.; Roschangar, F.; Vourloumis, D. *Angew. Chem. Int. Ed.* **1998**, *37*, 2014–2045.

173 Balog, A.; Meng, D.; Kamenecha, T.; Bertinato, P.; Su, D.-S.; Sorensen, E. J. *et al. Angew. Chem. Int. Ed. Engl.* **1996**, *35*, 2801–2803.

174 Yang, Z.; He, Y.; Vourloumis, D.; Vallberg, H.; Nicolaou, K. C. *Angew. Chem. Int. Ed. Engl.* **1997**, *36*, 166–168.
175 Schinzer, D.; Bauer, A.; Limberg, A.; Bohm, O. M.; Cordes, M. *Angew. Chem. Int. Ed. Engl.* **1997**, *36*, 523–524.
176 Su, D.-S.; Meng, D.; Bertinato, P.; Balog, A.; Sorensen, E. J.; Danishefsky, S. J. *et al. Angew. Chem. Int. Ed. Engl.* **1997**, *36*, 757–759.
177 (a) Funahashi, Y.; Kawamura, N.; Ishimaru, T. Japan Patent 08231551 A2960910, 1996; *Chem. Abstr.* **1997**, *126*, 6553. (b) Funahashi, Y.; Kawamura, N.; Ishimaru, T. Japan Patent 08231552, 1996; *Chem. Abstr.* **1996**, *125*, 326518.
178 Su, M. H.; Hosken, M. I.; Hotovec, B. J.; Johnston, T. L. U.S. Patent 5728727 A 980317, 1998; *Chem. Abstr.* **1998**, *128*, 239489.
179 Bode, H. B.; Zeeck, A. *J. Chem. Soc., Perkin Trans. 1* **2000**, 323.
180 Yuan, Y.; Men, H.; Lee, C. *J. Am. Chem. Soc.* **2004**, *126*, 14720.
181 (a) Mulzer, J.; Pichlmair, S.; Green, M. P.; Marques, M. M. B.; Martin, H. *J. Proc. Natl. Acad. Sci. U.S.A.* **2004**, *101*, 11980; (b) Sengoku, T.; Arimoto, H.; Uemura, D. *Chem. Commun.* **2004**, *1220*; (c) White, J. D.; Smits, H. *Org. Lett.* **2005**, *7*, 235.
182 Smith, III, A. B.; Mesaros, E. F.; Meyer, E. A. *J. Am. Chem. Soc.* **2005**, *127*, 6948–6949.
183 Wall, M. E.; Wani, M. C.; Cook, C. E.; Palmer, K. H.; Mcphail, A. T.; Sim, G. A. *J. Am. Chem. Soc.* **1966**, *88*, 3888.
184 Covey, J. M.; Jaxel, C.; Kohn, K. W.; Pommier, Y. *Cancer Res.* **1989**, *49*, 5016–5022.
185 Miao, Z.-H.; Player, A.; Shankavaram, U.; Wang, Y.-H.; Zimonjic, D. B.; Liao, Z.-Y. *et al. Cancer Res.* **2007**, *67*, 8752–8761.
186 Pommier, Y. *ACS Chem. Biol.* **2013**, *8*, 82–95.
187 (a) Giovanella, B. C.; Stehlin, J. S.; Wall, M. E.; Wani, M. C.; Nicholas, A. W.; Liu, L. F.; *et al. Science* **1989**, *246*, 1046. (b) Saltz, L. B.; Cox, J. V.; Blanke, C.; Rosen, L. S.; Fehrenbacher, L.; Moore, M. J.; *et al. New Engl. J. Med.* **2000**, *343*, 905. (c) Ozols, R. F. *Int. J. Gynecol. Cancer* **2000**, *10*, 33. (d) Vanhoefer, U.; Harstrick, A.; Achterrath, W.; Cao, S.; Seeber, S.; Rustum, Y. M. *J. Clin. Oncol.* **2001**, *19*, 1501.
188 (a) Cuendet, M.; Pezzuto, J. M. In *Modern Alkaloids: Structure, Isolation, Synthesis and Biology*; Fattorusso, E., Taglialatela-Scafati, O. (eds); Weinheim: Wiley-VCH, **2008**, Ch. 2, pp. 29–33. (b) Thomas, C. J.; Rahier, N. J.; Hecht, S. M. *Bioorg. Med. Chem.* **2004**, *12*, 1585. (c) Du, W. *Tetrahedron* **2003**, *59*, 8649. (d) Hutchinson, C. R. *Tetrahedron* **1981**, *37*, 1047.
189 (a) Kanazawa, A.; Muniz, M. N.; Baumlova, B.; Ljungdahl, N.; Greene, A. E. *Synlett* **2008**, *2275*. (b) Chavan, S. P.; Pathak, A. B.; Kalkote, U. R. *Synlett* **2007**, *2635*. (c) Peters, R.; Althaus, M.; Nagy, A.-L. *Org. Biomol. Chem.* **2006**, *4*, 498.

190 Stock, G.; Schultz, A. G. *J. Am. Chem. Soc.* **1971**, *93*, 4074.

191 Liu, G.-S.; Dong, Q.-L.; Yao, Y.-S.; Yao, Z.-J. *Org. Lett.* **2008**, *23*, 5393–5396.

192 Liu, G.-S.; Yao, Y.-S.; Xu, P.; Wang, S.-Z.; Yao, Z.-J. *Chem.-Eur. J.* **2011**, *17*, 10462–10469.

193 (a) Wang, J.; Soisson, S. M.; Young, K., Singh, S. B. *et al. Nature* **2006**, *441*, 358–61. (b) Singh, S. B.; Jayasuriya, H.; Ondeyka, J. G.; Herath, K. B.; Zhang, C.; Zink, D. L.; *et al. J. Am. Chem. Soc.* **2006**, *128*, 11916–11920. (c) Habich, D.; von Nussbaum, F. *ChemMedChem* **2006**, *1*, 951–954.

194 Manallack, D. T.; Crosby, I. T.; Khakham, Y.; Capuano, B. *Curr. Med. Chem.* **2008**, *15*, 705–710.

195 (a) Zou, Y.; Chen, C.-H.; Taylor, C. D.; Foxman, B. M.; Snider, B. B. *Org. Lett.* **2007**, *9*, 1825–1828. (b) Nicolaou, K. C.; Tang, Y.; Wang, J. *Chem. Commun.* **2007**, 1922–1923. (c) Li, P.; Payette, J. N.; Yamamoto, H. *J. Am. Chem. Soc.* **2007**, *129*, 9534–9535. (d) Lalic, G.; Corey, E. J. *Org. Lett.* **2007**, *9*, 4921–4923. (e) Tiefenbacher, K.; Mulzer, J. *Angew. Chem. Int. Ed.* **2007**, *46*, 8074–8075. (f) Nicolaou, K. C.; Pappo, D.; Tsang, K. Y.; Gibe, R.; Chen, D. Y.-K. *Angew. Chem. Int. Ed.* **2008**, *47*, 944–946. (g) Kim, C. H.; Jang, K. P.; Choi, S. Y.; Chung, Y. K.; Lee, E. *Angew. Chem. Int. Ed.* **2008**, *47*, 4009–4011. (h) Matsuo, J.-I.; Takeuchi, K.; Ishibashi, H. *Org. Lett.* **2008**, *10*, 4049–4052. (i) Ghosh, A. K.; Xi, K. *J. Org. Chem.* **2009**, *74*, 1163–1170.

196 (a) Nicolaou, K. C.; Li, A.; Edmonds, D. J. *Angew. Chem. Int. Ed.*, **2006**, *45*, 7086–7090; (b) Nicolaou, K. C., Edmonds, D. J.; Li, A.; Tria, G. S. *Angew. Chem. Int. Ed.* **2007**, *46*, 3942–3945; (c) Nicolaou, K. C.; Li, A.; Ellery, S. P.; Edmonds, D. J. *Angew. Chem. Int. Ed.* **2009**, *48*, 6293–6295; (d) Nicolaou, K. C.; Li, A.; Edmonds, D. J.; Ttria G. S.; Ellery, S. P. *J. Am. Chem. Soc.* **2009**, *131*, 16905–16918.

197 (a) Liu, J. S.; Zhu, Y. L.; Yu, C. M. *et al. Can. J. Chem.* **1986**, *64*, 837–839; (b) A. P. Kozikowski, W. Tuckmantel, *Acc. Chem. Res.* **1999**, *32*, 641–650.

198 Tudhope, S. R.; Bellamy, J. A.; Ball, A.; Rajasekar, D.; Azadi-Ardakani, M.; Meera, H. S. *et al. Org. Process Res. Dev.* **2012**, *16*, 635–642.

199 (a) Kozikowski, A. P.; Reddy E. R.; Miller, C. P. *J. Chem. Soc., Perkin Trans. 1*, **1990**, 195–197; (b) Kozikowski, A. P.; Xia, Y.; Reddy, E. R.; Tuckmantel, W.; Hanin, I.; Tang, X. C. *J. Org. Chem.* **1991**, *56*, 4636–4645; (c) Campiani, G.; Sun, L. Q.; Kozikowski, A. P.; Aagaard, P.; McKinney, M. *J. Org. Chem.* **1993**, *58*, 7660–7669.

200 Hecker, E. *Cancer Res.* **1968**, *28*, 2338 – 2348.

201 Zeichmeister, K.; Brandl, F.; Hoppe, W.; Hecker, E.; Opferkuch, H. J.; Adolf, W. *Tetrahedron Lett.* **1970**, *47*, 4075 – 4078.

202 Fujiwara, M.; Fujiwara, M.; Ijichi, K.; Tokuhisa, K.; Katsuura, K.; Shigeta, S.; *et al. Antimicrob. Agents Chemother.* **1996**, *40*, 271 – 273.

203 Lebwohl, M.; Swanson, N.; Anderson, L. L.; Melgaard, A.; Xu, Z.; Berman, B. N. *Engl. J. Med.* **2012**, *366*, 1010–1019.

204 (a) Siller, G.; Rosen, R.; Freeman, M.; Welburn, P.; Katsamas, J.; Ogbourne, S. M. *Aust. J. Dermatol.* **2010**, *51*, 99 – 105. (b) ClinicalTrials.gov, NCT00329121, http://clinicaltrials.gov/show/NCT00329121.

205 Jørgensen, L.; McKerrall, S. J.; Kuttruff, C. A.; Ungeheuer, F.; Felding, J.; Baran, S. *Science.* **2013**, *341*, 878–882.

206 McKerrall, S. J.; Jørgensen, L.; Kuttruff, C. A.; Ungeheuer, F.; Baran, S. *J. Am. Chem. Soc.* **2014**, *136*, 5799–5810.

11

Modern Technologies in Natural Product Synthesis

Zhu-Jun Yao and Shouyun Yu

School of Chemistry and Chemical Engineering, Nanjing University, Nanjing, China

The increasing need for more efficient synthetic technologies and sustainable processes can be seen as a major driving force for chemists to invent new synthetic methods. It also stimulates the creative rethinking of known concepts, which, in turn, will lead to the development of innovative chemistry. Subsequently, using these robust methods, more efficient, concise and "green" synthetic routes have been designed for the synthesis of natural products. Recently, some new and some old synthetic technologies, such as visible-light photochemistry, organic electrochemistry, flow chemistry, and their combinations, have emerged as powerful tools to promote useful transformations. In this chapter, we will highlight recent significant progress in total synthesis of natural products using these modern synthetic technologies.

11.1 Visible-Light Photochemistry

As early as the beginning of the twentieth century, synthetic chemists have attempted to develop new efficient processes using photochemistry inspired by nature's ability to convert solar energy to chemical energy. However, the lack of visible-light absorption by many organic molecules has limited the application of photochemical synthesis. This drawback can be addressed by using photosensitizers and photocatalysts. Although the use of visible-light photoredox catalysis to initiate organic transformations has evoked interest since the 1970s,[1] this synthetic strategy had its renaissance in 2008.[2] In 2008, two independent reports from the MacMillan[3a] and Yoon[3b] groups described the use of $Ru(bpy)_3^{2+}$ as a visible-light photoredox catalyst to promote an α-alkylation of aldehydes and a

Efficiency in Natural Product Total Synthesis, First Edition. Edited by Pei-Qiang Huang, Zhu-Jun Yao, and Richard P. Hsung.

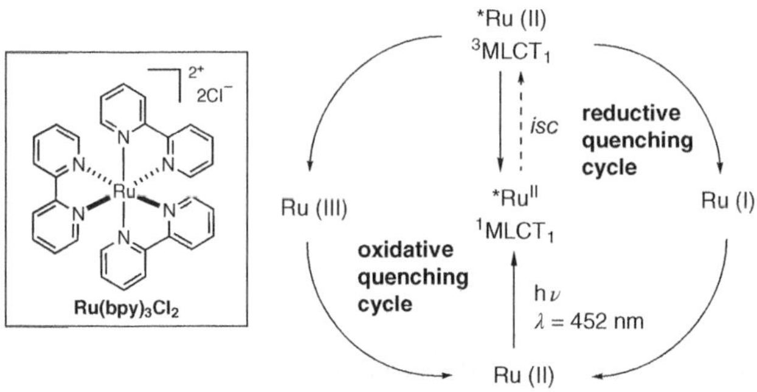

Scheme 11.1 General principle of visible-light photoredox catalysis.

[2 + 2] cycloaddition, respectively. Stephenson and coworkers[3c] also disclosed a photoredox reductive dehalogenation of activated alkyl halides using the same photocatalyst.

Mechanistically, photocatalysts, upon irradiation, are converted into their excited states, which are chemically more reactive because of the significantly altered electronic distribution. These excited photocatalysts can undergo energy-transfer processes and electron-transfer processes to initiate a radical or radical-polar cross-over reaction (Scheme 11.1).[4]

The efforts of these three research groups, as well as others, have initiated a renewed interest in visible-light photoredox catalysis, prompting to develop new organic reactions using this conceptually novel approach. For further reading and information, readers may refer to the comprehensive reviews on this subject.[2] This chapter first aims to highlight recent efforts on the total synthesis of natural products assisted by this photochemistry.

During their studies into the visible-light-mediated synthesis of indole alkaloid natural products using photoredox chemistry, Stephenson and coworkers serendipitously discovered an efficient method for the reductive dehalogenation of activated C–X bonds.[5] When 2-substituted indoles were used as a radical acceptor, C3–C3' bisindole motifs could be forged. They synthesized gliocladin C (**5**) in 10 steps from commercially available Boc-D-tryptophan methyl ester (**1**) at a 30% overall yield. The key reaction is the visible-light-promoted indole coupling reaction of bromopyrroloindoline **2** with 2-substituted indole **3** (Scheme 11.2).

Li and coworkers employed α,β-unsaturated carbonyl compounds as radical acceptors instead of indole derivatives, radical 1,4-addition could take place. Based on this reaction, an efficient route to drimentine F and indotetine A was developed (Scheme 11.3).[6] Optically active α,β-unsaturated ketone (**7**) was prepared from (+)-sclareolide (**6**), which was subjected to the radical 1,4-addition in the present of $Ir(ppy)_2(dtbbpy)PF_6$ under visible-light irradiation to give the key intermediate **9**. The intermediate **9** could be converted to drimentine

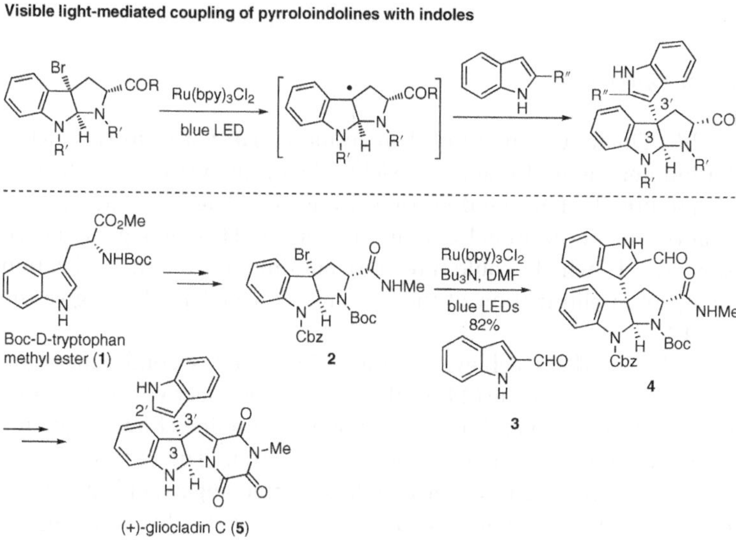

Scheme 11.2 Stephenson's total synthesis of gliocladin C.

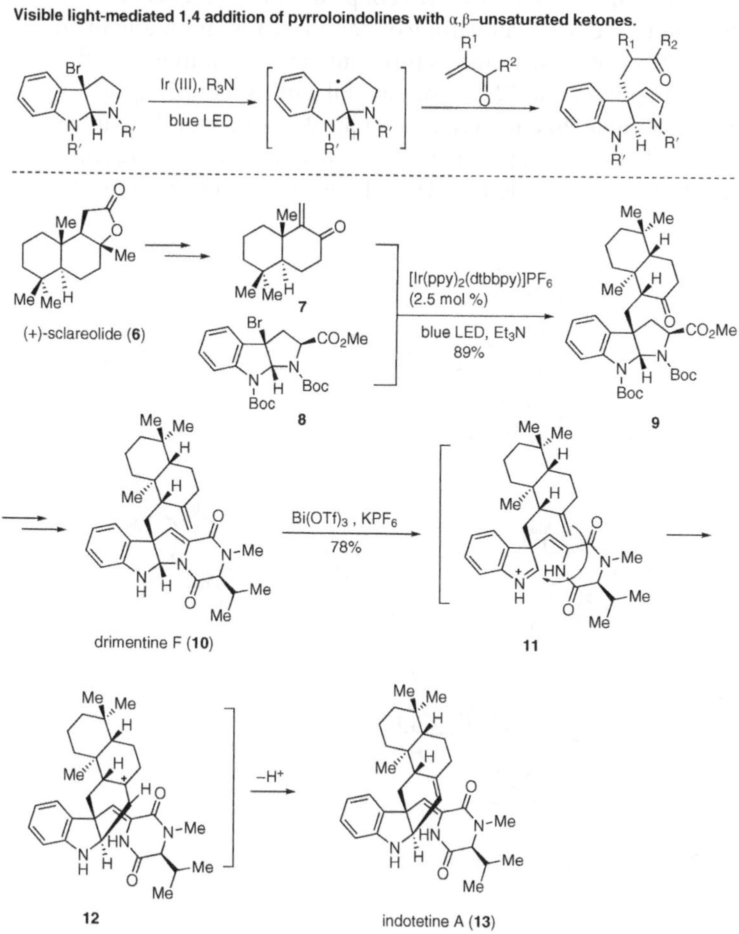

Scheme 11.3 Li's total synthesis of drimentine F and indotetine A.

F (**10**) smoothly. Finally treatment of drimentine F (**10**) with $Bi(OTf)_3/KPF_6$ could render indotertine A (**13**) at a 78% yield. The mechanism of transformation from drimentine F (**10**) to indotertine A was postulated in Scheme 11.3. Acidic activation of the germinal diamine moiety of **10** might generate an iminium species **11**, which could undergo an iminium-olefin cyclization followed by a proton elimination of the cationic intermediate **12** to give the final product **13**.

Overman and coworkers achieved a much-improved second-generation total synthesis of (−)-aplyviolene (**24**) enabled the stereoselective coupling of a tertiary carbon radical and a carbon electrophile to combine chiral fragments with formation of a new quaternary carbon stereocenter.[7] The radical precursor had to be considered carefully. The couplings employing a halide as a radical precursor were not accessible at a satisfactory yield. By contrast, an (*N*-acyloxy)phthalimide precursor appeared distinctly advantageous. Model studies showed that a solution of (*N*-acyloxy)-phthalimide **14** and methyl vinyl ketone **15** in a mixture of THF/H_2O was coupled in the presence of 1 mol% of $[Ru(bpy)_3]Cl_2$ and excess 1,4-dihydronicotinamide **16** with blue light to give adduct **18** at a 70% yield. This success prompted the group modified the total synthesis of (−)-aplyviolene (**24**). The pivotal radical coupling reaction of *N*-(acyloxy)phthalimide **20** and cyclopentenone **21** using modified anhydrous conditions provided coupled product **23** at a 61% yield. Following the established procedure, (−)-aplyviolene (**24**) could be completed (Scheme 11.4).

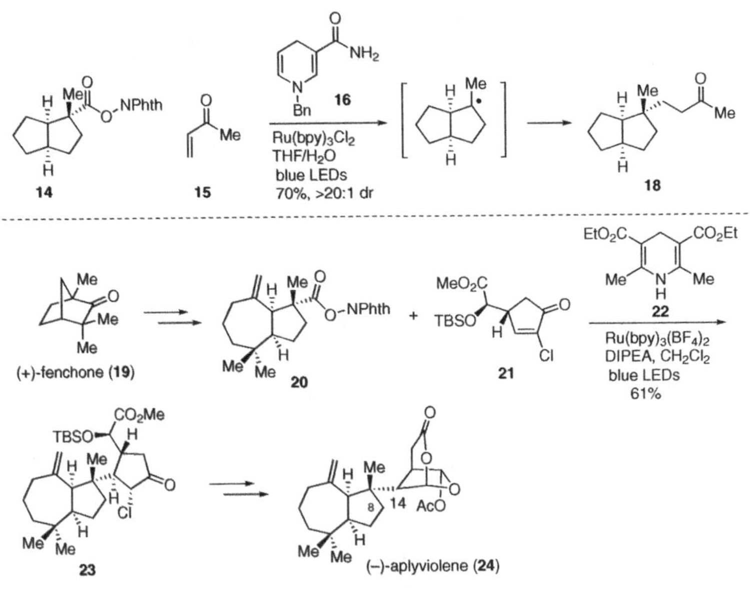

Scheme 11.4 Overman's total synthesis of (−)-aplyviolene.

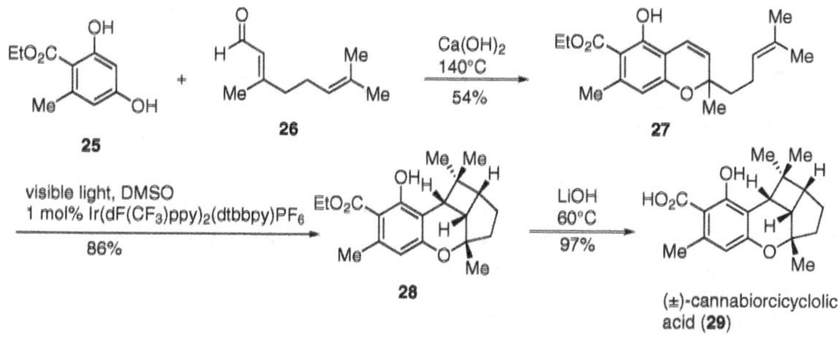

Scheme 11.5 Yoon's total synthesis of (±)-cannablorcicyclolic acid.

Yoon and coworkers recently finished a concise synthesis of (±)-cannablorcicyclolic acid (**29**) by exploring the visible-light-mediated intramolecular [2 + 2] cycloaddition as shown in Scheme 11.5.[8] The chromene precursor **27** was easily obtainable by base-catalyzed condensation of phenol **25** with citral (**26**) at a 54% yield. Photocycloaddition sensitized by 1 mol% $Ir(dF(CF_3)ppy)_2(dtbbpy)PF_6$ provided cyclobutane **28** at a 86% yield after 8 h of irradiation with a 23 W compact fluorescent lightbulb. Finally, saponification of the ethyl ester **28** afforded (±)-cannabiorcicycloic acid (**29**) at a 97% yield.

Another successful example of using visible-light-mediated pericyclic reactions is the total synthesis of prenylflavonoid natural products by Lei's group.[9] In this work, they demonstrated the first $Ru(bpy)_3Cl_2$-mediated regioselective Schenck ene reaction. Subsequently, they finished synthesis with kuwanons I and J, as well as brosimones A and B (Scheme 11.6).

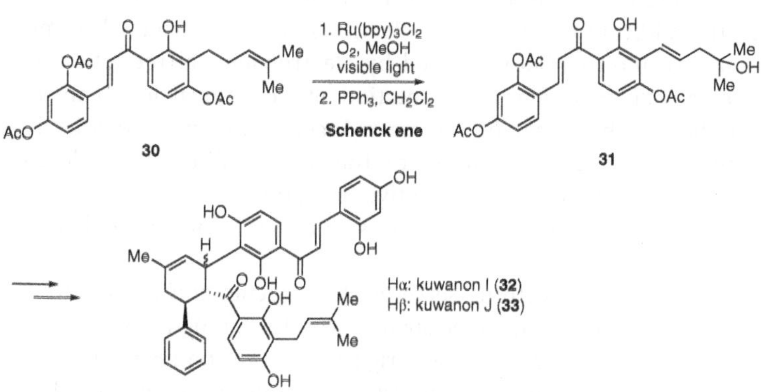

Scheme 11.6 Lei's total synthesis of kuwanons I and J.

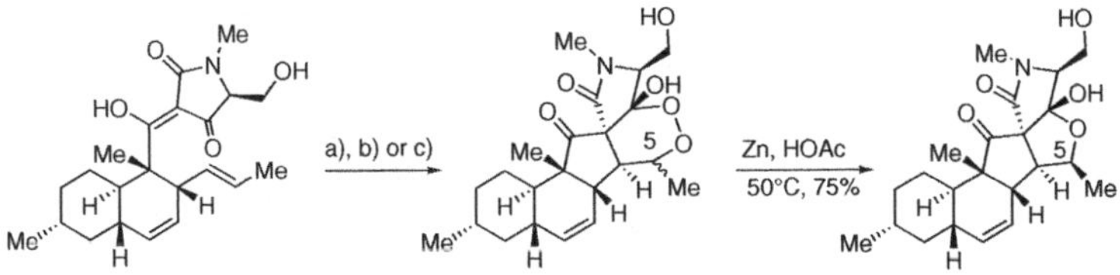

equisetin (**34**)

35

(+)-fusarisetin A (**36**) and its C5 epimer

a) $Mn(OAc)_3$ (0.1 equiv), air or 1 atm O_2, dr = 1.1:1 (at C5), 60%
b) $Ru(bpy)_3Cl_2$, Et_3N, air, visible light, dr = 2.0:1 (at C5), 68%
c) methylene blue, Et_3N, air, visible light, dr = 3.0:1 (at C5), 70%

Scheme 11.7 Gao's total synthesis of fusarisetin A.

Visible-light-mediated oxidation could be applied to total synthesis of natural products. As demonstrated by Gao and coworkers (Scheme 11.7), equisetin (**34**) was oxidized by molecular oxygen to peroxide **35**, which served as precursor for fusarisetin A (**36**). The oxidation step could be promoted by $Mn(OAc)_3$. Improved results were achieved when visible-light-mediated oxidations were employed using either a ruthenium complex or methylene blue as a photocatayst.[10]

11.2 Electrochemistry

Electrochemical synthesis, which has much in common with photochemical conversions, is an attractive method for the construction of organic compounds as it allows useful organic transformations, such as reductions, oxidations, C-C coupling reactions, and installation of heteroatom moieties.[11]

Electrochemistry offers the most straightforward method for the preparation of radical-ions, subsequent radical species, which can find numerous applications in organic synthesis. As shown in Scheme 11.8, molecules are activated by the addition or removal of electrons at an electrode. The most important reactive intermediates formed by electrochemical reduction and oxidation are radical ions. The fragmentation of the radical ions can lead typically to radicals, which may be further reduced to anions or oxidized to cations. Furthermore, organic compounds that are candidates for electrochemical conversion must have electroactive functionality (an electrophore), which is the analog of the chromophores necessary for photochemical activation.[11e]

In electrosynthesis, only electrons serve as the reagents and the electrodes employed are not consumed. Thus, large quantities of stoichiometric oxidants and reductants can be avoided and the waste originating from the reagents used is almost negligible. Electrochemistry complies with all the criteria of green and sustainable chemistry. Since renewable energy, like solar energy, will play a major role

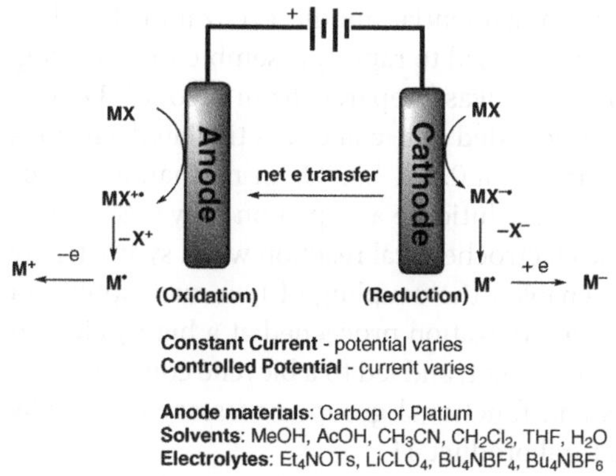

Scheme 11.8 General principle of organic electrochemistry.

in future industrial societies there will definitely be a surplus of electric energy that can best be used for the conversion of chemicals into valuable products. Unfortunately, despite years of impressive advances by electroorganic chemists, synthetic organic chemists do not routinely receive training in electrochemical methods, which means this method is not yet very common for the construction of larger organic compounds. Consequently, in the preparation of complex organic molecules, electrochemistry has still not been embraced as a standard technique by the broader synthetic community and is commonly considered the last option when all other conventional chemical methods have failed. For further reading, readers may refer to the comprehensive reviews on this subject to get familiar with this useful and powerful synthetic technology.[11] Recently, some encouraging results underline the potential of electroorganic synthesis and pave the way for the renaissance of this method. In this chapter, we will highlight recent significant progresses on total synthesis of natural products using electrochemistry.

One of most classical example employing electrochemistry in synthesizing natural products is Moeller's total synthesis of alliacol in 2003 (Scheme 11.9).[12]

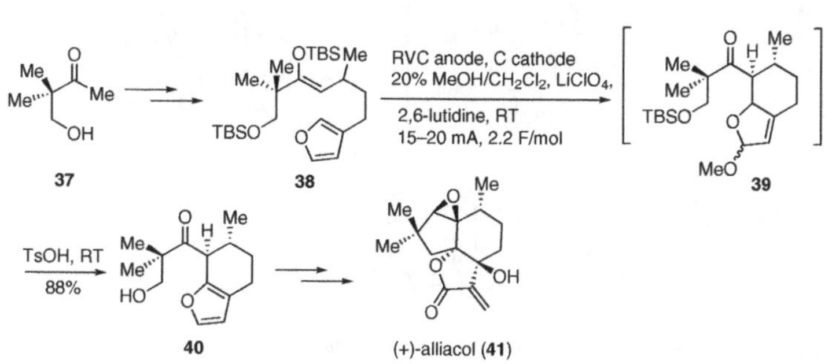

Scheme 11.9 Moeller's total synthesis of (+)-alliacol.

The key step of this synthesis was a sequential anodic cyclization-Friedel–Crafts alkylation of furan **38**, which was used to rapidly assemble the core ring skeleton of alliacol A. Furan **38**, which was prepared from alcohol **37**, was oxidized to give acetal **39** in an undivided cell using a reticulated vitreous carbon anode (RVC), a carbon cathode, a 0.4 M $LiClO_4$ in methanol/dichloromethane (1:4) electrolyte solution, 2,6-lutidine as a proton scavenger, and a constant current of 15–20 mA. The electrochemical reaction was a synthetically powerful Umpolung process that enabled the coupling of two nucleophiles (a silylenol ether and a furan ring). The cyclization proceeded at a high yield, did not require the use of specialized equipment, and led to a bicyclic compound **39** that contained all of the carbons and functional groups necessary for rapidly completing the synthesis of the natural product **41**.

Similarly, Trauner and coworkers adapted the same anodic oxidation strategy to finish a total synthesis of (−)-heptemerone B (**47**) and (−)-guanacastepene E (**48**) (Scheme 11.10).[13] In this event, anodic oxidation of the silyl enol ether **42**, under the conditions established by Moeller, gave tetracycle **46** at a 81% yield as a single isomer. This cyclization probably proceeds through oxidation of **42** to the corresponding radical cation **43**, followed by stereoselective intramolecular attack of the furan. Interception of the resulting carboxonium radical cation **44** by methanol, further oxidation, and finally desilylation of intermediary cation **45** furnishes acetal **46**. Acetal **46** could be rapidly converted to (−)-heptemerone B (**47**) and (−)-guanacastepene E (**48**).

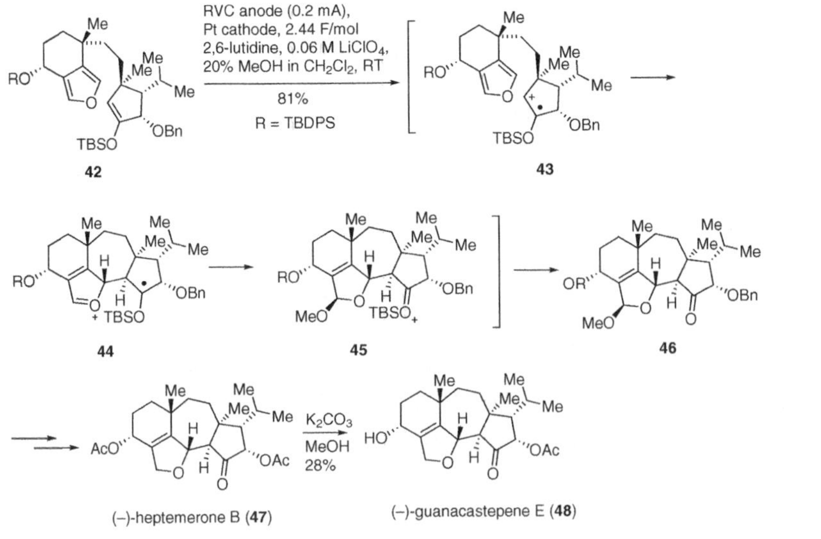

Scheme 11.10 Trauner's total synthesis of (−)-heptemerone B and (−)-guanacastepene E.

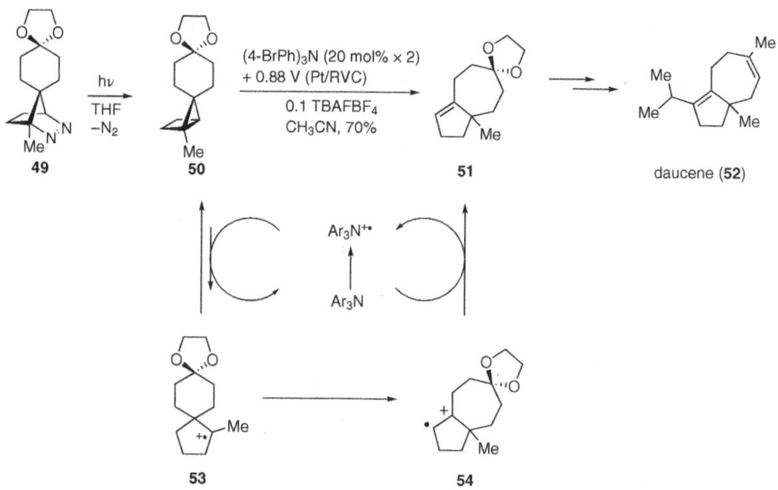

Scheme 11.11 Little's total synthesis of daucene.

Little and coworkers used an electrochemically mediated electron transfer to accomplish the key step of a total synthesis of a sesquiterpene daucene (**52**) (Scheme 11.11).[14] In the key step, the transformation of housane **50**, which was readily available using a photochemically induced N_2 extrusion from diazene precursor **49**, was converted to the bicyclic compound **51**. The rearrangement was achieved at a 70% yield in an undivided cell using tris(*p*-bromophenyl)amine as a mediator. The process was carried out at a potential of 0.88 V versus Ag/0.01 M $AgNO_3$ in order to oxidize the mediator rather than housane **50**. The sequence began with oxidation of tris(*p*-bromophenyl) amine at the anode. The resulting aminium cation radical, in turn, oxidized the housane **50** to the ring-opened cation radical **53**. The subsequent irreversible Wagner–Meerwein rearrangement of **53** generated the seven-membered ring **54**. The following step of **54** is reduced by tris(*p*-bromophenyl)amine to form the [5.3.0] adduct **51**.

Interestingly, during Yao's synthesis of azonazine, they found that regular chemical oxidation and electrochemical oxidation of the same substrate gave different products.[15] As shown in Scheme 11.12, treatment of cyclic dipeptide **55** with oxidant (diacetoxyiodo)benzene (PIDA) in 2,2,2-trifluoroethanol (TFE) led to oxidative C-C/C-O couplings to give two diastereomers **57a** and **57b** via benzofuranoindolines **56a** and **56b**, which could be further converted to proposed azonazine (**58a**) and enantiomer of revised azonazine (**59a**), respectively. When the dipeptide **55** was subjected to electrooxidation conditions, the carbocation generated by anode oxidation was intramolecularly captured by the indole nitrogen atom, instead of its 3'-position of the indole moiety. The 12-membered cyclic compound **59**

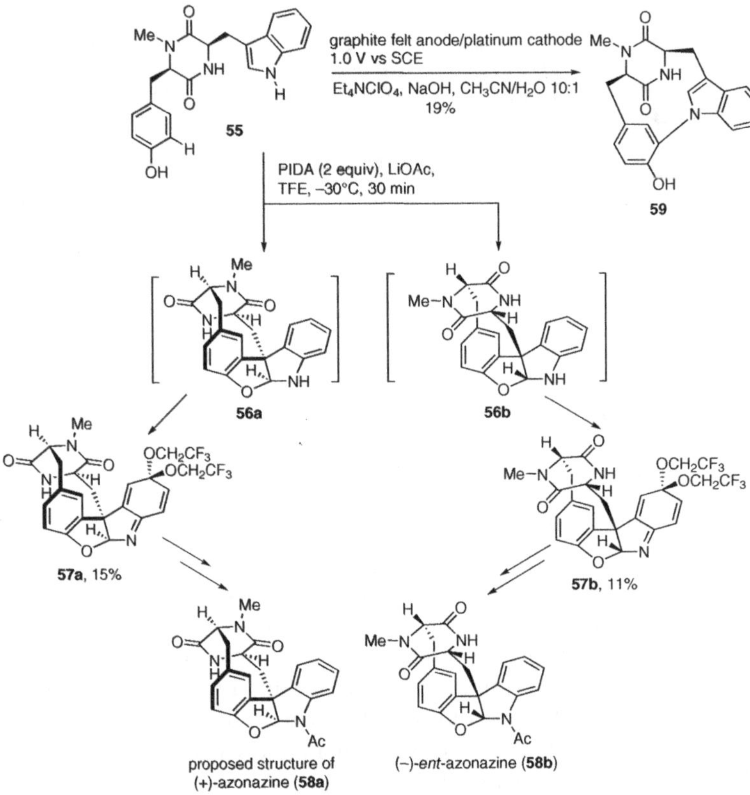

Scheme 11.12 Yao's total synthesis of *ent*-azonazine.

was provided. This example shows that electrochemical method can be complementary to regular chemical methods.

One of the newest examples is Baran's total synthesis of dixiamucin B (Scheme 11.13).[16] They planned to synthesize N – N-linked dimeric indole alkaloid dixiamucin B through late-stage N – N bond formation from xiamucin A. Early optimization attempts focused on the dimerization reaction of the readily available parent carbazole **60** using chemical oxidants. An exhaustive survey of chemical oxidants generally resulted in little to no substrate conversion. Finally, they turned to electrochemical oxidation to solve this problem. When carbazole **60** was subjected to electrochemical oxidation in DMF/MeOH – Bu_4NBr at a potential of +1.2 V versus Ag/AgCl for 18 h led to 60% of an isolated yield of dimer **61**. These conditions could be applied to dimerization of xiamucin A (**64**). Thus, anodic oxidation of **64** at a slight decreased potential (+1.15 V vs Ag/AgCl) for 4 h afforded dixiamucin B (**65**) at a 28% isolated yield, along with 13% recovered **64**; furthermore, bromoxiamycin (**66**) was isolated at a 17% yield.

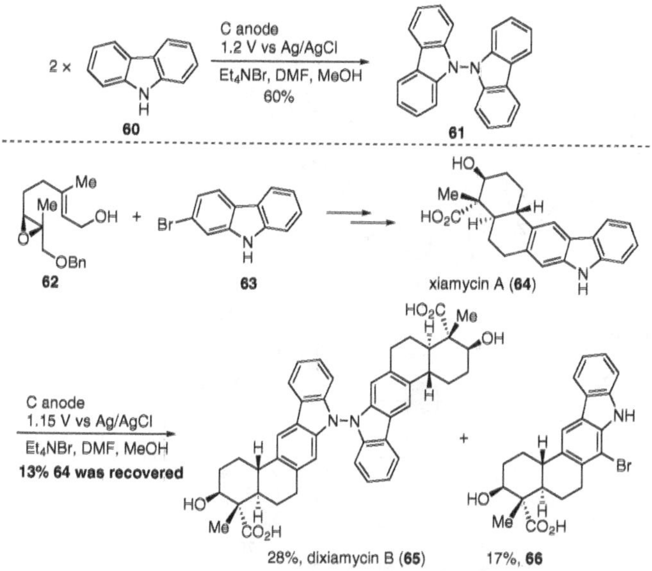

Scheme 11.13 Baran's total synthesis of dixiamucin B.

11.3 Flow Chemistry

Flow chemistry represents an important opportunity for contributing to the identification of an efficient and modern strategy to develop the necessary synthetic tools for greening chemical production.[17] This technology carries various highly interesting features, such as minimization of reaction time, optimization for time screening, waste minimization, safety improvement, process intensification and easy scale up, energy, and cost efficiency. Although the use of flow chemistry technologies for natural product synthesis is still in its early stage, benefits and opportunities can already be recognized. A comprehensive review contributed by Ley and coworkers summarized the recent progress in this subject up to 2013.[18] Flow reactor configurations and components can also be found in this review.

One of the most remarkable examples, which must be mentioned, is Oishi's total synthesis of the ladder-shaped polyether yessotoxin (**67**).[19] In work directed to the synthesis of this challenging target, Oishi and coworkers reported the successful application of reductive etherification of **68** to close K ring of yessotoxin (Scheme 11.14). They found that utilization of a micro-flow reactor was found to be very effective for this key step. When this crucial reaction was scaled up in batch mode, the yield of desired product **69** dramatically dropped to 19% due to the formation of the hydroxyketone by-product **70** as a pair of diastereomers at a 73% yield. When this reaction was

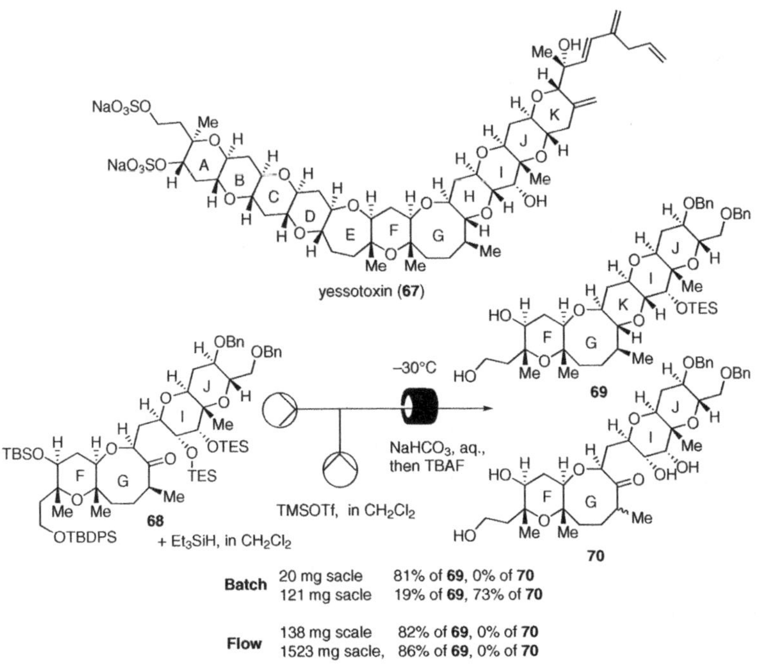

Scheme 11.14 Oishi's reductive etherification under continuous flow conditions.

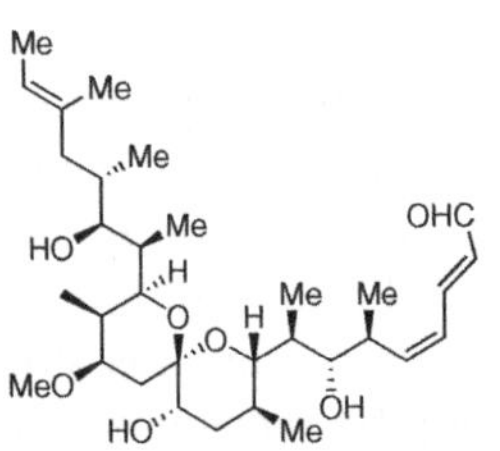

spirodienal A (71)

Scheme 11.15 The structure of spirodienal A.

subjected to continuous flow conditions, the isolated yield of **69** was more than 80% up to a 1.5-gram-scale. The key reason for this success is that precise control of the temperature and reaction time is possible in flow mode, which is difficult in batch mode.

Readers can refer to the previously mentioned review for more examples. Another remarkable example, which has not be covered in the review, is Ley's synthesis of the polyketide natural product spirodienal A (**71**) (Scheme 11.15).[20] In this synthesis, around 70% of the total synthesis steps are in flow chemistry mode. This synthesis therefore represents the current state-of-the-art of complex natural product construction using flow chemistry devices.

The synthesis began with 2,3-butane diacetal protected aldehyde **72**. All the steps toward the key intermediate homoallylic alcohol **76**, including the Wittig reaction, hydrogenation, protecting group switch, reduction, and crotylation, were in flow mode (Scheme 11.16).[20]

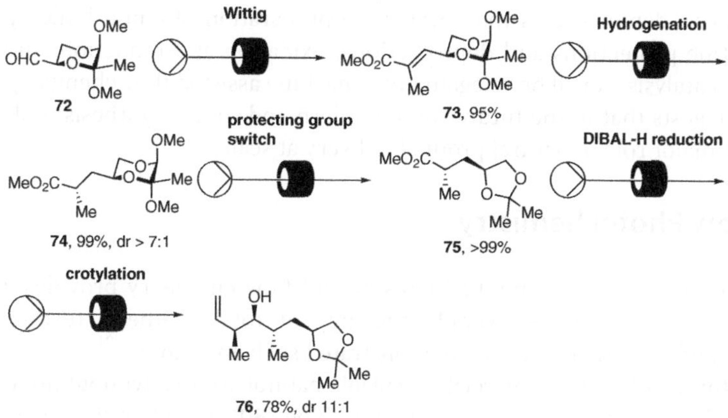

Scheme 11.16 Synthesis of homoallylic alcohol **76**.

Homoallylic alcohol **76** was used to prepare two advanced fragments, aldehyde **77** and bis-alkyne **78** (Scheme 11.17), which contain the same four contiguous stereocenters. Synthesis of advanced aldehyde fragment **77** was completed in 13 steps (9 in flow and 4 in batch) and a 11.6% overall yield from homoallylic alcohol **76**. The second advanced fragment, *bis*-alkyne **78**, was prepared in eight steps (7 in flow and 1 in batch) and at a 22% overall yield, also from homoallylic alcohol **78**. Aldehyde **77** and *bis*-alkyne **78** are among the most advanced substrates prepared in flow to date. Coupling of aldehyde **77** and bis-alkyne **78** and the subsequent steps were carried out in batch, which led to the completion of spirodienal A.[20]

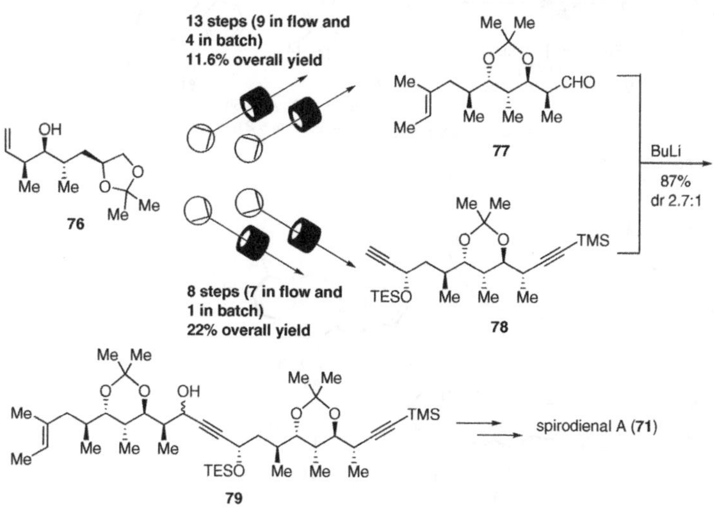

Scheme 11.17 Ley's total synthesis of spirodienal.

The synthesis demonstrates that large scale preparation of chiral building blocks, routine protections and deprotections, oxidation and reduction, and asymmetric catalysis can all be relegated to a machine assisted flow chemistry. It further suggests that in the future these modern and robust synthesis tools could play a major role in natural product delivery at scale.

11.4 Flow Photochemistry

Combination of visible-light photochemistry and flow chemistry provides a big opportunity to improve overall efficiency of total synthesis further. Stephenson and coworkers recently demonstrated such an example.[21]

Catharanthine (**80**) is a commercially available natural alkaloid without notable biologically activity. However, (+)-catharanthine (**80**) was identified as an ideal entry point for the synthesis of a number of structurally related and biologically important alkaloids, such as vinblastine alkaloids, through a common α-aminonitrile intermediate (**81**). Thus, a visible-light-mediated transformation from **80** to **81** has been investigated by Stephenson and coworkers. Visible-light irradiation of 80 in the presence of photocatalyst $Ir[dF(CF_3)ppy]_2(dtbbpy)PF_6$ and 2 equiv. of trimethylsilyl cyanide (TMSCN) provided the cyanated fragmentation product **81** at a 93% yield after 3h in batch mode. This transformation could be subjected to flow mode, which could give 96% yield in 2 min (Scheme 11.18). A possible mechanism has been proposed. The fragmentation

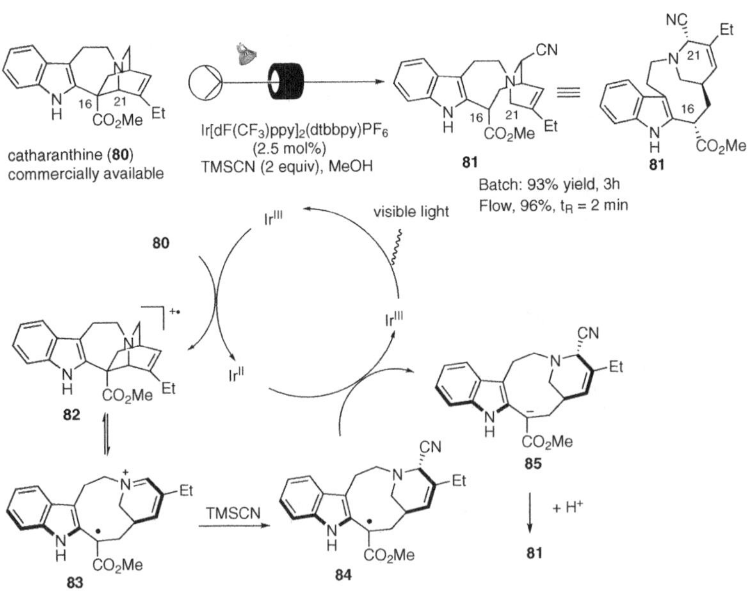

Scheme 11.18 Synthesis of bridged-ring system **81**.

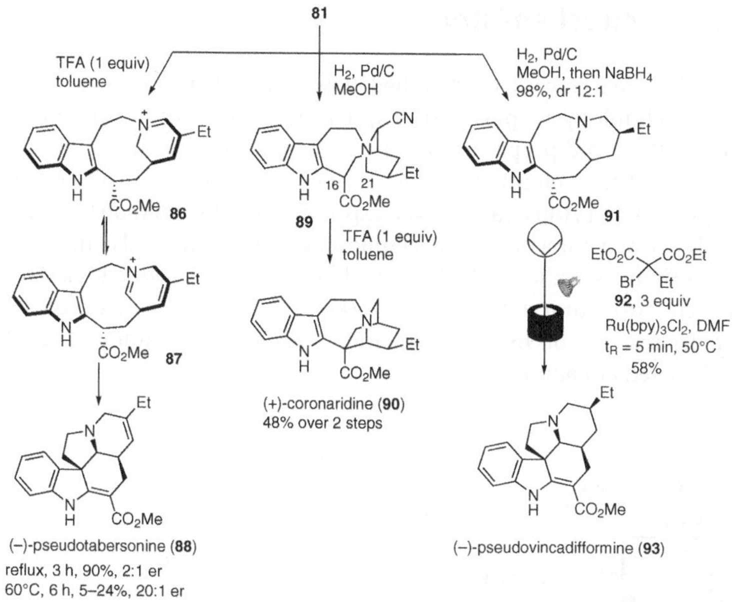

Scheme 11.19 Stephenson's synthesis of vinblastine alkaloids **88**, **90**, and **93**.

begins with oxidation of **80** by the excited state of photocatalyst. The resultant radical cation **82** undergoes fragmentation to provide the ring-opened radical cation **83**. Radical cation **83** is then trapped by TMSCN. Reduction and followed by protonation of **83** then affords the cyanated fragmentation product **81** and regenerates the Ir(III) species.

With abundant nitrile **81** in hands, its ability to undergo a unique fragmentation of its C16 – C21 bond, which has chiefly been exploited in the synthesis of vinblastine alkaloids, was studied (Scheme 11.19). Treatment of **81** with TFA in hot toluene gave pseudotabersonine (**88**). It was found that temperature has a huge effect on the yield and selectivity. When the reaction mixture was fluxed in toluene, a 90% yield was obtained but with disappointed enantioselectivity (2:1 *er*). The *er* value could be improved to 20:1 by performing the reaction at 60 °C but with inconsistent yields ranging from 5 to 24%. Hydrogenation of **81** was performed to yield **89** with high diastereoselectivity, which was subjected to the aminonitrile rearrangement conditions yielded the natural product (+)-coronaridine (**90**) as the sole product at a 48% yield over two steps. Hydrogenation of aminonitrile **81** with Pd/C followed by workup with sodium borohydride provided the tertiary amine **91** at a 98% yield with a 12:1 diastereomeric ratio in favor of the desired β-epimer. Exposure of the resultant amine to oxidative photoredox conditions in flow led to the formation of (−)-pseudovincadifformine (**93**) at a 58% yield using diethyl 2-bromo-2-methylmalonate (**92**) as the terminal oxidant.[21]

11.5 Flow Electrochemistry

Very recently, Ley and coworkers identified flow electrochemistry as a suitable synthetic technology to permit the quick generation of a compound library through the rapid preparation of protected cyclic α-methoxyamines (Scheme 11.20). These intermediates were then applied in a subsequent Pictet–Spengler reaction, thus leading to an expedient two-step method to access the biologically active natural product nazlinine and related unnatural congeners. The first step is Shono oxidation of *N*-Boc pyrrolidine 94 in a microfluidic electrolytic cell to give α-methoxylation product **95**. Then, a microwave-assisted Pictet–Spengler reaction of **95** with tryptamine (**96**) gave the natural product nazlinine and related unnatural congeners **97**.[22]

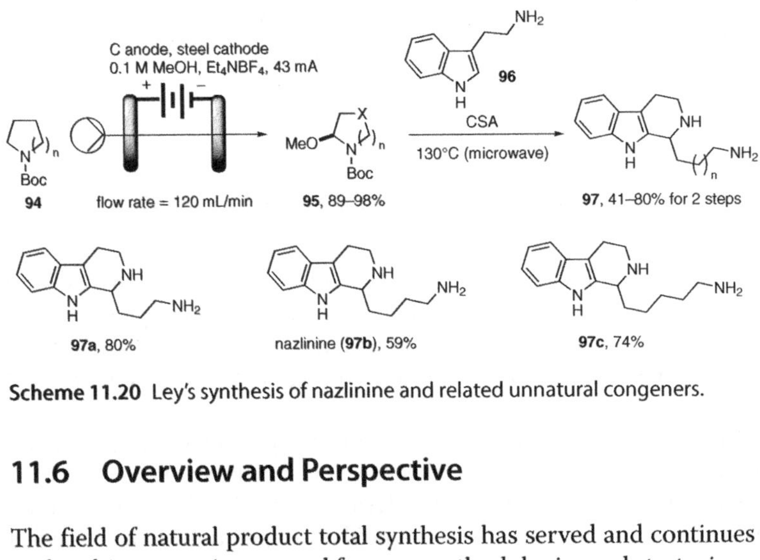

Scheme 11.20 Ley's synthesis of nazlinine and related unnatural congeners.

11.6 Overview and Perspective

The field of natural product total synthesis has served and continues to serve as the ultimate testing ground for new methodologies and strategies, as well as creation of new useful unnatural chemical structures, and the art and science of total synthesis targeting complex natural products has defined the frontiers of organic chemistry. For decades, synthetic chemists have been pursuing ideal total synthesis and have paid great attention to efficiency and selectivities of multi-step synthesis. Given that we are facing problems in supply of energy and resources, economy and sustainability of a route for a molecule must be considered nowadays. The synthetic technologies highlighted in this chapter bear some interesting features, such as minimization of reaction time, optimization for time screening, waste minimization, safety improvement, process intensification and easy scale up, energy, and cost efficiency, which complies with the criteria of green and economic chemistry. Consequently, in the areas of visible-light photochemistry, synthetic electrochemistry, and flow chemistry, there is still a significant demand for further investigations. The encouraging results introduced in this chapter underline the potential of these untraditional technologies in the total synthesis of natural products and pave the way for the greater advances of modern methodology in organic synthesis.

Acknowledgements

The authors thank National Natural Science Foundation (No. 21532002, 21778031) and Double First-class University Construction Plan of Nanjing University for the financial support.

References

1 For selected early examples on visible-light photoredox catalysis in organic synthesis, see: (a) Pac, C.; Ihama, M.; Yasuda, M.; Miyauchi, Y.; Sakurai, H. *J. Am. Chem. Soc.* **1981**, *103*, 6495. (b) Ishitani, O.; Pac, C.; Sakurai, H. *J. Org. Chem.* **1983**, *48*, 2941. (c) Hironaka, K.; Fukuzumi, S.; Tanaka, T. J. *Chem. Soc., Perkin Trans. 2* **1984**, 1705. (d) Kern, J.-M.; Sauvage, J.-P. *J. Chem. Soc., Chem. Commun.* **1987**, *546*.

2 Selected reviews on visible-light photoredox catalysis: (a) Prier, C. K.; Rankic, D. A.; MacMillan, D. W. C. *Chem. Rev.* **2013**, *113*, 5322. (b) Xuan, J.; Lu, L.-Q.; Chen, J.-R.; Xiao W.-J., *Eur. J. Org. Chem.* **2013**, *2071*. (c) Xi, Y.; Yi, H.; Lei, A. *Org. Biomol. Chem.* **2013**, *11*, 2387. (d) Shi, L.; Xia, W.; *Chem. Soc. Rev.* **2012**, *41*, 7687. (e) Xuan, J.; Xiao, W.-J. *Angew. Chem., Int. Ed.* **2012**, *51*, 6828. (f) Narayanam, J. M. R.; Stephenson, C. R. J. *Chem. Soc. Rev.* **2011**, *40*, 102. (g) Yoon, T. P.; Ischay, M. A.; Du, J. *Nature Chem.* **2010**, *2*, 527. (h) Zeitler, K. *Angew. Chem. Int. Ed.* **2009**, *48*, 9785.

3 (a) Nicewicz, D. A.; MacMillan, D.W. C. *Science* **2008**, *322*, 77. (b) Ischay, M. A.; Anzovino, M. E.; Du, J.; Yoon, T. P. *J. Am. Chem. Soc.* **2008**, *130*, 12886. (c) Narayanam, J. M. R.; Tucker, J. W.; Stephenson, C. R. J. *J. Am. Chem. Soc.* **2009**, *131*, 8756.

4 Tucker, J. W.; Stephenson, C. R. J. *J. Org. Chem.* **2012**, *77*, 1617.

5 Furst, L.; Narayanam, J. M. R.; Stephenson, C. R. *J. Angew. Chem. Int. Ed.* **2011**, *50*, 9655.

6 (a) Sun, Y.; Li, R.; Zhang, W.; Li, A. *Angew. Chem. Int. Ed.* **2013**, *52*, 9201. (b) Zhou, S.; Zhang, D.; Sun, Y.; Li, R.; Zhang, W.; Li, A. *Adv. Synth. Catal.* **2014**, *356*, 2867.

7 Schnermann, M. J.; Overman, L. E. *Angew. Chem. Int. Ed.* **2012**, *51*, 9576.

8 Lu, Z.; Yoon, T. P. *Angew. Chem. Int. Ed.* **2012**, *51*, 10329.

9 Han, J.; Li, X.; Guan, Y.; Zhao, W.; Wulff, W. D.; Lei, X. *Angew. Chem. Int. Ed.* **2014**, *53*, 9257.

10 (a) Yin, J.; Kong, L.; Wang, C.; Shi, Y.; Cai, S.; Gao, S. *Chem. Eur. J.* **2013**, *19*, 13040. (b) Yin, J.; Wang, C.; Kong, L.; Cai, S.; Gao, S. *Angew. Chem. Int. Ed.* **2012**, *51*, 7786.

11 (a) Francke, R.; Little, R. D. *Chem. Soc. Rev.* **2014**, *43*, 2492. (b) Frontana-Uribe, B. S.; Little, R. D.; Ibanez, J. G.; Palma, A.; Vasquez-Medrano, R. *Green Chem.* **2010**, *12*, 2099. (c) Sperry, J. B.;. Wright, *D. L Chem. Soc. Rev.* **2006**, *35*, 605. (d) Yoshida, J.; Kataoka, K.; Horcajada, R.; Nagaki, A.; *Chem. Rev.* **2009**, *108*, 2265. (e) Moeller, K. D. *Tetrahedron* **2000**, *56*, 9527. (f) Moeller, K. D. *Synlett* **2009**, *8*, 1208. (g) Utley, J. *Chem. Soc. Rev.* **1997**, *26*, 157.

12 (a) Mihelcic, J.; Moeller, K. D. *J. Am. Chem. Soc.* **2003**, *125*, 36. (b) Mihelcic, J.; Moeller, K. D. *J. Am. Chem. Soc.* **2004**, *126*, 9106.

13 (a) Hughes, C. C.; Miller, A. K.; Trauner, D. *Org. Lett.* **2005**, *7*, 3425. (b) Miller, A. K.; Hughes, C. C.; Kennedy-Smith, J. J.; Gradl, S. N.; Trauner, D. *J. Am. Chem. Soc.* **2006**, *128*, 17057.

14 Park, Y. S.; Little, R. D. *J. Org. Chem.* **2008**, *73*, 6807.

15 (a) Zhao, J.-C.; Yu, S.-M.; Qiu, H.-B.; Yao, Z.-J. *Tetrahedron* **2014**, *70*, 3197. (b) Zhao, J.-C.; Yu, S.-M.; Yao, Z.-J. *Org. Lett.* **2013**, *15*, 4300.

16 Rosen, B. R.; Werner, E.W.; Brien, A. G. O.; Baran, P. S. *J. Am. Chem. Soc.* **2014**, *136*, 5571.

17 (a) Mason, B. P.; Price, K. E.; Steinbacher, J. L.; Bogdan, A. R.; Mcquade, D. T. *Chem. Rev.* **2007**, *107*, 2300. (b) Webb, D.; Jamison, T. F. *Chem. Sci.* **2010**, *1*, 675. (c) Hartman, R. L.; McMullen, J. P.; Jensen, K. F. *Angew. Chem. Int. Ed.* **2011**, *50*, 7502. (d) Ley, S. V. *Chem. Rec.* **2012**, *12*, 378. (e) Wegner, J.; Ceylan, S.; Kirschning, A. *Adv. Synth. Catal.* **2012**, *354*, 17. (f) Zhao, D.; Ding, K. *ACS Catal.* **2013**, *3*, 928. (g) Tsubogo, T.; Ishiwata, T.; Kobayashi, S.; *Angew. Chem. Int. Ed.* **2013**, *52*, 6590. (h) Elvira, K. S.; Solvas, X. C.; Wootton, R. C. R.; de Mello, A. J. *Nat. Chem.* **2013**, *5*, 905. (i) Vaccaro, L.; Lanari, D.; Marrocchia, A.; Strappaveccia, G. *Green Chem.* **2014**, *16*, 3680.

18 Pastre, J. C.; Browne, D. L.; Ley, S. V. *Chem. Soc. Rev.* **2013**, *42*, 8801.

19 Oishi, T. *J. Synth. Org. Chem. Jpn.* **2012**, *70*, 1170.

20 Newton, S.; Carter, C. F.; Pearson, C. M.; de C. Alves, L.; Lange, H.; Thansandote, P.; Ley, S. V. *Angew. Chem. Int. Ed.* **2014**, *53*, 4915.

21 Beatty, J. W.; Stephenson, C. R. J. *J. Am. Chem. Soc.* **2014**, *136*, 10270.

22 Kabeshov, M. A.; Musio, B.; Murray, P. R. D.; Browne, D. L.; Ley, S. V. *Org. Lett.* **2014**, *16*, 4618.

12

Concluding Remarks and Perspectives

Pei-Qiang Huang[1], *Richard P. Hsung*[2], *Zhi-Xiong Ma*[2], *and Zhu-Jun Yao*[3]

[1] *Fujian Provincial Key Laboratory of Chemical Biology, Department of Chemistry and iChEM, Xiamen University, Xiamen, China*
[2] *Division of Pharmaceutical Sciences, School of Pharmacy, University of Wisconsin-Madison, Madison, WI, USA*
[3] *School of Chemistry and Chemical Engineering, Nanjing University, Nanjing, China*

The development of the field of natural product synthesis relies on support from both funding agencies and related industries, particularly the pharmaceutical industry. The development of drugs from natural products is undoubtedly a biomimetic approach in which total synthesis can play an important role. The total number of living species was estimated to range from 2 to 100 million.[1] According to Gordon M. Cragg and David J. Newman,[2] only 5–15 % of the approximately 250,000 species of higher plants have been systematically investigated. Hawksworth and Rossman have estimated that there may be as many as 1 million different fungal species. However, only about 100,000 of them have been described.[3,4] Marine natural products are a rich source of drug candidates. It has been estimated that in marine environments there may be 3.7×10^{30} microorganisms and more than 99% of all bacteria that cannot be cultured.[5] Hence, the opportunity for finding drug leads from natural sources remains immense. However, for total synthesis, even for very challenging natural products such as maitotoxin (**1**) that has a molecular weight of 3422 and contains 142 carbon atoms, 32 cyclic ether unities, and 28 hydroxyl groups (Figure 12.1), it is hard to obtain financial support.[6] Without further innovation, the field of natural product synthesis will find it difficult to play important roles in drug discovery and development. In addition to the problem of synthetic efficiency, which has been the focus of this book, the following aspects should be addressed in the future.

Efficiency in Natural Product Total Synthesis, First Edition. Edited by Pei-Qiang Huang, Zhu-Jun Yao, and Richard P. Hsung.

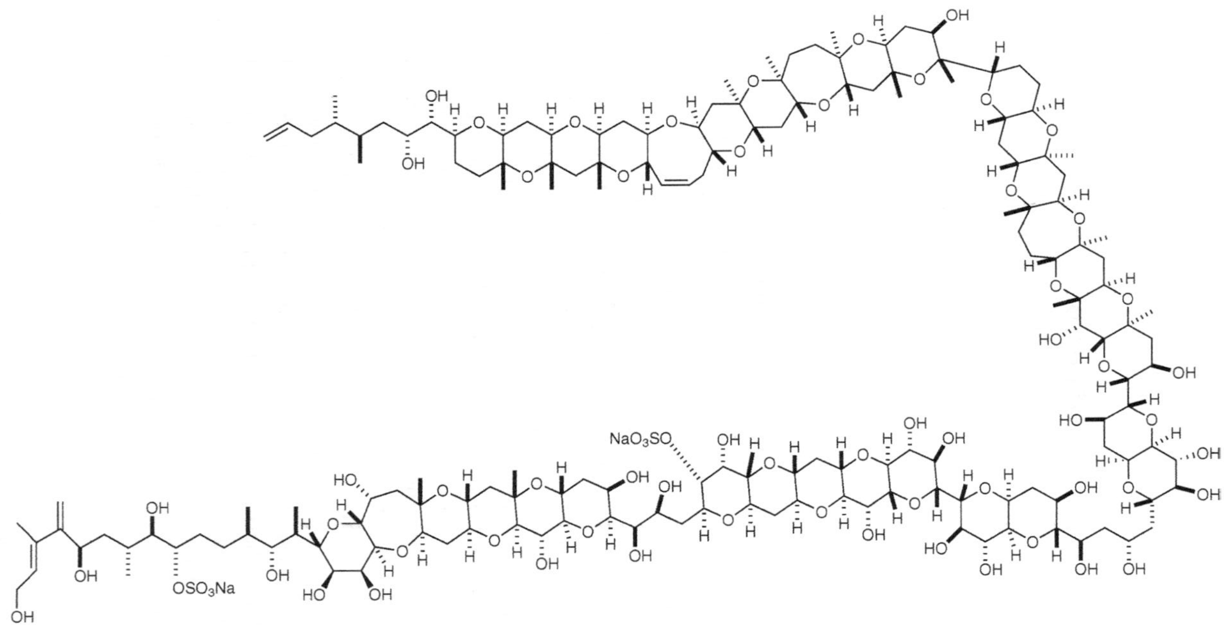

Figure 12.1 Structure of maitotoxin (**1**).

12.1 The Enantioselective Total Synthesis of Natural Products

In their review of 2012 entitled: "Is There No End to the Total Syntheses of Strychnine? Lessons Learned in Strategy and Tactics in Total Synthesis," Overman and coauthors selected 10 representative approaches to strychnine.[7] Among those accomplished during 2000–2011, only two are enantioselective total syntheses, suggesting that this remains a challenge in current organic synthesis. Moreover, most of the total syntheses of natural products reported in the last decade are racemic syntheses that lead to two enantiomers in equal amounts. However, most natural products exist in nature in only one enantiomer. Since different enantiomers display different, or even opposite bioactivities, enantioselective total synthesis should become an important criterion in the future and be pursued.

12.2 A Novel Model of Total Synthesis: The Combination of Chemical Synthesis with Synthetic Biology

Although the story about the re-discovery of Qinghaosu (artemisinin, **2**)[8a] and its industrial production through semisynthesis by Sanofi have been discussed in the introduction and in Chapter 10, the sustainable supply of Qinghaosu remains a challenge.[8] The continuing multidisciplinary endeavor to forge an environmentally benign, cost-effective, industrial production of Qinghaosu may also serve as a perspective for the field of natural product total synthesis.

Synthetic biology is a field with the goal to build novel biological systems.[9] In this emerging field, after years of success in replacing chemical synthesis by biology,[10] it has been recognized that synthetic biology should become a new incentive for interdisciplinarity, and "without the possibility of synthesis on the molecular level, efforts in reconstructing or engineering life are certainly bound to fail."[11] Thus "a cross-fertilization of the synthetic disciplines seems to be highly desirable."[11]

The semi-synthetic Artemisinin Project can serve as a model for the use of synthetic biology in the development of pharmaceutical agents.[12] The project was first funded in 2004 by the Bill and Melinda Gates Foundation, and later became a partnership with the University of California (Berkeley, USA), Amyris Inc. and the Institute for OneWorld Health (now known as PATH Drug Solutions).[12a] The breakthrough has been achieved by the collaborative work of the research groups directed by C. J. Paddon, D. McPhee, and J. D. Newman.[12c] By means of synthetic biology, strains of *Saccharomyces cerevisiae* (baker's yeast) were coaxed genetically to express *Artemisia annua* genes, which allowed the production in high levels of amorpha-4,11-diene (AD) and artemisinic acid (AA, 25 g/L), the precursors of artemisinin.[12b] Artemisinic acid was then converted to artemisinin by previously published chemical syntheses with modifications.[12c] Although synthetic biology-based biosynthesis is highly efficient in producing both AD and AA, the environmental and economic costs of this semisynthetic approach are still high because of the chemical transformations. On the other hand, dihydroartemisinic acid (DHAA) and AA are found in the waste from plant extraction. Thus, several synthetic organic research groups have pursued the development of environmental and economically effective chemical transformation of DHAA and AA.[13–17,17a]

12.2.1 Seeberger's One-Pot Photochemical Continuous-Flow Strategy

Conversion of AA and DHAA to artemisinin requires the use of photochemically generated singlet oxygen ($^1O_2$) on a large scale. Peter H. Seeberger and coworkers tackled this problem by developing a one-pot photochemical continuous-flow process, which allowed the conversion of DHAA to artemisinin at a 39% yield.[13a,b] More recently, they developed a divergent, multistep continuous synthesis and purification process that was capable of producing four anti-malarial artemisinins (APIs).[13c]

12.2.2 Wu's "Dark Singlet Oxygen" Strategy

In 2013, Yikang Wu's group developed a very simple and efficient method for the conversion of DHAA into artemisinin (**2**, at a 41% overall yield) (Scheme 12.1).[14] Their method features the use of the Na_2MoO_4/H_2O_2 to replace the almost unanimously employed photosensitized oxygenation to generate singlet oxygen. The synthesis was completed by stirring at ambient temperature without need for any special equipment.

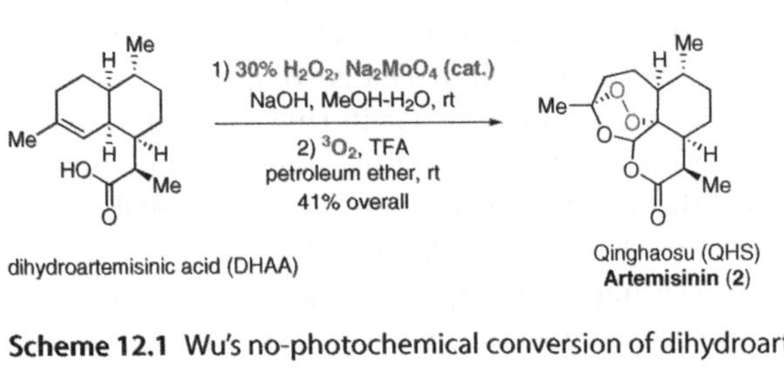

Scheme 12.1 Wu's no-photochemical conversion of dihydroartemisinic acid to artemisinin.

12.2.3 George's "Green" Photochemical Strategies

More recently, Michael W. George addressed the problems in current photochemical processes by applying the principles of green chemistry. His team developed two strategies.[15a] One used liquid CO_2 as the solvent and a recyclable heterogeneous photosensitizer/acid system consisting of porphyrin photosensitizer bound to amberlyst-15. The other one involved the photocatalytic oxygenation of DHAA in a mixed solvent of ethanol and water. The reaction was conducted at room temperature with dihydroartemisinic acid, O_2, and light to provide pure, crystalline artemisinin. The solvents, photocatalyst, and aqueous acid are recyclable. More recently, the same group reported the construction of a vortex reactor that used a rapidly rotating cylinder to generate Taylor vortices for continuous-flow thermal and photochemical reactions. This technique allowed the three-step one-pot transformation of dihydroartemisinic into artemisinin at a 50% yield.[15b]

12.2.4 A Novel Strategy Merging Synthetic Biology with Chemistry

Because none of the known methods has reduced the cost in artemisinin production, scientists of Ipca Laboratories Limited and Amyris Inc. recently developed a novel synthesis of artemisinin by merging synthetic biology with chemistry (Scheme 12.2).[16] This approach features the use of amorpha-4,11-diene (AD), an early-stage intermediate produced in two-fold higher molar

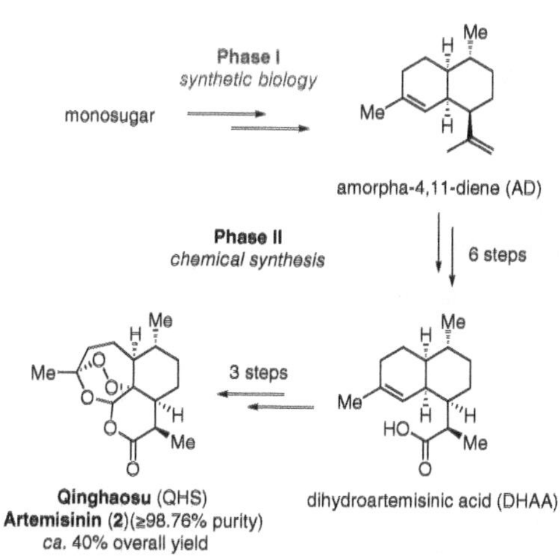

Scheme 12.2 Ipca Laboratories Limited and Amyris Inc.'s strategy merging synthetic biology with chemistry.

concentrations than AA during the production of AA by engineered yeast.[12] AD was next transformed via a six-step procedure to give DHAA, which was converted in three steps into artemisinin. Although this route relies on a lengthy nine-step chemical transformation and involves the use of toxic dimethyl sulfate, the authors claimed that only cheap and environmentally friendly reagents and conditions have been used. Moreover, the authors are optimistic that their procedure can bring the cost of artemisinin to less than US $100/kg.[16]

12.2.5 Zhang's Two-Step Catalytic Transformation of AA to Artemisinin: The End-Game?

In 2012, Wanbin Zhang of Shanghai Jiaotong University, patented a catalytic two-step method for the synthesis of artemisinin.[17a] Asymmetric hydrogenation of AA using (R,R_p)-RuPHOX – Ru as a chiral catalyst with 5 bar of H_2 yielded (R)-DHAA at an excellent yield and 99.7:0.3 d.r.[17b] The latter was converted into artemisinin via a peroxidation process, which avoids the disadvantages associated with the photochemical transformation (Scheme 12.3). This process is high yielding (>60%) and simple to operate. "The optimized procedure has been applied to a 10-kg production scale with a competitive price and shows potential for use in industry application prospects."[17c]

Although the synthesis of a natural product starting from an advanced intermediate obtained from fermentation has traditionally been defined as the semisynthesis,[18] the **synthetic biology–chemical synthesis** combination,

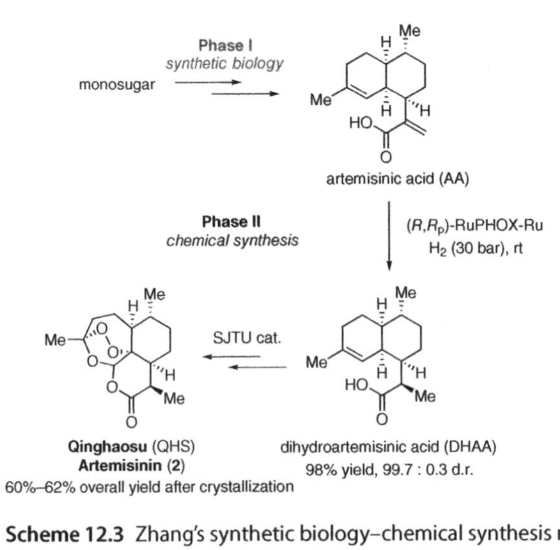

Scheme 12.3 Zhang's synthetic biology–chemical synthesis relay-total synthesis of QHS.

referred here as **relay-total synthesis**, may represent a future direction for the total synthesis of structurally complex natural products.

12.3 The Robot Chemist and the Generalized Automation of Small-Molecule Synthesis

In addition to the synthetic biology–chemical synthesis relay-total synthesis, as a long-term perspective, with the substantial development of flow chemistry syntheses of natural products[19] and computational chemistry,[20] the engagement of the robot chemist[21] in the field of the total synthesis of natural products could be expected.

In an attempt to develop generalized automation of many different types of small-molecule synthesis, Martin D. Burke from the University of Illinois at Urbana-Champaign reported the synthesis of 14 distinct classes of small molecules using the same fully automated process.[22] This approach relies on a common building block-based strategy and a common catch-and-release chromatographic purification protocol. As an illustration of this platform, the synthesis of structurally complex secodaphnane core (**3**) is outlined in Scheme 12.4. Because thousands of building blocks are commercially available, it is expected that numerous small molecules could be accessible through this platform. Moreover, this concept could inspire a more general and automated approach for the synthesis of natural products and natural product-like small molecules.

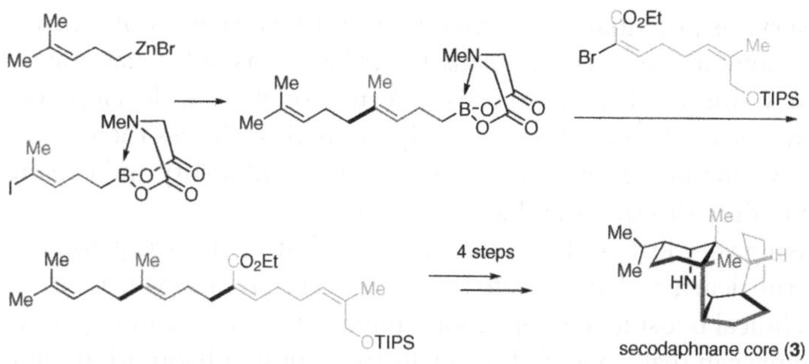

Scheme 12.4 Burke's strategy for the synthesis of 14 distinct classes of small molecules using the same fully automated process: the synthesis of secodaphnane core as an example.

12.4 A Synergistic Future with Academia and Industry Coming to the Same Table

The advancement of chemical, biological, and clinical sciences has allowed pharmaceutical industries to evolve into a prolific machinery that has led to solutions if not cures to countless diseases at an unprecedented rate in the past 50 years. However, as new technologies continue to emerge, and as we continue to benefit from progress made in drug discovery and development, the pharmaceutical industry has gravitated toward a model of business first and science at a distant second because the ongoing rate has been deemed not fast enough. Industries have become obsessed with crunching numbers at all cost and shuttling new drugs from pipelines to the market at an increasingly greater rate. Indeed, we may be faster and better than at the turn of twentieth century, but we remain at a snail's pace in drug discovery, relatively speaking. For instance, cancer is nowhere near being "cured" after the "War on Cancer" was declared in 1955 by the National Cancer Institute (NCI) in the USA.

Although greed is never a good thing to run a successful business, pharmaceutical industries should be allowed to make sufficient profit and garner ample revenue to keep this important machinery churning. Nevertheless, the engine for pharmaceutical industries is science and not number crunching for profit. Perhaps, unlike any other businesses, the business of drug discovery is extremely unique. Without real investment in sound and innovative science, how pharmaceutical industries could keep up this unsustainable model is anyone's guess.

While we are confident that pharmaceutical industries will resolve their own plight, it is the opinion here and of many others[23] that the silver lining of the current predicament can (if not already) bring invaluable opportunities to academia. Hundreds and thousands of new chemical structures have been generated from academic research in the last 30 years alone. These *de novo* chemical entities could serve as potential lead compounds in drug discovery. Academia can use these potential leads as platforms to strike an instant collaboration with the industry, and avail their expertise on methodological development and chemical synthesis. These efforts can help industries address their issues or woes in discovering new chemical spaces, which should fortify and expand the foundation of drug discovery and development.

In return, having very little to no experience in real drug discovery process, academia can truly learn and appreciate the art of real translations from basic science to the clinical bedside. Perhaps more importantly, and definitely more honestly, academia can really benefit financially from these collaborations; and we are not referring to benefits for personal pockets of researchers involved. We are referring to valuable dollars for conducting more experiments. Although financial support from industries has been limited on specific projects, if not also occasionally reserved for select individuals for the past five decades, a few dollars could go a long way in sustaining academic research, especially when the framework of a collaborative project is within the research interest of a given academic laboratory. This is especially true in the current toxic climate in federal funding on a global scale with the US leading the headline. Academic institutions around the world are desperately in need of support – the kind of support that will be instrumental for continuing their missions in training the next generation of scientists, who by the way will most likely be working in industry.

This pipeline of human resource cannot run dry – period! Or it will be the end of drug discovery and, worse, healthcare altogether.

Such a synergistic effort, if successful, could be rewarding to both academia and industry. One recent success is from the Baran laboratory at The Scripps Research Institute. They made fruitful progress through collaborations with an array of pharmaceutical companies that featured research interests in radical chemistry and natural product total synthesis.[23b] Another more general success story that comes to mind would be Eli Lilly's Open Innovation Drug Discovery Program in recent years. Along the same lines, we might also uphold the venture at Vanderbilt University as a poster child for drug discovery in academia that has been led by folks with a tremendous amount of industrial experience.

Once again, regrettably, these lines of thinking have largely remained as concepts on paper. True successes have been few and far in between. There is no doubt that vast conflicts exist between the two sides that are an ocean apart in ultimate goals: One is set for basic training and the celebrated purpose of education, while the other is striving for needed revenue and practical solutions that ultimately matter to life. We have not even begun to mention minute details in the daily operation of these two drastically polarized institutions: Academia prefers meticulous studies and quick publications, while industry focuses more on quick answers and protecting intellectual properties. There does not exist a standard working model that could serve as a true win-win relationship.

Thus, coming together to the same table to collaborate sounds easy on paper but has and will encounter insurmountable challenges. This then brings us back to natural product chemistry, which is the theme of this book and perhaps also the bias within our opinions.

Natural products have been widely utilized as folk medicines in China, South America, and other places around the world for centuries, and have a longstanding history of playing important roles for the modern pharmaceutical industry from quinine to more recent examples such as artemisinin and Taxol®. In terms of collaborative projects, natural products have been one of the most active and successful topics. More specifically, collaborations between academia and industry based on natural products have led to numerous successful drugs being released onto the market even just in the last few years.

So, what is – or rather who is – standing in the way of such synergistic pursuit and untapped potential? Why did the interest in natural products begin to fade as early as the 1990s after being the staple of drug discovery in the pharmaceutical industry for nearly a century?

It is not fair to blame all things on greed. While natural products have provided an incredibly rich array of structural motifs in contrast to flat and man-made small molecules, as well as inspirations for innovations, drawbacks are significant when considering drug discovery: (1) Typically low overall yields either from isolation processes using natural sources or via total syntheses in a laboratory have been an impediment for further biological studies and assessments; (2) the level of structural complexities in most natural products prevents generations of analogs or small libraries for necessary SAR study; and (3) tuning physical properties of natural products for seemingly trivial but important tasks such as solubility, formulation, and crystallinity to make them "drug-like" has been difficult.

The bottom line is efficiency and efficiency is the quintessential element that drives innovations and has led to nearly all discoveries, and not just drug discovery, since the days when the wheel was invented or somewhere around that time. The lack thereof can and will deter any industries from pursuing any scientific endeavors let alone natural product chemistry. Once again to be honest, as much as we have been striving to develop efficient chemical processes, the current state of natural product synthesis is a far cry from being truly efficient; and we are not even talking about ideal synthesis. On the flip side of this same coin, with so much to strive for and so much on the line, how can anyone in their right frame of mind declare natural product synthesis is dead, and that the discipline of organic chemistry has matured? Nevertheless, to re-engage pharmaceutical industry in natural product chemistry we must also improve our efficiency in their syntheses. We must keep on developing more creative approaches – or to "think outside the box" to borrow from an old cliché.

Fortuitously, prospects are changing these days with the onset of new technologies particularly in biological sciences such as genetic engineering. Biological synthesis, or so-called synthetic biology, has emerged to be a complementary and competitive tool to chemical synthesis.[24] A significant number of natural products can be produced via synthetic biology methods with much greater efficiency compared to their corresponding total synthesis. The interdisciplinary approach through a joint usage of the two methods in a well-designed sequence can prove to be an even more powerful tool for accessing natural products[25a] and their analogs.[25] Consequently, one can hope that such success can advocate more natural product-related projects in industries. Overall, it is very true that accepting and utilization of progress from other disciplines can create new opportunities in chemical synthesis toward a more efficient and productive direction.

Another exciting but somewhat crazy idea would be automated synthesis using robots, which is still at its very early stage with sporadic reports on partial successes. An ideal automated synthesis raises high requirements for both hardware and software. A programmed machine running a sequence of multistep continuous-flow reactions to deliver the desired molecule could liberate chemists from tedious bench work, but a software system designed to offer feasible new synthetic routes to a given molecule can be even more meaningful.[26,27] While conceptually such thinking is no longer new, its actual development is still at an archaic level – and immense challenges are lying ahead. To remain optimistic, if an AI player AlphaGO can learn to beat the best human player in the most complex brain game, the Go game, it could certainly be trained to analyze reaction conditions, substrate suitability, molecular complexities, and mechanistic pathways. Maybe, just maybe, the hope for an ultimate ideal synthesis of a chemical entity, be that a natural or unnatural product, lies within a machine.

Whatever solutions there may ultimately be, we know for sure that natural product synthesis is alive and kicking. However, to remain relevant for decades to come, we must be determined to strive for that next plateau because there is so much to be desired in chemical synthesis. At the same time, we recognize that with overwhelming challenges ahead, we must also learn to adapt and evolve. Just look at physics; it is far from being obsolete or matured, quite on the contrary. A collaborative endeavor between academia and industry that can embrace values and goals from both sides could very well be one solution that would aid such evolutionary process for natural product synthesis; and who knows? It may even benefit mankind.

Acknowledgements

PQH would like to thank Ms. Ling-Ling Li for her assistance in the preparation of this manuscript (in part). RPH would like to thank Laura and Edward Kremers Family Foundation for a generously endowed chair in natural products chemistry and also Vilas Trust and the University of Wisconsin for a Distinguished Achievement Professorship.

References

1 Li, W.-H.; Vederas, J. C. *Science* **2009**, *325*, 161.
2 Cragg, G. M.; Newman, D. J. *Pure Appl. Chem.* **2005**, *77*, 7.
3 Strobel, G.; Daisy, B.; Castillo, U.; Harper, J. *J. Nat. Prod.* **2004**, *67*, 257.
4 Hawksworth, D. C.; Rossman, A. Y. *Phytopathology* **1987**, *87*, 888.
5 Kennedy, J.; Marchesi, J.; Dobson, A. D. W. *Microb. Cell Fact.* **2008**, *7*, 27.
6 (a) Nicolaou, K. C.; Frederick, Michael O. *Angew. Chem. Int. Ed.* **2007**, *46*, 5278; (b) Murata, M.; Yasumoto, T. *Nat. Prod. Rep.* **2000**, *17*, 293; (c) Kishi, Y. *Pure Appl. Chem.* **1998**, *70*, 339; (d) Sasaki, M.; Murata, M. *J. Syn. Org. Chem. Jpn.* **1997**, *55*, 535; (e) Krämer, K. *Chem. World* 15 January, **2015**,

Available online at: www.chemistryworld.com/news/chemistrys-toughest-total-synthesis-challenge-put-on-hold-by-lack-of-funds/8152.article (accessed March 2018).

7 Cannon, J. S.; Overman, L. E. *Angew. Chem. Int. Ed.* **2012**, *51*, 4288.

8 (a) For a special issue dedicated to the 40th anniversary of the publication of the structure and relative stereochemistry of Qinghaosu (artemisinin) including Bill Gates' felicitation, see: *Kexue Tongbao* **2017**, *62 (*18*)*, 1903–2022; (b) Peplow, M. *Nature* **2016**, *530*, 389; (c) Njoroge, M.; Njuguna, N. M.; Mutai, P.; Ongarora, D. S. B.; Smith, P. W.; Chibale, K. *Chem. Rev.* **2014**, *114*, 11138.

9 For a recent review entitled: "Synthetic biology: the synthesis of biology," see: Ausländer, S.; Ausländer, D.; Fussenegger, M. *Angew. Chem. Int. Ed.* **2017**, *56*, 6396.

10 (a) Liu, X.; Ding, W.; Jiang, H. *Microb. Cell Fact.* **2017**, *16*, 125; (b) Breitling, R.; Takano, E. *Cold Spring Harb. Perspect. Biol.* **2016**, *8*, a023994; (c) Awan, A. R.; Shaw, W. M.; Ellis, T. *Adv. Drug Deliv. Rev.* **2016**, *105*, 96; (d) Mitchell, W. *Curr. Opin. Chem. Biol.* **2011**, *15*, 505.

11 Schwille, P. *Angew. Chem. Int. Ed.* **2013**, *52*, 2616.

12 (a) Paddon, C. J.; Keasling, J. D. *Nat. Rev. Microbiol.* **2014**, *15*, 355; (b) Benjamin, K. R.; Silva, I. R.; Cherubim, J. P.; McPhee, D.; Paddon, C. J. *J. Braz. Chem. Soc.* **2016**, *27*, 1339; (c) Ref. 159 in Chapter 10.

13 (a) Lévesque, F.; Seeberger, P. H. *Angew. Chem. Int. Ed.* **2012**, *51*, 1706; (b) Kopetzki, D.; Lévesque, F.; Seeberger, P. H. *Chem. Eur. J.* **2013**, *19*, 5450; (c) Gilmore, K.; Kopetzki, D.; Lee, J. W.; Horváth, Z.; McQuade, D. T.; Seidel-Morgenstern, A.; *Chem. Commun.* **2014**, *50*, 12652.

14 Chen, H.-J.; Han, W.-B.; Hao, H.-D.; Wu, Y. *Tetrahedron* **2013**, *69*, 1112.

15 (a) Amara, Z.; Bellamy, J. F. B.; Horvath, R.; Miller, S. J.; Beeby, A.; Burgard, A.; *et al. Nat. Chem.* **2015**, *7*, 489; (b) Lee, D. S.; Amara, Z.; Clark, C. A.; Xu, Z.; Kakimpa, B.; Morvan, H. P.; *Org. Process Res. Dev.* **2017**, *21*, 1042.

16 Singh, D.; McPhee, D.; Paddon, C. J.; Cherry, J.; Maurya, G.; Mahale, G.; *et al. Org. Process Res. Dev.* **2017**, *21*, 551.

17 (a) Zhang, W.; Liu, D.; Yuan, Q. *Method for preparing artemisinin through arteannuic acid.* CN102718773A, **2012**; (b) Li, J.; Shen, J, Xia C.; Xia, C.; Wang, Y. Z.; Liu, D. L.; *et al. Org. Lett.* **2016**, *18*, 2122. (c) Liu, D. L.; Zhang, W. B. *Kexue Tongbao* **2017**, *62*, 1997.

18 Gober, C. M.; Joullié, M. M. *J. Org. Chem.* **2016**, *81*, 10136.

19 For a recent review on the Flow chemistry syntheses of natural products, see: Pastre, J. C.; Browne, D. L.; Ley, S. V. *Chem. Soc. Rev.* **2013**, *42*, 8849.

20 (a) Skoraczynski, G.; Dittwald, P.; Miasojedow, B.; Szymkuć, S.; Gajewska, E. P. B. A. Grzybowski, B. A.; *et al. Scientific Rep.* **2017**, *7*:3582; (b) Szymkuć, S.; Gajewska, E. P.; Klucznik, T.; Molga, K.; Dittwald, P.; Startek, M.; *et al. Angew. Chem. Int. Ed.* **2016**, *55*, 5904; (c) Cadeddu, A.; Wylie, E. K.; Jurczak, J.; Wampler-Doty, M.; Grzybowski, B. A. *Angew. Chem. Int. Ed.* **2014**, *53*, 8108; (d) Arrieta, A.; de la Torre, M. C.; de Cózar, A.; Sierra, M. A.; Cossío, F. P. *Synlett* **2013**, *535*.

21 (a) Peplow, M. *Nature* **2014**, *512*, 20; (b) Steven V. Ley, S. V.; Fitzpatrick, D. E.; Ingham, R. J.; Myers, R. M. *Angew. Chem. Int. Ed.* **2015**, *54*, 3449.

22 Li, J.; Ballmer, S. G.; Gillis, E. P.; Fujii, S.; Schmidt, M. J.; Palazzolo, A. M. E.; *et al. Science* **2015**, *347*, 1221.

23 (a) Michaudel, Q.; Ishihara, Y.; Baran, P. S. *Acc. Chem. Res.* **2015**, *48*, 712; (b) Hanessian, S. *ACS Med. Chem. Lett.* **2016**, *7*, 6; (c) McGlacken, G. P.; Hayes, S.; Maguire, A. R.; O'Halloran, J.; Hodnett, K. *Chem. Eur. J.* **2018**, *24*, 499.

24 Ausländer, S.; Ausländer, D.; Fussenegger, M. *Angew. Chem. Int. Ed.* **2017**, *56*, 6396.

25 (a) Kirschning, A.; Hahn, F. *Angew. Chem. Int. Ed.* **2012**, *51*, 4012; (b) Maier, M. E. *Org. Biomol. Chem.* **2015**, *13*, 5302; (c) Bebbington, M. W. P. *Chem. Soc. Rev.* **2017**, *46*, 505.

26 Mascia, S.; Heider, P. L.; Zhang, H.; Lakerveld, R.; Benyahia, B.; Barton, P. I.; *et al. Angew. Chem. Int. Ed.* **2013**, *52*, 12359.

27 Fuller, P. E.; Gothard, C. M.; Gothard, N. A.; Weckiewicz, A.; Grzybowski, B. A. *Angew. Chem. Int. Ed.* **2012**, *51*, 7933.

Index

Efficiency in Natural Product Total Synthesis, First Edition. Edited by Pei-Qiang Huang, Zhu-Jun Yao, and Richard P. Hsung.

Printed and bound by CPI Group (UK) Ltd, Croydon, CR0 4YY
17/09/2026
14964286-0002